The Evolution of Artiodactyls

The Evolution
of Artiodactyls

Edited by

DONALD R. PROTHERO *and* SCOTT E. FOSS

THE JOHNS HOPKINS UNIVERSITY PRESS, *Baltimore*

9 8 7 6 5 4 3 2

The Johns Hopkins University Press
2715 North Charles Street
Baltimore, Maryland 21218-4363
www.press.jhu.edu

Frontispieces: Restoration of *Dromomeryx borealis* from Scott
(1913) and *Agriochoerus antiquus* skeleton from Scott (1940).

Library of Congress Cataloging-in-Publication Data

The evolution of artiodactyls / edited by Donald R. Prothero
and Scott E. Foss.
     p. cm.
    Includes bibliographical references and index.
    ISBN-13: 978-0-8018-8735-2 (hardcover : alk. paper)
    ISBN-10: 0-8018-8735-6 (hardcover : alk. paper)
    1. Artiodactyla, Fossil—Evolution.   2. Phylogeny.
  I. Prothero, Donald R.   II. Foss, Scott E, 1968–
QE882.U3E96   2007
569'.63—dc22   2007009445

Special discounts are available for bulk purchases of this book.
For more information, please contact Special Sales at 410-516-6936 or
specialsales@press.jhu.edu.

To our friends and colleagues
Dr. Jeremy Hooker and Dr. Alan Gentry
for their crucial contributions
to the study of fossil and living artiodactyls

CONTENTS

ACKNOWLEDGMENTS

We thank each of the individual authors, who invested a huge amount of time, research, and passion into each of their respective chapters. Second only to the authors, we thank the many specialists who have donated time as peer reviewers. They are usually acknowledged in each chapter as appropriate. Jessica Theodor contributed enormously to this project by coordinating many of the European contributions, and this book would not be as complete as it is without her tireless input and assistance. We thank Carl Buell for his stunning cover painting, which captures the essence of this book so well. We also thank the Johns Hopkins University Press editorial staff, especially Vincent Burke, for their ongoing enthusiasm and support toward the publication of this book.

CONTRIBUTORS

JEAN-RENAUD BOISSERIE, Département Histoire de la Terre, USM 0203, UMR 5143 CNRS, Unité Paléobiodiversité et Paléoenvironnement, Muséum National d'Histoire Naturelle, CP 38, 57 Rue Cuvier, Paris F-75005, France (jrboisserie@mnhn.fr)

EDWARD BYRD DAVIS, Department of Geological Sciences, University of Oregon, Eugene, OR 97403 (daviseb@berkeley.edu)

STÉPHANE DUCROCQ, Laboratoire de Géobiologie, Biochronologie et Paléontologie Humaine, UMR 6046 CNRS, Université de Poitiers, 40 Avenue du Recteur Pineau, F-86022 Poitiers cedex, France (stephane.ducrocq@univ-poitiers.fr)

JÖRG ERFURT, Institute for Geological Sciences and Geiseltalmuseum, Martin-Luther-Universität Halle-Wittenberg, Von-Seckendorff-Platz 3, D-06120 Halle (Saale), Germany (joerg.erfurt@geo.uni-halle.de)

SCOTT E. FOSS, Bureau of Land Management, Utah State Office, PO Box 45155, Salt Lake City, UT 84145 (scott_foss@blm.gov)

JONATHAN H. GEISLER, Department Geology/Geography and Georgia Southern Museum, Georgia Southern University, Statesboro, GA 30460-8149 (geislerj@georgiasouthern.edu)

COLIN P. GROVES, School of Archaeology and Anthropology, Building 14, Australian National University, Canberra, ACT 0200, Australia (colin.groves@anu.edu)

JOHN M. HARRIS, George C. Page Museum, 5801 Wilshire Boulevard, Los Angeles, CA 90036 (jharris@nhm.org)

JAMES G. HONEY, Department of Geological Sciences, University of Colorado, Boulder, CO 80309 (tiznat@hotmail.com)

CHRISTINE M. JANIS, Department of Ecology and Evolutionary Biology, Brown University, Providence, RI 02912 (christine_janis@brown.edu)

FABRICE LIHOREAU, Département de Paléontologie, FSEA, Université de N'Djaména, BP 1117, N'Djaména, Chad (flihoreau@intnet.td)

MATTHEW R. LITER, Department of Geology, Occidental College, Los Angeles, CA 90041 (liter@oxy.edu)

LIU LI-PING, Institute of Vertebrate Paleontology and Paleoanthropology, PO Box 643, Beijing 10044, China (liuliping@pa.ivpp.ac.cn)

JOSHUA A. LUDTKE, Department of Biology, San Diego State University, 5500 Campanile Drive, San Diego, CA 92182-4614 (joshualudtke@gmail.com)

JONATHAN D. MARCOT, Paleontology Section, University of Colorado Museum, UCB 265, Boulder, CO 80309 (jonathan.marcot@colorado.edu)

GRÉGOIRE MÉTAIS, Vertebrate Paleontology Section, Carnegie Museum of Natural History, 4400 Forbes Avenue, Pittsburgh, PA 15213 (akssyria3@hotmail.com)

DONALD R. PROTHERO, Department of Geology, Occidental College, Los Angeles, CA 90041 (prothero@oxy.edu)

GERTRUD E. RÖSSNER, Department of Earth and Environmental Sciences, Palaeontology Section, and Geobio-Center of the Ludwig-Maximilians-Universität München, Richard-Wagner-Strasse 10, D-80333 Munich, Germany (g.roessner@lrz.uni-muenchen.de)

NIKOS SOLOUNIAS, Department of Paleontology, American Museum of Natural History, Central Park West at 79th Street, New York, NY 10024; and Department of Anatomy, New York College of Osteopathic Medicine, Old Westbury, NY 11568-8000 (nsolouni@nyit.edu)

JAMES BOWIE STEVENS, PO Box 608, Terlingua, TX 79852 (stevensjb@hal.lamar.edu)

MARGARET SKEELS STEVENS, PO Box 608, Terlingua, TX 79852 (stevensjb@hal.lamar.edu)

JESSICA M. THEODOR, Department of Biological Sciences, University of Calgary, 2500 University Drive NW, Calgary, Alberta, Canada T2N 1N4 (jtheodor@ucalgary.ca)

MARK D. UHEN, PO Box 37012, Washington, DC 20013 (uhen@umich.edu)

INESSA VISLOBOKOVA, Paleontological Institute, Russian Academy of Sciences, ul. Profsoyuznaya 123, Moscow 117868, Russia (ivisl@paleo.ru)

The Evolution of Artiodactyls

SCOTT E. FOSS
AND DONALD R. PROTHERO

1

Introduction

THE LARGEST LIVING ORDER of terrestrial hoofed mammals is the even-toed hoofed mammals, or Artiodactyla. They are "even toed" or "cloven hoofed" because the axis of symmetry of the foot runs between the third and fourth toes, and so they usually have either two toes or four. Today, they are the most diverse and abundant ungulates on the planet, with over 190 living species, including pigs, peccaries, hippos, camels and llamas, deer, pronghorns, giraffes, sheep, goats, cattle, and dozens of species of antelope. Nearly every domesticated animal we eat is an artiodactyl (cattle, pigs, sheep, goats, and even deer), and they provide us with all of our milk (whether from a cow, goat, or camel) and wool (from either sheep or alpacas). Almost every large herbivore you might see in East Africa is an artiodactyl except for zebras, rhinos, and elephants. Artiodactyls are indeed a modern success story, yet they acquired that dominance gradually as the odd-toed perissodactyls (especially horses, rhinos, tapirs, and brontotheres) dominated in the Eocene and gradually were displaced while artiodactyls (especially the ruminants, such as camels, sheep, goats, and cattle), with their alternative mode of digestion, came to dominate the earth.

Our understanding of artiodactyl evolution has been revolutionized in the past few years. Advances in global information sharing, computer technology, and analytical techniques and new fossils from more varied localities around the world have all contributed to a rapidly increasing ability to understand complex evolutionary questions. Paleoecological and geochemical analyses of fossil material are elucidating the timing of hypsodont tooth development and herbivore exploitation of grasslands. Anatomical analyses of fossil and recent organisms are revealing information critical to our understanding of jaw mechanics, joint movement, and behavior. Descriptions of new species and revisions of outdated taxonomy are allowing previously poorly known groups to be better understood, thus permitting much more complete taxonomic classifications to be proposed. Large-scale phylogenetic analyses by both molecular biologists and functional morphologists are providing greater and greater resolution of the artiodactyl "tree of life."

It was 57 years ago that molecular systematists looked at molecular factors in milk casein and concluded that there was a close relationship between extant hippopotamids and whales (Boyden and Gemeroy, 1950; Irwin and Arnason, 1994; Graur and Higgins, 1994; Gatesy et al., 1996; and others). To paleontologists and vertebrate morphologists, this conjured images of a Gary Larson cartoon that offered a whale plus Holstein cow chimera. Initially, both the chimera and the molecular assertion were laughable to many of us. Over the next few years, however, molecular biologists conducted multiple experiments on other portions of somatic and mitochondrial DNA (summarized by Gatesy, 1998) and became convinced that, among extant taxa, whales and hippos are close relatives, closer in fact than hippos are to extant pigs or peccaries.

By 2001, paleontologists had largely agreed that whales and artiodactyls were closely related but still had not verified the seemingly radical conclusion of the molecular biologists that whales were nested deep within the Artiodactyla (Theodor, 2001). However, in the fall of 2001, paleontologists were shocked by the concurrent publication of two papers that each unveiled new whale fossils with traits previously thought to be unique to the artiodactyls, the paraxonic ("even-toed") foot and the double-pulley astragalus (Gingerich et al., 2001; Thewissen et al., 2001) (Fig. 1.1).

With the confirmation of these characters (previously considered unique to artiodactyls), the whales could not be ignored as the sister group (closest relative) of Artiodactyla. In 2003, Foss and Theodor described juvenile artiodactyl teeth that have accessory cuspules and morphology that was previously thought to be unique to whales (Foss and Theodor, 2003; Theodor and Foss, 2005) (Fig. 1.2). At the same time, Geisler and Uhen presented phylogenetic reconstructions with more complete character suites from morphological and stratigraphic data that showed whales nested within Artiodactyla, thus verifying the molecular cetartiodactyl (whale plus artiodactyl) phylogenetic hypothesis (Geisler and Uhen, 2003, 2005). Geisler et al. (this volume) goes one step further toward validating the concept of a monophyletic whale plus hippopotamus clade, Whippomorpha (Waddell et al., 1999) or Cetancodonta (Arnason et al., 2000). Thus, a monophyletic whale plus hippo clade is supported not only by molecular studies on extant organisms but also by morphological analysis of fossil organisms.

Rather than engage in a highly visible scientific disagreement with various personalities taking sides (as we have seen in the "birds are dinosaurs / no they are not" debate), paleontologists interested in whale and artiodactyl evolution, for the most part, have maintained a wait-and-see attitude while attempting to test the molecular results by performing independent morphological character analysis based not only on extant species (which is the limitation of molecular biol-

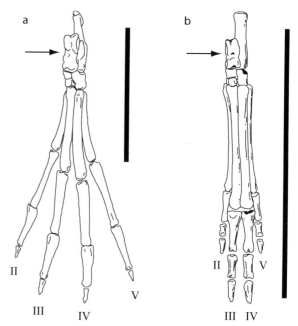

Fig. 1.1. Left pes (rear foot) of (A) *Rhodocetus* (early whale from Pakistan) and (B) *Diacodexis* (early artiodactyl from North America). The astragalus (ankle bone: marked by arrows) has a rotational articulation on both the proximal and distal surfaces. Until recently, this "double-pulley" astragalus was thought to be unique to the artiodactyls. The paraxonic foot, also diagnostic of whales and artiodactyls, is recognized by a sharing of load between digits III and IV. This two-toed condition is what has facilitated the development of the "cloven" hoof of most living artiodactyls. Images redrawn from Rose (1990) and Gingerich et al. (2001). Scale bars equal 10 cm.

ogy) but also on the fossils (O'Leary, 1999a, 1999b, 2001; O'Leary and Geisler, 1999; Geisler and Uhen, 2003, 2005). The result has been an aggressive reanalysis of both artiodactyl and whale systematics performed in an atmosphere of rigorous science rather than the judgment of the popular press. It took a century to verify Sir Richard Owen's assertion that birds are descended from dinosaurs; it has taken only a decade to accomplish the same level of consensus with regard to the origin of whales from artiodactyls. However, the search for the actual sister group of early fossil whales is only in its infancy. It is our hope that this book will contribute to that and to other inquiries by assembling the current state of artiodactyl taxonomy, systematics, and evolution and by setting the stage for the next series of questions regarding artiodactyl evolution.

We have assembled an international panel of experts as contributors to this volume, including scientists from the United States, Canada, France, Germany, China, Australia, Chad, and Russia. Family-level cetacean taxonomy is not considered in this volume. However, it has been discussed in numerous recent publications (including Thewissen, 1998; Prothero and Schoch, 2002; Arnason et al., 2004; May-Collado and Agnarsson, 2006). All of the living artiodactyl families are covered in this book. We are very pleased that many extinct artiodactyl families that have not received adequate scrutiny in the past are addressed in this volume (including the Raoellidae, Mixtotheriidae, Anoplotheriidae, Xiphodontidae, and many others). A few artiodactyl fami-

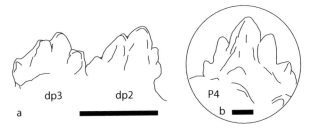

Fig. 1.2. (A) Eocene artiodactyl deciduous dentition (MNHN.EBA 327, *Cebochoerus lautricensis*) and (B) Eocene archaeocete whale dentition (USNM 14382, *Dorodon osiris*). Scale bar equals 1 cm.

lies are not fully covered in these chapters. This is largely because no specialist in this group was available to summarize them at this time or, in a few cases, there is little to report since the most recent revision. Among the tylopods only the Oromerycidae is omitted because they were fully summarized by Prothero (1998c), and among the cervoids, only the tiny (two genera) family Hoplitomerycidae is missing because little recent research has been conducted on these unusual multihorned, multitusked deer. Honey et al. (1998) provided a general overview of the Camelidae, so Honey (this volume) does not repeat that work here but instead provides a study on sexual dimorphism within the protolabine camels that sheds important light on their systematics. A detailed monograph of all the North American Camelidae, utilizing the excellent largely unpublished specimens in the Frick Collection, has yet to be undertaken.

Only a few years ago Prothero wrote, "Artiodactyla [is] possibly the best-substantiated order within the Mammalia" (1993: 176). This statement still applies when the concept is broadened to include whales. With the inclusion of a monophyletic Cetacea within the Artiodactyla, it would not be surprising, phylogenetically speaking, to find this to be one of the last publications to be titled "The Evolution of Artiodactyls." Future tomes may prefer the more inclusive title "The Evolution of Cetartiodactyla."

JONATHAN D. MARCOT

2

Molecular Phylogeny of Terrestrial Artiodactyls
Conflicts and Resolution

THE EVEN-TOED UNGULATES (order Artiodactyla) constitute the most diverse group of living medium- to large-bodied mammals. Artiodactyls have been of considerable importance throughout the Cenozoic, dominating many ancient ecosystems (especially in the Neogene) and flourishing in domestication (e.g., cattle [*Bos taurus*], swine [*Sus scrofa*], and sheep [*Ovis aries*]). As a result, considerable attention has been paid to numerous aspects of their evolution, including their taxonomic diversification, and the relationship of their ecology and morphology to ancient environments and climate. An accurate understanding of the evolutionary interrelationships, or phylogeny, of artiodactyls, especially with respect to fossil taxa, would undoubtedly contribute considerably to our understanding of their origin, diversification, and rise to the dominant position they hold in many ecosystems today. Phylogeny provides a more complete understanding of biodiversity, and its applications extend to studies of taxonomic (e.g., Smith, 1988; Magallón and Sanderson, 2001), morphological (e.g., Smith, 1994; Gaubert et al., 2005), molecular (e.g., Adkins et al., 1996; Ward et al., 1997), and behavioral (Caro et al., 2003; Stoner et al., 2003) evolution and the quality of the fossil record (e.g., Norell, 1992; Benton et al., 2000). However, any inferences drawn from such studies depend on the accuracy of our estimates of phylogenetic relationships, so the reliability of data used in phylogenetic analysis is of utmost importance.

Traditionally, phylogenetic relationships of most groups have been estimated using morphological data from both extant and fossil taxa (Wiens, 2004). However, the use of molecular data (particularly nucleotide or amino acid sequences) for phylogeny reconstruction has exploded in recent years. The results of some of the earliest molecular studies of mammalian phylogeny were highly discordant with traditional conceptions of phylogeny and classifications based on morphological data (e.g., Miyamoto and Goodman, 1986; Irwin et al., 1991; Graur and Higgins, 1994; see also Honeycutt and Adkins, 1993; Springer et al., 2004). Since these earliest studies, more and more taxa and genetic markers have been sampled, and as a result, some of these conflicts have evaporated while others remain strongly supported.

Workers have noted that apparent conflict between morphological and molecular results is often overstated (Hillis and Wiens, 2000). However, some workers assert that cases in which conflict is real serve to demonstrate the superiority of molecular data over morphological data for phylogeny reconstruction (Scotland et al., 2003). At the extreme, these workers have suggested that molecular phylogenetics has eliminated the role of morphology in phylogeny reconstruction except to corroborate the results of molecular analyses. Such assertions needlessly impose a false dichotomy between the two data types; there are many reasons that morphological data can be equally relevant, if not more so (Jenner, 2004; Wiens, 2004; Smith and Turner, 2005). However, they do raise substantive issues regarding the effectiveness and applicability of data that must be addressed prior to phylogenetic analysis. In particular, which data and methods will yield the most reliable estimates of phylogeny, so that they may be used confidently in subsequent biologic studies?

Artiodactyls provide several excellent examples of incongruence between traditional morphological and molecular data. Here, I review recent studies of terrestrial artiodactyl (i.e., noncetacean) molecular phylogeny, highlighting three such cases of conflict: (1) the relationships of the major clades of artiodactyls, (2) the relationships among the extant families of the suborder Ruminantia, and (3) the phylogeny of the family Bovidae. These examples illustrate multiple reasons that molecular and morphological data might yield different results and provide lessons that have the potential to guide future systematic research. I will not argue for the supremacy of one data type over another but instead follow others (e.g., Springer et al., 2004) and argue that the two types must be integrated for a complete understanding of mammal phylogeny that includes both extant and fossil taxa.

SUPERMATRIX ANALYSIS

To summarize recent molecular studies of artiodactyl phylogeny, I present a phylogenetic analysis of a "supermatrix," a single large data matrix consisting of several concatenated datasets of previously published molecular sequence data (e.g., Gatesy et al., 2002; Driskell et al., 2004; Philippe et al., 2005). Unlike alternatives such as supertree analyses (see Sanderson et al., 1998; Sanderson and Driskell, 2003; Bininda-Emonds, 2004), supermatrix analyses can assess how strongly data support particular relationships

(Gatesy et al., 2002) and can uphold relationships not indicated by any individual dataset (Gatesy et al., 1999a). I downloaded published DNA sequence data from GenBank (www.ncbi.nlm.nih.gov/gquery/gquery.fcgi) that represented the broadest possible coverage of terrestrial artiodactyls (e.g., Gatesy et al., 1999b; Hassanin and Douzery, 1999b; Matthee and Robinson, 1999; Gatesy and Arctander, 2000b; Matthee et al., 2001; Matthee and Davis, 2001; Hassanin and Douzery, 2003). I then concatenated these data into a single matrix of 216 taxa coded for 16 different genetic markers (see Appendix) spanning 13,989 nucleotide sites, 3,823 of which are parsimony informative. Of the roughly 220 extant terrestrial artiodactyl species, 198 (90%) were represented by at least one gene segment. The position of cetaceans within Artiodactyla is strongly supported (see below), so, because of computational limitations and the focus of this chapter on terrestrial artiodactyls, this analysis excluded all cetacean taxa. Instead, cetaceans were assumed to be the sister group to Hippopotamidae (Fig. 2.1).

Gene segments include two protein-coding mitochondrial genes (cytochrome *b* and cytochrome *c* oxidase subunit III), two mitochondrial ribosomal genes (12S and 16S rDNA), and 12 nuclear gene segments (κ-casein, β-casein, α-lactalbumin, aromatase cytochrome P450, lactoferrin promoter, β-spectrin nonerythrocytic 1, protein kinase C ι, stem cell factor, signal transducer and activator of transcription 5A, thyroglobulin, tumor necrosis factor, and thyrotropin). Most nuclear genes include segments of protein-coding exons and noncoding introns. Protein-coding regions (e.g., cytochrome *b* and κ-casein exon 4) were aligned by eye. Other nuclear datasets including intron regions were aligned according to published alignments (e.g., Matthee et al., 2001; Hassanin and Douzery, 2003). Mitochondrial ribosomal genes were aligned according to the secondary structure models of Springer and Douzery (1996) and Burk et al. (2002). Regions of ambiguous alignment were then excluded prior to analysis. The concatenated data matrix is available from the author on request. Not all taxa are represented for all genetic markers, but every taxon shares at least one sequence in common with at least one other taxon. In some instances (particularly mitochondrial genes such as cytochrome *b* and ribosomal DNA), several sequences were available for particular species, and all such sequences were merged. Resulting polymorphic sites within species were treated as true polymorphisms to accurately represent the genetic diversity of individual species.

The matrix analyzed here is the most densely sampled for terrestrial artiodactyls to date. Despite this dense taxon sampling, the matrix includes fewer genes than found in some previous analyses (e.g., Gatesy et al., 2002). Many genetic markers that have been useful for higher-level eutherian systematics (e.g., IRBP and vWF) were not included within this supermatrix because of sparse taxon sampling for terrestrial artiodactyls (e.g., fewer than 10 taxa). It is apparent, however (see results below), that these data are important to understanding the deepest nodes within artiodactyl phylogeny.

Fig. 2.1. Results of parsimony analysis of the artiodactyl supermatrix. The tree shown is a strict consensus of 21,100 most-parsimonious trees. Numbers above branches indicate bootstrap support percentages (see text). The mammalian Order Perissodactyla is used as an outgroup for rooting purposes. This tree represents the results of a single analysis but is subdivided for graphic purposes into (A) higher-level relationships within Artiodactyla, (B) relationships within the Families Suidae and Tayassuidae, (C) relationships within the Family Camelidae, (D) relationships among the Cervidae (note that *Cervus elaphus* has recently been shown to be polyphyletic; see Ludt et al., 2004; Pitra et al., 2004), (E) relationships within the Family Moschidae, (F) relationships within the bovid clade Boodontia, and (G) relationships within the bovid clade Aegodontia.

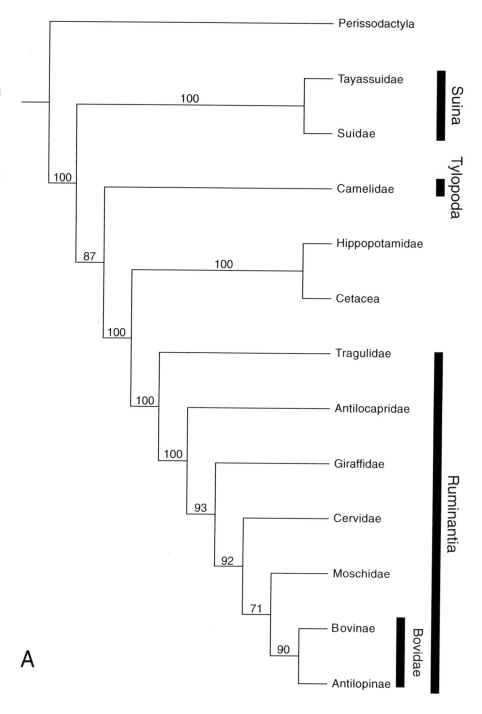

I performed a heuristic search using the supermatrix under maximum parsimony optimality criteria with PAUP*4β10 (Swofford, 2002) with the following settings: 1,000 searches using random taxon addition sequences, TBR branch swapping, parsimony-uninformative characters excluded prior to analysis, and the remaining characters treated as unordered and with equal weight. A strict consensus of 21,100 most-parsimonious trees of 27,216 steps is shown in Figure 2.1A–G. Although I divide the tree into several smaller trees in Figure 2.1, it represents the results of a single phylogenetic analysis of all taxa. Support for the resulting trees was assessed using nonparametric bootstrapping (Felsenstein,

1985), with 1,000 replicate heuristic searches using only parsimonious-informative characters. Bootstrap percentages greater than 50% are indicated in Figure 2.1.

Predictably, the results of this analysis are generally in accord with the studies from which the original data were derived, making the consensus tree a reasonable summary of previous molecular phylogenetic research, with some exceptions (see below). In general, the interrelationships among higher taxa (Fig. 2.1A) are all highly supported (bootstrap >70%). Species-level relationships are generally well resolved and supported (Figs. 2.1B–G), with notable exceptions among some clades within the family Bovidae.

Fig. 2.1. *Continued*

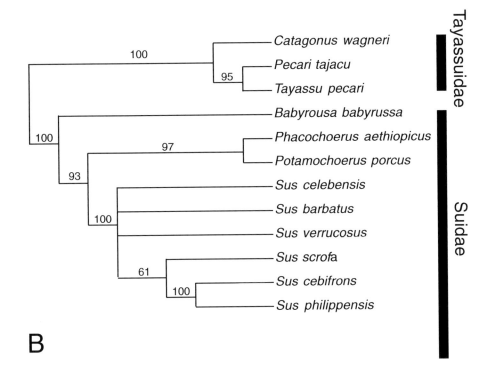

B

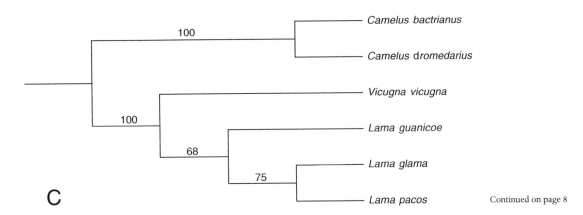

C

Continued on page 8

INTERRELATIONSHIPS OF MAJOR ARTIODACTYL CLADES

The mammalian order Artiodactyla has traditionally been composed of several major clades including Suidae (i.e., pigs; Fig. 2.1B), Tayassuidae (i.e., peccaries; Fig. 2.1B), Hippopotamidae (i.e., hippos), Camelidae (i.e., camels and llamas; Fig. 2.1C), and several families within the suborder Ruminantia (numerous ruminants; see below). The order Cetacea (i.e., whales) has also been variously allied with artiodactyls. In every case in which taxon sampling was sufficient, the monophyly of each of these major clades has been upheld by molecular analyses. However, recent molecular phylogenies differ from traditional classifications and morphological phylogenies in several key respects with respect to the interrelationships among these clades.

Whales Are Artiodactyls

The best-known controversy stemming from early molecular phylogenetic studies of mammalian orders is the position of whales with respect to the artiodactyl clade (see Gingerich, 2005). Traditional hypotheses consider whales to be ungulates outside crown-group Artiodactyla (e.g., Van Valen, 1966; Thewissen, 1994; Luckett and Hong, 1998; Geisler, 2001b). However, even some of the earliest phylogenetic studies using nucleotide sequences suggested that whales nest within Artiodactyla, (e.g., Irwin et al., 1991; Graur and Higgins, 1994; M. Smith et al., 1996; Springer et al., 1997), but these early analyses typically utilized few genes and taxa (artiodactyl or whale). Because of the apparent conflict with anatomical data, these early studies were intensely scrutinized and reanalyzed using more so-

Fig. 2.1. *Continued*

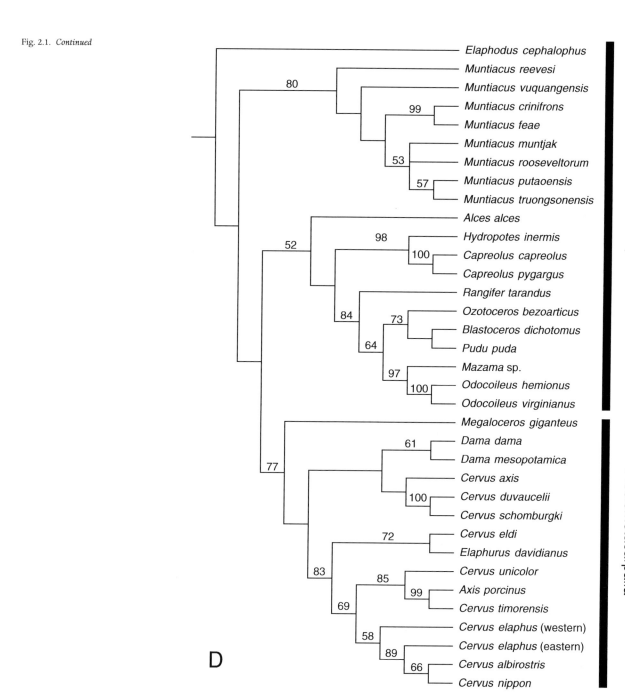

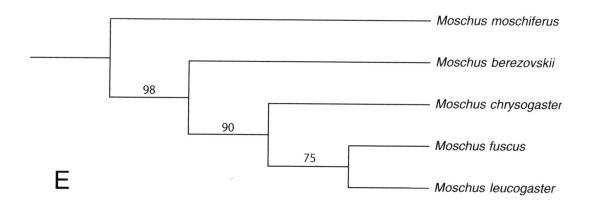

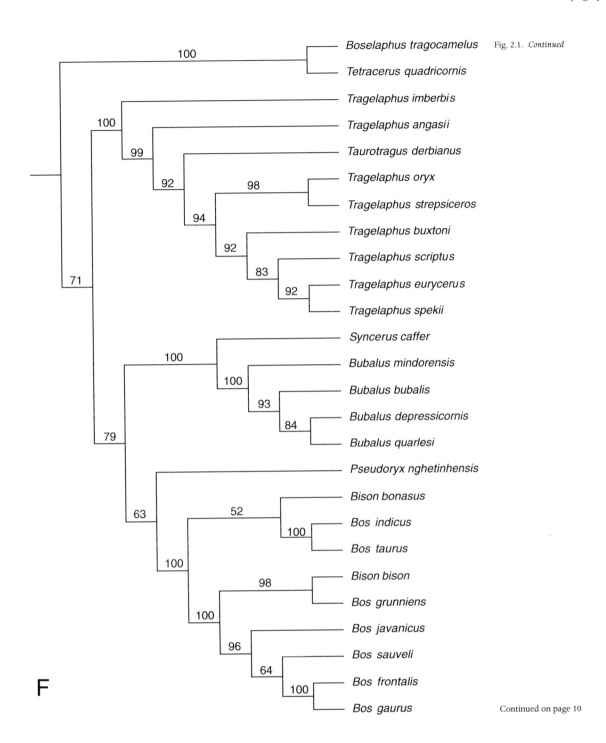

Fig. 2.1. *Continued*

F

Continued on page 10

phisticated phylogenetic methods. However, this only resulted in their further support (Adachi and Hasegawa, 1996; Hasegawa and Adachi, 1996), with additional taxon and genetic sampling suggesting that whales are the sister group to extant hippos (family Hippopotamidae) (e.g., Irwin and Árnason, 1994; Gatesy et al., 1996; Randi et al., 1996; Gatesy et al., 1997; Montgelard et al., 1997). In the past decade, further taxonomic sampling of these original datasets, as well as the addition of newer datasets, in particular large multigene sets of mitochondrial DNA and amino acid sequences (e.g., Ursing and Arnason, 1998; Ursing et al., 2000), sets of previously unsampled nuclear-encoded genes (e.g., Kleinei-

dam et al., 1999; Madsen et al., 2002; Amrine-Madsen et al., 2003) and large supermatrices of such sets (e.g., Gatesy et al., 1999a, 1999b; Madsen et al., 2001; Matthee et al., 2001; Murphy et al., 2001a; Scally et al., 2001; Gatesy et al., 2002) have produced better resolved and supported trees with whales nested within artiodactyls.

Short and long interspersed elements (SINEs and LINEs, respectively) provide additional support for the sister-taxon relationships of whales and hippos. SINEs and LINEs are insertions of apparently nonfunctional DNA, generally between 75 and 500 base pairs in length, that are copied and inserted into various locations throughout chromosomes

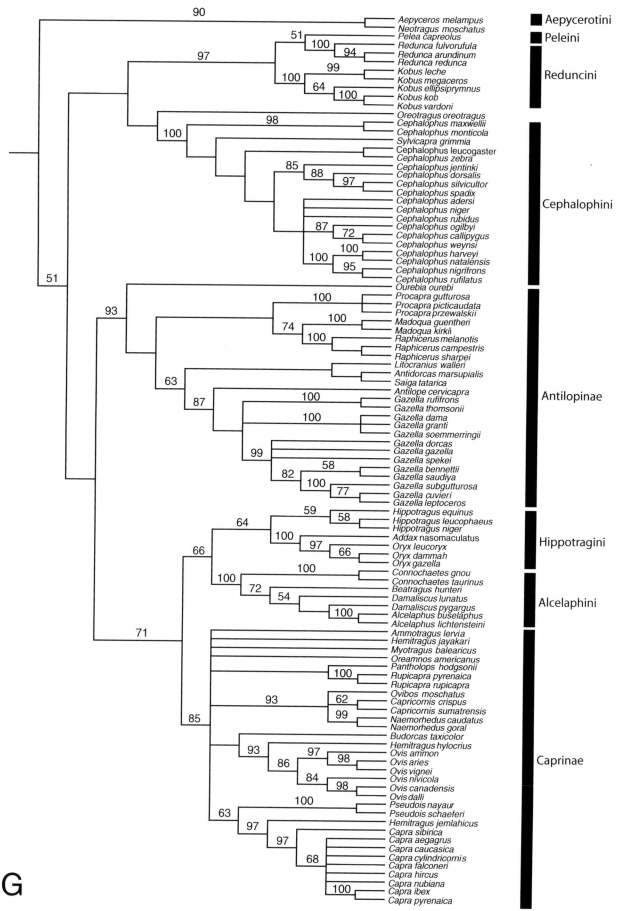

90 — Aepyceros melampus — **Aepycerotini**
Neotragus moschatus — **Peleini**
51 — Pelea capreolus
100 — Redunca fulvorufula
94 — Redunca arundinum
Redunca redunca
97 — 99 — Kobus leche
100 — Kobus megaceros
64 — Kobus ellipsiprymnus
100 — Kobus kob
Kobus vardoni — **Reduncini**
Oreotragus oreotragus
98 — Cephalophus maxwellii
100 — Cephalophus monticola
Sylvicapra grimmia
Cephalophus leucogaster
Cephalophus zebra
85 — 88 — Cephalophus jentinki
97 — Cephalophus dorsalis
Cephalophus silvicultor
Cephalophus spadix
Cephalophus adersi
Cephalophus niger
Cephalophus rubidus
87 — Cephalophus ogilbyi
72 — Cephalophus callipygus
Cephalophus weynsi
100 — Cephalophus harveyi
100 — Cephalophus natalensis
95 — Cephalophus nigrifrons
Cephalophus rufilatus — **Cephalophini**
Ourebia ourebi
100 — Procapra gutturosa
Procapra picticaudata
Procapra przewalskii
74 — 100 — Madoqua guentheri
Madoqua kirkii
100 — Raphicerus melanotis
Raphicerus campestris
Raphicerus sharpei
Litocranius walleri
Antidorcas marsupialis
63 — Saiga tatarica
87 — Antilope cervicapra
100 — Gazella rufifrons
Gazella thomsonii
100 — Gazella dama
Gazella granti
Gazella soemmerringii
99 — Gazella dorcas
Gazella gazella
Gazella spekei
82 — 58 — Gazella bennettii
Gazella saudiya
100 — Gazella subgutturosa
77 — Gazella cuvieri
Gazella leptoceros — **Antilopinae**
59 — 58 — Hippotragus equinus
Hippotragus leucophaeus
Hippotragus niger
64 — 100 — Addax nasomaculatus
97 — Oryx leucoryx
66 — Oryx dammah
Oryx gazella — **Hippotragini**
66 — 100 — Connochaetes gnou
Connochaetes taurinus
100 — 72 — Beatragus hunteri
54 — Damaliscus lunatus
Damaliscus pygargus
100 — Alcelaphus buselaphus
Alcelaphus lichtensteini — **Alcelaphini**
Ammotragus lervia
Hemitragus jayakari
Myotragus balearicus
Oreamnos americanus
Pantholops hodgsonii
100 — Rupicapra pyrenaica
Rupicapra rupicapra
93 — Ovibos moschatus
62 — Capricornis crispus
99 — Capricornis sumatrensis
Naemorhedus caudatus
Naemorhedus goral
Budorcas taxicolor
93 — Hemitragus hylocrius
86 — 97 — Ovis ammon
98 — Ovis aries
Ovis vignei
84 — Ovis nivicola
98 — Ovis canadensis
Ovis dalli
100 — Pseudois nayaur
Pseudois schaeferi
63 — Hemitragus jemlahicus
97 — Capra sibirica
97 — Capra aegagrus
Capra caucasica
68 — Capra cylindricornis
Capra falconeri
Capra hircus
Capra nubiana
100 — Capra ibex
Capra pyrenaica — **Caprinae**

85 — 71 —

G

Fig. 2.1. *Continued*

(Shedlock et al., 2000). Unlike nucleotide and amino acid substitutions, homoplasy among SINEs and LINES is thought to be exceedingly rare simply because of the improbability that identical SINEs will insert into exactly the same location on the same chromosome independently in different lineages (Shedlock and Okada, 2000; Nikaido et al., 2006). They also are generally thought to be irreversible (Shedlock and Okada, 2000; Shedlock et al., 2004; see also Hillis, 1999; Miyamoto, 1999). Early studies of interspersed elements showed that whales shared particular SINEs with some artiodactyls to the exclusion of other mammals (Buntjer et al., 1997), including camels and pigs (Shimamura et al., 1997; Shimamura et al., 1999). Additional SINEs were subsequently found to support the sister-taxon relationship between whales and hippos (Nikaido et al., 1999). Taken together, the extensive number of nucleotide sequences and the SINE data convincingly indicate that not only are whales related to artiodactyls but rather that whales *are* artiodactyls.

Polyphyly of Suiformes

A corollary of the apparent sister-taxon relationship between hippos and whales is that the suborder Suiformes, which includes hippos, the extant Suidae (pigs), and Tayassuidae (peccaries), is not monophyletic. This group is united by no fewer than 11 cranial, dental, and postcranial characters (Geisler, 2001b), and until recently the relationship between hippos and whales had little or no support from morphological data. Not only were there few morphological characters linking these two disparate clades, but the relationships also posited a gap in the fossil record of the lineage leading to hippos from the first appearance of whales in the early Eocene (about 50 million years ago [Ma]) to the first appearance of hippopotamids in the middle Miocene (about 15 Ma) (Boisserie et al., 2005b). However, this controversy inspired additional anatomical research that has now supported the conclusions that whales are artiodactyls, the whales and hippos are sister taxa, and Suiformes, as traditionally defined, is polyphyletic (e.g., Gingerich et al., 2001; Geisler and Uhen, 2003; Boisserie et al., 2005b). Subsequently, Geisler and Uhen (2005) have redefined Suiformes to exclude hippos.

Polyphyly of Selenodontia

Another relationship traditionally suggested on the basis of anatomical data was that of the Tylopoda (camels and their fossil allies) and Ruminantia into a clade known as Selenodontia (e.g., Gentry and Hooker, 1988) or Neoselenodontia (e.g., Webb and Taylor, 1980). This grouping is supported by several dental characters as well as the possession of a ruminating stomach with at least three chambers (Webb and Taylor, 1980) and continues to be found by phylogenetic analyses of morphological data (e.g., Gentry and Hooker, 1988; Geisler, 2001b; Theodor and Foss, 2005). In contrast, the sister-taxon relationship between extant Camelidae and Ruminantia has not been supported by any

phylogenetic analysis of molecular data. Instead, three hypotheses have been commonly supported with camels as the sister taxon to either (1) Suina (Suidae plus Tayassuidae) (Arnason et al., 2000; Arnason and Janke, 2002; Gatesy et al., 2002; Arnason et al., 2004; Geisler and Uhen, 2005), (2) a clade containing whales, hippos, and ruminants (Kleineidam et al., 1999; Matthee et al., 2001; Corneli, 2002; Lin et al., 2002; Maniou et al., 2004), or, most commonly, (3) all other extant artiodactyls (Graur and Higgins, 1994; Gatesy et al., 1996; Gatesy, 1997; Gatesy et al., 1999a, 1999b; Madsen et al., 2001; Murphy et al., 2001a, 2001b; Madsen et al., 2002; Amrine-Madsen et al., 2003; Waddell and Shelly, 2003; May-Collado and Agnarsson, 2006). Most sequence data support the last hypothesis (some of which were not included in this study), and it is also supported by SINE insertions (Nikaido et al., 1999; Shimamura et al., 1999).

The results of the supermatrix analysis in this study identify Suina as sister group to all other artiodactyls, with camelids nested one branch higher (Fig. 2.1A). This differs from the most commonly supported hypothesis that camels are the deepest divergence within artiodactyls, followed by Suina. This discrepancy is caused by the preponderance of data in the supermatrix that, in previous studies, have suggested such an arrangement (e.g., data from Matthee et al., 2001), and the omission of other data (e.g., Murphy et al., 2001a; Madsen et al., 2002; Amrine-Madsen et al., 2003) as a result of extremely sparse taxon sampling. It also occurs at the deepest node in artiodactyl phylogeny; such nodes are often difficult to reconstruct, especially when there are few extant taxa to be sampled, as is the case for camels (see below).

Thus, available molecular data suggest that camels and ruminants are not each other's closest relatives but, in fact, could not be further separated within the artiodactyl clade. This has profound implications for evolution of digestive physiology in artiodactyls (Langer, 2001), such as the complex, three (or more)-chambered stomach, which has been assumed to be shared between camels and ruminants (e.g., Webb and Taylor, 1980; Janis and Scott, 1987). This also throws into question the phylogenetic position of many extinct selenodont artiodactyls (e.g., oreodonts, protoceratids, oromerycids). For example, the family Protoceratidae has been variably considered closely related either to tylopods (e.g., Webb and Taylor, 1980) or to ruminants (e.g., Norris, 2000). If tylopods and ruminants were indeed sister taxa, this difference may have been slight, but the molecular results indicate that protoceratids might bear significantly more information with respect to the early radiation of selenodont artiodactyls, and their further study is strongly warranted.

INTERRELATIONSHIPS OF RUMINANT FAMILIES

Ruminants are by far the most diverse group of extant terrestrial artiodactyls. The majority of taxa fall into two large families: Bovidae (cattle, antelope, goats, and sheep) and Cervidae (deer). The four other extant ruminant families—Antilocapridae (the pronghorn antelope), Giraffidae (giraffes and okapis), Moschidae (musk deer), and Tragulidae (chevrotains and mouse-deer)—contain far fewer species today but were more diverse throughout much of the Neogene. Moreover, there are several extinct families.

Traditional assessment of anatomical data suggests that the Tragulidae are the sister taxon to the monophyletic Pecora (i.e., all other extant ruminant families). All molecular studies that have included tragulids have strongly supported this hypothesis (Graur and Higgins, 1994; Chikuni et al., 1995; Cronin et al., 1996; Gatesy et al., 1996; Gatesy, 1997; Gatesy et al., 1999a; Gatesy and Arctander, 2000b; Breukelman et al., 2001; Matthee et al., 2001; Wallis and Wallis, 2001; Gatesy et al., 2002; Hassanin and Douzery, 2003; Maniou et al., 2004). However, there has been considerable difficulty resolving the relationships among the extant pecoran families using either morphology or molecular data. An alarming number of possible arrangements of the pecoran families have been proposed on the basis of morphological data (see Janis and Scott, 1987; Kraus and Miyamoto, 1991; Gatesy et al., 1992; Hassanin and Douzery, 2003). In one of the earliest sequence-based phylogenetic analyses of pecorans, Kraus and Miyamoto (1991) found it equally difficult to obtain a well-resolved result using a large segment of mitochondrial DNA, with the results changing depending on how the data were analyzed. They attributed this difficulty to a taxonomic radiation in which the pecoran families rapidly diverged from one another, leaving little time for substitutions to accumulate along the branches separating them.

Analysis of the nuclear-encoded κ-casein gene provided some further resolution of Pecoran relationships (Chikuni et al., 1995; Cronin et al., 1996; Gatesy et al., 1996). Early analyses of these datasets suggested that Antilocapridae were the earliest divergence within pecorans. These studies did not resolve relationships among giraffes, deer, and bovids, but later datasets with more taxa and sequences (Gatesy et al., 1999a, 1999b; Gatesy and Arctander, 2000b; Matthee et al., 2001; Hassanin and Douzery, 2003) generate a consistent picture with the Giraffidae being the outgroup to a cervid–bovid clade (Fig. 2.1A). However, even some of the most recent studies are still equivocal as to whether giraffes are the sister taxon of antilocaprids or of all other ruminants to the exclusion of antilocaprids (e.g., Hassanin and Douzery, 2003). This difficulty may be caused in large part by long-branch attraction (see below), as these families include only one (Antilocapridae) and two extant species (Giraffidae).

Traditional classification divides the Cervidae into four extant subfamilies: Muntiacinae (including *Muntiacus* and *Elaphodus*), Cervinae (most Old World deer), Odocoileinae (mostly New World deer and moose), and Hydropotinae (only the antlerless *Hydropotes*). An early molecular study of mitochondrial rDNA indicated that muntiacines and cervines formed a clade, with odocoileines as their sister clade (Miyamoto et al., 1990). However, this study assumed

(following prior studies such as Groves and Grubb, 1987) that *Hydropotes* is the outgroup to the antlered cervids. Later studies with more complete taxon sampling and using a noncervid outgroup revealed *Hydropotes* to be nested within the cervid clade (Douzery and Randi, 1997; Polziehn and Strobeck, 1998; Randi et al., 1998b), suggesting that the lack of antlers in *Hydropotes* represents a secondary reversal. Other recent studies have supported two major clades of cervids that had been traditionally suggested on morphological grounds (e.g., Groves and Grubb, 1987) (Fig. 2.1D): Plesiometacarpalia (Cervinae + Muntiacinae), and Telemetacarpalia (Odocoileinae + Hydropotinae) (Randi et al., 1998b; Pitra et al., 2004; Gilbert et al., 2006). However, note that Telemetacarpalia traditionally did not include *Hydropotes*. The data for the most recent study were not available for the current supermatrix analysis, so Figure 2.1D differs from this recent conclusion.

The family Moschidae currently includes only a few extant species belonging to one genus, *Moschus* (Fig. 2.1E). Moschids historically have been allied with true deer (e.g., Simpson, 1945), although they have also been considered the sister group of all other pecoran taxa (Webb and Taylor, 1980). Initial molecular phylogenetic studies using mitochondrial data supported the sister-clade relationship between musk deer and true deer (e.g., Su et al., 1999). For some time, the cytochrome *b* data used in that study were the only available sequence data for *Moschus*. However, a recent and quite thorough analysis of ruminant phylogeny by Hassanin and Douzery (2003) using seven different genetic markers (three mitochondrial and four nuclear genes) suggested that moschids are, in fact, the sister group to bovids, an arrangement that had not been previously suggested based on *any* dataset. This arrangement was strongly supported (nonparametric bootstrap support >90%) by the combined dataset as well as separately by two of the individual nuclear gene segments. Hassanin and Douzery conducted several statistical analyses that suggested that although available morphological data had never supported the sister relationship between the Moschidae and Bovidae, they did not significantly contradict it. Their study stands out in the comprehensive manner in which they statistically test hypotheses of both morphological and molecular phylogenies.

PHYLOGENY OF THE BOVIDAE

Like the relationships among ruminant families, the phylogeny within the Bovidae is an example of little resolution by either morphology or molecules rather than conflict between those data. A split between the two major clades, Boodontia (e.g., cattle, buffalo, kudus; Fig. 2.1F) and Aegodontia (e.g., antelope, gazelle, goats, sheep; Fig. 2.1G), has traditionally been accepted based on morphological data. These, in turn, were divided into several tribes, the interrelationships of which are very poorly understood on the basis of morphological data (Gentry, 1992). Because of the notable recent diversity of bovids and their importance to humans, there have been a number of molecular phylogenetic studies of single or a few tribes (e.g., Miyamoto et al., 1989; Groves and Shields, 1996, 1997; Pitra et al., 1997; Ward et al., 1997; Hassanin et al., 1998; Ludwig and Fischer, 1998; Pitra et al., 1998; Hassanin and Douzery, 1999a; Rebholz and Harley, 1999; Gatesy and Arctander, 2000a; Lalueza-Fox et al., 2000; Birungi and Arctander, 2001; Hammond et al., 2001; van Vuuren and Robinson, 2001; Lalueza-Fox et al., 2002; Kuznetsova and Kholodova, 2003; Hassanin and Ropiquet, 2004; Ropiquet and Hassanin, 2004, 2005). Unfortunately, despite this attention, the interrelationships among these groups are poorly understood.

Early molecular phylogenetic studies using mitochondrial ribosomal DNA strongly support the monophyly of both Boodontia and Aegodontia as well as the monophyly of most of the bovid tribes (Allard et al., 1992; Gatesy et al., 1992). However, they offer very little support for any further interrelationships among the bovid tribes. As with pecoran ruminants, workers have attributed this pattern to the rapid radiation of these clades (Allard et al., 1992; Gatesy et al., 1992). Further sampling of these datasets links some tribes, such as the Peleini with the Reduncini, most of the Neotragini with the Antilopini, and the Hippotragini and the Alcelaphini with the Caprinae (Gatesy et al., 1997), but leaves many interrelationships unresolved (see Fig. 2.1G). To date, large mitochondrial (Hassanin and Douzery, 1999b; Matthee and Robinson, 1999) and nuclear (Matthee and Davis, 2001) DNA datasets have provided more support for some of these earlier findings but still offer little support for other interrelationships of many of the tribes of aegodont bovids. The continued lack of resolution in spite of thousands of base pairs of sequence data and dense taxon sampling suggests that the tribal radiation of bovids was indeed a very rapid one (e.g., Rokas et al., 2005). If so, then it may be more appropriate to represent bovid tribal phylogeny as a multifurcation (i.e., a "hard polytomy" *sensu* Maddison, 1989) rather than an arbitrarily resolved bifurcating tree. In this respect, the morphology and molecular data "agree" in their inability to resolve these relationships.

REASONS FOR CONFLICT

There is no doubt that many of the genetic markers used in molecular phylogenetic analyses have evolved under strong selective pressure, and the probability for functional convergence among taxa is nontrivial. However, in most of cases, hundreds to thousands of nucleotide sites support the phylogenetic results shown in Figure 2.1, from genes that span a variety of selective regimes (e.g., protein-coding exons, noncoding introns, ribosomal DNA) and from two different genomes (i.e., mitochondrial and nuclear). Moreover, the molecular phylogeny is also strongly supported by several SINE insertions that have an undoubtedly low probability of convergence. There is only one obvious reason why such numerous and disparate datasets should all converge on a single answer, and that is their shared phylogenetic history. To contend that the molecular phylogenetic relationships that are strongly supported by these numerous and disparate datasets do not reflect shared phylogenetic history requires an explanation of how such disparate datasets could all systematically be biased in nearly identical ways.

There are, therefore, two likely factors responsible for much of the observed conflict between traditional morphology-based phylogeny and recent molecular-based estimates: evolutionary convergence among morphological characters and rapid taxonomic origination. Evolutionary convergence among distantly related taxa will tend to mislead phylogenetic analysis (Wiens et al., 2003), grouping these taxa together on the basis of homoplasy (i.e., analogous structures) instead of shared history (i.e., homology). Character suites of different taxa may evolve convergently to be similar if they have experienced common selective pressures for similar function. A large proportion of characters used in morphological phylogenetics of artiodactyls are derived from the dentition and limbs (e.g., Webb and Taylor, 1980; Janis and Scott, 1987; Gentry and Hooker, 1988; Gentry, 1992). It is likely that these suites of characters have experienced strong and consistent functional selection for more efficient mastication and locomotion, respectively, throughout artiodactyl evolution. In particular, as Neogene terrestrial environments have become more open (i.e., less forested), many clades of mammals, including artiodactyls, have evolved more lophate and higher-crowned dentitions (Jernvall et al., 1996; Fortelius and Solounias, 2000; Janis et al., 2000; Williams and Kay, 2001; Fortelius et al., 2002; Jernvall and Fortelius, 2002; Strömberg, 2006) and longer limbs with fewer skeletal elements (e.g., Janis and Scott, 1987; Janis et al., 2002). These environmental changes exerted similar selection pressures on most ungulates, and it is not surprising that they responded with similar morphological adaptations.

The second potential reason for the observed conflict between morphological and molecular phylogenies is that some artiodactyl clades, particularly the pecoran ruminants and, within this group, bovids (Kraus and Miyamoto, 1991; Allard et al., 1992; Gentry, 2000a) probably underwent very rapid radiations in which taxonomic origination rates were unusually high. Such rapid cladogenesis allows little time for evolutionary transitions to accumulate along intervening branches between successive origination events (Kraus and Miyamoto, 1991). This results in clades that are characterized by very few unambiguous synapomorphies (e.g., ruminant families; Janis and Scott, 1987). There is no doubt that such a phenomenon will affect both morphological and molecular data in the same way (Kraus and Miyamoto, 1991; Rokas et al., 2005). However, given the immense numbers of nucleotides within mammalian genomes, there are orders of magnitude more opportunities for evolutionary change within molecular data than there are for discrete morphological characters, which generally number in the hundreds. This means that it may be possible to tease these rapid divergences apart, given very large sequence databases (although this is not necessarily the case), as has been done for pecoran ruminants.

Alone or together, evolutionary convergence and rapid radiations are a perfect recipe for long-branch attraction, a particularly challenging problem for most methods of phylogenetic reconstruction (Felsenstein, 1978; Huelsenbeck and Hillis, 1993). This occurs when there are many evolutionary steps (e.g., nucleotide substitutions) along branches leading to terminal taxa but few along internal branches that unite them. Some evolutionary changes along these long branches undoubtedly yield character states (e.g., morphology or nucleotide substitutions) shared between taxa simply by chance. If there are a sufficient number, these taxa will be grouped by these chance similarities rather than by any close relationship. This is most likely the case for pecoran ruminants, in which families diverged quickly from one another, followed by long branches along which considerable morphological convergence has occurred. Long-branch attraction is exacerbated in many artiodactyl clades by the extinction of most of the constituent taxa, leaving only a few extant taxa (e.g., one in Antilocapridae and two in Giraffidae). This is particularly a problem for molecular phylogenetics, for which additional extant taxa cannot be sampled simply because they do not exist. In these cases, incorporation of fossil taxa can "break up" these long branches (Gauthier et al., 1988; Huelsenbeck, 1991) in morphology-based analyses, further underscoring the utility of morphological data and the critical role of fossil taxa in phylogenetic analysis.

RESOLUTION

Despite the emphasis on cases of conflict within this chapter, I wish to stress that there are even more cases in which molecular phylogenetic results have corroborated previous morphology-based clades. For example, molecular data corroborate the monophyly of all artiodactyl families, despite several families having few diagnostic synapomorphies (Janis and Scott, 1987). These examples further reinforce the obvious fact that both morphological and molecular data are taken from the same organisms and at some level record the same phylogenetic history.

For future morphological phylogenetic analysis, the problem of evolutionary convergence among artiodactyls remains worrisome, yet several approaches should serve to mediate its effect. Careful character analysis (e.g., Scott and Janis, 1993; Wiens, 2001; Gaubert et al., 2005) should help define characters more accurately and explicitly and identify potentially homoplastic characters prior to phylogenetic analysis. Furthermore, discrete morphological characters should be augmented with new types of data such as morphometric data or stratigraphic distributions. Although rarely used today primarily for methodological reasons, these data represent a wealth of information that can be informative for phylogenetic analysis. Also, probabilistic methods of phylogenetic reconstruction (e.g., maximum-likelihood or Bayesian methods) should be used, as they are less easily misled by convergence than traditionally parsimony-based methods. This is because they make use of the probability of character distributions among taxa to determine relationships (e.g., Wagner, 1998, 2000; Lewis, 2001) rather than relying on synapomorphies. These methods are still in their infancy, but as they become further developed, they will be increasingly important tools for phylogenetic reconstruction with morphological data. Last, when conflict is encountered, statistical tests should be employed to determine the strength of the relative phylogenetic signals of both the morphological and molecular data (e.g., Hassanin and Douzery, 2003; Gaubert et al., 2005).

A broader approach would be to combine morphological and molecular data in a single analysis. These are of particular importance for the integration of extant taxa with molecular data and fossil taxa. The simultaneous analysis approach concatenates morphological and molecular data in a single matrix for phylogenetic analysis (i.e., the supermatrix approach above), allowing the strengths of each dataset to be considered simultaneously. Here, again, probabilistic methods will be of considerable utility in that they are able to provide simultaneous and commensurate fits of phylogeny to disparate datasets (e.g., morphological, molecular, or stratigraphic [Huelsenbeck and Rannala, 1997; Wagner, 1998, 2000]). The simultaneous analysis approach has been used to integrate morphological and molecular data in several previous studies of artiodactyl phylogeny (Montgelard et al., 1998; O'Leary, 1999a, 1999b, 2001; Gatesy et al., 2002; Geisler and Uhen, 2005). Alternatively, well-supported molecular phylogenies can be used as "scaffolds" or "backbone constraints" (Springer et al., 2001; Sánchez-Villagra et al., 2003; Springer et al., 2004; Teeling et al., 2005) in which relationships among extant taxa that are robustly supported by molecular data are fixed as a template onto which other taxa, especially fossil taxa, are then optimally placed. As molecular phylogenies become more robustly supported, the simultaneous analysis and molecular scaffold approaches will likely converge on the same result.

Note that, in at least some cases, conflicting phylogenetic results have had a largely positive result. They have compelled molecular systematists and comparative anatomists alike to critically reconsider existing data and sparked research to collect even more. The most obvious example of this is illustrated by the relationships of whales. The controversy following the early molecular phylogenetic studies that indicated that whales nest within artiodactyls resulted in a flurry of molecular phylogenetic research to the point that this clade has become one of the most intensely sampled of all mammals. This research has resulted in strongly supported phylogenies that could be used to make predictions about the anatomy of expected ancestral taxa (Gatesy et al., 1996). This increased effort was mirrored by paleontologists, who continued to collect fossil specimens of transitional forms, and by comparative anatomists, who began reexamining other fossil taxa. Together, these efforts resulted in the discovery of early whales with the diagnostic artiodactyl synapomorphy, the double-trochleated astragalus (Gingerich et al., 2001; Thewissen et al., 2001b) that had been predicted by earlier molecular phylogenetic studies (Gatesy et al., 1996), and other morphological support for the sister-taxon relationship between whales and hippos (Geisler and Uhen, 2003; Boisserie et al., 2005b). These subsequent studies provide an encouraging example of how molecular and rigorous anatomical systematics can be combined to achieve complete consensus. Other parts of the artiodactyl tree await such intense and integrative study.

Too often morphological and molecular data are pitted against one another, and conflicts between them are emphasized. Suggestions that molecular data alone will be sufficient to determine the complete tree of life (e.g., Scotland et al., 2003) naïvely underestimate the importance that fossil taxa have to our understanding of phylogeny and biology, in general. On the other hand, the vast contribution that has been made by molecular systematics cannot be minimized. Molecular phylogenies have corroborated many longstanding hypotheses. They have also given comparative anatomists and paleontologists a fresh perspective and firm foundation from which to study their organisms. The further development of approaches and phylogenetic methods that can even-handedly evaluate morphological and molecular data is crucial to a complete understanding of phylogenetic relationships and any study that relies on an accurate assessment thereof.

APPENDIX 2.1: SUMMARY OF MOLECULAR DATA USED IN SUPERMATRIX ANALYSIS

First row below gene names (Species) indicates the number of taxa sampled for each. First column following taxon names (*n*) indicates number of genes sampled for each. Gene abbreviations are as follows: cyt*b*, cytochrome *b*; 12S, 12S rDNA; 16S, 16S rDNA; cox III, cytochrome *c* oxidase subunit III; κc, κ-casein; βc, β-casein; αlac, α-lactalbumin; *CYP19*, aromatase cytochrome P450; lf, lactoferrin promoter; βsn1, β-spectrin nonerythrocytic 1; pkcι, protein kinase C ι; scf, stem cell factor; st, signal transducer and activator of transcription 5A; tg, thyroglobulin; tnf, tumor necrosis factor; and tt, thyrotropin. Cells marked with a dot (•) indicate that gene was sampled for that taxon.

ACKNOWLEDGMENTS

I thank K. Sears, P. Wagner, and three anonymous reviewers, whose comments on earlier versions greatly improved this chapter.

	n	cytb	12S	16S	cox III	κc	βc	αlac	CYP19	Lf	βsn1	pkcι	scf	st	tg	tnf	tt
Species		207	152	109	52	88	38	26	26	40	50	50	25	25	25	24	24
Hippidion saldiasi	1	•															
Equus caballus	12	•	•	•	•	•					•	•	•	•	•	•	•
Equus przewalskii	1	•															
Equus asinus	4	•	•	•	•												
Equus burchellii	2	•	•														
Equus grevyi	2	•	•														
Equus hemionus	2		•	•													
Equus kiang	1		•														
Equus zebra	1		•														
Tapirus indicus	3	•	•	•													
Tapirus bairdii	1		•														
Tapirus pinchaque	1		•														
Tapirus terrestris	4	•	•	•	•												
Ceratotherium simum	4	•	•	•	•												
Diceros bicornis	2	•	•														
Coelodonta antiquitatis	2	•	•														
Dicerorhinus sumatrensis	2	•	•														
Rhinoceros sondaicus	2	•	•														
Rhinoceros unicornis	4	•	•	•	•												
Catagonus wagneri	2	•					•										
Pecari tajacu	12	•	•	•			•	•			•	•		•	•		
Tayassu pecari	2	•					•										
Babyrousa babyrussa	4	•	•	•				•									
Phacochoerus aethiopicus	1	•															
Phacochoerus africanus	3	•	•	•													
Sus celebensis	1	•															
Sus barbatus	1	•															
Sus verrucosus	1	•															
Sus scrofa	14	•	•	•				•			•			•	•	•	
Sus cebifrons	1	•															
Sus philippensis	1	•															
Camelus bactrianus	11	•	•	•		•					•	•		•	•	•	
Camelus dromedarius	6	•	•	•		•		•									
Vicugna vicugna	1	•															

continued

Appendix 2.1. Continued

	n	cytb	12S	16S	cox III	κc	βc	αlac	CYP19	Lf	βsn1	pkcι	scf	st	tg	tnf	tt
Lama guanicoe	5	•	•	•				•	•								
Lama glama	11	•	•	•		•					•	•	•	•	•	•	•
Lama pacos	4	•	•	•	•												
Hexaprotodon liberiensis	12	•	•	•		•	•				•	•	•	•	•	•	
Hippopotamus amphibius	8	•	•	•	•		•	•		•							
Tragulus javanicus	6	•	•			•		•	•	•							
Tragulus meminna	6					•					•	•	•	•	•		
Tragulus napu	4	•	•	•			•										
Antilocapra americana	15	•	•	•		•		•	•		•	•	•	•	•	•	
Giraffa camelopardalis	15	•	•	•		•	•	•	•		•	•	•	•	•	•	
Okapia johnstoni	14	•	•	•		•		•	•		•	•	•	•	•	•	
Elaphodus cephalophus	3	•	•	•													
Muntiacus reevesi	15	•	•	•	•			•			•	•	•	•	•	•	
Muntiacus vuquangensis	2	•		•													
Muntiacus crinifrons	4	•	•	•													
Muntiacus feae	2	•		•													
Muntiacus muntjak	4	•	•	•	•												
Muntiacus putaoensis	1		•														
Muntiacus truongsonensis	1		•														
Muntiacus rooseveltorum	1		•														
Alces alces	4	•	•			•	•										
Hydropotes inermis	6	•	•	•				•	•								
Capreolus capreolus	7	•	•	•		•		•	•								
Capreolus pygargus	1	•															
Rangifer tarandus	14	•	•	•		•		•			•	•	•	•	•	•	•
Ozotoceros bezoarticus	1	•															
Blastoceros dichotomus	1	•															
Pudu puda	1	•															
Mazama sp.	3	•	•			•											
Odocoileus hemionus	13	•	•			•		•			•	•	•	•	•	•	•
Odocoileus virginianus	4	•	•	•		•											
Megaloceros giganteus	1	•															
Dama dama	1	•															
Dama mesopotamica	1	•															
Cervus axis	2	•		•													
Cervus duvaucelii	2	•					•										
Cervus schomburgki	1	•															
Cervus eldi	3	•	•	•													
Elaphurus davidianus	2	•					•										
Cervus unicolor	4	•	•	•		•											
Axis porcinus	3	•	•	•													
Cervus timorensis	1	•															
Cervus elaphus2	1	•															
Cervus elaphus	8	•	•			•		•	•		•	•					
Cervus albirostris	2	•	•														
Cervus nippon	4	•	•		•	•											
Moschus moschiferus	7	•	•	•		•		•	•								
Moschus berezovskii	2	•	•														
Moschus chrysogaster	3	•	•	•													
Moschus fuscus	2	•	•														
Moschus leucogaster	1	•															
Boselaphus tragocamelus	15	•				•	•	•			•	•	•	•	•	•	•
Tetracerus quadricornis	5	•				•			•								
Tragelaphus imberbis	15	•	•	•				•	•		•	•	•	•	•	•	•
Tragelaphus angasii	5	•							•								
Taurotragus derbianus	1	•															
Tragelaphus oryx	9	•	•	•			•		•		•						
Tragelaphus strepsiceros	4	•		•					•								
Tragelaphus buxtoni	2	•		•													
Tragelaphus scriptus	4	•	•	•					•								
Tragelaphus eurycerus	5	•	•	•					•								
Tragelaphus spekii	4	•	•	•					•								
Syncerus caffer	9	•	•	•	•	•	•				•	•	•				

Appendix 2.1. Continued

	n	cytb	12S	16S	cox III	κc	βc	αlac	CYP19	Lf	βsn1	pkcι	scf	st	tg	tnf	tt
Bubalus mindorensis	1	•															
Bubalus bubalis	9	•	•	•	•	•	•	•	•	•							
Bubalus depressicornis	5	•					•			•							
Bubalus quarlesi	1	•															
Pseudoryx nghetinhensis	5	•	•			•	•			•							
Bison bonasus	2	•				•											
Bos indicus	12	•	•	•	•	•					•	•	•	•	•	•	•
Bos taurus	11	•	•	•	•	•	•	•	•	•	•	•					
Bison bison	5	•	•	•		•				•							
Bos grunniens	8	•	•	•	•	•		•	•	•							
Bos javanicus	2	•				•											
Bos sauveli	2	•								•							
Bos frontalis	2	•	•														
Bos gaurus	2	•				•											
Aepyceros melampus	16	•	•	•	•	•	•	•				•	•				
Neotragus moschatus	8	•	•	•	•	•				•		•	•				
Pelea capreolus	8	•	•	•	•	•	•					•	•				
Redunca fulvorufula	10	•	•	•	•	•	•	•					•				
Redunca arundinum	3	•	•	•													
Redunca redunca	6	•	•	•	•							•	•				
Kobus leche	6	•	•	•	•							•	•				
Kobus megaceros	3	•	•	•													
Kobus ellipsiprymnus	12	•	•	•	•		•					•		•	•	•	•
Kobus kob	3	•	•	•													
Kobus vardoni	3	•	•	•													
Oreotragus oreotragus	7	•	•	•	•				•			•					
Cephalophus maxwellii	4	•	•	•			•										
Cephalophus monticola	5	•	•		•							•	•				
Sylvicapra grimmia	6	•	•	•	•							•	•				
Cephalophus leucogaster	2	•	•														
Cephalophus zebra	2	•	•														
Cephalophus jentinki	2	•	•														
Cephalophus dorsalis	7	•	•			•	•	•	•								
Cephalophus silvicultor	2	•	•														
Cephalophus spadix	2	•	•														
Cephalophus adersi	2	•	•														
Cephalophus ogilbyi	2	•	•														
Cephalophus callipygus	2	•	•														
Cephalophus weynsi	2	•	•														
Cephalophus harveyi	2	•	•														
Cephalophus natalensis	3	•	•		•												
Cephalophus nigrifrons	2	•	•														
Cephalophus rufilatus	2	•	•														
Cephalophus niger	2	•	•														
Cephalophus rubidus	2	•	•														
Ourebia ourebi	6	•	•	•	•							•	•				
Procapra gutturosa	3	•	•	•													
Procapra picticaudata	3	•	•	•													
Procapra przewalskii	2	•	•														
Madoqua guentheri	2	•			•												
Madoqua kirkii	11	•	•	•		•						•	•	•	•	•	•
Raphicerus melanotis	7	•	•	•	•	•						•	•				
Raphicerus campestris	8	•	•	•	•	•	•					•	•				
Raphicerus sharpei	6	•	•	•		•						•	•				
Litocranius walleri	7	•	•	•	•	•						•	•				
Antidorcas marsupialis	8	•	•	•	•	•	•					•	•				
Saiga tatarica	4	•	•	•			•										
Antilope cervicapra	5	•	•	•	•		•										
Gazella bennettii	2	•			•												
Gazella saudiya	2	•			•												
Gazella subgutturosa	4	•		•	•												
Gazella cuvieri	2	•			•												
Gazella leptoceros	2	•			•												

continued

Appendix 2.1. Continued

	n	cytb	12S	16S	cox III	κc	βc	αlac	CYP19	Lf	βsn1	pkcт	scf	st	tg	tnf	tt
Gazella dorcas	2	•			•												
Gazella gazella	2	•			•												
Gazella spekei	2	•			•												
Gazella dama	4	•			•						•	•					
Gazella granti	8	•	•		•	•	•	•	•	•							
Gazella soemmerringii	2	•			•												
Gazella rufifrons	2	•			•												
Gazella thomsonii	12	•	•	•	•	•					•	•	•	•	•	•	•
Hippotragus equinus	6	•	•	•		•					•	•					
Hippotragus leucophaeus	1	•															
Hippotragus niger	15	•	•	•		•	•	•	•			•	•	•	•	•		•
Addax nasomaculatus	3	•	•	•													
Oryx leucoryx	3	•	•	•													
Oryx dammah	6	•	•	•		•					•	•					
Oryx gazella	7	•	•	•		•					•	•					
Connochaetes gnou	7	•	•	•		•	•				•	•					
Connochaetes taurinus	3	•	•	•													
Beatragus hunteri	6	•	•	•		•					•	•					
Damaliscus lunatus	13	•	•	•	•	•					•	•	•	•	•	•	•
Damaliscus pygargus	7	•	•	•		•		•	•	•							
Alcelaphus buselaphus	6	•	•	•		•					•	•					
Alcelaphus lichtensteini	4	•	•	•			•										
Ammotragus lervia	3	•	•			•											
Budorcas taxicolor	2	•				•											
Hemitragus hylocrius	1	•															
Ovis ammon	3	•	•	•													
Ovis aries	15	•		•	•	•	•	•	•	•	•	•	•	•	•		•
Ovis vignei	2	•	•														
Ovis nivicola	2	•	•														
Ovis canadensis	3	•	•	•													
Ovis dalli	4	•	•	•		•											
Pseudois nayaur	4	•	•	•		•											
Pseudois schaeferi	2	•	•														
Hemitragus jemlahicus	4	•	•	•		•											
Capra sibirica	4	•	•	•		•											
Capra aegagrus	3	•	•			•											
Capra caucasica	1	•															
Capra cylindricornis	3	•	•	•													
Capra falconeri	4	•	•	•		•											
Capra hircus	15	•	•	•	•	•		•			•	•	•	•	•		•
Capra ibex	3	•	•			•											
Capra pyrenaica	1	•															
Capra nubiana	2	•				•											
Capricornis crispus	4	•	•	•		•											
Capricornis sumatrensis	2	•				•											
Naemorhedus caudatus	1	•															
Naemorhedus goral	4	•	•	•		•											
Ovibos moschatus	12	•	•	•		•	•				•	•	•	•	•	•	•
Hemitragus jayakari	1	•															
Myotragus balearicus	2	•	•														
Oreamnos americanus	5	•	•	•		•	•										
Pantholops hodgsonii	5	•	•	•		•	•										
Rupicapra pyrenaica	1	•															
Rupicapra rupicapra	2	•				•											

JONATHAN H. GEISLER,
JESSICA M. THEODOR,
MARK D. UHEN,
AND SCOTT E. FOSS

3

Phylogenetic Relationships of Cetaceans to Terrestrial Artiodactyls

TEN YEARS AGO, including a chapter on cetaceans in a volume on artiodactyls was unthinkable. At that time, most studies had found that among the extant mammalian orders, cetaceans were closest to Perissodactyla, and artiodactyls were the most basal branch of an "ungulate" clade (e.g., Novacek, 1986; Prothero et al., 1988; Thewissen, 1994). The prevailing view was that the Mesonychia, an extinct group of carnivorous and / or omnivorous hoofed mammals, was the sister group to Cetacea (Van Valen, 1966; Thewissen, 1994), and McKenna (1975) recognized this relationship by placing both taxa in Mirorder Cete. The grouping of cetaceans with mesonychians was predominantly supported by dental characters (see O'Leary, 1998a) but also by a few basicranial features (Geisler and Luo, 1998; Luo and Gingerich, 1999).

As recently as 1999, most of us were quite opposed to the idea that cetaceans were artiodactyls, but one by one, we changed our views as new evidence came in. The evidence that convinced us can be divided into two groups. First was the overwhelming molecular evidence that placed Cetacea not only within Artiodactyla but specifically as the extant sister group to Hippopotamidae (e.g., Irwin and Arnason, 1994; Gatesy et al., 1996, 2002). Initially all of us were skeptical of these new data, and this is reflected in some of our publications from that time (Theodor and Mahoney, 1998; Geisler and Luo, 1998; O'Leary and Geisler, 1999). However, as different molecular laboratories independently discovered genes that supported the hippo plus whale signal (e.g., Gatesy et al., 1996; Gatesy, 1997; Montgelard et al., 1997; Ursing and Arnason, 1998; Nikaido et al., 1999), it became apparent that this clade was strongly supported and that any study that included these molecular data, either with or without morphological data, would come to the same conclusion.

Although molecular data robustly supported a close relationship between whales and hippos, apart from a few behavioral and integument characters (see Gatesy, 1997), there was no morphological support for this hypothesis. In fact, the paleontological and skeletal data available at the time distinctly contradicted it (Geisler and Luo, 1998; O'Leary and Geisler, 1999; Geisler, 2001a). All this changed in 2001, when in the same week two groups announced the discovery of the ankles of protocetid (Gingerich et al., 2001) and pakicetid (Thewissen et al., 2001b) whales. In our opinion, these fossil discoveries are the most convincing evidence for the inclusion of cetaceans within Artiodactyla. The astragali and calcanei of these early cetaceans bear distinct characters that, prior to these discoveries, had only been seen in artiodactyls. When these and other new paleontological data were incorporated into existing morphological data matrices, not only did cetaceans fall inside Artiodactyla but, like the molecular topology, they were close to hippopotamids (Geisler and Uhen, 2003, 2005; Boisserie et al., 2005b; but see Thewissen et al., 2001b; Theodor and Foss, 2005).

EVIDENCE THAT CETACEANS ARE ARTIODACTYLS

Although Gingerich et al. (2001) and Thewissen et al. (2001b) described the artiodactyl features in the ankles of early cetaceans, we think it is appropriate to briefly review those and other characters with a similar taxonomic distribution in this chapter. The astragalus is arguably the most distinctive bone in artiodactyls, and it has a characteristic "double-pulley" shape (Schaeffer, 1947). The proximal "pulley" is a trochleated articular surface for the tibia, and the distal "pulley" is a trochleated articular surface for the navicular (Fig. 3.1). Having both ends of the astragalus concave

transversely also occurs in perissodactyls and mesonychids; however, in artiodactyls, the proximal and distal ends are much more similar in size, shape, depth, and the degree of rotation they allow at their respective joints (Thewissen and Madar, 1999). Part of the symmetry in artiodactyl astragali can be attributed to the presence of a large articular surface for the cuboid that faces distolaterally (Schaeffer, 1947), a trait that also occurs in early cetacean astragali (Gingerich et al., 2001; Thewissen et al., 2001b) (Fig. 3.1). A cuboid facet occurs on the astragalus of some other mammals (e.g., mesonychids, the tapir *Heptodon*), but it is much smaller and faces more laterally than distally (Geisler, 2001a). Similarly, the sustentacular and ectal facets of the artiodactyl astragalus are distinctive. The sustentacular facet is in a far lateral position, reflecting an ankle that is transversely compact, and the ectal facet faces primarily laterally and lacks a supporting lateral process (Schaeffer, 1947). These features occur in pakicetids (Thewissen et al., 2001b) as well as in *Rodhocetus* and *Artiocetus* (Gingerich et al., 2001). The final artiodactyl character in the ankles of archaic cetaceans is a distinct step between the articular facets for the calcaneus and the astragalus on the cuboid (Schaeffer, 1947; Thewissen and Madar, 1999; Geisler, 2001a; Gingerich et al., 2001) (Fig. 3.1).

Although the ankle is the most important source of synapomorphies that group terrestrial artiodactyls and whales, similarities are also found in other anatomical regions. In the skull, the alisphenoid canal is absent in all cetaceans and all terrestrial artiodactyls (except for *Cainotherium*), and the P4 in archaic cetaceans and most artiodactyls has an entocingulum that encircles the protocone (Geisler, 2001a). Two soft tissue characters are synapomorphies of the cetacean and artiodactyl clade: (1) two, instead of three, primary bronchi of the lungs (Slijper, 1979) and (2)

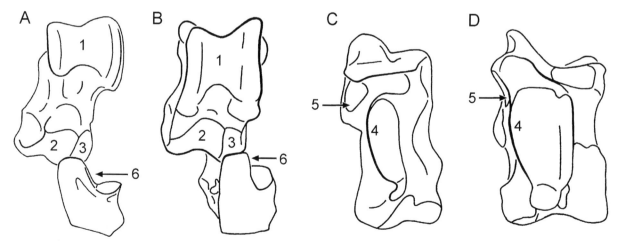

Fig. 3.1. Ankle characters supporting the inclusion of Cetacea within Artiodactyla. (A) Left astragalus and cuboid of the cetacean *Artiocetus clavus* in dorsal view. (B) Right astragalus and cuboid of the terrestrial artiodactyl *Archaeotherium mortoni* (AMNH FM 1277) in dorsal view. (C) Left astragalus of *Artiocetus* in plantar view. (D) Right astragalus of *Archaeotherium* in plantar view. Ankle bones of *Archaeotherium* reversed to ease comparison. Derived features shared by cetaceans and terrestrial artiodactyls include (1) a deeply trochleated articular facet for the tibia (also shared with Perissodactyla), (2) a trochleated navicular facet, (3) a large cuboid facet that faces distally, (4) a sustentacular facet in the far lateral position, and (5) absence of the lateral process and ectal facet facing at least partially laterally. Drawings A and C are based on Gingerich et al. (2001), and B and D are modified from Geisler (2001a). Abbreviation: AMNH FM, American Museum of Natural History (New York, NY) Frick Catalog Number.

sparse cavernous tissue in the penis with erection via relaxation of a retractor muscle (Slijper, 1966).

A PROVISIONAL PHYLOGENY OF ARTIODACTYLA

Our current attempt to resolve the phylogeny of Artiodactyla, including Cetacea, is based on a matrix that combines our independent investigations into this problem conducted over the past 10 years (Theodor, 1996; Geisler and Luo, 1998; Geisler, 2001a; Geisler and Uhen, 2003, 2005; Theodor and Foss, 2005). Although many of the characters employed in the present study developed from work conducted in the 1990s, the current data matrix was created by joining our matrices that were recently published in a *Journal of Mammalian Evolution* Festschrift volume for Dr. William Clemens (Geisler and Uhen, 2005; Theodor and Foss, 2005).

All characters in our previous two studies (Geisler and Uhen, 2005; Theodor and Foss, 2005) were used in the present study, although in some cases characters were combined or modified to accommodate differing descriptions or methods of coding. The combined matrix includes 217 morphological characters: 80 from Geisler and Uhen (2005), 32 from Theodor and Foss (2005), 99 characters shared by Geisler and Uhen (2005) and Theodor and Foss (2005), and 6 new characters (available at personal.georgiasouthern .edu/~geislerj/). The stratigraphic data from Geisler and Uhen (2005) were incorporated, as were the molecular data in the matrix of Gatesy et al. (2002). All combined, the matrix includes 150 unordered morphological characters, 67 ordered morphological characters, 46 transposon characters, 37,084 nucleotide characters, and 738 amino acid characters. Of the 38,040 characters, 8,250 (22%) are parsimony informative. In cases where we had come to incompatible character codings for the 100 characters shared by our previous studies, the following authors took the lead in resolving discrepancies: Foss (bunodont artiodactyls), Geisler (noncetartiodactyl taxa), Theodor (selenodont artiodactyls), and Uhen (cetaceans). When a discrepancy was not caused by a coding error and we had previously examined different specimens, the taxon was coded as polymorphic.

In combining such diverse types of data, we are following the "total evidence" (i.e., simultaneous analysis) approach to developing and testing phylogenetic hypotheses (Kluge, 1989; Nixon and Carpenter, 1996). The most challenging component of our analysis involved the inclusion of stratigraphic data. All previous computer-assisted phylogenetic analyses that have included stratigraphic data (e.g., Uhen, 1998; Bodenbender and Fisher, 2001; Bloch et al., 2001) have used the following heuristic method for finding the most parsimonious trees: (1) parsimony searches are conducted without stratigraphic data; (2) the matrix and the most parsimonious trees from the search are opened in MacClade (Maddison and Maddison, 2000); (3) a MacClade stratigraphic character is added; and (4) the shortest trees are subjected to manual branch swapping and MacClade searches. As acknowledged by opponents (Sumrall and

Brochu, 2003) and proponents (Fisher and Bodenbender, 2003) of stratocladistics alike, even though this is the only published methodology for stratocladistic analysis, it is not effective at finding all most-parsimonious trees.

The present study uses a new method to allow stratigraphic data to be incorporated into automated parsimony searches as executed by PAUP4.0β10 (Swofford, 2002). Our new method builds on the use of a step matrix in calculating the Manhattan Stratigraphic Metric as developed by Siddall (1998) and then modified by Pol and Norell (2001). As with the MacClade stratigraphic character, each distinct chronostratigraphic interval is given a separate character state; in our matrix the Cenozoic is divided into 13 states. The stratigraphic character is ordered and irreversible, and it can be viewed at personal.georgiasouthern .edu/~geislerj/. In theory, our step matrix should be the same as the irreversible character option in PAUP; however, the number of steps for the irreversible character did not always equal the number of steps as a MacClade stratigraphic character.

Although a step matrix has clear advantages over the manual method described above, it cannot completely replace manual searches in MacClade because PAUP does not directly consider ancestors, which can shorten the length of some phylogenetic hypotheses. The present matrix was not subjected to manual swapping in MaClade because autapomorphies for ingroup taxa were not systematically incorporated into our character list, thus making it impossible to differentiate descendants from their immediate ancestors. We used PAUP4.0β10 to conduct analyses of our entire data matrix. Our first set of analyses included 1,000 heuristic searches with a random taxon addition sequence and TBR (tree bisection and reconnection) branch swapping. Only 20 trees were saved per search to reduce computational time. This was followed by TBR branch swapping on all most-parsimonious trees from the 1,000 heuristic searches, without a limit on the number of trees saved.

Our phylogenetic analysis, which included morphological, molecular, and stratigraphic data, yielded 16,472 most-parsimonious trees, each 33,815 steps in length. Despite the large number of trees, the strict consensus is fairly well resolved with 89% of nodes resolved (Fig. 3.2). The strict consensus includes monophyletic Perissodactyla, Mesonychia, Mesonychidae, Hapalodectidae, Cetacea, Odontoceti, Autoceta, Hippopotamidae, Ruminantia, Agriochoeridae, Xiphodontoidea, Camelidae, Suidae, Suoidea, Entelodontidae, and an artiodactyl clade that includes Cetacea but not mesonychians (Fig. 3.2: node 1). Surprisingly the widely recognized families Anthracotheriidae, Tayassuidae, Protoceratidae, Oromerycidae, and the Superfamily Oreodontoidea were each found to be paraphyletic. We suspect this is an artifact of our character sampling, which focused on higher-level artiodactyl relationships, and expect that monophyly of these groups would be obtained if more synapomorphies of these groups were included.

As in Geisler and Uhen (2003, 2005) but unlike Theodor and Foss (2005), our analysis supported a clade including

Table 3.1. Character list for phylogenetic relationships of cetaceans to terrestrial artiodactyls

VASCULAR

1. *Medial edge of ectotympanic bulla.* Not notched (0); has notch and sulcus for the internal carotid artery (1) (modified from Webb and Taylor, 1980; Geisler, 2001b: GU1).

2. *Transpromontorial sulcus for the internal carotid artery.* Present, forms anteroposterior groove on promontorium, medial to the fenestrae rotunda and ovalis (0); absent (1) (Cifelli, 1982b; Thewissen and Domning, 1992: GU2, TF74).

3. *Internal carotid foramen.* Absent or confluent with piriform fenestra (0); present at basisphenoid/basioccipital suture with lateral wall of foramen formed by both of these bones and thus separated from the piriform fenestra (1) (Geisler and Luo, 1998: GU3, TF73).

4. *Sulcus on promontorium for proximal stapedial artery.* Present, forms a groove that branches from the transpromontorial sulcus antero-medial to the fenestra rotunda and extends to the medial edge of the fenestra vestibuli (0); absent (1) (Cifelli, 1982b; Thewissen and Domning, 1992: GU4, TF75).

5. *Foramen for ramus superior of stapedial artery.* Present (0); absent (1) (modified from Novacek, 1986; Thewissen and Domning, 1992: GU5, TF76).

6. *Position of foramen for ramus superior of stapedial artery.* Lateral to epitympanic recess (0); anterolateral to epitympanic recess, adjacent to ventrally convex portion of the tegmen tympani (1). Cannot be scored for taxa that lack this foramen (modified from Geisler and Luo, 1998; O'Leary and Geisler, 1999: GU6).

7. *Size of postglenoid foramen (ordered).* Large, much larger than fenestra vestibuli of petrosal (0); small, slightly larger than or equal in size to the fenestra vestibuli (1); absent (2) (modified from Geisler and Luo, 1998; O'Leary and Geisler, 1999: GU7, TF68).

8. *Position of postglenoid foramen.* Enclosed entirely by the squamosal (0); situated on petrosal/squamosal suture, and if bulla present, a secondary ventral opening between the bulla and the squamosal may form (1) (modified from Geisler and Luo, 1998; O'Leary and Geisler, 1999: GU8).

9. *Mastoid foramen.* Present, skull in posterior view (0); absent (1) (see MacPhee, 1994: GU9, TF89).

10. *Posttemporal canal (for arteria diploetica magna, also called percranial foramen).* Present, occurs at petrosal/squamosal suture with skull in posterior view, and the canal continues within the petrosal/squamosal suture (0); absent (1) (Wible, 1990; MacPhee, 1994: GU10, TF91).

OTIC REGION AND SURROUNDING FEATURES

11. *Subarcuate fossa of petrosal.* Present and deep, is enclosed by a bony arch of a semi-circular canal (0); present but is not deep (1); absent (2) (Novacek, 1986; Norris, 1999, 2000: GU11: TF90).

12. *Shape of tegmen tympani (ordered).* Uninflated, forms lamina lateral to facial nerve canal (0); inflated, forms barrel-shaped ossification lateral to the facial nerve canal (1); hyperinflated, transverse width of tegmen tympani greater than or equal to width of promontorium (2) (modified from Cifelli, 1982b; Geisler and Luo, 1998; Luo and Gingerich, 1999; O'Leary and Geisler, 1999: GU12, TF70).

13. *Anterior process of petrosal.* Absent (0); present, anterior edge of tegmen tympani far anterior to edge of promontorium (1) (Geisler and Luo, 1998; Luo and Gingerich, 1999: GU13, TF71).

14. *Fossa for tensor tympani muscle.* Shallow anteroposteriorly elongate fossa (0); circular pit, no groove (1); circular pit with deep tubular anterior groove (2); long narrow groove between tegmen tympani and promontorium (3) (Geisler and Luo, 1998; Luo and Gingerich, 1999: GU14, TF72).

15. *Stylomastoid foramen.* Complete, ectotympanic contacts both the tympanohyoid and the petrosal (0); incomplete, ectotympanic contacts tympanohyoid laterally and petrosal medially, in some cases ectotympanic separated from petrosal by a narrow fissure (1) (modi-

fied from Geisler and Luo, 1998; O'Leary and Geisler, 1999; Luo and Gingerich, 1999: GU15, TF86).

16. *Articulation of pars cochlearis with basisphenoid/basioccipital.* Present (0); absent (1) (Thewissen and Domning, 1992: GU16, TF77).

17. *Ectotympanic.* Simple ring (0); medial edge expanded into bulla (1). Cannot be scored for taxa in which the ectotympanic is not preserved (derived from Novacek, 1977; MacPhee, 1981: GU17, TF78).

18. *Ectotympanic bulla.* Contains middle middle ear space only (0); medial side is pachyosteosclerotic and forms an involucrum (1); houses middle ear space and highly cancellous bone (2) (Gentry and Hooker, 1988; Thewissen, 1994; Luo, 1998: TF79, GU18 and 19).

19. *Pterygoid sinus fossa (ordered).* Absent or so small that it is not clearly differentiated from the middle ear cavity (0); present and extends slightly anterior to the middle ear cavity (1); present and large, forms a large trough that extends toward the interorbital region (2) (GU196).

20. *Lateral furrow of tympanic bulla.* Absent (0); present, forms a groove on the lateral surface of the ectotympanic bulla anterior to the base of the sigmoid process (1) (Geisler and Luo, 1998; Luo and Gingerich, 1999: GU20).

21. *Ventral inflation of ectotympanic bulla (ordered).* Absent, ventral edge of bulla dorsal to ventral edge of occipital condyles (0); intermediate, edge of bulla at same level as occipital condyles (1); present, ventral edge of bulla ventral to occipital condyles (2) (Geisler, 2001b: GU21).

22. *Posterior extension of bulla.* Absent, stylohyoid does not rest in notch on posterior edge of bulla (0); present, bulla expanded around stylohyoid forming notch on posterior edge of bulla (1); bulla extends posterior to stylohyoid medially (2); bulla extends posterior to stylohyoid laterally (3); dorsal end of stylohyoid completely enveloped or nearly so by bulla (4) (modified from Gentry and Hooker, 1988; Geisler, 2001b: GU22, TF80).

23. *Orientation of stylohyoid.* Ventral or ventrolateral, may rest in notch on the posterior edge of a tympanic bulla (0); anteroventral, may rest in longitudinal furrow on ventral surface of a tympanic bulla (1) (Geisler, 2001b: GU23, TF81).

24. *Articulation of ectotympanic bulla to squamosal (ordered).* Broad articulation with medial base of postglenoid process (0); broad contact with the entoglenoid process (1); narrow contact with a transversely narrow entoglenoid process (2); contact absent (3) (Geisler and Luo, 1998; Luo and Gingerich, 1999: GU24).

25. *Contact between exoccipital and ectotympanic bulla.* Absent (0), present (1) (Geisler and Luo, 1998; Luo and Gingerich, 1999: GU25).

26. *Sigmoid process (homologous to anterior crus of tympanic ring).* Absent (0); present, forms transverse plate that projects dorsolaterally from the anterior crus of the ectotympanic ring and forms the anterior wall of the external auditory meatus (1) (modified from Thewissen, 1994; Geisler and Luo, 1998; Luo and Gingerich, 1999: GU26, TF82).

27. *Morphology of sigmoid process.* Thin and transverse plate (0); broad and flaring, base of the sigmoid process forms dorsoventral ridge on lateral surface of ectotympanic bulla (1) (Geisler and Luo, 1998; Luo and Gingerich, 1999: GU27).

28. *Ectotympanic part of the meatal tube (ordered).* Absent (0); present but short, length of tube <30% the maximum width of the bulla (1); present and long, length >60% maximum width of bulla (2) (Geisler and Luo, 1998: GU28, TF83).

29. *Basioccipital crests (falcate processes).* Absent (0); present, form ventrolaterally flaring basioccipital processes (1) (derived from Barnes, 1984; modified from Thewissen, 1994; Geisler and Luo, 1998: GU29, TF92).

30. *Hypoglossal foramen (ordered).* Closer to occipital condyle (0); equidistant from occipital condyle and jugular foramen (1); closer to

Table 3.1. **Continued**

jugular foramen (2); confluent with the jugular foramen (3) (modified from Thewissen et al., 2001b: GU193: TF84).

31. Condyloid foramen. Absent (0); present, separate from hypoglossal foramen (1) (Geisler and Luo, 1998: TF85).

GLENOID, POSTGLENOID, AND TEMPORAL FEATURES

32. *Paroccipital process* (ordered). Short, in posterior view distal end terminates dorsal to ventral edge of occipital condyle (0); intermediate size, extends just ventral to ventral edge of condyle (1); elongate, terminates far ventral to occipital condyle (2) (Geisler, 2001b: GU30).

33. *Posterior edge of squamosal* (ordered) Flat (0); sharply upturned (1); sharply upturned and bears dorsally projecting process (2) (modified from Gentry and Hooker, 1988: GU31).

34. *Lambdoidal crest.* Present (0); absent or forms low ridge (1) (Geisler, 2001b: GU32).

35. *Minimum width of intertemporal region.* Wide, width is more than half the maximum width of the palate, teeth included (0); narrow, intertemporal width narrower than half the palatal width (1) (Thewissen et al., 2001b: GU187).

36. *Sagittal crest size* (ordered). Absent or barely present, dorsoventral thickness of crest < 7% of dorsoventral height of braincase (measured from ventral edges of condyles to dorsalmost point of supraoccipital) (0); small, 10% < sagittal crest thickness < 15% of braincase height (1); substantial, 20% < sagittal crest thickness < 33% of braincase height (2); dorsally expanded, 39% < sagittal crest thickness < 52% of braincase height (3) (Geisler, 2001b: GU33).

37. *Sagittal crest shape.* Single crest (0); a single crest posteriorly but bifurcates into two crests in the intertemporal region, the crests diverge anteriorly and lead to the postorbital processes, or homologous regions (1); two crests along entire intertemporal region, one on each side of the sagittal plane (2) (Theodor and Foss, 2005: T1999: TF60). Cannot be scored for taxa that lack a sagittal crest.

38. *Dorsal edge of braincase, relative to occlusal plane* (ordered). In lateral view, slopes posterodorsally (0); approximately level relative to upper toothrow (1); curves posteroventrally (2) (Geisler, 2001b: GU34).

39. *Facial nerve sulcus distal to stylomastoid foramen.* Absent (0); anterior wall of sulcus formed by squamosal (1); anterior wall formed by mastoid process of petrosal (2); anterior wall formed by meatal tube of ectotympanic (3) (modified from Geisler and Luo, 1998; O'Leary and Geisler, 1999: GU35).

40. *Length of mastoid process of petrosal* (ordered). Ventral portion absent (0); ventral portion short, <70% of the anteroposterior length of promontorium (1); elongate, >100% length of promontorium (2); hypertrophied, >200% length of promontorium (3) (modified from Geisler and Luo, 1996; Luo and Marsh, 1996; Geisler and Luo, 1998; Geisler, 2001b: GU36).

41. *Lateral exposure of mastoid process of petrosal* (ordered). Present between exoccipital and squamosal (0); constricted, dorsal part of exposure forms lamina (1); absent (2) (modified from Geisler and Luo, 1996; Luo and Marsh, 1996: GU37).

42. *Mastoid process of petrosal.* Exposed externally on posterior face of braincase as a triangle between the lambdoidal crest of the squamosal dorsolaterally, the exoccipital ventrally, and the supraoccipital medially (0); not exposed posteriorly, lambdoidal crest of squamosal in continuous contact with exoccipital and supraoccipital (1) (Geisler, 2001b: GU38, TF88).

43. *Angle of suture of squamosal with petrosal or exoccipital, skull in ventral view* (ordered). Very large, forms a 147° angle with the sagittal plane (0); large, forms an angle between 127° and 125° (1); angle between 111° and 105° (2), angle < 100° (3) (Geisler, 2001b: GU39).

44. *Length of external auditory meatus of the squamosal* (ordered). Very short or absent, length < 4% of half the basicranial width (0); short, length between 19% and 23% (1); intermediate length be-

tween 29% and 36% (2); long, length between 41% and 45% (3); very long, length > 52% (4) (Geisler, 2001b: GU40).

45. *Postglenoid process orientation.* Forms transversely oriented and ventrally projecting ridge (0); ventrally projecting prong, roughly oval in coronal section (1) (Geisler, 2001b: GU41).

46. *Postglenoid process shape.* Smooth caudally (0); indented by external auditory meatus (1) (Thewissen et al., 2001b: TF67).

47. *Glenoid fossa position.* Medially bordered by crest, or elevated out of plane of basicranium (0); medially continuous with middle ear cavity (1) (Thewissen et al., 2001b: TF69).

48. *Glenoid fossa* (ordered) Concave longitudinally (0); flat longitudinally (1); convex longitudinally (2) (Geisler, 2001b: GU42).

49. *Posterolateral border of glenoid fossa.* Slightly downturned ventrally or flat (0); conspicuously upturned with concave surface facing posteroventrally (1) (Geisler, 2001b: GU43).

50. *Foramen ovale.* Anteromedial to glenoid fossa (0); medial to glenoid fossa (1) (derived from Zhou et al., 1995; Geisler and Luo, 1998: GU45).

51. *Ectopterygoid process of the alisphenoid.* Present (0); absent (1) (Novacek, 1986: GU189, TF62).

52. *Dorsoventral thickness of zygomatic process of the squamosal* (ordered). Small, 7% < dorsoventral thickness of zygomatic process < 15% of dorsoventral height of braincase (measured from ventral edges of condyles to dorsalmost point of supraoccipital) (0); intermediate, 17% < dorsoventral thickness of zygomatic process < 28% of dorsoventral height of braincase (1); dorsoventrally deep, 32% < dorsoventral thickness of zygomatic process < 40% of dorsoventral height of braincase (2) (Geisler, 2001b: GU47).

53. *Zygomatic portion of jugal.* Directed posterolaterally (0); directed posteriorly (1) (Geisler, 2001b: GU48).

54. *Preglenoid process.* Absent (0); present, forms transverse, ventrally projecting ridge at anterior edge of glenoid fossa (1) (modified from Thewissen, 1994; Geisler and Luo, 1998: GU44, TF66).

55. *Posterior edge of foramen ovale.* Formed by the alisphenoid (0); formed by the petrosal and or tympanic bulla (1); formed by squamosal (2); formed by squamosal and alisphenoid (3) (Geisler, 2001b: GU46, TF65).

ORBITAL MOSAIC AND FORAMINA

56. *Alisphenoid canal (alar canal).* Absent (0); present (1) (Novacek, 1986; Thewissen and Domning, 1992: GU49, TF64).

57. *Foramen rotundum.* Absent, maxillary division of trigeminal nerve exits skull through the sphenorbital fissure (0); present (1) (Novacek, 1986; Thewissen and Domning, 1992: GU50, TF63).

58. *Contact of frontal and maxilla in orbit.* Absent (0); present (1) (Novacek, 1986; Thewissen and Domning, 1992: GU52, TF55).

59. *Contact of frontal and alisphenoid.* Present (0); absent, separated by orbitosphenoid or parietal (1) (Prothero, 1993: GU51).

ORBITAL POSITION AND SURROUNDING FEATURES

60. *Supraorbital horns* (ordered). Absent (0); present, unbranched (1); present, branched (2) (Janis and Scott, 1987; Scott and Janis, 1993: GU53).

61. *Supraorbital process.* Absent, region over orbit does not project lateral from sagittal plane (0); present, laterally elongate and tabular (1) (derived from Barnes, 1984; Geisler and Luo, 1998: GU54).

62. *Foramen for frontal diploic vein.* Absent (0); present (1) (Thewissen and Domning, 1992: GU186, TF59).

63. *Postorbital process of jugal* (ordered). Absent (0); present but does not contact frontal (1); present and contacts postorbital process of frontal forming postorbital bar (2) (modified from Gentry and Hooker, 1988: GU55, TF56).

64. *Ventral edge of orbit flared laterally.* Absent (0); present (1) (Geisler, 2001b: GU56).

65. *Lacrimal tubercle.* Absent (0); present, situated on anterior edge of orbit adjacent to the lacrimal foramen (1) (Novacek, 1986: GU58, TF54).

continued

Table 3.1. **Continued**

66. *Position of anterior edge of orbit relative to toothrow* (ordered). Over P^4 or P^4/M^1 division (0); over M^1 or M^1/M^2 division (1); over M^2 or M^2/M^3 division (2); over or posterior to M^3 (3) (Geisler, 2001b: GU57).

67. *Lacrimal foramina* (ordered). Two (0); one (1); one and highly reduced (2); absent (3) (modified from Gentry and Hooker, 1988; Geisler, 2001b: GU59).

68. *Facial portion of lacrimal relative to the orbit* (ordered). Restricted to orbital rim, 15% < maximum anteroposterior length of facial portion of lacrimal < 30% of the maximum anteroposterior diameter of the orbit (0); small facial portion present, 30% < length of facial portion of lacrimal < 67% of the orbital diameter (1); moderate facial portion present, 70% < length of facial portion of lacrimal < 93% of the orbital diameter (2); large facial portion present, 100% < length of facial portion of lacrimal < 180% of the orbital diameter (3) (Geisler, 2001b: GU60).

69. *Nasolacrimal contact on face.* Absent (0); present (1) (Thewissen et al., 2001b: TF52).

70. *Antorbital pit in lacrimal.* Absent (0); present (1) (Janis and Scott, 1987; Scott and Janis, 1993: GU61).

71. *Fenestra in rostrum at junction of lacrimal, nasal, and maxilla.* Absent (0); present (1) (Scott and Janis, 1993: GU62).

72. *Position of anterior edge of Jugal* (ordered). Anterior to or over P^4 (0); over M^1 or M^1/M^2 division (1); over M^2 or M^2/M^3 division (2); over or posterior to M^3 (3) (Geisler, 2001b: GU63).

FACE AND PALATE

73. *Elongation of face* (ordered). Absent and face short, face (defined as part of skull anterior to anterior edge of orbit) < 85% of the remaining posterior part of the skull (defined as part between anterior edge of orbit and posterior edge of occipital condyle) (0); long, 90% < face < 170% remaining part of skull (1); elongate, 190% < face < 230% remaining part of skull (2) (Geisler, 2001b: GU64).

74. *Anterior opening of infraorbital canal* (ordered). Over M^1 or P^4 (0); at level between P^3 and P^4 (1); anterior or over P^3 (2) (Geisler, 2001b: GU65, TF57).

75. **GU66:** *Lateral surface of maxilla.* Flat or slightly concave or convex (0); highly concave (1) (Geisler, 2001b: GU66).

76. *Posterior edge of nasals.* Terminate anterior to orbit (0); extended posteriorly, terminate posterior to the anterior edge of the orbit (1) (Geisler, 2001b: GU67, TF58).

77. *Palatine fissures* (ordered). Enlarged, transverse distance between lateral edges of palatine fissures > 53% of the width of the palate in the same transverse plane (0); small, transverse distance between lateral edges of palatine fissures < 48% of the width of the palate in the same transverse plane (1); absent (2) (Geisler, 2001b: GU68).

78. *Palate.* Flat (0); vaulted, portion along sagittal plane well dorsal to lateral edge (1) (Geisler, 2001b: GU69).

79. *Embrasure pits on palate.* Absent (0); present, situated medial to the toothrow, accommodate the cusps of the lower dentition when the mouth is closed (1) (modified from Thewissen, 1994; Geisler and Luo, 1998: GU70, TF51).

80. *Floor of nasopharyngeal duct.* Not ossified (0); ossified (1) (Thewissen, 1994; Thewissen et al., 2001b: GU188, TF61).

81. *Posterior margin of external nares* (ordered). Anterior to or over the canines (0); immediately anterior to P^1 (1); between P^1 and P^2 (2); posterior to P^2 (3) (Geisler and Luo, 1998: GU71).

MANDIBLE

82. *Angular process of mandible.* No dorsal hook (0); dorsal hook present (1) (Gentry and Hooker, 1988: GU72).

83. *Mandibular foramen.* Small, maximum height of opening 25% or less the height of the mandible at M_3 (0); enlarged and continuous with a large posterior fossa, maximum height > 50% the height of the mandible at M_3 (1) (modified from Thewissen, 1994; Geisler and Luo, 1998: GU74, TF44).

84. *Elongation of coronoid process* (ordered). Absent, 50% < dorsal height of coronoid process < 90% of the width of the coronoid process (measurement taken at posterior base of coronoid process, immediately anterior to mandibular condyle of process) (0); moderate, 110% < dorsal height < 180% of its width (1); substantial, 200% < dorsal height < 270% of its width (2) (Geisler, 2001b: GU75).

85. *Height of coronoid process* (ordered). Low, 150% < height of coronoid (measured from ventral edge of mandible to dorsal edge of coronoid) < 190% of the depth of the mandible at M_3 (0); high, 210% < height of coronoid < 310% of the depth of the mandible at M_3 (1); very high, 320% < height of coronoid < 440% of the depth of the mandible at M_3 (2) (Geisler, 2001b: GU76).

86. *Height of dentary condyle* (ordered). Low, 60% < height of condyle (measured from ventral edge of mandible to dorsal edge of condyle but excluding any portion of the mandible that extends ventrally below the edge of the mandible at M_3) < 140% of the depth of the mandible at M_3 (0); moderately elevated, 160% < height of condyle < 230% of the depth of the mandible (1); well elevated, 240% < height of condyle < 300% of the depth of the mandible (2) (Geisler, 2001b: GU78).

87. *Angle of mandible.* Distal end at same level as ventral edge of dentary below molars (0); forms distinct flange that projects posteroventrally well below ventral edge of dentary below molars but that flange is in the same parasagittal plane as the ramus (1); projects ventrolaterally (2); projects caudally or ventrocaudally (3) (modified from Gentry and Hooker, 1988: TF46, GU73).

88. *Ramus of mandible.* Approximately same dorsoventral thickness from M_1 to M_3 (0); deepens posteriorly from M_1 to M_3 (1) (modified from Gentry and Hooker, 1988: GU79, TF48).

89. *Mandibular symphysis* (ordered). Unfused and no contact between left and right halves (0); contact and suture present (1); fused (2) (Pickford, 1983: GU80, TF45).

INCISORS AND CANINES

90. *Lower incisors.* Apex of cusp pointed or narrower than base (0); spatulate, apex of cusp wider than base (1); peg-shaped, width of base equal to width of tip of tooth (2); tusklike (3); large peg with basal flare (occasionally massively spatulate) (4) (modified from Geisler, 2001b; Theodor and Foss, 2005: TF22, GU83).

91. *Upper incisors.* Present and arranged in transverse arc (0); incisors aligned longitudinally with intervening diastemata (1); incisors absent (2) (modified from Prothero et al., 1988; Thewissen, 1994: TF47, GU81 and 82).

92. *Size of I^3.* Similar in size to I^{1-2} (0); I^3 distinctly larger than I^{1-2} (1) (Theodor and Foss, 2005: TF1).

93. *Orientation of upper canines.* Vertically oriented (0); laterally splayed (1) (Theodor and Foss, 2005: TF2).

94. *Roots of upper canine.* Single-rooted (0); double-rooted (1) (Theodor and Foss, 2005: TF3).

95. *Elongation and transverse compression of upper canines.* Absent (0); present (1). If sexually dimorphic, score for males only (modified from Webb and Taylor, 1980: GU84).

96. *Lower canine size* (ordered). Larger than incisors (0); approximately same size as incisors (1); smaller than incisors (2) (Geisler, 2001b: GU85, TF23).

97. *Lower canine shape.* Oval in cross section (0); triangular or D-shaped and pointing anteriorly (if D-shaped, rounded portion directed anteriorly) (1) (modified from Gentry and Hooker, 1988: GU86, TF24).

98. *Lower canine.* Consists of a distinct crown and root (0); hypsodont, no clear boundary between crown and root (1) (Pickford, 1983: GU87).

PREMOLARS

99. *P^1* (ordered). Absent (0); present, one-rooted (1); present, two-rooted (2) (Zhou et al., 1995; O'Leary, 1998a: GU88, TF4).

Table 3.1. **Continued**

100. *P₁*. Present, has a single low cusp that is transversely compressed (0); caniniform, single cusp high and pointed (1); molariform, two main cusps in trigonid followed by compressed talonid basin (2); absent (3) (modified from Gentry and Hooker, 1988: TF25, GU89 and 90).

101. *Morphology of dP₂*. Simple, conical (0); trenchant tooth with small talonid (1); with central conid and anterior and posterior accessory cuspids and small talonid with central cusp (protoconulid), total number of cuspids is three to four (2) (Theodor and Foss, 2005: TF41).

102. *P³ roots*. Three (0); two (1) (Zhou et al., 1995: GU91, TF5).

103. *Length of P₃*. Less than or equal to M/1 length (0); 120% M/1 length < P/3 length < 150% (1) (Gentry and Hooker, 1988: TF27).

104. *Morphology of dP³*. Simple with metacone absent (0); three distinct cusps forming triangle in occlusal view, with single cusp anteriorly and molariform posteriorly (1); buccolingually compressed, major cusp with accessory denticles arranged mesiodistally (2) (Theodor and Foss, 2005: TF20).

105. *P₃ metaconid*. Absent (0); present (1) (Thewissen and Domning, 1992: GU92, TF26).

106. *Morphology of dP₃*. Simple, conical (0); trenchant tooth with small talonid (i.e., metaconid) (1); with central conid and anterior and posterior accessory cuspids and small talonid (2) (Theodor and Foss, 2005: TF1).

107. *P⁴ protocone*. Present (0); absent (1) (Thewissen, 1994: GU93, TF6).

108. *P⁴ paracone*. Equal or subequal to height of paracone of M¹ (0); greater than twice the height of M¹ paracone (1) (Thewissen, 1994: GU94, TF7).

109. *P⁴ metacone*. Absent (0); present (1) (Geisler, 2001b: GU95, TF8).

110. *P⁴ entocingulum*. Absent or very small (0); present, partially or completely surrounds the base of protocone (1) (Geisler, 2001b: GU96, TF9).

111. *Morphology of dP⁴*. Premolariform (0); molariform (1) (Theodor and Foss, 2005: TF21).

112. *P₄ metaconid*. Absent (0); present (1) (Thewissen and Domning, 1992: GU97, TF28).

113. *Deciduous P₄*. Resembles M₁ (0); additional cusp on paracristid, six-cusped (1); elongate, buccolingually compressed with accessory cuspules (2) (derived from Gentry and Hooker, 1988; Luckett and Hong, 1998; Theodor and Foss, 2005: TF43, GU98).

MOLARS

114. *M³* (ordered). Present, larger than M² (0); present, approximately equal (1); reduced, maximum mesiodistal length <60% the length of M² (2); absent (3) (modified from Zhou et al., 1995; Geisler and Luo, 1998: GU103, TF17).

115. *Molars*. Have stylar shelves (0); stylar shelves absent (1) (O'Leary and Geisler, 1999; GU107).

116. *Ectocingula on upper molars*. Present (0); absent (1) (O'Leary, 1998a: GU108, TF18).

117. *M¹ parastyle* (ordered). Absent (0); weak (1); moderate to strong (2) (Zhou et al., 1995; O'Leary, 1998a: GU99, TF10).

118. *Parastyle or preparacrista position*. Lingual position, in line with or lingual to a line that connects the paracone or metacone (0); labial position, labial to the line that connects the paracone and metacone (1) (Geisler, 2001b: GU106).

119. *Mesostyle* (ordered). Absent (0); present but low (1); present and high (2) (modfied from Thewissen and Domning, 1992).

120. *Mesostyle on upper molars*. Not connected to centocrista (0); mesostyle connected and forms an open "V" shape (1); mesostyle connected and forms an a narrow "U" shape (2) (Theodor and Foss, 2005: TF16). Taxa that lack a mesostyle would be coded as "–."

121. *M² metacone* (ordered). Distinct cusp, subequal to paracone (0); distinct cusp, approximately half the size of the paracone (1); highly reduced, indistinct from paracone (2) (Zhou et al., 1995; O'Leary, 1998a: GU101, TF11).

122. *Paraconule of upper molars* (ordered). Present (0); reduced (1); absent (2) (O'Leary, 1998a: GU110, TF13).

123. *M² metaconule* (ordered). Absent (0); similar in size to paraconule (1); approaching size of protocone (2) (modified from Thewissen et al., 2001b: TF14).

124. *M² hypocone* (ordered). Absent (0); present and small (1); present and similar in size to protocone (2) (modified from Thewissen et al., 2001b: TF15).

125. *Shape of lingual cusps on upper molars*. Conical (0); crescent-shaped; postprotocrista and premetaconule crista well developed and labially directed (1) (Theodor and Foss, 2005: TF19).

126. *M₃ hypoconulid* (ordered). Long, protrudes as separate distal lobe (0); reduced, does not protrude substantially beyond rest of talonid (1); absent (2) (Thewissen, 1994: GU105, TF38).

127. *Protoloph*. Absent (0); present (1) (Hooker, 1989: GU112).

128. *Metaloph*. Absent (0); present (1) (Hooker, 1989: GU113).

129. *Lingual cingulid on molars*. Poorly defined or absent (0); continuous to mesial to distal extreme (1) (O'Leary, 1998a: GU114, TF39).

130. Lower molar paraconids. Present (0); absent (1) (O'Leary and Geisler, 1999: TF30).

131. *Lower molar paraconid or paracristid position*. Cusp lingual or crest winds lingually (0); cusp anterior or crest straight mesodistally on lingual margin (1) (O'Leary, 1998a: GU115, TF31).

132. *Crest connecting entoconid and hypoconid to the exclusion of the hypoconulid on lower molars (hypolophid)*. Absent (0); present (1) (Gentry and Hooker, 1988: GU117).

133. *Talonid basins*. Broad, hypoconid and entoconid present (0); compressed, with hypoconid displaced lingually and centered on the width of the tooth, entoconid absent (1) (modified from Zhou et al., 1995; description based on O'Leary and Rose, 1995a; Geisler and Luo, 1998: GU123).

134. Entoconid on M/1 and M/2. Present (0); absent (1) (Thewissen et al., 2001b: TF35).

135. M/1 and M/2 metaconid and entoconid. Cuspate (0); elongate, form mesiodistal crest (1) (Thewissen et al., 2001b: TF36).

136. *M₁ and M₂ hypoconulid*. Absent (0); present (1) (Gentry and Hooker, 1988: GU116, TF34).

137. *Metaconids on lower molars*. Most molars have metaconids (0); most molars lack metaconids or occasionally present as swelling on lingual side of protoconid (1) (modified from Zhou et al., 1995).

138. *Metastylid of lower molars*. Absent (0); present (1) (Gentry and Hooker, 1988: GU118).

139. *Entoconulid of lower molars*. Absent (0); present (1) (Gentry and Hooker, 1988: GU119).

140. *Molar trigonid*. Subequal to height of talonid (0); closer to twice height of talonid or greater (1) (O'Leary, 1998a: GU120, TF33).

141. Elongate shearing facets on molars. Absent (0); present, extending below gum line (1) (O'Leary and Geisler, 1999: TF40).

142. *Loph formation on anterior aspect of lower teeth*. Absent (0); present (1) (Hooker, 1989: GU121).

143. *Reentrant grooves* (ordered). Proximal (0); absent (1); distal (2) (Thewissen, 1994; O'Leary, 1998a: GU122, TF37).

OCCIPITAL CONDYLE AND VERTEBRAL

144. *Occipital condyles*. Broadly rounded in lateral view (0); V-shaped in lateral view, in posterior view the condyle is divided into a dorsal and a ventral half by a transverse ridge (1) (Geisler, 2001b: GU124, TF93).

145. *Anteroventral border of occipital condyle*. Tapers medially (0); flared laterally and ventrally to form stop for ventral movement of the cranium (1) (Geisler, 2001b: GU125).

146. *Odontoid process of axis*. Forms anteriorly pointed peg (0); spoutlike, dorsal surface forms concave trough (1); bears central dorsal ridge that separates two spoutlike troughs (2) (modified from Webb and Taylor, 1980; Geisler, 2001b: GU126, TF94).

continued

Table 3.1. **Continued**

147. *Atlantoid facet of axis vertebra.* Restricted below neural arch or extends slightly dorsal to the base of the neural pedicle (0); extended dorsally at least halfway up neural arch (1) (modified from Webb and Taylor, 1980: GU127, TF95).

148. *Cervical vertebrae* (ordered). Short, length shorter than centra of anterior thoracics (0); long, length of centrum greater than or equal to the centra of the anterior thoracics (1); very long, length closer to twice the length of the anterior thoracics (2) (derived from Gingerich et al., 1995: GU128, TF96).

149. *Arterial canal for vertebral artery in cervical vertebrae 3–6.* Posterior openings exterior to neural canal (0); inside neural canal (1) (Gentry and Hooker, 1988: GU129, TF97).

150. *Revolute zygapophyses of lumbar vertebrae.* Absent, zygapophyses are flat or slightly curved (0); present (1) (Thewissen et al., 2001b: GU192, TF98).

151. *Articulation between sacral vertebrae and illium of pelvis* (ordered). Broad area of articulation between pelvis and S1 and possibly S2 (0); narrow articulation of pelvis with end of transverse process of S1 (1); articulation absent (2) (Geisler and Luo, 1998: GU130, TF101).

152. *Number of sacral vertebrae* (ordered). One (0); two or three (1); four (2); five or six (3). Cannot be scored for taxa that lack articulation of vertebral column to illium (Thewissen and Domning, 1992; Gingerich et al., 1995: GU131, TF99).

153. *First sacral vertebra.* Articulates with or is fused to the second sacral vertebra via pleuropophyses on the transverse processes (0); does not articulate with the second via pleuropophyses (1) (Geisler and Uhen, 2005: GU195).

FORELIMB

154. *Supraspinatus fossa of the scapula.* Large, portion on neck faces laterally and is equal to or larger than the infraspinatus fossa (0); small, portion on neck faces anterolaterally and is smaller than the infraspinatus fossa (1) (Geisler, 2001b: GU133, TF103).

155. *Direction of acromion process of scapula* (ordered). Ventral (0); slightly anteroventral (1); anteroventral (2); anterior (3) (Geisler and Uhen, 2005; GU194).

156. *Acromion process.* Long (0); short (1) (modified from O'Leary and Geisler, 1999: GU132).

157. *Clavicle.* Present (0); absent (1) (Theodor and Foss, 2005: TF100).

158. *Greater tuberosity of humerus.* Low, not above head of humerus (0); greater tuberosity enlarged above head (1); greater tuberosity arched over bicipital groove of humerus (2) (Theodor and Foss, 2005: TF104).

159. *Deltoid tuberosity of the humerus.* Present (0); absent (1) (Theodor, 1996; Thewissen et al., 2001b: GU190, TF105).

160. *Entepicondyle of humerus.* Wide, width 50% or greater than the width of the ulnar and radial articulation facets (0); narrow, 25% or less than the width of the ulnar and radial articulation facets (1) (derived from O'Leary and Rose, 1995b; Geisler and Luo, 1998: GU134, TF106).

161. *Entepicondylar foramen.* Present (0); absent (1) (Thewissen and Domning, 1992: GU135, TF107).

162. *Distal articular surface of humerus.* Restricted by medial edge of trochlea (0); expanded medially past trochlear edge to form convex surface (1) (Gentry and Hooker, 1988: GU136).

163. *Distal humerus intercondylar ridge between capitulum and epicondyle.* Absent (0); present (1) (modified from Gentry and Hooker, 1988: GU137).

164. Supratrochlear foramen of humerus. Absent (0); present (1) (Theodor and Foss, 2005: TF108).

165. *Metacarpal I.* Present (0); absent (1) (Theodor and Foss, 2005: TF117).

166. *Length of olecranon process* (ordered). Short, <10% of total ulnar length (0); long, >20% of ulnar length (1) (derived from O'Leary and Rose, 1995b; O'Leary and Geisler, 1999: GU138, TF109).

167. *Posterior edge of ulna* (ordered). Convex posteriorly (0); straight (1); concave posteriorly (2) (derived from O'Leary and Rose, 1995b: GU139).

168. *Radius and ulna* (ordered). Completely separate (0); fused distally (1); fused completely (2) (Webb and Taylor, 1980: GU140, TF112).

169. *Proximal end of radius* (ordered). Single fossa for edge of trochlea and capitulum of humerus (0); two fossae, for the medial edge of the trochlea and the capitulum (1); three fossae, same as state "0" but with additional fossa for the lateral lip of the humeral articulation surface (2) (Geisler and Luo, 1998: GU141, TF110).

170. *Distal articulation surface of radius* (ordered). Single concave fossa (0); split into scaphoid and lunate fossae (1); three fossae for carpals (2) (derived from O'Leary and Rose, 1995b; Geisler and Luo, 1998; Theodor and Foss, 2005: TF111, GU142).

171. *Centrale.* Present (0); absent (1) (Thewissen, 1994: GU143, TF113).

172. *Lunar position.* Rests equally on magnum and unciform in anterior view (0); lunar rests primarily on unciform in anterior view (1) (Theodor and Foss, 2005; TF115).

173. *Magnum and trapezoid.* Separate (0); fused (1) (Webb and Taylor, 1980: GU144, TF114).

174. *Manus* (ordered). Second digit is longest (0); mesaxonic, axis of symmetry of foot passes along center of digit III (1); paraxonic, axis lies between digits III and IV (2) (O'Leary and Geisler, 1999: GU145, TF116).

175. *Second metacarpal contact with magnum.* Present (0); absent, excluded by proximal end of metacarpal III (1) (Geisler, 2001b: GU146).

176. *Second digit of forelimb* (ordered). Long, distal end of third phalanx terminates distal to distal end of second phalanx of third digits (0); reduced, distal end of third phalanx terminates proximal to distal end of second phalanx of third digit (1); highly reduced, metacarpal forms proximal splint or nodule (2); absent (3) (Geisler, 2001b: GU147). The length of the fifth digit was not included because its variation closely matches that of the second digit. However, these digits were not combined because of some variation in outgroup taxa.

177. *Relative widths of second and fifth metacarpals* (ordered). Sum of minimum widths of second and fifth metacarpals ≥170% minimum width of third metacarpal (0); constricted, 170% > metacarpal widths > 92% minimum width of third metacarpal (1); highly compressed, 92% of third metacarpal ≥ second and fifth metacarpal (2) (modified from Geisler, 2001b; Theodor and Foss, 2005).

178. *Fifth metacarpal.* Present (0); absent (1).

HINDLIMB

179. *Greater trochanter of femur* (ordered). Below level of head of femur (0); approximately same level as head of femur (1); elevated dorsally well beyond head of femur (2) (derived from O'Leary and Rose, 1995b: GU151, TF120).

180. *Third trochanter of femur* (ordered). Present (0); highly reduced (1); absent (2) (Luckett and Hong, 1998; O'Leary and Geisler, 1999: GU152, TF121).

181. *Patellar articulation surface on femur.* Wide (0); narrow (1) (O'Leary and Geisler, 1999: GU153).

182. *Ridges bordering patellar facet.* Equal in height (0); medial border of patellar facet projects beyond (i.e., anteriorly) the lateral border of the facet (1) (Theodor and Foss, 2005: TF122).

183. *Fibula* (ordered). Complete (0); incomplete but has both the proximal and distal end (1), reduced to a distal splint (2) (Webb and Taylor, 1980; Theodor and Foss, 2005 TF155).

184. *Proximal end of astragalus* (ordered). Nearly flat to slightly concave (0); well grooved, but depth of trochlea <25% its width (1); deeply grooved, depth >30% its width (2) (derived from Schaeffer, 1947; O'Leary and Geisler, 1999: GU156, TF126).

185. *Astragalar canal.* Present (0); absent (1) (Shoshani, 1986; Thewissen and Domning, 1992: TF125, GU157).

Table 3.1. Continued

186. *Astragalar head.* Dorsoplantarly mildly convex (0); strongly convex (1) (Thewissen and Madar, 1999: TF128).

187. *Distal calcaneal facet of astragalus.* Indistinct from sustentacular facet (0); distinct (1) (Thewissen and Madar, 1999: TF132).

188. *Tibia and fibula* (ordered). Separate (0); fused proximally (1); fused proximally and distally (2) (Webb and Taylor, 1980: GU154).

189. *Navicular facet of astragalus* (ordered). Convex (0); flat (1); concave mediolaterally (2) (Schaeffer, 1947; Thewissen and Domning, 1992; Geisler and Luo, 1998: GU158, TF127).

190. *Astragalonavicular joint.* Lacks defined plane of rotation (0); rotates in dorsoplantar plane (1); oblique plane (2) (modified after Thewissen and Madar, 1999: TF129).

191. *Distal end of astragalus contacts cuboid* (ordered). Contact absent (0); contact present, articulating facet on astragalus forms a steep angle with the parasagittal plane (1); contact present and large, facet almost forms a right angle with the parasagittal plane (2) (Geisler, 2001b: GU159, TF130).

192. *Long axes of proximal and distal articulating surfaces of astragalus.* If extrapolated, form angle that is obtuse and opens medially (0); parallel, no angle formed (1) (modified from Gentry and Hooker, 1988: GU160, TF134).

193. *Proximal half of lateral surface of astragalus.* Concave (0); flat (1) (modified from Gentry and Hooker, 1988: GU161).

194. *Lateral process of astragalus.* Present, ectal facet of the astragalus faces in the plantar direction, and its distal end points laterally (0); absent, ectal facet faces laterally, and its long axis is parasagittal (1) (Schaeffer, 1947: GU162, TF133).

195. *Sustentacular facet of the astragalus.* Narrow and medially positioned, lateral margin of sustentacular facet of the astragalus well medial to the lateral margin of the trochlea (0); wide and laterally positioned, lateral margin in line with the lateral margin of the trochlea (1) (derived from Schaeffer, 1947; Geisler and Luo, 1998: GU163, TF131).

196. *Sustentacular facet* (ordered). Completely separated from navicular/cuboid facet (0); medial edge of sustentacular facet continuous (1); completely continuous with cuboid/navicular facet (2) (Geisler, 2001b: GU164).

197. *Fibular facet of calcaneum.* Absent (0); flat or simply concave (1); convex (2); concave anteriorly and convex posteriorly (3) (modified from Webb and Taylor, 1980; Thewissen and Madar, 1999: TF124, GU191).

198. *Articulation of calcaneus and cuboid.* Flat, proximal articulating surface of the cuboid in one plane and corresponding surface of the calcaneus faces distally (0); sharply angled and curved, proximal surface of the cuboid has a distinct step between the facets for the calcaneus and astragalus (1) (Thewissen and Madar, 1999; Geisler, 2001b: GU165, TF135).

199. *Cuboid and navicular.* Unfused (0); fused (1) (Webb and Taylor, 1980: GU166, TF136).

200. *Cubonavicular and ectocuneiform.* Separate (0); fused (1) (Webb and Taylor, 1980: GU167).

201. *Ectocuneiform and mesocuneiform.* Separate (0); fused (1) (Webb and Taylor, 1980: GU168, TF137).

202. *Mesocuneiform.* Clearly visible in dorsal (anterior if digitigrade) view (0); completely hidden by the ectocuneiform in dorsal view, or extreme medial edge of mesocuneiform partially visible (1) (derived from Gingerich et al., 2001: GU185).

203. *Pes* (ordered). Mesaxonic, axis of symmetry of foot passes along center of the third digit (0); paraxonic, axis lies between digits 3 and 4 (1); axis passes along center of digit 4 (2) (derived from Gingerich et al., 1990; Thewissen, 1994; O'Leary and Geisler, 1999: GU169, TF138).

204. *First metatarsal* (ordered). Unreduced, length > 50% length of third metatarsal (0); reduced, length < 50% length of third metatarsal (1); highly reduced, metatarsal forms nodule or small splint or is absent (2) (O'Leary and Geisler, 1999: GU170, TF139).

205. *Second digit of hindlimb.* Long, distal end of third phalanx terminates distal to distal end of second phalanx of third digit (0); reduced, distal end of third phalanx terminates proximal to distal end of second phalanx of third digit (1); highly reduced, forms nodule or small splint (2); absent (3); reduced, distal end of phalanx terminates proximal to the distal end of the second phalanx of the third digit because the second metatarsal is 50% of the length of the third metatarsal (4) (Geisler, 2001b: GU171). The length of the fifth digit was not included because its variation closely matches that of the second digit. However, these digits were not combined because of some variation in outgroup taxa.

206. *Relative widths of second and fifth metatarsals* (ordered). Sum of minimum widths of second and fifth metatarsals ≥ 110% minimum width of third metatarsal (0); constricted, 110% > metatarsal widths > 53% minimum width of third metatarsal (1); highly compressed, 53% of third metatarsal ≥ second and fifth metatarsal (2) (modified from Geisler, 2001b; Theodor and Foss, 2005).

207. *Fifth metatarsal.* Present (0); absent (1).

208. *Fusion of third and fourth metatarsals.* Absent (0); present (1) (Webb and Taylor, 1980: GU177, TF141).

209. *Ventral side of distal phalanges of foot and manus.* Distinctly concave (0); flat (1) (O'Leary and Geisler, 1999: GU179, TF119).

210. *Elongation of third metatarsal* (ordered) Absent, 20% < length of third metatarsal < 39% of the length of the femur (0); slight elongation, 47% < length of third metatarsal < 54% of the length of the femur (1); substantial elongation, 63% < length of third metatarsal < 95% of the length of the femur (2) (Geisler, 2001b: GU175).

211. *Keels on distal ends of the metapodials* (ordered) Absent (0); present, restricted to distal and plantar surfaces (1); present and extended onto dorsal surface (or anterior surface in a digitigrade stance) (2) (Webb and Taylor, 1980: GU176).

212. *Anterior surface of distal ends of third and fourth metatarsals.* Unfused (0); fused, fusion forms distal end to prominent gully between third and fourth metatarsals (1) (Janis and Scott, 1987; Scott and Janis, 1993: GU178).

213. *Distal phalanges of foot in dorsal view.* Phalanx compressed transversely (0); broad transversely, each phalanx is bilateral with central anteroposterior axis (1); broad transversely, each phalanx is asymmetrical (2) (O'Leary and Geisler, 1999: GU180).

INTEGUMENT

214. *Hair.* Abundant to common on body (0); almost completely absent (1) (Gatesy, 1997; O'Leary and Geisler, 1999: GU181).

215. *Sebaceous glands.* Present (0); absent (1) (Gatesy, 1997; O'Leary and Geisler, 1999: GU182).

OTHER SOFT TISSUE

216. *Cavernous tissue of penis.* Abundant (0); sparse (1) (derived from Slijper, 1936; Thewissen, 1994: GU183).

217. *Primary bronchi of lungs.* Two (0); three, two on the right and one on the left (1) (Thewissen, 1994: GU184).

STRATIGRAPHIC CHARACTER

218. *Stratigraphic character.* Early Paleocene (0); Late Paleocene (1); Early Eocene (2); Middle Eocene (3); Late Eocene (4); Early Oligocene (5); Late Oligocene (6); Early Miocene (7); Middle Miocene (8); Late Miocene (9); Early Pliocene (A); Late Pliocene (B); Early Pleistocene (C); Middle Pleistocene (D); Late Pleistocene (E); Recent (F).

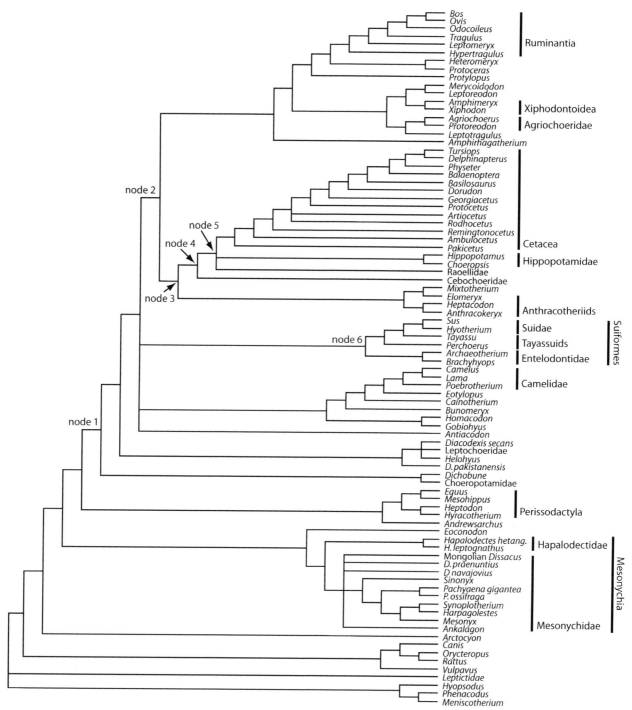

Fig. 3.2. Provisional phylogeny of Artiodactyla, including Cetacea. Shown here is a strict consensus of 16,472 most-parsimonious trees for our data set of molecular, morphologic, and stratigraphic data. Each most-parsimonious tree is 33,815 steps in length. Node 1, Artiodactyla (also known as Cetartiodactyla); Node 2, Cetruminantia; and remaining nodes discussed in the text.

Cetacea, Hippopotamidae, and Raoellidae (Fig. 3.2: node 5). Anthracotheriidae, which had been thought to be a para-phyletic stem group to Hippopotamidae (Colbert, 1935; Gentry and Hooker, 1988; Boisserie et al., 2005b), is in a more basal position. In 8,239 of our most parsimonious trees we found a novel arrangement for the hippo, whale, and raoellid clade; Hippopotamidae and Raoellidae are sis-ter groups. The hippopotamid and raoellid grouping is sup-

ported by the presence of a hypocone in both families, yet this depends on the postulated homology of these cusps. Al-though a true cingular hypocone occurs in raoellids, as in-dicated by its joint presence with a metaconule in some taxa (Kumar and Sahni, 1985), it is possible that the "hypocone" in hippos is actually an enlarged metaconule, such as occurs in several other artiodactyl groups (e.g., ruminants and camelids). In the remaining 8,233 trees, Raoellidae is the sole sister group to Cetacea, a result found in some of the most parsimonious trees recovered by Geisler and Uhen (2005). This hypothesis is supported by the paracone on P4 being much higher than the paracones on the molars (Geisler and Uhen, 2003).

The next node down in our consensus tree (Fig. 3.2: node 4) includes Cetacea, Hippopotamidae, Raoellidae, and *Cebochoerus,* a result also found in some of the most parsimonious trees of Geisler and Uhen (2005). This clade is supported by four unequivocal synapomorphies, of which the most intriguing is a dP3 that is buccolingually compressed and has a central cusp flanked by accessory denticles. In a previous study, two of us speculated that this morphology in artiodactyls was homologous to that seen in basilosaurid cetaceans (Theodor and Foss, 2005), and the result here corroborates this hypothesis. However, it should be noted that hippopotamids were coded as autapomorphic for this character based on their unique dP3 morphology. In the extant pygmy hippo, *Choeropsis liberiensis,* the dP3 is a mesiodistally elongate, molariform tooth (see Fig. 3.4 in Luckett and Hong, 1998). In a future study, we intend to better represent the complexity of the hippopotamid deciduous P3 by developing multiple characters for the morphology of this tooth instead of just one.

The second putative synapomorphy for the clade that includes Cetacea, Hippopotamidae, and *Cebochoerus* (Fig. 3.2: node 4) is the absence of a mastoid foramen. Hippopotamids were coded as "?" because of a problem in establishing homology. Both extant species of hippos have a foramen that pierces the posterior aspect of the skull along the supraoccipital and squamosal suture. This foramen is homologous to either the mastoid foramen, which lies medial to the mastoid process of the petrosal, or to the posttemporal canal, which lies lateral to the mastoid process (Wible, 1987; Geisler and Luo, 1998). Unfortunately, the primary difference between these foramina disappears in hippos and whales, where the mastoid process is greatly reduced and not exposed in posterior view. The third synapomorphy optimized for this clade is a reduction (e.g., raoellids) or loss of the molar paraconules (e.g., hippopotamids and cetaceans), and the fourth is contact of maxilla and frontal in the orbit, although the latter character could not be scored in most extinct taxa.

Although our previous studies have supported the inclusion of anthracotheres in a clade of suiform artiodactyls (Geisler, 2001a; Geisler and Uhen, 2003, 2005; Theodor and Foss, 2005), the most parsimonious trees for our merged matrix have a very different topology; anthracotheres and *Mixtotherium* are the sister group to the whale plus hippo plus raoellid plus cebochoerid clade (Fig. 3.2: node 3). Four basicranial characters support this grouping: postglenoid foramen greatly reduced or absent, postglenoid foramen situated on petrosal / squamosal suture (when present), enlarged tegmen tympani of the petrosal, and contact between petrosal and basioccipital. Although a postglenoid foramen is absent in *Cebochoerus* and hippopotamids, a small foramen occurs in the anthracothere *Elomeryx.* As in

cetaceans, the foramen in anthracotheres is on the petrosal / squamosal suture, and there is also a secondary postglenoid foramen between the squamosal and the tympanic bulla (Geisler and Luo, 1998) (Fig. 3.3). Anthracotheres, cetaceans, and hippos all share an enlarged tegmen tympani of the petrosal (Fig. 3.4). Previously this character was cited as evidence for a close relationship between cetaceans and mesonychids (Luo and Gingerich, 1999), but here that feature in mesonychids is interpreted as convergent because the tegmen tympani of most other artiodactyls is fairly thin and is not pachyostotic. The final basicranial character has a more mixed distribution: contact between the petrosal and basioccipital occurs in pakicetids (Gingerich and Russell, 1981) and anthracotheres, but that contact is lost in hippopotamids (pers. obs.) and later cetaceans (Luo and Gingerich, 1999). Other characters that support the clade that includes whales and anthracotheres are a reduced mastoid process of the petrosal that is not exposed on the posterior aspect of the skull and a zygomatic arch that is directed posterolaterally. Reduction of the mastoid process also occurs in pigs, peccaries, and entelodonts, although that similarity is interpreted here as convergent.

Cetruminantia (Fig. 3.2: node 2), which includes cetaceans, hippopotamids, and ruminants, receives little morphological support from the present study despite being supported by nucleotide sequences (Gatesy et al., 1996; Mathee et al., 2001; Gatesy et al., 2002) and more complex molecular characters such as SINES and multiple-basepair deletions (Gatesy et al., 1996; Shimamura et al., 1997, 1999; Nikaido et al., 1999; Geisler, 2001b). Among morphological characters, Cetruminantia is diagnosed only by the loss of molar hypocones, a character that is plagued by homology issues. Based on our coding of this cusp, the hypocone reappears at least once in cetruminants because it occurs in raoellids and also hippopotamids. Although our analysis supports Cetruminantia, it is clear from previous cladistic analyses

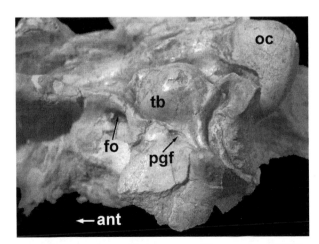

Fig. 3.3. Ventrolateral view of the basicranium of *Elomeryx armatus* (AMNH FM 582). Note the small postglenoid foramen that is situated between the tympanic bulla and the squamosal, which is also found in early cetaceans and hippopotamids. Abbreviations: AMNH FM, American Museum of Natural History (New York, NY) Frick Catalog Number; ant, anterior; fo, foramen ovale; oc, occipital condyle; pgf, postglenoid foramen; tb, tympanic bulla.

Fig. 3.4. Left petrosals of (A) an anthracotheriid (AMNH FM 511) and (B) an undescribed protocetid cetacean (CIS P1514) in ventral view. (C) Reconstruction of the skull of the protocetid cetacean *Georgiacetus vogtlensis*, showing the orientation and position of the petrosal (shaded in gray) in the artiodactyl skull. The right side of the reconstruction shows the tympanic bulla in place, which ventrally covers the petrosal. Drawing A is a composite based on the right and left sides of AMNH FM 511. Note the enlarged tegmen tympani in A and B, a trait also seen in hippopotamids. Scale bars in A and B are 1 cm in length, and C is not to scale. Abbreviations: AMNH FM, American Museum of Natural History (New York, NY) Frick Catalog Number; ant, anterior; ap, anterior process; br, break at base of mastoid process; CIS, Cranbrook Institute of Science; fc, fenestra cochleae; fv, fenestra vestibuli; lat, lateral; pr, promontorium; tt, tegmen tympani.

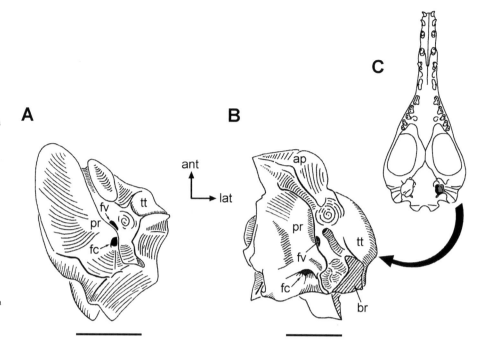

that morphology alone consistently supports a clade of cud-chewing artiodactyls (ruminants and camelids) (Geisler, 2001a; Geisler and Uhen, 2003; Theodor and Foss, 2005).

Although this chapter focuses on the phylogenetic position of Cetacea, the provisional phylogeny presented here has important implications for other parts of artiodactyl phylogeny. As noted by Geisler and Uhen (2005), the relationships of most artiodactyls with selenodont teeth have become even more problematic with the inclusion of molecular data that strongly contradict a close relationship between Camelidae and Ruminantia, the two extant groups of selenodont artiodactyls. In our most parsimonious trees, most extinct selenodont groups are more closely related to ruminants than to camelids. This includes Protoceratidae (even though its monophyly was not supported), Xiphodontidae, and Amphimerycidae. The phylogeny of oromerycids is most perplexing; it is paraphyletic with *Protylopus* being closely related to ruminants and *Eotylopus* being closely related to camelids. Cainotheriidae appears to be more closely related to camelids than to ruminants, as was found by Geisler (2001a).

Based on both molecular (Gatesy et al., 1996; Shimamura et al., 1997, 1999; Nikaido et al., 1999; Mathee et al., 2001; Gatesy et al., 2002) and morphological (Geisler and Uhen, 2003; Boisserie et al., 2005b) data that supported hippopotamids being more closely related to cetaceans than to pigs and peccaries, Geisler and Uhen (2005) redefined Suiformes as "the clade that includes Suidae, Tayassuidae, Entelodontidae, and Anthracotheriidae, but excludes Camelidae, Hippopotamidae, Cetacea, Raoellidae, Cebochoeridae, Mixtotheriidae, Cainotheriidae, and Oreodontoidea." In the most parsimonious trees of the present study, entelodonts remain in the suiform clade, but anthracotheres are the sister group to a clade that includes whales, hippos,

and raoellids. Boisserie et al. (2005b) also supported the exclusion of Anthracotheriidae from Suiformes (*sensu* Geisler and Uhen, 2005), although their position for anthracotheres differs from ours. Given the growing consensus that anthracotheres are not suiform artiodactyls, we redefine Suiformes as the clade that includes Suidae, Tayassuidae, and Entelodontidae but excludes Anthracotheriidae, Camelidae, Hippopotamidae, Cetacea, Raoellidae, Cebochoeridae, Mixtotheriidae, Cainotheriidae, and Oreodontoidea (Fig. 3.2: node 6). This more exclusive Suiformes is diagnosed by two unequivocal synapomorphies: fused mandibular symphysis and presence of a P4 metacone. Three other characters are probable synapomorphies of Suiformes but have an ambiguous optimization in trees that have *Antiacodon* as the sister group to Suiformes: the posterior edge of the squamosal bears a dorsally projecting process, the distal articular surface of the humerus extends past the trochlear edge, and a ridge is present between the capitulum and the epicondyle on the humerus.

CONCLUSION

Over the past 15 years there has been a dramatic effort to sequence the genes of artiodactyls (e.g., Gatesy et al., 1996; Mathee et al., 2001; Gatesy et al., 2002), and these sequences have led to a phylogeny of artiodactyls that at its higher levels is radically different from that based on anatomy (e.g., Matthew, 1934). The most unexpected aspect of the molecular phylogeny was the recognition that Cetacea is a deeply nested member of Artiodactyla. The implications of this and other novel aspects of the molecular tree are still being investigated. Although the positions of most extant taxa are fairly constrained by molecular characters, the positions of many extinct artiodactyl clades remain uncertain. The phylogeny depicted in this chapter is our best guess as to where many of these taxa go; however, we take the paraphyly of several widely recognized families (e.g., Protoceratidae) as evidence that the data matrix on which it is based does not yet adequately represent published morphological data. Given the vast body of literature on artiodactyls, a complete representation will probably not be reached soon; however, we do have a sense of where additional efforts should be made in order to delineate the higher-level clades within Artiodactyla.

Although there are many more artiodactyl taxa that could be coded for our characters and incorporated into our phylogenetic hypothesis, we think that the exemplars we have chosen are adequate to resolve most of the higher-level clades. One notable exception to this involves anthraco-theriids. Boisserie et al. (2005b) included eight anthracotheres in their study and found the group to be paraphyletic with respect to Hippopotamidae, and we think that in future analyses of our matrix, some of the more derived and hippo-like anthracotheriids (e.g., *Merycopotamus*) should be included. Although adding taxa is important, we think that more effort should be directed to adding new information for the taxa we have already sampled (Fig. 3.2). This can be done in two ways. In joining our previous matrices, we focused on resolving discrepancies and developing a single synthetic character list. There is a substantial amount of missing data that developed because our previous studies sampled different taxa and characters. Many of these missing entries could fairly easily be filled in. The second way to add new data for our exemplar taxa is to incorporate new characters. We suspect that the addition of dental characters would help recover monophyly of some of the extinct families, which could then have a ripple effect on the higher-level phylogeny within Artiodactyla. Developing a matrix that adequately addresses both lower-level and higher-level artiodactyl phylogeny simultaneously is a daunting task, one that will require a concerted effort by experts on artio-dactyl systematics for several years to come.

ACKNOWLEDGMENTS

We acknowledge Philip Gingerich, Zhe-Xi Luo, and Hans Thewissen, who reviewed this chapter and greatly improved it through their suggestions. Bolortsetseg Minjin took the photograph in Figure 3.3 and took several photographs of anthracothere petrosals, from which Figure 3.4A was drawn. We thank Donald Prothero for the opportunity to contribute to this volume.

JESSICA M. THEODOR,
JÖRG ERFURT,
AND GRÉGOIRE MÉTAIS

The Earliest Artiodactyls

Diacodexeidae, Dichobunidae,
Homacodontidae, Leptochoeridae,
and Raoellidae

THIS CHAPTER DEALS WITH THE OLDEST, most primitive members of Artiodactyla, often referred to the Dichobunoidea, probably best regarded as a paraphyletic group of stem lineages. The members of the group are, however, of considerable importance in understanding the origin and early evolution of Artiodactyla, even with the currently incomplete Asian record. We attempt here to provide a useful summary for these groups.

STRATIGRAPHY

In North America, the first appearance of artiodactyls, *Diacodexis ilicis* in the earliest Wasatchian NALMA (North American Land Mammal Age), is a hallmark of the Eocene. Artiodactyls are rare in early Eocene faunas but are much more diverse by the middle Eocene, implying a taxonomic radiation. Artiodactyls are important marker taxa in the biostratigraphic zonation of the Paleogene of North America, and the biostratigraphy of the early Tertiary has recently been extensively reviewed (Prothero and Emry, 2004; Robinson et al., 2004).

Artiodactyls are also important for biostratigraphy in Europe (see Table 4.1) and for the definition of the European Land Mammal Ages (ELMA). The oldest, most primitive diacodexeid is the early Eocene *Diacodexis antunesi* from Silveirinha (Portugal, Mammal Paleogene [MP] level 7). It co-occurs with more specialized forms of *Diacodexis,* implying that diversification of the genus occurred earlier. Diacodexeids survived until the base of the middle Eocene, when dichobunids appeared and radiated rapidly until the late Eocene. In the Oligocene, other families such as cainotheriids provide biostratigraphic information (Erfurt and Métais, this volume), and

Table 4.1. Biostratigraphy of artiodactyl sites in the West European Tertiary

MN/MP		France; Austria (A); Italy (I)	Switzerland; Germany (D); Portugal (P)	Great Britain; Belgium (B); Spain (ES)
MN	13	I: Pirro; Brisighella; Grivelli		**ES: El Arquillo 1,** Milagros, Venta del Moro
MN	12	Cucuron. I: Baccinello V1; Casteani		**ES: Los Mansuelos;** Concud; Los Mansuetos
MN	11	Andante	D: Dorn-Dürkheim	**ES: Crevillente 2,** Piera
MN	10			**ES: Masia del Barbo;** La Roma 2
MN	9	A: Inzersdorf	D: Eppelsheim, Höwenegg	**ES: Can Llobateres;** El Lugarejo; Santita; Los Valles de Fuentidueña
MN	7/8	**La Grive M**	D: Oehningen; Steinheim	ES: Hostalets
MN	6	**Sansan.** A: Göriach	D: Georgensgmünd; Goldberg (Ries); Stätzling; Steinberg (Ries)	ES: Arroyo de Val-Barranca; Manchones
MN	5	**Pontlevoy;** Beaugency; Tavers	Käpfnach; Veltheim. D: Reisensburg; Sandelzhausen; Randecker Maar (in part)	ES: Puente de Vallecas; Tarazona
MN	4	**La Romieu;** Artenay; Bézian; La Romieu; Montréal-du-Gers. A: Oberdorf	D: Petersbuch 2; Erketshofen; Langenau; Oggenhausen 2?; Rauscheröd	ES: Artesilla; Buñol; Els Casots; Sant Mamet; Tarazona
MN	3	Chilleurs-aux-Bois; Chitenay; Espira-du-Conflent; Neuville; Pontigné	**D: Wintershof-West;** Schnaitheim 1; Öllingen; Stubersheim 3	ES: Moratilla;
MN	2	**Montaigu;** Barbotan-Les-Thermes; Langy; Laugnac; Montaigu-leBlin; St.-Gérand-le-Puy		ES: Cetina de Aragón; Loranca; Navarrete del Rio
MN	1	**Paulhiac;** Aarau		
MP	30	**Coderet**	D: Ehrenstein 4	
MP	29		**Rickenbach**	
MP	28b	*Pech du Fraysse*[8]	D: Gaimersheim 1	
MP	28a	*Pech Desse*[8]		
MP	27	Sarèle	Mümliswyl-Hardberg. **D: Boningen;** Gaimersheim 2	
MP	26	*Mas de Pauffié; Puycelci*[6]		
MP	25	*Le Garouillas;* Authezat; Rabastens; Saint-André[9]; Tauriac[9]	Bumbach	ES: Carrascosa del Campo
MP	24	Itzac	**D: Heimersheim**	ES: Can Quaranta
MP	23	*Itardies;* Pech Crabit 1		
MP	22	**Villebramar;** La Plante 2; *Pinchenat*[9]; *Ravet;* Ronzon;	D: Herrlingen 1; Möhren 13; Veringen[9]	ES: Calaf; Montalban 3c, 8
MP	21	**Soumailles;** *Aubrelong 1*	D: Möhren 19, 20	Bouldnor [upper Hamstead Member][7]; B: Hoogbutsel
MP	20	**St-Capraise-d'Eymet;** Baby 2; *Tabarly;* Vermeils[9]; Villeneuve-la-Comptal	D: Frohnstetten	Bouldnor [lower Hamstead Member][7]; Whitecliff Bay 2 [Bembridge Marls Member][7]
MP	19	*Escamps; Coânac 1;* Montmartre 1[4]; Mas S[tes] Puelles; Mormoiron; Pont d'Assou; *Rosières 1-4*	Obergösgen; Mt. Eclépens Gare 2[3]; Eclépens C[3], E[3]; Entreroches[3]. D: Möhren 6; Neuhausen; Pappenheim; Rixheim[9]. P: Côja	Headon Hill[7]; [Bembridge Limestone beds 4-18]; Binstead. ES: San Cugat de Gavadons
MP	18	**La Débruge;** Gargas[9]; *Gousnat;* Les Ondes; Montmartre 2[4]; *Ste Néboule; Saindou D*[9]	Cinq-Sous[3]. D: Ehrenstein 1-3, 6; Eselsberg[9]; Gösgen-Kanal; Herrlingen 3	Headon Hill [Hatherwood Limestone]. ES: Zambrana[2]
MP	17b	*Perrière;* Argenteuil[9]; *Celarié*[9]; *Malpérié; Rosières 5;* Verrerie des Roches[9]		Headon Hill [How Ledge Limestone; Hordle Rodent bed][3] Both belong to the upper Totland Bay Member = lower Headon beds in part]
MP	17a	*Fons 4; Aubrelong 2;* Baby 1; Fons 1-7; Hippolyte-de-Caton[9] (Euzèt); *La Bouffie; Lavergne*[3]; *Lebratières*[9]; Les Clapiès; Les Pradigues; Les Sorcières[9]; *Mouillac*[9]; Salesmes; Souvignargues; St.	Mt. Eclépens B[3]; Verrerie des Roche[9]	Hordle (Hordwell) Mammal bed [lower Totland Bay Member]. ES: Sossís; Roc de Santa
MP	16	**Robiac;** Caylux[9]; Grisolles; La Milette[9]; *Larnagol*[9]; Lautrec[6]; *Le Bretou;* Le Castrais; Montespieu[9]; Paris Gare du Nord; Sicardens[9]	Les Alleveys (= St. Loup)[3]; Moutier[9]; Mt. Eclépens A[3]; Mt. Eclépens-Gare (in part)[3]. D: Heidenheim; Herrlingen 4	Barton [Barton Clay Fm. beds C-F][3]; Creechbarrow. ES: Caenes[1]; Villamayor[9]

continued

Table 4.1. Continued

MN/MP		France; Austria (A); Italy (I)	Switzerland; Germany (D); Portugal (P)	Great Britain; Belgium (B); Spain (ES)
MP	15	**La Livinière 2**		ES : Pontils 26; 38; Sant Jaume de Frontanyà 1
MP	14	Arcis le Ponsart; Issel; *Laprade;* Le Guépelle; Lissieu	**Egerkingen á+â, Bolus** (= abberant Fazies), blaue Mergel; Chamblon[9]	ES: Capella; Sant Jaume de Frontanyà 2, 3
MP	13b	Eygalayes[9]	Egerkingen γ[9], Huppersand[9]. D: Gt [Oberkohle] site Ce V[9]	
MP	13a	Aumelas; Bouxwiller; Calcaire Grossier; Grabels[9]; Mas Gentil[9]; La Défense; St-Maximin; St. Martin-de-Londres[9]	**D: Gt [obere Mittelkohle]** sites Cecilie IV, Leo I-III, XXVI, XXXV- XXXVII, XLI; Gt [oberes Hauptmittel] site VII, Cecilie I-III; Eckfeld	
MP	12		**D: Gt [untere Mittelkohle]** sites VI, XVIII, XXII, XLIII	
MP	11	Argenton[5]	**D: Gt [Unterkohle]** site XIV; Messel (D)	
MP	10b	Prémontré[4]; Rouzilhac[9]		
MP	10a	**Grauves;** Chavot; Cuis; Epernay[9]; Mancy; Mas de Gimel; Monthelon; St. Agnan		ES: Corsa I; II; III[1]; El Pueyo Güell 1-3[1]; Laguarres[1]; La Roca; Les Badies; Los Saleres; Montderoda; Repeu del Güaita[1]
MP	8-9	**Avenay;** Condé-en-Brie; Mutigny; Sézanne-Broyes;		Abbey Wood; Herne Bay; Harwich [London Clay]
MP	7	Le Quesnoy (Creil); Pourcy; Rians	Silveirinha (P)	**B: Dormaal.** Kyson [Suffolk Pebble Beds]
MP	6	**Cernay**		Beltinge
MP	1-5	?Menat	Walbeck (D)	**B: Hainin**

Emended from Russell et al. (1982), Schmidt-Kittler (1987), Sudre et al. (1992), and BiochroM'97 according to [1]Antunes et al. (1997), [2]Astibia et al. (2000), [3]Hooker and Weidmann (2000: Fig. 75, 79), [4]Mertz et al. (2000: Fig. 3). [5]Sudre and Lecomte (2000), [6]Astruc et al. (2003; Fig. 4), [7]Hooker et al. (2004: Fig. 3), [8]Blondel (2005), [9]tentative rating by J. Erfurt based on artiodactyls. Miocene localities emended from Heizmann (1983) and Gentry et al. (1999). For fossil complexes, only those sites are indicated that are mentioned in the text. Italic = Quercy localities; bold = reference localities; Eg = Egerkingen, Gt = Geiseltal (Haubold and Hellmund, 1997). Lithological units are in brackets. Entreroches and Eclépens belong to Mormont. *Bach, Lamandine,* and Saint Saturnin are estimated from MP 17–20, *Mémerlein* from MP 18–19.

artiodactyls provide important index fossils until the Pleistocene. Their presence in the European Paleogene record is increasingly calibrated by radioisotope methods. The first absolute dates for the Geiseltalian ELMA (Franzen and Haubold, 1986) for the Eckfeld basalts (Germany) yield an age of 44.3 Ma based on the ^{40}Ar/^{39}Ar method (Mertz et al., 2000: 270), which fits with the upper limit of the nannofossil biochrons NP 15b and CP 13b (Berggren et al., 1995: Fig. 2) and establishes an upper limit for this ELMA. The lower boundary falls between the chrons C22n/C21r, yielding a time span from 49 to 44 Ma for the reference levels MP 11 to MP 13. This ELMA is characterized by the dichobunid genera *Eygalayodon, Lutzia, Parahexacodus, Messelobunodon, Buxobune,* and *Neufferia.* Further important markers are provided by the investigations of the basaltic flows and trachyte tuff from Enspel and Kärlich (Germany) by Mertz et al. (2007). They indicate a time interval of 24.9–24.5 Ma for reference level MP 28 (Enspel) and a maximum age of 25.5 Ma for the interval MP 28 to MP 30 (Kärlich).

Because the Asian record of fossil mammals has not been extensively investigated, the biochronology of continental rocks and faunas is much less precise than that in Europe and particularly North America. The Asian Land Mammal Ages (ALMA), inspired by the NALMA system, are based on fossiliferous formations restricted to Mongolia and northern China. Correlations with mammal faunas from Central or South Asia have proven difficult because of Paleogene provincialism. The first artiodactyl in Asia is either the pu-

tative diacodexeid *Tsaganohyus pecus* from the base of the Bumban Member of the Naran Bulak Formation in Mongolia (Kondrashov et al., 2004), correlative to the Wasatchian and probably the upper part of Clarkforkian NALMAs (Beard and Dawson, 1999) or the enigmatic suiform *Wutuhyus primiveris* from the early Eocene Wutu Formation, Shandong Province, coastal China (Tong and Wang, 2006). We have tried to provide the most recent stratigraphic and age updates of the fossil taxa discussed below. Question marks by the ages of some Asian fossiliferous localities indicate that the age is contested or currently under investigation. Figures 4.1 and 4.2 illustrate the taxa discussed below.

BIOGEOGRAPHY

Diacodexeid and dichobunid genera, except for *Diacodexis, Bunophorus,* and possibly *Simpsonodus,* are restricted to a single continent. These taxa have their origins in the earliest Eocene, when faunal exchange between North America and Europe was temporarily possible along the Greenland–Iceland–Faeroes Ridge or via Spitzbergen (McKenna, 1972, 1975; Collinson and Hooker, 2003). In the middle Eocene, Europe was isolated to the south by the Tethys Sea, to the west by the Atlantic, and to the east by connections between the Atlantic (North Sea) and Tethys through the Black Sea, and between the Arctic Sea and Tethys by Lake Ob and the Turgai Strait. The only connection in the north to America

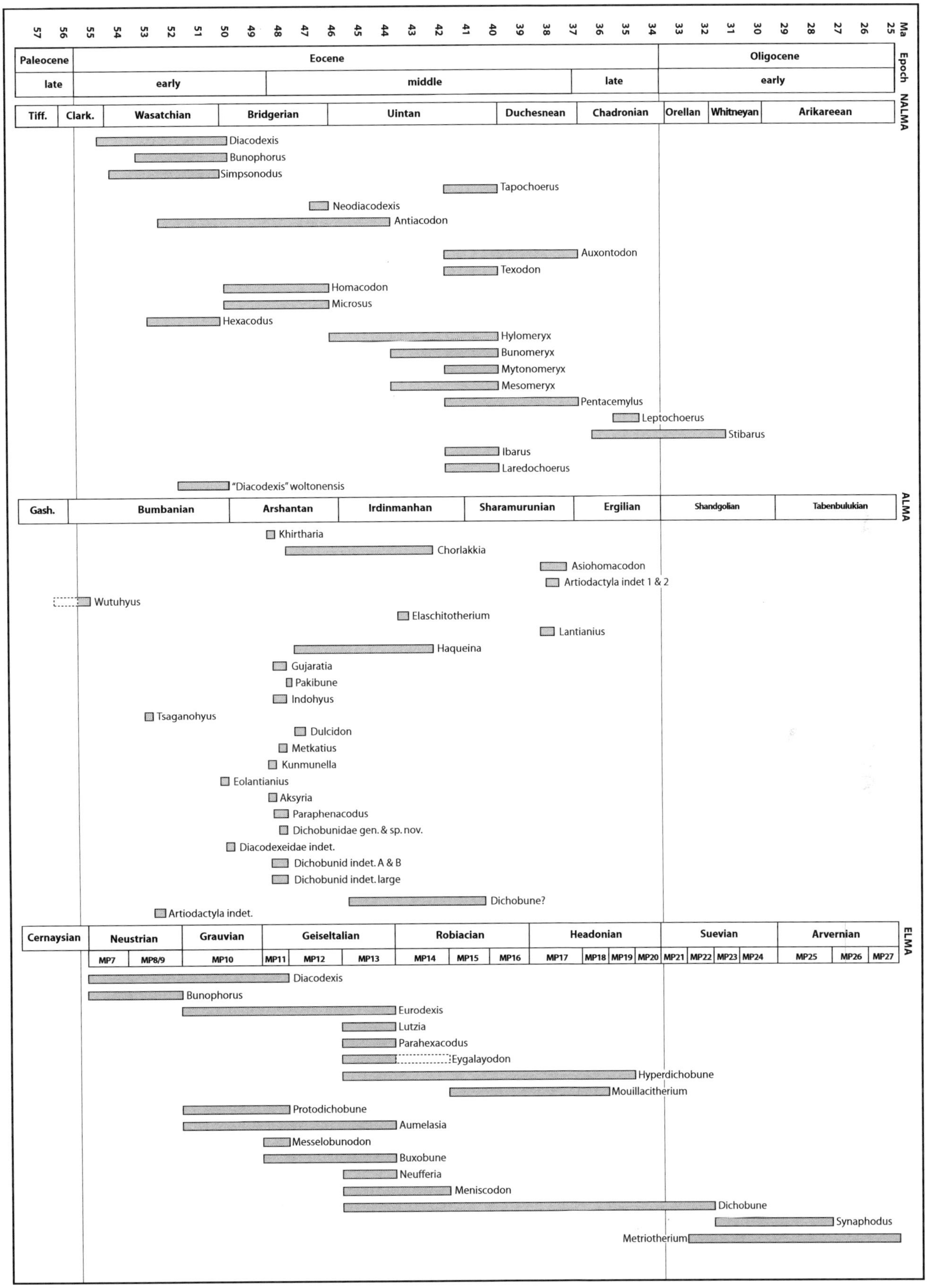

Figure 4.1. Occurrence of early artiodactyls in the Holarctic MP levels and reference localities according to Schmidt-Kittler (1987) and Escarguel et al. (1997).

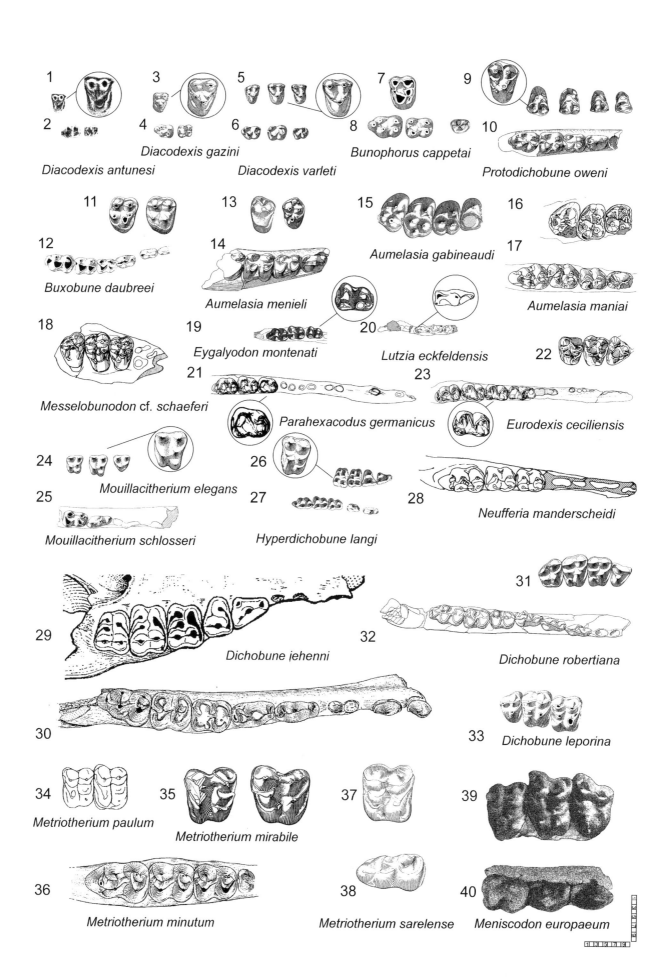

1
2
3
4
Diacodexis antunesi
Diacodexis gazini

5
6
Diacodexis varleti

7
8
Bunophorus cappetai

9
10
Protodichobune oweni

11
12
Buxobune daubreei

13
14
Aumelasia menieli

15
Aumelasia gabineaudi

16
17
Aumelasia maniai

18
Messelobunodon cf. *schaeferi*

19
Eygalyodon montenati

20
Lutzia eckfeldensis

21
Parahexacodus germanicus

22

23
Eurodexis ceciliensis

24
25
Mouillacitherium elegans
Mouillacitherium schlosseri

26
27
Hyperdichobune langi

28
Neufferia manderscheidi

29
Dichobune jehenni

30

31
32
Dichobune robertiana

33
Dichobune leporina

34
Metriotherium paulum

35
Metriotherium mirabile

36
Metriotherium minutum

37
38
Metriotherium sarelense

39
40
Meniscodon europaeum

was open during the Oligocene and temporarily interrupted by sea level changes (Tarling, 1982). Endemic taxa diversified and evolved in situ during the middle Eocene, and they experimented with several unique dental and postcranial adaptations. Diacodexeids and dichobunids are restricted to western and central Europe and were formerly regarded as endemic. However, as our knowledge of epicontinental marine barriers changes, it is important to reevaluate the European record, which can be locally rich but poor overall, and to compare with diversity patterns observed in North America and Asia. The apparent lack of an intensive artiodactyl faunal exchange between Europe and other areas may have been caused by interruptions of the Northern land bridge or by filters during the middle Eocene (Godinot, 1982).

There is little direct evidence of any interchange between North America and Europe after the Wasatchian, when *Diacodexis, Bunophorus,* and homacodonts appear. However, Stucky (1998) noted that the strong resemblances between dichobunids and homacodonts suggest either high homoplasy or repeated immigration not seen in other groups of mammals. Although numerous groups of Asian origin arrived in North America throughout the Eocene, there is little evidence of later Asian immigrants among the North American basal artiodactyls, but this may be related to the relatively poor Paleocene and early Eocene Asian record. Asia has long been thought to be the continent where (cet)artiodactyls originated (e.g., Pilgrim, 1940; but see Gingerich et al., 1986, 1989).

The Asian landmass is vast, and our knowledge derives from few spots (often poorly dated), mostly in the Indo-Pakistani region, Central Asia, and Southeast Asia. Northern Asia, notably the most eastern part of Russia, which may have been a migration route between Asia and North America during most of the Paleogene, has only recently yielded any evidence. The best record of early middle Eocene

artiodactyl faunas comes from the Indian Subcontinent. The distinctive faunas recovered from the Kuldana and Subathu formations are morphologically homogeneous, suggesting local evolution during the short Eocene record there. The hypothesis that the Indian Subcontinent may have been a distinct faunal province during at least the late early and early middle Eocene is corroborated by data from other faunal elements (e.g., Thewissen et al., 2001a). However, the geographical extent of this province is unclear and may have also included southeast Asia (Métais et al., in press). Additional data from other areas in Asia are needed to test biogeographic hypotheses.

PALEOECOLOGY

Based on dental morphology (Janis, 1990), diacodexeids and dichobunids can be divided into two groups: smaller forms with tribosphenic, crescentiform molars (European *Diacodexis,* eurodexeines, hyperdichobunines, lantianiines, some homacodontids) and larger taxa with quadrituber-cular, flatter, more square-shaped molars. The trigonid/talonid height difference of the first group suggests a crushing adaptation, as in insectivorous or frugivorous mammals, but gut contents are unknown. The second group includes omnivorous and/or herbivorous dentitions such as in dichobunines. The fossil gut content of *Messelobunodon* is consistent with that interpretation (Richter, 1981, 1987) and indicates an ecological differentiation of the two groups. The latter is also postcranially documented in different ecomorphs (or better "Wuchstyp": Erfurt and Altner, 2003); smaller taxa such as *Diacodexis* and *Eurodexis* and larger forms by dichobunines (e.g., *Aumelasia* and *Dichobune*). Both groups show long tails, hindlimbs longer than forelimbs, and unfused zeugopodials. The first group resembles small and slender antelopes with longer tails and occurred in both families. The second group is stockier and characterized by

Figure 4.2. (*opposite*) Selected dentitions of European diacodexeids and dichobunids. Dentitions are approximately scaled only; those in circles are magnified (2×). Bold specimens are types. (1) *Diacodexis antunesi* CEPUNL SV3 74 right M2 (Estravis and Russell, 1989: pl. II/2). (2) *Diacodexis antunesi* CEPUNL **SV3 338,** SV3 129 left m3 (mirrored), right m2 (Estravis and Russell, 1989: pl. I/2a, 3a). (3) *Diacodexis gazini* USTL RI 168 left M2 (mirrored) (Sudre et al., 1983: Fig. 1a). (4) *Diacodexis gazini* USTL RI 162, **RI 164,** right m3, left m2 (mirrored) (Sudre et al., 1983: Fig. 1b,c). (5) *Diacodexis varleti* MNHN CB 2461, **CB 973,** CB 1165 right M3, M2, M1 (Sudre et al., 1983: Fig. 5a–c). (6) *Diacodexis varleti* MNHN CB 258, CB 1276 CB 255 right m3, m2, m1 (Sudre et al., 1983: Fig. 5h–j). (7) *Bunophorus cappettai* MNHN **MU 6183,** left M1 or M2 (Sudre et al., 1983: Fig. 15b). (8) *Bunophorus cappettai* MNHN AV 4666, MU 5561, MU 5577 right m3, m2, p4 (Sudre et al., 1983: Fig. 15c–f). (9) *Protodichobune oweni* coll. P. Louis GR 117, GR 160, GR 131, GR 126 right M1, left M2, right dP4, left P4 (Sudre et al., 1983: Fig. 10c–f). (10) *Protodichobune oweni* MNHN **AI 5232,** right mandible with p4–m3 (Sudre et al., 1983: Fig. 11b). (11) *Buxobune daubreei* USTL Bchs 430, **Bchs 516,** right upper molars, M2? (Sudre, 1978: pl. I/1,2). (12) *Buxobune daubreei* USTL Bchs 569, right mandible with p2–m3 (Sudre, 1978: pl. I/3). (13) *Aumelasia menieli* coll. Monganaste Mt 15730, MB 15731 right M2 (mirrored), left M1 (Sudre et al., 1983: Fig. 13a,b). (14) *Aumelasia menieli* coll. P. Louis **MA 1L,** right mandible with p4–(m3) (Sudre et al., 1983: Fig. 14b). (15) *Aumelasia gabineaudi* USTL **AUM 145,** left maxilla with P4–(M3) (Sudre, 1980: Fig. 2). (16) *Aumelasia maniai* GMH **LII-52,** left maxilla with (M1), M2, (M3), Erfurt and Haubold (1989: Fig. 2a). (17) *Aumelasia maniai* GMH **LII-52,** left mandible with (p4)–m3, Erfurt and Haubold (1989: Fig. 2b). (18) *Messelobunodon* cf. *schaeferi* GMH XIV-1763, right maxilla with M1–M3 (Erfurt, 1988: pl. 3b). (19) *Eygalayodon montenati* USTL **EYG 06,** left mandible with m1–m3 (Sudre and Marandat, 1993: Fig. 2). (20) *Lutzia eckfeldensis* MNHM **PW 1991/43–LS,** left mandible with p4–(m2) (Franzen, 1994: Fig. 1b). (21) *Parahexacodus germanicus* GMH **XXXVII-198,** left mandibula with p1, m1–m3 (Erfurt, 1988: pl. 5b). (22) *Eurodexis ceciliensis* GMH **Ce IV-305,** left maxilla with M1–M3 (Erfurt and Haubold, 1989: Fig. 3b). (23) *Eurodexis ceciliensis* GMH Leo I-3875, left mandibula with dp4, m1–m3 (Erfurt and Haubold, 1989: Fig. 3a). (24) *Mouillacitherium elegans* USTL MPR 1022, 1026, 1019, left M1–M3 (combined) (Sudre, 1978: pl. III/6). (25) *Mouillacitherium schlosseri* BSPG **1879 XV-138,** right mandible with p4, m1 (Sudre, 1978: pl. IV/5). (26) *Hyperdichobune langi* FSL 2542, right P3–M2 (combined) (Sudre, 1972: Fig. 3a). (27) *Hyperdichobune langi* FSL Li 1009 - 1011, left p3, right p4 (mirrored), left mandibula m1–m3 (Sudre, 1972: Fig. 4a–c). (28) *Neufferia manderscheidi* MNHM **PW 1994/54-LS,** left mandible with m1–m3 (Franzen, 1994: Fig. 2b). (29) *Dichobune jehenni* LPVP **Vil 1973–14,** skull with P3–M3 (Brunet and Sudre, 1980: Fig. 4). (30) *Dichobune jehenni* LPVP Vil 1970-234, right mandible with c–m3 (Brunet and Sudre, 1980: Fig. 6). (31) *Dichobune robertiana* FSL 2559, right P4–M3 (combined) (Sudre, 1972: Fig. 2). (32) *Dichobune robertiana* USTL SMF 108, right mandible with p2–m3 (Sudre, 1988a: Fig. 1b). (33) *Dichobune leporina* USTL ESC 1222, right maxilla with M1–3 (Sudre, 1988a, 1988b: Fig. 4b). (34) *Metriotherium paulum* MNB **Q.A. 127,** left M1–M2 (Stehlin, 1906: Fig. 42). (35) *Metriotherium mirabile* USTL GAR 2051, right dP4 and M1 (Sudre, 1995: Fig. 1). (36) *Metriotherium minutum* LPVP **Vil 1974-162,** right mandible with m1–3 (Brunet and Sudre, 1980: Fig. 9). (37) *Metriotherium sarelense* USTL **SAR-Ar-01,** right M1 (Astruc et al., 2003: Fig. 17). (38) *Metriotherium sarelense* USTL SAR-Ar-03, right m3 (Astruc et al., 2003: Fig. 18). (39) *Meniscodon europaeum* MNB **Eg 518,** left maxilla with M1–M3. (40) *Meniscodon europaeum* MNB Eh 794, right mandible with p3–m1.

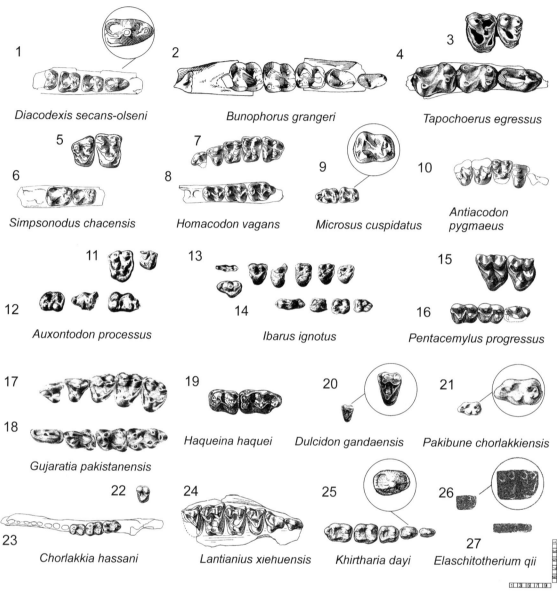

1 *Diacodexis secans-olseni*

2 *Bunophorus grangeri*

3 4 *Tapochoerus egressus*

5 6 *Simpsonodus chacensis*

7 8 *Homacodon vagans*

9 *Microsus cuspidatus*

10 *Antiacodon pygmaeus*

11 12 *Auxontodon processus*

13 14 *Ibarus ignotus*

15 16 *Pentacemylus progressus*

17 18 *Gujaratia pakistanensis*

19 *Haqueina haquei*

20 *Dulcidon gandaensis*

21 *Pakibune chorlakkiensis*

22 23 *Chorlakkia hassani*

24 *Lantianius xiehuensis*

25 *Khirtharia dayi*

26 27 *Elaschitotherium qii*

Figure 4.3: Selected dentitions of North American and Asian diacodexeids and dichobunids. Dentitions are approximately scaled only; those in circles are magnified (2×). Boldface specimens are types. North American taxa above, Asian taxa below. (1) *Diacodexis secans-olseni* AMNH 14937 p4-m3 (Sinclair, 1914: Fig. 26). (2) *Bunophorus grangeri* **AMNH 15516 p3–m3** (p3 reversed, Sinclair, 1914: Fig. 1). (3) *Tapochoerus egressus* LACM(CIT) 5230 left ?M2, M3 (McKenna, 1959: pl. 37a). (4) *Tapochoerus egressus* LACM(CIT) 1588, p4–m2 (McKenna, 1959: pl. 37b). (5) *Simpsonodus chacensis* AMNH 15510 right M2–M3 (Sinclair, 1914: Fig. 28). (6) *Simpsondus chacensis* AMNH 15513 right m1–m2 (Sinclair, 1914: Fig. 27b). (7) *Homacodon vagans* AMNH 12695, left P4–M3 (Sinclair, 1914: Fig. 20). (8) *Homacodon vagans* AMNH 12139 m1–3 (Sinclair, 1914: Fig. 20). (9) *Microsus cuspidatus* AMNH 12143 left m2–m3 (Sinclair, 1914: Fig. 22). (10) *Antiacodon pygmaeus* AMNH 12043 right P4–M3 (Sinclair, 1914: Fig. 21). (11) *Auxontodon processus* SMNH 1654.449 right M1 or M2, SMNH P1589.10 right M3 (Storer, 1984: Fig. 8d–e). (12) *Auxontodon processus* SMNH ROM 35.026 right m1 or m2, SMNH P1654.456 left m3, **SMNH P1654.457** right m3 (Storer, 1984: Fig. 8f–h). (13) *Ibarus ignotus* SMNH P1654.458 left P2; SMNH P1654.459 left P3; SMNH 1654.463 right dP4; SMNH 1654.465 right P4; SMNH 1654.468 right M1 or M2; SMNH 1654.469 right M1 or M2; SMNH 1654.471 right M3 (Storer, 1984: Fig. 9a–f). (14) *Ibarus ignotus* SMNH P1589.11 left p3; SMNH P1654.474 left p4; **SMNH P1589.12 left m1 or m2;** SMNH P1654.479 right m3 (Storer, 1984: Fig. 9h–k). (15) *Pentacemylus progressus* **CM 11865 left M1–2** (Peterson, 1931: Fig. 9). (16) *Pentacemylus progressus* **CM 11865 left p4–m2** (Peterson, 1931: Fig. 9). (17) *Gujaratia pakistanensis* **HGSP 300 5003 left maxilla with P3–M3** (Thewissen et al., 1983, Fig. 1). (18) *Gujaratia pakistanensis* **HGSP 300 5003 left mandible with p4–m3** (Thewissen et al., 1983, Fig. 1). (19) *Haqueina haquei* **BSPG 1956 II 10 left m2–3** (Dehm and Oettingen-Spielberg, 1958: Fig. 5). (20) *Dulcidon gandaensis* **BSPG 1956 II 1 left M2** (Dehm and Oettingen-Spielberg, 1958: Fig. 3). (21) *Pakibune chorlakkiensis* **GSP-UM 259 right m3** (Thewissen et al., 1987: Fig. 2k). (22) *Chorlakkia hassani* GSP-UM 1437 left M2 (Thewissen et al., 1987: Fig. 3f). (23) *Chorlakkia hassani* **GSP-UM 66 left m1–3** (Thewissen et al., 1987: Fig. 1a). (24) *Lantianius xiehuensis* **IVPP-V.2933 right maxilla** with P2–M3 (Chow, 1964: Fig. 1a). (25) *Khirtharia dayi* GSP-UM 1412, 1510, 115, 1404, 87, composed P3–M3 (Thewissen et al., 1987: Fig. 5e). (26) *Elaschitotherium qii* **IVPP 12759.1 right M1–2** (Métais et al., 2004: Fig. 2a). (27) *Elaschitotherium qii* IVPP 12759.44 left m1–3 (Métais et al., 2004: Fig. 2f).

a still longer tail, resembling carnivores (coatis and raccoons). This morph is unknown in diacodexeids but is represented in European dichobunines, cebochoerids, and choeropotamids (Erfurt and Métais, this volume) and suggests a suid-like habit.

SYSTEMATICS

The lists of fossil localities are mainly compiled from Sudre (1978), Sudre et al. (1983), Erfurt and Haubold (1989), Sudre and Ginsburg (1993), and Sudre (1997) for Europe; Walsh (1996), Stucky (1998), Robinson et al. (2004), and Prothero and Emry (2004) for North America; Russell and Zhai (1987), Thewissen et al. (1987, 2001a), and Bajpai et al. (2005) for Asia. The type locality is listed first, and all other localities are in alphabetical order. Geographical affiliation of the European sites is indicated in Table 4.1. In the morphological descriptions, capital letters indicate elements of the upper dentition, and lower-case letters elements of the lower dentition. All measurements are in millimeters (L for length, W for width). If no specimen is indicated, all dimensions represent average values based on the given literature. Museums are indicated by abbreviations.[1]

[1] Abbreviations: AMNH FM, American Museum of Natural History, Frick Mammal Collection (U.S.A.); ANSP, Academy of Natural Sciences, Philadelphia (U.S.A.); BMNH, Natural History Museum London (Great Britain); BSPG, Bayerische Staatssammlung für Paläontologie und historische Geologie München (Germany); CEPUNL, Centro de Estratigrafia e Paleobiologie da Universidade Nova de Lisboa (Portugal); CM, Carnegie Museum of Natural History (U.S.A.); FSL, Faculté des Sciences, Université Claude Bernard Lyon (France); GMH, Geiseltalmuseum, Martin-Luther-Universität Halle (Germany); GSP, Geological Survey of Pakistan, Islamabad (Pakistan): HGSP, Howard University collection, GSP-UM, University of Michigan collection; HLMD, Hessisches Landesmuseum Darmstadt (Germany); IVPP, Institute of Vertebrate Paleontology and Paleoanthropology, Beijing (China); LACM (CIT), Natural History Museum of Los Angeles County, California Institute of Technology collection (U.S.A.); LGBPH, Laboratoire de Géobiologie, Biochronologie et Paléontologie Humaine, Poitiers; LPVP, Laboratoire de Paléontologie Vertébrés Poitiers (France); ISM, Illinois State Museum (U.S.A.); LUVP, Lucknow University, Geological Museum, Lucknow (India); MCZ, Museum of Comparative Zoology, Harvard (U.S.A.); MGL, Muséum Géologique de Lausanne (Switzerland); MNHL, Musée d'Histoire naturelle de la ville de Lyon (France); MNHM, Naturhistorisches Museum Mainz und Landessammlung für Naturkunde Rheinland-Pfalz (Germany); MNHN, Institut de Paléontologie, Muséum National d'Histoire Naturelle Paris (France); NMB, Naturhistorisches Museum Basel (Switzerland); NHME, Natural History Museum of Egée University (Turkey); NMMP, National Museum of Myanmar, Paleontology, Yangon (Myanmar); ONG, Oil and Natural Gas Commission, Dehra Dun (India); PIN, Paleontological Institute, Russian Academy of Sciences, Moscow (Russia); PIT, Paleobiological Institute, Georgian Academy of Sciences, Tbilisi (Georgia); ROM, Royal Ontario Museum (Canada); SMF, Senckenberg Museum Frankfurt (Germany); SMNH, Royal Saskatchewan Museum of Natural History (Canada); SMNS, Staatliches Museum für Naturkunde Stuttgart (Germany); TMM, Texas Memorial Museum (U.S.A.); UM, University of Michigan (U.S.A.); USNM, National Museum of Natural History, Smithsonian Institution V (U.S.A.); USTL, Institut des Sciences de l'évolution Université Montpellier II (France); VLP/K, Panjab University, Chandigarh (India); YPM, Yale Peabody Museum (U.S.A.); ZIN, Zoological Institute, Russian Academy of Sciences, Saint-Petersburg (Russia).

Order ARTIODACTYLA Owen, 1848
Superfamily DICHOBUNOIDEA Gill, 1872

We unite the diacodexeid and dichobunid artiodactyls in the Dichobunoidea. Simpson (1945), Romer (1966), Sudre et al. (1983), and McKenna and Bell (1997) have included Diacodexeidae in the broadly conceived family Dichobunidae. We follow Gazin (1955), Krishtalka and Stucky (1985), and Theodor et al. (2005) and separate Diacodexeidae and Dichobunidae. The following subfamilies are used, partly emended from McKenna and Bell (1997) and Hooker and Weidmann (2000):

Dichobunoidea Gill, 1872
    Diacodexeidae (Gazin, 1955)
    Dichobunidae Turner, 1849
        Dichobuninae Turner, 1849
        Hyperdichobuninae Gentry and Hooker, 1988
        Eurodexeinae Erfurt and Sudre, 1996
        Lantianiinae Métais et al. 2004
    Homacodontidae Marsh, 1894
    Leptochoeridae Marsh, 1894
    Helohyidae Marsh, 1877
    Cebochoeridae Lydekker, 1883 (see next chapter)
    Raoellidae Sahni et al., 1981

The Asian Eocene taxa are temporarily placed into three distinct families that, except for raoellids, are known in Europe and/or North America. Allocations of these poorly known taxa are provisional and will evolve as additional material becomes available. The systematic scheme proposed here does not pretend to reflect phylogeny; only the Raoellidae is sufficiently distinctive to justify its familial rank. The suprafamilial affinities of raoellids are problematic: they have been considered as stem Bunodontia (e.g., McKenna and Bell, 1997) or stem Selenodontia (e.g., Thewissen et al., 2001a). We have opted to leave them *incertae sedis* pending additional fossil data.

Family DIACODEXEIDAE Gazin, 1955

Diacodexeids are small artiodactyls with short snouts. The tribosphenic upper molars bear smaller para- and metaconules. The main cusps are bunodont or crescentiform. The hypocone is usually absent. Lower molars bear moderately elevated trigonids. The shallow entocristid leaves the talonid basin open lingually. Posthypocristid and postentocristid are not directly transversely connected. If present, the hypoconulid connects the posthypocristid and postentocristid. p2–p4 tend to become elongate, with one main cusp and a low talonid; they may also bear a parastylid or a small paraconid. p4 usually bears a lingual metaconid.

Genus *Diacodexis* Cope, 1882

Type Species and Type Specimen. *D. secans* Cope, 1882, AMNH 4899, left and right mandible with p3–m3; Wind River Fm., Wind River Basin (Wyoming). Bighorn Basin (includes *D. metsiacus, D. kelleyi, D. primus* as "lineage segments" [Krishtalka and Stucky, 1985]; junior synonyms of *D. secans: D. laticuneus, D. olseni, D. brachystomus,* and *D. metsiacus).*

Included North American Species. *D. ilicis* Gingerich, 1989, Early Wasatchian, Wa0 and Sandcouleean subage, Sand Coulee, Bighorn Basin (Wyoming); *D. minutus* Krishtalka and Stucky, 1985, Late Wasatchian (Lostcabinian) and early Bridgerian (Gardnerbuttean), Wind River Basin; *D. gracilis* Krishtalka Stucky, and Bakker, 1985 (in Krishtalka and Stucky, 1985), Bighorn Basin; *D.* cf. *secans* Piceance and San Juan Basins (Krishtalka and Stucky, 1986); *Diacodexis* sp., early Bridgerian, South Pass, Wyoming (Gunnell and Bartels, 2001).

Included European Species. *D. gazini* Godinot, 1978, Rians; *D. varleti* Sudre et al., 1983:, Condé-en-Brie, Avenay, Mutigny, Pourcy; *D. antunesi* Estravis and Russell, 1989, Silveirinha; *D. gigasei* R. Smith et al., 1996, Dormaal; *D. corsaensis* Checa, 2004, Corsa inferior, Barranc del Guesot; *D.* cf. *varleti,* Cuis, Grauves, Mancy, Saint-Agnan, Sézanne-Broyes (Sudre et al., 1983: 302), Premontré and Geiseltal site XIV (Erfurt and Sudre, 1996: 376); *D.* indet., Abbey Wood, Kyson (Sudre et al., 1983: 306) ; *D.* sp., Les Saleres and other Spanish localities (Antunes et al., 1997: 344), Creil (De Ploëg et al., 1998: 109)

Included Asian Species. *D.* sp.: Andarak 2 (early Eocene), Kyrgyzstan (Averianov, 1996).

Diagnosis. Upper molars triangular, central protocone, generally lacking hypocone and mesostyle; para- and metaconule well developed; protocone, paracone, and metacone bulbous; hypoconulid on posterior ridge of talonid between entoconid and hypoconid; metaconid moderately inflated; paraconid distinct but reduced, especially on m2–3 (modified from Stucky, 1998).

Dimensions. m2: L = 3.7–5.3 mm (Sudre et al., 1983: Tables 1 and 2; Estravis and Russell, 1989: Table 1; Stucky, 1998).

Discussion. The genus *Diacodexis* may be paraphyletic, as the name has been widely applied to the earliest known artiodactyls on each continent (Gentry and Hooker, 1988; Sudre and Erfurt, 1996; Stucky, 1998; Métais et al., 2004; Theodor et al., 2005), and it is clear that a revision is needed. A number of North American species have been synonymized with *D. secans,* representing morphologically differentiated temporal groups within one lineage ("lineage segments," Krishtalka and Stucky, 1985).

Van Valen (1971) noted the occurrence of *Diacodexis* in Europe based on material from Dormaal. Godinot (1978) described the first European species, *D. gazini* from Rians. Most materials are isolated, poorly preserved teeth; postcranial fragments are rare, and articulated skeletons have not been found. The oldest, most primitive species is *D. antunesi* from Silveirinha in Portugal. In contrast to the oldest American species (*D. ilicis*), *D. antunesi* is smaller: the length of m2 ranges from 3.2 to 3.4 mm (Estravis and Russell, 1989:

Table 1). R. Smith et al. (1996: Fig. 1) proposed that *D. antunesi* gave rise to both *D. gazini* and *D. gigasei.* Smith suggested that North American forms of *Diacodexis* evolved from *D. gigasei. D. varleti* could also be related to *D. antunesi.* All European materials display a plesiomorphic crescentiform, tribosphenic pattern in the upper molars. Unlike North American species, such as *D. secans,* there are no bunodont European dentitions.

An isolated P4 (ZIN 34351) and an M1 (ZIN 34352) from the early Eocene of Andarak 2 (Kyrgyzstan) have been referred to the Asian *Diacodexis* sp. The upper molar is distinctive from other diacodexeids in its derived features: the apparent reduction of its paraconule into a strong preprotocrista and the prominent distolingual cingulum, which tends to form a shelf but not a true hypocone. The prominent cingular shelf on upper molars may be a morphological precursor of the hypocone characterizing European dichobunids. Averianov (1996) suggested possible affinities with *D. antunesi* from Silveirinha based on size, but the limited material referable to *Diacodexis* sp. from Andarak 2 limits its further investigation.

Genus *Bunophorus* Sinclair, 1914

Type Species and Type Specimen. *B. etsagicus* Cope, 1882, AMNH FM 4698, left mandible with p3–m3; Wind River Basin.

Included North American Species. *B. macropternus* Cope, 1882, Bighorn, Green River, Wind River Basins, Wyoming; *B. sinclairi* Guthrie, 1966, Bighorn, Bridger, Washakie, Wind River Basins, Wyoming; Huerfano Basin, Piceance Basin, Colorado; *B. grangeri* Sinclair, 1914 (including *Antiacodon crassus, Wasatchia dorseyana,* and *Wasatchia lysitensis,* according to Stucky, 1998): Bighorn, Piceance, Wind River Basins; San Juan Basin, New Mexico; *B. pattersoni* Krishtalka and Stucky, 1986, Huerfano, Piceance Basins; *B. robustus* Sinclair, 1914, Bighorn Basin; Sand Wash Basin (Colorado); *Bunophorus* sp. and *B.* cf. *sinclairi,* South Pass (Gunnell and Bartels, 2001).

Included European Species. *"B." cappettai* Sudre et al., 1983: Mutigny, Avenay, Pourcy.

Diagnosis. P3/3–M3/3 more robust than in other diacodexeids; proportionately more inflated P4/4, molar cusps and conules; p4 paraconid more reduced; P4 lacking metacone; M1–2 more nearly square, conular wings reduced or absent; included species larger than all diacodexeids except *Simpsonodus* sp. (Krishtalka and Stucky, 1986) and *Diacodexis* (some specimens in lineage segment *D. s.-secans* Krishtalka and Stucky, 1985 (Stucky and Krishtalka, 1990: 150).

Dimensions. m2 (North America), L = 5.8–9.0 mm (Stucky, 1998); m2 (Europe), L = 6 mm (Sudre et al., 1983: 340).

Discussion. *Bunophorus* is known from Europe by isolated teeth of one species. According to Stucky and Krishtalka (1990: 166), materials from Avenay are referable to

Bunophorus; specimens from Mutigny belong to *Simpson-odus.* We refer all European material to *"B." cappettai* as a paraphyletic form until more material is found, as it is larger and has more bulbous teeth with a stronger metaconule than *Protodichobune.* The protocone is mesiolingually shifted and close to the paraconule. In the bulbous lower molars, the trigonid is slightly higher than the talonid. The distally placed, shallow entoconid leaves the talonid open lingually. In contrast to *Diacodexis* and *Protodichobune,* the p4 is stout and bears a strong metaconid in the middle of the lingual side. In contrast to *Buxobune,* there is no lingual cingulum on the upper molars. According to Escarguel et al. (1997: 448), the oldest form comes from Pourcy, and the specimens from Avenay and Mutigny are slightly younger.

Genus *Tapochoerus* McKenna, 1959

Type Species and Type Specimen. *Tapochoerus egressus* Stock, 1934 (=*Hyopsodus egressus*) LACM (CIT) 1590 right p4–m3; Simi Valley, California.

Included Species. *T. mcmillini* Walsh, 2000, San Diego Co., California.

Diagnosis. Teeth more bunodont than selenodont; hypocone on M1 and/or M2; lacks mesostyle and ectoloph ribbing on M1 and/or M2; metaconules large, not displaced, with five posterior crests when unworn; p2 double-rooted and isolated by strong diastemata; p3 trenchant, lacks metaconid; p4 trenchant with distinct entoconid, metaconid separate from protoconid high on crown; lower molars elongate, hypoconulids similar to *Hexacodus;* m1 paraconid strong, separated from metaconid, but more narrowly than in *Antiacodon* or *Auxontodon;* m2 paraconid separation variable; m3 with single lingual trigonid cusp (modified from McKenna, 1959).

Dimensions. m2: L = 7.4–8.0 mm, W = 5.1–5.3 mm, from LACM (CIT) 1292 (McKenna, 1959).

Discussion. Although it was originally considered to be a homacodontine (McKenna, 1959), West and Atkins (1970) placed *Tapochoerus* in the former subfamily Diacodexeinae with *Neodiacodexis, Diacodexis,* and *Bunophorus.* Storer (1984) argued that *Neodiacodexis, Auxontodon,* and *Antiacodon* should be grouped together as Antiacodontinae (within Dichobunidae), citing the squared-off upper molars, strong cingulae, large hypocones, well-developed mesostyles, and strong para- and metaconules. Stucky (1998) added to this list *Sarcolemur* and *Tapochoerus* on the basis of the cristid obliqua rising up the posterior wall to the apex of the metaconid. However, *Tapochoerus* clearly lacks the mesostyle cited by Storer (1984) as one of the features uniting the Antiacodontinae, and two of the three features that potentially unite Antiacodontinae cited by Walsh (2000) are variable and often absent in *Tapochoerus.* Because the characters cited to support Antiacodontinae should be tested in a thorough phylogenetic analysis, we retain the earlier assignment to the diacodexeids.

Genus *Simpsonodus* Krishtalka and Stucky, 1986

Type Species and Type Specimen. *Simpsonodus chacensis* (Cope, 1875b) (= *Pantolestes chacensis*); neotype AMNH 48694, right p3–m3, left p4-m3, left M1–3; San Juan Basin. Bighorn, Piceance Basins (holotype lost, Matthew, 1899).

Included Species. *Simpsonodus* sp.: Piceance, Bighorn Basins (Krishtalka and Stucky, 1986).

Diagnosis. Larger than contemporaneous *Diacodexis;* upper molars more quadrate, lowers with less expanded molar metaconid (excepting *D.* cf. *secans* from the San Juan Basin), higher trigonid, longer talonid. Larger p4 talonid basin, more gracile p4, molar cusps and crowns not bulbous, higher molar trigonids and deeper talonid basins than *Bunophorus* (modified from Krishtalka and Stucky, 1986).

Dimensions. m2: L = 5.5 mm; W = 3.4 mm (Krishtalka and Stucky, 1986: Table 2).

Discussion. Krishtalka and Stucky (1986) originally included *Wasatchia lysitensis* in *Simpsonodus,* but Stucky (1998) placed *W. lysitensis* in *Bunophorus. Simpsonodus* is often confused with *Diacodexis* because they often occur in the same deposits. Although Krishtalka and Stucky (1986) discussed the possible presence of *Simpsonodus* in European deposits, more specimens are needed to confirm it.

Genus *Neodiacodexis* Atkins, 1970 (in West and Atkins, 1970)

Type Species and Type Specimen. *Neodiacodexis emryi* West and Atkins, 1970, AMNH 56054 (P4–M2); Green River Basin, Wyoming.

Diagnosis. P4 with two large sharp cusps; M1 and M2 tritubercular with central protocone, paracone, and metacone equal in size, conical sharp and high; strong conular wings, conules well developed, relatively independent of major cusps; metaconule not hypertrophied; mesostyle present (modified from West, 1984; Stucky, 1998).

Dimensions. m2: L = 6.0 mm; W = 7.5 mm (West and Atkins, 1970).

Discussion. For the reasons outlined above for *Tapochoerus,* we follow McKenna and Bell (1997), placing *Neodiacodexis* among the diacodexeids pending testing of the validity of Antiacodontinae.

Genus *Gujaratia* Bajpai et al., 2005

Type Species and Type Specimen. *G. pakistanensis* (Thewissen et al., 1983), HGSP 300 5003, left maxilla with P3–M3 associated with a left mandible with p4-m3; Barbora Banda I, Kuldana Fm., North-West Frontier Province (NWFP), Pakistan. (This new genus includes *"Diacodexis" pakistanensis* and the type material designated by Thewissen et al., 1983.)

Included Species. *G. indica* Bajpai et al., 2005: Vastan lignite mine, Taluka Mangrol, Surat District, State of Gujarat (India).

Diagnosis. Small size; short C/P1, P1/P2, P2/P3 diastemata; well-developed lingual cusp on P3 and P4; triangular upper molars with continuous cingulum, interrupted on the lingual face of the protocone, particularly thick mesially and distally; lacks hypocone; paracone and metacone slightly crested anteriorly and posteriorly, medially situated conules barely crested. p1–4 simple, triangular in lateral view; distinct paraconid on p3/p4; m1–3 bear small paraconids; entoconid small; hypoconulid incorporated in posthypocristid; trigonid transversely narrower than the talonid in m1–2; m3 with large third posterior hypoconulid lobe. *Gujaratia* (= *"Diacodexis" pakistanensis*) retains a clavicle and has five digits in the manus and four in the pes. Postcranial features are described in detail and discussed in Thewissen and Hussain (1990) (adapted from Thewissen et al., 1983; Bajpai et al., 2005).

Dimensions. m2 (*G. pakistanensis*): L = 3.7 mm; W = 3.1 mm (Thewissen et al., 1983: Fig. 6).

Discussion. The generic assignment of *"D." pakistanensis* from Pakistan has long been questioned (Gentry and Hooker, 1988; Sudre and Erfurt, 1996; Stucky, 1998; Métais et al., 2004; Theodor et al., 2005). Bajpai et al. (2005) included all the specimens from the Kuldana Fm. (Pakistan) previously referred to *Diacodexis pakistanensis* within their new genus *Gujaratia*. This generic distinction is justified in regard to the distinct dental and postcranial morphology of specimens from the Indian Subcontinent. *G. indica* Bajpai et al., 2005 is based on a lower jaw from the Vastan lignite mine, Taluka Mangrol (Surat District, India), described by Bajpai et al. (2005) and was diagnosed as larger than *G. pakistanensis*.

Although *G. pakistanensis* is based on a larger sample that shows intraspecific variation in dental features (Thewissen et al., 1983), *G. indica* shows several dental characters that are distinctive from the material described from Pakistan. Those include the cingulids and paraconid being more developed on p3 and p4, the stronger paraconid on m1–2, and the trigonid of m3 appearing transversely wider in *G. indica* than in *G. pakistanensis*.

The age of *Gujaratia* (= *"D." pakistanensis*) is important because this taxon is often considered to be more primitive than the earliest European and North American species of *Diacodexis*. The continental red beds of northern Pakistan that yielded mammal assemblages are probably diachronous, and the age of the different localities of Punjab and North West Frontier Province (NWFP) is still debated. The type specimen was found in the Mami Khel Clay (lower part of the Kuldana Fm.), at the locality Barbora Banda I (NWFP), a locality Leinders et al. (1999) and Thewissen et al. (2001a) considered to be earliest Eocene on the basis of comparisons with other mammal assemblages of the Kuldana Fm. Gingerich (2003), based on benthic foraminifera and the pattern of transgressions and regressions, suggested an early

middle Eocene age for all the mammal faunas from the Kuldana Fm. in Pakistan as well as those from the Subathu Fm. in India (Sahni and Jolly, 1993). The dentition and postcranial material of *Gujaratia* (Thewissen and Hussain, 1990) are clearly distinct from those of North American *Diacodexis* and, to a lesser degree, from the European species known only by dental remains. Thewissen et al. (1983) stressed that North American and European species of *Diacodexis* differ in the greater lingual length of upper molars, the greater development of the paraconid on lower molars, and m3 morphology. However, it appears unlikely that North American *Diacodexis* was derived from *Gujaratia*. We here consider this form as the most primitive artiodactyl of the Indian Subcontinent, probably close to the ancestry of the radiation of artiodactyls in that region.

Family DICHOBUNIDAE Turner, 1849
Dichobunidae *Incertae Sedis* from Asia

Most of the Asian forms reviewed below are based on limited dental material, making any systematic allocation particularly difficult. Most workers have tried to attach these forms to early artiodactyls known in Europe or North America, but these provisory allocations may not reflect the diversity of Asian artiodactyls during the early Paleogene, and more data will likely change their placement. The term "dichobunoid" employed below should be understood in the sense of stem Asian artiodactyls, but their relationship to contemporaneous forms from Europe and North America is still poorly understood. This section aims to gather information about these Asian dichobunoids often scattered in the literature and to encourage workers to improve the fossil record of artiodactyls in Asia.

Genus *Haqueina* Dehm and Oettingen-Spielberg, 1958

Type Species and Type Specimen. *H. haquei* Dehm and Oettingen-Spielberg, 1958, BSPG 1956 II 10, left m2–3; Ganda Kas 21, Kuldana Fm., Pakistan.

Included Species. *H. haichinensis* Vislobokova, 2004b: Khaichin-Ula II, Khaichin Fm., Mongolia.

Diagnosis. P2 elongate, with small posterolingual cingular shelf; P3 and P4 with lingual protocone, well-developed cingulae; upper molars retain a small paraconule, strong cingulae; anterior lower premolars separated by diastemata; p4 slightly inflated, lacks paraconid; bunodont to bunolophodont lower molars, absence of paraconid and faint crest joining protoconid and metaconid anteriorly; distinct hypolophid connecting hypoconid and entoconid on m2, attenuated on m3; hypoconulid small and medially situated on m2, larger and twinned on m3.

Dimensions. m2 (holotype): L = 7.8 mm; W = 5.7 mm (Dehm and Oettingen-Spielberg, 1958).

Discussion. This genus has long been known only by the type specimen collected in the Ganda Kas area, and its affinities are ambiguous. *Haqueina* is comparable in size and morphologically close to *Chorlakkia* (Gingerich et al., 1979) but differs in its more lophodont dental pattern and centrally located hypoconulid not twinned with the entoconid on m2, although this cusp is twinned on the third lobe of m3. Thewissen et al. (2001a) claimed that *Haqueina* cannot be allied with dichobunids because it lacks a paraconid and has a somewhat connate protoconid and metaconid, although these features also characterize *Chorlakkia*. The definition of the genus was substantially improved with the report of additional dental material (including upper dentition) of *Haqueina* from the middle Eocene of Mongolia (Vislobokova, 2004b). The molars of *H. haichinensis* do not show a clear tendency toward lophodonty as in the raoellids *Khirtharia* and *Metkatius*. Likewise, the upper molars referred to *Haqueina* by Vislobokova (2004b) are closer to a lower stage of bunoselenodont molar pattern than the bilophodont pattern that characterizes the Raoellidae. Thus, it is still unclear whether *Haqueina* should be included within raoellids. We favor its placement close to *Chorlakkia* within the *Dichobunidae,* although this allocation should be seen as provisory.

Genus *Dulcidon* Van Valen, 1965

Type Species and Type Specimen. *D. gandaensis* Van Valen, 1965, BSPG München 1956 II 1, left ?M1; locality 1, Ganda Kas area, Kuldana Fm. (Pakistan) (first figured in Dehm and Oettingen-Spielberg, 1958, under *"Promioclaenus? gandaensis"*).

Diagnosis. Upper molars with winged conules; paracone and metacone equal in size; postprotocrista stronger than preprotocrista, which extends labially toward weak stylar shelf lingual to paracone; labial cingulum moderately developed, thicker below the paracone; precingulum weak, postcingulum very faint except at the level of the protocone, where it forms an apparent shelf (adapted from Van Valen, 1965).

Dimensions. ?M1 (holotype): L = 4.1 mm; W = 4.8 mm (Dehm and Oettingen-Spielberg, 1958).

Discussion. The identification of *Dulcidon,* based on a unique poorly preserved, fairly worn upper molar, is uncertain, as the taxon has been thought to be a hyopsodontid (Dehm and Oettingen-Spielberg, 1958), a new genus of insectivore paroxyclaenid (Van Valen, 1965), and a dichobunid artiodactyl (Gingerich and Russell, 1981; Thewissen et al., 1987). Thewissen et al. (1987) suggested that additional material might prove it to be congeneric with *Chorlakkia* because of the overall resemblance. Although it is close in size to *C. hassani,* the type of *Dulcidon* differs in its more triangular occlusal outline and the distinct configuration of cingulae on upper molars. The basic tribosphenic pattern and triangular outline of the type specimen might suggest close affinities with *Gujaratia pakistanensis.* Only additional material, including lower dental elements, could shed light on the uncertain status of *D. gandaensis.*

Genus *Paraphenacodus* Gabunia, 1971

Type Species and Type Specimen. *P. solivagus* Gabunia 1971, PIT 3–21a, right m1; Chakpaktas Svita and lower Obayla Subsvita, Zaysan Depression (eastern Kazakhstan).

Diagnosis. Cheek teeth low-crowned and bunodont; lacks paraconid; strong mesial cingulid; talonid cuspids bulbous; entoconid and hypoconid subequal in size, connected by weak hypolophid, smaller hypoconulid on m1–2 connected to hypoconid by a faint crest; prominent labial cingulid (adapted from Gabunia, 1971).

Dimensions. m1 (holotype) L = 5.9 mm; W = 4.3 mm (Gabunia, 1971).

Discussion. *Paraphenacodus* is known from one lower molar because the other heavily worn lower molar was labeled as "Dichobunoidea indet" in Gabunia (1977). The two specimens probably belong to the same species, among the largest of the Asian dichobunoids. Gabunia (1971) tentatively referred these lower molars from the Zaysan Basin to phenacodontid condylarths, but the absence of a paraconid and the overall morphology do not exclude *Paraphenacodus* from artiodactyl affinities. Crests are poorly developed, and cusps bulbous, the paraconid is lost, the faint cristid obliqua seems to rise to the protoconid fairly labially, and the hypoconulid is medially placed on the m1 figured by Gabunia (1964a). These supposedly derived features may indicate either condylarth affinities or the occurrence of a bunodont lineage of dichobunid in Central Asia showing a dental pattern comparable to that of contemporaneous bunodont forms (*Buxobune* in Europe or *Ibarus* in North America).

Genus *Aksyiria* Gabunia, 1973

Type Species and Type Specimen. *A. oligostus* Gabunia, 1973; PIT No. 25, right M1 or M2; Chakpaktas Svita and lower Obayla Subsvita, Zaysan Depression (eastern Kazakhstan).

Diagnosis. Subquadrate in outline; basic tribosphenic pattern with voluminous protocone; poorly expressed crests; conules large and fairly crescentic; labial cingulum thick and continuous; lacks lingual cingulum and distinct hypocone on the distolingual cingular ledge (adapted from Gabunia, 1973).

Dimensions. M1 or M2 (holotype): L = 4.5 mm; W = 5.6 mm (Gabunia, 1973).

Discussion. Initially thought by Gabunia (1973) to be a small dichobunid close to *Diacodexis,* but the affinities of *Aksyiria* are still enigmatic. Thewissen et al. (1983) suggested that this unique molar might be a deciduous tooth, and Sudre (1978) claimed some resemblance with an indeterminate dichobunid from La Livinière (MP 15). In the absence of further fossil evidence, *Aksyiria* is provisionally reassigned to the heterogeneous group of Asian dichobunids.

Genus *Chorlakkia* Gingerich et al., 1979

Type Species and Type Specimen.   *C. hassani* Gingerich et al., 1979, GSP-UM 77082, left mandible with m1–3; near Chorlakki village Kuldana Fm. (NWFP, Pakistan).

Included Species.   *D. valerii* Vislobokova, 2004a: Khaichin-Ula II locality, Khaichin Fm., Mongolia.

Diagnosis.   Small artiodactyls with bunodont dentition. Upper molars tribosphenic-like with bulbous cusps; hypocone absent; cingulae moderately developed, absent lingual to the protocone. Lower molars with subequal protoconid and metaconid; no apparent paraconid; talonid broadly basined, presence of twinned entoconid and smaller hypoconulid, the latter roughly medially located; hypoconulid larger on m3, forming a distinct lobe (adapted from Gingerich et al., 1979).

Dimensions.   m1 (holotype of *C. hassani*): L = 3.6 mm; W = 2.5 mm (Gingerich et al., 1979).

Discussion.   After *Gujaratia pakistanensis*, *Chorlakkia* is the smallest artiodactyl from the Eocene of Indo-Pakistan. It is clearly more bunodont than the Barbora Banda form but does not show any trend to a bunolophodont pattern as raoellids do. Upper molars (Thewissen et al., 1987) resemble those of *G. pakistanensis* but are more bunodont. Affinities of *Chorlakkia* are not well established, although close relationships with *Gujaratia* are likely. *Chorlakkia* has also been recognized in the middle Eocene of Mongolia (Vislobokova, 2004a) and thus would not be endemic to the Indian Subcontinent.

Genus *Pakibune* Thewissen et al., 1987

Type Species and Type Specimen.   *P. chorlakkiensis* Thewissen et al., 1987, GSP-UM 259, right m3; near Chorlakki Village, Kuldana Fm. (NWFP, Pakistan).

Diagnosis.   Moderate-sized artiodactyl larger than *Gujaratia pakistanensis*, *Chorlakkia*, and *Dulcidon*; protoconid slightly larger than metaconid; paraconid well developed, mesiolingually located; metacristid much weaker than paracristid, which is mesially arched, connected to the paraconid; cristid obliqua faint, medially oriented; hypoconid larger than entoconid; hypoconulid present, connected to both entoconid and hypoconid by weak crests; mesial and labial cingulids well developed, the latter being interrupted below the hypoconulid; m3 hypoconulid strong, medially situated (adapted from Thewissen et al., 1987).

Dimensions.   m3 (holotype): L = 6.0 mm; W = 3.2 mm.

Discussion.   An m2 is also referred to this genus (Thewissen et al., 1987), but elements of the upper dentition are still unknown. *Pakibune* differs from other supposed dichobunids from the Indian Subcontinent by its larger size, relatively greater development of cingulids, and strong, fairly arched paracristid on the trigonid. The molar morphology resembles lowers of *Haqueina*, especially in the development of cingulids and general configuration of crests, but *Haqueina* has a more reduced paraconid. The overall morphology sets it closer to *Gujaratia pakistanensis* than to *Chorlakkia* (Thewissen et al., 1987).

Dichobunidae indet. A and B (cited by Thewissen et al., 1983, 1987)

Geographic Distribution.   Chorlakki locality, Kuldana Fm. (NWFP, Pakistan).

Remarks.   Thewissen et al. (1987) figured a single fragmentary m3 equivalent in size to that of *Pakibune* but differing in its reduced entoconid, weaker crests and cingulids. An M3 was also questionably referred to the Dichobunidae; this unique molar displays a reduced metaconid that led Thewissen et al. (1987) to suggest hyopsodontid affinities.

"Dichobunid indet. large" (West, 1980b; Thewissen et al., 2001a)

Geographic Distribution.   Ganda Kas area, Kuldana Fm. (Punjab, Pakistan).

Remarks.   The existence of this large dichobunid is based on a partial upper molar (HGSP 1974D) described and referred to the helohyid *Gobiohyus* by West (1980b). Thewissen et al. (2001a) added a damaged left P4 from the same area that could match with the upper molar and suggested possible affinities of those specimens with large North American dichobunids such as *Bunophorus* and *Wasatchia*.

Artiodactyla indet. 1 Tsubamoto et al., 2005

Geographic Distribution.   Pondaung Fm. (Myanmar).

Discussion.   The specimen (NMMP-KU 1556), questionably designated as a left M2 and referred to "Artiodactyla indeterminate 1" by Tsubamoto et al. (2005), likely belongs to a new genus of Asian dichobunid. The diagnostic characters of this upper molar, such as the distinct hypocone emerging from a thick posterolingual cingulum and the central position of the protocone associated with weak cresting, are common dental features present in European dichobunids. This tooth obviously belongs to a more derived dichobunid than those from the Indo-Pakistan region and is equivalent in its "morphological grade" to the early representatives of the European genus *Dichobune* (*D. robertiana*). North American "homacodonts" and leptochoerids from the Uintan are much smaller and morphologically distinct from the Burmese form. Even if the content of the family and its monophyly are disputable, the new specimen from Myanmar could be confidently assigned to the "dichobunoid complex" (see Métais et al., 2004).

Genus *Wutuhyus* Tong and Wang, 2006

Type Species and Type Specimen.   *W. primiveris* Tong and Wang 2006, IVPP V 10720, left mandible preserving p2–m3; Wutu Fm. (Shandong Province, China).

Dimensions.   m1 (holotype): L = 4.1 mm; W = 3.3 mm.

Discussion.   The premolar morphology is simple, p1 is leaf-like and separated from p2 by a slight diastema, p4 is inflated. The molars are bunodont; there is a marked gradient of molar size from m1 to m3; cuspids are all strikingly inflated; the trigonid occupies about two thirds of the total length of the molar with a prominent paraconid situated on the same longitudinal line as the metaconid. The talonid consists of three bulbous cuspids, including a medially located hypoconulid. The third lobe of m3 is duplicated in two equally developed cuspids on the same transverse row. Tong and Wang (1998) questionably assigned this specimen to entelodontids, and finally referred it to indeterminate suiform in the formal description. If this apparently new genus proves to belong to Artiodactyla (upper dentition and tarsal evidence would be helpful), its extreme bunodonty suggests early differentiation of a bunodont lineage of artiodactyls in Asia; supporting a Paleocene origin for the order. This speculative hypothesis would corroborate the phylogeny proposed by Gentry and Hooker (1988) and could resolve issues related to the relatively derived postcranial features of *Diacodexis*. The Wutu fauna is thought to be contemporaneous with Wasatchian faunas of North America (Tong and Wang 2006), but correlation with the Clarkforkian LMA has also been suggested (Beard and Dawson, 1999). If correct, this inferred age makes *Wutuhyus* the earliest occurrence of artiodactyls in Asia.

Further Preliminary Determinations

Dichobunoidea Indet.   Pondaung Fm. (Myanmar); Métais (2006).

?*Dichobune* sp.   Locality 7, Rencun Member, Heti Fm., Shanxi Province (China); Zdansky (1930), Métais et al. (in press).

***Dichobune* sp.**   Upper part of Lushi Fm. (Henan Province, China); Chow et al. (1973).

Subfamily DICHOBUNINAE Turner, 1849

Larger dichobunids with unspecialized, bulbous dentition and slightly elongated snouts. Upper molars tend to bear three broad distal cusps and a hypocone. The metaconule becomes larger than the paraconule. P3 displays a protocone, p4 a metaconid. The first lower premolar is premolariform. The dentition is generalized but more bunodont than diacodexeids. The presence of a hypocone is variable in less differentiated genera (*Protodichobune* as well as *Diacodexis*) ranging from absence to a cuspule to a full hypocone (Sudre et al., 1983; Estravis and Russell, 1989). This character pattern may reflect variability in the earliest Eocene that became functionally important and taxonomically significant in the middle Eocene. Dichobunines may have their origin in bunodont lineages of *Diacodexis,* via *Protodichobune* or similar forms in the earliest Eocene.

Genus *Dichobune* Cuvier, 1822

Type Species and Type Specimen.   *D. leporina* Cuvier, 1822, MNHN GY-655, mandible with p2, p3, m3; Montmarte (Gypsum of Paris). Aubrelong 1, Baby, Bembridge Limestone, Coânac, Escamps, Frohnstetten, Hoogbutsel, La Débruge, Mas S^tes Puelles, Eclépens-Gare, Perrière, Pont d'Assou, Rosières 2, 4; Ste Néboule; Saindou D; Villeneuve-la-Comptal.

Included Species.   *D. robertiana* Gervais, 1848–52: Calcaire Grossier. Aumelas, La Défense, St. Martin-de-Londres (Sudre, 1988a: 412); *D. fraasi* Schlosser, 1902: Eselsberg near Ulm; *D. sigei,* Sudre, 1978: Lavergne, Le Bretou; *D. jehennei* Brunet and Sudre, 1980: Villebramar, La Plante 2, Pinchenat; *D.* cf. *robertiana:* Egerkingen, Bolus, Grauer Ton (Stehlin, 1906: 617); Lissieu (Sudre, 1988a: 414); *D.* cf. *fraasi:* Gösgen, Hoogbutsel (Sudre, 1978: 31); *D.* sp. or indet. (summarized): St.-Maximin, Tabarly (Sudre, 1978, 1997).

Dimensions.   m2: L = 7.3 mm; p2–m3 (holotype *D. leporina*): L = 46 mm (Sudre, 1978: Table 2, 26); skull (*D. jehenni*): L = 130 mm (Brunet and Sudre, 1980: 128).

Diagnosis.   Medium-sized dichobunids with slightly elongated skull. Neurocranium rounded, orbits large and subcentral. Mastoid well-developed, tympanic bulla unossified. Upper molars quadratic, bunodont, buccal cusps crested. M1 and M2 with strong hypocone and preprotocrista, including a paraconule. Postprotocrista absent. Lower molars with rounded lingual cuspids, slightly crested buccal cuspids; talonid basin is lingually open, lacks a preentocristid. m1 and m2 distal cingulid bears a hypoconulid. All premolars display a paraconid and hypoconid; p4 a lingual metaconid; p2 mesial and distal diastemata (emended from Sudre, 1978: 24).

Discussion.   The genus is well represented in Western Europe by dentitions (Sudre, 1978; Brunet and Sudre, 1980). In addition, cranial remains are described by Dechaseaux (1974). The mastoid of *D. leporina* is covered by a thin temporal process (Stehlin, 1906: 606). The third and fourth metapodials are longer than the second and fifth. Zeugopodials, metapodials, and carpals are unfused. MC III articulates with the magnum, and MC II with the trapezoid. The oldest and smallest species is *D. robertiana* from Aumelas, which with *D.* cf. *robertiana* from Egerkingen belongs to one lineage (Sudre, 1988a). A further lineage is represented with *D. sigei* as ancestor of *D. leporina*. The ancestor of the genus is probably an unknown dichobunid similar to *Messelobunodon*. *D. langi* Rütimeyer, 1891, *D. nobilis* Stehlin, 1906, *D. spinifera* Stehlin, 1906, and another undetermined form were transferred to *Hyperdichobune* by Stehlin (1910a).

Genus *Synaphodus* Pomel, 1848

Type Species and Type Specimen.   *S. brachygnathus* Pomel, 1848, MNHN LIM 265, left mandible with alveoli for c-p4, m1-m3, molars poorly preserved; Authezat.

Dimensions.   m2: L = 12 mm; c-m2: L = 90 mm (MNHN LIM 265).

Diagnosis.   Large dichobunid similar to *Metriotherium*, but with more bunodont lower dentition. Canine large; premolars with paraconid; lacks p2/p3 diastema; p4 longer than in *Metriotherium*, bears a strong metaconid in the centre of the lingual side (emended from Lavocat, 1951).

Discussion.   The type specimen is not mentioned in Pomel (1848). The genus is probably referred to "*Cyclognathus gergovianus*" from the old collection of Abbé Croizet. Based on similarities of *Synaphodus* with *Metriotherium*, some authors (Viret, 1961) synonymized it with *M. mirabile*. Sudre (1978: 48) restored the genus because of differences in the degree of molarization compared with *Metriotherium*. He saw the origin of *S. brachygnathus* near *Dichobune leporina*. The upper dentition is unknown, and the taxonomic and biostratigraphic status remains unclear.

Genus *Protodichobune* Lemoine, 1878

Type Species and Type Specimen.   *P. oweni* Lemoine, 1878, MNHN Al 5232, mandible with p4–m3; *Teredina*, sands near Epernay, Chavot, Cuis, Grauves, Mancy, Mas de Gimel, Monthelon, Saint-Agnan.

Included Species.   *P.* cf. *oweni:* Geiseltal site XIV (Erfurt and Sudre, 1996: 377).

Diagnosis.   Small dichobunid artiodactyls with bunodont dentition, lacking diastemata. Upper molars triangular, tritubercular, conules relatively large. M1 and M2, paracone and metacone equal in size, not crested. M2 is the largest molar, may bear a mesostyle. M3 paracone reduced; p1 incisiform; p4 with paraconid, lacks metaconid. Trigonid only slightly higher than talonid. Paraconid separated from metaconid, which is the highest conid. Wide entoconid is distally situated and connected with the hypoconulid by a postentocristid (emended from Sudre et al., 1983: 313).

Dimensions.   m2: L = 6.1 mm (Sudre et al., 1983: 316); p4–m3 (type *P. oweni*): L = 24 mm.

Discussion.   A second species of *Protodichobune*, described as *P. lyddekkeri* (Lemoine, 1891: 287), is now regarded as the genus *Cuisitherium* (Sudre et al., 1983: 351). Apart from the type specimen of *P. oweni*, the genus is known from isolated teeth. The high variability among French collections caused Sudre et al. (1983) to introduce two morphotypes. Type A shows bunodont upper molars with rounded cusps, strong cingulae, and a mesostyle, which sometimes occurs on M2. Type B is more crescentiform, displays a trapezoidal outline, and lacks a mesostyle. These types are based on upper teeth, and the taxonomic state of these morphotypes remains problematic.

Genus *Metriotherium* Filhol, 1882

Type Species and Type Specimen.   *M. mirabile* Filhol, 1882, MNHN Qu 84, left mandible with dentition alveoli for p1-2, p3–m3 from Quercy (without indication), Authezat, Le Garouillas, Rabastens, Saint-André, Tauriac; Mümliswyl-

Hardberg; Carrascosa del Campo (Lacomba and Morales, 1987).

Included Species.   *M. paulum* Stehlin, 1906: Bach; *M. minutum* Brunet and Sudre, 1980: Villebramar, Montalban, and La Plante; *M. sarelense* Sudre, 2003, in Astruc et al. (2003: 646): Sarèle; *M.* cf. *sarelense:* Puycelci (Astruc et al., 2003: 644).

Dimensions.   m2: L = 13 mm; p4-m3 (*M. mirabile*): L = 55 mm (Brunet and Sudre, 1980: Table 3; Sudre, 1995: 208; Astruc et al., 2003: 646).

Diagnosis.   Large dichobunids with selenodont dentition. Upper molars with oblique buccal side as a result of the lingually shifted position of paracone and metacone. Lacks paraconules; strong hypocone, distally crested. Cuspids crested, trigonid slightly higher than talonid. A lophid-like cristid runs transversely to hypoconid from entoconid (emended from Astruc et al., 2003: 645).

Discussion.   *Metriotherium* is the sole dichobunid genus in the late Eocene of Europe. The size difference between the younger, smaller *M. sarelense* from Sarèle (MP 27) and the larger, older *M. mirabile* from Carrascosa del Campo (MP 25) could indicate either two different lineages or sexual dimorphism. Unfortunately, most of the species are represented only by isolated teeth, and the genus is in need of revision. The ancestor of *Metriotherium* could be seen in preliminary determined forms of *Dichobune* (see above).

Genus *Meniscodon* Rütimeyer, 1888

Type Species and Type Specimen.   *M. picteti* Rütimeyer, 1891, NMB EG 513, isolated M2; Egerkingen blaue Mergel (synonymy see Erfurt and Sudre, 1995a: 867). Egerkingen Huppersand, grauer Mergel, Bouxwiller, Geiseltal site Leo III, XLI.

Included Species.   *M. europaeum* (Rütimeyer, 1888): Egerkingen Bolus, Huppersand, Arcis le Ponsart, Bouxwiller, Calcaire Grossier, La Défense, Lissieu.

Diagnosis.   Medium-sized dichobunines with subquadratic M1 and M2. Upper molars broader than long, with a buccal entoflex and parastyle, lack mesostyles. M3 subtriangular in outline. All main cusps triangularly arranged. Paraconule close to protocone, smaller than metaconule. Lower molars subquadratic; trigonid only slightly higher than talonid. Fused paraconid and metaconid. Premolars short; p4 bears a metaconid.

Dimensions.   m2: L = 11.2 mm (Erfurt and Sudre, 1995a: Table 1).

Discussion.   In contrast to Stehlin (1906: 654), some of the Swiss *Meniscodon* specimens are from Egerkingen. They belong to *M. europaeum*, which is distinguishable from *M. picteti*. The sudden occurrence of *Meniscodon* was regarded as evidence of an immigration event at the beginning of the middle Eocene in Europe (Sudre, 1972: 131). However, rapid evolution in size during one or two MP levels could account for this pattern, with origins of the genus in Europe. This is supported by smaller *M. picteti* in Buchswiller and at Geiseltal (Erfurt and Sudre, 1995a), where this lineage occurs in MP 13. Apart from one calcaneus from Geiseltal, only dental fragments are known, indicating that *Meniscodon* was rare in comparison to other artiodactyls in all fossil assemblages. The ancestor is still unknown but should lie in the clade of bunodont diacodexeids with incipient hypocones.

Genus *Buxobune* Sudre, 1978

Type Species and Type Specimen. *B. daubreei* Sudre, 1978, NMB Bchs 516, isolated M1; Bouxwiller.

Included Species. *B.* aff. *daubreei:* Geiseltal site XIV (Erfurt and Haubold, 1989).

Dimensions. m2: L = 5.6 mm (Sudre, 1978: Table 1; Erfurt, 1988).

Diagnosis. Larger dichobunine artiodactyls with bunodont dentition. Metaconule much stronger than paraconule; buccal cusps slightly crested; a cingulum surrounds the buccal and lingual sides of the molars and protocone; lacks mesostyle. Trigonid not significantly higher than talonid. Paraconid closely appressed to metaconid; hypoconid and protoconid slightly crested. Entoconid round, higher than in *Diacodexis*. m1 and m2 bear a broad precingulid and weak hypoconulid. Lower premolars stout; p4 not longer than m1. p4 with lingual metaconid, small hypoconid in distal prolongation of main cusp (emended from Sudre, 1978: 21).

Dimensions. L m2 = 5.6 mm (Sudre, 1978: Table 1).

Discussion. This genus represents a further lineage of larger bunodont dichobunines, which differs from the similar-sized *Aumelasia* in stronger metaconules and the lack of mesostyles. With tetratubercular upper molars, *Buxobune* resembles less differentiated choeropotamids such as *Haplobunodon*. According to Sudre (1978), *Protodichobune* is the direct ancestor of *Buxobune*. The contemporaneous occurrence of *Buxobune* and first choeropotamids in MP 11 means it cannot be ancestral, but a common ancestor in the group around *Protodichobune* is likely. The latter is indicated by *B.* aff. *daubreei,* which occurs in the Unterkohle of Geiseltal (MP 11) and connects the type specimen from Bouxwiller (MP 12) with the findings of *P. oweni* from MP 10.

Genus *Aumelasia* Sudre, 1980

Type Species and Type Specimen. *A. gabineaudi* Sudre, 1980, USTL AUM 145, maxilla with P4-M3; *Aumelas.*

Included Species. *A. menieli* Sudre et al., 1983: Mancy, Cuis, Mont-Bernon, Monthelon; *A. maniai* Erfurt and Haubold, 1989: Geiseltal site LII; *A.* cf. *gabineaudi:* Messel (Franzen, 1988: 309); *A.* aff. *menieli:* Geiseltal site XIV (Erfurt and Haubold, 1989: 137).

Diagnosis. Bunodont dichobunines, larger than *Protodichobune*. Upper molars: protocone mesially shifted; distolingually situated metaconule only slightly larger than paraconule; M2 is longest and broadest tooth, bears a strong mesostyle. Lower molars with broad crown base; reduced paraconid; large hypoconid; entoconid high, mesially located. m2 much longer than m1 (emended from Sudre et al., 1983: 327).

Dimensions. m2: L = 7.2 mm; P4–M2 (type *A. gabineaudi*): L = 22 mm; skull (*A.* aff. *gabineaudi*): L = 92 mm; postcranial data see Marcot (this volume) and Franzen (1988: Table 1).

Discussion. *Aumelasia* can be considered as descendant of *Protodichobune* type A (Sudre et al., 1983). The genus displays a size increase from *A. menieli, A. gabineaudi, A.* aff. *gabineaudi* to *A. maniai* and an increasing in differentiation of p3 and p4 (*A. maniai* bears tiny entoconids distolingual to the protoconid). In addition, there are incipient hypoconids on p2–p4. The most complete specimens of the genus belong to *A.* cf. *gabineaudi* from Messel with nearly complete skeletons (Franzen, 1988). *Aumelasia* is cursorially adapted, with longer hindlimbs than forelimbs. Preserved gut contents, which contain seeds and fruits (Franzen and Richter, 1988) support the interpretation of *Aumelasia* as a browser and mixed feeder. Another morphotype is *A.* aff. *menieli* from the Unterkohle of Geiseltal, which is less differentiated than *A. maniai* and *A. gabineaudi*. The specimens from the Unterkohle and Messel likely belong to another species.

Genus *Messelobunodon* Franzen, 1980

Type Species and Type Specimen. *M. schaefer,* Franzen, 1981, SMF-ME 501, nearly complete skeleton with dentition; Messel. *Messelobunodon* was first mentioned by Franzen and Krumbiegel (1980) with the description of *"M." ceciliensis* from Geiseltal. Because of a delay in printing, Franzen's diagnosis was published in 1981. According to IUZN § 68a (Kraus, 2000), the nomenclature is retained as *Messelobunodon,* Franzen, 1980 based on the primary definition of the type species *M. schaeferi* (Franzen, 1981: 315).

Included Species. *M.* aff. *schaeferi,* Geiseltal site XIV (Erfurt and Haubold, 1989).

Diagnosis. Medium sized dichobunines with buccally rounded upper molars, lacking styles and hypocone. M1 and M2 subrectangular, M3 triangular. The snout is elongated by diastemata mesially and distal to P2/p2. P3 bears a distolingually shifted protocone. Lower molars with isolated paraconid; trigonid higher than talonid. Third lobe on m3 short; weak hypoconulid. p2–3 with paraconid and parastylid; p4 shorter than p3 and p2, lacks metaconid. P1/p1 caniniform, not elongated; canines incisiform.

Dimensions. m2: L = 5.3 mm, skull (type *M. schaeferi*) L = 106 mm (Franzen, 1981: Tables 1, 2); body length including tail about 60 cm.

Discussion. *Messelobunodon* is closely related to bunodont forms of *Protodichobune* (Franzen, 1981). In contrast to other bunodont Messel artiodactyls, *M. schaeferi* bears an elongated snout and lacks caniniform premolars and molarized P4 (see diagnosis of choeropotamids, Marcot, this volume). It is the single European dichobunine for which the postcranial skeleton and gut content are known. These animals are slender, their hind limbs are elongate, and the tail is very long. All zeugopodials are unfused; the fibula is very thin. The cuboid and navicular are unfused. The paraxonic metatarsus includes four unfused bones, of which MT III and MT IV are elongated. The forelimb contains five metacarpals, and MC I and MC V are about half of the length of the equal-sized MC III–IV. All terminal phalanges are clawlike but not fissured. From the gut contents (Richter, 1981, 1987), the food consisted mostly of mushrooms, fruits, and leaves.

Another smaller specimen from Messel, also described as *Messelobunodon* (Franzen, 1983), was later referred to as *Eurodexis* (Erfurt and Sudre, 1996). In contrast to the holotype, there is a hypocone on the upper molars, p2 bears a hypoconid, the femur displays a third trochanter, and the P1 is premolariform. Isolated tooth fragments from the Unterkohle of Geiseltal are more similar to *M. schaeferi* (Erfurt and Haubold, 1989), and an isolated astragalus, which resembles the holotype. The state of *M. n. sp.* (level of Grauves; BiochroM'97: 784) needs revision.

Genus *Neufferia* Franzen, 1994

Type Species and Type Specimen. *N. manderscheidi* Franzen, 1994, MNHM PW 1994/545-LS, fragmentary left lower jaw with m1–m3; Eckfeld (Haupt-Turbidit).

Diagnosis. Bunodont dentition with sharp crests, short premolars. M3 with mesostyle and hypocone. m1 and m2, preentocristid mesiobucally oriented toward the middle of prehypocristid; hypoconid and entoconid divided by a deep valley, without hypoconulid. (Franzen, 1994: 196).

Dimensions. m2: L = 7.3 mm; m1–m3 (type) L = 23.8 mm (Franzen, 1994: 197).

Discussion. In contrast to *Aumelasia*, *Neufferia* bears an additional lophid-like cristid on the lower molars. This cristid runs from the mesiobuccal base of the entoconid to the lingual side of the prehypocristid and divides the talonid in a mesial and a distal portion. This differentiation of the talonid is also found in the choeropotamid *Amphirhagatherium weigelti* from Geiseltal, which lacks a hypocone. This cristid seems functionally uncorrelated with a mesostyle and evolved independently in the two families.

Subfamily EURODEXEINAE Erfurt and Sudre, 1996

On the basis of newly discovered small crescentiform teeth of artiodactyls in German and French localities, Erfurt and Sudre (1996) described Eurodexeinae to explain differ-

ent parallel middle Eocene lineages of small artiodactyls in Europe and North America. They followed the restriction of homacodontines to less-differentiated, older taxa by Gentry and Hooker (1988). In contrast to dichobunines, eurodexeine molars are tribosphenic and crescentiform. They lack a hypocone, or this cone is very small ("in statu nascendi," Franzen, 1981: 315). The snout is elongated by diastemata. M3 is smaller and more triangular than M2. In the lower molars, the trigonid is much higher than the talonid and bears a strong metaconid. On the talonid, posthypocristid and postentocristid are connected, and the talonid basin is lingually open. p1 is caniniform (*Parahexacodus*) and resembles choeropotamids, which are known with *Choeropotamus* from MP 18 (La Débruge). Thus, eurodexeines could be ancestral to choeropotamids. Their radiation must have taken place earlier than MP 13 because of the contemporaneous occurrence of eurodexeine and choeropotamid (e.g., *Amphirhagatherium*) at Geiseltal. Eurodexeines evolved from a lineage of diacodexeids with crescentiform teeth, older than *Diacodexis* from MP 10.

Genus *Eygalayodon* Sudre and Marandat, 1993

Type Species and Type Specimen. *E. montenati* Sudre and Marandat, 1993, USTL EYG O6, left mandible with m1–m3; Eygalayes.

Included Species. *E. isavenaensi* Checa, 2004: Güell 2, El Pueyo, Les Radies, Montderoda, La Roca, Corsà II, III; *E. sp.*: St-Maximin (Sudre, 1997: 761).

Diagnosis. Very small dichobunids. Lower molars: trigonid higher than talonid; paraconid close to metaconid, reduced from m1 to m3; posterior wall of trigonid very straight; protoconid weakly crescentiform; entoconid rounded, conical, very distal, separated from trigonid by a deep notch; posthypocristid short and transverse. Hypoconulid very strong, situated at beginning of a strong cingulum, which regularly descends toward the labial side. m2 larger than m1; m3 relatively short, bears a narrow hypoconulid (according to Sudre and Marandat, 1993).

Dimensions. m2: L = 4.1 mm.

Discussion. *Eygalayodon* was the first European genus with small crescentiform teeth similar to North American homacodontines and was originally placed in this subfamily (Sudre and Marandat, 1993). *Eygalayodon* sp., based on an isolated m3 from St.-Maximin, supports the concept of eurodexeines as a European group of small crescentiform dichobunids and shows different degrees of bunodonty. Despite the similar size to *Lutzia*, *E.* sp. is placed into *Eygalayodon* because of differences in the height of trigonid and talonid. The age of Eygalayes was regarded as MP 14 by Sudre and Marandat (1993: Fig. 3). The fact that *E.* sp., *Eurodexis,* and *Lutzia* belong to MP 13 indicates a slightly older age. Therefore, Erfurt and Sudre (1996: 378) used MP 13 as an upper limit for the subfamily.

Genus *Lutzia* Franzen, 1994

Type Species and Type Specimen. *L. eckfeldensis* Franzen, 1994, MNHM PW 1991/43–LS, left lower jaw with p4-m1, (m2); Eckfeld (Haupt-Turbidit).

Diagnosis. Very small dichobunids with bunodont dentition. Lower molars elongate with lingually open trigonid and talonid. Postprotocristid and postmetacristid connected and transversely oriented as well as the posthypocristid and postentocristid. Prehypocristid elongated and longitudinally oriented to the middle of the connection between protoconid and metaconid. Buccal side of hypoconid is bent (ectoflexid). p4 significantly longer than m1, bears a metaconid, lacks hypoconulid. Buccal side of the tooth bent (according to Franzen, 1994: 196).

Dimensions. m2 unknown, m1: L = 2.9 mm (Franzen, 1994: 193).

Discussion. *Lutzia* is the smallest dichobunid. The worn dentition of the holotype indicates that the given measurements belong to an adult specimen. The m1 is shorter than that of *Diacodexis antunesi* and broader than that of *D. gazini*. All American forms are larger but have shorter premolars. Because of the isolated position of the paraconid on the trigonid of *Diacodexis,* this larger genus is less differentiated than the smaller *Lutzia.* There, the paracristid is lacking, and the close position (or fusion?) of paraconid and metaconid indicates an advanced state. *Lutzia* is regarded as a late development of the dichobunid lineage. This supports the idea of a radiation of small European artiodactyls with primitive dentitions in contrast to larger, cursorially adapted bunodont species.

Genus *Eurodexis* Erfurt and Sudre, 1996

Type Species and Type Specimen. *E. ceciliensis* (Franzen and Krumbiegel, 1980) GMH Ce IV–305, skull with upper dentition; Geiseltal (site Cecilie IV) obere Mittelkohle.

Included Species. *E. russell,* Sudre and Erfurt, 1996: Premontré; *E.* cf. *russelli:* Cuis, Grauves, Mancy, St. Agnan (Sudre and Erfurt, 1996: 404); *E.* sp.: Messel (Erfurt and Sudre, 1996: 378).

Diagnosis. Small dichobunids with triangular, evenly sized upper molars, M3 slightly smaller than M2. M1 and M2 bear a hypocone but no mesostyle. P1/p1 are separated by a mesial diastema from the canines and a distal diastema from P2/p2. All canines small, C and p1 premolariform. Lower canine incisiform, closely appressed to incisors. Trigonid much higher than talonid; paraconid weakly separated from metaconid by a shallow notch. Talonid basin is lingually closed by a preentocristid. Postentocristid often connected to posthypocristid. A hypoconulid is situated on a short postcingulid.

Dimensions. m2: L = 5.1 mm, skull (type *E. ceciliensis*) L = 100 mm (Erfurt and Sudre, 1996: 379).

Discussion. The oldest specimens of *Eurodexis* were found in the French early Eocene and referred to *Diacodexis* (*"D. russelli"*). In contrast to *Diacodexis,* the upper molars are lingually more expanded; the conules are larger and more rounded. On the lower molars, the conules are also more rounded, the paraconid is nearly fused with the metaconid, and the hypoconulid is smaller. As in the North American lineage of *Diacodexis* (Krishtalka and Stucky, 1985), there are transitions between the typical characters of *Eurodexis* and *Diacodexis.* These similarities suggest a close phylogenetic relationship between the genera, and the roots of *Eurodexis* lie in the European stock of *Diacodexis.*

The type specimen of *Eurodexis* was formerly attributed to *Messelobunodon ceciliensis.* In contrast to the bunodont dentition of the type species of *Messelobunodon* (*M. schaeferi*), *Eurodexis* represents smaller crescentiform artiodactyls. Postcranial characters are preserved in *Eurodexis* sp. from Messel (SMF-ME 1001), formerly described as *Messelobunodon* (Franzen, 1983). Compared with *M. schaeferi* (SMF-ME 501), this form bears a third femoral trochanter and is smaller.

Genus *Parahexacodus* Erfurt and Sudre, 1996

Type Species and Type Specimen. *P. germanicus* Erfurt and Sudre, 1996, GMH XXXVII-198, lower jaw with p1, m1–m3; Geiseltal (site XXXVII) obere Mittelkohle.

Diagnosis. Tribosphenic dichobunids with caniniform p1 and smaller size than *Eurodexis.* The lower canine is premolariform, smaller than p1. p4 bears a metaconid. Paraconid and metaconid are clearly separated. Trigonid is much higher than the lingually closed talonid. In contrast to *Eurodexis,* there is a transverse connection between posthypocristid and postentocristid. A hypoconulid is lingually situated on a broad postcingulid. The third lobe of m3 is stout.

Dimensions. m2: L = 4.7 mm (Erfurt and Sudre, 1996: 381); i3–m3: L = 45 mm.

Discussion. The lower molars of *Parahexacodus* and *Eurodexis* resemble those of *Microsus, Homacodon,* and *Hexacodus.* This is probably a parallel development from similar ecological adaptation, but p1 is unknown in American diacodexeids. The main differences among the genera are in the premolar dentition: form of p1, presence of p4 metaconid. In addition, the talonid structure is slightly different from the size of the entoconid. *Parahexacodus* and *Eurodexis* are found in the same stratigraphic levels, but the interpretation of first premolar variation as sexual dimorphism is refuted by differences in talonid structure, and the corpus mandibulae seems to be stronger in *Parahexacodus.*

Subfamily HYPERDICHOBUNINAE Gentry and Hooker, 1988

Hyperdichobunines are a diverse group of small dichobunids with the tendency to broaden the distal molar side with the hypocone (especially on M1 and M2). The mesial molar side is relatively narrow because of a small or absent paraconule. Premolars and molars are molarized and crescentiform. The snout is elongated by elongated premolars and diastemata. The taxonomy of hyperdichobunines is complicated by a lack of complete dentition and postcranial material. The patchy fossil record produced several preliminary described forms, whose taxonomic state and phylogenetic relationships remain unclear. A recent review by Hooker and Weidmann (2000) clarified the stratigraphic relationships of some older collections within the Swiss Mormont fossil complex.

Genus *Hyperdichobune* Stehlin, 1910

Type Species and Type Specimen.   *H. spinifera* (Stehlin, 1906) NMB Mt 147, right maxilla with (dP3), dP4-M2; Entreroches.

Included Species.   *H. nobilis* (Stehlin, 1906) Stehlin, 1910a: Egerkingen Grauer Ton. Bolus, Lissieu; *H. spectabilis* Stehlin, 1910: Mt. Entreroches. Eclépens C ; *H. langi* (Rütimeyer, 1891) Sudre, 1972: Egerkingen Bolus, Lissieu, Arcis-le-Ponsart; *H. hammeli* Sudre, 1978: Bouxwiller, La Défense, Eckfeld; *H.* cf. *hammeli:* St.-Maximin (Sudre, 1997: 761); *H.* sp. (summarized): Egerkingen, Arcis-le-Ponsart, Aumelas, Lissieu (Sudre, 1978), Eclépens B, ?Creechbarrow (Hooker and Weidmann, 2000: 102).

Diagnosis.   Very small hyperdichobunines with increasing size from first to third molar. All cusps are acute and bear sharp crests. The paraconule is weak. P3 and P4 molarized, bearing protocone, paracone, and metacone. p3 and p4 bear a lingual metaconid and a strong paraconid (emended from Sudre, 1978: 35).

Dimensions.   m2: L = 3.9 mm (Sudre, 1972: Table 2; Hooker and Weidmann, 2000: Tables 18, 19).

Discussion.   *Hyperdichobune* occurs in the late middle Eocene in several localities with mostly isolated teeth. *H. hammeli* from Buchswiller (MP 13) the oldest species, is regarded as ancestral to *H. langi* (Sudre 1978: 36). Unspecified forms from Lissieu and Aumelas indicate that the origin of *Hyperdichobune* is older than MP 13, probably derived from forms such as *Protodichobune oweni* type B (Sudre et al., 1983: 318) in the early Eocene. At the beginning of the late Eocene, there are already different lineages evolved (*H. langi*, *H. nobilis*), which lead to the Oligocene *H. spinifera* and *H. spectabilis*. Both species occur in Mormont, and their diagnoses were mainly based on size differences (Stehlin, 1910: 1099). Hooker and Weidmann (2000: 101) follow Stehlin's concept in separation of both forms, of which *H. spectabilis* is probably more advanced because of its larger size and higher molarization of the premolars with higher protocones.

Genus *Mouillacitherium* Filhol, 1882

Type Species and Type Specimen.   *M. elegans* Filhol, 1882, MNHN Qu 17506 left maxilla with P3–M3; Mouillac.

Aubrelong 2; Bach; Caylus; Eclépens B; Fons 1, 4; La Bouffie; Lavergne; Le Bretou; Les Clapiers; Les Pradigues (J. Sudre, Montpellier, pers. comm., 2005), Malpérié; Perrière; Robiac; Souvignargues; ?Creechbarrow.

Included Species.   *M. cartieri* (Rütimeyer, 1891) Stehlin, 1906: Egerkingen grauer Ton; Laprade; *M. schlosseri* Sudre, 1978: Ste. Néboule; *M.* cf. *elegans:* Lebratières (Sudre, 1978: 213); *M.* sp.: Egerkingen, Bolus (Stehlin, 1906: 636).

Diagnosis.   Small hyperdichobunines with elongated skull, rounded neurocranium, small ossified tympanic bulla, thin zygomatic arch, strong occipital and sagittal crests and reduced mastoid. Mandible is elongated, ventrally concave and bears a broad ramus angularis with high and slender coronoid process and high mandibular condyles. Upper premolars very sharp, separated by diastemata of different length. Molars similar to *Hyperdichobune,* but cusps are less elevated. Paraconule is mostly absent, a weak mesostyle may be present. Strong connection between protocone and metaconule. Lower premolars elongate, with diastemata mesial and distal to p2. P4 triangular, simple. P3 elongated. Canines not enlarged (emended from Sudre, 1978: 41).

Dimensions.   L m2 = 4.2 mm (Sudre, 1978: Table 4; Hooker and Weidmann, 2000: Table 18).

Discussion.   According to Sudre (1978: Fig. 7), *Mouillacitherium* forms a lineage that starts at the level of MP 14 with *M. cartieri* from Egerkingen via *M. elegans* and ends with *M. schlosseri* at MP 18 (Escarguel, 1997: Table 2). *M. elegans* occurs in MP 16 and MP 17, displaying increasing molar and premolar complexity and a size decrease. The origin of the genus is thought to be in early diacodexeid forms of the early Eocene. Pearson (1927) and Dechaseaux (1969) described the skull of *Mouillacitherium,* reconstructed with an estimated length of 7 cm (Dechaseaux, 1974: Fig. 12).

Subfamily LANTIANIINAE Métais et al., 2004
Genus *Lantianius* Chow, 1964

Type and Unique Species and Type Specimen.   *L. xiehuensis* Chow, 1964, IVPP-V.2933 right P2–M3; Kangwangou section, Bailuyuan Fm., Xie-hu, Lantian County (Shaanxi Province, China).

Diagnosis.   Premolars fairly elongated; protocone very tiny and posterior on P2, well developed and centrally placed on P3; P4 with strong lingual cingulum, distinct parastyle. Molars increase in size from M1 to M3, with subtriangular outline; cingulae are marked, continuous lingual cingulum; incipient hypocone located just posterior to protocone on M1-2; paraconule nearly merged with paracrista; parastyle prominent, labially projected (adapted from Chow, 1964).

Dimensions.   M1 (holotype): L = 5.0 mm; W = 6.6 mm (Chow, 1964).

Discussion.   This genus has been referred to several orders of mammals, and its ordinal status is in question until significant additional material is found. Chow (1964) concluded that *Lantianius* is a specialized form of adapid-like primates. In his review of Asian early primates, Gingerich (1976) noted that the morphology of its premolars matches better with dichobunoid artiodactyls. McKenna and Bell (1997) included *Lantianius* in their extended concept of oxyclaenids, and Métais et al. (2004) tentatively considered it as part of the Lantianiinae, a new subfamily of Asian dichobunids. The most compelling evidence to include *Lantianius* within artiodactyls rests on the morphology of the premolars and the fact that no features of the molars can convincingly exclude it from belonging to a poorly known group of dichobunoid artiodactyls. Additional material (especially the lower dentition) is necessary to clarify the status of this genus.

Genus *Eolantianius* Averianov, 1996

Type and Unique Species and Type Specimen. *E. russelli* Averianov, 1996, ZIN 34357, right M1; Andarak 2 (Kyrgyzstan).

Diagnosis. Subquadrate low-crowned upper molars with complete, robust lingual cingulum and well-developed hypocone. Paraconule and postparaconule crista on M1–3 virtually absent. Distinct parastylar area can be recognized on the type, well developed on M3 (adapted from Averianov, 1996).

Dimensions. M1 (holotype): L = 4.1 mm; W = 4.8 mm (Averianov, 1996).

Discussion. Initially defined as a small diacodexeid by Averianov (1996), this genus was removed to the Lantianiinae (Métais et al., 2004). *Eolantianius* was tentatively included in the Lantianiine on the basis of its strong cingulum and the development of the hypocone on upper molars. The m2 tentatively referred to the species displays a distinct hypolophid connecting hypoconid and entoconid, but the paraconid is apparently lost. The affinities of *Eolantianius* should be clarified when additional material is found.

Genus *Elaschitotherium* Métais et al., 2004

Type and Unique Species and Type Specimen. *E. qii* Métais et al., 2004, IVPP-12759.1, right maxilla with M1–2; Shanghuang D (Jiangsu Province, China).

Diagnosis. Transversely widened upper molars; thick lingual cingulum; well-differentiated hypocone on M1–2, reduced on M3. Lower molar with tiny paraconid, well basined talonid, and strong postcingulid on m1–2 (adapted from Métais et al., 2004).

Dimensions. M1 (holotype): L = 3.1 mm; W = 3.8 mm (Métais et al., 2004).

Discussion. *Elaschitotherium* is based on a significant sample of dental material extracted from the matrix of the Shanghuang fissure fillings. It seems to include several species currently under study. *Elaschitotherium* combines a unique set of dental characters, but its intermediate morphology prevents reference to any group of Paleogene artiodactyls. It is more advanced than diacodexeids but more primitive than European dichobunids or North American homacodonts. The development of the lingual cingulum allied with the presence of a distinct hypocone is unknown in the forms of the Indo-Pakistani area, and *Elaschitotherium* probably represents a new lineage of early selenodont artiodactyls in Asia. The artiodactyls from Shanghuang provide an important window for investigating selenodont diversity in the middle Eocene of eastern Asia.

Family HOMACODONTIDAE Marsh, 1874

We include Bunomerycidae *sensu* Gentry and Hooker, 1988 in Homacodontidae. The Homacodontidae are paraphyletic, but until a thorough phylogenetic analysis allows better resolution, we retain it here.

Storer (1984), Stucky (1998), and Walsh (2000) placed *Antiacodon*, *Auxontodon*, *Tapochoerus*, and *Neodiacodexis* in Antiacodontinae, either in the family Dichobunidae or *incertae sedis*. The potential autapomorphies for this grouping include (1) strong cristid obliqua extending to the metaconid apex, (2) paraconid larger than or subequal to the metaconid on m1, and (3) weak accessory cusp on the anterior cingulum (Walsh, 2000). The only phylogenetic studies that attempted to test the relationships of this genus within basal artiodactyls (Theodor and Foss, 2005; Geisler et al., this volume) included only *Antiacodon*, and it appeared as the sister taxon to *Bunomeryx* and to Suiformes, respectively. Although this conflict does not provide a test of monophyly of Antiacodontinae, the potential sister-group relationship with *Bunomeryx* and the weakness of at least two of the three potential apomorphies in *Tapochoerus* lead us to retain *Antiacodon* and *Auxontodon* in Homacodontidae pending further analysis.

Homacodontids are small, bunoselenodont artiodactyls with a posterolingually enlarged metaconule, small paraconule, and anterior protocone. The hypocone is variably present in M1–2 and lost in M3. Mesostyles may be present or absent. c1 and p1 are similar in size. Paraconid is reduced, talonid equal in size to trigonid. Hypoconulid on postcingulum (Scott, 1891; Peterson, 1919; West, 1984; Stucky, 1998).

Genus *Homacodon* Marsh, 1872a

Type Species and Type Specimen. *Homacodon vagans* Marsh, 1872a, YPM 13129, skull and jaws, Henry's Fork (Bridger D).

Included Species. Stucky (1998) briefly describes three unnamed new species from Twin Buttes Mbr., Bridger Fm. (?*Homacodon* n. sp. A), Bridger B (*Homacodon* n. sp. B), and Cathedral Bluffs Tongue of the Wasatch Fm. (*Homacodon* n. sp. C).

Diagnosis. m1–3 paraconid small, closely appressed to metaconid. P3/p3, P4/p4 gracile; p4 with double crest on posterior of protoconid; crests on lower molars stronger than *Hexacodus;* posthypocristid well-developed, at 45° angle to hypoconulid; hypoconulid medial on postcingulid. Uppers more crescentic than *Hexacodus,* lack mesostyle. M1–2 with strong cingular hypocone (modified from Stucky 1998).

Dimensions.   m2: L = 4.0–5.8 mm (Stucky, 1998).

Discussion.   Stucky (1998) suggests that there are at least four undescribed morphospecies that belong to *Homacodon,* all sharing a p4 that is gracile, with a doubled ridge from the protoconid to a lateral ridge on the talonid, and lacking a metaconid.

Genus *Antiacodon* Marsh, 1872b

Type Species and Type Specimen.   *Antiacodon venustus* Marsh, 1872b, YPM 11765 m3; Green River Basin (Wyoming), southern California.

Included Species.   *A. vanvaleni* Guthrie, 1971, from Bighorn, Huerfano, and Wind River Basins (Wyoming); *A. pygmaeus* Cope, 1872 (including *Nanomeryx caudatus*): Bighorn, Huerfano, and Green River Basins (Utah and Wyoming); *A. furcatus* (Cope, 1873c): Green River Basin (Wyoming); *Antiacodon* sp. undescribed species (R. K. Stucky, Denver Museum of Science and Nature, and L. Krishtalka, Natural History Museum and Biodiversity Research Center at the University of Kansas, unpublished data): Bighorn and Wind River Basins (Stucky 1998); *Antiacodon* sp. small and large: South Pass (Gunnell and Bartels, 2001).

Diagnosis.   Well-developed hypocone on M1 and M2, may be double; small cusp anterior to protocone; small distinct mesostyle; prominent conules. p4 with metaconid. Metaconid reduced; paraconid and metaconid closely appressed but distinct (modified from West, 1984).

Dimensions.   m2: L = 4.1–5.5 mm (Stucky, 1998).

Discussion.   Lucas (1983a) placed *Nanomeryx caudatus* in *Homacodon,* but Stucky (1998) concluded that the diastema in the type of *N. caudatus* showed that it should be synonymized with *Antiacodon.* Stucky (1998) also notes that in Lostcabinian and Gardnerbuttean *Antiacodon,* the paraconid and metaconid are equal in size, and the hypocone and mesostyle are absent or barely present, but in Bridgerian taxa, the paraconid is larger than the metaconid. Walsh (2000) referred *Sarcolemur furcatus* to *Antiacodon* because the main character cited to retain *Sarcolemur* (Stucky, 1998), the presence of an m1–2 hypoconulid on the postcingulid, is also found in specimens of *Antiacodon.*

Genus *Auxontodon* Gazin, 1958

Type Species and Type Specimen.   *Auxontodon pattersoni* Gazin, 1958, MCZ 9316, Left mandibular ramus with p3–m1, part of m3; Uinta Basin (Utah); Wind River Basin.

Included Species.   *A. processus* Storer, 1984: Swift Current Creek (Saskatchewan, Canada).

Diagnosis.   Inferior margin of lower jaw strongly convex in anteroposterior profile. Canine incisiform, small. Enlarged p1; p1–p2 diastema marked; p2–4 anteroposteriorly elongate. Cheek teeth resemble *Antiacodon.* Paraconid larger than metaconid; cristid obliqua rises to apex of metaconid as in *Antiacodon.* Hypoconulid located on posterior cingulid. Referred upper molars with continuous cingulum; crescentiform cusps are nearly equal in size and height. Central protocone with small anterior and posterior cuspules on cingulum. Labial margin of tooth has strong stylar ridges; mesostyle high, strongly connected to paracone and metacone (modified from Gazin, 1958; Black, 1978b).

Dimensions.   m2: L = 6.7–7.1 mm (Storer, 1984).

Discussion.   Stucky (1998) referred two additional specimens to *Auxontodon,* a dP4 from Texas (United States) originally placed in *Texodon meridianus* (West, 1982), and an upper molar from the Wagon Bed Fm., Wyoming (Black, 1978b).

Genus *Texodon* West, 1982

Type Species and Type Specimen.   *Texodon meridianus* West, 1982, TMM 41672-62, M2–3; Big Bend Region, Devil's Graveyard Fm. (Texas).

Diagnosis.   Known only from type. Medium-sized bunodont artiodactyl; M2 with strong hypocone, M3 lacks hypocone; M3 smaller than M2; M2 metaconule small, slightly larger than paraconule; parastyles small; strong cingulae; no mesostyle; M3 protocone more central, M2 more anterior (modified from West, 1982; Stucky, 1998).

Dimensions.   m2: L = 5.0 mm (West, 1982).

Genus *Microsus* Leidy, 1870

Type Species and Type Specimen.   *Microsus cuspidatus* Leidy, 1870, USNM 1178 (formerly ANS 10260), lower molars; Green River Basin (Wyoming).

Included Species.   Microsus sp.: South Pass, Green River, Uinta Basins (Stucky, 1998; Gunnell and Bartels, 2001).

Diagnosis.   All cusps relatively higher, less bunodont, more crescentic than Homacodon. p4 with metaconid. Paraconid on m2–3 is vestigial or absent resulting in a compressed trigonid. m1 may retain a relatively well-separated paraconid and metaconid of equal size. M1–2 hypocone present, mesostyle may be incipient or absent (modified from Stucky, 1998).

Dimensions.   m2: L = 4.0–4.9 mm (Stucky, 1998).

Genus *Hexacodus* Gazin, 1952

Type Species and Type Specimen.   *Hexacodus pelodes* Gazin, 1952, USNM 19215, p4–m2; Fossil Basin. Also known from Green River Basins (Wyoming), Bridger, Piceance, Washakie, and Wind River Basins.

Included North American Species. *H. uintensis* Gazin, 1952: Fossil Basin (Wyoming); *Hexacodus* sp.: Fossil Basin (Stucky, 1998); *Hexacodus* sp. nov.: South Pass (Gunnell and Bartels, 2001).

Diagnosis. P3, P4/p4 robust, broad relative to length; P3 short, slightly longer than P4/p4. p4 simple with weak to absent metaconid. m2–3 paraconid and metaconid elevated, higher than protoconid, joined or closely appressed. Paraconid smaller than metaconid (equal on m1 in *H. uintensis*); metaconid not inflated. m1–2 hypoconulid on posterior cingulid connected to posthypocristid. Entoconid is tall, lingual to hypoconid. M1–2 metaconule posterolingually expanded; protocone apex anterior to the midline of the tooth. Preparaconule crista and preprotocrista form distinct shearing crest. Hypocone very small, developed as an elevated ridge on cingulum. Mesostyle on M1–3 weak to absent (modified from Stucky, 1998).

Dimensions. m2: L = 4.4–4.8 mm (Stucky, 1998).

Discussion. *Hexacodus* may represent a paraphyletic grouping (Stucky, 1998); with *H. pelodes* more closely related to *Microsus* and *Homacodon* than to *H. uintensis,* based on twinning of the para- and metaconid in *H. pelodes.* Stucky (1998) suggests that *Hexacodus* may represent the earliest North American selenodont because of the enlarged metaconule, strong posthypocristid, position of the hypoconulid on the cingulum, and the anterior position of the protocone.

Genus Bunomeryx Wortman, 1898

Type Species and Type Specimen. *Bunomeryx montanus* Wortman, 1898, AMNH 2071, rostral partial skull with P2–M3, and mandible with p4–m3; Uinta Basin.

Included North American Species. Monospecific (=*B. elegans* Wortman, 1898, fide Gazin, 1955; Stucky, 1998).

Diagnosis. Cheek teeth more crescentic, p4 more gracile than *Hylomeryx.* M1 with well-developed hypocone, M2 hypocone reduced; para-, meso-, and metastyles strong; paracone and metacone labially ribbed. Posthypocristid directed posteriorly behind entoconid (modified from Stucky, 1998).

Dimensions. m2: L = 5.4 mm (Gazin, 1955).

Discussion. Norris (1999), in a study of the ear region of *Bunomeryx,* noted that many of the dental characters in use to group homacodontid taxa are likely plesiomorphic. However, the ear region of *Bunomeryx* shows derived features of the endocranial surface of the periotic, notably a deep mastoid fossa and an expanded mesoventral crista promontorii medioventralis, both of which unite *Bunomeryx* with the camelids and are not found among suiforms or ruminants. However, this region has not yet been explored for any other homacodontids, diacodexeids, or dichobunids.

Genus Hylomeryx Peterson, 1919 (includes Sphenomeryx Peterson, 1919)

Type Species and Type Specimen. *Hylomeryx annectens* Peterson, 1919, CM 2335, anterior skull and lower jaws; Uinta Basin, also from the Washakie Basin.

Included Species. *H. quadricuspis* (Peterson, 1919): Uinta, Sand Wash Basins; *Hylomeryx* sp.: Big Bend Region (Texas), questionable form.

Diagnosis. Para-, meso-, and metastyle smaller than *Bunomeryx,* better developed than *Homacodon, Microsus.* Paracone and metacone conical, bunodont, with trend to labial ribbing as in *Bunomeryx, Hypertragulus,* and North American tylopods. Hypocone on M1–2. Paraconule distinct on M1, small on M2 protoloph, absent on M3. P4/p4 robust with strong p4 metastylid. Posthypocristid directed toward entoconid as in *Homacodon* and *Microsus,* not behind entoconid as in *Bunomeryx* (modified from Stucky, 1998).

Dimensions. L m2 = 4.7–5.8 mm (Stucky, 1998).

Genus Pentacemylus Peterson, 1931

Type Species and Type Specimen. *Pentacemylus progressus* Peterson, 1931, CM 11865, M1–2, partial p4, m1–2; Uinta Basin.

Included Species. *P. leotensis* Gazin, 1955: Uinta Basin.

Diagnosis. More crescentic cusps than *Bunomeryx;* lacks hypocone on M1–2; well-developed paraconule. Paraconid better defined on p3 and larger on p4 than in *Bunomeryx* (modified from Peterson, 1931; Stucky, 1998).

Dimensions. m2: L = 6.5 mm, W = 4 mm (Peterson, 1931).

Genus Mytonomeryx Gazin, 1955

Type Species and Type Specimen. *Mytonomeryx scotti* Gazin, 1955, USNM 20401, skull, jaws, partial skeleton; Uinta Basin.

Diagnosis. Elongate rostrum; marked P1/P2, P2/P3, p1/p2, and p2/p3 diastemata; M1–M2 quadrate with well-developed hypocone. Upper molars moderately selenodont with prominent mesostyle (modified from Gazin, 1955).

Dimensions. m2: L = 6.0 mm (Gazin, 1955).

Discussion. *Mytonomeryx,* as noted by Stucky (1998), bears a striking dental resemblance to *Bunomeryx* but bears an elongated rostrum. More detailed comparisons to *Bunomeryx* based on the ear region could help interpret the data from the ear region of *Bunomeryx* (Norris, 1999) if the crushing of the type specimen permits.

Genus Mesomeryx Gazin, 1955

Type Species and Type Specimen. *Mesomeryx grangeri* Peterson, 1919, CM 3189, maxillary fragment with P3–M2; Uinta Basin.

Diagnosis. P3 anteroposteriorly shortened; protocone anteriorly positioned. Bunoselenodont molars lacking mesostyle and hypocone; M1 with cingular flexure in hypocone position. Paraconule absent; preprotocrista continuous with preparaconule crista; paracone and metacone labially ribbed; metaconule crescentic (modified from Stucky, 1998).

Dimensions. m2 unknown; M2: L = 5 mm (Peterson, 1919).

Discussion. Stock (1934b) suggested that *Mesomeryx* might be ancestral to the hypertragulids. Stucky (1998) argued that the loss of the hypocone, expansion of the M1 metaconule, and well-developed centrocristae lacking mesostyles unite *Mesomeryx, Pentacemylus,* and the Ruminantia. However, new postcranial material from the Uinta Basin (Rasmussen et al., 1999) does not show any postcranial apomorphies of Ruminantia, such as the fused navicular and cuboid.

Genus *Limeryx* Métais et al., 2005

Type Species and Type Specimen. *Limeryx chimaera* Métais et al., 2005, IVPP V12760.1, upper molar (probably M2); Shanghuang D (Jiangsu Province, China).

Diagnosis. Upper molars with thick buccal cingulum, parastyle, mesostyle, and centrocrista well developed and salient labially, paraconule and metaconule distinct, metaconule subcrescentic and not fully lingual in position, and protocone relatively central in position. Lower molars retain straight hypolophid and minute paraconid, and the entoconid exhibits a distinct *Zhailimeryx* fold (modified from Métais et al., 2005).

Dimensions. m2 (paratype, IVPP V12760.2): L = 7.5; M2: L = 7.5 mm (Métais et al., 2005).

Discussion. The material referable to this genus also includes the *?Dichobune* sp. pointed out by Zdansky (1930) from the locality 7, Rencun Member, Heti Fm., Shanxi Province (China).
Lower molars show a *Zhailimeryx* fold, which is also present in several Eocene early ruminants from Asia. *Limeryx* is temporarily included in the Homacodontidae because of its relatively derived preruminant molar pattern, but additional data (dental and postcranial) would be necessary to test this systematic scheme. *Limeryx* most probably belongs to an Asian group of buselenodont dichobunoids so far poorly documented in Asia.

Genus *Asiohomacodon* Tsubamoto et al., 2003

Type Species and Type Specimen. *A. myanmarensis* Tsubamoto et al., 2003, NMMP 0713, fragmentary maxilla with left P4–M3; Kd2 locality near Bahin Village, Pondaung Fm. (central Myanmar).

Diagnosis. Simple bunoselenodont molars with small paraconule, continuous lingual cingulum; metaconule smaller than protocone; styles salient labially; lower molars fully selenodont and low-crowned; external cuspids strongly crescent-shaped; posterior cingulid fairly thick; hypoconulid (on m1–2) and paraconid absent; enamel noticeably wrinkled.

Dimensions. M1 (holotype): L = 7.1 mm; W = 8.0 mm (Tsubamoto et al., 2003).

Discussion. *Asiohomacodon* is the most selenodont dichobunoid known in the middle Eocene of Asia, and its allocation to the North American Homacodontinae is provisional. The fossil record of early selenodonts in Asia is particularly poor, and the sudden emergence of ruminants in the late middle Eocene remains a largely unresolved issue. As noted by Tsubamoto et al. (2003), *Asiohomacodon* is reminiscent of the anthracotheriid *Atopotherium* from the late Eocene of Thailand and of North American agriochoerids. Despite their similar basic structure, the lower molars of *Asiohomacodon* differ from those of early agriochoerids in retaining a weak *Dorcatherium* fold, whereas all agriochoerids exhibit a well-developed metastylid. Allocation of *Asiohomacodon* to homacodontids may prove to be incorrect because such a large size and highly derived selenodont pattern are unknown within the North American homacodontids.

Genus *Tsaganohyus* Kondrashov et al., 2004

Type Species and Type Specimen. *T. pecus* Kondrashov et al., 2004, PIN 3104-479, left dentary fragment with dp4-m1 and alveoli of c-p3 ; Tsagan-Khushu, Bumban Member, Naran Bulak Fm. (Mongolia).

Diagnosis. Markedly bunodont molars; paraconid reduced, medially located, and connected to the protoconid and metaconid by faint crests; entoconid and hypoconid roughly conical and equal in size; hypoconulid medially situated, emerges from the posterior cingulid, approximately the same size as the paraconid; a small entoconulid occurs anterolingual to entoconid; the symphysis of the mandible extends back to p3 (adapted from Kondrashov et al., 2004).

Dimensions. m1 (holotype): L = 3.1 mm; W = 2.0 mm (Kondrashov et al., 2004).

Discussion. Artiodactyls are much rarer in the early-middle Eocene of northern Asia than southern Asia. This is particularly true in Mongolia, which has otherwise yielded most of the reference fauna to establish Asian Land Mammal Ages. For example, Bumbanian, Arshantan, and Irdinmanhan faunas consist mostly of perissodactyls and condylarths, but very few recognizable dichobunoids have been reported. Dashzeveg (1982) pointed out indeterminate artiodactyls from the Bumban Member of the Naran Bulak Fm. (Mongolia), but these remains have not been figured. Kondrashov et al. (2004) claimed close affinities between *Tsaganohyus* and North American homacodonts. Several features on this single specimen may cast some doubt on the ordinal identity of this genus, as the dp4 is clearly not trilobed, and the extreme bunodonty of the molar does not support close relationships with homacodonts. However, the bunodont molars of *Tsaganohyus* may be reminiscent of *Wutuhyus* figured in Tong and Wang (1998, 2006), which is comparable in age and size. As *Wutuhyus* is poorly known, and given the ambiguous and fragmentary nature of the material from the Naran Bulak Fm., it is premature to add another interpretation.

Family LEPTOCHOERIDAE Marsh, 1894

This family seems to be monophyletic, but its position in Artiodactyla is not well understood. In most recent phylogenetic analyses (Gentry and Hooker, 1988; Theodor and Foss, 2005; Geisler et al., this volume) they group with a species of *Diacodexis*. They are known mainly from dental remains; although Marsh (1894b) reported postcranial remains, the specimens appear to have been lost. Leptochoerids are rare, and more data are needed to understand the placement of the family within Artiodactyla and relationships within it.

Genus *Leptochoerus* Leidy, 1856 (including *Laopithecus* Marsh, 1875)

Type Species and Type Specimen. *Leptochoerus spectabilis,* Leidy, 1856, ANSP 15593, left mandibular fragment with m1–2; White River Fm. (Wyoming, Colorado). Also known from Brule Fm. (South Dakota).

Included North American Species. *L. elegans* (Macdonald, 1955): Brule Fm. (South Dakota); *L. supremus,* Macdonald, 1955: Brule Fm. (South Dakota); *L. emilyae,* Edwards, 1976: White River Fm. (Colorado); *Leptochoerus* sp.: White River Fm. (Wyoming), Brule Fm., Rosebud Fm., Sharps Fm. (South Dakota), Brule Fm. (Nebraska), Renova Fm. (Montana).

Diagnosis. Skull with short rostrum, more massive, bunodont lower molars than *Stibarus,* which do not have a subcrescentic appearance when little worn; paraconids lacking, p2, P2, and p3 may be tall, single-cusped, or low tricusped, elongate teeth. M1–3 triangular, but lingual edge is blunted off (Edwards, 1976: 103).

Dimensions. m1: L = 5.6–6.2 mm (Stucky, 1998).

Discussion. Storer (1984) noted that all unworn leptochoerid lower molars showed evidence of reduced paraconids that were subsequently worn. This likely applies to *Leptochoerus,* where the paraconids are absent or vestigial and easily worn (Westgate, 1994).

Genus *Stibarus* Cope, 1873a (including *Menotherium* Cope, 1873b, and *Nanochoerus* Macdonald, 1955)

Type Species and Type Specimen. *Stibarus obtusilobus* Cope, 1873a, AMNH 6784, mandibular fragment with p3; White River Fm. (Colorado, Wyoming), Brule Fm. (South Dakota, Nebraska), Chadron Fm. (Nebraska).

Included North American Species. *S. quadricuspis* (Hatcher, 1901): White River Fm. (Wyoming, Colorado); ?*S. montanus* Matthew, 1903: Renova Fm. (Montana); *S. yoderensis* Macdonald, 1955: Yoder Local Fauna (Wyoming).

Diagnosis. Skull with pinched rostrum. Massive, bunodont lower molars that appear subcrescentic when little worn, with reduced paraconids. The p2, P2, and p3 are low and tricuspid, the medial cusp being largest. M1–3 are triangular with a sharp lingual apex (Edwards, 1976: 100).

Dimensions. m1: L = 3.1–4.0 mm (Stucky, 1998).

Genus *Ibarus* Storer, 1984

Type Species and Type Specimen. *Ibarus ignotus* Storer, 1984, SMNH P1589.12, left m1 or m2; Swift Current Creek (Saskatchewan, Canada).

Included North American Species. *I.* cf. *ignotus:* Mission Valley and Santiago Formations (San Diego, California) (Walsh, 1996: 86).

Diagnosis. Cheek teeth low-crowned; P3/3–p4 elongate, narrow relative to molars; molar series not massive as in *Leptochoerus;* M3/3 relatively unreduced; M1–3 conules bulbous, not strongly crescentic; m1–3 hypoconulid, accessory cusp, entoconid subequal, set in arc; m3 talonid shorter than in Oligocene taxa (Storer, 1984: 76).

Dimensions. m1 or m2: L = 4.4–4.6 mm (Storer, 1984).

Discussion. The genus is very similar to *Stibarus,* with the exception of the larger m3 (Storer, 1984; Westgate, 1994). Storer (1984) suggested that this genus might be ancestral to *Stibarus* and *Leptochoerus,* but Westgate (1994) argued that *Laredochoerus* was a more likely ancestor for *Leptochoerus.*

Genus *Laredochoerus* Westgate, 1994

Type Species and Type Specimen. *Laredochoerus edwardsi* Westgate, 1994, TMM 42486-1130, partial mandible with p4–m3; TMM locality 42486 Laredo (Texas).

Included North American Species. Monospecific.

Diagnosis. Progressive size decrease from p4 to m3; m3 about two-thirds the length of m1 or m2; m1 or 2 with small paraconid; m1 or 2 protoconid, metaconid, and hypoconid large and subequal; entoconid and hypoconulid small and subequal; talonid and trigonid subequal in length; crescentic entoconid. M3 with well-developed paraconule, weak hypocone, lacks metaconule (after Westgate, 1994; Stucky, 1998).

Dimensions. m2: L = 5.5 mm; W = 6.1 mm (Westgate, 1994).

Discussion. Westgate (1994) suggested that *Laredochoerus* might be ancestral to *Leptochoerus,* based on the robust p4 morphology and the reduced m3.

"Diacodexis" woltonensis Krishtalka and Stucky, 1985

Type Species and Type Specimen. *"Diacodexis" woltonensis* Krishtalka and Stucky, 1985, CM 43478, right p4–m3; Wind River Basin. Big Bend Region, Green River Basin (Wyoming), Huerfano Basin.

Diagnosis. Differs from other leptochoerids: m1 smaller than m2; p4–m2 less inflated; more prominent paraconids on lower molars. Differs from *Diacodexis:* m1–2 nearly square; p4–m2 inflated; inflated proto- and hypoconids; hyperinflated metaconid; small talonid basin with well-developed notch and uncompressed trigonid (modified from Stucky, 1998).

Dimensions. m2: L = 4.21 mm, W = 3.76 mm (Krishtalka and Stucky, 1985).

Discussion. Stucky (1998) notes that material from the Huerfano Basin and the New Fork–Big Sandy area may represent an additional species within this unnamed new genus.

Family RAOELLIDAE Sahni et al., 1981

The family is based in *Raoella* Sahni and Khare, 1971, which was synonymized with *Indohyus* Ranga Rao, 1971 by Thewissen et al. (1987). Most of the genera now included within the Raoellidae (Sahni et al., 1981) were originally referred to Dichobunidae or Helohyidae and to Suiformes *incertae sedis* in McKenna and Bell (1997). Two groups can be distinguished within the family: (1) small, bunodont and slightly lophodont (*Khirtharia* and *Metkatius*), and (2) *Kunmunella* and *Indohyus,* both larger, with a marked lophodont pattern. It is worth noting that *Indohyus* and *Kunmunella* are much more abundant in the Subathu Fm. at Kalakot, whereas *Khirtharia* is very rare. This difference in taxonomic composition may reflect different ecological conditions, although it is difficult to estimate because faunas from the Subathu Fm. (India) are slightly younger than those from Pakistan (Thewissen et al., 2001a).

The family consists of extremely bunodont forms showing various degrees of lophodonty; the premolars are generally simple and trenchant; molars have reduced or absent paraconule on uppers, loss of the paraconid on lowers. They can be more or less easily distinguished from the dichobunids *Haqueina, Chorlakkia,* and *Pakibune* but are dentally very distinct from *Gujaratia.* The issue of whether the posterolingual cusp is a metaconule or a hypocone remains open, as no intermediate forms are known in the fossil record. However, though variously developed on the upper molars, the distolingual cingulum is always present, and the posterolingual cusp is most likely an enlarged metaconule. Thewissen et al. (1987) suggested that, though endemic, raoellids may be closely related to the European dacrytheriid *Tapirulus,* based on the lophodont trend observed in the molars of *Tapirulus* (Sudre, 1978), but this hypothesis remains highly speculative. The current evidence would rather suggest a local radiation of raoellids, probably from an ancestry close to *Gujaratia.* Raoellids are known mainly in the Eocene of the Indian Subcontinent, although their occurrence in the Pondaung Fm. of Myanmar is likely (see below).

Genus Khirtharia Pilgrim, 1940

Type Species and Type Specimen. *K. dayi* Pilgrim, 1940, BMNH M15796, fragmentary dentary with left M2; Lammidhan, Kuldana Fm. (Kala Chitta Hills Area, Pakistan).

Included Species. *K. inflatus* (Ranga Rao, 1972) proposed by Kumar and Sahni (1985): Kalakot area, Subathu Fm. (Jammu and Kashmir, India); *K. aurea* Thewissen et al., 2001a: Chorgali Fm. (= basal member of Kuldana Fm., according to Gingerich, 2003), Pakistan.

Diagnosis. Upper molars showing four main cusps and crests very attenuated; subtriangular M3 (Pilgrim, 1940).

Dimensions. m2 (holotype of *K. dayi*): L = 9.5 mm; W = 8.5 mm (Pilgrim, 1940).

Discussion. First described as a helohyid by Pilgrim (1940), the genus was removed to the Dichobunidae together with the newly described genera *Haqueina* and *Pilgrimella* by Dehm and Oettingen-Spielberg (1958), until Coombs and Coombs (1977b) suggested that *Khirtharia* might not be an artiodactyl. Ranga Rao (1972) reported a new dichobunid *Bunodentus inflatus* from the Kalakot area in Indian Kashmir, considered a junior synonym of *Khirtharia* by West (1980b), who reassigned *Khirtharia* to Helohyidae. Pilgrim (1940) stressed the extreme bunodonty of his small sample (the genus is based on a single m2; additional material consists of m3, and M1–2) and diagnosed this taxon mainly on the lower molars bearing a "doubtful paraconid and the lack of hypoconulid." He based its assignment to artiodactyls on a double-pulley astragalus found in the same horizon in another locality of the Kala Chitta Hills. *Khirtharia* is the most bunodont raoellid, with incipient transverse lophs on the molars. Other discriminating features include the marked crenulation of enamel and the presence of a labial cingulid on lower molars. *Khirtharia* is reported from both Kuldana and Subathu formations and includes three species (*K. dayi, K. inflata,* and *K. aurea*).

Genus Indohyus Ranga Rao, 1971

Type Species and Type Specimen. *I. indirae* Ranga Rao, 1971, ONG/K/1, lower jaw with left p3–m3; Sindkhatuti, upper Subathu Fm. (Indian Jammu and Kashmir).

Included Species. *I. indirae* Ranga Rao, 1971: Kalakot area, Subathu Fm. (Jammu and Kashmir, India); *I. major* Thewissen et al., 1987: Chorlakki, Kuldana Fm., Pakistan.

Diagnosis. Medium-sized raoellid with bunolophodont dentition; P4 is triangular, three rooted; upper molars are subquadrate, four cusped, metaconule is the smallest cusp; a faint parastyle is present; well-developed cingulae.

Lower molars elongated and surrounded (except lingually) by a distinct cingulid; no paraconid; small hypoconulid medially located on m1–2, widely developed on m3 (adapted from Ranga Rao, 1971; Kumar and Sahni, 1985).

Dimensions.   m2 (holotype of *I. indirae*): L = 9.5 mm; W = 6.5 mm (Ranga Rao, 1971).

Discussion.   The current definition of *Indohyus* includes specimens described as *Raoella* and *Kunmunella* in Sahni and Khare (1971). The upper dentition of this genus was not recognized before Kumar and Sahni (1985) described and figured upper molars previously referred to *Raoella*. In their original description, Sahni and Khare (1971) suspected that the type and then unique specimen (maxilla with P4–M3) of *Raoella* might be the upper dentition of *I. indirae* Ranga Rao, 1971. *Indohyus* is the best known raoellid; numerous partial jaws have been described from the type locality, Kalakot, where it is the most abundant artiodactyl.

Genus *Kunmunella* Sahni and Khare, 1971

Type Species and Type Specimen. *K. kalakotensis* (Ranga Rao, 1971), LUVP 15004, right m3; Sindkhatuti, upper Subathu Fm. (Jammu and Kashmir, India).

Included Species.   *K. transversa,* Kumar and Sahni, 1985: Sindkhatuti, upper Subathu Fm. (Indian Kashmir), Chorgali Fm. (= basal member of the Kuldana Fm., according to Gingerich, 2003), Pakistan.

Diagnosis.   Bunodont molars that tend to bilophodonty; paracone, protocone larger than metacone and metaconule, this trend accentuated from M1 to M3; paraconule tends to merge into the lophed paracrista; no styles, slight cingulum interrupted lingual to the protocone. Four trenchant premolars; P4 with prominent lingual cusp connected by a loph to main acute cusp (adapted from Sahni and Khare, 1971; Kumar and Sahni, 1985).

Dimensions.   M2 (*K. kalakotensis*): L = 8.6 mm; W = 10.5 mm (Kumar and Sahni, 1985).

Discussion.   Initially regarded as a dichobunid by Sahni and Khare (1971), it was included within raoellids by Sahni et al. (1981). Defined on the basis of a single M3, the definition and specific content of the genus have been emended by Kumar and Sahni (1985) and Thewissen et al. (1987). *Kunmunella* has often been confused with *Indohyus*, although the former genus is generally more lophodont than the latter.

Genus *Metkatius* Kumar and Sahni, 1985

Type Species and Type Specimen.   *M. kashmiriensis* Kumar and Sahni, 1985, VLP/K 562 anterior portion of skeleton including fragmentary skull with lower and upper dentitions, vertebrae, and forelimb; West Babbian Gala Section, upper Subathu Fm. (Jammu and Kashmir, India).

Diagnosis.   Medium-sized raoellid characterized by elongation of molars, intermediate in bunodonty between *Khirtharia* and *Indohyus*. No trace of hypoconulid on m1–2. The lower m2 is approximately 40% smaller than m3; protoconid larger than metaconid, and transverse lophs moderately developed on lower molars (adapted from Kumar and Sahni, 1985).

Dimensions.   M2 (holotype): L = 6.5 mm; W = 3.5 mm (Kumar and Sahni, 1985).

Discussion.   This genus was proposed by Kumar and Sahni (1985), and it is based on a partial skeleton and badly damaged skull that prevents thorough observation of the dental morphology. However, the size range and the known dental features of *Metkatius* warrant the validity of the genus. *Metkatius* is about 40% smaller than the smallest specimens of *Kunmunella,* and its transverse lophids on lower molars are poorly developed and more obliquely oriented.

Artiodactyla indet. 2 Tsubamoto et al., 2005

Specimens.   NMMP 1765, a right M2; NMMP 1742, a right M3; Pk5 locality, Pondaung Fm. (central Myanmar).

Geographic Distribution.   Pondaung Fm. (Myanmar).

Stratigraphic Range.   Sharamurunian ALMA.

Dimensions.   M2: L = 9.7 mm, W = 10.7 mm; M3: L = 10.8 mm, W = 10.7 mm (Tsubamoto et al., 2005).

Discussion.   Upper molars referred to "Artiodactyla indeterminate 2" are reminiscent of raoellids' molar pattern although they do not display a really characteristic bilophodont molar (Sahni et al., 1981). The morphology of NMMP-KU 1742 is somewhat unusual within artiodactyls. The metaconule ("fourth cusp" for Tsubamoto et al., 2005) is strongly reduced, giving the tooth a subtriangular outline, and three small distinct accessory cusps emerge from a distolabially situated cingulum. Moreover, the metacone is noticeably reduced, and the cingulum surrounding the tooth appears stronger than that on the corresponding M2. The morphology of this molar is difficult to analyze, as it is not found in any raoellids so far reported, which usually display a less marked reduction of the posterior lobe of their M3. Although the M3 is usually variable in shape, such a distinct M3 morphology is unknown in material from Pakistan and India, or even in any Eocene artiodactyls, and it constitutes an obvious apomorphy of this new form. If raoellids are indeed present in the Pondaung Fm., it significantly extends the geographic and chronological ranges of the family. This occurrence of raoellids in Myanmar argues for the existence of a distinct South Asian faunal province during the middle Eocene. Although the evidence is still thin, it is expected that the accentuation of latitudinal climatic zoning may have implied vegetative and thus faunal distinction among South Asia, Central Asia, and northern China.

CONCLUSION

The sudden appearance of basal artiodactyls in Europe and North America has led most authors to conclude

Artiodactyla is of Asian origin. The limited Asian record has been supplemented by new taxa in recent years, which has improved and complicated our understanding of artiodactyl (and whale) evolution. In spite of the rich fossil record of early artiodactyls, many groups are in need of revision, and a more complete phylogenetic analysis of these basal groups is needed to better place them in a phylogenetic framework and to understand their relationships to more derived artiodactyl groups.

ACKNOWLEDGMENTS

J.E. and G.M. are especially grateful to Dr. Jean Sudre (Montpellier) for his useful hints and information and to Drs. J. J. Hooker (NHM, London), C. Sagne, P. Tassy, and D. E. Russell (MNHN, Paris), R.J. Emry (NMNH, Washington DC), P. D. Gingerich (UM, Ann Arbor), and M. R. Dawson and K. C. Beard (CMNH, Pittsburgh) for access to collections in their care. We thank these individuals as well as Dr. Jens L. Franzen for their permission to reproduce drawings of their publications. G.M. is grateful to the Foundation Singer-Polignac and the Carnegie Museum of Natural History for their financial support. J.T. thanks R. K. Stucky (DMNH), J. Meng and C. Norris (AMNH), R. W. Purdy (NMNH), and M. A. Turner (YPM) for access to collections in their care. R. K. Stucky, S. Walsh, and one anonymous reviewer provided helpful comments on this chapter. Last but not least, we thank the editors, who encouraged us to write this chapter.

JÖRG ERFURT
AND GRÉGOIRE MÉTAIS

5

Endemic European Paleogene Artiodactyls

Cebochoeridae, Choeropotamidae,
Mixtotheriidae, Cainotheriidae,
Anoplotheriidae, Xiphodontidae,
and Amphimerycidae

THE SEVEN FAMILIES treated in this chapter share the characteristic of being geographically restricted to Europe, which was intermittently isolated from the rest of Asia during the Eocene. Soon after the contact between Europe and North America (via Greenland) was interrupted during the late early Eocene, artiodactyls underwent a radiation of forms that experimented with various ecological specializations in the context of an apparent endemism that lasted until the end of the Eocene and the famous *Grande Coupure*. The incursion of Asian taxa during the earliest Oligocene is probably related to the extinction of several groups of these endemic forms, although cainotheriids persisted until the early Miocene. Most of these Eocene artiodactyls endemic to Europe were described during the nineteenth century through critical monographs. As a result, several genera have long taxonomic histories that have sometimes become problematic because of poor preservation or the loss of original type materials. In addition, there are many forms with only preliminary identifications, and whose phylogenetic positions are still unclear. Here, we have tried to gather taxonomic, stratigraphic, and geographic information scattered in several seminal works including Stehlin (1908, 1910a), Sudre (1978), Hooker (1986), Erfurt and Haubold (1989), Hooker and Weidmann (2000), and Hooker and Thomas (2001). We have also added some information about the paleoecology of these forms as well as a discussion of their phylogenetic relationships that will certainly require further analyses as new fossil evidence appears.

SYSTEMATICS

This section provides an overview of the diversity of Paleogene artiodactyls stemming from a European radiation. The specific phylogeny of these taxa has not been taken up here and is treated elsewhere in this volume. Details dealing with methods of compilation and stratigraphy used are in accordance with Chapter 4 (Theodor et al., this volume). The lists of fossil localities are mainly compiled from Stehlin (1908, 1910a), Sudre (1978), Hooker (1986), Erfurt and Haubold (1989), Antunes et al. (1997), and Hooker and Weidmann (2000).

Family CEBOCHOERIDAE Lydekker, 1883

Characteristics.   Small to medium-sized artiodactyls that display a short and robust cerebral portion of the skull. The lower jaws deepen posteriorly in advanced forms. The external acoustic meatus is tubular as in pigs, but it is shorter and opens between the retroglenoid process and the retrotympanic process of the squamosal. Upper molars are bunodont with four quadratically arranged bulbous cusps. The lingual cusps point buccally, and the buccal cusps may be crested. Paraconule small or absent, mesostyle generally absent. The posterolingual metaconule is as large as the protocone. P4 is triangular; P1–P3 are elongated mesiodistally. First upper and lower premolars are enlarged, mostly caniniform, and separated from both P2/p2 and C1/c1 by diastemata. Lower canines are incisiform and close to the incisors. Lower molars are transversely narrow and only slightly crested. The paraconid is separated from the metaconid, and the hypoconulid is posteriorly displaced. The p4 is shorter than p2–3 and lacks a paraconid.

Included Genera.   *Cebochoerus, Acotherulum, Moiachoerus,* "*Gervachoerus.*"

Choeromorus Gervais, 1848–1852 was added to *Cebochoerus,* and *Leptacotherulum* Filhol, 1877 to *Acotherulum* (Sudre, 1978: 51, 52; Hooker, 1986: 389).

Comments.   Apart from a size increase, there are two main trends in the evolution of cebochoerids: the strengthening of the masticatory apparatus by a shorter, stronger, and more robust mandibular ramus and the elongation of the first premolars and upper canines. The first observed trend may be linked to grubbing adaptations and/or harder food, and the second may be related to sexual dimorphism. Large upper canines associated with caniniform teeth also have a grubbing role in pigs and hippos. Another similarity shared with these extant groups is the presence of additional cusps on the molar occlusal surface, especially on m3, which tends toward a polybunodont pattern, as in the modern pig *Sus.* In contrast with extant pigs and hippos, the cebochoerid skull is short, and the dentition is not interrupted by diastemata. There are additional differences in the cranial anatomy, especially in the ear region, which is amastoid (Pearson, 1927), unlike the condition seen in modern ruminants.

Based on dental features such as metaconule development and molarization of the cheek teeth, Erfurt and Sudre

(1996: Fig. 3) regarded cebochoerids as descendants of unknown diacodexeids of the European lower Eocene. *Cuisitherium* indicates that the choeropotamid clade had already evolved in that time and that this group probably evolved from European diacodexeids close to *Protodichobune.* The quadrangular arrangement of molar cusps seems to be an archaic character, which also occurs in such other "ungulates" as *Phosphatherium,* an early proboscidean afrothere (Gheerbrant et al., 2005). Theodor and Foss (2005) recently demonstrated that the second and third deciduous premolars of cebochoerids possess accessory denticles, which are very similar to deciduous and adult premolars in archaeocete whales. Until additional new material is collected, the origin of cebochoerids and their phylogenetic relationships with early whales will remain unclear.

Genus *Cebochoerus* Gervais, 1852

Type Species and Type Specimen.   *C. anceps* Gervais, 1852, FSL 6796, left maxilla with P4–M3; La Débruge. *C. lacustris* was chosen as lectotype species by Gervais (1859), Stehlin (1908), and Depéret (1917) because the type specimen of the older *C. anceps* had been regarded as lost. Despite the pending formal application to the ICZN, *C. lacustris* is not listed as type species here because of the rediscovery of FSL 6796 by Hooker (1986: 390).

Included Species.   *C. lacustris* Gervais, 1856: Souvignargues, Lamandine, Mémerlein. The materials from Fons 6 and Sossís (Sudre, 1978: 75) very likely belong to *C. helveticus; C. helveticus* (Pictet and Humbert, 1869): Eclépens (Tunnel?), Barton (Bed C), Eclépens Gare, Euzèt, Lamandine, Lautrec, Le Bretou, Le Castrais, Robiac, Sossís (including *C. minor* Gervais, 1876 according to Hooker and Weidmann [2000: 78] and probably *C.* sp. [Biochrom, 1997: 787]); *C. ruetimeyeri* Stehlin, 1908: Egerkingen a, Egerkingen grauer Ton; *C. robiacensis* Depéret, 1917: Robiac, Creechbarrow, Lautrec, Le Castrais, Caylux (Astruc et al., 2003); *C. fontensis* Sudre, 1978: Fons 4. Fons 5, 6, 7 ; *C.* cf. *minor:* Aubrelong 2 (Sudre, 1978); *C.* cf. *ruetimeyeri:* Aumelas (Sudre, 1978); *C.* aff. *fontensis:* Euzèt, Entreroches (Hooker and Weidmann, 2000: 78).

Diagnosis.   Medium-sized to large animals. Two-rooted and elongated upper canines. P1 is two rooted and larger than P2. The p1 is caniniform, and c1 is incisiform. Upper cheek teeth bear strong mesial and distal cingula and crests. Upper molars: large paraconule; crown height lingually higher than or equal to buccal height; fused lingual roots; postprotocrista restricted to distal wall of protocone; premetaconule crista often thickened and joining the connection of postprotocrista and premetacrista; protolophule may or may not join paracone. P4 protocone lingual to paracone. P3 protocone lingual or distolingual to paracone, and outline may or may not narrow mesially. P2 protocone present or absent. Lower molars with angles of protoconid and hypoconid crests acute. Supraorbital foramina above level of M2 or M3. In adults, the horizontal mandibular ramus is about twice as deep below m3 as below p2 (according to Hooker, 1986: 390).

Dimensions.   m2: L = 8.2 mm, W = 6.5 mm; L snout to distal orbits (*C. minor*) 80 mm (Hooker, 1986: pl. 31).

Discussion.   *Cebochoerus* represents a cebochoerid lineage with more robust forms and greater molarization. On the molars, there are additional crests and conules intercalated between the main cusps, which foreshadow a polybunodont pattern. The large number of forms listed above illustrates the difficulties of recognizing infrageneric entities, which are accentuated by several transitional characters shared with the genus *Acotherulum*. For example, *Cebochoerus* aff. *fontensis* is morphologically similar to *A. quercyi*, and the latter differs only in size from *A. saturninum*. The genus first appeared in the MP 13 with *C.* cf. *ruetimeyeri* from Aumelas (Sudre, 1978: 83), which is more primitive than *C. ruetimeyeri*. A cladistic approach (Hooker, 1986: Table 34) indicates that *C. ruetimeyeri* (MP 14) is the most basal form, and two clades are recognized: the clade *C. robiacensis–C. fontensis–C.* aff. *fontensis* and the clade *C. helveticus / minor–C. lacustris*. *C.* aff. *fontensis* is from Entreroches, a locality that is slightly younger than Escamps and belongs to MP 19 (Hooker and Weidmann, 2000: 129). The second clade begins with the Bartonian *C. helveticus* (MP 16) and ends with *C. lacustris*. Although the stratigraphic position of Lamandine and Mémerlein is uncertain, this clade probably terminated in MP 17 because of the stratigraphic position of the type locality Souvignargues.

Genus *Acotherulum* Gervais, 1852

Type Species and Type Specimen.   *A. saturninum* Gervais, 1850, MNHN LDB 107, right maxilla with dP2–M1; La Débruge, Baby, Bach, Escamps, Lamandine, Perrière, Pont d'Assou, Rosières 1, Ste Néboule (including *Leptacotherulum cadurcence* Filhol, 1877 from Mouillac [Hooker, 1986: 399]).

Included Species.   *A. campichii* (Pictet, 1857): Eclépens (Gare?), Eclépens A, Le Castrais, Creechbarrow, Grisolles, Le Bretou, Eclépens A, Robiac (including *Metadichobune* Filhol, 1877 [Hooker, 1986: 399]); *A. pumilum* (Stehlin, 1908): Mormont (Eclépens B?), Gousnat, Lamandine, Hordle Cliff (Rodent Bed, Headon Hill Fm.; Harrison et al., 1995), Perrière, Sossís; *A. quercyi* (Stehlin, 1908): Bach, Headon Hill (Bembridge Limestone), Entreroches, Soumailles; *A.* cf. *quercyi*: Soumailles, Hoogbutsel (Sudre, 1978); *A.* sp.: Sossís (Cuesta et al., 2006).

Diagnosis.   Small to medium-sized with large and caniniform first premolars. c1 incisiform. Upper cheek teeth bear weak mesial and distal cingula and crests. Upper molars: small paraconule, crown height lingually higher than buccally, fused or unfused lingual roots, postprotocrista restricted to distal wall of protocone, protolophule, which joins the paracone. P4 protocone mesiolingual to paracone. P3 protocone lingual to paracone and outline narrowing mesially. P2 lacks protocone. Lower molars with angles of

protoconid and hypoconid crests right-angled to obtuse. Supraorbital foramina above level of M2. Adult horizontal mandibular ramus scarcely deeper below m3 than below p2 (according to Hooker, 1986: 399).

Dimensions.   m2: L = 6.2 mm (Sudre, 1978: Tables 5, 6); m1 (*A. pumilum*): L = 4.4 mm, W = 2.6; P4–M3 (*A. campichii*): L = 21.8 mm.

Discussion.   *Acotherulum* represents a lineage of medium-sized cebochoerids in the upper Eocene and Oligocene, characterized by weaker molarization of the premolars than in *Cebochoerus*. There are fewer additional cusps and cuspids on the molars. The oldest form is *A. campichii* from Eclépens-Gare (MP 16); the youngest are *A. quercyi* from Quercy (Bach) with uncertain stratigraphic range and *A.* cf. *quercyi* from Soumailles (MP 21). Hooker (1986: 403) advocated close phylogenetic relationships between these species via *A. saturninum*. *A. pumilum* is regarded as sister taxon of *A. saturninum* because of synapomorphies such as the high paraconid on p4 and narrow lower cheek teeth.

Genus *Moiachoerus* Golpe-Posse, 1972

Type Species and Type Specimen.   *M. simpsoni* Golpe-Posse, 1972, IPS 1685, right maxilla with P4–M1; San Cugat de Gavadons.

Diagnosis.   Medium-sized cebochoerid close in size to *Acotherulum saturninum*. P4 is anteroposteriorly compressed and displays a concave anterior border. Upper molars are less transversely developed than in *A. saturninum*. Lower molars bear a hypoconulid (according to Sudre, 1978: 82).

Dimensions.   m2: L = 5.8 mm (Sudre, 1978: Table 7).

Discussion.   *Moiachoerus* is a monospecific genus, and it is only known from its type locality, whose age is estimated as being between the level of La Débruge and Hoogbutsel (Golpe-Posse, 1972: 12). It occurs with the xiphodontids *Dichodon* cf. *frohnstettensis* and *D. cervinus*. Because of the poor fossil record, this genus remains problematic. Sudre (1978) pointed out that *M. simpsoni* differs from *Acotherulum* by the morphology of its P4. Hooker (1986: 399) tentatively included *M. simpsoni* in *Acotherulum*, but *Moiachoerus* is not mentioned in Hooker and Weidmann (2000). Although close phylogenetic relationships between *Moiachoerus* and *Acotherulum* or *Cebochoerus* are very likely, additional material would be necessary to improve the definition of this genus.

Genus "*Gervachoerus*" Sudre, 1978

Type Species and Type Specimen.   *Cebochoerus minor* Gervais, 1876, MNHN Qu60b, left maxilla with M1–M3; Lamandine (see discussion), Barton Cliff (Bed C), Mas Stes. Puelles, Sossís, Villeneuve-la-Comptal.

Included Species.   "*G.*" *dawsoni* Sudre, 1978: Bouxwiller; "*G.*" *jaegeri* Sudre, 1978: Bouxwiller, Geiseltal Cecilie III; "*G.*" *suillus* (Gervais) 1859: Calcaire Grossier, Capella;

"G." cf. *suillus*: Egerkingen Bolus, Arcis le Ponsart, Capella, Lissieu (Sudre, 1978); "G." cf. *jaegeri*: St.-Maximin (Sudre, 1997).

Diagnosis.   Small cebochoerids with elongated skull. Mandibles slender; height underneath p2 less than half of the height underneath m3. Lower canine incisiform and adjacent to incisors. P1/p1 premolariform and isolated by diastemata. Upper molars buccally rounded, without styles. Medivallum broad, without thickened premetaconule crista. Postprotocrista weak but joins the premetaconule crista at least on M3. Lower molars: trigonid is slightly higher or equal to the talonid, hypoconulid weak to absent on m1 and m2. Lower premolars are elongated with decreasing length from p2 to p4. On p4, mesial cristid of the main cusp without separated paraconid or parastylid.

Dimensions.   m2: L = 5.3 mm; "G." *jaegeri* i1–m3: L = 58 mm; skull: L = 110 mm, H = 50 mm; body: L = 750 mm, H = 250 mm.

Discussion.   The type species for the subgenus *Gervachoerus* is *Cebochoerus minor* (Sudre, 1978: 57), which is a junior synonym of *C. helveticus*. The latter species was placed in the nominate subgenus *Cebochoerus* by Sudre (1978). Based on a nearly complete skeleton from Geiseltal, Erfurt and Haubold (1989: 143) followed Sudre's concept to separate *C. dawsoni* and *C. jaegeri* and regarded "*Gervachoerus*" as an independent genus. They based "*Gervachoerus*" on the type specimen "*G.*" *dawsoni* from Bouxwiller USTL BX 66816, a right maxilla with P2–(M1), M2, M3. The taxonomy of these small cebochoerids remains unclear, and a new genus name should be used (J. J. Hooker, BMNH, pers. comm., 2006).

"G." *dawsoni* is the smallest and morphologically most primitive form. Postcranial evidence is preserved only for the contemporaneous "*G.*" *jaegeri* from Cecilie III at Geiseltal (Oberes Hauptmittel, MP 13). In GMH Ce III-4227, a very long tail is preserved, which represents about 30% of the total body length. The estimated body length is 70–80 cm, and the shoulder height is about 30 cm. The habit (posture) is somewhat squatter than in *Diacodexis,* but the hindlimb is elongated as well. Metapodials, zeugopodials, and basipodials remain unfused, and the third and fourth metapodials are slightly longer than second and fifth ones. The third phalanges are slender and morphologically intermediate between claws and hooves. Their distal side is concave. The origin of "*Gervachoerus*" is unknown, but bunodont forms of the lower Eocene such as *Protodichobune* have been hypothesized as a possible ancestral group. Phylogenetic relationships between *Cebochoerus* and "*Gervachoerus*" remain poorly understood, as both genera appear suddenly in the fossil record at the onset of the Middle Eocene.

Family CHOEROPOTAMIDAE Owen, 1845

Characteristics.   Bunoselenodont artiodactyls with a diastema separating p1/p2 and incisiform lower canine. The p1 is mostly caniniform and occludes behind the upper canine. Molars are bunodont to bunoselenodont, with an enlarged metaconule, no hypocone, and a small paraconule. The astragalus bears a distally extensive sustentacular facet and a trochleated head, which is only slightly bent with respect to its long axis (Hooker and Weidmann, 2000: 80).

Included Genera.   *Choeropotamus, Tapirulus, Rhagatherium, ?Diplopus, Thaumastognathus, Amphirhagatherium, Haplobunodon, Lophiobunodon, Masillabune, Cuisitherium, Parabunodon, Hallebune.*

Comments.   Some choeropotamids were formerly regarded as anthracotheriids (Stehlin, 1908) or close to cebochoerids (*Choeropotamus*). Pilgrim (1941) proposed the family Haplobunodontidae for *Haplobunodon, Lophiobunodon,* and *Rhagatherium,* and Sudre (1978) also included *Amphirhagatherium* and *Anthracobunodon.* Sudre (1978) listed a set of characters for the family including brachyodont and bunoselenodont dentition with molarization of P4/p4, elongation of the snout, and loss of P1/p1. Although *Choeropotamus* was regarded as the single genus of the Choeropotamidae, *Masillabune, Parabunodon,* and *Hallebune* were successively included within an enlarged haplobunodontid concept. Based on a cladistic analysis of dental, cranial, and postcranial characters, Gentry and Hooker (1988) united Haplobunodontidae and *Cuisitherium* into the Choeropotamidae, but this concept is far from being unanimously accepted (e.g., McKenna and Bell, 1997), and further analyses are needed to improve both the definition and outline of these two families.

Genus *Choeropotamus* Cuvier, 1822

Type Species and Type Specimen.   *C. parisiensis* Cuvier, 1822, MNHN GY 283, skull with upper right P1, P3-M3 and upper left P1, P2, P4-M3; Montmartre (undifferentiated), Escamps, Frohnstetten, Headon Hill (Bembridge Limestone Fm., beds 9–18), La Débruge, Mas Stes. Puelles, Neuhausen, St.-Capraise-d'Eymet, Vermeils, Villeneuve-la-Comptal, White Cliff Bay 2 (Bembridge Marls Member).

Included Species.   *C. affinis* Gervais, 1852: Mormoiron, La Débruge; *C. depereti* Stehlin, 1908: St. Hippolyte-de-Caton (Euzèt), Argenteuil?, Hordle, Lamandine, Malpérié, Eclépens (Gare?); *C. sudrei* Casanovas-Cladellas, 1975a: Roc de Santa. Fons 1, 4, 6 ; *C. lautricensis* (Noulet, 1870): Le Castrais, Robiac; ?*C.* sp.: Creechbarrow (Hooker, 1986).

Diagnosis.   Larger choeropotamids with low, broad skulls. The cerebral cranium is elongated, and zygomatic arches are laterally projected. The mastoid is not exposed posterolaterally (amastoid condition). Long mandibular symphysis extending back to the level of p2/p3, and long diastema between p1 and p2. Teeth bear thick wrinkled enamel and strong cingula on molars and premolars. p1 caniniform. Lower molars display metastylid, mesoconid, and entostylid. The m3 has two main cusps on the hypoconulid lobe, and a wrinkled crest mesial to the hypoconulid lobe occurs in the talonid basin. Upper molars bear an accessory cusp in the middle of the central valley. P4 with metacone on a short postprotocrista; P3 lacks a protocone (adapted from Viret, 1961: 897; Hooker and Thomas, 2001: 838).

Dimensions. m2: L = 14 mm (Cuvier, 1822: 261); m1–m3: L = 28–62 mm (Sudre, 1978: 87, 88).

Discussion. Species of the genus increase in size from *C. lautricensis* (MP 16), *C sudrei*, *C. depereti* (MP 17), *C. affinis* (MP 18), to *C. parisiensis* (MP 19/20). It is still difficult to provide a clear specific differentiation because of a relatively stable morphology and size overlap. For example, Hooker and Weidmann (2000) pointed out that the material from Mormont could belong to either *C. lautricensis* or *C. depereti*. The phylogenetic origin of *Choeropotamus* remains unclear. Several authors have mentioned similarities to cebochoerids (Stehlin, 1908) or inferred hypothetical relationships with large bunodont artiodactyls from Asia such as *Bunodentus inflatus* (Sudre, 1978: 89). But the fossil record of artiodactyls in Asia remains very scarce, and there is no clear evidence to support this hypothesis; the morphological resemblances are probably the result of convergent evolution. Recent cladistic analyses suggested a close phylogenetic relationship between *Choeropotamus* and *Thaumastognathus*, together as sister group of *Haplobunodon* (Hooker and Thomas, 2001).

Genus *Tapirulus* Gervais, 1850

Type Species and Type Specimen. *T. hyracinus* Gervais, 1850, MNHN LDB 206, left mandible with m2, m3; La Débruge, Coânac, Ehrenstein 1 (A), Frohnstetten, Hoogbutsel, Lamandine, Quercy (undifferentiated), Rosières 2, 4, Ste. Néboule.

Included Species. *T. majori* Stehlin, 1910: Egerkingen Huppersand, Mas Gentil; *T. depereti* Stehlin, 1910: Egerkingen α, Lissieu; *T. schlosseri* Stehlin, 1910: Larnagol, Lebratières, Le Bretou, Eclépens (Gare?), Robiac, Quercy (undifferentiated); *T. perrierensis* Sudre, 1978: Perrière, Malpérié, La Bouffie, Lavergne, Le Pradigues, Malpérié, Mouillac, Quercy (undifferentiated); *T. cf. majori:* Bouxwiller; Grabels (Sudre, 1978); *T. cf./aff. schlosseri:* Le Bretou; Robiac (Sudre, 1978); *T. sp.* (summarized): Aubrelong 1, Ravet (Sudre, 1978); Eclépens B (Hooker and Weidmann, 2000); Whitecliff Bay 2 (Bembridge Marls), Yarmouth-Bouldnor Cliff (upper Hamstead Member) (Hooker et al., 2004).

Diagnosis. Smaller choeropotamids with flat skull, elongated cranium, and reduced mastoid. Large brain but neopallium only slightly differentiated. Dentition bilophodont (tapiroid); diastemata and mesostyles are lacking. The distal half of upper and lower molars (the metastyle and hypoconulid, respectively) tend to be transversely larger than the mesial border. Premolars are elongated and sharp edged.

Dimensions. m2: L = 6.4 mm; W = 3.6 mm (Sudre, 1978: Table 9).

Discussion. *Tapirulus* was successively included in the Anoplotheriidae (Viret, 1961), Dacrytheriidae (Sudre, 1978), or as a tylopod *incertae sedis* (Hooker and Weidmann, 2000). Later, Hooker and Thomas (2001) placed this genus in the newly defined choeropotamids. The most striking characteristic of *Tapirulus* is its bilophodont tooth pattern, which contrasts with the dental pattern seen in the other choeropotamids. Likewise, the molar wear facets are mesially larger on trigon and trigonid and distally larger on the metastyle and hypoconid. In other choeropotamids, the facets on the opposite cusp sides are larger. In *T. major* from the Geiseltal, the lophs are poorly expressed; in later species they are higher (*T. perrierensis*). Furthermore, larger parastyles appear (*T. cf. schlosseri*), and the teeth became larger and broader. Therefore, the attribution to choeropotamids should be regarded tentatively. In addition, *Tapirulus* crosses the Eocene/Oligocene boundary with undetermined forms reported from Aubrelong and Ravet.

Genus *Rhagatherium* Pictet, 1857

Type Species and Type Specimen. *R. valdense* Pictet, 1857, syntypes MGL LM.768 skull with left P1, (P4–M1), right (I3)–P1, P4–M3; LM.767 mandible with left p1–p4, m3, and right p1–m3; Entreroches, Headon Hill (Bembridge Limestone), Hordle (Totland Bay Member, Headon Hill Fm.).

Included Species. *R. kowalevskyi* Stehlin, 1908: Egerkingen α, Huppersand, Bolus; *R. sp.*: Geiseltal Ce V (Erfurt, 1995).

Diagnosis. Medium-sized brachybunodont choeropotamids with molarized P4/p4. P4 displays three main cusps: metacone, paracone (only slightly higher than metacone), and protocone. Parastyle large, preprotocrista with paraconule. p4 with distinct paraconid and entoconid; the metaconid is attached or only slightly separated from the protoconid. Lower molars with accessory cusps along the posthypocristid and fused posthypocristid and postentocristid.

Dimensions. m2: L = 6.6 mm, W = 5.3 mm; L (type) I3–M3 = 67 mm (Pictet, 1855–1857: 47).

Discussion. Hooker and Thomas (2001) regarded this genus, which occurs only in MP 19, as monospecific. *R. kowalevskyi* Stehlin, 1908, from Egerkingen Huppersand and Bolus, was referred to the Mixtotheriidae because of the style of selenodonty and the position of the P4 cuspule. Indications of similar characters are also seen in choeropotamids such as *Hallebune*. Therefore, former citations of *R. kowalevskyi* from Geiseltal (obere Mittelkohle) have been emended as belonging to that genus (Erfurt and Sudre, 1995b). Until a comprehensive study of the Egerkingen material considering the different strata is undertaken, the species *R. kowalevskyi* is provisionally included in this genus. The poorly preserved *R. aegyptiacum* Andrews, 1906 resembles *Anthracotherium* and has no relation to *R. valdense* (Sudre, 1978: 93). The only similar form is *R. sp.* from Geiseltal Ce V (obere Oberkohle). Altogether, this single fragmentary mandible is smaller but displays a similar pattern of the premolars. On p4, the metaconid is also distally shifted, but this tooth is shorter than that of *R. valdense.*

Genus ?*Diplopus* Kowalevsky, 1873

Type Species and Type Specimen.   *D. aymardi* Kowalevsky 1873, syntypes BMNH.29739 right scapula; BMNH.30160 distal right humerus; BMNH.29745 left ulna; BMNH.29747 right tibia; BMNH.30179 distal fibula; BMNH.30119 proximal right metacarpal III; BMNH.30177 right calcaneum; BMNH.29721 partial right hind foot with articulated navicular, ectocuneiform, cuboid, and fragmentary metatarsals III and IV; BMNH.30118 right metatarsal IV; Hordle (Totland Bay Member, Headon Hill Fm.).

Included Species.   ?*D.* sp.: Eclépens-Gare (Hooker and Weidmann, 2000).

Diagnosis.   Large didactyl choeropotamids similar in size to reindeer. Limbs slender and cursorially adapted with unfused zeugopodials. First, second, and fifth metatarsals reduced; second and fifth metacarpals much shorter than the third and fourth. Tarsal bones (including the three cuneiforms) are unfused. The dentition was very likely selenodont with elongated parastyles in the last premolars and molars.

Dimensions.   Ulna: L = 270 mm; Tibia: L = 280 mm (Kowalevsky, 1873: 36, 40).

Discussion.   The type series includes only postcranial material, which probably belongs to several individuals. This genus is tentatively attributed to choeropotamids on the basis of a single P3, which clearly indicates a choeropotamid pattern (Hooker and Weidman, 2000: 83). This tooth was found in the Totland Bay Member of Hordle (Hampshire), which is MP 17 in age. The stratigraphic range of *Diplopus* is still unclear because of remaining uncertainties on the age of Eclépens-Gare (Hooker and Weidmann, 2000: Fig. 79). Compared to Geiseltalian choeropotamids (*Masillabune martini, M. franzeni,* and *Amphirhagatherium weigelti*), *D. aymardi* appears more specialized in the elongation of the forelimb and reduction of the second and fifth metapodials. If we assume that all syntypes belong to one individual, the length of MC IV plus ulna is about 93% of the length of MT IV plus tibia, whereas this ratio is about 70–75% in the older species from Geiseltal. Until associated postcrania and teeth are found, the systematic position of this taxon will remain unclear.

Genus *Thaumastognathus* Filhol, 1890

Type Species and Type Specimen.   *T. quercyi* Filhol, 1890, MNHN Qu 17505, left mandible with A1 i1-i3, p1-m2; Quercy (undifferentiated), Euzèt, Eclépens A, Gare, Robiac.

Diagnosis.   Large choeropotamids with elongated snout. p1/p2 diastema is about one-fourth of the total mandibular length. Dentition bunoselenodont; p1 caniniform; p3 and

p4 are not molarized. Mandibular ramus is more or less equal in height from p3 to m3.

Dimensions.   m2: L = 10 mm, W = 8 mm (Filhol, 1890: 37); mandible: L to p1 alveolus 140 mm, height of ramus ascendens = 60 mm, p1/p2 diastema: L = 34 mm.

Discussion.   Hooker and Thomas (2001: 840) reidentified a fragmentary maxilla (*Haplobunodon* sp. in Sudre, 1978: pl. 11/13) and a nearly complete mandibular ramus (UM.3080) from Euzèt as belonging to *Thaumastognathus*. Furthermore, an isolated M1 from Mormont belongs to this genus, which was formerly regarded as *Haplobunodon* sp. by Stehlin (1908: pl. 13/4).

Genus *Amphirhagatherium* Depéret, 1908

Type Species and Type Specimen.   *A. fronstettense* (Kowalevsky, 1874), lectotype (Hooker and Thomas, 2001: 836) SMNS 44059, left mandibular ramus with p4–m3; Frohnstetten, Headon Hill (Bembridge Limestone Fm., beds 9-18), Isle of Wight (undifferentiated, Bembridge Marls). *A. frohnstettense* is an unjustified emendation of the species name by Stehlin (1908: 791).

Included Species.   *A. ruetimeyeri* (Pavlov, 1900): Mormont (undifferentiated), Quercy (undifferentiated); *A. weigelti* (Heller, 1934): Geiseltal Ce III, Geiseltal sites VI, XVIII, XXII, XLIII (untere Mittelkohle); XXXV, XLI, Leo I, Leo III, Ce IV, Ce VI (obere Mittelkohle); Ce I, Ce II, Ce III (oberes Hauptmittel); *A. louisi* (Sudre, 1978): Grisolles; *A. neumarkensis* (Erfurt and Haubold, 1989): Geiseltal XVIII (untere Mittelkohle); *A. edwardsi* Hooker and Thomas, 2001: Isle of Wight (Headon Hill, upper Hatherwood Limestone Member).

Diagnosis.   Medium-sized bunodont/bunoselenodont choeropotamids with the tendency toward molarization of P4/p4 and P3, and weak parastyles. In advanced species, the P4 bears a paracone, and protocone and metacone are almost equal in size. p4 bears a paraconid, entoconid, and distal metaconid in advanced species. Paraconid and protoconid are separated. Upper molars lack parastyles; in some species there are large mesostyles and a protolophule, which links to the premetaconule crista. Lower molars without paraconid, entoconid distally situated, crest of the metaconid mesiobucally oriented, hypolophid attached to the cristid obliqua.

Dimensions.   m2: L = 7.9 mm, W = 6.1 mm; *A. weigelti* i1–m3: L = 60 mm, skull: L = 120 mm, H = 50 mm, body: L = 700 mm, H = 300 mm.

Discussion.   *Amphirhagatherium* includes the genus *Anthracobunodon* Heller, 1934 (Hooker and Thomas, 2001) with the species *A. weigelti, A. neumarkensis,* and *A. louisi.* As a result, the stratigraphic range of *Amphirhagatherium* extends from MP 12 to MP 20. The poorly preserved *A. ruetimeyeri* is regarded as a *nomen dubium* (Hooker and Weidmann, 2000), probably synonymous with *A. fronstettense.*

The newly described *A. edwardsi* bears only slightly crested conical cusps on the upper molars. The mesostyle is rounded at its base and is linked to both the postparacrista and premetacrista. The premetaconule crista is bifurcated: one branch runs buccally to reach the base of the metacone; the other branch runs mesially and joins a crest descending from the protoconid. The most advanced species is *A. fronstettense*.

Genus *Haplobunodon* Depéret, 1908

Type Species and Type Specimen. *H. lydekkeri* Stehlin, 1908, BMNH 29851, skull with nearly complete upper dentition; Hordle Cliff (Totland Bay Member). BMNH 29851 was originally described as the holotype, but BMNH 29713 (left and right mandible with p2–m3) is almost certainly the same individual and also part of the holotype (Hooker and Thomas, 2001). *H. ruetimeyeri* (Pavlov, 1900) is referred to *Amphirhagatherium*.

Included Species. *H. solodurense* Stehlin, 1908: Egerkingen α, β, Huppersand, Bolus; *H. muelleri* (Rütimeyer, 1862): Egerkingen Huppersand, α, ?β, Lissieu; *H. venatorum* Hooker, 1986: Creechbarrow (Creechbarrow Limestone Fm.); *H. meridionale* Sudre, 1997: St. Maximin; *H.* cf. *muelleri*: Geiseltal site XXII (untere Mittelkohle). XXXV, XXXVI, Ce IV (obere Mittelkohle) (Erfurt, 1988); *H.* sp.: level of Robiac (Biochrom, 1997: 787).

Diagnosis. P1/p1 premolariform to caniniform. No diastema between p2 and p3. P3/P4 with outline as an isosceles triangle and without metacone. Upper molars semibunodont with mesostyles less than half the height of the paracone; paraconule smaller than protocone and not separated from it by deep fissure; premetaconule crista short, often joined to protocone. M3 distal edge straight. p3/p4 without hypoconid and p4 often with large paraconid. Lower penultimate molars with postcristid joining hypoconulid usually independently of entoconid. Mandibular ramus increasing in thickness under p1 to m3 (adapted from Hooker, 1986).

Dimensions. m2 (type) L = 8.2 mm, W = 6.7 mm; p2–m3: L = 52 mm; m1–m3: L = 30 mm.

Discussion. After the definition of *Haplobunodon* by Depéret (1908), Stehlin (1908) introduced *H. solodurense* and *H. muelleri* from Egerkingen, which are different in size and in the shape of their mandibular ramus. He stressed the possibility that the *Haplobunodon* material might be heterogeneous and that their taxonomic status was preliminary. *H. ruetimeyeri* (Pavlov, 1900) is the third species mentioned by Stehlin from an undifferentiated site of Mormont. This species, which is based only on a fragmentary upper maxilla with two molariform teeth, displays fewer similarities with the first two Egerkingen species than to *Amphirhagatherium fronstettense* and is, therefore, referred to this genus. *H.* cf. *muelleri* indicates a second choeropotamid lineage at Geiseltal, which should be regarded as a distinct species. *H.*

cf. *muelleri* can be differentiated from the older *Amphirhagatherium neumarkensis* as well as from the partly contemporaneous *A. weigelti* in having less differentiated P3 and P4, which displays only a small, mediodistally shifted protocone. In contrast to *H. muelleri* from Egerkingen, the molars are less crested. This Geiseltal form is similar in the P2–P4 structure to *H. lydekkeri* from Hordle Cliff (BMNH.29851) but is more primitive in having premolariform p1, smaller size, lacking parastyles, more bulbous cusps, and a more triangular M3.

Genus *Lophiobunodon* Depéret, 1908

Type Species and Type Specimen. *L. minervoisensis* Depéret, 1908, FSL.3016, left maxilla with P1–M3; La Livinière.

Included Species. *L. rhodanicum* Depéret, 1908: Lissieu; *L.* sp.: Rouzilhac (J. Erfurt, unpublished data).

Diagnosis. Small lophodont choeropotamids with slender lower jaws and diastemata in front of and behind p2. The p1 is premolariform, p2 and p3 are elongated, and the p4 bears a distolingually situated metaconid. In lower molars, the entoconid is bulbous and as high as the hypoconid. The hypoconulid on m1 and m2 is small. P4 displays only a paracone on the buccal side. Upper molars: parastyles stronger than the mesostyles, metastyles weak, paraconule close to protocone, protocone round and lacks a postprotocrista, metaconule crescentiform, preparaconule crista, and premetaconule crista elongated (lophodont), protocone round and displays no postprotocrista.

Dimensions. m2: L = 4.5 mm, W = 3.6 mm; P1–M3 (syntype): L = 31.5 mm.

Discussion. In his original description of *L. minervoisensis*, Depéret (1908: 160) mentioned *L. rhodanicum*, a smaller species from Lissieu. According to Richard (1942: 144), the Lissieu material is lost. Sudre and Lecomte (2000: Fig. 5) figured this type specimen (FSL 2627) and saw similarities with *Rhagatherium*. Because of taxonomic problems of the latter genus, we tentatively keep *L. rhodanicum* here. *Lophiobunodon* represents a very early lineage of lophodont artiodactyls in the middle Eocene and occupies an isolated position within the choeropotamid clade. Because of the relatively high specialization, earlier forms of the genus would be expected at least at the end of the early Eocene.

Genus *Masillabune* Tobien, 1980

Type Species and Type Specimen. *M. martini* Tobien, 1980, HLMD, skeleton with nearly complete skull and dentition; Messel. The holotype belongs to a private collector but is accessible in the HLMD.

Included Species. *M. franzeni* Erfurt and Haubold, 1989: Geiseltal site VII (Oberes Hauptmittel), XXVI (obere Mittelkohle).

Diagnosis. Medium-sized slightly bunoselenodont choeropotamids with short cranium. Angular process of the mandible rounded. Upper dentition: weak mesostyle and weak parastyle, only one diastema between C1/P1. The molars are less crested than in *Amphirhagatherium*. Lower dentition: c1 incisiform, p1 caniniform, p4 with small metaconid. In lower molars, paraconid and metaconid are joined, but well spaced.

Dimensions.   m2 (holotype): L = 7 mm; mandible: L = 80 mm.

Discussion.   Unlike other choeropotamids such as *Rhagatherium* and *Amphirhagatherium*, *Masillabune martini* has a relatively robust skull. The mandible is slender, not continuously increasing in height under the cheek teeth. The hindlimbs are only slightly longer than the forelimbs (see Fig. 5.4). *M. franzeni* from Geiseltal (MP 13) displays a similar robust outline of the skull as the type species from Messel (MP 11). In contrast, the limb proportions are more similar to those of *Amphirhagatherium weigelti*. Although the dentition is more bulbous in *M. franzeni*, close phylogenetic relationships with *Amphirhagatherium* have been advocated (Hooker and Thomas, 2001: 838). Unfortunately, the diagnostically important p1 is not preserved in the Geiseltal form for a definitive decision.

Genus *Cuisitherium* Sudre, Russell, Louis, and Savage, 1983

Type Species and Type Specimen.   *C. lydekkeri* (Lemoine, 1891), MNHN Al 5236, left mandibular ramus m1–(m3); Teredina sands near Epernay, Cuis, Grauves, Mancy, Mas de Gimel, Monthelon, Saint-Agnan.

Included Species.   *C.* sp.: Avenay (Sudre et al., 1983).

Diagnosis.   Small choeropotamids with bulbous bunodont dentition. Upper molars: paracone and metacone only slightly anteroposteriorly crested; parastyles and mesostyles lacking; protocone mesially shifted with attached paraconule; metaconule larger than paraconule, slightly crested, and in internal position; precingulum and postcingulum connected or only shortly interrupted near the protocone. Lower molars with elevated trigonid; paraconid mostly absent; trigonid mesially open, preprotocristid not connected with paraconid or metaconid. Postcingulid with central hypoconulid (on m3 as large as hypoconid), m2 much longer than m1 (adapted from Sudre et al., 1983).

Dimensions.   m2: L = 5.8 mm, W = 4.2 mm; m1–m3 (type): L = 17.8 mm (Sudre et al., 1983: 351).

Discussion.   The genus *Cuisitherium* was first described as a dacrytheriid in order to stress differences with more bunodont dichobunids such as *Protodichobune* (Sudre et al., 1983). The holotype, which was first mentioned as *P. lydekkeri* (Lemoine, 1891), is more advanced than *Protodichobune* or *Diacodexis* in its more strongly crested buccal cusps, a larger and more mesially shifted entoconid, and the nearly fused paraconid and metaconid. Recently, Hooker and Thomas (2001) included *Cuisitherium* within

the Choeropotamidae because of its similar selenodonty and proximity of paraconule and protocone in the upper molars compared with *Haplobunodon* and *Lophiobunodon*. The main argument for this systematic position is incomplete because the antemolar dentition of *Cuisitherium* is poorly known. If this genus is a primitive choeropotamid, the radiation of this family should have taken place at the end of the early Eocene.

Genus *Parabunodon* Ducrocq and Sen, 1991

Type Species and Type Specimen.   *P. anatolicum* Ducrocq and Sen, 1991, NHME EC15, fragmentary skull with right P4–M3 and left M1–M3; Eski Celtek (Turkey).

Diagnosis.   Medium-sized; M3 smaller than M2; trapezoidal shaped molars with a crescentiform metaconule; mesostyle rather internally situated; parastyle weak; no internal cingulum; metacone with flattened external wall; external wall of the labial cusps strongly canted inward; rather strong P4 with elongated external cusp and vertically notched on its labial side; crenulated enamel (Ducrocq and Sen, 1991: 13).

Dimensions.   M1–M3: L = 27.0 mm (Ducrocq and Sen, 1991: 16).

Discussion.   This monospecific genus is regarded as choeropotamid because of its large, blunt, and bunodont premolars and molars, but it may also be related to raoellids from the Indian Subcontinent. In contrast to most of the other choeropotamids, the main cusps and the mesostyles are not crested. In this regard *"Haplobunodon* cf. *muelleri"* and *Hallebune* from Geiseltal are similar. In *Hallebune*, the protocone is also mesially shifted, and the M3 is of triangular shape. *Parabunodon* is more advanced in having a broad lingual lobe and mesially shifted protocone on P4. On the upper molars, the metaconules are stronger and in a more lingual position. Therefore, close phylogenetic relationships between *Hallebune* and *Parabunodon* are unlikely. *Parabunodon* would be the single choeropotamid genus outside of western Europe. The early Lutetian or Ypresian age of the type locality Eski Celtek is still poorly constrained, and it is mostly based on the occurrence of the embrithopod *Palaeomasia* (Russell et al., 1982). If *Parabunodon* is a choeropotamid, and the size development of western European taxa is similar, it suggests a younger age for this locality.

Genus *Hallebune* Erfurt and Sudre, 1995b

Type Species and Type Specimen.   *H. krumbiegeli* Erfurt and Sudre, 1995b, GMH Ce IV-3925, fragmentary skull with right P3–M3 and left P3–(M3); Geiseltal Cecilie IV, Geiseltal Leo III (both sites: obere Mittelkohle).

Diagnosis.   Small brachybunodont choeropotamids with molarized P3 and P4, which display crested protocones and paracones with closely attached small metacones. P3 with parastyle and distally shifted protocone. P4 triangular, wider than long, protocone simple. Upper molars rectangular in outline, with bulbous cones; they lack hypocones, parastyles, or mesostyles. Metaconule similar in size to the paracone and larger than the paraconule. M2 larger than M1 (adapted from Erfurt and Sudre, 1995b: 87).

Dimensions. m2: L = 5.1 mm, W = 3.9 mm; M1–M3 (type): L = 13.8 mm.

Discussion. *Hallebune* is known only by the type species and represented at two localities within Geiseltal (obere Mittelkohle). Except for the poorly known *Cuisitherium*, *Hallebune* is the smallest and most primitive choeropotamid genus. The lower premolars are elongated and bear a paraconid and hypoconid, and the p4 an additional metaconid. The lower molars are less crescentiform than in *Cuisitherium*. The upper molars of *Hallebune* resemble those of *Protodichobune* but are more derived in their more rectangular outline and the more lingual position and larger size of the metaconule. Therefore, *Hallebune* is regarded as a stem taxon within the choeropotamid clade that is different from the similar sized *Rhagatherium* in having less differentiated molars lacking any additional conules. In the cladistic analysis of Hooker and Thomas (2001), it is indicated as the sister group to *Rhagatherium*.

Family MIXTOTHERIIDAE Pearson, 1927

Characteristics. Roof of the skull relatively low, muzzle short and wide, orbits medially situated, incomplete postorbital bar, frontal bones enlarged, prominent sagittal crest connected posteriorly to marked transverse occipital crests, no exposure of the pars mastoidea between the exoccipital and the squamosal, relatively small tympanic bulla; upper molars brachydont, bunoselenodont, subtriangular to subtrapezoidal in occlusal view, paraconule reduced and included in the preprotocrista, parastyle and mesostyle rounded labially; P4 triangular in outline, molariform but without paraconule, P3 and P2 with a well-developed lingual cusp, P1 narrow and adjacent to a prominent canine; C and I3 are separated by a short diastema. Dentary posteriorly deep; prominent lower canine possibly separated from p1 by a short diastema; p2 and p3 are elongated and show three main cuspids situated on the same line, p4 tends to be molariform with a distinct paraconid, a strong metaconid lingual to the protoconid, and a distinct posterior heel bearing the hypoconid and entoconid; m1–2 typically show a distal cingulid rounded lingually and extending up at the rear of the entoconid, complete postcristid, the postprotocristid meets a faint postmetacristid, a weak parastylid occurs mesiolingually, third lobe of m3 transversely compressed (adapted from Viret, 1961).

Included Genus. *Mixtotherium*.

Comments. Simpson (1945) considered mixtotheriines as a subfamily of the Cebochoeridae based on resemblances with the skull of *Cebochoerus*. Gentry and Hooker (1988: Fig. 9.8) placed *Mixtotherium* as the sister group of

two typically North American families, the Merycoidodontidae and Agriochoeridae, whereas in their PAUP-generated cladogram (Fig. 9.7), mixtotheriids lie at the base of a heterogeneous clade including anoplotherioids, xiphodontids, protoceratids, amphimerycids, ruminants, and cameloids. Recently, Hooker and Weidmann (2000) allied *Mixtotherium* with the Cainotherioidea on the basis of the tendency toward deepening of the dentary and molarization of P4 and the retention of a complete postcristid on the lower molars (contrary to the condition observed in the Anoplotheriidae, Xiphodontidae, and Amphimerycidae). These authors also proposed to remove some specimens formerly allocated to the species *Robiacina lavergnensis* into the genus *Mixtotherium*. The clade Cainotherioidea investigated by Hooker and Weidmann (2000) is characterized by several apomorphies: dentary posteriorly deep, trend toward the molarization of P4, and retention of a complete postcristid on lower molars. As mentioned by the authors, the second feature also characterizes the xiphodontid *Dichodon;* the first feature is unknown in several taxa, and its value as an apomorphy of the clade should be considered to be provisory.

Genus *Mixtotherium* Filhol, 1880

Type Species and Type Specimen. *M. cuspidatum* Filhol, 1880, MNHM M.41, fragmentary maxilla with right C-M3; Quercy (undifferentiated), Celarié, La Bouffie, Perrière.

Included Species. *M. gresslyi* Rütimeyer, 1891: Egerkingen α, Lissieu (according to Hooker [1986] *M. priscum* Stehlin, 1908 is included here from Egerkingen Huppersand, Bolus, α; Chamblon [Stehlin, 1908] and Lissieu [Sudre, 1978]); *M. depressum* (Filhol, 1884): Quercy (undifferentiated); *M. quercyi* (Filhol, 1884): Quercy (undifferentiated); *M. leenhardti* Stehlin, 1908: Quercy (undifferentiated); *M. infans* Stehlin, 1910: Egerkingen α; *M. lavergnensis* (Sudre, 1977): Lavergne, Eclépens-Gare (includes *Robiacina lavergnensis* Sudre, 1977 and *R. weidmanni* Sudre, 1978 [Hooker and Weidmann, 2000]); *M.* aff./cf. *gresslyi*: Creechbarrow (Hooker, 1986: 383); La Défense (Sudre and Ginsburg, 1993: 165).

Diagnosis. See Family.

Dimensions. M2 (holotype): L = 10.2 mm; W = 10.6 mm.

Discussion. The type species *M. cuspidatum* is known from several skulls and a partial mandible figured in Stehlin (1908). Dechaseaux (1974) provided a reliable reconstruction of the skull and mandible on the basis of additional material of a juvenile individual. As noted by these authors, the posterior enlargement of the mandible resembles that of extant hyracoids, suggesting similar ecological niches. The postcranial morphology of *Mixtotherium* remains poorly documented. Sudre and Ginsburg (1993) described several astragali from La Défense (MP 13) showing a strong asymmetry of the mesial pulley, which is not in line with the distal one. The sudden appearance of *Mixtotherium* during the Geiseltalian-Robiacian transition (MP 13–MP 14) cannot be

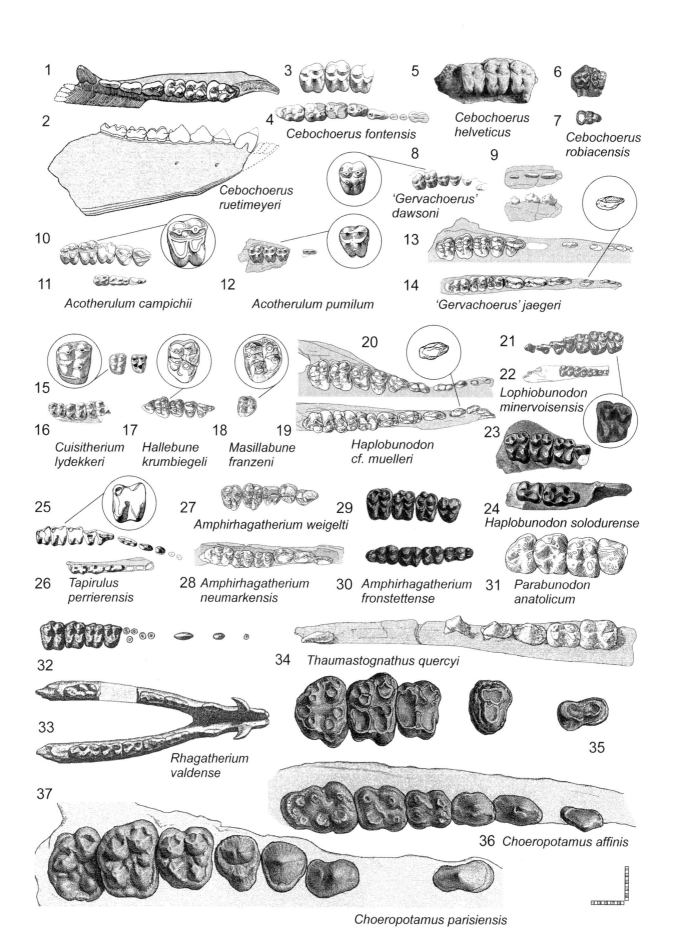

1
2
Cebochoerus
ruetimeyeri

3
4
Cebochoerus fontensis

5
Cebochoerus
helveticus

6

7
Cebochoerus
robiacensis

8
'Gervachoerus'
dawsoni

9

10
11
Acotherulum campichii

12
Acotherulum pumilum

13

14
'Gervachoerus' jaegeri

15
16   17
Cuisitherium
lydekkeri

18
19
Hallebune
krumbiegeli

Masillabune
franzeni

20
Haplobunodon
cf. muelleri

21

22
Lophiobunodon
minervoisensis

23

24
Haplobunodon solodurense

25
26 Tapirulus
perrierensis

27
Amphirhagatherium weigelti

28 Amphirhagatherium
neumarkensis

29

30 Amphirhagatherium
fronstettense

31 Parabunodon
anatolicum

32

33
Rhagatherium
valdense

34 Thaumastognathus quercyi

35

36 Choeropotamus affinis

37

Choeropotamus parisiensis

connected to any taxa known in the different levels of the Geiseltal or earlier in the Eocene. Likewise, artiodactyls reported from the few other MP 13 localities (Bouxwiller, and localities from the southern France) do not display special affinities with *Mixtotherium.* Hooker and Thomas (2001: 840) transferred *Rhagatherium kowalevskyi* from the former haplobunodontids to mixtotheriids, suggesting close affinities between choeropotamids and mixtotheriids.

Family CAINOTHERIIDAE Camp and Van der Hoof, 1940

Characteristics. Complete postorbital bar, enlarged bullae; the morphology of the lower jaw is variable within the family but tends to become massive and deepen posteriorly in some genera; diastemata within the premolar row are present in the Eocene species and tend to be reduced or lost (*Cainotherium*) in the Oligocene and Miocene forms. Lower canine incisiform, anterior premolars narrow, tendency toward molarization of p4, crowns of cheek teeth moderately high, lingual mesiostylid (or mediostylid) on lower molars; P1 and P2 without anterior lobe, P3 labially concave, with a lingual cusp, P4 subrectangular, upper molars with a W-shaped ectoloph, crescent-shaped lingual cusps, and retaining conules, trend toward a protocone distolingually placed, along with a displacement of the paraconule in the mesiolingual corner of the molar; limbs tetradactyl with a reduction of lateral digits (II and V), forelimb shorter than hindlimb, long tail.

Included Genera. *Cainotherium, Caenomeryx, Plesiomeryx, Oxacron, Paroxacron.*

Comments. According to Hooker and Weidmann (2000), this family and the monogeneric Mixtotheriidae are now referred to the Cainotherioidea (Camp and Van der Hoof, 1940). The former authors also removed *Robiacina* from the Anoplotheriidae to the Cainotheriidae as first

proposed by Sudre (1969, 1978). Cainotheriids have the longest range among artiodactyls in Europe, as they first appeared in the late middle Eocene with *Robiacina,* and the last representatives of the family are middle Miocene with *Cainotherium huerzeleri,* up to 20 million years of evolution in Europe.

Genus *Robiacina* Sudre, 1969

Type Species and Type Specimen. *R. minuta* Sudre, 1969, USTL RBN 5066, right M1; Robiac-Nord, Le Bretou.

Included Species. (*R. lavergnensis* Sudre, 1977 and *R. weidmanni* Sudre, 1978; see *Mixtotherium;*) *R. quercyi* Sudre, 1977; La Bouffie, Eclépens B? (Hooker and Weidmann, 2000: 88), Les Pradigues, Les Sorcières; *R.* sp.: Sant Jaume de Frontanyà 1,3 (Antunes et al., 1997).

Diagnosis. Very small artiodactyl with triangular upper molars, conules retained, W-shaped ectoloph, flat labial wall of the paracone and metacone, the latter being more lingually situated, styles developed, paraconule conical but connected to the parastyle, P3 with a lingual cusp, P4 triangular with a strong bunodont lingual cusp connected to the parastyle; lower molars with crescent-shaped labial cuspids, the cristid obliqua reaches the metaconid, trigonal crescent mesiodistally narrow, large valley between the metaconid and entoconid that opens lingually, p4 and to a lesser degree p3 tend to be molariform (adapted from Sudre, 1969, 1977).

Dimensions. M2? (holotype of *R. minuta*): L = 3.0 mm; W = 3.25 mm (Sudre, 1969).

Discussion. Sudre (1969, 1978) regarded this genus to be close to the ancestry of cainotheres but placed this genus within the Anoplotheriidae (and proposed a new and monogeneric subfamily, the Robiacinae). Following Sudre's initial opinion, Hooker and Weidmann (2000) removed *Robiacina* to the Cainotheriidae in their tentative phylogeny of cain-

Fig. 5.1. (*opposite*) Cebochoerid and choeropotamid dentitions. Dentitions occlusal and approximately scaled; those in circles are magnified (2x). Bold specimens are types. (1) *Cebochoerus ruetimeyeri* **NMB Ef 210,** right mandible with p1–m3 (Stehlin, 1908: Fig. CIII a). (2) *Cebochoerus ruetimeyeri* **NMB Ef 210,** right mandible with p1–m3, lateral (Stehlin, 1908: Fig. CIII b). (3) *Cebochoerus fontensis* USTL F4 214 (mirrored), left maxilla with M1–M3 (Sudre, 1978: pl. 7/1). (4) *Cebochoerus fontensis* **USTL F5 07,** left mandible with p3–m3 (Sudre, 1978: pl. 8/2). (5) *Cebochoerus helveticus (minor)* **MNHN Qu60b,** left maxilla with M1–M3 (Hooker, 1986: pl. 31/2). (6) *Cebochoerus robiacensis* **NMB Rb 52,** right maxilla with M2–M3 (Stehlin, 1908: Fig. CIX). (7) *Cebochoerus robiacensis* **NMB Rb 70,** left m3 (Stehlin, 1908: Fig. CIX). (8) "*Gervachoerus*" *dawsoni* **USTL Bx 66816,** right maxilla with P2–M3 (Sudre, 1978: pl. 5/1). (9) "*Gervachoerus*" *dawsoni* USTL Bx 66817 (mirrored), right mandible with p1–p3 (Sudre, 1978: pl. 5/2). (10) *Acotherulum campichii* USTL RBN 5447 (mirrored), left maxilla with P3–M3 (Sudre, 1978: pl. 6/3). (11) *Acotherulum campichii* USTL RBN 2313, right mandible with dp3–m1 (Sudre, 1978: pl. 6/3). (12) *Acotherulum pumilum* USTL PRR 1465, right maxilla with dP2, dP4–M2 (Sudre, 1978: pl. 9/3). (13) "*Gervachoerus*" *jaegeri* GMH Ce III-4220, right maxilla with P1, P3–M3 (Erfurt, 1988: pl. 7b). (14) "*Gervachoerus*" *jaegeri* GMH Ce III-4220, left mandible with I1–M3 (Erfurt, 1988: pl. 7c). (15) *Cuisitherium lydekkeri* coll. Louis Cuis 1-DE, isolated M2; coll. Dégremont C13-L, isolated M1; Sudre et al. (1983: Fig. 17c,d). (16) *Cuisitherium lydekkeri* **MNHN Al 5236,** left mandible with m1–(m3); Sudre et al. (1983: Fig. 18a). (17) *Hallebune krumbiegeli* **GMH Ce IV-3925,** right maxilla with P3–(M3); Erfurt and Sudre (1995b: Fig. 1). (18) *Masillabune franzeni* **GMH VII-58,** left M2 isolated drawn (Erfurt, 1988: pl. 19b). (19) *Haplobunodon* cf. *muelleri* GMH XXII/553, right mandible with i1–m3; Erfurt (1988: pl. 11c). (20) *Haplobunodon* cf. *muelleri* GMH XXII/553, right maxilla with (C1)–M3; Erfurt (1988: pl. 11b). (21) *Lophiobunodon minervoisensis* **FSL 3016,** left maxilla with P1–M3; A. Franke, Halle. (22) *Lophiobunodon minervoisensis* **FSL without number,** right mandible maxilla with (dp3)–m2; A. Franke, Halle. (23) *Haplobunodon solodurense* **NMB Ef 43** right maxilla with (P4)–M3; Stehlin (1908: pl. 13/3). (24) *Haplobunodon solodurense* NMB Ef 63 right mandible with m2–m3; Stehlin (1908: pl. 13/31). (25) *Tapirulus perrierensis* coll. F.S. Marseille PQ 689, right maxilla with C1–M3; Sudre (1978: pl. 12/1). (26) *Tapirulus perrierensis* **USTL PRR 1449,** right mandible with p3–m1; Sudre (1978: pl. 12/3b). (27) *Amphirhagatherium weigelti* **GMH Ce III-4225,** right maxilla with P3–M3; Erfurt (1988: pl. 14b). (28) *Amphirhagatherium neumarkensis* GMH XVIII/369, right mandible with p2–m3; Erfurt (1988: pl. 18b). (29) *Amphirhagatherium fronstettense* **SNMS 44056, 44058** and unknown specimen, combined P4–M3; Kowalevsky (1874: pl. VIII/58). (30) *Amphirhagatherium fronstettense* **SNMS 44059** (mirrored), left p4–m3; Kowalevsky (1874: pl. VIII/59). (31) *Parabunodon anatolicum* **NHME EC15,** right maxilla with P4–M3; Ducrocq and Sen (1991: Fig. 2b). (32) *Rhagatherium valdense* **MGL LM.768,** right maxilla with (I3)–M3; Pictet (1855–1857: pl. III/1b). (33) *Rhagatherium valdense* **MGL LM.767,** right mandible with p1–m3; Pictet (1855–1857: pl. III/6). (34) *Thaumastognathus quercyi* **MNHN without number,** left mandible with p1–m2; Filhol (1890: 33). (35) *Choeropotamus affinis* **MGL without number,** right maxilla with P3–M3; Gervais (1848–1852: pl. XXXI/2–4). (36) *Choeropotamus affinis* **MGL without number,** left mandible with p2–m3; Gervais (1848–1852: pl. XXXI/6). (37) *Choeropotamus parisiensis* **MNHN without number** detail of maxilla with right P1, P2–M3; Gervais (1848–1852: pl. XXXII/1).

otheroids and placed *Mixtotherium* as the sister group of the family. *Robiacina* remains the most primitive and the earliest cainotheriid known.

First described from Robiac (MP 16), the genus is now reported in the early Robiacian of San Jaume de Frontanya (Moyà-Solà and Köhler, 1993) and possibly from Laprade (Sudre et al., 1990). The origin of *Robiacina* remains obscure, although it may be close to the Geiseltalian Eurodexinae (Erfurt and Sudre, 1996). It is worth noting that the important radiation that occurred by the Geiseltalian/Robiacian transition includes both hypoconiferous (*Hyperdichobune, Pseudamphimeryx*) and nonhypoconiferous taxa (*Robiacina, Dichodon*). The important radiation of selenodont artiodactyls in Europe, including the Cainotheriidae (even if they diversified only by the late Eocene, MP 18–19), is thought to be linked to climatic changes that occurred by the late Lutetian in Europe (Collinson et al., 1981).

Subfamily OXACRONINAE Hürzeler, 1936

The upper molars of the Oxacroninae are subtriangular, and they do not show the derived "Cainotherian plan" (Stehlin, 1906): the protocone is in a subcentral position, and the mesostyle is looplike instead of being pinched as in the Cainotheriinae. Short diastemata are usually present. The muzzle is short and generally narrow, and the orbits are large with respect to the small size of the skull. As stressed by Hooker and Weidmann (2000), the distinction of *Oxacron* from *Paroxacron* is not clear, and a revision of these genera may prove useful.

Genus *Oxacron* Filhol, 1884

Type Species and Type Specimen. *O. courtoisi* Gervais, 1848-1852, MNHN, maxilla with P3–M3; Quercy (undifferentiated), Escamps, Hoogbutsel, La Débruge, Rosières 2, 4, Saindou D.

Included Species. *O. quinquedentatus* (Filhol, 1877): Quercy (undifferentiated); *O.* cf. *courtoisi*: Coânac (Sudre, 1978: 206).

Diagnosis. P1 caniniform and separated from P2 by a short diastema, protocone of upper molars centrally located; diastema between p2 and p3 (adapted from Hürzeler, 1936).

Dimensions. M2 (*O. courtoisi*): L = 3.1 mm; W = 4.0 mm.

Genus *Paroxacron* Hürzeler, 1936

Type Species and Type Specimen. *P. valdense* (Stehlin, 1906), lectotype (Hooker and Weidmann, 2000: 89) NMB Mt. 230, cranial fragment with right P2–M3; Entreroches (undifferentiated), Eclépens C, Village (Cinq-Sous).

Included Species. *P. bergeri* Heissig, 1978: Mouillac; *P.* sp. (summarized): Escamps (Legendre, 1980); Aubrelong 1 (Sudre, 1978: 203).

Diagnosis. P1/1 and P2/2 elongated and premolariform; diastemata between P1 and P2, p1 and p2, p2 and p3

generally very short or absent, anterior lobe of P3 elongated, upper molars with a mesostyle notch (adapted from Hürzeler, 1936).

Dimensions. M2 (*P. valdense*): L = 2.9 mm; W = 4.3 mm.

Discussion. Hooker and Weidmann (2000) summarized the complex systematic history of these two genera, which may represent a single genus. In any case, there is obviously taxonomic oversplitting within the Cainotheriidae; our understanding of generic characters and their variability are further obstructed by the paucity of Eocene fossil samples in comparison with the abundant fossil material collected in the Oligocene and Miocene. Interestingly, the range of *Paroxacron* extends over the *Grande Coupure*, and the genus may have persisted into the late Oligocene (Blondel, 2005).

Subfamily CAINOTHERIINAE Camp and Van der Hoof, 1940

The upper molars of the Cainotheriinae are subquadrate, more crested than those of oxacronines, and they show the typical "Cainotherian plan" (Stehlin, 1906): the protocone is placed in the distolingual corner of the molar, and the mesostyle is pinched. The preprotocrista and the postparaconule crista tend to meet in the center of the molar. Diastemata are absent. The muzzle is elongated, and the maxilla extends higher and forms with the frontal a subhorizontal line in lateral view.

Genus *Cainotherium* Bravard, 1828

Type Species and Type Specimen. *C. renggeri* (von Meyer, 1837), private collection of Rengger (whereabouts unclear), fragmentary lower jaw from Aarau, unreferenced specimen.

Included Species. *C. laticurvatum* Geoffroy Saint-Hilaire, 1833: Langy (includes *C. minimum* Bravard, 1835, *C. elegans* Pomel, 1846, *C. metopias* Pomel, 1851 [Heizmann, 1999: Table 21.1]); *C. commune* Bravard, 1835: St-.Gérand-le-Puy, Escamps; *C. miocaenicum* Crusafont-Pairó et al., 1955: Sant Mamet. For further French and Spanish localities see Heizmann (1983: 823); *C. bavaricum* Berger, 1959: Wintershof-West, Schnaitheim 1, Stubersheim 3, Oggenhausen 2, Randecker Maar, Petersbuch 2; *C. lintillae* Baudelot and Crouzel, 1974: Espira-du-Conflent (includes *C. laticurvatum ligericum* Ginsburg et al., 1985 [Heizmann, 1999: Table 21.1]); *C. huerzeleri* Heizmann, 1983: Steinberg, Goldberg; *C.* cf. *laticurvatum*: Öllingen (Heizmann, 1983: 820); *C.* aff./cf. *bavaricum*: Petersbuch 2, Randecker Maar (Heizmann, 1983: 814, 817); *C.* sp.: Langenau 1 (Heizmann, 1983: 816).

Diagnosis. Upper molars quadrangular, cusps markedly crescent shaped, crests developed, mesostyle pinched; diastemata very short or absent; P1, P2, and P3 relatively elongated, P1 and P2 without lingual extension; the nasal is long and rectangular, the premaxilla reaches the frontal, presence of vacuities in the maxilla below the orbit, a long narrow crescent-shaped ethmoidal slit variably present, choanae are anteriorly short or large; the mandible is more slender than in other cainotheriines.

Dimensions.   M2 (*C. laticurvatum*): L = 7.7 mm; W = 8.0 mm.

Discussion.   Heizmann (1983, 1999) provided the last overview of the genus and suspected that cainotheres may have occupied the same ecological niche as present-day hares. The geographic distribution of *Cainotherium* extends from Portugal to central Europe; its specific diversity and time range are the greatest within the family. Berger (1959) reported *C. commune* from Escamps (MP 19), which represents the oldest and only occurrence of the genus in the Eocene. This determination seems to be questionable, however, because the same species was first described from MN2 and is also known from other localities, which belong to MP 30 (Biochrom, 1997: 794).

Genus *Caenomeryx* Hürzeler, 1936

Type Species and Type Specimen.   *C. filholi* (Lydekker, 1885), BMNH-M1399, complete skull; Quercy (undifferentiated).

Included Species.   *C. procommunis* Hürzeler, 1936: Quercy (undifferentiated); *C.* cf. *procommunis*: level of Le Garouillas (Biochrom, 1997: 791); *C.* sp.: level of Pech du Fraysse (Biochrom, 1997: 791).

Diagnosis.   Upper molars wider (transversely) than long, premolars mesiodistally short, short diastema between C and P1, diastema between P1 and P2 very short or missing, P3 with a strong lingual cusp, P2 with a smaller lingual cusp, P1 with a very faint lingual lobe, P4 is subtriangular; the muzzle is high and narrow at its anterior extremity but broad at the level of orbits; the premaxilla does not reach the frontal, the nasal bones broaden behind the posterior extremities of the premaxilla, choanae are large, the mandible is massive, deepens rapidly posteriorly, and the coronoid process is smooth (adapted from Hürzeler, 1936).

Dimensions.   M2 (*C. procommunis*): L = 7.0 mm; W = 8.7 mm.

Discussion.   *Caenomeryx* became diversified in MP23 (Blondel, 2005) and is recorded up into the Oligocene. However, the specific diversity of the genus is poor for the upper part of Oligocene. Much work is still needed to determine intraspecific variation and then to estimate specific diversity within the genus.

Genus *Plesiomeryx* Gervais, 1873

Type Species and Type Specimen.   *P. cadurcensis* Gervais, 1873, MNHN QU 1772, lower jaw with right p2–m3; Mouillac.

Included Species.   *P. huerzeleri* Berger, 1959: Gaimersheim; *P.* cf. *cadurcensis*: levels of Itardies, Le Garouillas,

Pech du Fraysse (Biochrom, 1997: 791); *P.* indet.: level of Mas de Pauffié (Biochrom, 1997: 791).

Diagnosis.   Upper molar showing an intermediate morphological stage between oxacronines and *Cainotherium* in the achievement of the "Cainotherien plan" (Stehlin, 1906), paraconule smaller than the protocone, premolars mesiodistally short, long diastema between C and P1 and between P1 and P2, P3 with a developed lingual cusp, P2 with a small lingual cusp, P4 is subquadrate; the nasal is long, narrow, straight, and high, the choanae are large, the mandible tends to be more slender than in other genera (adapted from Hürzeler, 1936).

Dimensions.   M2 (*P. huerzeleri*): L = 4.3 mm; W = 5.4 mm.

Discussion.   Sudre (1995) demonstrated the presence of high morphological diversity in the genus at Le Garouillas (MP 25) and suspected the occurrence of several lineages (or populations) that may have evolved concurrently in the same ecological zone. That would imply several "ghost lineages" in the early Oligocene that remain unknown to date. *Plesiomeryx* is present in MP 21 when the diversity of the family is minimal, the Oligocene radiation of the family occurring later by mammal levels MP 22 and particularly MP 23 with the appearance of *Caenomeryx* and *Cainotherium*. *Plesiomeryx* is thus the first post–*Grande Coupure* Cainotheriinae.

Family ANOPLOTHERIIDAE Bonaparte, 1850

These forms show dental and postcranial specializations without any equivalent in the contemporaneous groups of artiodactyls known from other Holarctic landmasses. Based on the morphology of the astragalus, Heissig (1993) evidenced affinities with North American oreodonts, suggesting a common origin of both groups in Central Asia. Anoplotheriids (such as xiphodontids and amphimerycids) are strictly restricted to the Paleogene of Europe, where they radiated roughly in the same period as the North American selenodonts. The data on the early Asian selenodonts are still very incomplete (see Foss and Prothero, this volume) and make the phylogeny of the European Selenodontia still ambiguous.

Subfamily DACRYTHERIINAE Depéret, 1917

Characteristics.   Medium-sized artiodactyls with a complete dentition; diastemata in the anterior dentition variably present; upper molars with a centrally placed conical lingual protocone, a subconical paraconule connected to the parastyle, metaconule generally crescent shaped, paracone and metacone included in the W-shaped ectoloph; premolars moderately molarized and retaining a crescent-shaped paraconule, canine undifferentiated, lower premolars elongated and trenchant. Paroccipital apophysis formed by both the mastoid and the exoccipital.

Included Genera.   *Dacrytherium, Catodontherium*.

Comments. Sudre (1978) followed Depéret (1917) in considering the Dacrytheriidae as a distinct family including *Dacrytherium, Catodontherium, Leptotheridium, Tapirulus,* and possibly *Cuisitherium*. However, Sudre and Lecomte (2000) cast some doubts on the familial affinities of *Cuisitherium,* and Hooker and Thomas (2001) moved *Cuisitherium* and *Tapirulus* to the Choeropotamidae. Stehlin (1910a) suggested that xiphodontids might be closely related to dacrytheriids. Following that logic, Hooker and Weidmann (2000) moved *Leptotheridium* from the Dacrytheriidae to the Xiphodontidae and placed the dacrytheriine *Dacrytherium* and *Catodontherium* as a subfamily of the Anoplotheriidae. Xiphodontids and anoplotheriids are now included in the broad concept of Anoplotherioidea following the phylogeny of Gentry and Hooker (1988). We have espoused this systematic scheme below. Dacrytheriids are common in Bouxwiller (MP 13) but are unknown in the levels of Geiseltal suggesting peculiar ecological conditions in that area during the early middle Eocene (e.g., Franzen, 2003).

Genus *Dacrytherium* Filhol, 1876

Type Species and Type Specimen. *D. cayluxi* Filhol, 1876 (junior synonym of *D. ovinum* Owen, 1857 from the Hordle Cliff [Totland Bay Member, Headon Hill Fm.] based on BMNH.29174a, left and right lower jaws with i3–m3). Recently, Cuesta et al. (2006) indicated *D. ovinum* from Sossís.

Included Species. *D. elegans* (Filhol, 1884): Lamandine, Creechbarrow, Eclépens (A, Gare), Lautrec, Le Bretou, Le Castrais, Les Alleveys, Robiac; *D. priscum* Stehlin, 1910: Egerkingen Huppersand or Bolus, Lissieu?; *D. saturnini* Stehlin, 1910: La Débruge, Ste. Néboule, Saindou D; *D.* cf. *elegans:* Egerkingen (very common in a, less frequent in Bolus, Huppersand and β), Chamblon (Stehlin, 1910a); Bouxwiller, Le Bretou (Sudre, 1978); *D.* cf. *saturninii:* Gousnat (Sudre, 1978: 210) ; *D.* sp.: Repeu del Güaita (Antunes et al., 1997: 345).

Diagnosis. Roof of the cranium low, muzzle elongated, transversely developed, and with a wide preorbital vacuity, lacks postorbital bar, pars mastoidea exposed laterally between the exoccipital and squamosal; upper incisors triangular in section, canine undifferentiated, P1–3 moderately elongated with a weak development of the lingual lobe, P4 triangular with a crescentic lingual cusp, trapezoidal and upper molars with five cusps, paraconule connected to a prominent parastyle, labial side of the paracone and metacone are slightly ribbed and noticeably excavated, loop-shaped mesostyle; lower canine incisiform, p1, p2, p3 narrow and trenchant, p4 with a metaconid distolingual to the protoconid, and a weak and bifurcated paraconid; lower molars with two labial crescent-shaped cuspids and three lingual cuspids (small lingual mediostylid), postcristid and paracristid extending lingually (adapted from Depéret, 1917; Sudre, 1978).

Dimensions. M2 (holotype *D. cayluxi*): L = 10.3 mm; W = 7.2 mm.

Discussion. Sudre (1978) proposed two distinct lineages throughout the evolution of the genus: *D. elegans–D. saturnini* on one hand, and *D. priscum–D. ovinum* on the other hand. The most recent species of the two lineages tend to show an accentuated W-shaped ectoloph on the upper molars and a larger size.

Genus *Catodontherium* Depéret, 1908

Type Species and Type Specimen. *C. robiacense* (Depéret, 1906), Lectotype (Sudre, 1969: 127) FSL-4799, right lower jaw with p2–m3; lower Calcaires de Fons (Robiac), Le Bretou, Eclépens (A, Gare), Les Alleveys, Quercy (undifferentiated).

Included Species. *C. fallax* Stehlin, 1910: Egerkingen Huppersand?, α, β, Bolus, Chamblon, Lissieu; *C. buxgovianum* Stehlin, 1910: Egerkingen α, β, Bolus, Huppersand, Lissieu; *C. paquieri* Stehlin, 1910: Le Castrais (type probably lost); ?*C. argentonicum* Stehlin, 1910: Argenton; *C.* cf. *fallax* and *C.* cf. *argentonicum:* Bouxwiller (Sudre, 1978: 205); *C.* sp.: Sant Jaume de Frontanyà 1, Sossís (Antunes et al., 1997: 346).

Diagnosis. Lower premolars more elongated (except p4) than in *Dacrytherium,* lower molars lacking the lingual mediostylid, third lobe of m3 is double cusped; upper molars trapezoidal in outline; labial side of the paracone and metacone are slightly ribbed, styles weakly salient labially, strong mesial cingulum.

Dimensions. m2 (holotype of *C. robiacense*): L = 17.6 mm; W = 10.1 mm.

Discussion. This genus is not as well known as *Dacrytherium.* ?*Catodontherium argentonicum* from Argenton (MP 11) might be the earliest occurrence of the family (Stehlin, 1910a: 926). This taxon is still poorly documented, but its molars seem noticeably more bunodont than those of *Catodontherium* and *Dacrytherium* from MP 13.

Subfamily ANOPLOTHERIINAE Bonaparte, 1850

Characteristics. Medium to large-sized animal with a complete dentition without any or limited diastemata; upper molars retaining a crescent-shaped paraconule, undifferentiated canine, and lower premolars elongated and trenchant.

Included Genera. *Anoplotherium, Diplobune, Ephelcomenus, Robiatherium.*

Comments. Anoplotheriids suddenly appear in the mammal level MP 18 (like the Cainotheriidae), and *Diplobune* and *Ephelcomenus* persisted in the Oligocene. The Anoplotheriidae are highly distinctive because of their derived postcranial anatomy. In some, such as *Diplobune,* the fifth metacarpal and metatarsal are reduced to a small nodule, and there is a short but complete second digit, leaving a tridactyl foot unique among artiodactyls. Early descriptions conflicted on the interpretation of this foot morphology, especially in the forefoot, where some workers contended that the first digit was retained (Cuvier, 1822; Blainville, 1849; Schlosser, 1883; Sudre, 1983; Abush-Siewert, 1989). Apart from their foot morphology, anoplotheres are relatively heavily built and short-limbed.

Genus *Anoplotherium* Cuvier, 1804

Type Species and Type Specimen. *A. commune* Cuvier, 1804, MNHN GY 29, part of a maxilla with left P1–M3; "Gypse de Montmartre." Bembridge limestome, Frohnstetten, La Débruge, Mas Stes. Puelles, Neuhausen, Perrière, St.-Capraise-d'Eymet, Villeneuve-la-Comptal (includes partly *A. secundarium* Cuvier, 1894 [Stehlin, 1910a: 941]).

Included Species. *A. pompeckji* Dietrich, 1922: Eselsberg; *A. latipes* Gervais, 1852: Quercy (undifferentiated), La Débruge; *A. laurillardi* Pomel, 1851: Quercy (undifferentiated), La Débruge, Les Ondes, Montmartre; Obergösgen, Rixheim; *A.* cf. *commune*: Côja (Telles-Antunes, 1986); *A.* sp. or indet. (summarized): Entreroches (Hooker and Weidmann, 2000); La Débruge, Rosières 1, 2, 4, Saindou D (Sudre, 1978); Obergösgen (Stehlin, 1910a).

Diagnosis. Medium-sized to large animal with a long tail, skull low, and muzzle elongated on the same line as the top of the cranium.

Dimensions. M2 (holotype of *A. commune*); L = 36.3 mm; W = 34.5 mm.

Discussion. Several complete individuals of this animal were recovered from the Gypsum quarries of Montmartre early in the nineteenth century. Cuvier united all specimens with a complete and continuous dentition and a premolariform canine under the name *Anoplotherium*. However, these specimens turned out to belong to several genera (including *Diplobune* and probably *Ephelcomenus*), and only the large and didactyl species *A. commune* initially described by Cuvier (1804–1805) remains valid. Subsequent discoveries of *Anoplotherium* outside the Paris Basin by Gervais (1850) and Filhol (1877) provided additional data on specific diversity of anoplotheriids during the late Eocene.

The morphology of the limbs has been the focus of several studies from Cuvier to Heissig (1993). *Anoplotherium* is one of the rare Eocene artiodactyls for which specific characters rest mostly on the postcranial morphology (instead of dentition). Anoplotheriids are characterized by a high mobility of the elbow joint, especially between the radius and ulna, as well as the characteristic distal pulley of the humerus that displays a flat pulley lacking keels or longitudinal grooves, allowing a high range of motion between the radius and humerus.

Genus *Diplobune* Rütimeyer, 1862

Type Species and Type Specimen. *D. minor* (Filhol, 1877), MNHN Qu 395, left mandible with i1-2, Al c-p1, p2-3, m1-3; Quercy (undifferentiated), Calaf, Itardies, Veringen (dental and postcranial details in Sudre, 1974).

Included Species. *D. quercyi* Filhol, 1877: Quercy (undifferentiated), Eselsberg, Escamps; *D. bavarica* Fraas, 1870: Pappenheim; *D. secundaria* (Cuvier, 1822): "Gypse de Montmartre," Baby, Côja, Eselsberg, Frohnstetten, Heidenheim, La Débruge, Obergösgen, Villeneuve-la-Comptal; *D.* cf. *secundaria*: Entreroches (Stehlin, 1910a); *D.* sp. or indet. (summarized): Aubrelong 1, Rosières 4 (Sudre, 1978).

Diagnosis. Medium-sized; tridactyl forelimb, mandible increasing in height backward, the articulation jaw / cranium is high, transversely elongated, and not oblique, the coronoid apophysis is wide to moderately wide and curved backward. Upper incisors are separated by short diastemata, I1 are large and procumbent and curved, I2 and I3 are smaller and vertically implanted in the premaxilla, C undifferentiated, transversely compressed and ribbed, P1 canine-like but larger, P2 and P3 moderately elongated with a posterolingual heel, P4 subtriangular with a salient parastyle, upper molar with five cusps and prominent styles; rounded and procumbent i1 and i2, i3 are subtriangular and vertically implanted in the maxilla, p4 short, lower molar with a duplicated metaconid and a talonid open lingually anterior to the entoconid.

Dimensions. m2 (*D. minor*); L = 10.3 mm; W = 7.2 mm.

Discussion. Sudre (1974, 1988a) provided an extensive review of the type species thanks to new fossils, especially postcranial material from Itardies (MP 23). Gervais (1850) hypothesized the presence of palmate limbs in *Diplobune* and postulated it was a better swimmer than modern hippos. Kowalevsky (1874) supposed that the divergent digit II was not functional. However, the similar length of the three segments of the limb as well as the morphology of the tarsus would indicate a walking semidigitigrade locomotion. *Diplobune* probably inhabited forested and swampy environments.

Genus *Ephelcomenus* Hürzeler, 1938

Type Species and Type Specimen. *E. filholi* (Lydekker, 1889), BMNH M.41, fragmentary maxilla with left I2–P4; Bach.

Diagnosis. C prominent and separated from P1 by a small diastema. Upper premolars short and moderately developed transversely. Molars increase in size from M1 to M3, subtriangular upper molars with a protocone larger and more centrally situated than in other anoplotheriids, mesostyle voluminous.

Dimensions. M2: L = 16.3 mm; W = 20.7 mm (Beaumont, 1963).

Discussion. The postcranial features described by Hürzeler (1938) suggest a high mobility of the elbow, and morphology of the humerus is reminiscent of that of carnivores. *Ephelcomenus* probably inhabited forested environments of western Europe.

Genus *Robiatherium* Sudre, 1988b

Type Species and Type Specimen. *R. cournovense* (Sudre, 1969), USTL RBN 2257, left M3: Robiac, Le Bretou, Quercy (undifferentiated).

Diagnosis. Small anoplotheriid, trapezoidal upper molars increasing in size from M1 to M3, protocone subconical and centrally situated, postparaconule crista absent, labial side of the paracone and metacone concave and without any rib, styles salient labially forming the characteristic W-shaped ectoloph, mesial cingulum developed.

Dimensions. M3 (holotype): L = 10.8 mm; W = 9.5 mm (Sudre, 1988b).

Discussion. Based on dental material formerly assigned to the xiphodontid *Paraxiphodon*, this genus constitutes the earliest occurrence of anoplotheriids. Sudre (1988b) considered *Robiatherium* as the most primitive anoplotheriine, although it is not as well known as other upper Ludian representatives of the subfamily.

Family XIPHODONTIDAE Flower, 1884

Characteristics. Small to medium-sized artiodactyls, undifferentiated canine, elongated premolars, trend toward a molarized P4, upper molars bearing four or five crescent-shaped cusps; lower molars tetraselenodont with transversely compressed lingual cuspids, and markedly crescent-shaped labial cuspids, the labial crescent of the trigonid being strongly compressed mesiodistally.

Included Genera. *Xiphodon, Dichodon, Haplomeryx, Paraxiphodon, Leptotheridium.*

Comments. Xiphodontids are common, very small to medium-sized mammals in the middle and late Eocene of Europe. As in the case of *Anoplotherium*, Cuvier first recognized the genus *Xiphodon*, and relatively complete fossil material was collected in the upper Ludian Gypsum of Montmartre early in the nineteenth century. There are four easily distinguished genera, the relatively small *Haplomeryx* and *Dichodon*, mostly Robiacian in age (even if some species referred to this genus were known in the late Eocene), and the larger and mostly Headonian aged *Xiphodon* and *Paraxiphodon*. Recently, Hooker and Weidmann (2000) proposed inclusion of *Leptotheridium*, formerly referred to the dacrytheriines, because of overall dental resemblance to the Robiacian forms.

Genus *Xiphodon* Cuvier, 1822

Type Species and Type Specimen. *X. gracile* Cuvier, 1822, MNHN GY 263, 270-275 two articulated hindlimbs with fragmentary tarsals, MT III-IV and phalanges; "Gypse de Montmartre," Baby, Bembridge limestone, Escamps, La Débruge, Mas Stes. Puelles, Perrière, Rosières 4, Saindou D, San Cugat de Gavadons, St.-Capraise-d'Eymet, Villeneuve-la-Comptal.

Included Species. *X. castrensis* Kowalevsky, 1873: Le Castrais, Eclépens (A, Gare), La Milette, Le Bretou, Montespieu, Quercy (undifferentiated), Robiac, Sicardens; *X. intermedium* Stehlin, 1910: Quercy (undifferentiated), Euzèt, Fons 4, St. Hippolyte?, Roc de Santa, Sossís; *X.* cf. *intermedium:* Fons 5, Roc de Santa (Sudre, 1978: 208); *X.* sp.: Laguarres (Antunes et al., 1997: 346).

Diagnosis. Medium-sized artiodactyls; elongated skull with large tympanic bulla, and mastoid condition (the periotic bone is visible), orbit open posteriorly, muzzle relatively elongated, large palatine foramen extending from I3 to P1; upper molars subquadrate increasing noticeably in size from M1 to M3; four main crescent-shaped cusps with paraconule and metaconule connected to the salient parastyle and metastyle, respectively, protocone forming a slightly rotated and isolated crescent with a very short preprotocrista, ectoloph markedly W-shaped, mesial cingulum variably extended lingually; premolars elongated and undifferentiated, no diastema. Slim and elongated limbs showing a strong reduction of lateral digits so that only two were apparently functional during locomotion; cuboid and navicular unfused (adapted from Sudre, 1978).

Dimensions. M2 (*X. gracile*): L = 11.5 mm; W = 10.3 mm.

Discussion. *Xiphodon* was described by Cuvier (1822) from the famous Ludian gypsum layers of Montmartre early in the nineteenth century and represents the only xiphodontid genus for which postcranial anatomy is adequately documented (Zittel, 1893; Viret, 1961). According to Dechaseaux (1967), *Xiphodon* shows endocranial and external cranial similarities with North American tylopods, such as an elongated skull and neck, long legs and high-suspended body, a zeugopod more elongated than the stylopod; forelimbs have reduced lateral digits. Norris (1999) linked North American bunomerycids with European xiphodontids based on the morphology of the mastoid fossa, the expansion of the ventral margin of the petrosal, and its contribution to the roof of the petrobasilar canal. *X. castrense* and *X. intermedium* are mostly differentiated by size, and the lack of associated dentitions for each species prevents any definitive taxonomic definition. Antunes et al. (1997: 346) reported the earliest occurrence of the genus in the late early Eocene (MP 10), but this allegation is based on very poor material and needs additional fossil evidence for verification.

Genus *Dichodon* Owen, 1848

Type Species and Type Specimen. *D. cuspidatus* Owen, 1848, BMNH, right maxilla with I3–C?, dP2–M2, mandible with i1–c?, p1–m2; Hordle Cliff (lower Headon Beds).

Included Species. *D. cervinus* (Owen, 1841): Binstead (Seagrove Bay Member, Headon Hill Fm.), Eclépens B, Euzèt, Fons 4, La Bouffie, Roc de Santa, San Cugat de Gavadons, Sossís; *D. frohnstettensis* Meyer, 1852: Frohnstetten, Escamps; *D. simplex* Kowalevsky, 1874: Egerkingen Bolus, Huppersand; *D. cartieri* Rütimeyer, 1891: Egerkingen α, β; *D.*

subtilis Stehlin, 1910: Eclépens-Gare or Les Alleveys, Eclépens A; *D. ruetimeyeri* Stehlin, 1910: Egerkingen Huppersand and/or Bolus; *D. lugdunensis* Sudre, 1972: Lissieu; *D. stehlini* Sudre, 1973: La Débruge; *D. vidalenci* Sudre, 1988: Le Bretou; *D. biroi* Hooker and Weidmann, 2000: Eclépens A, Eclépens-Gare, ?Creechbarrow; *D.* cf. *frohnstettensis:* Larnagol, La Débruge, Rosières 4, Saint Saturnin, San Cugat de Gavadons (Sudre, 1978); *D.* cf. *cervinum:* Moutier (Stehlin, 1910a); Celarié, Fons 4, Perrière, Roc de Santa, Salesmes (Sudre, 1978); Creechbarrow (Hooker, 1986: 413); *D.* sp. or indet.: Egerkingen Huppersand, Verrerie de Roches (Stehlin, 1910a); Aubrelong 2, Gousnat, Le Bretou (Sudre, 1978); Creechbarrow (Hooker, 1986: 414).

Diagnosis. Small to medium-sized artiodactyls; complete dentition with reduced and undifferentiated canine; premolars tend to become markedly elongated and thinner except P4/4, which show a strong tendency toward molarization (p4 is trilobed), especially in the more recent species of the genus, lacks diastema; upper molars subquadrate and four-cusped, tend to become transversely compressed in some species, preprotocrista very short (adapted from Sudre, 1978).

Dimensions. M2 (*D. lugdunensis*): L = 7.0 mm; W = 7.7 mm.

Discussion. *Dichodon* is known from several species as early as MP 14, and the genus persists into the latest Eocene. Some species are based on very scarce dental material (Stehlin, 1910a; Sudre, 1973), and there might be some taxonomic oversplitting (Hooker and Weidmann, 2000). The tetraselenodont pattern of the upper molars is an advanced stage of dental evolution; the sister taxon of this genus is probably *Haplomeryx*, but their derived dental features and sudden appearance in the earliest Robiacian make any assessment of their relationships and origin difficult. This animal reached a large size (e.g., *D. stehlini*) unlike *Haplomeryx* (Sudre, 1973), which remained in the small to very small size range. Dentally, *Haplomeryx* and *Dichodon* are very close to one another. Dechaseaux (1965) has provided a detailed description of the cranium of *Dichodon*, showing an elongated muzzle. She suggested that the absence of diastemata with marked elongation of premolars is related to the muzzle morphology, which is reminiscent of that of North American tylopods, especially *Poebrotherium*.

Genus *Haplomeryx* Schlosser, 1886

Type Species and Type Specimen. *H. zitteli* Schlosser, 1886, MNHN, fragmentary left maxilla with M1-3; Escamps, Gousnat, Rosières 2, Saindou D, Ste. Néboule.

Included Species. *H. picteti* Stehlin, 1910: Eclépens-Gare, Le Bretou, Malpérié, Perrière, Robiac; *H. egerkingensis* Stehlin, 1910: Egerkingen α, Huppersand, ?Eclépens B; *H. euzètensis* Depéret, 1917: Euzèt, Aubrelong 2, Clapies, Eclépens B, La Bouffie, Lavergne, Malpérié, Perrière, Pradigues, Roc de Santa, Sossís; *H.?* *obliquus* (Cuvier, 1822):

Montmartre; Larnagol; *H.* cf. *picteti:* Lavergne, Le Bretou, Robiac (Sudre, 1978); *H.* cf. *euzètensis:* Fons 4 (Sudre, 1978: 208); *H.* sp. (summarized): Egerkingen α (Stehlin, 1910a); Bembridge Marls (Sudre, 1978).

Diagnosis. Very small artiodactyls; upper molars with four crescent-shaped cusps (a small paraconule is present in *H. egerkingensis*), W-shaped ectoloph, looplike mesostyle, labial wall of the paracone and metacone strongly excavated, conical protocone connected to a strong parastyle but protocone without V-shaped postprotocrista; P4 triangular with a crescent-shaped lingual cusp; lower premolar elongated (excepted p4), diastema between p2 and the anteriorly situated tooth which might be a canine (p1 may be lost) (adapted from Sudre, 1978).

Dimensions. M2 (*H. zitteli*): L = 4.9 mm; W = 5.1 mm.

Discussion. The biochronologic range of *Haplomeryx* is difficult to estimate with certainty. The familial affinity of this genus is based on the close dental resemblance to *Dichodon*, from which it differs in having a more prominent parastyle. Similarly, the different species tend to increase in size through time, although *Haplomeryx* remained very small. The cranium and postcranial features of *Haplomeryx* are unknown. According to Sudre (1988b: 146), *H. picteti*, *H. Euzètensis,* and *H. zitteli* belong to one clade showing a trend toward tooth enlargement. Hooker and Weidmann (2000: 96) regard these taxa as subspecies of *H. zitteli*. The status of *H.? obliquus* is problematic as the single specimen from the Paris Gypsum referable to this species was tentatively referred to *Haplomeryx* by Stehlin (1910a: 1063).

Genus *Leptotheridium* Stehlin, 1910a

Type Species and Type Specimen. *L. lugeoni* Stehlin, 1910a, MGL-8989, left maxillary fragment with P3-M2; Mormont (undifferentiated), Aubrelong 2, Eclépens (A, Gare?), Euzèt, La Bouffie, Lebratières, Malpérié, Perrière, Roc de Santa, Rosières 5, Sossís.

Included Species. *L. traguloides* Stehlin, 1910: Egerkingen α, β, Huppersand (synonymized with *L. lugeoni* and *L.* cf. *traguloides* from Mormont [Stehlin, 1910a] and Aubrelong 2 [Sudre, 1978] by Hooker and Weidmann [2000: 95]); *L.* cf. *traguloides:* Le Bretou, Robiac (Sudre, 1978) ; *L.* aff./cf. *lugeoni:* Fons 1, 2, 4, 6; Lavergne (Sudre, 1978) ; *L.* sp.: Sant Jaume de Frontanyà 1, 3 (Antunes et al., 1997), Caenes (Cuesta, 1998), Villamayor (Cuesta and Jiménez, 2000).

Diagnosis. Selenodonty weakly developed with crests poorly expressed, in particular weak postprotocrista and postparaconule crista on upper molars; P4 relatively long; upper molars with broadly looped mesostyle and prominent paraconule; premolars nonmolariform; p1 absent; lower canine trilobed; diastemata between lower canine and p2 and between p2 and p3 (from Hooker and Weidmann, 2000).

Dimensions. M2 (*L. lugeoni*): L = 7.1 mm; W = 6.8 mm (Stehlin, 1910a).

Discussion. Stehlin (1910a) placed *Leptotheridium* close to *Dacrytherium, Catodontherium,* and *Ephelcomenus.* Hooker and Weidmann (2000) considered the resemblances with dacrytheriines as plesiomorphies common to anoplotheroids (*sensu* Gentry and Hooker, 1988). Also, these authors listed a series of dental synapomorphic features shared with xiphodontids. Following that logic, they moved *Leptotheridium* from dacrytheriines to the Xiphodontidae.

Genus *Paraxiphodon* Sudre, 1978

Type Species and Type Specimen. *P. teulonensis* Sudre, 1978, USTL F4-287, left lower jaw with p2–m3, USTL F4-287, alveoli of p1; Fons 4. Fons 1, 5, 6.

Included Species. *P. cournovensis* Sudre, 1978: Robiac-Nord.

Diagnosis. Medium-sized artiodactyls close in size to *X. gracile;* upper molars transversely reduced; styles salient and bulbous, mesostyle looplike and symmetric; premolars elongated, p1 is one rooted, strongly reduced, and separated from p2 by a diastema (adapted from Sudre, 1978).

Dimensions. M2 (*P. teulonensis*): L = 11.7 mm; W = 10.3 mm (Sudre, 1978: Fig. 18).

Discussion. Sudre (1978) utilized the abundant material from Robiac and Fons to differentiate this genus, which differs from *Xiphodon* by the longitudinal elongation of upper molars, the symmetry of the mesostyle, and the reduction of p1. As mentioned by Sudre (1978), the lower molars of the two genera can easily be confused, and only the lower premolars are really distinctive. The stratigraphic range of *Paraxiphodon* extends into the lower Oligocene (Jehenne, 1969).

Family AMPHIMERYCIDAE Stehlin, 1910a

Characteristics. Small artiodactyls with elongated muzzle; vast orbit largely open posteriorly; complete dental formula; elongated premolars separated or not by diastemata, incisiform lower canine, and three-lobed p1, p2, p3; upper molars subtriangular and transversely developed, five crescent-shaped cusps, remains of hypocone variably present in all species. They are exclusively known in Europe from MP 14 to MP 20 (Robiacian and Headonian ELMA); their presence in the earliest Oligocene is poorly documented (adapted from Stehlin, 1910a).

Included Genera. *Amphimeryx, Pseudamphimeryx.*

Comments. Pomel (1851) first mentioned the presence of a cubonavicular tarsal bone in *Amphimeryx collatarsus* (see *A. murinus* below). This led Matthew (1929b), Simpson (1945), and McKenna and Bell (1997) to include this small European family within the Ruminantia. Yet, Webb and Taylor (1980) followed Viret (1961) in excluding the Amphimerycidae from the Ruminantia pending further evidence. The retention of p1 and their very distinctive molar morphology clearly set amphimerycids apart from early ru-

minants known in the Eocene of Asia and North America. The recent description by Hooker and Weidmann (2000) of a lower jaw of *Amphimeryx murinus* from Entreroches (MP 19) clearly shows a procumbent and rather incisiform lower canine that is adjacent to i3. However, the morphology of the premolars and molars of amphimerycids is very distinctive from that of the earliest Asian ruminants (Webb and Taylor, 1980; Vislobokova, 1998; Guo et al., 2000) in the following characters: marked elongation allied to a transverse compression of p1, p2, and p3, their shape consisting of three rounded cuspids aligned in a single plan. These premolars are usually separated by small diastemata, except p1, which seems to be invariably adjacent to the lower canine. The morphology of the fourth premolar is fairly variable within the Tragulina, and it is often an important element to differentiate the different genera. The p4 of amphimerycids is not comparable to any premolar pattern observed in early ruminants. Similarly, the upper molars of amphimerycids are triangular in occlusal outline and consist of five crescent-shaped cusps (protocone somewhat bunodont in *Pseudamphimeryx*) with a marked W-shaped ectoloph. The lower molars have a very basic pattern within the advanced Selenodontia with crescent-shaped labial cuspids and a variable transverse compression of the lingual cuspids. As stressed by several authors (e.g., Viret, 1961), the dental morphology of amphimerycids is more reminiscent of those of xiphodontids or dacrytheriids than of that of any other Eocene taxon of early ruminants from Asia or North America.

Genus *Amphimeryx* Pomel, 1848

Type Species and Type Specimen. *A. murinus* (Cuvier, 1822), MNHN GY 676, right mandible with (dp4), m1-m3; "Gypse de Montmartre," Baby, Entreroches, Escamps, Cinq-Sous, Rosières 2, 4, Saindou D, Tabarly (includes *A. collatarsus* Pomel, 1849 from La Débruge and Mormont [Sudre, 1978]).

Included Species. *A. riparius* Aymard, 1855: Ronzon; *A.* cf. *murinus:* Coânac, Gösgen-Canal, Ste. Néboule (Sudre, 1978) ; *A.* sp.: Ste. Néboule (Sudre, 1978).

Diagnosis. Cranium described in detail by Dechaseaux (1974); complete dentition, incisiform and procumbent lower canine, premolars elongated and transversely compressed, p2 isolated from p1 and p3 by diastemata, upper molars with five crescent-shaped cusps, protocone connected to the parastyle, W-shaped ectoloph, lack of mesial cingulum; lower molars with strongly crescentic labial cuspids, the trigonid crescent being more compressed mesiodistally, subconical lingual cuspids; lower border of the mandible is straight, and the angular process is slightly rounded posteriorly.

Dimensions. m2 (*A. murinus* from Montmartre): L = 4.5 mm; W = 3.5 mm (Sudre, 1978).

Discussion. Pomel (1851) reported postcranial elements of *A. murinus* from La Débruge, and Sudre (1978) described a complete posterior limb of *A. murinus* from Es-camps. The skull anatomy including endocranial features was described by Pearson (1927) and Dechaseaux (1974). Unfortunately, the diagnostically important upper incisors were not mentioned. The limbs consist of partially fused and elongated Mt III and IV with coalescent and reduced lateral digits II and V and, most importantly, a fused cubonavicular tarsal bone. *Amphimeryx* extends from the late Eocene to the earliest Oligocene (thus, over the *Grande Coupure*). The sudden appearance of *Amphimeryx* in the MP 18 led Sudre (1978: 162) to state that an immigration event is likely, although the family is known as early as MP 14. As for xiphodontids, the origin of amphimerycids in the early middle Eocene of Europe is fairly unclear. Much work is needed to increase the number of Geiseltalian aged localities in order to get a better sampling outside of that provided by the rich fossiliferous localities of the Geiseltal (Germany).

Genus *Pseudamphimeryx* Stehlin, 1910a

Type Species and Type Specimen. *P. renevieri* Pictet and Humbert, 1869; lectotype (Sudre, 1978) MGL 9056: right maxilla with P4–M3; "Station d'Eclépens," Aubrelong 2; Eclépens A?, B; Euzèt; Fons 4; Gösgen-Canal; Gousnat; La Bouffie; Lavergne; Le Bretou; Les Alleveys; Les Clapies; Les Pradigues; Malpérié; Moutier; Perrière; Robiac; Rosières 5; Salesmes; Verrerie des Roche. Type designated and illustrated in Stehlin (1910a: pl.18/27).

Included Species. *P. schlosseri* (Rütimeyer, 1891): Egerkingen α, β, Huppersand, Bolus, ?Arcis le Ponsart, Lissieu; *P. valdensis* Stehlin, 1910a: Eclépens-Gare; *P. pavloviae* Stehlin, 1910a: Larnagol, Eclépens A, B, Gare, La Bouffie, Le Bretou, Les Clapies, Les Pradigues, Malpérié, Perrière, Quercy (undifferentiated), Robiac; *P. hantonensis* Forster-Cooper, 1928: Hordwell; *P. salesmei* Sudre, 1978: Salesmes.

Diagnosis. Cranium described in detail by Dechaseaux (1974); complete dentition, subtriangular and five cusped upper molars with an incipiently to moderately crescentic protocone that is connected to the mesial cingulum, W-shaped ectoloph, premolars elongated, diastema between P1 and P2; incisiform lower canine, p1 adjacent to the lower canine and separated from p2 by a diastema, lower molars with crescentic labial cuspids, the trigonid crescent being more compressed mesiodistally, subconical lingual cuspids; lower border of the mandible is convex, the angular process is massive and posteriorly rounded, and the articular joint is lower than in *Amphimeryx*.

Dimensions. m2 (*P. renevieri* from Le Bretou): L = 3.9 mm; W = 3.1 mm (Sudre, 1978).

Discussion. Stehlin (1910a) distinguished five species of *Pseudamphimeryx*, extending from MP14 to MP17. Sudre (1978) proposed a new species from Salesmes (MP17) and doubted the validity of the poorly known species *P. decedens* Stehlin, 1910a, which was tentatively assigned to an inde-terminate ruminant. The specific distinction has proven delicate because of the apparent wide intraspecific size range, and the size differences result in oversplitting between contemporaneous species that are morphologically very close. Sudre and Blondel (1995) reported a fused cubonavicular in *Pseudamphimeryx salesmi* from Salesmes. Likewise, Hooker and Weidmann (2000) cited the derived morphology of an isolated astragalus to conclude that a fused cubonavicular was present in *P. renevieri* from Euzèt (MP 17). However, the morphology of the tarsus in the early Robiacian species *P. schlosseri* remains uncertain, and pending further fossil evidence, the fused cubonavicular should not be considered as a diagnostic character of the whole family (see above). Among the fossils from Egerkingen (MP 14) referred to *P. schlosseri,* some upper molars bear a small hypocone emerging from a thin lingual cingulum extending below the metaconule. This incipiently hypoconiferous pattern is also visible in the holotype of *P. hantonensis* from the Hordwell Barton Beds (MP 17a) and was interpreted as an aberrant morphology resulting from geographic isolation of southern England during the middle Eocene (Sudre, 1978).

STRATIGRAPHY

Families discussed in this chapter are mainly known from the middle and upper Eocene of Europe (Fig. 5.3). *Cuisitherium* from the early Eocene of Cuis and Avenay in France (MP 8–10), the oldest choeropotamid taxon (Hooker and Weidmann, 2000), consists of isolated dental material whose attribution to choeropotamids remains questionable. Except for some uncertain indications of *Lophiobunodon, Dacrytherium,* and *Xiphodon, Masillabune* from Messel (MP 11) is the oldest western European artiodactyl known by a nearly completely preserved skeleton. This genus displays the typical choeropotamid character set such as bunodont dentition, caniniform p1, and incisiform c1. The different levels of the Geiseltal that characterize the early Eocene in Europe show other primitive species (e.g., *Amphirhagatherium neumarkensis*) suggesting a choeropotamid radiation in the lower Geiseltalian, contemporaneously with cebochoerids. Both families reached their highest diversity in the Robiacian and Headonian.

The smallest and most primitive cebochoerid is *"Gervachoerus" dawsoni* from Bouxwiller (MP 13). *"G."dawsoni* and the contemporaneous and larger *"G."jaegeri* already display the typical cebochoerid upper molar pattern with four blunt, quadrangularly arranged bunodont main cusps. In the Geiseltal area, *"G."jaegeri* is found in the "Oberes Hauptmittel," which is somewhat younger than the MP 13 reference level "obere Mittelkohle" (Franzen and Haubold, 1986: Fig. 2). Although expected in the early Eocene, cebochoerids are currently unknown in these levels.

Most choeropotamids, cebochoerids, mixtotheriids, and xiphodontids became extinct at the *Grande Coupure* or just after the early Oligocene, which is consistent with the concept of the *Grande Coupure* as a contemporaneous significant faunal change in Europe driven by climate change

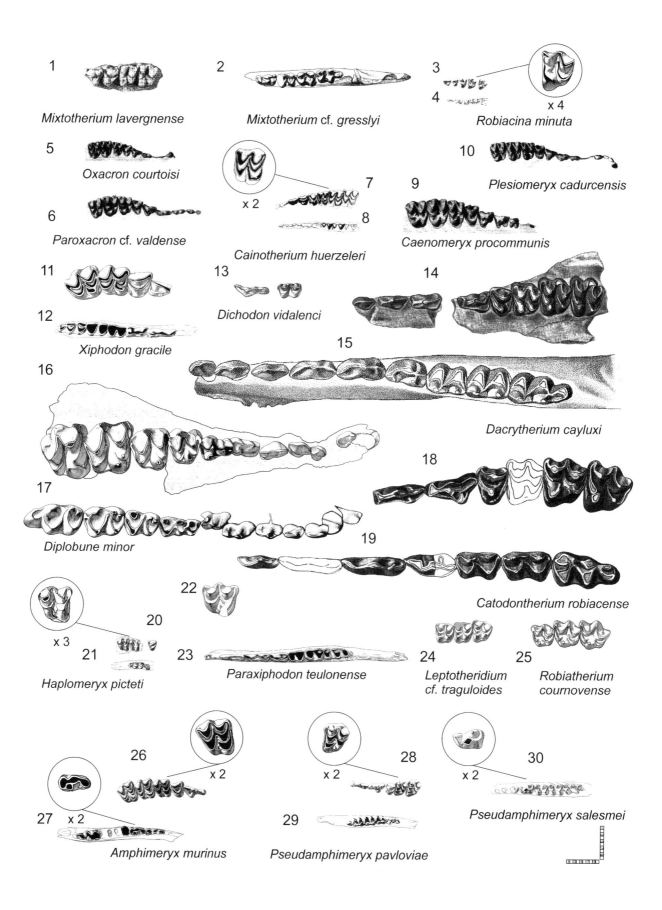

1   *Mixtotherium lavergnense*

2   *Mixtotherium* cf. *gresslyi*

3
4   x 4   *Robiacina minuta*

5   *Oxacron courtoisi*

6   *Paroxacron* cf. *valdense*

x 2   7
8   *Cainotherium huerzeleri*

9   *Caenomeryx procommunis*

10   *Plesiomeryx cadurcensis*

11
12   *Xiphodon gracile*

13   *Dichodon vidalenci*

14

15   *Dacrytherium cayluxi*

16

17   *Diplobune minor*

18   *Catodontherium robiacense*

19

20
21   x 3   *Haplomeryx picteti*

22

23   *Paraxiphodon teulonense*

24   *Leptotheridium* cf. *traguloides*

25   *Robiatherium cournovense*

26
27   x 2   *Amphimeryx murinus*

28   x 2

29   *Pseudamphimeryx pavloviae*

30   x 2   *Pseudamphimeryx salesmei*

and/or competition with modern taxa such as tragulids, anthracotheriids, or ruminants dispersing (arriving) from Asia. Within choeropotamids, only *Tapirulus* (Ravet, Aubrelong; MP 21) and another undescribed species from MP 22 at Möhren and Herrlingen (K. Heissig, pers. comm., 2005) in Germany survived the Eocene–Oligocene transition. The systematic position of *Diplopus* is questionable. *Acotherulum* cf. *quercyi* from Soumailles (southwestern France) is the youngest cebochoerid (MP 21), and its upper stratigraphic occurrence is not well known. The extinction of both families is not exactly coincident with the Eocene–Oligocene boundary and would be congruent with the earliest Oligocene age for the *Grande Coupure* (Hooker et al., 2004).

Cebochoerids and choeropotamids have not yet been recorded outside Europe, although Ducrocq and Sen (1991) reported a new choeropotamid in the middle Eocene of Turkey. According to Hooker and Thomas (2001: Fig.1), the earliest choeropotamids such as *Cuisitherium* (MP 10), *Masillabune* (MP 11), or *Amphirhagatherium* (MP12) are restricted to north of the line joining southwest Switzerland, Paris Basin, and southern Great Britain. *Tapirulus* and *Lophiobunodon* would have dispersed south of this line later, although *Lophiobunodon* has so far been recorded only from Lissieu (southern France, MP 14). Further species of "northern" genera occur later in southern Europe, including the different Quercy localities. This distribution would be consistent with a basal radiation of the clades in central Europe and subsequent dispersal events to the north (British Isles) and south (southern French localities, Spain), probably as far as Asia Minor during the middle Eocene.

Mixtotheriids, cainotheriids, and probably anoplotheriids diversified from the base of the Robiacian (MP 14) and were the most variant during the late Eocene. Anoplotheriids are already present in MP 13b with *Dacrytherium priscum*. This genus is further mentioned with an undetermined species from Repeu del Güaita (MP 10). *Catodontherium* should first occur in Argenton (MP 11). Because of poor preservation, it is not clear whether these groups were present earlier in the middle Eocene. Until now, anoplotheres

and mixtotheres have not been recorded in these levels (e.g., in the Geiseltal), but an ecological bias that would have favored bunodont forms living in more forested habitats cannot be ruled out for these groups. The stratigraphic range of *Mixtotherium* is limited to the Eocene, whereas anoplotheriids survived in the lower Oligocene (Legendre et al., 1995).

Cainotheriids, xiphodontids, and amphimerycids are known from the base of the Robiacian LMA (MP 14), and both groups diversified during the late Eocene. The stratigraphic range of the cainotheriids is more extensive, as *Cainotherium* is known in the early Miocene. Their first recorded appearence is from the late middle Eocene of San Jaume de Frontanya (MP 14–15) with the genus *Robiacina*, also well documented from different middle Eocene localities (MP 16) of southern France. Two phases of diversification can be noted in the evolutionary history of cainotheres: MP 18–19 (Oxacroninae) and MP 22 (essentially Cainotheriinae with some relict lineage of oxacronines; highest diversity of the family; this timing extends over the *Grande Coupure* and is in accordance with the diversity pattern of anoplotheriids. Most other groups of European Selenodontia, which were common in the late Eocene, such as xiphodontids and amphimerycids, became extinct during the Eocene–Oligocene transition. The global cooling and subsequent increase in aridity by the beginning of the Oligocene (Prothero, 1994) gave rise to cainotheriines that show a more crested and selenodont molar pattern. The sudden appearance of oxacronines in the late Eocene, and the lack of any possible sister taxon in the middle-late Eocene, led several authors to suggest an invasion from Eastern Europe or Asia.

PALEOECOLOGY

Paleoecological interpretations depend, at the least, on having autochthonous conditions and complete skeletons. For the families discussed in this chapter, such conditions are given in the localities Messel and Geiseltal. Therefore, this section is mostly focused on cebochoerids and choero-

Fig. 5.2. (*opposite*) Mixtotheriid, cainotheriid, anoplotheriid, xiphodontid, and amphimerycid dentitions. Dentitions occlusal and approximately scaled; those in circles are magnified (2x). Bold specimens are types. (1) *Mixtotherium lavergnense* USTL Lav 3, right maxilla with P3-M2 (Hooker and Weidmann, 2000: Fig. 51c). (2) *Mixtotherium* cf. *gresslyi* MNHN DF 620 left mandible with p4–m3 (Sudre and Ginsburg, 1993: Fig. 2a). (3) *Robiacina minuta* USTL RBN 5438-5442, left combined P3–M3 (Sudre, 1978: pl. 13/1). (4) *Robiacina minuta* USTL RBN 5443-5447, left combined p2–m2 (Sudre, 1978: pl. 13/2). (5) *Oxacron courtoisi* NMB Q.S. 525, right maxilla (C), P1–M3 (Hürzeler, 1936: Fig. 55). (6) *Paroxacron* cf. *valdense* NMB Q.S. 486 (mirrored), left maxilla (I3), C–M3 (Hürzeler, 1936: Fig. 56). (7) *Cainotherium huerzeleri* BSPGM 1970 XVIII 7035, C-P1; 7102, P2; **6916 P3-M3**, left combinated C–M3 (Heizmann, 1983: Fig. 9a). (8) *Cainotherium huerzeleri* BSPGM 1970 XVIII 7034, i2–p2; **6941, p3–m3**, left combinated i2–m3 (Heizmann, 1983: Fig. 4a). (9) *Caenomeryx procommunis* NMB Q.S. 480, right maxilla (C), P1–M3 (Hürzeler, 1936: Fig. 54). (10) *Plesiomeryx cadurcensis* NMB Q.S. 485, right maxilla (I2), I3–M3 (Hürzeler, 1936: Fig. 53). (11) *Xiphodon gracile* USTL Esc 1214, right maxilla with P3–M2 (Sudre, 1978: pl. 15/1). (12) *Xiphodon gracile* USTL Esc 1213, right mandible with p3–m3 (Sudre, 1978: pl. 15/5). (13) *Dichodon vidalenci* **Musée de Montauban BRB 2950** isolated P2; BRB 74 isolated m1 (Sudre, 1988b: Fig. 13). (14) *Dacrytherium cayluxi* NMB Q.A. 275, 282 combinated left maxilla with C–M3 (Stehlin 1908: Fig. 68). (15) *Dacrytherium cayluxi* **NMB Q.A. 324** right mandible with i3–m3 (Stehlin 1908: Fig. 73). (16) *Diplobune minor* USTL ltd 45, right maxilla with I1–P2, dP3–4, M1–M3 (Sudre, 1974: Fig. 1a). (17) *Diplobune minor* USTL ltd 41, right mandible with i1–p3, dp4, m1–3 (Sudre, 1974: Fig. 1b). (18) *Catodontherium robiacense* MGL LM 906, MGL LM 935, MGL LM 936, MGL LM 912, combined and partly mirrored P2–M3 (Stehlin 1910a: Fig. 84). (19) *Catodontherium robiacense* MGL LM 921, MGL LM 905, NMB Mt 185, MGL LM 923, combinated and partly mirrored p1–m3 (Stehlin 1910a: Fig. 88). (20) *Haplomeryx picteti* USTL BRT 28, right maxilla with M1–M2 combined with BRT 44, P4 (Sudre, 1988b: Fig. 12a,b). (21) *Haplomeryx picteti* USTL BRT 40, right mandible with (m2), m3 (Sudre, 1988b: Fig. 12c). (22) *Paraxiphodon teulonense* USTL F4 288, upper molar (Sudre, 1978: pl. 15/4). (23) *Paraxiphodon teulonense* **USTL F4 287,** left mandible with p2–m3 (Sudre, 1978: pl. 15/6a). (24) *Leptotheridium* cf. *traguloides* USTL BRT 95, left maxilla M1–3 (Sudre, 1988b: Fig. 5). (25) *Robiatherium cournovense* **USTL ACQ 10,** left maxilla M1–3 (Sudre, 1988b: Fig. 9a). (26) *Amphimeryx murinus* MHNL Gg 496, right maxilla with P3–M3 (Sudre, 1978: pl. 19/3). (27) *Amphimeryx murinus* USTL Esc 1206, left mandible with p3–p4, (m1), m3 (Sudre, 1978: pl. 20/2a). (28) *Pseudamphimeryx pavloviae* USTL PRR 1052, left maxilla with P2–M2 (Sudre, 1978: pl. 18/6). (29) *Pseudamphimeryx pavloviae* BR2-83, left mandible with p4–m3 (Sudre, 1978: pl. 17/3a). (30) *Pseudamphimeryx salesmei* MNHN Sal 336, left mandible with (p3), p4, m1–3 (Sudre, 1978: pl. 20/3).

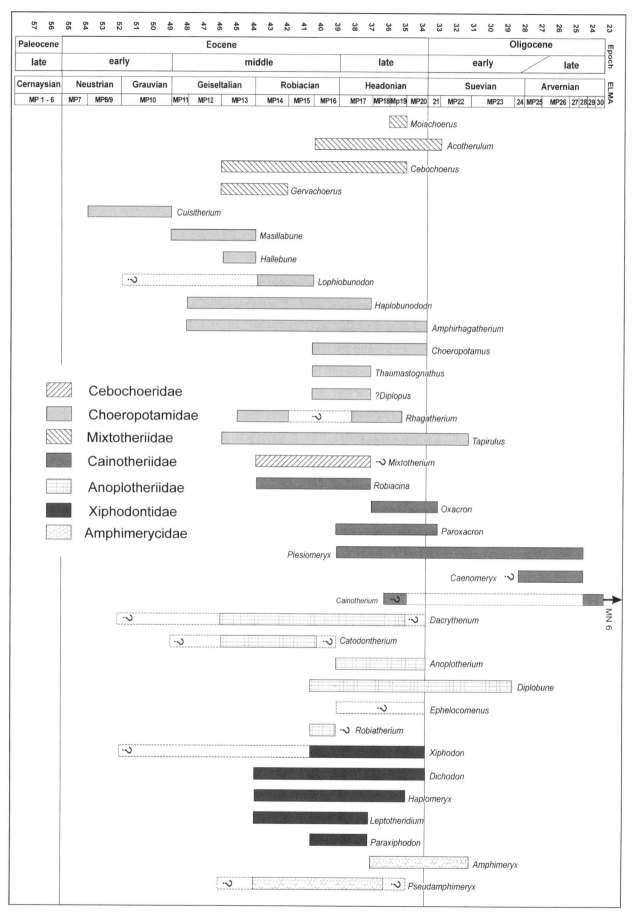

Fig. 5.3. Stratigraphic occurrence of taxa discussed in the text. Emended after Sudre (1978: Fig. 13), Hooker (1986: Fig. 62), Erfurt and Haubold (1989), Hooker and Weidmann (2000), and Blondel (2005). MP levels and reference localities are according to Schmidt-Kittler (1987) and Escarguel et al. (1997). Compare also Table 4.1 (Theodor et al., this volume). *Parabunodon* is not indicated here because of the uncertain age of the type locality.

potamids from these sites. In addition, postcranial materials of cainotheriids are briefly considered. Several species are known by relatively complete skeletons or soft tissues with accompanying paleobotanical and taphonomic records for ecological interpretations: *"Gervachoerus" jaegeri, Amphirhagatherium (Anthracobunodon) weigelti, Haplobunodon* cf. *muelleri, Masillabune franzeni, Masillabune martini,* and *Caenotherium* sp. These species display distinct patterns in their habits and probably behavior (e.g., locomotion, feeding, adaptation), which are inferred from skeletal and dental characteristics. Different ecological groups have been identified.

Group 1

Very small, brachybunodont species known only from dental remains and including species such as *"Gervachoerus" dawsoni* or *Hallebune krumbiegeli* comprise this group. In contrast to similarly sized diacodexeids and dichobunids, the snout is short, and the teeth are bunodont. Broad molar surfaces with round and blunt conules indicate grinding or squeezing of food, not cutting (Janis, 1990). This distinguishes smaller cebochoerids and choeropotamids from artiodactyls with crescentiform molars and significantly higher trigonids than talonids. Group 1 was mostly frugivorous, consuming lesser amounts of animal diet such as insects or tougher vegetation. This group is typical of the initial radiation in both families at the end of the early Eocene and survived until the middle Eocene.

Group 2

Medium-sized bunodont artiodactyls with generalized limbs (Fig. 5.4) identify this group, which is represented postcranially by the cebochoerid *"Gervachoerus" jaegeri* and the choeropotamid *Haplobunodon* cf. *muelleri* from Geiseltal. A blunt tooth pattern and a tendency to increase in size are typical. The hindlimb is only slightly elongated. *"G." jaegeri* bears an unusually long tail, which is about 30% of body length. The stylopodials (humerus and femur) and zeugopodials (radius, ulna, tibia, and fibula) are of about the same length in the fore- and hindlimbs (Fig. 5.4), unlike the more elongate diacodexeid tibia (Rose, 2001: Fig. 15). Metapodials II and V are only slightly shorter than metapodials III and IV. Upper Eocene cebochoerids (*Cebochoerus*) have a short snout, the mandibular ramus becomes very strong, and p1 is caniniform. The latter indicates the intake of harder food. Group 2 probably became extinct at the end of the Eocene.

Group 3

Medium-sized bunoselenodont artiodactyls with elongated hind limbs (Fig. 5.4) represent this group. In all species, the locomotion is cursorially specialized: the hindlimb is elongated, and the tail is reduced. In the Eocene, slender choeropotamids (e.g., younger species of *Amphirhagatherium*) belong to this group, which display a tendency to elongate the skull and molarize the premolars. In the

Oligocene, Group 3 is represented by cainiotheriids, which still display unfused zeugopodials and the ability of supination (Hürzeler, 1936). In the Eocene species, a long snout and dental wear indicate selective browsers. We assume their diet was comparable to that of the dichobunid *Messelobunodon* (Richter, 1987) and included mostly mushrooms, fruits, and leaves. Leaves contain a relatively small amount of energy and are not the main food component in recent small mammals. Leaves became more important in specially adapted animals with larger bodies or slower locomotion. Mixtotheriids, dacrythriids, xiphodontids, and amphimerycids can also be included in this ecological group, although for most of them the postcranial morphology remains poorly documented. Xiphodontids were probably the ecological counterpart of North American camelids with their cursorial adaptation, reduced lateral metapodials, and sharpness of phalanges. However, it is not possible to demonstrate whether or not they adopted a pacing mode of locomotion. The singular postcranial adaptations of anoplotheres suggest that they occupied a specific ecological niche with perhaps tree-climbing or branch-walking activities (Sudre, 1983). Amphimerycids occupied the same ecological niche as the contemporaneous traguloids from Asia and North America. Modern tragulids and moschids probably constitute a reliable approximation of the ecological characteristics of amphimerycids with a diet relying on fruits, tender shoots, and occasional insects. Geologic data suggest that these taxa inhabited gallery forests and more open savanna woodlands for larger forms of the ungulate guild.

Group 4

Large bunodont and polybunodont artiodactyls are typical for this group. These mostly upper Eocene choeropotamids have elongated skulls with longer diastemata and caniniform p1. The molars may display additional cusps and cuspids as on the m3 of *Choeropotamus*. The wear facets of the molars indicate grinding in the tooth surface, and few on the surface between consecutive molars. Unlike *Tapirulus*, wear facets are larger distal to the trigonid (*Thaumastognathus* BMNH.M61216). Mesially, there is only a smaller facet on the preparaconule crest. These artiodactyls are regarded as unspecialized omnivores and display increasing body mass and robust limbs. This group survived the *Grande Coupure* and persisted with short-tailed piglike animals until today.

Group 5

Small lophodont artiodactyls comprise this group. The lophodonty of *Lophiobunodon* and *Tapirulus* is unique among Paleogene artiodactyls and represents a convergence with perissodactyls. These taxa may have been directly in competition with larger and more highly specialized perissodactyls. They were probably relying on soft ground vegetation, and their postcranial characteristics remain poorly

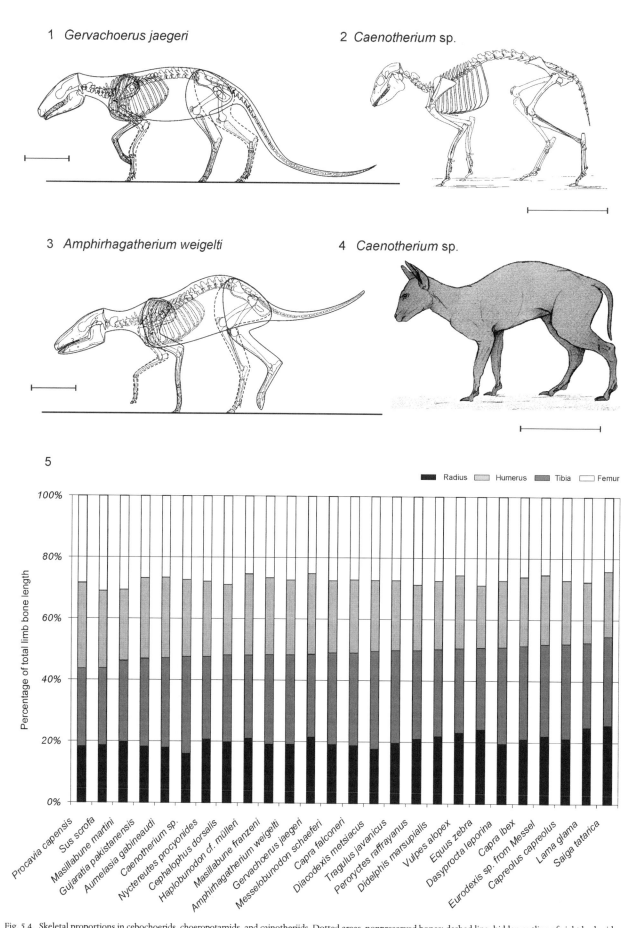

1 *Gervachoerus jaegeri*

2 *Caenotherium* sp.

3 *Amphirhagatherium weigelti*

4 *Caenotherium* sp.

5

Radius ▣ Humerus ▣ Tibia ▢ Femur

Percentage of total limb bone length

Procavia capensis
Sus scrofa
Masillabune martini
Gujaratia pakistanensis
Aumelasia gabineaudi
Caenotherium sp.
Nyctereutes procyonides
Cephalophus dorsalis
Haplobunodon cf. *mülleri*
Masillabune franzeni
Amphirhagatherium weigelti
Gervachoerus jaegeri
Messelobunodon schaeferi
Capra falconeri
Diacodexis metsiacus
Tragulus javanicus
Peroryctes raffrayanus
Didelphis marsupialis
Vulpes alopex
Equus zebra
Dasyprocta leporina
Capra ibex
Eurodexis sp. from Messel
Capreolus capreolus
Lama glama
Saiga tatarica

Fig. 5.4. Skeletal proportions in cebochoerids, choeropotamids, and cainotheriids. Dotted areas, nonpreserved bones; dashed line, hidden outline of right body side. The bone measurements and the reconstruction of *Cainotherium* are taken from Hürzeler (1936: Figs. 50, 51). The diagram includes the relative lengths of humerus, radius, femur, and tibia as percentages (measurements: J.E.) Species ordered according to the elongation of the distal limb bones. Scale bars equal 100 mm.

known. These lophodont forms of artiodactyls were never as diversified as the selenodont forms, and they became extinct with the *Grande Coupure*.

Figure 5.4 compares the habit of animals in Group 2 (*"Gervachoerus"* cf. *jaegeri*) and Group 3 (*Amphirhagatherium weigelti*) according to measurements of their limbs, vertebrae, and skulls. The different habits are reflected in length differences in the hindlimb and vertebral column. A general tendency is the increasing zeugopodial length from Eocene artiodactyls to modern camelids and ruminants. The increased length in the latter is mostly regarded as a terrestrial cursorial adaptation but is also seen in aquatic pinnipeds. The modern *Sus* represents the generalized "Wuchstyp" (Erfurt and Altner, 2003) with long and strong stylopodials. Most of the Eocene artiodactyls have similarly elongated zeugopodials, but they bear different stylopodials. Among Eocene taxa, only *Eurodexis* sp. from Messel (SMF ME-1001) displays a cursorial adaptation similar to modern cervids, but this specimen is juvenile and could be a result of allometric growth trajectories. Erfurt and Altner (2003) reconstructed *Amphirhagatherium weigelti* as a slender and agile mammal that lived in small groups on the ground in the less densely covered areas of the Eocene Geiseltal and fed on leaves and fruits. Body size was about 70 cm (including the tail), and the body mass totaled about 4 kg (Erfurt, 2000). It is supposed that this habit was only slightly different in all other artiodactyls of Group 3. The Oligocene Group 3 is represented by the relatively short-tailed *Cainotherium*. Their ecotype is often compared with that of modern hares because of the possible supination of radius and ulna, the pelvis structure, and the clawlike phalanges (Hürzeler, 1936). Earlier reconstructions suggest that they had the habit of modern agoutis (Hürzeler, 1936: Fig. 52), but later reconstructions are closer to extant chevrotains (Heizmann, 1999: Fig. 21.2).

The postcranial peculiarities of Anoplotheriidae (the divergent digit 2 in the tarsus, robustness and large size of the scapula and humerus) led to many speculations about their habitat and paleoecology, including aquatic habits and a digging behavior. It seems that the ecomorph of anoplotheres has no modern analogue, and the functional role of their divergent digit 2 in the pes has been variously interpreted (Cuvier, 1822; Kowalevsky, 1874; Sudre, 1983).

CONCLUSION

The restriction of cebochoerids and choeropotamids to the middle and late Eocene of Europe is consistent with the isolated geographic position of this continent during that time. The origin of both families can be seen in early Eocene European diacodexeids and dichobunids, which developed into endemic taxa and became extinct with the *Grande Coupure* at the beginning of the Oligocene as modern artiodactyls invaded from Asia. Descriptions of new materials have increased our understanding of both families in the last 10 years. Some taxa are known postcranially and represent an extinct ecological type consisting of small,

doglike, mostly frugivorous browsers with long tails as well as heavier, medium-sized omnivorous animals with elongated snouts. Some taxa need systematic revision because of improvements in lithostratigraphic information about fossil complexes in France, Germany, Great Britain, and Switzerland.

The record of cainotheres is restricted to Europe so far, and their highly distinctive molar pattern (Stehlin, 1910a; Hürzeler, 1936) suggests a European origin from a still unidentified group of middle Eocene dichobunoids. Sudre (1977) showed that the middle Eocene genus *Robiacina* shares important synapomorphies of the later well-defined Cainotheriidae, including the specific dental wear pattern and related mechanisms of the jaw. These features likely appeared in an unknown group of nonhypoconiferous dichobunoids in the Geiseltalian or even earlier. Eurodexeinae are still poorly documented, but they exhibit a mixture of primitive and derived dental pattern that might be related to the emergence of Selenodontia in Europe. The apparent low generic diversity of Robiacian cainotheriids may indicate that a climatic stimulus and/or the rupture of a geographic barrier may be invoked to explain their sudden appearance and rapid diversification in MP 18–19.

Xiphodontids show an unexpected tetraselenodont molar pattern, although primitive species of *Haplomeryx* retained a small paraconule (Sudre, 1973). The Xiphodontidae appeared abruptly in the earliest Robiacian (MP 14) with *Haplomeryx* and *Dichodon*. Their excavated, looplike mesostyle suggests a close relationship with dacrytheriids as suggested by early authors and supported by Gentry and Hooker (1988). This family was formerly classified as Tylopoda (Simpson, 1945; McKenna and Bell, 1997; Hooker and Weidmann, 2000). However, dental and postcranial characters exhibited by *Xiphodon* have been alternately considered to be either more primitive or more derived than those of the North American camelid *Poebrotherium* (Matthew, 1929a; Dechaseaux, 1963, 1965, 1967). Almost all recent molecular phylogenies place the Tylopoda as the first offshoot of the Cetartiodactyla (e.g., Gatesy and O'Leary, 2001), a surprising result, as it does not fit with the fossil record of terrestrial artiodactyls. Given our limited fossil record, this molecular hypothesis is certainly worth testing with additional morphological data. The first appearance of camelids in North America (late Uintan, Honey et al., 1998) roughly corresponds to the earliest occurrence of xiphodontids in Europe. The origin of camelids is not clearly established (Webb and Taylor, 1980; Janis, 1993), but a North American origin of the family is most frequently advanced (Golz, 1976; Prothero, 1996). The derived morphology of xiphodonts suggests that this small European family is not closely related to North American tylopods but is more likely a convergence (Viret, 1961), which was caused by the isolation of Europe during the middle and late Eocene. The sudden emergence of xiphodontids during the Geiseltalian/Robiacian transition remains enigmatic, as they appeared fairly derived with respect to the group (incompletely) documented in the lower part of the Geiseltalian ELMA. As for other groups of European selenodonts, we assume that xiphodontids may have emerged from a still unknown group of Geiseltalian dichobunoid probably close to the Eurodexeinae (Erfurt and Sudre, 1996). Their first radiation is roughly coincident (MP 14) with the appearence of agriochoerids, oromerycids, and protoceratids in the middle Uintan (Ui2, Uinta B Stucky, 1990; Robinson et al., 2004) of North America. The appearance of North American selenodonts is thought to be closely linked to the global cooling event that characterized the late Uintan (although for hypertragulids and leptomerycids, an immigration event from Asia is most likely [Vislobokova, 1998]). This global cooling is also correlated with the radiation of protoselenodont groups in Europe, including amphimerycids and xiphodontids.

Amphimerycids form a homogeneous and small family of rabbit-like selenodont artiodactyls, and because of the fusion of the cuboid and navicular tarsal bones, they have often been included in the Ruminantia (e.g., Hooker and Weidmann, 2000). The sudden appearance of amphimerycids led various authors to suggest an allochthonous origin for the family. This hypothesis also served to explain the rapid emergence of xiphodontids, cainotheriids, or anoplotheriids during the Robiacian. In the case of amphimerycids, the real significance of ruminant features has been a matter of debate for decades, and the issue of whether or not they should be included within the Ruminantia is still debated. The presence of a remnant hypocone in the earliest and most primitive species of *Pseudamphimeryx* may indicate that amphimerycids evolved from an unknown European group of hypoconiferous artiodactyls. Such an endemic evolution of mammals is corroborated with the relative geographic isolation of western Europe during the middle Eocene. The morphological evidence for the origin of amphimerycids is thin, but we suspect that this family probably evolved from an unknown autochthonous group of small hypoconiferous forms. Again, this ancestry may lie close to the Eurodexeinae (see above). This interpretation

may be invoked to explain the emergence of hyperdichobunines, xiphodontids, or *Robiacina* later in the Robiacian. A dichobunoid ancestry of European selenodonts would be in accordance with the concept of Merycotheria proposed by Gentry and Hooker (1988).

It is worth noting that the Geiseltalian/Robiacian transition is characterized by the sudden appearance of several groups of selenodont artiodactyls endemic to Europe, documented from Egerkingen and subcontemporaneous localities. However, there is a relative bias in the knowledge of these early middle Eocene mammal communities, especially concerning the levels MP 11–12, for which our knowledge rests exclusively on the Geiseltal assemblages, which certainly do not reflect the diversity of habitats, taxa, and adaptations during the early–middle Eocene transition in Europe. Much fieldwork is still needed in Europe to investigate the early Geiseltalian, thus clarifying important issues related to the origin and early phylogeny of these European selenodont artiodactyls.

ACKNOWLEDGMENTS

We are grateful to J. Sudre (Murviel-lès-Montpellier) for insightful remarks that much improved our chapter; J. Theodor (University of Calgary) for proofreading parts dedicated to the stratigraphy, ecology, and systematics of cebochoerids and choeropotamids; and M. R. Dawson (CMNH) for her useful comments on the text. G.M. is grateful to the vertebrate paleontology section of the Carnegie Museum of Natural History (Pittsburgh), and the Singer-Polignac Fondation (Paris) for their financial support. We also thank J. Hooker (BMNH) for helpful discussions and S. Ducrocq (LGBPH), P. Tassy (MNHN), K. Heissig (BSPG), and E. J. Heizmann (SMNS), who provided information on holotypes in their care. Two anonymous reviewers provided useful comments on the chapter.

6

Family Helohyidae

ELOHYIDS, LIKE OTHER EARLY ARTIODACTYLS, were initially small in size (similar to extant tragulids) but later became the first truly large artiodactyls. As the forests of the later Paleocene and early Eocene gave way to more open plains, ungulate taxa in general began to increase in size and diversity. The helohyids acquired an omnivorous diet that may have included vegetation, fruit, seeds, and insects initially, and perhaps carcass scavenging in the later achaenodonts. Helohyids and other artiodactyls (dichobunids) were a minor component of early and middle Eocene mammalian faunas that were dominated by a diverse array of arboreal forms, notably primates (Rose, 1981; Gunnell, 1997). Artiodactyl diversity increased dramatically during the late–middle Eocene and was coincident with a decline in primate diversity, which has been interpreted by some to indicate a shift from heavily forested to more open habitats (Stucky and Krishtalka, 1990; Stucky, 1990, 1992, 1998).

Although a relatively small group, the Helohyidae are important because they reflect some of the earliest derivations in morphology, habitat, and perhaps behavior that reflect a move to an omnivorous diet. This shift was accomplished independent of the later suiform groups (including entelodonts, suids, tayassuids, and sanitheres).

SYSTEMATIC PALEONTOLOGY

Class MAMMALIA Linnaeus, 1758

Order ARTIODACTYLA Owen, 1848: 131

Family HELOHYIDAE Marsh, 1877: 364

Family Diagnosis.   Bunodont molars somewhat rounded, cusps conical; upper molars with a small hypocone (but lost in later forms), mesostyle absent (Stucky,

1998); lower molars possessing a small hypoconulid except m3, where the hypoconulid is fully developed, large, and sometimes expanded into three distinct rounded subdivisions. The premolars are large and crowded; later forms lose the first premolar altogether. Foss et al. (2001) showed that the first premolar (p1) is present in a juvenile *Achaenodon* before being lost in the adult as a result of canine enlargement.

Discussion. Gazin (1955) offered clear reasoning for erecting a subfamily (Helohyinae Gazin, 1955: 37) that reflected this group's association with the Homacodontinae. Both subfamilies were assigned to the Dichobunidae. The Helohyidae is unique and warrants McKenna and Bell's (1997) interpretation that the group be given family level status. However, I fully concur with Gazin's reasoning for assigning the name Helohyidae for the group: "The name (Helohyidae Marsh, 1877) is selected rather than the original (Achaenodontidae Zittel, 1893) as it is based on the older familial designation. . . . Moreover, this name does not carry the implication of including the entelodonts as originally defined, or of being a subfamily of the Entelodontidae as later assigned, as does Zittel's Achaenodontidae" (Gazin, 1955: 37).

Genus *Helohyus* Marsh, 1872

Type Species. *Helohyus plicodon* Marsh, 1872.

Diagnosis. *Helohyus* is characterized by double-rooted upper premolars (the existence or character of P1 is still unknown); P4 has a tall conical paracone, conical protocone, and strong, but incomplete, postcingulum. The upper molars have four major cusps and multiple less-developed intermediate cusps, including both a paraconule and hypocone. The postcingulum is well developed. The lower dentition has four premolars, including a single-rooted p1; the double-rooted p3–4 have small postcingula; the lower p4 has a tiny metaconid. The lower molars are fully developed and include a paraconid. The hypoconulid heel on m3 is large and often subdivided into multiple blunt tubercles.

Discussion. Marsh (1872b) first proposed *Thinotherium* for the taxon but found it preoccupied by *Thinotherium* Cope, 1870 (a taxon formerly assigned to Tayassuidae). The genus remains poorly known, with only dentitions and a partial skull being reported. *Ithygrammodon* Osborn et al. (1878) is not directly comparable to *Helohyus* and was considered by Stucky (1998) to be *nomen dubium*.

Included Species. *H. plicodon* Marsh, 1872; *H. lentus* (Marsh, 1871); *H.* new species Stucky, 1998.

Genus *Achaenodon* Cope, 1873

Type Species. *Achaenodon insolens* Cope, 1873.

Diagnosis. The teeth of *Achaenodon* are large and bunodont. Contrary to early descriptions, the dental formula of the lower jaw reflects the complete artiodactyl 3-1-4-3 pattern. Foss et al. (2001) showed that the first lower premolar,

present in juvenile forms, is crowded out by the development and eruption of the canine and is lost in the adult form. The crowding in the tooth rows reflects both the enlargement of the teeth and the relative shortening of the muzzle. It is possible that this is the case for the upper tooth row as well, although there are no known juvenile skulls that may be examined in order to verify this.

The P2 is single rooted, P3 double rooted; both have crenulate but not quite serrated ridges on the mesial and distal edges. The P4 has a tall protocone and paracone with a significantly smaller metacone situated on its distal flank. The upper molars are square and sometimes simplified (by loss of the hypocone and protoconule) to four large conical cusps surrounded by robust pre- and postcingula and greatly thickened enamel. *A. fremdi* is striking because it has retained both the protoconule and the hypocone (Lucas et al., 2004), which are reduced in other achaenodonts.

The skull has large jugal arches that are expanded laterally, giving the animal a wide appearance (the skull is nearly as wide as it is long). The sagittal crest is tall and meets a large, laterally expanded, nuchal crest. The expanded sagittal and nuchal crests, together with the widened jugal arches, facilitate an enormous amount of origin area for biting and chewing musculature.

The forelimb is short, metacarpals are unfused, and the manus has four digits. The hindlimb is slightly longer and has unfused metacarpals. Little more is known about the postcrania of this genus (Osborn, 1895; Walsh, 2000; Foss et al., 2001).

Discussion. *Achaenodon* was approximately the size of a black bear and was the earliest artiodactyl to surpass the 40-kg mark. Foss et al. (2001) estimated 200 kg based on ankle and long bone size, whereas Townsend (2004) estimated its weight at 285 kg based on molar area regression. It remained the only megafaunal artiodactyl in North America for the duration of middle and early-late Eocene (from approximately 43 to 39 million years ago). Subsequent artiodactyls didn't reach that size again in North America until the Chadronian entelodonts (e.g., *Archaeotherium crassum*) evolved large size (approximately 35 Ma). It is the convergent artifacts of the increase to large size that have caused so many investigators to associate achaenodonts with entelodonts. This is especially notable in convergently massive tooth morphology and thick enamel. The similarities are likely a combination of the effects of both allometric scaling and evolutionary convergence resulting from a similarity in ecological niche. Other artiodactyls have similarly been compared to achaenodonts when they reach large size, including large North American anthracotheres (*Elomeryx*, *Heptacodon*) and enormous Asian suids (*Tetraconodon*, *Sivachoerus*).

Protelotherium Osborn (1895: 105) is an incompletely described hindlimb that is referable to *Achaenodon*.

Included Species. *A. insolens* Cope, 1873; *A. robustus* Osborn, 1883 (Fig. 6.1); *A. uintense* Osborn, 1895; *A. fremdi* Lucas et al., 2004.

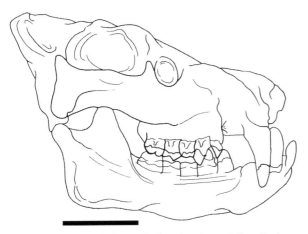

Fig. 6.1. YPM.PU 10033, holotype of *Achaenodon robustus*. Collected in the Washakie Basin, Sweetwater County, Wyoming, by the Princeton Expedition of 1878.

Genus Parahyus Marsh, 1876

Type Species.   *Parahyus vagus* Marsh, 1876.

Diagnosis.   Dentition similar to *Achaenodon,* albeit smaller, except that p2 appears to be single rooted, and the overall molar tooth row is longer relative to the depth of the jaw (Gazin, 1955). Both the upper and lower molars are generally more inflated than in *Helohyus;* the lower molars have a largely reduced (or absent) hypoconulid (Gazin, 1955).

Discussion.   Intermediate in size between *Helohyus* and *Achaenodon, Parahyus* is closely referable to *Achaenodon* and, if it were not for the smaller size, might be attributable to the larger genus. The length of the tooth row indicates that the animal had a relatively longer snout than *Achaenodon,* and it will be interesting to note, when more comparable material is reported, whether the animal had a complete dentition (including a first premolar).

Included Species.   *P. vagus* Marsh, 1876.

Genus Simojovelhyus Ferrusquía-Villafranca, 2006

Type Species.   *Simojovelhyus pocitosense* Ferrusquía-Villafranca, 2006: 989.

Diagnosis.   From Unit C of a measured, but as yet unnamed, late Oligocene strata in northwestern Chiapas, Mexico, *Simojovelhyus* is the latest known member of the Helohyidae (Ferrusquía-Villafranca, 2006).

The morphology of the trigonid is derived relative to other helohyids, including a well-defined precingulum and parallel crests ("anterolingual cristid" and "anteromesocristid") that are not present in other helohyids. The paraconid, normally closely appressed to the metaconid in other helohyid genera, is not readily discernable and appears to be lost in this lineage. The trilobed hypoconulid heel of m3, characteristic of most helohyids, is present.

Discussion.   The youngest helohyid in North America is *Achaenodon fremdi,* found in the John Day region of Oregon (Hancock Mammal Quarry, upper Clarno Formation), which is dated at slightly older than 39.5 Ma (Lucas et al., 2004; Foss et al., 2004). The Hancock quarry is situated near the Uintan/Duchesneyan NALMA boundary and is characterized by a mix of late Uintan North American endemic fauna and Asian immigrant taxa. *Simojovelhyus* is reported to be from strata between 26 and 28 Ma, suggesting a gap in the helohyid lineage of 12–14 million years. This large time gap, and the fact that it is currently the earliest known Paleogene mammal from middle America, confer enormous paleobiogeographic implications (Ferrusquía-Villafranca, 2006).

The holotype and only known specimen of *Simojovelhyus* (IHNE 6784) is currently missing, but it is represented by "plastotype" casts (IHNE 6784bis and IGM-1902G) (Ferrusquía-Villafranca, 2006).

Included Species.   *Simojovelhyus pocitosense,* Ferrusquía-Villafranca, 2006.

Genus Dyscritochoerus Gazin, 1956

Type Species.   *Dyscritochoerus lapointensis* Gazin, 1956: 26.

Diagnosis.   The p4 has a high protoconid, anterior and posterior ridges, and a posterior cingulum. The first lower molar (m1) has a paraconid separate from the metaconid, and Peterson (1934: 376) reported, "From the base of the proto- and metaconids there are low but discernable ridges which extend backward, connecting with the base of the hypoconid." This appears to be the result of a strong buccally oriented cristid obliqua that parallels the presence of a mesially oriented hypocristid (a feature that is unknown in the Entelodontidae).

Discussion.   This taxon is known from only a single dentary fragment with the left p4–m1. Peterson (1934) recognized the specimen (CM 11912) as either *Lophiohyus* or *Helohyus* and noted that the diagnostic features of the m1 (specifically the disassociation of the para- and metaconid) were unlike those of entelodonts. Peterson (1934: 376) tentatively associated the specimen with *Helohyus* based on size but recognized that it might be unique: "That this fragmentary specimen pertains to a distinct genus is entirely probable, but I refrain from using it as a type."

Gazin (1956: 26) concurred with Peterson's assertion that the specimen was too fragmentary to represent a new taxon; then in the following sentence, he introduced a generic and specific name: "I agree with Peterson that the Lapoint specimen [CM 11912] is probably inadequate as a type. . . . I propose the new name *Dyscritochoerus lapointensis.*" Stucky (1998) concurred with Gazin's assignment of a new genus.

McKenna and Bell (1997: 412) assigned *Dyscritochoerus* to the entelodont *Brachyhyops.* However, the above-mentioned unique characters, along with the lack of a raised trigonid and a lingually oriented hypoconulid on m1, all suggest that *Dyscritochoerus* is not an entelodont.

Gazin noted that the specimen matched *Helohyus lentus* most closely in size (Gazin, 1956). Only the presence of a hypocristid (separate from the cristid obliqua) makes this specimen significantly different from *H. lentus*. Marsh (1871: 39) originally assigned *H. lentus* to *Elotherium* before later reassigning the specimen to *Helohyus* (King, 1878: 404).

Because the specimen (CM 11912) has been missing since before 1956, none of the later authors (including Gazin, Stucky, McKenna, Bell, or this author) has ever actually viewed it.

RELATED TAXA

Coombs and Coombs (1977a, 1977b, 1982) allied both *Gobiohyus* and anthracotheres with helohyids. Stucky (1998) rejected these associations at the broad level. Previously associated Asian taxa should be excluded from the Helohyidae, including not only the Raoellidae (*Indohyus, Kirtharia, Kunmunella, Haqueina,* and *Raoella*) but perhaps also *Gobiohyus* and *Pakkokuhyus*.

The evolutionary relationships between and among the Helohyidae, Raoellidae, and Anthracotheridae (and also the Cebochoeridae and Hippopotamidae) warrant further study, as the results will be of tantamount importance in resolving the position of Cetacea within the Artiodactyla.

Also of interest will be the resolution of the various North American homacodontid Dichobunidae (including *Hexacodus, Antiacodon,* and *Homacodon*) in order to better understand the origin of the Helohyidae in North America. Stucky (1998) cited similarities between North American *Hexacodus* (a homacodont) and *Helohyus* and went on to suggest that the homacodonts might have also provided the origin to tylopods and ruminants. Finally, the evolutionary positions of Asian *Gobiohyus* and *Pakkokuhyus* need to be reconciled relative to similar North American helohyid taxa.

ACKNOWLEDGMENTS

I thank K. E. Beth Townsend for her insight and discussions regarding middle and late Eocene fauna in the Uinta Basin and for offering improvements to the late version of this chapter. Margery Coombs, Lia Vella, and Jessica Theodor offered comments to improve the early version. I am indebted to Jude Higgins for producing the original illustration for Figure 6.1 at extremely short notice. Finally, I dedicate this chapter to Bill Turnbull for his enthusiastic assistance and discussions about the little-studied achaenodont group and for introducing me to the fauna of the Washakie Basin.

FABRICE LIHOREAU
AND STÉPHANE DUCROCQ

7

Family Anthracotheriidae

ANTHRACOTHERES ARE A FOSSIL GROUP of Tertiary "suiform"
artiodactyls that ranged from small, terrier-sized animals to beasts approaching a hippopotamus in size. They occurred in Eurasia, North America, and
Africa from the middle Eocene until the late Pliocene and were represented by 37
genera (following McKenna and Bell, 1997). This group, known since Cuvier (1822),
included many more taxa several decades ago, and several Eurasian forms need to be
revised in a phylogenetic context that takes into account the entire family diversity.
Recent efforts have led to a phylogenetic hypothesis in which Anthracotheriidae
are the stem group of Hippopotamidae (Boisserie et al., 2005a, 2005b; Boisserie and
Lihoreau, 2006), but we do not consider the latter family in this chapter. However,
these new results suggest that anthracotheres might not be extinct and that the extant representatives of the clade might indeed be called hippos. Furthermore, a clade
Anthracotheriidae-Hippopotamidae has recently been proposed as the sister group
of Cetacea, which is congruent with molecular data linking modern whales and hippos (Boisserie et al., 2005a, and references therein).

Anthracotheres have often been early members of large-scale mammal dispersal
events, thus suggesting early phases of connections between landmasses. They are
therefore very useful fossils for reconstructing the paleogeographic and paleoenvironmental history of areas where they lived. Although first thought to be exclusively
restricted to swampy environments, recent investigations of anthracotheres have revealed them to be much more diversified in their habitats and diets (Lihoreau, 2003).
Their extinction has been commonly related to competition with hippopotamids
(e.g., Pickford, 1983, 1991: 1523), but the diachronism of their extinction on different
landmasses, together with their contemporaneous occurrence with hippopotamids

in Asia and Africa over several million years, leads to a reconsideration of the causes of their demise.

ANTHRACOTHERIID DIVERSITY

For most authors, the origin of Anthracotheriidae appears to root within Asian Helohyidae, a group of primitive artiodactyls (e.g., Coombs and Coombs, 1977b; Ducrocq et al., 1997a; Foss, this volume). Although no transitional form between the families is known, *Siamotherium* is considered to be the most primitive anthracothere (Suteethorn et al., 1988), the one that displays the basic morphotype of the family. Anthracotheriidae can thus be characterized by an association of derived characters: amastoid skull, brachyodont and bunoselenodont molars, upper and lower molars increasing in size posteriorly, enlarged m3 and M3, the anterior rim of the orbit above M1 in primitive forms and more posteriorly located in derived taxa, a complete dental formula, upper and lower fourth premolar never molariform, upper molars with four or five cusps (with a paraconule), no hypocone, but well-developed distolingual metaconule in the hypocone position, postprotocrista never reaching the metaconule, mesostyle primitively pinched in and becoming looplike, no paraconid on lower molars, distinct mesiolingual metacristid, V-shaped hypolophid when present, prehypocristid that ascend the distal trigonid wall, cingulum spur on the distal face of m1–m2 (commonly misnamed hypoconulid), well-developed m3 hypoconulid, unfused metacarpal and metatarsal bones, manus primitively with five digits but reduced to four (with digit I lost) in some lineages and pes with four toes (Fig. 7.1). Incisors and canines are sexually dimorphic in all anthracotheriid subfamilies.

In order to establish more precisely an original hypothesis of relationships within the Anthracotheriidae, a phylogenetic analysis of 51 dental and cranial morphological characters (see Appendix 7.1 for character description and data matrix) observed on 27 taxa has been conducted using PAUP 4.0β10 (Swofford, 2002). We selected representatives of the family characterized by their good preservation, their abundant fossil record, and their diversified geographic and stratigraphic range. Three subfamilies (Anthracotheriinae, Microbunodontinae nov., and Bothriodontinae) can be distinguished in the present classification (Fig. 7.2). We also used the cladogram of Figure 7.2 to distinguish the characteristics of each subfamily and genus. *Siamotherium* Suteethorn et al., 1988 (late Eocene of Thailand) is the closest genus to the ancestral condition within the family (see below). We therefore consider this form to be the best outgroup because of its abundant and well-preserved material.

Family ANTHRACOTHERIIDAE Leidy, 1869

Genus *Siamotherium* Suteethorn et al., 1988

Type Species.   *Siamotherium krabiense* Suteethorn et al., 1988; late Eocene of Thailand.

Range.   Late middle and late Eocene of Thailand and Myanmar.

Characteristics.   *Siamotherium* is considered to be the most primitive anthracotheriid genus (Suteethorn et al., 1988; Ducrocq et al., 1997a; Ducrocq, 1999). It differs from helohyids in the derived features listed above for anthracotheriids and likely represents the stem group of all anthracotheres. First described from the late Eocene of Thailand and later from the late middle Eocene of Myanmar (*S. pondaungensis* Ducrocq et al., 2000), *Siamotherium* has yielded numerous cranial, dental, and postcranial remains, which represent most of its skeleton. *Siamotherium pondaungensis* has been synonymized with the helohyid *Pakkokuhyus lahirii* by Tsubamoto et al. (2002) mainly on the basis of matching size, but we consider here *S. pondaungensis* as an anthracotheriid. Indeed, the structural arguments used by Tsubamoto et al. (2002) to refer *S. pondaungensis* to a helohyid (bunodont and conical cusps, enlarged metaconule, and vestigial styles) are features also seen in primitive anthracotheres. On the other hand, helohyids have their upper molar lingual margin rounded and not square as in *S. pondaungensis*, their paraconule is reduced (it is almost as big as

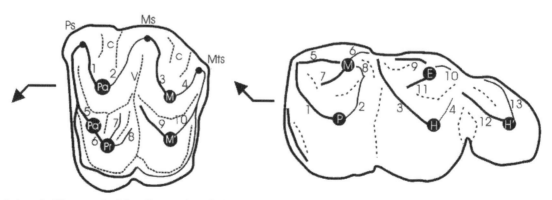

Fig. 7.1. Anthracotheriid upper molar (left) and lower molar (right) dental terminology. Left upper molar: Pr, protocone; Pa', paraconule; Pa, paracone; M, metacone; M', metaconule; Ps, parastyle; Ms, mesostyle; Mts, metastyle; V, transverse valley; C, rib; 1, preparacrista; 2, postparacrista; 3, premetacrista; 4, postmetacrista; 5, preparacristule; 6, preprotocrista; 7, postprotocrista; 8, lingual postprotocrista; 9, premetacristule; 10, postmetacristule. Left lower molar: P, protoconid; H, hypoconid; H', hypoconulid; M, metaconid; E, entoconid; 1, preprotocristid; 2, postprotocristid; 3, prehypocristid; 4, posthypocristid; 5, mesiolingual metacristid; 6, distolingual metacristid; 7, premetacristid; 8, postmetacristid; 9, mesiolingual entocristid; 10, distolingual entocristid; 11, preentocristid; 12, prehypocristulid; 13, posthypocristulid. The angled arrows point to the mesiolingual border of the tooth.

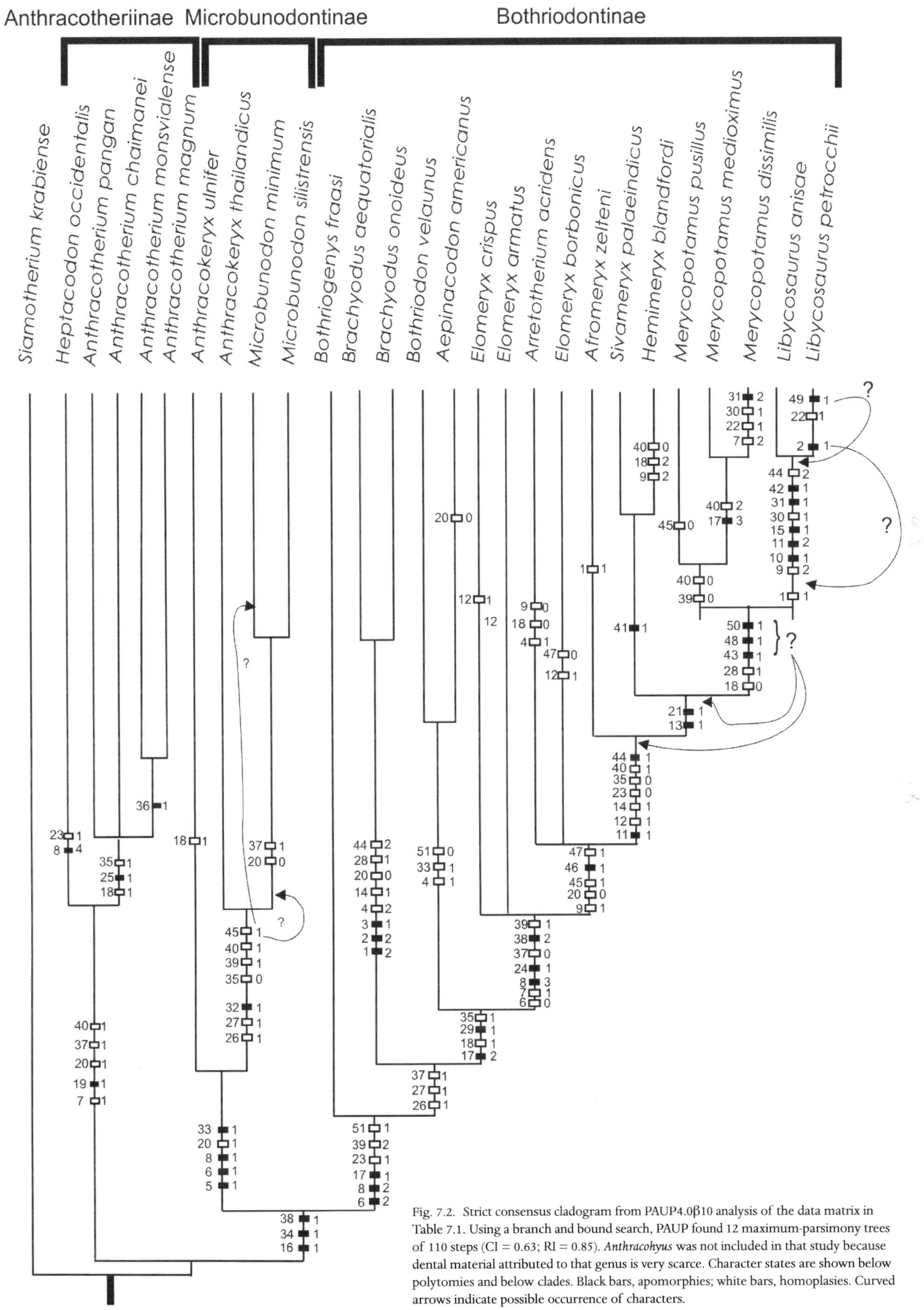

Fig. 7.2. Strict consensus cladogram from PAUP4.0β10 analysis of the data matrix in Table 7.1. Using a branch and bound search, PAUP found 12 maximum-parsimony trees of 110 steps (CI = 0.63; RI = 0.85). *Anthracohyus* was not included in that study because dental material attributed to that genus is very scarce. Character states are shown below polytomies and below clades. Black bars, apomorphies; white bars, homoplasies. Curved arrows indicate possible occurrence of characters.

the metacone in *S. pondaungensis*), and they generally have a small hypocone on the postcingulum (there is no trace of such a cusp in anthracotheres). *Siamotherium* has upper and lower tooth rows without diastemata, incipient selenodonty on molars, a short snout, its orbits above the M1, simple triangular lower premolars (generally without accessory cusps), small canines, weakly developed styles, mesostyle sometimes absent, and slender limbs (the estimated weight for *Siamotherium* is about 30 kg).

Genus *Anthracohyus* Pilgrim and Cotter, 1916

Type and Only Included Species. *Anthracohyus choeroides* Pilgrim and Cotter, 1916; late middle Eocene of Myanmar.

Range. Late middle and late Eocene of Thailand and Myanmar.

Characteristics. Medium-sized anthracothere with five-cusped upper molars that have a distal lobe buccally narrower than the mesial one, a rather straight centrocrista, and no para- or mesostyle. Tsubamoto et al. (2002) recently synonymized *Anthracohyus* with *Anthracotherium,* but we consider *Anthracohyus* to be a valid genus because it can be distinguished from *Anthracotherium* by the features listed above.

No lower teeth can be referred with certainty to *Anthracohyus,* and those attributed to that genus by Pilgrim and Cotter (1916) cannot be distinguished from the dental material of *Anthracotherium* (see that genus).

Only one species (*A. choeroides*) is recognized from the late middle Eocene of Myanmar, and it is represented by very few isolated teeth (Pilgrim and Cotter, 1916; Aung Naing Soe, 2004). The species *A. slavonicus* Heissig, 1990, from the Eocene of Slovenia, is very likely a cebochoerid and mainly differs from anthracotheriids by the presence of mediodistal accessory cusps on upper molars (Ducrocq, 1999). We do not agree with Heissig (2001) that *Anthracohyus* should be included in the Achaenodontidae because the latter is characterized by the presence of a hypocone, which *Anthracohyus* lacks.

Subfamily ANTHRACOTHERIINAE Leidy, 1869

Characteristics. Small to large-sized forms, upper molars with an accessory cusp (protostyle) on the mesial cingulum, rather bunodont cheek teeth compared with other subfamilies, a strong vertical lower canine with a circular section and a distal wear facet, no diastema between the lower canine and p1, and a symphysis with an elliptical cross section (Fig. 7.3A,D,G,J). Microbunodontinae nov. subfam. differ by their smaller size, long upper canine laterally compressed in males, small lower canine that mesially wears against I3, and the transverse constriction of their mandible at c–p1 diastema. Bothriodontinae differ by their upper molars with the mesostyle invaded by the transverse valley, selenodont cusps, ventrally concave symphysis, and diastema between c and p1.

Genus *Heptacodon* Marsh, 1894a

Synonyms. *Anthracotherium*, in part; *Octacodon* Marsh, 1894b. See detailed data on occurrences in Kron and Manning (1998).

Type Species. *Heptacodon curtus* Marsh, 1894; early Oligocene (Whitneyan) of North America.

Range. Middle Eocene (late Uintan) to early Oliocene (late Whitneyan) of North America.

Characteristics. Medium to large-sized anthracotheriine with a short rostrum, fused mandibular symphysis, and canines set far laterally with an elliptical transverse section, displaying an inflated buccal face (Macdonald, 1956).

Holroyd (2002) distinguished four species of this North American genus: *H. curtus* Marsh, 1894 (early Oligocene-Whitneyan of South Dakota), *H. occidentale* Osborn and Wortman, 1894 (early Oligocene-Orellan of South Dakota), *H. pellionis* Storer, 1983 (late middle Eocene-Duchesnean of Saskatchewan, Oregon, Texas, and Utah), and *H. yeguaensis* Holroyd, 2002 (late middle Eocene-Duchesnean of Texas). However, fragmentary remains attributed to *Heptacodon* by Hanson (1996) have been recovered from the Hancock Quarry local fauna in Oregon that has been dated from the late Uintan (middle Eocene) by Lucas et al. (2004). The specimens from Oregon might, therefore, represent the oldest record for that genus in North America. *Heptacodon* is the earliest known anthracothere and the only representative of the Anthracotheriinae in the New World.

Genus *Anthracotherium* Cuvier, 1822

Selected Synonymy. *Bugtitherium* Pilgrim, 1908; *Anthracothema* Pilgrim, 1928; *Heothema* Tang, 1978; *Huananothema* Tang, 1978.

Type Species. *Anthracotherium magnum* Cuvier, 1822; early late Oligocene of Europe.

Range. Late middle Eocene to late Oligocene of Asia and Europe.

Characteristics. Medium to large-sized anthracotheriine, which has an m3 with an entoconulid, m1–2 with developed distal cingular spur, accesory cusps on the upper molar mesial cingulum, and a large upwardly oriented canine with a rounded section.

Fifteen species of *Anthracotherium* have been recognized: *A. pangan* and *A. crassum* Pilgrim and Cotter, 1916 (late middle Eocene, Myanmar); *A. chaimanei* Ducrocq, 1999 (late Eocene, Thailand); *A. monsvialense* De Zigno, 1888, and *A. alsaticum* Cuvier, 1822 (early Oligocene [MP 21–22], western Europe); *A. bimonsvialensemagnum* Golpe-Posse, 1971 (early Oligocene [MP 23], western Europe); *A. seckbachense* Kinkelin, 1884 (early Oligocene [MP 22–25?], western Europe); *A. illyricum* Teller, 1886, and *A. bumbachense* Stehlin, 1910 (late early Oligocene [MP25], central and western Europe); *A. magnum* Cuvier, 1822 (early late Oligocene [MP 25–26], western Europe); *A. cuvieri* Gaudry, 1873 (early late

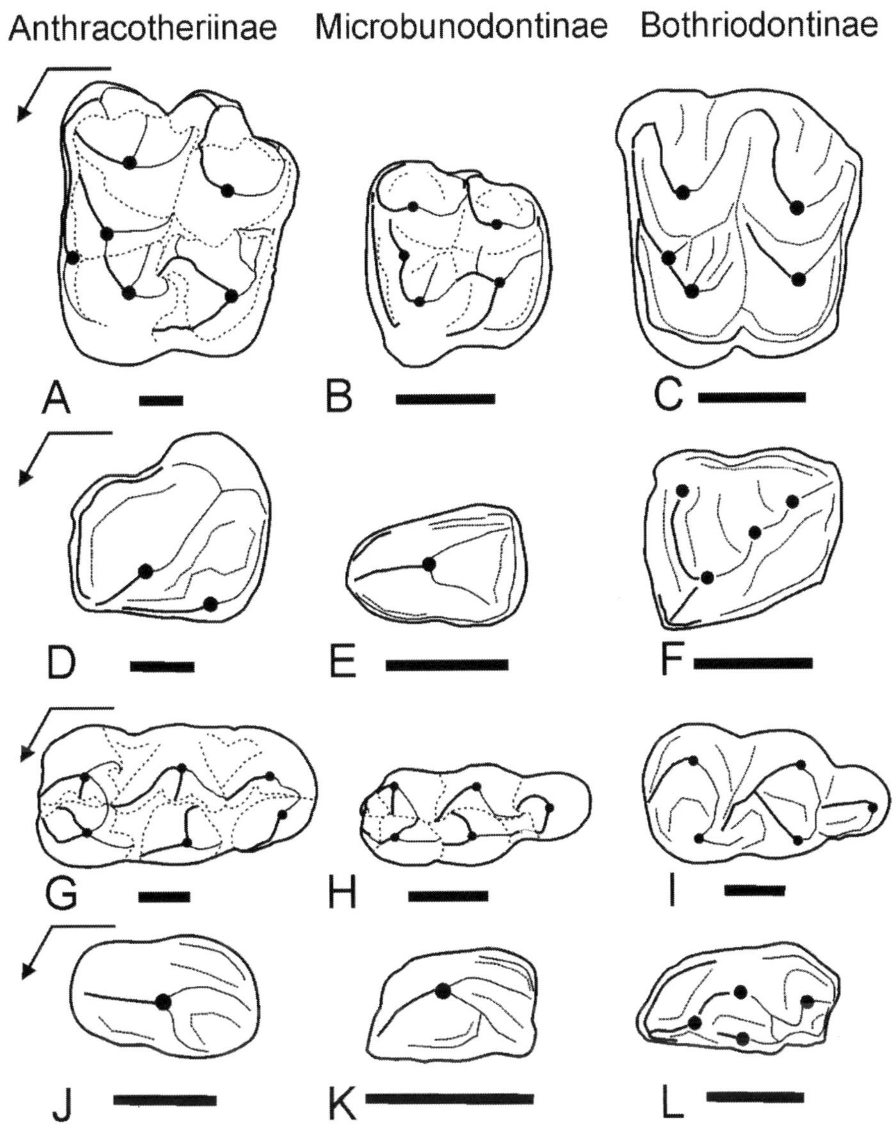

Anthracotheriinae   Microbunodontinae   Bothriodontinae

Fig. 7.3. Dental occlusal pattern of the three anthracotheriid subfamilies. All teeth are oriented with lateral (buccal) at top, and anterior (mesial) at left. A, left M3 *Anthracotherium*; B, left M3 *Microbunodon*; C, left M3 *Sivameryx*; D, left P3 *Anthracotherium*; E, left P3 *Microbunodon*; F, left P3 *Merycopotamus*; G, right m3 *Anthracotherium*; H, right m3 *Microbunodon*; I, right m3 *Brachyodus*; J, right p4 *Heptacodon*; K, right p4 *Microbunodon*; L, right p4 *Merycopotamus*; Scale bars equal 1 cm. The angled arrows point to the mesiolingual border of the tooth. Scale bars equal 1 cm.

Oligocene [MP 26], western Europe); *A. hippoideum* Rütimeyer, 1857 (late Oligocene [MP 27], western Europe); *A. valdense* Kowalevski, 1876 (late Oligocene [MP 29], western Europe); *A. bugtiense* Pilgrim, 1907 (late Oligocene, Indian Subcontinent); *A. kwablianicum* Gabunia, 1964 (late Oligocene, eastern Europe). Six species previously attributed to the genera *Heothema* and *Huananothema* from the late Eocene of China (Tang, 1978; Zhao, 1981, 1983) might, in fact belong to *Anthracotherium* (Ducrocq, 1999). Finally, the genus has been reported from the Eocene of Timor Island in Indonesia (von Koenigswald, 1967), but the dental morphology of the specimen and the geological history of that area suggest that the fossil from Timor is not autochthonous (Ducrocq, 1996).

The genus *Anthracotherium* flourished and diversified in the Oligocene of the Old World. However, several species, especially those described during the nineteenth century, for which intraspecific variation was not taken into account,

should likely be synonymized on the basis of very similar dental morphology.

Genus *Prominatherium* Teller, 1884

Synonyms. *Anthracotherium*, in part.

Type and Only Included Species. *Anthracotherium dalmatinum* (von Meyer, 1854); late Eocene of Croatia.

Range. Late Eocene of southern Europe (Croatia).

Characteristics. Small-sized anthracotheriine known from little material but regarded as a smaller version of *Anthracotherium*. Upper molars are mesiodistally compressed with strongly developed para- and mesostyles and the transverse valley not obstructed by crests.

Hellmund and Heissig (1994) have pointed out that *Prominatherium* is morphologically very close to *Anthracotherium illyricum* and mainly differs from it in size and skull structure (longer skull with regularly rounded zygomatic arch and oblique glenoid facet), the latter feature being very variable.

Subfamily MICROBUNODONTINAE nov.

Characteristics.   Small-sized anthracotheres with slender limbs, long upper canine laterally compressed in males, mesial wear facet on the small lower canine caused by contact with I3, transverse constriction of mandible at c–p1 diastema with, in some forms, a chin apophysis (Fig. 7.3B,E,H,K). Anthracotheriinae differ by their larger size, upper molars with an accessory cusp on the mesial cingulum, their mandible not transversely constricted at c–p1 diastema, their symphysis with an elliptical section, and their stronger lower canine that distally wears against their upper canine. Bothriodontinae differ by their upper canine premolariform (in primitive genera) or strong with a circular cross section (in derived genera) and by their upper molars invaded by the transverse valley.

In the proposed phylogenetic hypothesis (Fig. 7.2), the clade Microbunodontinae is excluded from the Anthracotheriinae and Bothriodontinae (*sensu* Scott, 1940). The creation of a new taxon of subfamilial rank for this monophyletic clade including the genera *Microbunodon* and *Anthracokeryx* is therefore warranted. These genera were not involved in the previous phylogenetic studies of the Anthracotheriidae family by Scott (1940) or Kron and Manning (1998) and were until now considered as Anthracotheriinae.

Genus *Anthracokeryx* Pilgrim and Cotter, 1916

Selected Synonymy.   *Anthracotherium*, in part; *Anthracohyus*, in part; *Anthracosenex* Zdansky, 1930; ?*Probrachyodus* Xu, 1962.

Type Species.   *Anthracokeryx birmanicus* Pilgrim and Cotter, 1916; late middle Eocene of Myanmar.

Range.   Late middle Eocene to late Eocene of Asia.

Characteristics.   Small-sized microbunodontine characterized by upper molars with paracone and metacone that tend to be buccally flattened, incipient selenodont protocone and metaconule, mesial and distal crests of P3 oriented in the longitudinal axis of the tooth, lower molars with a flattened lingual face, and p3 longer and taller than p4.

We consider that five species can be attributed to *Anthracokeryx*: *A. tenuis* Pilgrim and Cotter, 1916, *A. ulnifer* Pilgrim, 1928, and *A. birmanicus* Pilgrim and Cotter, 1916 (late middle Eocene, southeast Asia); *A gungkangensis* Qiu, 1977 (= *A. kwangsiensis* Qiu, 1977) and *A. sinensis* Zdansky, 1930 (middle-late Eocene, China). Compared to other species of *Anthracokeryx* in which the anterior part of the mandible is known, *A. thailandicus* Ducrocq, 1999 (late Eocene, southeast Asia) appears to be more derived, suggesting that this form might represent a separate genus (Fig. 7.2). *Anthracokeryx ulnifer* is known by a very well-preserved skull associated with a lower jaw that allows the accurate description of

the genus. The validity of *A. tenuis,* however, is dubious, and it might be synonymized to *A. ulnifer.*

Tsubamoto et al. (2002) included most species of *Anthracokeryx* (except *A. sinensis*) in the genus *Anthracotherium* on the basis of what they perceived as very similar dental morphology. However, we retain *Anthracokeryx* as a valid genus, because it is characterized by the set of features (listed above) that are never seen in *Anthracotherium*. The molar morphology of *Anthracosenex* Zdansky, 1930 does not differ from that of *Anthracokeryx,* and we thus consider that the latter can be included in *Anthracokeryx.*

Genus *Microbunodon* Depéret, 1908

Synonyms.   *Anthracotherium,* in part; *Sus,* in part; *Choeromeryx,* in part; *Microselenodon* Depéret, 1908; *Rhagatherium,* in part. See complete list of synonymy in Lihoreau et al. (2004a).

Type Species.   *Microbunodon minimum* (Cuvier, 1822); late Oligocene of France.

Range.   Late Oligocene to late Miocene of Eurasia.

Characteristics.   Small-sized microbunodontine (weight estimated by astragalus measurement to be around 20 kg), characterized by a fused mandibular symphysis with a crested ventral edge, marked sexual dimorphism with a bladelike upper canine in males, and loss of third metacristule.

This long-ranging Eurasian genus includes three species: *M. minimum* (Cuvier, 1822) (late Oligocene, France, Germany, Switzerland, Austria, Turkey); *M. silistrensis* Pentland, 1828 (early to middle Miocene, Indian Subcontinent); *M. milaensis* Lihoreau et al., 2004a (late Miocene, Indian Subcontinent).

Subfamily BOTHRIODONTINAE Scott, 1940

Characteristics.   Small to large-sized forms, small premolariform upper canine in primitive genera (strong with a circular cross section in derived genera), low and premolariform lower canine with an elliptical cross section and mesial and distal keels in primitive genera (large and high, with a labially concave cross section and a distal keel in derived genera), upper molar mesostyle invaded by the transverse valley, crescentic cusps, upper molars with five or four cusps (paraconule lost in some representatives), m1 and m2 cingulum spur reduced, m3 hypoconulid forming a transversely compressed loop, ventrally concave symphysis, and manus with four or five digits (Fig. 7.3C,F,I,L). Anthracotheriinae differ by their bunodont cusps, the presence of an accessory cusp on the mesial cingulum of upper molars and on the distolingual border of the P3, the absence of diastema between c and p1, and by their symphysis with convex dorsal and ventral borders. Microbunodontinae differ by their transversely compressed upper canine, their lower canine that mesially wears against the I3, and by their mandible with a transverse constriction at c–p1 diastema.

Genus *Ulausuodon* Hu, 1963

Type and Only Included Species. *Ulausuodon parvus* Hu, 1963; late middle Eocene of Mongolia.

Range. Late middle Eocene of Mongolia.

Characteristics. Small-sized bothriodontine, five-cusped upper molars anteroposteriorly compressed with wide transverse valley, paraconule in line with para- and protocone, and shallow mandible with a slightly developed angular process.

The attributed material displays characteristics that can be found in primitive members of *Bothriodon* (see features of that genus below), but more complete specimens are needed to infer precise relationships.

Genus *Elomeryx* Marsh, 1894b

Synonyms. *Hyopotamus,* in part; *Xiphodon,* in part; *Brachyodus,* in part; *Ancodus,* in part; *Microbunodon,* in part; *Bunobrachyodus,* in part; *Heptacodon,* in part; *Bothriodon,* in part. See complete list of synonymy in Hellmund (1991).

Type Species. *Elomeryx crispus* (Gervais, 1849); late Eocene (MP 18) of France.

Range. Late middle Eocene to early Miocene of Europe, Asia, and North America.

Characteristics. Small to medium-sized bothriodontine with marked sexual dimorphism on upper canines, serrated posterior edge of upper canines in males, short rostrum, five-cusped upper molars, looplike mesostyle, premolar row without diastemata (very short diastema in *E. cluai*), and presence of accessory cusps on premolars in derived species.

Four species are recognized: *E. armatus* Marsh, 1894b from the late Eocene (middle Chadronian) to early-late Oligocene (late early Arikareean) of North America (detailed data on occurrences in Kron and Manning, 1998); *E. cluai* (Depéret, 1906) from the late early Oligocene of western Europe; *E. crispus* (Gervais, 1849) from the late Eocene to early Oligocene of Europe; and *E. borbonicus* (Geais, 1934) from the late Oligocene to early Miocene of western Europe. There is also strong evidence (see Ducrocq and Lihoreau, in press) for *Elomeryx* to have occurred in the Eocene and Oligocene of Asia (Georgia, Pakistan, China, and Japan). The proposed phylogeny (Fig. 7.2) shows that *Elomeryx* is paraphyletic and suggests that this genus might be the stem group of several lineages of anthracotheres on different landmasses. A detailed review of the North American and Eurasian forms attributed to *Elomeryx* is therefore needed.

Genus *Bothriogenys* Schmidt, 1913

Synonymy. *Ancodon,* in part, *Brachyodus,* in part. See complete list of synonymy in Ducrocq (1997).

Type Species. *Bothriogenys fraasi* (Schmidt, 1913); late Eocene to early Oligocene of Egypt.

Range. Late Eocene to late Oligocene of Thailand, China, Egypt, and Ethiopia.

Characteristics. Small to medium-sized anthracothere, no angular process on the shallow mandible, complete dental formula, no incisor enlargement, simple and elongated premolars, small canines, five-cusped upper molars with flattened parastyle and mesostyle, lower molars with short preprotocristid and prehypocristid.

There are five recognized species in this genus: *B. fraasi, B. rugulosus* (Schmidt, 1913), *B. andrewsi* (Schmidt, 1913), and *B. gorringei* (Andrews and Beadnell, 1902) all from the late Eocene to early Oligocene of Egypt, and *B. orientalis* Ducrocq, 1997, from the late Eocene of Thailand. This genus also likely occurred in the late Eocene of South China (Ducrocq and Lihoreau, 2006). The North African forms might have originated from Asian ancestors (Ducrocq, 1997). Black (1978a) suggested that the upper molar attributed to *Rhagatherium aegypticum* by Andrews (1906) might, in fact, belong to a small species of *Bothriogenys,* but the morphology of the tooth (especially the bunodonty and the very weakly developed styles) makes that hypothesis questionable.

Genus *Qatraniodon* Ducrocq, 1997

Synonymy. *Ancodon,* in part; *Brachyodus,* in part; *Bothriogenys,* in part. See complete list of synonymy in Ducrocq (1997).

Type and Only Included Species. *Qatraniodon parvus* (Andrews, 1906); late Eocene of Egypt.

Range. Late Eocene of Egypt.

Characteristics. *Qatraniodon* was long included in the genus *Bothriogenys,* but it can be distinguished from the latter by the following features: smaller size, comparatively longer and more slender lower molars with narrower mesial end of the crown, straight and mesiolingually oriented preprotocristid, and no premetacristid.

Qatraniodon is known so far only from a fragmentary lower jaw from the early Oligocene of Egypt (Ducrocq, 1997).

Genus *Bothriodon* Aymard, 1846

Synonyms. *Hyopotamus,* in part; *Ancodon,* in part; *Ancodus,* in part; *Anthracotherium,* in part.

Type Species. *Bothriodon velaunus* (von Meyer, 1832); early Oligocene of western Europe.

Range. Late Eocene to early Oligocene of Eurasia and North America.

Characteristics. Middle-sized bothriodontine with very elongated rostrum, mesiodistally compressed five-cusped upper molars, looplike mesostyle, numerous diastemata between c/C and p1/P1, p1/P1 and p2/P2, and p2 and p3, fused symphysis, premolariform canine and large i2 compared to other incisors, and internal choanae displaced anteriorly to the level of M3.

Several taxa were described in the nineteenth century, but most authors consider as valid only the following five species: *B. advena* Russell, 1978 from the late Eocene (early Chadronian) and *B. rostratus* (Scott, 1894b) from the early Oligocene (Orellan) of North America (detailed data on occurrences in Kron and Manning, 1998); *B. velaunus* (von Meyer, 1832), *B. leptorynchus* Aymard, 1846, and *B. aymardi* (Pomel, 1847a) from the early Oligocene of western Europe. Isolated teeth have also been reported from the late Eocene of southern China and attributed to *Bothriodon* (Chow, 1957; Xu, 1961).

Genus *Bakalovia* Nikolov and Heissig, 1985

Synonyms.   *Elomeryx*, in part. See complete list of synonymy in Hellmund (1991).

Type Species.   *Bakalovia palaeopontica* (Nikolov, 1967); late Eocene of Bulgaria.

Range.   Late Eocene of Bulgaria.

Characteristics.   Medium-sized bothriodontine morphologically very close to *Elomeryx* and differing from it only in the lack of junction between the posthypocristid and the postentocristid on lower molars.

According to Hellmund (1991), two species are known: *B. palaeopontica* and *B. astica* (Nikolov, 1967), both from the late Eocene of southeastern Europe.

Genus *Aepinacodon* Troxell, 1921

Synonyms.   *Hyopotamus*, in part; *Ancodus*, in part; *Ancodon*, in part; *Bothriodon*; in part. See detailed data on occurrences in Kron and Manning (1998).

Type Species.   *Aepinacodon deflectus* (Marsh, 1890); late Eocene (late Chadronian) of North America.

Range.   Late Eocene (early to late Chadronian) of North America.

Characteristics.   Large bothriodontine with elongated rostrum, premolariform canine, large and curved incisors, internal choanae displaced anteriorly to the level of or just behind M3, square five-cusped upper molars, looplike mesostyle, and fused symphysis.

Two species have been referred to this genus: *A. americanus* (Leidy, 1856b) and *A. deflectus* (Marsh, 1890), both from the late Eocene (Chadronian) of North America. The genus has also been reported by Eaton et al. (1999) from the late middle Eocene (Duchesnean) of Utah. According to the hypothesized relationships of anthracotheres presented in Fig. 7.2, *Aepinacodon* might be the sister group of *Bothriodon*.

Genus *Arretotherium* Douglass, 1901

See detailed data on occurrences in Kron and Manning (1998).

Type Species.   *Arretotherium acridens* Douglass, 1901; early Miocene (early late Arikareean) of Montana, North America.

Range.   Late Oligocene (early early Arikareean) to early Miocene (early Hemingfordian) of North America.

Characteristics.   Medium-sized bothriodontine with laterally developed orbits that are somewhat elevated above the cranial roof, short snout, four-cusped upper molars, spatulate lower incisors with i2 larger than the other ones, long diastema between p1 and p2, deep angular process of the mandible, and marked sexual dimorphism on canines.

According to Kron and Manning (1998), three species are known in the early Miocene and possibly in the late Oligocene of North America: *A. acridens* Douglass, 1901, *A. fricki* Macdonald and Schultz, 1956, and *A. leptodus* (Matthew, 1909b). *Arretotherium* likely appears to have been derived from *Elomeryx* according to its phylogenetic position in Fig. 7.2 and following Macdonald (1956) and Macdonald and Martin (1987).

Genus *Kukusepasutanka* Macdonald, 1956

See detailed data on occurrences in Kron and Manning (1998).

Type and Only Included Species.   *Kukusepasutanka schultzi* Macdonald, 1956; early late Oligocene (early early Arikareean) of Montana, North America.

Range.   Early late Oligocene (early early Arikareean) of Montana, North America.

Characteristics.   Large bothriodontine with a rostrum tubular in cross section, pentacuspidate upper molars with strongly developed meso- and metastyles, P4 and upper molars mesiodistally compressed, P2 and P3 with a well-developed distolingual heel, and P1 possibly lost.

Genus *Telmatodon* Pilgrim, 1907

Synonyms.   *Ancodus*, in part; *Brachyodus*, in part; *Gonotelma*, in part. See complete list of synonymy in Pickford (1987).

Type Species.   *Telmatodon bugtiensis* Pilgrim, 1907; late Oligocene to early Miocene of Pakistan.

Range.   Late Oligocene to early Miocene of Pakistan.

Characteristics.   Large bothriodontine, four-cusped upper molars, lingual face of the protocone displaying a gentle slope compared to other bothriodontines, and lingual postprotocrista reaching the lingual margin of the tooth.

Two species are recognized: *T. bugtiensis* and *T. orientale* Forster-Cooper, 1924, both from Pakistan. Lower molars of *Telmatodon* do not significantly differ in size or in morphology from those of *Hemimeryx*, and a revision of some of the material attributed to *Telmatodon* is thus necessary in order to confirm the validity of the genus.

Genus *Parabrachyodus* Forster-Cooper, 1915

Synonyms. *Anthracotherium*, in part; *Hyopotamus*, in part; *Brachyodus*, in part; *Telmatodon*, in part. See complete list of synonymy in Pickford (1987).

Type and Only Included Species. *Parabrachyodus hyopotamoides* (Lydekker, 1883b); late Oligocene or early Miocene of Pakistan.

Range. Late Oligocene or early Miocene of Pakistan and India.

Characteristics. Large bothriodontine with narrow muzzle, pentacuspidate upper molars, accessory cusps on upper premolars, looplike mesostyle, parastyle formed by preparacrista and not by buccal cingula (in contrast to *Brachyodus*).

The genus has also been cited in the Oligocene site of Benara in Georgia (Russell and Zhai, 1987), but the material is very scarce, and this attribution is uncertain.

Genus *Brachyodus* Depéret, 1895

Synonyms. *Masritherium* Fourtau, 1920; *Bothriogenys*, in part; *Anthracotherium*, in part; *Hyopotamus*, in part; *Sihongotherium* Liu and Zhang, 1993. See complete list of synonymy in Pickford (1991).

Type Species. *Brachyodus onoideus* (Gervais, 1859); early Miocene of France, Switzerland, Germany, Austria, Portugal, and Greece.

Range. Early Miocene to early middle Miocene of Europe, Africa, and Asia.

Characteristics. Large bothriodontine with long and wide snout and shallow mandible, weakly developed angular process, long diastema between canine and premolars, pentacuspidate upper molars, P3 with a quadratic occlusal outline, two mesial crests on upper premolars, no premetacristid on lower molars, canines premolariform, and tusklike central upper and lateral lower incisors, reduction in incisor number, and anterior position (between C–P1) of the main palatine foramen.

Four species are recognized: *B. depereti* (Fourteau, 1918) and *B. mogharensis* Pickford, 1991, both from the late early Miocene of Egypt; *B. aequatorialis* McInnes, 1951 from the early Miocene of East and North Africa, and *B. onoideus* (Gervais, 1859) from the early Miocene of Europe.

Brachyodus has been recognized in the early middle Miocene of eastern Asia, but specific attribution is not yet established (Liu and Zhang, 1993; Ducrocq et al., 2003). *Brachyodus depereti* might also occur in the early-middle Miocene of South Africa (Pickford, 2003), but the morphology of the scarce material leads us to question the attribution to that taxon.

Genus *Afromeryx* Pickford, 1991

Synonyms. *Brachyodus*, in part; *Bothriogenys*, in part; *Ancodus*, in part; *Hyoboops*, in part, *Masritherium*, in part; *Gelasmodon*, in part. See complete list of synonymy in Pickford (1991).

Type Species. *Afromeryx zelteni* Pickford, 1991; early middle Miocene of Libya.

Range. Late early Miocene to middle Miocene of Africa and Arabian Peninsula.

Characteristics. Small to medium-sized bothiodontine with four-cusped upper molars; short diastemata between C and p1/P1, no diastema between i3 and c, lower premolars low crowned with pustulate mesial and distal crest[10], mesial crests of buccal cusps of lower molars do not reach lingual margin of tooth, upper incisors separated by small gaps, incisive foramen very large, markedly sexually dimorphic canines; p1 single rooted, symphysis reaches back to level of p1–2, genial spine strongly developed, talonid of m3 centrally placed and only slightly obliquely oriented (Pickford, 1991).

Afromeryx has been recovered only in Africa, with two species: *A. zelteni* from the late early Miocene to middle Miocene of Libya, Kenya, and the Sultanate of Oman (Pickford, 1991; Roger et al., 1994) and *Afromeryx africanus* (Andrews, 1899), which is endemic to the late early Miocene site of Moghara in Egypt (Pickford, 1991; Miller, 1999).

Genus *Sivameryx* Lydekker, 1883

Synonyms. *Choeromeryx*; *Hyopotamus*, in part; *Hyoboops*, in part; *Merycops*; *Ancodon*, in part; *Hemimeryx*, in part; *Brachyodus*, in part; *Rhagatherium*, in part; *Gonotelma*, in part. See complete list of synonymy in Pickford (1991) and Lihoreau (2003).

Type Species. *Sivameryx palaeindicus* (Lydekker, 1877); late middle Miocene to early middle Miocene of Pakistan.

Range. Late early Miocene to middle Miocene of Africa and the Indian Subcontinent.

Characteristics. Medium-sized bothriodontine with five-cusped upper molars and markedly reduced paraconule, upper molars longer than wide, looplike mesostyle, cingulum developed on the lingual side of the protocone, two distal crests run down from the protocone apex, p1 two rooted, preprotocristid and prehypocristid reaching the lingual border of lower molars, four distinct crests run down from the metaconid apex, m3 hypoconulid in line with buccal cusps, symphysis long and shallow and reaches back to level of p1.

Two species are known: *S. palaeindicus* from the late early Miocene to middle Miocene of the Indian Subcontinent and East and North Africa and *S. moneyi* (Fourteau, 1918) from the late early Miocene of Egypt.

The species *S. africanus* (Andrews, 1914) should be included in *S. palaeindicus* because of the lack of significant morphological and biometric differences between them (Lihoreau, 2003). Similarly, *Gonotelma shabhazi* Pilgrim, 1908, displays a very reduced paraconule that we consider to be within the range of size variation of *Sivameryx*. In the same way, the strong distolingual crest of the protocone in *Gonotelma*, which is also known in *Telmatodon* (see below), *Sivameryx*, *Afromeryx*, and *Elomeryx*, is not a feature that sets *Gonotelma* phylogenetically closer to *Telmatodon* (*contra* Pickford, 1987) than to *Sivameryx*. The structure, morphology, and size of the lower molars attributed to *Gonotelma* correspond to those of *Sivameryx palaeindicus*, discovered in the same quarries that yielded *Gonotelma* remains.

Genus *Hemimeryx* Lydekker, 1883

Synonyms.   *Ancodus*, in part; *Gelasmodon*. See complete list of synonymy in Lihoreau (2003).

Type and Only Included Species.   *Hemimeryx blandfordi* Lydekker, 1883; early middle Miocene of Pakistan.

Range.   Late early to early middle Miocene of Pakistan.

Characteristics.   Large bothriodontine, four-cusped upper molars with postprotocrista not running into the transverse valley, looplike mesostyle, upper molars longer than wide, numerous accessory cusps on lower premolar crests, all lower premolars with a lingual accessory cusp, m3 hypoconulid in line with buccal cusps, very shallow and long mandibular symphysis.

"*Hemimeryx*" *pusillus* Lydekker, 1885 was attributed to *Merycopotamus* because of its symphysis and lower molar morphology and mandibular depth (Lihoreau et al., 2004b).

Genus *Merycopotamus* Falconer and Cautley, 1847

Synonyms.   *Hippopotamus*, in part; *Hemimeryx*, in part. See complete list of synonymy in Lihoreau et al. (2007).

Type Species.   *Merycopotamus dissimilis* (Falconer and Cautley, 1836); late Miocene and early Pliocene of India.

Range.   Middle Miocene to late Pliocene of the Indian Subcontinent, late Miocene of Iraq and Thailand (Lihoreau et al., 2007).

Characteristics.   Medium-sized bothriodontine, canines markedly sexually dimorphic, oblique facial crest that reaches a marked facial tuberosity, large canine fossa, four-cusped upper molars wider than long with looplike mesostyle (divided in derived species), no second distal protocrista on upper molars, one or two accessory cusps on distobuccal crest of upper premolar, one accessory cusp on mesiolingual crest of lower premolars, markedly oblique mandibular symphysis with concave ventral border, no lower molar premetacristid, preprotocristid and prehypocristid reaching the lingual border of lower molars, m3 hypoconulid in line with buccal cusps, strong angular process of the mandible, deep vascular groove (= gonial angle), and manus with four digits.

Three species are recognized in Asia only: *M. nanus* (ex *pusillus*) Falconer, 1868 from the middle Miocene (see Lihoreau et al., 2007), *M. medioximus* Lihoreau et al., 2004b, from the early late Miocene and *M. dissimilis* from the late Miocene to early Pliocene (Steensma and Hussain, 1992).

Genus *Libycosaurus* Bonarelli, 1947

Synonyms.   *Anthracotherium*, in part; *Gelasmodon*, in part; *Merycopotamus*, in part. See complete list of synonymy in Pickford (1991) and Lihoreau (2003).

Type Species.   *Libycosaurus petrocchii* Bonarelli, 1947; late Miocene of Libya.

Range.   Late middle Miocene to late Miocene of Libya, Tunisia, Algeria, Chad, and Uganda.

Characteristics.   Large bothriodontine, facial tuberosity situated anteriorly to the facial crest and not in contact with it, marked anterior palatine depression, main anterior palatine foramina open at canine level, more than two accessory cusps on the distobuccal crest of upper premolars and on the mesiolingual crest of lower premolars, five upper premolars, four-cusped upper molars, looplike mesostyle undivided, lower incisors with very long root, variable number of lower incisors, preprotocristid and prehypocristid reaching the lingual border of lower molars, strong angular process of the mandible, no vascular groove, metapodials and phalanges medially widened and dorsoventrally flattened, and manus with four digits.

This exclusively African genus is known from three species: *L. algeriensis* Ducrocq et al., 2001, from the early late Miocene of North Africa, *L. anisae* (Black, 1972) from the early late Miocene of North and East Africa, and *L. petrocchii* Bonarelli, 1947, from the late Miocene of North and Central Africa. It should be noted that *Libycosaurus* is the only anthracothere that has five upper premolars (extremely rare in Tertiary mammals), a feature interpreted as a single event that occurred in a single population of limited size that transmitted this characteristic to all its descendants (Lihoreau et al., 2006).

Pickford (2006) does not consider *L. algeriensis* to be a valid species. Although *L. algeriensis* has been described only from isolated teeth, the morphology of its p4 and the significantly smaller size of its M3 and m3 lead us to consider this species valid until more material is discovered.

PROBLEMATIC TAXA

Atopotherium Ducrocq et al., 1996, is a small-sized monospecific genus from the late Eocene of Thailand characterized by strongly selenodont lower molars and wide lower premolars. The absence of a distal basin on the p4, the distolingually open lower molars lacking a distal cingular spur (= "hypoconulid"), and the selenodont structure of the lower molars make the attribution of *Atopotherium* to the anthracotheriids questionable (Ducrocq et al., 1996). Rather, these features are more suggestive of those of North American oreodonts. Upper teeth need to be recovered to assess the familial affinities of *Atopotherium* more precisely.

Anthracochoerus Dal Piaz, 1930, is known from two species (*A. stehlini* and *A. fabianii*) from the lower Oligocene of Italy and is characterized by four-cusped bunoselenodont upper molars, without paraconule. This morphology is unusual for Paleogene anthracotheres, which otherwise all display a paraconule. Tetracuspidy in general appeared in the Miocene and served as a character linking the derived bothriodontines *Arretotherium, Afromeryx, Hemimeryx, Telmatodon, Merycopotamus,* and *Libycosaurus.* However, the phylogeny proposed in Figure 7.2 indicates that tetracuspidy might have been a convergent character within the family.

MAIN EVOLUTIONARY TRENDS

The great diversity that can be observed in Anthracotheriidae results from several main evolutionary trends within each subfamily (Fig. 7.4). Anthracotheriinae increase in body size, a trend that is generally associated with stronger and larger canines and incisors. The somewhat elongated muzzle observed in derived forms is correlated with the occurrence of diastemata. The Microbunodonti-

nae had increasingly transversely compressed upper canines, which became bladelike. This feature can be correlated with a deepening and fusion of the mandibular symphysis in later taxa. Unlike Anthracotheriinae, Microbunodontinae do not show an increase in body size, and their limbs stay slender. The Bothriodontinae are represented by several lineages and show more diversified trends. Molars become more selenodont (elongation of the main crests, looplike upper molar styles and m3 hypoconulid, and loss of the paraconule, although this feature seems to occur several times in the subfamily). Premolars increase in complexity (development of cingulum and accessory cusps). In some lineages (for example, *Bothriodon, Aepinacodon, Hemimeryx*), the skull lengthens and tapers anteriorly, and it has long diastemata. In other taxa (for example, *Libycosaurus*), the skull becomes wider anteriorly and displays a general elevation of orbits, nares and external auditory meatus, while the mandible deepens and widens, and the canine tends to have a continuous growth (for example, *Merycopotamus*). This evolution led to animals with a general hippo-like morphology. Bothriodontines lost the first digit on the manus at least once.

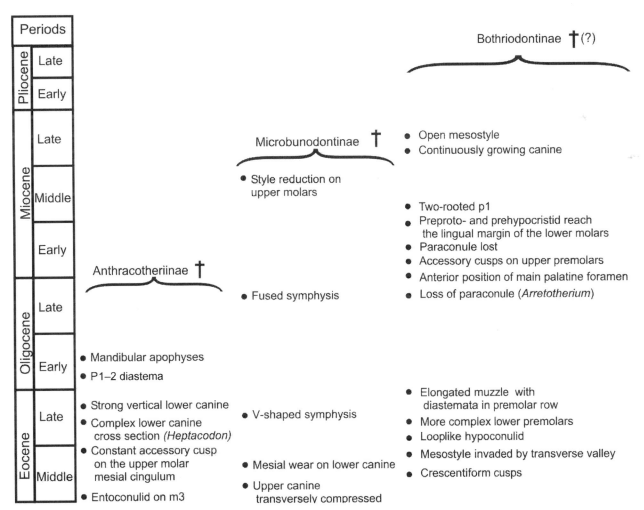

Fig. 7.4. Main evolutionary trends of each anthracotheriid subfamily in a chronological context.

BIOGEOGRAPHY AND DIVERSIFICATION

The oldest representatives of the family are known from the late middle Eocene of South Asia (Pilgrim, 1928; Colbert, 1938b; Ducrocq et al., 2000; Tsubamoto et al., 2002) and North America (Holroyd, 2002). At that time, the three subfamilies were already present in the fossil record and very likely had an Asian origin (in agreement with Pilgrim, 1941; Xu, 1962; Coombs and Coombs, 1977b; Ducrocq, 1994b; Ducrocq et al., 1997a; Beard, 1998; Ducrocq, 1999; Ducrocq et al., 2000). In addition, the family was already diverse at its first appearance, suggesting that the anthracotheriids had already differentiated from other artiodactyls by the beginning of the middle Eocene (Fig. 7.5). Helohyids

have been considered to be the sister group of anthracotheres (Coombs and Coombs, 1977b). This hypothesis, however, needs to be revised because characters used to ally these taxa define the selenodont artiodactyls (Kron and Manning, 1998; Stucky, 1998).

The biogeographic history of Anthracotheriidae was characterized very early by several dispersal events. *Heptacodon*, the oldest representative of the family in North America (and maybe worldwide; Hanson, 1996; Lucas et al., 2004), probably originated from an Asian ancestor that might have reached the New World through Beringia, as several other contemporary groups of mammals did (Beard and Wang, 1991; Ducrocq, 1994b; Sanmartin et al., 2001). *Bothriogenys* was likely the first anthracotheriid to enter Africa, thus providing evidence of a continental bridge

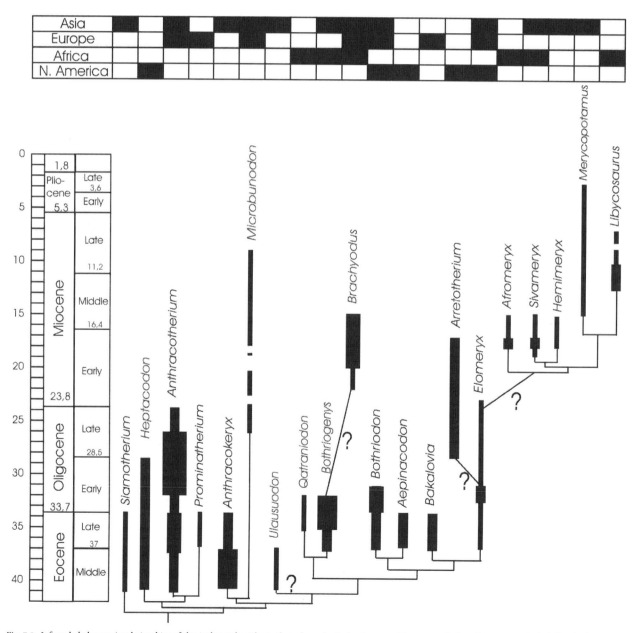

Fig. 7.5. Inferred phylogenetic relationships of the Anthracotheriidae in their chronological and geographic context. *Anthracohyus* was not included in the figure because dental material attributed to that genus is very scarce.

between Asia and Africa at least since the late Eocene (Ducrocq, 1997; Marivaux et al., 2000).

The Eocene–Oligocene boundary was characterized by a major faunal turnover, mainly in Europe (Stehlin, 1910b). However, *Anthracotherium* and *Elomeryx,* among other taxa, migrated to western Europe from Asia during the late Eocene (Sudre, 1978; Fejfar, 1987; Ducrocq, 1995). A second dispersal event into Europe involved *Bothriodon* at the basal Oligocene (Hooker, 1992), whereas this genus entered North America from Asia earlier, during the late Eocene (Fig. 7.5). The last dispersal event from Asia to North America was that of *Elomeryx* at the basal Oligocene. Following that, anthracotheres no longer reached the New World, and North American anthracotheres became isolated from the Old World and evolved from the Eocene and basal Oligocene stock.

Contemporaneously in Africa, the first Eocene immigrants led to an African bothriodontine lineage that evolved there until the basal Miocene, when Africa was reconnected to Eurasia. This lineage gave rise to *Brachyodus,* which entered Europe and Asia (China and Thailand) during the Miocene (Liu and Zhang, 1993; Ducrocq et al., 2003).

The genus *Microbunodon* testifies to exchanges between Europe and Asia during the late Oligocene, although it did not occur in the Asian fossil record until the middle Miocene (Lihoreau et al., 2004a). Likewise, *Elomeryx,* which evolved independently in Asia and Europe, showed limited exchanges between those areas during the late Oligocene. Asian species of *Elomeryx* gave rise to an advanced bothriodontine lineage on the Indian Subcontinent.

Anthracotheres disappeared from Europe and North America at the beginning of the Miocene. They survived only in Asia, where they constituted a pool from which Africa was repopulated: (1) during the Proboscidean Datum Event (around 18 million years), *Sivameryx* and *Afromeryx* entered Africa and then became extinct along with the last African representatives of *Brachyodus;* and (2) around 15–13 million years, a *Merycopotamus*-like stock entered Africa and gave rise to *Libycosaurus.* During one of these dispersal events, a bothriodontine likely settled in Africa and evolved into the hippopotamid family (see Boisserie et al., 2005a, 2005b). The Anthracotheriidae disappeared from Africa at the very end of the Miocene. During the middle to late Miocene in Asia, anthracotheres became increasingly specialized, depending on environmental shifts, until they eventually disappeared at the end of the Pliocene on the Indian Subcontinent, around 2.5 million years (Steensma and Hussain, 1992), when the family became extinct.

ECOLOGICAL ADAPTATIONS

For a long time, anthracotheres were believed to have been mainly semiaquatic, hippo-like, with heavily built forms, because the first described specimens (notably *Anthracotherium* itself) were large, massive species, recovered from lignite deposits indicating a swampy habitat (e.g., Cuvier, 1822; Falconer and Cautley, 1836; Rütimeyer, 1857).

We still understand very little about the locomotor adaptations of anthracotheriids, as there have been no functional studies of their postcrania. Although little is still known about the adaptations of the Paleogene forms, the quality and diversity of new fossil remains and recent techniques of investigation (dental micro- and mesowear, paleoenvironmental analyses) have demonstrated that anthracotheres were adapted to a wide diversity of environments and diets (Lihoreau, 2003). Small forms such as *Siamotherium, Microbunodon,* and *Anthracokeryx* (about 20–25 kg) very likely lived in forested environments. Micro- and mesowear analyses conducted on the teeth of *Microbunodon* have revealed that this genus probably had a folivorous–frugivorous diet (Lihoreau, 2003). *Microbunodon* was a lightly built animal with slender limbs, a short snout, and was characterized by marked sexual dimorphism in its bladelike upper canines (Fig. 7.6A). In its diet and mode of life, *Microbunodon* can be compared with extant Moschidae and Tragulidae. *Siamotherium,* from Southeast Asia, was morphologically very close to *Microbunodon* and probably had a very similar ecology.

On the other hand, several medium to large forms (with a body weight about 150 to 1,000 kg), such as *Arretotherium, Merycopotamus,* and *Libycosaurus,* were adapted to an aquatic habitat and were likely gregarious, reminiscent of the aquatic part of the hippopotamus lifestyle (Fig. 7.6B). They had many features for standing in water: orbits elevated above the cranial roof, external nares on the top of the snout, and highly placed auditory meatus. They also had massive limbs, high bone density, and strong sexual dimorphism. Some other taxa (*Bothriodon, Aepinacodon*) were large animals. They were characterized by extremely long skulls that were narrow anteriorly, with diastemata in their tooth rows, markedly spatulate incisors of different sizes, deep maxillae, shallow mandibles, and strongly selenodont teeth (Fig. 7.6C). Janis (1995) proposed that these features suggest a dominantly folivorous diet. However, the most recent known species of *Merycopotamus* shows a very deep lower jaw with a wide symphysis, incisors of similar size arranged in a straight row, and a low maxilla with a strong facial tuberosity. This morphology actually characterizes grazers (Janis, 1995), and that diet has been confirmed by dental microwear studies (Lihoreau, 2003). Some anthracotheriid taxa (e.g., *Elomeryx*) do not show marked morphological specializations, and their mode of life and diet are still unclear. They may represent generalist species and long-term survivors from which new species could readapt and colonize other habitats.

POSSIBLE CAUSES OF ANTHRACOTHERE EXTINCTION

Anthracotheres had diachronic extinctions on various continents: first in Europe in the early Miocene, then in North America at the end of the early Miocene, in Africa at the end of the Miocene, and finally, in Asia during the late Pliocene (Fig. 7.5). Several authors have suggested for the African and

Fig. 7.6. Lateral views of anthracotheriid skulls showing the main known ecological adaptations. A, *Microbunodon*; B, *Libycosaurus*; C, *Aepinacodon*. Scale bars equal 10 cm.

Asian events an extinction of anthracotheres caused by competition with hippos, but this hypothesis can be rejected because of the contemporaneous occurrence of both groups in Africa (for at least 8 million years) and in Asia (for at least 4 million years). Two groups of mammals competing for an ecological niche would not have lived together for so long.

During their history, anthracotheres dispersed several times, which forced them to constantly adapt to changing environments. Ongoing migrations of rather generalized forms between different geographic zones might have been a crucial factor in the survival and readaptation of the family. On the other hand, the last representatives of Anthracotheriidae in North America, Africa, and Asia were very specialized to an aquatic way of life and were isolated on distinct continents. Moreover, their distributions were locally constrained by the extension of the hydrographic system they lived in. Aridification (desertification) of the environment might have led to a shrinking of their habitats. Indeed, each anthracotheriid final extinction can be correlated with a major environmental shift, such as the early emingfordian–late Hemingfordian transition (drying and cooling trend) in North America (Tedford et al., 1987), the Messinian event (desiccation of the Mediterranean Sea) in the Mediterranean basin (Adams et al., 1977), and aridification in Asia (Patnaik, 2003) correlated with an Arctic icecap appearance (An et al., 2001). Consequently, we can infer that a low generic diversity at the end of the Miocene, marked specialization to constrained environments, and geographic isolation are the most likely causes for the extinction of anthracotheres (*sensu stricto*). However, the anthracothere–hippo sister group relationships demonstrated by Boisserie et al. (2005a, 2005b) and Boisserie and Lihoreau (2006) show that anthracotheres would not be extinct but still living in Africa. Consequently, the extinction of anthracotheres would only be on the regional level (see Boisserie, this volume). This would also imply that the subfamily Bothriodontinae should be considered paraphyletic with respect to hippos.

CONCLUSION

The anthracotheres were characterized by their ecological diversity, their distribution, and their persistence through the Tertiary, and they thus represent a key group that contributes to a better understanding of mammal communities. Moreover, they can be regarded as a pioneer group because of their ability to colonize new areas very early in large-scale main dispersal events. An attempt to study the ecological adaptations of some anthracotheres (morphofunctional and microwear analyses, see Lihoreau, 2003) demonstrated that they were diversified in their habitat and diet. Morphological specializations linked to various ecological niches suggest that anthracotheres could potentially become a valuable environmental marker for most of Tertiary deposits.

Nevertheless, the overview of the family presented here suggests the need for revisions of several genera and for a more thorough investigation of the origins of anthracotheres. In addition, the lack of data in the Paleogene North America, Asia, and Africa, points to a period during which the evolutionary history of anthracotheriids is still poorly known. Efforts to investigate the Oligocene deposits in these areas will probably lead to the recovery of taxa that will document important phases of exchanges and diversifications for all subfamilies. Filling the gaps of the anthracotheriid fossil record could also shed light on the phylogenetic history of the yet unknown group of anthracotheres from which hippopotamids might have originated.

APPENDIX 7.1: DESCRIPTION OF CHARACTERS USED IN THE CLADISTIC ANALYSIS

Anterior Teeth

Character 1. Number of lower incisors: three (0); variable from three to two (1); two (2). The number of permanent incisors is used without referring to tooth identification (Dineur, 1981). State 1 corresponds to a variable number of lower incisors within a fossil population (Black, 1972; Gaziry, 1987b).

Character 2. Number of upper incisors: three of equal size (0); three with a reduced I3, which represents 60% or even less of the diameter of the others (1); two (2).

Characters 1 and 2 are coded differently for the analyzed taxa (Table 7.1) and thus are not linked.

Character 3. Lower incisor morphology: not caniniform (0); at least one caniniform lower incisor (1). We consider that a strong incisor in which the crown is squeezed at its apex, like a fang, is caniniform.

Character 4. Relative dimensions of lower incisors: all of equal size (0); i2 larger (1); i3 larger (2). We consider that one incisor is larger when its crown is at least twice as high as the others.

Character 5. Wear on lower canine: distal wear facet caused by the contact with upper canine (0); mesial wear facet caused by the contact with I3 (1).

Character 6. Upper canine morphology: strong with subcircular cross section (0); strong and laterally compressed (1); premolariform (2). A premolariform canine has its unworn crown as high as the unworn premolar crown. In that case the canine displays a subrectangular occlusal outline with a mesiodistal maximal length.

Character 7. Lower canine in males: premolariform (0); large (1); ever-growing (2). These character states have only been observed on male specimens because lower canines are sexually dimorphic in size in many anthracotheriid species (e.g., Hellmund, 1991).

Character 8. Lower canine cross section at cervix: subcircular (0); elliptical with a rounded mesial margin and a distal keel (1); elliptical with a mesial and a distal crest (2); elliptical with a concave buccal margin and a distal keel (3).

Cheek Teeth

Character 9. Accessory cusps on the mesial crest of lower premolars: none (0); only one (1); several (2).

Character 10. Presence of five upper premolars: no (0); yes (1). This character does not correspond to the number of premolariform teeth but to the number of teeth present in adult (functional M3) between the upper canine and the first upper molar (Lihoreau et al., 2006).

Character 11. Distolabial crests of upper premolars: simple (0); with a maximum of two accessory cusps (1); with more than two accessory cusps (2). This character applies to the entire premolar row and not to a particular premolar because the number of accessory cusps varies with the position of the premolar in the dental row. The number of ac-

Table 7.1. Data matrix of anthracotheres used in the cladistic analysis

	12345	11111 67890	11112 12345	22222 67890	22223 12345	33333 67890	33334 12345	44444 67890	44445 12345	5 67890	1
S. krabiense	000??	00?00	00000	00000	00000	00000	00000	00000	000?0	0?00?	0
M. minimum	00001	10100	00000	10000	00000	11000	01110	01111	00001	00000	0
M. silistrensis	0?001	?0100	000?0	10000	00000	11000	01110	01111	0????	?????	?
A. magnum	00000	01000	0000?	00111	00001	00000	00001	11001	0000?	???0?	0
A. monsvialense	00000	01000	0000?	00111	00001	00000	00001	1100?	0000?	??00?	0
A. chaimanei	0?00?	0?000	00?0?	00111	00001	00000	000?1	0?00?	00000	?00??	0
A. pangan	???0?	01?0?	00000	00111	00001	00000	???0?	00???	0????	?????	?
A. thailandicus	0?001	?0100	0000?	10001	00000	11000	01110	00011	0?00?	?0???	0
A. ulnifer	00001	10100	00000	10101	00000	00000	00111	00100	00000	00000	0
H. occidentalis	0?000	01400	00000	00011	00100	00000	?0000	0100?	00000	0000?	0
B. fraasi	00??0	20200	0000?	11000	00100	00000	00010	00120	00?00	?000?	1
B. aequatorialis	22120	20200	0001?	11000	00100	11100	00010	01120	00020	00000	1
B. onoideus	22120	20200	0001?	11000	00100	11100	00010	01120	00020	??000	1
A. americanus	00010	20200	00000	12100	00100	11010	00111	01120	00000	0000?	0
B. velaunus	00010	20200	00000	12100	00100	11010	00111	01120	000?0	00000	0
E. crispus	00000	01300	01000	12100	00110	11010	00011	00210	00000	00000	1
E. armatus	00000	01300	00000	12100	00110	11010	00011	00210	00000	00000	1
E. borbonicus	00000	01310	01000	12100	00110	11010	00011	00210	0000?	?000?	?
A. acridens	00010	01300	0000?	12000	00110	11010	00011	00210	00001	?1?00	1
A. zelteni	10000	01310	1101?	12100	00010	11010	00010	00211	00?11	?1?0?	1
S. palaeindicus	0?000	?1310	11110	12100	10010	11010	00010	00211	1????	?????	?
H. blandfordi	??0??	?132?	?11??	?2200	10010	11010	00010	00210	1????	?????	?
M. pusillus	00000	01310	11110	12000	10010	11110	00010	00200	00110	11101	1
M. dissimilis	00000	02310	11110	13000	11010	11111	20010	00202	00111	11101	1
M. medioximus	00000	01310	11110	13000	11010	11110	00010	00202	00111	11101	1
L. anisae	1?000	01321	21111	12000	10010	11111	10010	00211	011??	??11?	1
L. petrocchii	11000	01321	21112	12000	11010	11111	10010	00211	01121	11111	1

Note: "?" indicates missing data. All character states are treated as unordered and of equal weight.

cessory cusps is always observed on the premolar where it is maximal.

Character 12. Presence of a lingual accessory cusp on p4 (Fig. 7.3B): no (0); yes (1). An accessory lingual cusp is defined as a tubercle issued from the cingulum or from a cristid of the main cusp that is taller than half its height, with a distinct apex.

Character 13. p1 roots: one (0); two (1). The p1 is considered to have two roots when at least half of the root height is divided.

Character 14. Number of mesial crests on P1–P3 (Fig. 7.3C): one (0); two (1).

Character 15. Number of P4 roots: three (0); two (1); one (2). All observed P4 possess three pulp canals, and states 1 and 2 indicate root fusion(s), not disappearance.

Character 16. Accessory cusp on distolingual margin of P3: one (0); none (1).

Character 17. Upper molar mesostyle: simple (0); V-shaped and invaded by the transverse valley (1); looplike (2); divided in two (3).

Character 18. Number of postprotocristae: one (0); two (1).

Character 19. Accessory cusp on upper molar mesial cingulum: no (0); yes (1). Some anthracotheriid species have a slightly inflated mesial cingulum in front of the protocone. These are not considered as accessory cusps because they do not possess a differentiated basis with a distinct apex and are not in front of the paraconule where the cingulum is less salient.

Character 20. Number of cristules issued from the metaconule: two (0); three (1). The development of a third metacristule might be linked to the position of the tooth in the molar row (preferentially on the M3). It is considered (taken into account) only if it joins the summit and the basis of the metaconule on each upper molar.

Character 21. Preprotocristids and prehypocristids on lower molars: do not reach the lingual margin of the tooth (0); reach the lingual margin (1).

Character 22. Hypoconulid on m3: looplike (0); single cusp (1).

Character 23. Postentocristid on lower molars: does not reach the posthypocristid and leaves the longitudinal valley open (0); reaches the posthypocristid and closes the longitudinal valley (1).

Character 24. Dimensions of lingual and labial cusps on lower molars: equal (0); different (1). State 1 is defined by labial cusps twice as large at their basis as the lingual cusps.

Character 25. Entoconulid on m3: absent (0); present (1).

Character 26. Number of cristids issued from the hypoconid: three (0); two (1).

Character 27. Position of the preentocristid on lower molars: reaches the hypoconid summit (0); reaches the prehypocristid (1).

Character 28. Premetacristid on lower molars: present (0); absent (1).

Character 29. Mesial part of looplike hypoconulid: open (0); pinched (1).

Character 30. Entoconid fold (see Lihoreau et al., 2004b) on lower molars: absent (0); present (1).

Mandible

Character 31. Ventral vascular groove on mandible: slightly marked (0); absent (1); strongly marked (2). When the ventral border is straight we considered that there is no vascular groove. A deep vascular groove is characterized by a sudden change in the curvature of the mandibular notch.

Character 32. Morphology of mandibular symphysis cross section: U-shaped (0); V-shaped (1).

Character 33. Transverse constriction of mandible at Cp1 diastema: no (0); yes (1).

Character 34. Cp1 diastema: absent (0); present (1).

Character 35. p1–p2 diastema: absent (0); present (1).

Character 36. Lateral mandibular tuberosity: absent (0); present (1).

Character 37. Dentary bone fusion at the symphysis: no (0); yes (1). This fusion is frequent in older individuals in some species. There is fusion when it can be observed in all the available specimens of one species.

Character 38. Morphology of the symphysis in sagittal section: elliptic (0); dorsally concave (1); ventrally concave (2).

Character 39. Maximal thickness of the symphysis in sagittal section: in the middle part (0); in the anterior part (1); in the posterior part (2).

Character 40. Number and position of main external mandibular foramina: only one foramen, below the anterior part of the premolar row (0); two foramina, one below the anterior part of the premolar row, the other below its posterior part (1); one foramen, below the posterior part of the premolar row (2). This character corresponds to the position of the largest external mandibular foramen considered as the main one. When two foramina of equal size are present, they are taken into account (State 1). The position of the main external mandibular foramen varies slightly within a population; in that case, the state of the character corresponds to a mean position within a species.

Character 41. Tuberosity on the dorsal border of the mandible at c–p1 diastema: no (0); yes (1).

Skull

Character 42. Palatine depression between the canines: no (0); yes (1).

Character 43. Canine fossa: short (0); long (1). This character is estimated from the length of the canine fossa (from the canine protuberance to the infraorbital foramen) relative to the cranium length: if the fossa extends more than 50% of the muzzle length (from the anterior border of the orbit to the anterior border of the maxillary), it is considered long. Among all the studied species, two groups are observed: those with a small canine fossa (<30%) and those with a large canine fossa (>70%).

Character 44. Aperture of the main palatine foramen: between M3 and P3 (0); between P2 and P1 (1); between P1 and C (2). The main palatine foramen is anterior and larger than the other foramina, which are lingually lined with the dental row. Its position does not vary within a species.

Character 45. Morphology of the frontonasal suture: V-shaped (0); rounded or straight (1).

Character 46. Lachrymal extension: separated from the nasal by the frontal (0); in contact with the nasal (1).

Character 47. Supraorbital foramina on the frontal: one (0); several (1).

Character 48. Facial crest: horizontal (0); oblique (1).

Character 49. Anterior border of premaxillary in lateral view: concave (0); convex (1).

Character 50. Postglenoid foramen position: posterior to the styloid process of the tympanic bulla (0); anterior to the styloid process of the tympanic bulla (1).

Character 51. Opening of internal choanes: at M3 (0); behind M3 (1).

ACKNOWLEDGMENTS

8

Family Hippopotamidae

THE ORIGIN OF MOST FOSSIL and extant artiodactyl families is found in the Paleogene of Eurasia and America. Africa, in contrast, harbored only some later developments, though sometimes far from negligible, such as the African bovid late Neogene radiation. There is, however, a remarkable exception to that pattern: the Hippopotamidae. For this strictly Neogene family, Africa is at the same time its place of emergence and its main center of evolution. Hippopotamids occurred first in middle Miocene deposits from Kenya (Pickford, 1983; Behrensmeyer et al., 2002), but the first well-known species, already exhibiting the morphology characteristic of their family (Fig. 8.1), were recorded in deposits younger than 7.5 Ma (Weston, 2003; Boisserie et al., 2005c). Hippopotamid unique morphology complicated the recognition of their affinities within artiodactyls, but their subsequent evolution also offers a rare record of early diversification within an artiodactyl distinct anatomical scheme.

Another distinctive trait of hippopotamids is ecological: they are the only extant large herbivorous amphibious mammals, a condition only seen in some Anthracotheriidae and very few other fossil mammals. Among the two extant hippopotamid species, *Choeropsis liberiensis* from western Africa is unfortunately secretive, and its habits are poorly known. The larger *Hippopotamus amphibius* from Sub-Saharan Africa is much more frequent, and its ecology has been extensively studied. This species has a strong impact on hydrography, on chemical and organic features of water, and on composition of periaquatic vegetation (Verheyen, 1954; Laws, 1968; Field, 1970; Lock, 1972; Mackie, 1976; Kingdon, 1979; Ogen-Odoi and Dilworth, 1987; Eltringham, 1993a; McCarthy et al., 1998; Eltringham, 1999; Wolanski and Gereta, 1999; Deocampo, 2002; Grey and Harper, 2002). This impact on both aquatic and terrestrial communities should by itself motivate extensive studies of fossil hippopotamids,

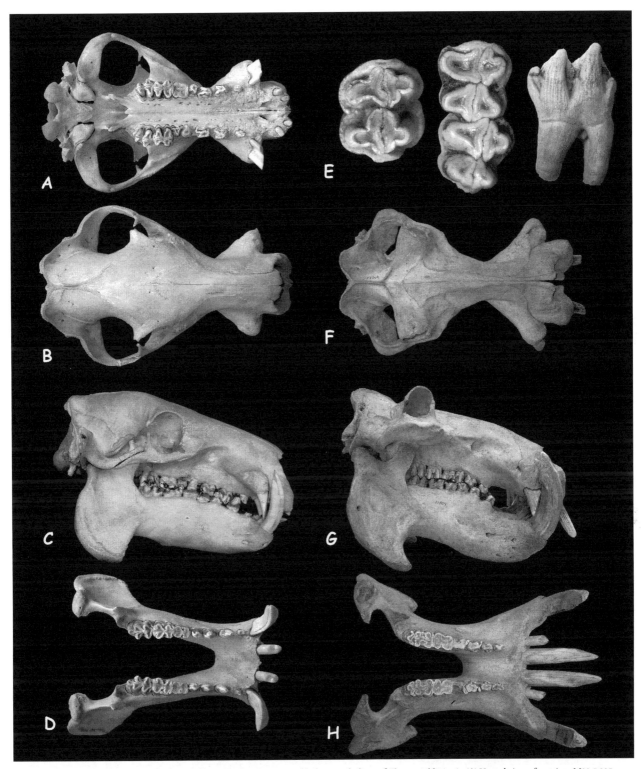

Fig. 8.1. Craniodental morphology of extant Hippopotamidae. Craniomandibular morphology of *Choeropsis liberiensis.* (A) Ventral view of cranium M09.5.005. A, preserved at the Laboratoire de Géobiologie, UMR CNRS 6046, Poitiers (LGBPH). (B) Dorsal view of cranium M09.5.005.A. (C) Right lateral view of skull M09.5.005. (D) Dorsal view of mandible M09.5.005.B. Dental morphology, (E) from left to right, is occlusal view of left M2 (LGBPH) of *Hippopotamus amphibius;* occlusal view of left M2–M3 (on M09.5.005.A) of *Choeropsis liberiensis;* lingual view of left m2 (LGBPH) of *Hippopotamus amphibius.* Craniomandibular morphology of *Hippopotamus amphibius:* (F) dorsal view of cranium MVZ 77804 preserved at the Museum of Vertebrate Zoology, Berkeley; (G) right lateral view of skull MVZ 77804; (H) dorsal view of mandible MVZ 124268. All specimens reduced to sizes facilitating comparisons (basal lengths of M09.5.005.A and MVZ 77804: 32 cm and 65 cm, respectively).

potential sources of data on wet habitats, periaquatic vegetation, and interbasin hydrographic connections (Jablonski, 2003; Boisserie et al., 2005d).

However, despite abundant fossil records in Africa and southern Asia, hippopotamid evolution has received little attention. Beyond size-related collection biases and other circumstances, the lack of changes in hippopotamid cheek teeth (Coryndon, 1977; Boisserie, 2005) probably accounted for most of this weak interest. Consequently, most species were attributed to the genus *Hexaprotodon* without actual phylogenetic significance (as shown by Stuenes, 1989; Harris, 1991; Weston, 2003; Boisserie, 2005) until a recent phylogenetic revision of Hippopotamidae was attempted (Boisserie, 2005) and resulted in taxonomic adjustments. These are discussed here and summarized in Table 8.1 and in the appendices (Appendix 8.1 is a proposed classification of the family; Appendix 8.2 lists generic diagnoses). Kahlke (1990) delineated a broad paleobiogeographic picture of the family, largely used to frame this chapter.

The reader may refer to Coryndon (1978a) and Faure (1983) for detailed historical accounts of "hippo evolutionary studies" since the eighteenth century. Along with these works, I would like to indicate further critical contributions to this field, with the proviso that, of course, such a list cannot be exhaustive: Hooijer (1950), Coryndon (1977), Gèze (1980), Pickford (1983), Stuenes (1989), Kahlke (1990), Harris (1991), Mazza (1995), Harrison (1997), Weston (1997, 2003), and Boisserie (2005).

A CONTROVERSIAL EMERGENCE

The Hippopotamids and the Paraphyly of Artiodactyla

Paleontologists have been arguing for 150 years about the closest relatives of Hippopotamidae, generally supporting either anthracotheriids or suoids (suids, tayassuids, and allies). This debate, at first glance rather anecdotic for nonspecialists, is in fact the centerpiece to a quite important issue: the relationships between cetaceans and artiodactyls (Gatesy and O'Leary, 2001). Among critical advances on this question: (1) the most recent reviews of fossil evidence support an emergence of Hippopotamidae within Anthracotheriidae (Gentry and Hooker, 1988; Boisserie et al., 2005a, 2005b; see also Fig. 8.4); (2) some morphological data and analyses now favor close relationships between Hippopotamidae and Cetacea (Gingerich et al., 2001; Naylor and Adams, 2001; Geisler and Uhen, 2003; Boisserie et al., 2005b); (3) molecular results frequently indicate closer affinities of cetancodonts (hippopotamids plus cetaceans; Arnason et al., 2000) with ruminants than with suoids (Gatesy et al., 1996; Gatesy, 1997, 1998; Nikaido et al., 1999; Amrine-Madsen et al., 2003; Arnason et al., 2004). This implies that Anthracotheriidae Leidy, 1869 and Artiodactyla Owen, 1848 are paraphyletic taxa, whereas Suiformes Simpson, 1945 is polyphyletic (see Appendix 8.1).

Despite those advances, deriving the hippopotamid triangular to trefoliate cusp pattern (Fig. 8.1E) from the apparently more complex anthracotheriid selenodont pattern (Colbert, 1935a, 1935b; Boisserie et al., 2005a, 2005b) still challenges classic views on tooth evolution. Teeth are often seen as reliable elements for reconstructing phylogenies, and selenodont dental crest patterns are considered much more derived than what is observed in hippopotamid cheek teeth. Tooth development genetics may indicate a new way to examine this question. Kangas et al. (2004) showed how minor genotype changes induce major alterations of mammal cheek tooth crest development. In mice and humans, those perturbations are also linked to depletion of hair and sebaceaous glands (Mikkola and Thesleff, 2003). Interestingly, hippos retain little hair and no sebaceous glands, a supposed aquatic adaptation (Luck and Wright, 1964). This track could be explored by looking for correlations between dental changes and changes in molecular factors of epithelial growth in artiodactyls.

Kenyapotaminae: Dawn of Hippopotamidae?

Another dilemma regarding hippo origins is the exact nature of the oldest hippopotamids. Mostly known by isolated teeth, with few craniodental fragments (Fig. 8.2A) or postcranial elements, their general shape and paleoecology are for now impossible to portray. Most of this material, dated between 10 Ma and 8.5 Ma and attributed to *Kenypotamus coryndoni*, comes from central Kenya (Pickford, 1983; Nakaya et al., 1984, 1986), Tunisia (Pickford, 1990), and maybe Ethiopia (Geraads et al., 2002). Older remains (from 16 Ma to 14 Ma) were found in Kenya (Pickford, 1983; Behrensmeyer et al., 2002) and designated *Kenyapotamus ternani*.

Kenyapotaminae were close in size to the extant *Choeropsis liberiensis* (Pickford, 1983). Their particularly brachyodont molars exhibit a cusp pattern reminiscent of later hippopotamids as well as features reminiscent of other "suiformes." Some fragmentary canine and incisor remains and postcranials can be readily attributed to hippopotamids. However, in the absence of associated mandibular symphyseal or cranial remains, only a thorough comparison of *Kenyapotamus* with a wide range of bunodont to bunoselenodont artiodactyls will definitely confirm its initial attribution and will make it possible to decipher its actual relationships with later hippopotamids.

The Sudden Advent of Hippopotaminae

Hippopotamids with anatomically modern cheek teeth (or Hippopotaminae, excluding *Kenyapotamus*) burst into the fossil record in the uppermost Miocene and number among the most frequently collected mammals. Although modalities of this sudden eruption and dominance in African wet ecosystems remain unclear, it must be noted that they correlate well in time and in estimated diets with

Table 8.1. Spatiotemporal occurrence, taxonomic authorities, and synonymy of genera and species in Hippopotamidae

Genera	Species and main synonyms	Key[1]	Temporal range[2]	Areas
Kenyapotamus	*coryndoni* Pickford, 1983		l. Miocene	Baringo Basin, central Tunisia, Afar (?)
Pickford, 1983	*ternani* Pickford, 1983		m. Miocene	Kenyan Rift Valley
Archaeopotamus[3]	*lothagamensis* (Weston, 2000)	***lot.***	l. Miocene	Turkana Basin
Boisserie, 2005	aff. *lothagamensis* Gentry (1999)	***aff. lot.***	l. Miocene	Abu Dhabi
	harvardi (Coryndon, 1977)	***har.***	l. Miocene to e. Pliocene	Turkana Basin
	aff. *harvardi* Kent (1942)	***aff. har.***	l. Pliocene	Rift Nyanza
Saotherium[3]	*mingoz* (Boisserie et al., 2003)	***min.***	e. Pliocene	Djourab
Boisserie, 2005	cf. *mingoz* Boisserie et al. (2003)	***cf. min.***	e. Pliocene	Djourab
Choeropsis[3]	*liberiensis* (Morton, 1844)	***lib.***	extant	Central western African coast
Leidy, 1853				
Hexaprotodon[3]	*sivalensis* Falconer and Cautley, 1836	***siv.***	l. Miocene – l. Pliocene (?)	Siwaliks, Yunnan (?)
Falconer and	*duboisi* (Hooijer, 1950)[4]	***dub.***	Pleistocene	Punjab
Cautley, 1836	*namadicus* (Falconer and Cautley, 1847)[4]	***nam.***	m. Pleistocene (?)	Narmada
	palaeindicus (Falconer and Cautley, 1847)[4]	***pal.***	m. Pleistocene	Narmada
	synahleyus (Deraniyagala, 1935)[4]	***syn.***	m. Pleistocene	Sri Lanka
	sivajavanicus (Hooijer, 1950)[4]	***sij.***	e. Pleistocene	Java
	soloensis (Hooijer, 1950)[4]	***sol.***	m. to l. Pleistocene	Java
	koenigswaldi (Hooijer, 1950)[4]	***koe.***	e. Pleistocene	Java
	iravaticus Falconer and Cautley, 1847	***ira.***	l. Pliocene (?)	Burmese Siwaliks
	= sp. indet. (?) Boisserie (2005)			
	bruneti Boisserie and White, 2004	***bru.***	l. Pliocene	Afar
	garyam Boisserie et al., 2005c	***gar.***	l. Miocene	Djourab
aff. Hippopotamus[3]	*protamphibius* Arambourg, 1944a	***pro.***	e. (?) Pliocene – e. Pleistocene	Turkana Basin
provisional attribution	= *shungurensis* (Gèze, 1985) (?)		m. Pliocene – e. Pleistocene	Turkana Basin
awaiting revision	cf. *protamphibius* Weston (2003)	***cf. pro.***	e. Pliocene	Turkana Basin
of the Plio-	*aethiopicus* Coryndon and Coppens, 1975	***aet.***	l. Pliocene to e. Pleistocene	Turkana Basin
Pleistocene material	*karumensis* (Coryndon, 1977)	***kar.***	l. Pliocene to e. Pleistocene	Turkana Basin
from Afar and	*afarensis* (Gèze, 1985)	***afa.***	m. Pliocene to l. Pliocene	Afar
Turkana following	= *Trilobophorus afarensis* Harrison			
Boisserie (2005)	(1997), Boisserie (2005)			
	coryndoni (Gèze, 1985)	***cor.***	m. Pliocene to l. Pliocene	Afar
	dulu (Boisserie, 2004)	***dul.***	e. Pliocene	Afar
Hippopotamus	*amphibius* Linné, 1758	***amp.***	e. (?) Pleistocene to extant	Africa, western Europe and Asia
Linné, 1758	= *incognitus* Faure, 1984[5]		m. to l. Pleistocene	Western Europe
	kaisensis Hopwood, 1926	***kai.***	e. (?) to l. Pliocene	Western Rift, Turkana Basin (?)
	gorgops Dietrich, 1928	***gor.***	e. Pleistocene to m. Pleistocene	Africa, Jordan Valley (?)
	= *behemoth* Faure, 1986 (?)	***beh.***	e. Pleistocene	Jordan Valley
	= *sirensis* Pomel, 1890 (?)		e. (?) to m. Pleistocene	Northwestern Maghreb
	antiquus[5]	***ant.***	e. to m. Pleistocene	Western and southeastern Europe
	= *major* Desmarest, 1822		e. Pleistocene	
	= *georgicus* Vekua, 1959 (?)	***geo.***	e. Pleistocene	Caucasus
	tiberinus Mazza, 1991[5]	***tib.***	e. to l. Pleistocene	Western Europe
	pentlandi Von Meyer, 1832	***pen.***	m. to l. Pleistocene	Sicily, Malta
	melitensis Major, 1902	***mel.***	l. (?) Pleistocene	Malta
	creutzburgi Boekschoten and Sondaar, 1966	***cre.***	e. to m. Pleistocene	Crete
	lemerlei Grandidier in Milne-Edward, 1868	***lem.***	Holocene	Madagascar
	madagascariensis Guldberg, 1883	***mad.***	Holocene	Madagascar
	laloumena Faure and Guérin, 1990	***lal.***	Holocene	Madagascar
Phanourios	*minutus* (Cuvier, 1824)	***mit.***	l. Pleistocene–Holocene	Cyprus
Boekschoten and	= *minor* (Desmarest, 1822)			
Sondaar, 1972				

continued

Table 8.1. Continued

Genera	Species and main synonyms	Key[1]	Temporal range[2]	Areas
Incertae sedis[3]	*hipponensis* (Gaudry, 1876)	**hip.**	e. Pliocene (?)	Orania
provisionally noted:	*imagunculus* (Hopwood, 1926)	**ima.**	l. Pliocene	Kaiso, western Rift
Hexaprotodon ?	cf. *imagunculus* – Pavlakis (1990),	**ima.**	e. to l. Pliocene	Western Rift
following Boisserie	Boisserie (2005)			
(2005)	*pantanellii* (Joleaud, 1920)	**pan.**	l. Miocene	Toscana
	siculus (Hooijer, 1946)	**sic.**	l. Miocene	Sicily
	crusafonti (Aguirre, 1963)	**cru.**	l. Miocene	Western Mediterranean area
	= *primaevus* Crusafont et al., 1964			
	sahabiensis Gaziry, 1987	**sah.**	l. Miocene	Sirt Basin

[1]Key refers to abbreviations used in caption to Fig. 8.4.
[2]l. = late; e. = early; m. = middle; (?) = uncertainty on synonymy or occurrence.
[3]Genus previously identified as *Hexaprotodon* following Coryndon's (1977) revision.
[4]Considered as subspecies of *Hex. sivalensis* by Hooijer (1950).
[5]Possibly a subspecies of *Hip. amphibius* following Kahlke (1997, 2001).

the main expansion of C_4 plants (mostly grass) on this continent (Cerling et al., 2003; Boisserie et al., 2005d).

Hippopotaminae exhibit the following unique character association (Figs. 8.1–8.3): prolonged to permanently growing incisors and canines; procumbent lower incisors; tetracuspidate cheek teeth; triangular to trefoliate molar wear pattern; wide and tall muzzle; intercanine position of main palatal foramina; large temporal fossae; poorly compressed postglenoid area of the braincase; wide and subvertical occipital plate; thick and wide mandibular symphysis; shallow mandibular constriction; ventrally extended angular process; digits II and V slightly reduced. They also share evolutionary trends occurring independently in most lineages: canine process expansion; angular process expansion; muzzle elongation; orbit and sagittal crest elevation; incisor differentiation and losses; premolar row shortening and simplification; increase of molar crown height.

The following parts summarize the evolutionary history of the Hippopotaminae. The reader can refer to Fig. 8.4, Table 8.1, and the Appendices for taxonomic authorities, spatiotemporal distributions, and phylogenetic relationships.

MIO-PLIOCENE DIVERSIFICATION IN AFRICA

After 7.5 Ma, African hippopotamines diversified into mutually exclusive lineages, resulting in locally disparate evolution. This situation probably resulted from a combination of hippopotamid semiaquatic habits and a likely fragmentation of watersheds. Consequently, Hippopotamidae achieved the greatest diversity and disparity in Africa.

Genus *Choeropsis*

This genus includes only one species, the extant *C. liberiensis* (Fig. 8.1A–E), one of the smallest hippopotamids (maximal weight 275 kg) (Kingdon, 1997). Secretive and solitary, it inhabits Atlantic coastal rainforests of western Africa (Eltringham, 1993b, 1999). Morphologically, it is the only

hippopotamid displaying orbits clearly below the cranial roof and a down-turned sagittal crest (Fig. 8.1C). It also exhibits a narrow but short mandibular symphysis, bearing only two incisors (Fig. 8.1D), a developed palatine nasal spine (Fig. 8.1A), and other unique features (Boisserie, 2005). Although *Choeropsis* is unknown in the fossil record, this unique mosaic of characters suggests that it differentiated a long time ago (Harrison, 1997; Boisserie, 2005).

Choeropsis also differs from *Hip. amphibius* by primitive features (e.g., lack of lachrymal-nasal contact) and by more terrestrial habits and a more diverse diet (Kingdon, 1979; Eltringham, 1999). Hence, its behavior was proposed as a model for fossil species sharing those primitive features (Coryndon, 1977, 1978a). However, a radically different interpretation may question the validity of this model. Like its extant relative, *Choeropsis* presents physiological adaptations to aquatic environments: a sun-sensitive skin and ability to seal nasal and auditory external apertures while diving. Its more terrestrial habits and some of its purported archaic cranial features could therefore be secondary adaptations to the shaded rainforest. For example, the low position of the orbits and sagittal crest on the cranium of *Choeropsis* could be interpreted as adaptations to better penetration of the rainforest dense vegetation. This habitat was probably marginal for hippopotamids, as most fossil hippopotamuses were accumulated in more open environments.

Genus *Saotherium*

The lower Pliocene of Chad recently produced hippopotamids characterized by (Figs. 8.2D and 8.3A) a strong angle of the cranial roof above the orbit rostral edges, an elongated braincase with a rounded transverse section, and a low occipital plate (Boisserie et al., 2003; Boisserie, 2005). This new genus *Saotherium* includes *S. mingoz* from Kollé (between 5 Ma and 4 Ma; Brunet et al., 1998) and *S.* cf. *mingoz* from Kossom Bougoudi (ca. 5 Ma; Brunet and M.P.F.T., 2000) The former differs from the latter by a smaller size and relatively shorter premolar rows.

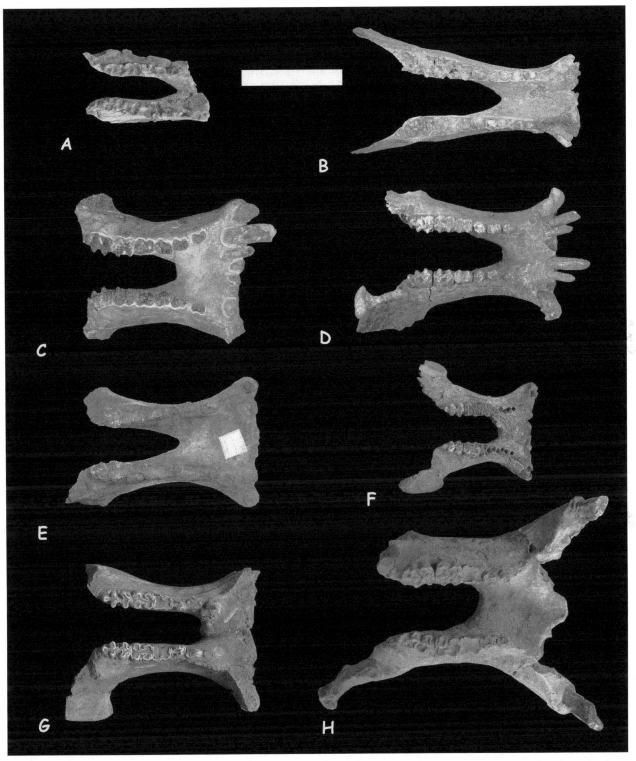

Fig. 8.2. Mandibles (dorsal views) of fossil Hippopotamidae: (A) KNM SH 15857, *Kenyapotamus coryndoni,* Samburu Hills, Kenya (Nakaya et al., 1986), preserved at the National Museums of Kenya (NMK), Nairobi; (B) M49464, *Archaepotamus* aff. *lothagamensis,* Abu Dhabi, United Arab Emirates (Gentry, 1999a), preserved at Natural History Museum (NHM), London; (C) BOU-VP-8/79 (holotype), *Hexaprotodon bruneti,* Bouri, Ethiopia (Boisserie and White, 2004), preserved at the National Museum of Ethiopia (NME), Addis Abeba; (D), KL19-99-049, *Saotherium mingoz,* Kollé, Chad (Boisserie et al., 2003), preserved at the Centre National d'Appui à la Recherche (CNAR), N'Djaména; (E) 16316, *Hexaprotodon sivalensis,* Siwalik Hills, India/Pakistan (Falconer and Cautley, 1847), preserved at the NHM; (F) P997-5 (holotype), aff. *Hippopotamus aethiopicus,* Omo Shungura, Ethiopia (Coryndon and Coppens, 1975), preserved at the NME; (G) Omo 75-69-2798, aff. *Hippopotamus protamphibus,* Omo Shungura, Ethiopia (Gèze, 1980), preserved at the NME; (H) KNM ER 1308* (paratype), aff. *Hippopotamus karumensis,* Koobi Fora, Kenya (Harris, 1991), preserved at the NMK. *Image reversed for comparison purposes; scale bar equals 20 cm.

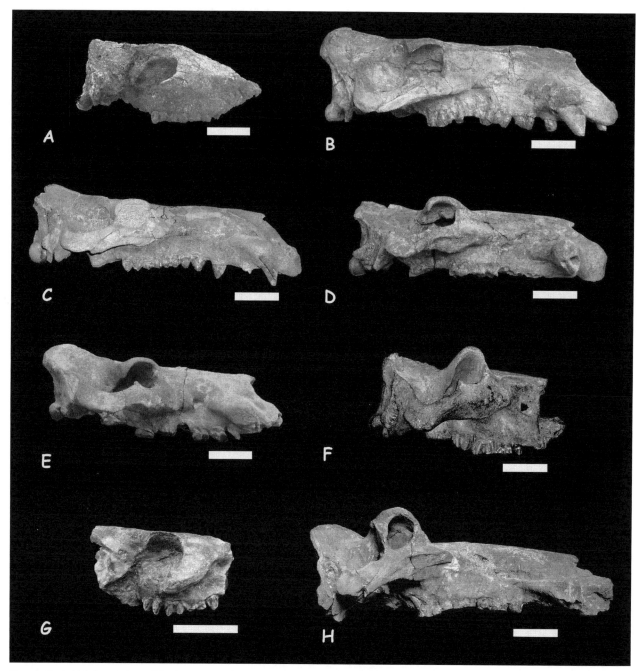

Fig. 8.3. Crania (right lateral views) of fossil Hippopotamidae: (A) KL09-98-049* (holotype), *Saotherium mingoz,* Kollé, Chad (Boisserie et al., 2003), preserved at the CNAR; (B) KNM LT 4 (holotype), *Archaepotamus harvardi,* Lothagam, Kenya (Weston, 2003), preserved at the NMK; (C) A.L. 126-69, aff. *Hippopotamus afarensis,* Hadar, Ethiopia (Gèze, 1980), preserved at the NME; (D) KNM WT 16588, Nachukui, Kenya (Harris et al., 1988), preserved at the NMK; (E) 2269 (holotype), *Hexaprotodon sivalensis,* Siwalik Hills, India/Pakistan (Falconer and Cautley, 1847), preserved at the NHM; (F) 36824, *Hexaprotodon palaeindicus,* Siwalik Hills, India/Pakistan (Falconer and Cautley, 1847), preserved at the NHM; (G) M14801, *Hexaprotodon? imagunculus,* Kazinga Channel, Uganda (Cooke and Coryndon, 1970), cast preserved at the NHM; (H) BOU-VP-2/89,* *Hippopotamus* cf. *gorgops,* Bouri, Ethiopia (Boisserie and Gilbert, in press), preserved at the NME. *Image reversed for comparison purposes; scale bars equal 10 cm.

As in *Choeropsis, Saotherium* exhibits a mosaic of plesiomorphies (e.g., low orbits, Fig. 8.3A) and apomorphies (e.g., cranial roof angle), indicating an early Pliocene lineage endemic to central Africa (Fig. 8.4; Boisserie, 2005). The shape of the braincase and the relatively large orbits of *Saotherium* (Fig. 8.3A) could also suggest affinities with *C. liberiensis,* but for now the lack of fossil data makes uncertain any phylogenetic placement of these taxa.

Genus *Archaeopotamus*

This genus includes the oldest hippopotamines, found in eastern Africa and on the Arabic Peninsula. It is mainly characterized by a mandibular symphysis that is relatively elongated, bearing six slightly differentiated incisors; premolar rows (P2–P4 and p2–p4) almost as long as molar rows; angular processes not laterally everted (Fig. 8.2B). According

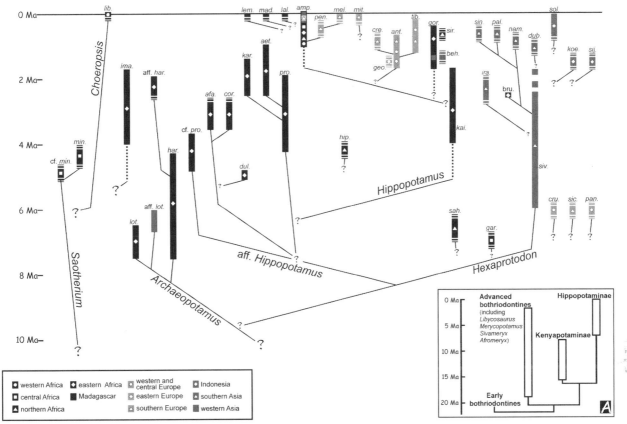

Fig. 8.4. Phylogenetic affinities and spatiotemporal placement of hippopotamine species. Possible relationships of Hippopotaminae within Hippopotamidae and Anthracotheriidae (after Boisserie and Lihoreau, 2006; see Lihoreau and Ducrocq (this volume) for discussion of anthracotheriid taxa). Abbreviations (see also Table 8.1 for further information on these taxa): *aet.*, aff. *Hip. aethiopicus; afa.*, aff. *Hip. afarensis;* aff. *har., A.* aff. *harvardi;* aff. *lot., A.* aff. *lothagamensis; amp., Hip. amphibius; ant., Hip. antiquus; beh., Hip. behemoth; bru., Hex. bruneti; cf. min., S. cf. mingoz; cf. pro.,* aff. *Hip. cf. protamphibius; cor.,* aff. *Hip. coryndoni; cre., Hip. creutzburgi; cru., Hex.? crusafonti; dub., Hex. duboisi; dul.,* aff. *Hip. dulu; gar., Hex. garyam; geo., Hip. georgicus; gor., Hip. gorgops; har., A. harvardi; hip., Hex.? hipponensis ; ima., Hex. ? imagunculus; ira., Hex. iravaticus; kai., Hip. kaisensis ; kar.,* aff. *Hip. karumensis; koe., Hex. koenigswaldi; lal., Hip. laloumena; lem., Hip. lemerlei; lib., C. liberiensis; lot., A. lothagamensis; mad., Hip. madagascariensis; mel., Hip. melitensis; min., S. mingoz; mit., P. minutus; nam., Hex. namadicus; pal., Hex. palaeindicus ; pan., Hex.? pantanellii; pen., Hip. pentlandi; pro.,* aff. *Hip. protamphibius; sah., Hex.? sahabiensis; sic., Hex.? siculus ; sij., Hex. sivajavanicus; sir., Hip. sirensis ; siv., Hex. sivalensis ; sol., Hex. soloensis; syn., Hex. synahleyus; tib., Hip. tiberinus.*

to Boisserie (2005), four forms share this morphology, which is often considered as archaic.

Lothagam, Kenya, produced abundant remains of *A. harvardi* aged between about 7.5 Ma and almost 4 Ma (Weston, 2000; McDougall and Feibel, 2003; Weston, 2003). The report of this species in Tanzania (Harrison, 1997), based on fragmentary remains, is uncertain. *A. harvardi* is somewhat smaller than the extant common hippo. Its cranium (Fig. 8.3B) exhibits orbits just below or at the level of the cranial roof, a low sagittal crest, and a relatively tall occipital plate. Its diet was dominated by C_4 plants (Cerling et al., 2003). The small-sized *A. lothagamensis* from the lowest levels of Lothagam (Weston, 2003) and the relatively larger *A.* aff. *lothagamensis* (Fig. 8.2B) from Abu Dhabi (Gentry, 1999a) are only documented by mandibular remains showing a particularly narrow symphysis. Finally, a mandible from the Plio-Pleistocene of Rawi, Kenya, strongly recalls *A. harvardi* by the proportions of its symphysis and premolar row (Boisserie, 2005). Altogether, this evidence suggests that this lineage, potentially the sister group of all later hippopotamids

described below (Fig. 8.4), had in fact a particularly long history in eastern Africa.

Genus aff. *Hippopotamus*

In the northern part of the eastern African Rift, *Archaeopotamus* was replaced during the early Pliocene by endemic forms, found in the Pliocene and early Pleistocene formations surrounding the lake Turkana (Arambourg, 1944a, 1947; Coryndon and Coppens, 1975; Coryndon, 1976; Gèze, 1980, 1985; Harris et al., 1988; Harris, 1991; Weston, 2003; Harris et al., 2004) and in the Pliocene deposits of the Afar region in Ethiopia (Gèze, 1980, 1985; Boisserie, 2004; Boisserie and Haile-Selassie, in press) All these forms exhibit features reminiscent of *Hippopotamus.* These derived features recall conditions seen in *Hippopotamus,* but Plio-Pleistocene Turkana and Pliocene Afar hippopotamines exhibit notably a primitive enamel pattern on the lower canine (more or less expressed, parallel grooving) differing from the strong, convergent ridges diagnostic of *Hippopotamus.*

Understanding the exact relationships among all these different species and between them and *Hippopotamus,* which is likely to lead to taxonomic revisions, requires a thorough analysis of the particularly abundant remains in both Turkana and Afar basins. In the meantime, they were grouped under the provisory identification aff. *Hippopotamus* in order to acknowledge the similarities and differences of these hippopotamines with *Hippopotamus* (Boisserie, 2005).

In the Turkana Basin, at least three species can be attributed to aff. *Hippopotamus* (Table 8.1). All exhibit expanded canine processes, elevated orbits and sagittal crest, and small premolars more similar to the condition seen in *Hippopotamus* than in *Archaeopotamus* (Figs. 8.2F–H and 8.3D). In the most frequent species, aff. *Hip. protamphibius,* the lachrymonasal contact seen in *Hippopotamus* occurs sporadically. Contrarily to *Hippopotamus,* Plio-Pleistocene Turkana hippopotamines exhibit a primitive lower canine enamel pattern as well as incisor differentiation and reduction, i1–i1 diastema increase, and shallow mandibular symphysis that differentiate them from other hippopotamines. Within such a restricted geographic area, these animals reached a substantial diversity (Table 8.1; Figs. 8.2F–H, 8.3D, and 8.4). Variations seen within each of the described species, especially in aff. *Hip. protamphibius* (Figs. 8.2G and 8.3D) and aff. *Hip. karumensis* (Fig. 8.2H), suggest a complex history and call for a complete revision of this group, incorporating recently discovered material.

These situations and uncertainties are paralleled in Afar, Ethiopia. A local evolution apparently occurred from early to late Pliocene, involving at least three different forms (Table 8.1; Fig. 8.3C). These hipppotamines, with little extended canine processes and little elevated orbits and sagittal crests, nevertheless exhibit features more comparable to *Hippopotamus* than to *Archaeopotamus,* e.g., the mandibular symphysis section. In some Hadar specimens, the lachrymal area, initially described as distinctive from that of all other hippopotamids by Gèze (1980, 1985), who created the monospecific genus *Trilobophorus* on this basis, appears in fact similar to that of *Hippopotamus.* For this reason, this genus *Trilobophorus* Gèze, 1980 is for now included within aff. *Hippopotamus.* Ongoing studies of new material from Middle Awash, Gona, and poorly known material from Hadar should allow reconstruction of the evolution of these Afar hipppotamids and clarification of their affinities among each other and with species from the Turkana Basin.

Small Hippopotamines from the Western Rift

Plio-Pleistocene Ugandan localities delivered generally fragmentary small-sized remains related to *Hexaprotodon? imagunculus.* Craniomandibular remains from Kazinga Channel described by Cooke and Coryndon (1970) retain primitive features such as low orbits (Fig. 8.3G) but also derived features such as a wide mandibular symphysis (Cooke and Coryndon, 1970: Table 29, Pl. 14A). Consequently, successive authors considered *Hexaprotodon? imagunculus* as a valid

taxon or heterogeneous in turn (Hopwood, 1926; Misonne, 1952; Krommenhoek, 1969; Cooke and Coryndon, 1970; Krommenhoek, 1971; Pavlakis, 1990; Faure, 1994). For now, Pavlakis (1990) suggested keeping the name *imagunculus* for the specimens from Kaiso (the type locality) and to use cf. *imagunculus* for other western Rift small-sized hippopotamines.

Additional Records

These strongly suggest that large portions of the history of African endemic Hippopotaminae remain to be unveiled. For example, the abundant late Miocene to Pleistocene collection from the Baringo Basin, central Kenya, only published in a preliminary report (Coryndon, 1978b), requires thorough description. The late Miocene dental remains from Lemudong'o, Kenya (Boisserie, in press) and the early Pliocene remains from the Manonga Valley, Tanzania (Harrison, 1997) wait for further discoveries to be better understood. Finally, the remains from Langebaanweg, South Africa, need both description and additional discoveries. Langebaanweg hippopotamines are particularly important, this Pliocene assemblage being the only significant one in southern Africa as well as the only one for which an exclusive C_3 diet was reported until now (Franz-Odendaal et al., 2002).

THE FIRST WAVE OUTSIDE AFRICA (TERMINAL MIOCENE)

Perhaps linked to the "Messinian Crisis," the first migration of Hippopotamidae outside Africa took place around 6 Ma. They settled briefly in Europe but much more durably in tropical Asia.

Miocene Species in Southern Europe

The first hippopotamid described in European upper Miocene deposits is *Hexaprotodon? pantanellii* (Pantanelli, 1878; Joleaud, 1920), based on very limited dental material from Casino, Italy. A fragmentary portion of a mandibular symphysis indicates a likely hexaprotodont condition. *Hexaprotodon? siculus* was created on more abundant material from Gravitelli, Sicily (Seguenza, 1902, 1907; Hooijer, 1946) but unfortunately lost. Several traits obviously indicate a relatively archaic hippopotamid: deep distal groove on the upper canines, low molar with a triangular wear pattern, and possible hexaprotodont condition.

The best known species is *Hexaprotodon? crusafonti.* This species was unearthed in various Spanish latest Miocene localities (Aguirre, 1963; Crusafont et al., 1964; Aguirre et al., 1973; Lacomba et al., 1986) and possibly also occurs in the early Pliocene of southern France (Faure and Méon, 1984). The best preserved material, from La Portera, Spain, includes a partial mandible bearing four incisors (Lacomba et al., 1986).

Van der Made (1999c) was in favor of merging those three species within *pantanellii,* whereas Weston (2000) related the Spanish material to narrow-muzzled species, i.e., *Archaeopotamus.*

The Asian *Hexaprotodon*

The genus *Hexaprotodon* was initially created by Falconer and Cautley (1836) for the material from the Siwalik Hills. The distinctness of this Asian lineage was confirmed by later authors (Lydekker, 1884; Colbert, 1935a, 1938b; Hooijer, 1950). Hooijer (1950) merged almost all forms in *Hex. sivalensis* but defined eight subspecies considered as species in this chapter.

Indian Subcontinent

Fossil hippopotamids from the Siwaliks of India, Pakistan, Nepal (Corvinus and Rimal, 2001), and maybe China (Chow, 1961) may have occurred as early as 5.9 Ma (Barry et al., 2002). Asian Mio-Pliocene hippopotamid remains are usually attributed to the medium-sized *Hexaprotodon sivalensis* (Figs. 8.2E and 8.3E), a species mainly characterized by a very high and robust mandibular symphysis, relatively short despite its limited canine processes; lower incisors equivalent in size; a double-rooted P1; lack of lachrymonasal contact; orbits and sagittal crests moderately elevated. Four more derived forms occurred during the Pleistocene: *Hex. duboisi* from the Siwaliks of Punjab; *Hex. namadicus* and *Hex. palaeindicus* (Fig. 8.3F) from the middle Pleistocene Narmada Hill deposits in central India; and *Hex. sinahleyus* from Sri Lanka (Deraniyagala, 1969). Following Lydekker (1884), Asian Pleistocene hippopotamids derived from *Hex. sivalensis* by orbit and sagittal crest elevation (Fig. 8.3F), reduction of i2, increase of i3 diameter (particularly in *palaeindicus* and *sinahleyus*), and increase of molar crown height and complexity.

Indonesian Archipelago

Hexaprotodon reached those islands around 1.5 Ma and survived until the late Pleistocene (Bergh et al., 2001). Hooijer (1950) described three forms: *Hex. sivajavanicus* and *Hex. koenigswaldi* from the lower Pleistocene, both small-sized with relatively high orbits and short symphysis, differing mainly on i3 size, respectively smaller and larger than i2; and the moderately sized *Hex. soloensis* from middle to late Pleistocene, with very elevated orbits, and i3 tending to be equal or superior in size to i1.

Myanmar

Hex. iravaticus, probably Pleistocene, was initially based on a juvenile fragmentary mandible (Falconer and Cautley, 1847; Colbert, 1935a, 1938b).

Other specimens from Myanmar could indicate a species somewhat smaller than *Hex. sivalensis* with rather developed canine processes. The cranium also exhibits a lachry-mal relatively closer to the nasal, whereas the molars remain relatively simple compared to other Pleistocene forms.

Hexaprotodon in Africa

Fossil evidence indicates that the origin of Eurasian *Hexaprotodon* should be looked for in Africa in levels older than 6 Ma. A possible precursor was identified in Chad at about 7 Ma (Vignaud et al., 2002). Although retaining more primitive premolars than *Hex. sivalensis, Hex. garyam* exhibits similarly high and robust mandible, wide occipital plate, and moderately raised sagittal crest (Boisserie et al., 2005c).

Three other assemblages based on fragmentary evidence could play a role in understanding early hippopotamid exchanges between Africa and Eurasia: *Hexaprotodon? sahabiensis* from the late Miocene of Sahabi, Libya (Gaziry, 1987a); the Pliocene *Hexaprotodon? hipponensis* from Pont du Vivier, Algeria (Gaudry, 1876; Pomel, 1890; Arambourg, 1944b); and the material from the middle Pliocene of Wadi el Natron, Egypt (Andrews, 1902; James and Slaughter, 1974).

Finally, the best established African occurrence of *Hexaprotodon* is *Hex. bruneti* from the Afar, Ethiopia. Aged at 2.5 Ma, this species is a likely Asian migrant that exhibits a high and robust mandible bearing an i3 three times larger than i1, even more hypertrophied than in advanced Asian species (Boisserie and White, 2004).

HIPPOPOTAMUS: AFRICAN NEW DEAL AND CONQUEST OF EUROPE

Hippopotamus emerged during the African diversification of Hippopotaminae. For long, it remained rare within hippopotamid faunas. The Plio-Pleistocene climatic changes apparently favored this ubiquitous genus. Maybe more tolerant to colder temperatures, it quickly expanded after 2 Ma in Africa, replacing previously dominant local genera, and became the first hippopotamid to cross the northern limits of the tropical/Mediterranean world.

Origin of *Hippopotamus*

Hippopotamus is robust, tetraprotodont, notably characterized by a shallow distal groove on the upper canine; strong convergent ridges on the lower canine enamel; relatively high-crowned molars; expanded canine processes; and wide lachrymonasal contact (Fig. 8.1E–H). Despite this distinctive character association, the first appearance datum of the earliest *Hippopotamus,* the relatively brachyodont *Hip. kaisensis* from Pliocene/early Pleistocene of the western Rift, is not well established. Its emergence from aff. *Hippopotamus* could be supported by Turkana specimens as old as 4 Ma (Harris et al., 1988; Feibel et al., 1989), whereas Faure (1994) indicated a possible occurrence of *Hip. kaisensis* about 5 Ma in the Nkondo Formation, Uganda. Given the scarcity of Pliocene remains, the possibility of confusion with other hippopotamids (Pavlakis, 1990; Gentry, 1999a) obscures the origin of *Hippopotamus* for now.

New Deal in Africa

More advanced than *Hip. kaisensis, Hip. gorgops* successfully exploited most of the continent during the basal Pleistocene. This species was initially described at Olduvai, between 1.9 Ma and 0.6 Ma (Dietrich, 1926, 1928; Coryndon, 1971). *Hip. gorgops* or closely related forms were notably reported in the Turkana Basin, possibly as early as 2.5 Ma and up to 0.7 Ma (Gèze, 1980; Harris et al., 1988; Harris, 1991); at Olorgesailie, Kenya, between 1.0 Ma and 0.6 Ma (Isaac, 1977); at Melka Kunturé, Ethiopia, between 1.6 Ma and 0.8 Ma (Gèze, 1980); at about 1 Ma at Buia, Eritrea (Martínez-Navarro et al., 2004) and Bouri, Ethiopia (Boisserie and Gilbert, in press); at Cornelia, South Africa, middle Pleistocene (Hooijer, 1958); and possibly in the Pleistocene of central and northern Africa (in Algeria and Morocco, possibly under the name *Hip. sirensis* Pomel, 1890). This species, larger than *Hip. amphibius,* is characterized by (Fig. 8.3H) its particularly elevated orbits and sagittal crest, its elongated muzzle, and its large and tall molars, which are more simple than in *Hip. amphibius.* Because of a lack of comparative studies, it is not possible to decide whether or not several distinctive forms currently included in *Hip. gorgops* should be recognized as separate species. In any case, these hippopotamids rapidly replaced all the large local species in what can be qualified as an early Pleistocene African "hippopotamid turnover." Finally, during the middle Pleistocene, *Hip. gorgops* was itself supplanted by a species with lower orbits and sagittal crest, shorter muzzle, but a more complicated molar wear pattern: the extant *Hip. amphibius* (Fig. 8.1E–H).

Beyond Africa Again: The European and Western Asian *Hippopotamus*

Western Europe

With three to four species, *Hippopotamus* was present in southern western Europe during most of the Pleistocene, successively colonizing northern latitudes (as north as Leeds, England [Faure, 1981]) when the climate was more favorable during interglacial intervals.

Hip. antiquus occurred first in the basal Pleistocene of Italy, about 1.8 Ma. This is the largest known hippopotamid, showing clear trends toward muzzle development and orbit and sagittal crest elevation. Its autopods appear relatively short and robust (Faure, 1983). Around 1 Ma, *Hip. antiquus* is known from Spain to Germany, being particularly well recorded at Untermassfeld (Kahlke, 1997, 2001). During the early middle Pleistocene, it also occurred in England but shortly afterward disappeared from Europe.

Hip. tiberinus, more advanced in term of muzzle elongation and orbit and sagittal crest elevation, was recognized by Mazza (1991) in southern and central Europe from the late early to the late Pleistocene. Equivalent in size to *Hip. amphibius,* it compares best with the most advanced *Hip. gorgops* from Olduvai, Tanzania. However, Mazza (1995) favored a parallel evolution from *Hip. antiquus* instead of a direct migration from an African stock.

A middle to late Pleistocene species chiefly identified in England, France, and Germany, *Hip. incognitus* was proposed by Faure (1983, 1984, 1985). Supposedly intermediate in size to *Hip. antiquus* and *Hip. amphibius,* it was rejected by other authors (Petronio, 1986; Mazza, 1991; Caloi and Palombo, 1994; Mazza, 1995) as based on material belonging to the latter and/or the former.

Hip. amphibius was the last continental European hippopotamid, perhaps first occurring in the middle Pleistocene (Caloi and Palombo, 1996). It is nevertheless known with certainty only from the last interglacial period around 0.12 Ma (Mazza, 1995) and then progressively retreated to southern Europe until it disappeared with the advent of the last glacial age.

The evolution and systematics of European species remain disputed. Kahlke (1997, 2001) considers all European forms as subspecies of *Hip. amphibius,* showing an early to middle Pleistocene evolution from *"amphibius*-like" forms (early *Hip. amphibius antiquus*) to high-orbits/elongated-muzzle forms (*Hip. amphibius tiberinus*). This could have happened in parallel with *Hip. gorgops* or through a genetic continuum between Africa and Europe (including *gorgops* as a subspecies of *Hip. amphibius*). In any case, the cause of this development and of the late replacement of advanced subspecies by an unchanged basal stock of *Hip. amphibius* remains poorly understood.

Eastern Europe

Hippopotamus is comparatively much less frequent in the Balkans and beyond. For the lower Pleistocene, Kahlke (1987) reported *Hippopotamus* sp. in Croatia, Romania, Greece, Moldavia, and *Hip. antiquus* in Hungary and Slovenia. The most eastern occurrence of European *Hippopotamus* is that of the lower Pleistocene *Hip. georgicus* at Akhalkalaki in Georgia (Vekua, 1959, 1986). It was described as a particularly large form, with an enlarged digit III of the manus. In fact, given the available evidence, it is better interpreted as *Hip. antiquus* (Faure, 1983; Kahlke, 1987).

Western Asia

The oldest occurrence of *Hippopotamus* in the Levant is paradoxically younger than in western Europe, suggesting possible migration routes through Gibraltar and/or the Siculo-Maltese archipelago. Two species were recorded at 'Ubeidiya, Israel, around 1.4 Ma (Tchernov, 1987): *Hip. gorgops,* represented by three postcranial elements, and the endemic *Hip. behemoth* (Faure, 1986; Horwitz and Tchernov, 1990). The latter is known by isolated teeth and postcranials, indicating a size slightly larger and metapodials relatively longer than in *Hip. amphibius,* but Martínez-Navarro et al.

(2004) noted that these elements do not really depart from the range of variation of *Hip. gorgops.*

During the middle and upper Pleistocene, *Hip. amphibius* occurred essentially in coastal areas of Israel, Lebanon, and Syria, with a late Pleistocene interruption (Horwitz and Tchernov, 1990). The Holocene Levantine *Hip. amphibius* was probably the last hippopotamid known outside Africa before it definitely retired to the Nile delta after 3,500 BP (Horwitz and Tchernov, 1990), then south to the Sahara after the seventeenth century (Manlius, 2000).

BEYOND THE SEAS: INSULAR HIPPOPOTAMIDS

The Pleistocene and Holocene hippopotamids from Mediterranean islands and Madagascar were probably close relatives of continental *Hippopotamus.* However, they were subjected to similar evolutionary pressures, resulting in parallel adaptations to insular microhabitats. For this reason, they are presented separately.

Mediterranean Islands

Most Pleistocene Mediterranean Sea corridors were not really an obstacle for hippopotamids: about 30 km wide to reach Crete, and probably nil between the continent and the Siculo-Maltese archipelago (Reyment, 1983). However, the semiaquatic habits of hippopotamids probably played a major role for colonizing Cyprus, as well as Madagascar, which were maybe 100 km (Reyment, 1983) and at least 270 km, respectively, in low stand level. Paradoxically, Mediterranean insular hippopotamids had to deal with rather arid and rocky environments. Along with size reduction, they developed more rigid, less graviportal limbs for a "low-gear locomotion" adapted to irregular nonmuddy slopes (Sondaar, 1977; Caloi and Palombo, 1994; Spaan, 1996).

Siculo-Maltese Archipelago

Two species are known, apparently both recorded in Sicily and Malta. The middle to late Pleistocene *Hip. pentlandi,* slightly smaller than *Hip. amphibius,* exhibited high orbits and a narrow elongate muzzle. It could be closely related to *Hip. amphibius* (Capasso Barbato and Petronio, 1983a; but see Faure, 1983). The much smaller *Hip. melitensis* may have occurred during the late Pleistocene, and could have been derived from *Hip. pentlandi* (Capasso Barbato and Petronio, 1983b; Faure, 1983; Petronio, 1986).

Crete

Hip. creutzburgi was an early Pleistocene to middle Pleistocene species, equivalent in size to *Hip. melitensis.* It may have been derived from *Hip. antiquus* (Petronio, 1986; Spaan, 1996). It exhibits trends toward lophodonty, reduced canine processes, and more gracile and more erected limbs with reduced lateral flexion (Caloi and Palombo, 1996). In terms of

adaptations to insularity, this species ranks between the Siculo-Maltese forms and the following.

Cyprus

The most derived insular hippopotamid is found in deposits from Pleistocene to 9,000 BP (Simmons, 1988). *Phanourios minutus,* slightly smaller than *Choeropsis,* is notably characterized by brachyodont lophodont molars; lack of P4; and postcranials derived in a way similar to *Hip. creutzburgi* (Boekschoten and Sondaar, 1972; Houtemaker and Sondaar, 1979). Given its advanced morphology, its affinities remain uncertain. Its disappearance is attributed either to hunting by *Homo sapiens* (Simmons and Reese, 1993) or to Holocene climatic changes (Diamond, 1992; Bromage et al., 2001).

Madagascar

Madagascar harbored three Holocene hippopotamids. The first two were thoroughly revised by Stuenes (1989). They were intermediate in size between *Hip. amphibius* and *Choeropsis. Hip. lemerlei* differs from *Hip. madagascariensis* by more pronounced sexual dimorphism; a narrower and longer muzzle; higher orbits; more apical incisor occlusions; a reduced upper canine distal groove; and less everted angular processes. Ecologically, Stuenes (1989) indicated that *lemerlei,* inhabiting coastal lowlands, was probably more aquatic than the highland dweller *madagascariensis.*

Hip. laloumena, a third species described by Faure and Guérin (1990), is closer in size and morphology to *Hip. amphibius.* Its occurrence could indicate several invasions of Madagascar by one or several continental form(s) over the Mozambique Channel (currently at least 420 km wide). Those hippopotamids disappeared around 1,000 BP, maybe in relation to human settlement on the island (MacPhee and Burney, 1991).

CONCLUSION: SEVENTEEN MAJOR QUESTIONS ABOUT HIPPO EVOLUTION

Many important pieces of hippopotamid evolutionary history remain poorly understood. In approximate chronological order, these are: (1) the exact relationships between Hippopotamidae and their purported stem group, Anthracotheriidae; (2) the evolution of the hippopotamid molar scheme, related to issue 1; (3) the exact nature of *Kenyapotamus* and its relationships with Hippopotaminae; (4) the taphonomical, biological, and ecological background of the sudden appearance of Hippopotaminae; (5) the basal relationships among the main lineages of Hippopotaminae; (6) the evolutionary history of *Choeropsis;* (7) the route(s) of the first hippopotamid migration toward Eurasia; (8) the links between the "Messinian crisis" and this dispersal; (9) the relationships between Miocene Mediterranean hippopotamids and the rest of the family; (10) the relationships within *Hexaprotodon;* (11) the emergence and earliest evolution of *Hippopotamus;* (12) the ecological differences between *Hippopotamus* and the other hippopotamines, notably regarding temperature and sea water; (13) the relationships among species of *Hippopotamus;* (14) the migratory route(s) of *Hippopotamus* toward Europe; (15) the reasons and modalities of the parallel, apparently correlated development of high orbits, high sagittal crest, and elongated muzzle within at least four lineages (endemic Pliocene Turkana hippopotamids; southern Asian *Hexaprotodon,* European *Hippopotamus,* African *Hippopotamus*); (16) the phylogenetic relationships between continental and insular hippopotamids; and (17) the morphological and paleoecological evolution of hippopotamids in response to island environments and constraints.

Advances on each of these issues and on many others of local importance will require increasing field work and further studies of extant collections. They would bring a lot to our perception of artiodactyl evolution in general, would facilitate the use of hippopotamids for paleoecological and paleobiogeographic reconstructions, and, of course, would be extremely satisfying for the sake of knowing better these astonishing semiaquatic mammals.

APPENDIX 8.1: A CLASSIFICATION OF HIPPOPOTAMIDAE

In order to avoid extensive taxonomic discussions that may result from the phylogenetic hypotheses linking hippopotamids with anthracotheriids and cetaceans, the classification of hippopotamid genera proposed below follows a monophyletic hierarchy. Hippopotamoidea Gray, 1821, anterior to Anthracotherioidea Leidy, 1869, and Ancodonta Matthew, 1929, is used here to designate the clade including Hippopotamidae and Anthracotheriidae according to the definition of Gentry and Hooker (1988).

Mammalia Linné, 1758
  Cetartiodactyla Montgelard et al., 1997
    Cetancodonta Arnason et al., 2000
      Hippopotamoidea Gray, 1821 sensu Gentry and
        Hooker (1988)
        Hippopotamidae Gray, 1821
          Kenyapotaminae Pickford, 1983
          Kenyapotamus Pickford, 1983
          Hippopotaminae Gray, 1821
          Hippopotamus Linné, 1758
          Hexaprotodon Falconer and Cautley, 1836
          Choeropsis Leidy, 1853
          Phanourios Boekschoten and Sondaar, 1972
          Archaeopotamus Boisserie, 2005
          Saotherium Boisserie, 2005

APPENDIX 8.2: DIAGNOSES OF HIPPOPOTAMID GENERA

Genus *Kenyapotamus* Pickford, 1983

Diagnosis. A genus of Hippopotamidae in which the molars are bunodont and rather tayassuid-like with strong

lingual cingula; P4 with two main cusps (one labial, the other lingual) with strong cingulum except labially; P3 with large distolingual cusp, outer enamel surface projects further rootward than lingual enamel, and with pustular enamel tubercles; P1 with two fused roots; lower molars more suid-like than uppers with median accessory cusps strongly joined to hyproconids by a large crest; lower premolars triangular in side view, posterior ones with strong labial folds of enamel on the distal edge; m3 with talonid; M3 without talon; lower incisors cylindrical, permanently growing; talus with navicular facet smaller than cuboid facet (from Pickford, 1983).

Genus *Hippopotamus* Linné, 1758

Emended Diagnosis. Tetraprotodont having the following apomorphies: skull with an elongated muzzle; upper canines with a longitudinal and shallow posterior groove, narrow and covered with enamel; lower canines with strong convergent enamel ridges; deep and widely open notch on the orbital anterior border; limbs short and robust with very large quadridigitigrade feet. This genus displays many other features that are derived within the family but that may be seen in other hippos: antorbital process of the frontal short to absent and a long contact between the maxillary bone and the lachrymal bone; high orbits; short globular braincase, with strong postorbital constriction; mandibular symphysis globular in sagittal section, without incisor alveolar process overhanging frontally; canine processes well developed laterally and frontally; molars high-crowned, compact, and relatively long mesiodistally (modified from Gèze, 1980; Harris, 1991).

Genus *Hexaprotodon* Falconer and Cautley, 1836

Emended Diagnosis. Hexaprotodont; characterized by a very high robust mandibular symphysis, relatively short in spite of its canine processes, which are not particularly extended laterally; dorsal plane of symphysis very inclined; thick incisor alveolar process, frontally projected; some relatively small differences between the incisor diameters, the i2 usually being the smallest; laterally everted but not hook-like gonial angle; orbit having a well-developed supraorbital process and a deep but narrow notch at its anterior border; thick zygomatic arches; elevated sagittal crest on a transversally compressed braincase. Some constant features of this genus appear to be primitive: the strong double-rooted P1 the quadrangular lachrymal, separated from the nasal bone by a long maxillary process of the frontal (from Boisserie, 2005).

Genus *Choeropsis* Leidy, 1853

Emended Diagnosis. Small-sized genus, distinct from all the other known Hippopotamidae by its downwardly bent nasal anterior apex, which clearly passes the premaxillae-nasal contact anteriorly; orbits clearly below the cranial roof; strong posterior nasal spine of the palatine; large and elongated tympanic bulla, which is apically rounded and without marked muscular process; presence of a lateral notch on the basilar part of the basioccipital, immediately posterior to the muscular tubercles; down-turned sagittal crest. *Choeropsis* shares many characters with the most primitive Hippopotamidae: weak extension of the canine processes (both lower and upper); facial crest regularly convex in ventral view, gradually sloping from the zygomatic arch toward the maxilla; slender zygomatic arch in ventral view; lachrymal separated from the nasal by a long maxillary process of the frontal; orbit anterior to the level of contact between M2–M3 seen in lateral view; weak supraorbital apophyses; braincase elongated and transversally rounded. *Choeropsis* also presents some derived features convergent on several other Hippopotamidae: diprotodont mandibular symphysis short and upright; gonial angle laterally everted; P4 and p4 generally without accessory cusps (from Boisserie, 2005).

Genus *Phanourios* Boekschoten and Sondaar, 1972

Diagnosis. A small member of the Hippopotamidae with a simple molar pattern that became lophodont by wear. The upper fourth premolar is lacking, leaving a diastema between the third premolar and the first molar in the adult animal (from Boekschoten and Sondaar, 1972).

Genus *Archaeopotamus* Boisserie, 2005

Diagnosis. Hexaprotodont, characterized by having a very elongate mandibular symphysis relative to its width. This symphysis also bears an incisor alveolar process strongly projected frontally, very procumbent incisors, and canine processes poorly extended laterally and not extended anteriorly. The length of the lower premolar row approaches the length of the molar row. The horizontal ramus height is low compared to its length but tends to increase posteriorly. The

gonial angle of the ascending ramus is not laterally everted (from Boisserie, 2005).

Genus *Saotherium* Boisserie, 2005

Diagnosis. Hexaprotodont with the following apomorphies: cranial roof showing an antorbital angle in lateral view; skull very high above the molars; slender mandibular symphysis in the sagittal plane. Also exhibiting these plesiomorphic or convergent features: orbits below the cranial roof; slender zygomatic arches; cylindrical braincase; slender and low sagittal crest; laterally developed occipital plate; maxillary process of the frontal separating the nasal and the lachrymal bones; short extension of the canine processes; lingual border of the lower cheek tooth alveolar process lower than the labial border (from Boisserie, 2005).

ACKNOWLEDGMENTS

My research as a hippopotamid specialist since 1998 was made possible and continually supported by Michel Brunet, F. Clark Howell, Patrick Vignaud, and Tim D. White as well as by all the field workers who dared to excavate and carry fossil hippopotamids. I thank John Harris, Terry Harrison, and John Barry for their reviews that greatly improved this chapter. I thank the people and financial support of the Ministère Français de l'Education Nationale et de la Recherche (UMR CNRS 6046, Université de Poitiers), the University of California (Human Evolution Research Center, Department of Integrative Biology and Museum of Vertebrate Zoology), the Mission Paléoanthropologique Franco-Tchadienne, the Middle Awash research program, the Fondation Fyssen (postdoctoral research grant and research grant), the Ministère des Affaires Etrangères (program Lavoisier and SCAC, French Embassy in Ethiopia), the Fondation Singer-Polignac (postdoctoral research grant), and the NSF-HOMINID program Revealing Hominid Origins Initiative.

9

Family Entelodontidae

THE ENTELODONTIDAE ARE A FAMILY of bunodont artiodactyls that arose about 38 million years ago in Asia during the late Eocene (Zhang et al., 1983) and became extinct about 19 million years ago during the early Miocene (Hunt, 1990; Foss and Fremd, 1998). Whereas many of the diagnostic features of the family (jugal flange, elongate snout, and mandibular tubercles) are highly derived, most of the postcrania and dentition are morphologically primitive with respect to other artiodactyls. The phylogenetic ancestor of the Entelodontidae is unknown, but it seems likely that it shares a common ancestry with the suids, probably originating in Asia.

Possibly because entelodonts have often been (incorrectly) allied with achaenodonts (descendants of helohyids: Gazin, 1955), there has been a lack of investigation into their origin. Romer (1966) suggested that entelodonts descended from an Asian group, such as choeropotamids or cebochoerids. Gazin (1955) suggested that early helohyids were ancestral to entelodonts (but independent of achaenodonts). Coombs and Coombs (1977a) assigned Helohyidae to the Anthracotherioidea but were uncertain about the placement of cebochoerids, choeropotamids, or entelodonts. With the exception of Scott (1898b, 1937), there is virtually no additional literature that discusses the origin of the family Entelodontidae, including the monographs that are devoted to the group.

Entelodonts, often dubbed "big pigs," are correctly associated with living pigs and peccaries, whereas most early artiodactyls that are often considered "piglike" are not related to true suids (Foss, 2001; Geisler and Uhen, 2005; Theodor and Foss, 2005; Geisler et al., this volume). Although not true pigs themselves, the superfamily Entelodontoidea is phylogenetically similar (sister group) to the Suoidea. The larger question of the origin of Suiformes is still a topic for investigation.

WHAT IS AN ENTELODONT?

At least 56 species of entelodonts have been assigned to 23 genera over the past 155 years. Dental characters (especially molar tooth morphology), although diagnostic of the family in general, are especially conservative and not always diagnostic of individual species. On the other hand, characters of the skull and jaw (diastema spacing, forehead rugosity, jugal flange size, mandibular tubercle shape) are highly variable within individual populations of entelodonts and offer only limited species-level recognition (although they may be of general assistance at the generic level). Postcranial and basal skull morphology is helpful in separating genera but is rarely available.

Over the past 155 years, the Entelodontidae have undergone revisions at every taxonomic level. Specimens have been assigned to the Perissodactyla (Marsh, 1874); names have been incorrectly attributed (including Pomel, 1847a, 1847b; Leidy, 1851a; Alston, 1876; Marsh, 1893) and have been found to be *nomina nuda* (Palmer, 1904; Simpson, 1945); and oversplitting has occurred to a great degree (see Troxell, 1920). These characteristics of existing taxonomic work must be understood, analyzed, and corrected before a valid phylogeny of the Entelodontidae may be constructed. This chapter outlines a generic-level taxonomy of the entelodonts and suggests assignments based on character traits that are recognizable across multiple taxa and on anatomical characters (such as dental traits) that are more likely to be identified on even fragmentary specimens.

The Entelodontidae are characterized by "complete" bunodont dentition (3-1-4-3/3-1-4-3); thus, the name entelodont, where the Greek *enteles* means "complete" or "perfect" and the Greek *odontos* means "tooth" or "teeth." They also possess the "typical" artiodactyl double-pulley astragalus, elongate limbs with didactyl feet exhibiting paraxony (weight is distributed equally over toes III and IV), a lateroventrally oriented dependent process of the zygomatic arch that is composed exclusively of the jugal bone, and at least one, but usually two, bilateral pairs of mandibular tubercles on the ventral surface of the mandible. Although mandibular tubercles or jugal flanges are seen in other taxa (some *Hippopotamus* skulls have small mandibular tubercles; edentates, many marsupials, and some rodents have a jugal flange), the jugal flanges and mandibular tubercles observed in the Entelodontidae are a unique phylogenetic derivation.

"LUMPING" AND "SPLITTING" IN ENTELODONT SYSTEMATICS

The question of how to assign subtly different entelodont taxa to different genera remains. Some authors have argued that *Entelodon* is separate from *Archaeotherium* by transcontinental geography only. Other genera of North American entelodonts have been assigned or supported based on seemingly inadequate criteria. Many North American genera have been synonymized with either *Daeodon* (*Dinohyus*, *Ammodon*, and *Boochoerus* as subjective synonyms)

or *Archaeotherium* (*Megachoerus*, *Pelonax*, and *Scaptohyus* as subgenera; *Choerodon* as a subjective synonym; and *Arctodon* as a *nomen nudum*).

The various species within these genera (especially *Archaeotherium*) have been erected based on characters of tooth wear and presumed behavior (*A. ramosum*, *A.* [*S.*] *altidens*), geography (*A. caninus*), overall size (*A. latidens*, *A. wanlessi*), and morphological characters (most others). The characters that support these species have historically involved the size, shape, and thickness of the jugal flanges and mandibular tubercles. In fact, "*Megachoerus*" is characterized by the large size of its jugal flanges, and "*Pelonax*" by extremely large mandibular tubercles on the ventral portion of the jaw. Even though these two subgenera co-occur in the Whitneyan beds of the White River Group, "*Pelonax*" cannot be reliably distinguished from "*Megachoerus*" because the skull of "*Pelonax*" is unknown.

Understanding entelodont systematics is difficult because so many characters that have been used to support taxa (flange/tubercle shape, overall size, geography, premolar diastema spacing, behavior) are variable both within populations as well as across the family. Most of these characters are probably affected by individual variation, ontogeny, and possibly nutrition and sexual dimorphism. Better morphological characters, such as molar cusp arrangement, are often too conservative to distinguish individuals from anything other than an overall family assignment. Skull features, such as a rugose forehead or separate temporal lines, are widely variable within populations (especially apparent in the "Pig Dig" sample from Badlands National Park and housed at SDSMT).

Brunet (1973) separated the two main species of *Entelodon* (*E. deguilhemi* and *E. magnus*) based on overall size but admitted that this is troublesome. In North America, where there are more known specimens of entelodonts, separating taxa based on size is even more confounding.

Effinger (1987) made an attempt to separate entelodont genera based on bivariate plots of teeth. He concluded that three entelodont taxa (*Archaeotherium mortoni*, *A. wanlessi*, and *A. latidens*) clustered as three distinct size groups. Attempting to follow Effinger's lead, Foss (2001) discovered that adding more individuals to his plots simply filled in the gaps and made separating species by size appear purely arbitrary. Additionally, specimens from single quarry samples (such as the "Pig Dig" in Badlands National Park, South Dakota, or the *Trigonias*/Horsetal Creek assemblage of northeastern Colorado) are widely variable with regard to size within a single population (Foss and Lucas, 1997; Foss, 2001).

SYSTEMATIC PAELONTOLOGY
Class MAMMALIA Linnaeus, 1758
Order ARTIODACTYLA Owen, 1848: 131
Suborder SUIFORMES Jaeckel, 1911: 233

Discussion. The inclusion of Cetacea within the Artiodactyla has necessitated a revision of the Suiformes. Geisler and Uhen (2005) included Suidae, Tayassuidae, Entelodontidae, and Anthracotheriidae and specifically excluded Hippopotamidae, Raoellidae, Cebochoeridae, Mixtotheriidae, Cainotheriidae, Oreodontoidea, Camelidae, and Cetacea. Geisler et al. (this volume) exclude Anthracotheriidae as well. The latest data support a monophyletic Suiformes with four families: Suidae, Tayassuidae, Sanitheriidae, and Entelodontidae.

Infraorder SUINA Gray, 1868: 20

Discussion.   Simpson (1945) included the Entelodontidae in Matthew's (1920) Palaeodonta. The infraorder Palaeodonta includes most of the early bunodont artiodactyls and is largely polyphyletic but may be rescued as a useful taxonomic concept with the reassignment of certain groups. Here, the Entelodontidae are reassigned to the infraorder Suina (Gray, 1868), which in part agrees with the classification proposed by Romer (1966) and fully agrees with the results of Theodor and Foss (2005), Geisler and Uhen (2005), and Geisler et al. (this volume). Thus, the Suina contain two distinct superfamilies, the Suoidea and the Entelodontoidea.

Superfamily ENTELODONTOIDEA (Colbert, 1938: 105)

Discussion.   The Entelodontoidea originally contained the Entelodontidae and Achaenodontidae. Following the reassignment of the Achaenodontidae to a helohyid relationship (Gazin, 1955; Coombs and Coombs, 1977a), the group contains only one family and is essentially synonymous with Lydekker's Entelodontidae. However, it serves here to denote the entelodonts as a distinctly separate grade within the Suina.

Family ENTELODONTIDAE Lydekker, 1883: 146

Family Diagnosis.   Teeth bunodont, muzzle long, cranium short, limbs elongated, feet didactyl (Peterson, 1909: 43). Also, large temporal foramen; complete ancestral artiodactyl dentition (3-1-4-3/3-1-4-3) with thick enamel; a well-developed precingulum and six cusps on the upper M1 and M2; an elevated trigonid and twinned paraconid-metaconid on the lower molars; co-ossified radius and ulna (*contra* Marsh, 1873); unfused foot bones; robust postorbital bar; inferior enlargement of the jugal bone resulting in a lateroventrally oriented pendulous process; at least one but usually two pairs of tubercles on the ventral margin of the mandible; and a low mandibular condyle at or below the level of the tooth row. Although the rostrum is characteristically long, the length varies by both age and group. The earliest entelodonts, *Eoentelodon,* from Asia, and *Brachyhyops,* from North America, have shorter snouts relative to later entelodonts.

Discussion.   Lydekker's original (1883b) assignment also included *Achaenodon* and *Tetraconodon,* which have sub-

sequently been reassigned to the Helohyidae and Suidae, respectively. Elotheridae Alston, 1878 (for 1876) (in part) is a synonym of Entelodontidae and is discussed by Marsh (1894d: 408) and Scott (1898b: 322).

The subfamily Entelodontinae (Osborn, 1909: 61) contained entelodonts and achaenodonts. With the reassignment of the achaenodonts to the Helohyidae (Gazin, 1955; Coombs and Coombs, 1977a), it is essentially synonymous with Entelodontidae.

Genus *Eoentelodon* Chow, 1958: 30

Type Species.   *Eoentelodon yunanense* Chow, 1958: 30.

Diagnosis.   Small entelodont (less than half the size of *Archaeotherium,* but only slightly smaller than *Brachyhyops*), p4 relatively large, m3 with large hypoconulid, mandible deep with a bony tubercle projecting from the inferior border of the ramus beneath p4 (Fig. 9.1).

Discussion.   This is the earliest known entelodont, and all of the characters that are discernible on the type specimen support its placement in the Entelodontidae: elevated trigonid, fused mandibular symphysis, twinned or paired parametaconid on m2.

Given the small amount of comparative material, little exists to separate this specimen from *Brachyhyops* except for the presence of a p2–p3 diastema, which is common to all entelodonts except for *Brachyhyops. Eoentelodon* and *Brachyhyops* share a few morphological characters that are unlike all other entelodonts: p4 is the largest lower premolar (p3 is the largest in all other entelodonts), and the hypoconulid is strong on m2 (it is weak or absent in all other entelodont genera).

Often cited as the oldest member of the Entelodontidae (Chow, 1958; Wilson, 1971b), *Eoentelodon* is part of the lower Lunan Basin fauna of China (Lumeiyi Formation), which,

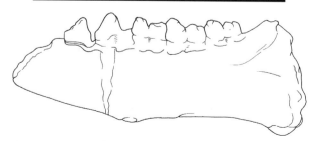

Fig. 9.1. GM.CPP 0051, holotype of *Eoentelodon yunanense.* Left ramus of mandible with p3–m3. Scale bar equals 10 cm.

according to Zhang et al. (1983), is late Eocene in age and faunally equivalent to the Irdin Manha fauna of Inner Mongolia.

Included Species. *Eoentelodon yunanense* Chow, 1958: 30.

Genus *Entelodon* (Aymard, 1846: 227–242)

Type Species. *Entelodon magnus* (Aymard, 1846 [1848?]: 227–242).

Diagnosis. Aymard (1846) reported: dental system of 3-1-7/3-1-7 (Aymard did not differentiate between molars and premolars); teeth marked by longitudinally intersecting grooves; upper and lower molars with anterior and posterior cingula; principal molar large (P4, p4); incisors increasing in size from 1 to 3 (upper and lower) and have small cingula on the inside (lingual) surface; three double-rooted molars just posterior to the upper canine; first lower molar (p1) single rooted, whereas first upper molar (P1) separated on each side by a space; third molar (P3) narrow relative to neighboring genera ("genres voisins") and is the tallest in the series. The principal inferior molar (p4) with conical crown in front (anterior) and a transverse ridge connecting two hills in back (ridge between metaconid and hypoconid); last three molars (M1–M3, m1–m3) divided by a transverse valley into two hills with three blunt points (cusps) in the upper jaw and two in the lower jaw (upper molars with six cusps, lower molars with four cusps); uppers (M1–M3) trapezoidal in shape with M2 stronger than M3; lower molars longer than large (great anterior–posterior length), and the last left devoid of a talon or hill (m3 lacks a hypoconulid) (Aymard, 1846:227–242; Peterson, 1909:45–46).

Discussion. Most of the characters noted above are not unique to *Entelodon* and, in fact, are general to the entire family. Characters unique to *Entelodon* include a hypoconulid on m3 and the loss of an entocristid on m2.

Scott (1898b) preferred *Elotherium* as the generic name for both the European and North American early Oligocene entelodonts. Peterson (1909) recognized the European varieties as *Entelodon* and the North American varieties as *Archaeotherium*. This convention of calling the Oligocene European forms *Entelodon* and the North American Oligocene forms *Archaeotherium* was accepted by Troxell (1920: 248).

Brunet (1979: 75) discussed the generic differences between *Entelodon* and *Archaeotherium* and concluded that they are insufficient to support distinction at the generic level. Scott and Jepsen (1940) came to the same conclusion but chose not to synonymize the two genera because the European form was incomplete. Brunet (1979) had more material to compare but still incomplete skeletal material; much of the anatomy of the European form remains unknown, and its affinity to *Archaeotherium* remains a topic for continuing investigation.

Entelodon or *Elotherium?* There has been longstanding uncertainty as to which generic name, *Entelodon* Aymard, 1846 (1848?), or *Elotherium* Pomel, 1847, was assigned first,

and therefore, which name is to be regarded as the senior synonym for the European genus of entelodonts, and which name has priority at the family level (i.e., Entelodontidae Lydekker, 1883, or Elotheridae Alston, 1878).

Bush (1903) notes that although all reprints of Pomel's 1846 description appear to have been printed in 1848, the original volume (Vol. XII of the *Annals of the Puy Society*) was dated 1846. A letter to Bush from Baron de Vinols, then the president of the Society of Agriculture, Science, Arts, and Commerce at Puy, states: "Ce qui a pu faire confondre la date de 1846 avec celle de 1848 c'est que cette dernière date est sur la couverture du livre et la date de 1846 est dans l'intérieur [Those who have confused the date of 1846 with that of 1848, it is the latter date that is on the outside cover of the book, and the date of 1846 is on the interior] (Bush, 1903: 97–98).

In a discussion of this uncertainty of generic priority, Peterson (1909) points out that although Aymard's (1846) description was likely written in 1846 (or earlier), it could not have been published until late 1847 or 1848 because on page 247 (or page 23 of the reprint) is a footnote that correctly cites page 385 of a separate volume that was published late in 1847 (Peterson, 1909: 43). Peterson concludes: "Pomel's description [1847] of *Elotherium* did probably appear before that of Aymard on *Entelodon*, but, inasmuch as the type of the former was rather inadequate, no illustrations were published, and the type has since been lost (Peterson [1909:43] notes that the British Museum, which acquired Pomel's collection, has no record of the type of *Elotherium*), the present writer is of the opinion that the later name should be used, as both text and figures are clear" (Peterson, 1909: 41).

Simpson (1945) summarized Peterson's (1909) conclusion and added that the name *Entelodon* is more common in recent literature. McKenna and Bell (1997) agree and quote Simpson (1945) verbatim: "The name *Elotherium* probably has priority and is often used (necessitating changes of subfamily, family, and superfamily names as well), but recently *Entelodon* is more common in the literature." As Peterson (1909: 144) has shown: "It is not certain that *Elotherium* is prior, and the type specimen was inadequate, poorly described, unfigured, and is lost. No one but Pomel ever saw it, and it is fair to say that his genus was not recognizably established. *Entelodon* was firmly established, and the name can legitimately continue in use" (Peterson, 1909: 144; *op. cit.* McKenna and Bell, 1997: 412).

For the past 100 years, all of the authors who have visited the question of which generic name should assume priority have agreed that *Entelodon* should be given priority over *Elotherium*, and hence the familial assignment of Entelodontidae should be considered senior to its synonym, Elotheridae. If the name *Elotherium* had been in print before *Entelodon*, it would seem that a strict reading of nomenclatural priority would favor the former name (regardless of the inadequate nature of the description and the absence of a holotype). Because of the potential instability of the nomenclature at the family level (recognition of *Elotherium* would then necessitate the reinstatement of Elotheridae Alston,

1878, which assumes a senior status to Entelodontidae Lydekker, 1883), and based on the arguments presented above (especially Peterson's), it may be prudent to petition the ICZN to formally assign *Entelodon* as the senior synonym of the genus so that continued speculation on the implied and actual dates of publication of these mid-nineteenth century reports will become unnecessary.

Included Species. *E. magnus* (Aymard, 1846); *E. ronzonii* (Aymard, 1846); *E. aymardi* (Pomel, 1853); *E. verdeaui* (Delfortrie, 1874); *E. antiquus* Repelin, 1919; *E. deguilhemi* Repelin, 1919; *E. dirus* Matthew and Granger, 1923; *E. gobiensis* (Trofimov, 1952).

Genus *Brachyhyops* Colbert, 1937

Type Species. *Brachyhyops wyomingensis* Colbert, 1937.

Diagnosis. 3-1-4-3 upper dentition (3?-1-4-3 lower dentition [Wilson, 1971b]), cranium broad with separate temporal crests, orbit closed by a strong postorbital bar, short snout (postorbital length is slightly longer than the preorbital length of the skull), jugal arch extended ventrally (as in *Archaeotherium*), glenoid processes shallow and level with the toothrow, paroccipital process short, and pterygoids weak (Colbert, 1937, 1938a) (Fig. 9.2).

Discussion. The assignment of this entelodont genus has not always been as unambiguous as it is today. In his original description, Colbert (1937) suggested that *Brachyhyops* may be descended from *Parahyus,* which is similar in size but is considerably older. Multiple discoveries since then have shown *Brachyhyops* to be unquestionably an entelodont (Foss, 2001). Colbert's (1938a) description is much more complete and compares the genus to *Choeropotamus, Helohyus, Achaenodon,* and *Parahyus.* Colbert's conclusion that *Brachyhyops* is "possibly descended" from *Parahyus* and that the two genera are "probably" related to the Choeropotamidae contrasts with Scott's (1937) conclusion that *Brachyhyops* and *Parahyus* should be assigned to the Entelodontoidea with *Brachyhyops* in the entelodont family proper. Both authors recognized the difficulty of classifying early bunodont artiodactyls with so few specimens, and given the lack of comparative material at the time, their conclusions were not radically different.

Fig. 9.2. CM 12048, holotype of *Brachyhyops wyomingensis.* From the late Eocene of Wyoming, the skull is vertically crushed and is missing all teeth. Scale bar equals 10 cm.

With the addition of referred material (Russell, 1980a; and unpublished Canadian Museum of Nature [CMN] specimens, including 12048, 12079, and 11989), the cusp arrangement of both the upper and lower molars of *Brachyhyops* may be inferred. It is not known whether the symphysis of *B. wyomingensis* was fused, although it is fused in all other entelodonts (including *B. viensis*). It is also unknown whether *Brachyhyops* had osseous tubercles along the ventral margin of the mandibular ramus.

Brachyhyops may have occurred as early as the Duchesnean if artiodactyl trackways reported in the Baca formation of west-central New Mexico are, in fact, referable to *Brachyhyops* (Schrodt, 1980; Lucas, 1983a, 1983b).

Colbert (1938a) assigned *Brachyhyops* and the similar-sized *Parahyus* to the Choeropotamidae. Scott (1945), disagreeing with his own 1937 conclusion and agreeing with Colbert (1938a), assigned *Brachyhyops* to the "?Choeropotamidae." Later, Gazin (1955) assigned *Parahyus* and *Achaenodon* to the Dichobunidae. Romer (1966) assigned *Parahyus* and *Achaenodon* to the family Achaenodontidae and assigned the Choeropotamidae (with "?*Brachyhyops*") to the superfamily Entelodontoidea. Finally, McKenna and Bell (1997) correctly recognized *Brachyhyops* as an entelodont (Entelodontidae), removed Choeropotamidae from the Entelodontoidea, and assigned *Parahyus* and *Achaenodon* to the Helohyidae. Stucky (1998) validates this latest assignment. In short, achaenodonts (including *Parahyus*) are not entelodonts and should no longer be confused as such, whereas *Brachyhyops* is, without question, an entelodont.

Included Species. *B. wyomingensis* Colbert, 1937: 473; *B. viensis* Russell, 1980a: 3.

Genus *Cypretherium* Foss, new genus

Type Species. *Cypretherium coarctatum* (Cope, 1889: 629).

Comments. The genus is named for the type area in the Chadronian beds of the Cypress Hills Formation of Saskatchewan, Canada. The holotype specimen (CMN 6260) is an incomplete mandibular ramus (Fig. 9.3). However, the animal is well represented by multiple specimens that have been excavated from the Hunter Quarry, also of the Cypress Hills Formation (illustrated in Russell, 1980b).

Diagnosis. "Premolars in a series uninterrupted by diastema except a very short one between p3 and p4" (Cope, 1889: 629) (Russell, 1980b, notes diastema between other premolars in referred material.) "The molars are peculiar in having the two anterior cusps elevated above the three posterior ones" (Cope, 1889: 629). (The two anterior cusps are the protoconid on the buccal side and a twinned paraconid plus metaconid on the lingual side. This trait is similar to both *Eoentelodon* and *Brachyhyops.*) "The posterior, or 5th tubercle, is well developed on the m3" (Cope, 1889: 629). Cope's "5th tubercle" is the hypoconulid, which is quite large. On m3 the hypoconulid is as large as the entoconid and hypoconid, thereby making a three-cusped talonid. This trait is common to both *Eoentelodon* and *Brachyhyops.*

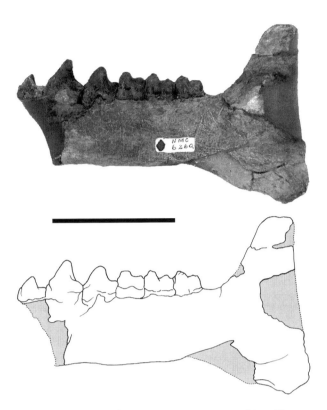

Fig. 9.3. NMC 6260, holotype of *Cypretherium coarctatum*. Left jaw with p2–m3. Scale bar equals 10 cm.

Using referred material (including ROM 11628, here assigned as a plesiotype; Fig. 9.4), Russell (1980b: 6) added upper molars with protocone higher than hypocone, frontal bones with a crenulated dorsal surface (rugose forehead) and with "strong superciliary ridges, which are distinct from superior orbital rim" (supraorbital furrow), double sagittal crest with a median sulcus, "the two crests diverge posteriorly to form the flaring and overhanging lambdoidal crest," jugals extend lateroventrally (as in all Entelodontidae), rounded at distal margin, long fused mandibular symphysis, prominent process on the anterior margin of the mandible (anterior tubercles), posterior mandibular tubercles absent.

Discussion. Russell (1980b) provided a good overview of the entelodont material from the Cypress Hills For-

Fig. 9.4. ROM 11628, complete skull and jaws of *Cypretherium coarctatum* from the Cypress Hills Formation of Saskatchewan. Scale bar equals 10 cm.

mation of Saskatchewan. All of the specimens of *C. coarctatum* that he discussed and referred to are housed at the Royal Ontario Museum (ROM) or Canadian Museum of Nature (CMN). Additional referred specimens of *C. coarctatum* may include Texas Memorial Museum (TMM) 40283-95, a single molar; Field Museum of Natural History Paleo-Mammal Collection (FMNH.PM) 98, a dentary fragment with p4–m2 (Wilson, 1971b); and specimens United States National Museum (USNM) 521245 and 521246, from the Chadron Formation of Flagstaff Rim, Wyoming (Emry, 1973; Lucas and Emry, 2004).

Troxell (1920: 249) reassigned *A. coarctatum* to *Entelodon*: "*Entelodon coarctatus* Cope is undoubtedly quite distinct from any other American species, and from the observations which follow it is apparent that if any of our entelodonts are closely related to *E. magnus* of France, it is this specimen described so long ago by Professor Cope." Troxell's "observations" included: a small diastema between the premolars, single-rooted first molar (this must be p1), crowns of (pre)molars wrinkled and compressed with cutting edges, large second molar (p2), and similarly sized lower molars.

Sinclair (1922:56) disagreed with Troxell and suggested: "In view of the comparisons it is a fair presumption that the Cypress Hills form, from its resemblance to the South Dakota and Nebraska skulls in the matter of crowded lower premolar dentition and other features, should properly be referred to *Archaeotherium* rather than to *Entelodon*."

Scott and Jepsen (1940: 429) then reassigned the species to *Entelodon*: "the fossil is too incomplete for any definite decision of this question. It should, in my opinion, be tentatively assigned to the European genus, until it can be shown that it belongs with the American forms."

Russell (1980b) described many nearly perfect skulls from the Cypress Hills Formation and conclusively demonstrated the distinctiveness of "*A. coarctatum*" from *Entelodon*. It is interesting that Russell stopped there, as the morphology of *Cypretherium* is similar to both *Archaeotherium* (long snout, jugal flanges, etc.) and *Brachyhyops* (highly rugose forehead, fully divided temporal crests, short diastema between the premolars, etc.). The assemblage of characters, although reminiscent of both *Archaeotherium* and *Brachyhyops*, is unique to the Entelodontidae, but the presence of anterior tubercles and absence of posterior tubercles on the inferior margin of the dentary is wholly unique to the family. *Cypretherium* can thus be supported as a separate genus.

Scott and Jepsen (1940: 429) predicted the unique condition of the mandibular ramus when they said, "this species may not have had the conspicuous tubercles on the lower jaw which characterize all the other American forms. If they were present, however, they must have had a different position from that seen in the other species." Scott and Jepsen were, as yet, unaware of the condition of the mandible of *Brachyhyops* and *Eoentelodon*, which, like *A. coarctatum*, have only one pair of tubercles. The difference is that whereas the former taxa have only posterior tubercles, *A. coarctatum* has only anterior tubercles.

Included Species. *C. coarctatum* Cope, 1889; unnamed species from the Chambers Tuff Formation of the Vieja Group, Trans-Pecos, Texas (Wilson, 1971b: 13); unnamed species from the Chadron Formation of Flagstaff Rim, Wyoming (Emry, 1973; Lucas and Emry, 2004).

Genus *Archaeotherium* Leidy, 1850

Type Species. *Archaeotherium mortoni* Leidy, 1850: 90.

Diagnosis. Elongate face, P3 narrower than P4, P3 convex externally (labially), base of enamel elevated above the gum line, P3 double rooted with a single cusp and a slight posterior cingulum, P4 triple rooted (two labial, one lingual) with a cuboidal shape, triangular area between P3 and P4 that allowed occlusion of p4, palatine process of maxillary bone convex, infraorbital foramen at the junction of the maxillary and jugal bones (Leidy, 1850).

Discussion. Leidy (1850: 93) originally identified the teeth of ANSP 10609 as P1 and P2 and, "posterior to the second premolar, the sockets alone for the three fangs of the third premolar" (Fig. 9.5). The teeth are actually P3 and P4, and the "sockets" represent the former location of M1. Leidy (1852) corrected this error.

The characters identified above represent a thorough description of ANSP 10609. Although the morphology is unquestionably entelodont, there are no traits that can be identified as species specific, and the generalized location of the horizon and locality prevents even a topotype from being assigned.

Leidy (1852), however, offered USNM 146 as a plesiotype (Fig. 9.6). This specimen, although a juvenile, has enough adult characters to adequately support a species description. Whereas both specimens (holotype and plesiotype) are from the White River Group, it is not certain that they are from the same formation (although circumstantial evidence supports an inferred origin in the late Eocene Chadron Formation for each specimen). The species *A. mortoni* must be supported by characters observed on the plesiotype, and although it may be correct to designate the species name as a *nomen dubium* (based on the incomplete nature of the holotype), it was Leidy's intention to support the species on both specimens, and because of the longstanding use of the taxonomic name, there would be little utility in suppressing the name. If confusion persists, it may be necessary to petition the ICZN to suppress the holotype (ANSP 10609) in favor of reassigning the plesiotype (USNM 146) as neotype.

It is necessary, however, to support the genus solely on the characters of the plesiotype, USNM 146, which agrees generally with the holotype ANSP 10609 while offering additional characters that may be used to support the genus. Derived characters that appear to be unique to *Archaeotherium* appear to be the loss of an ectoflexus on M2 (unique to the Entelodontidae) and the loss of a hypocone on M3 (unique to *Archaeotherium*). Leidy (1853) also used a third specimen, USNM 117, to support his description of the genus.

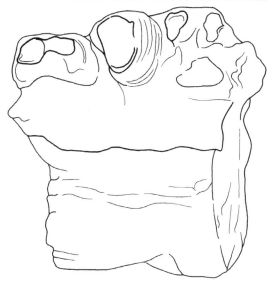

Fig. 9.5. ANSP 10609, holotype of *Archaeotherium mortoni*. Portion of left rostrom with P3 and P4 from the lowest (late Eocene) White River Group, Nebraska or Wyoming. Scale bar equals 10 cm.

Over the past 150 years many different taxa have been assigned to *Archaeotherium*; many have been synonymized with previously described taxa, whereas others have been suppressed as *nomina dubia*. However, the genus remains in need of thorough examination and revision. There are currently three distinct morphotypes within the genus *Archaeotherium*: the standard taxa (including the type species *A. mortoni*); Subgenus A, which includes enormous taxa with massively enlarged jugal flanges ("*Megachoerus*") and huge knob-like mandibular tubercles ("*Pelonax*"); and Subgenus B, which includes an unusually robust form, unlike any other known entelodont, known only from the upper Turtle Cove Member of the John Day Formation in eastern Oregon (Fig. 9.7). Subgenus A and subgenus B almost certainly belong to separate grades, but until the included taxa can be better characterized, it is premature to assign these to distinct genera.

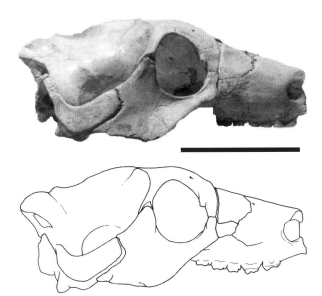

Fig. 9.6. USNM 146, plesiotype of *Archaeotherium mortoni*. Portion of juvenile skull from the lowest (late Eocene) White River Group, South Dakota. Scale bar equals 10 cm.

Included Species. *A. mortoni* Leidy, 1850; *A. crassum* (Marsh, 1873a); *A. marshi* Troxell, 1920; *A. scotti* Sinclair, 1921; *A. wanlessi* Sinclair, 1921; *A. palustris* Schlaikjer, 1935.

Subgenus A. *A. ramosum* (Cope, 1874a); *A. zygomaticus* Troxell, 1920; *A. latidens* (Troxell, 1920); *A. caninus* (Troxell, 1920); *A. altidens* (Sinclair, 1921); *A. praecursor* (Scott and Jepsen, 1940); *A. lemleyi* (Macdonald, 1951); *A. trippensis* Skinner et al., 1968.

Subgenus B. *A. calkinsi* (Sinclair, 1905).

Genus *Daeodon* Cope, 1878

Type Species. *Daeodon shoshonensis* Cope, 1878.

Diagnosis. Peterson (1909) reports the major characters of *Daeodon* as reduced median incisors, square P3, P4 with two cusps, long M3 with large meta- and hypocones, elevated trigonid, diastemata between all premolars, jugal flange relatively smaller than *Archaeotherium*, posterior jugal bone forms buttress at the anterior glenoid of the jaw, small anterior and large knoblike posterior tubercles on the inferior margin of the jaw, smooth caudal slope to the slightly dependent angle of the jaw, short chin, fibula co-ossified with the tibia, trapezium absent, and metatarsal V "sometimes" absent (Peterson, 1909: 67) (Fig. 9.8).

In addition to Peterson's (1909) characterization of *Daeodon*, the following characters may be unique to the genus: on the manus a distally enlarged lunate, poorly contacting magnum and unciform on the dorsal surface, and absence of a trapezium. On the pes, the cuboid and metatarsal do not come into articular contact (Peterson, 1909: 117; Foss and Fremd, 1998: 65–67).

Discussion. Lucas et al. (1998: 431) argued for a synonymy of the "giant" entelodont taxa, *Ammodon* and *Dino-*

hyus, with *Daeodon* and, although they maintained five species, suggested "we find it difficult to evaluate the validity of these taxa and offer the tentative, conservative conclusion that they represent a single species." Foss and Fremd (1998: 66–67) reassigned *Boochoerus* to *Daeodon*. After noting the differences between the postcranial skeleton of *D. hollandi* and *D. humerosum*, they concluded: "it is impossible to tell whether this difference [robustness] is attributable to individual or population level variation or if it is evidence of sexual dimorphism. We believe, however, that the differences outlined above do not justify separation at the generic level" (Foss and Fremd, 1998: 66–67). The difficulties in separating intraspecific from specific level variation in entelodonts, exacerbated by the paucity of comparative material, makes direct comparison among the various species of *Daeodon* nearly impossible, so it is difficult to characterize any single species of *Daeodon*, except for *D. hollandi*.

Daeodon or Dinohyus? In 1878 Cope named a new genus and species of ungulate (*Daeodon shoshonensis*) that he subsequently (Cope 1881: 397, 1884b: 713, 1887: 1063) considered a member of the Menodontidae (Perissodactyla, Brontotheridae). Peterson (1909: 63) and Matthew (1909a: 108) were the first authors to recognize this genus as an entelodont. The holotype specimen, *Daeodon shoshonensis* (AMNH 7387), consists of only the symphyseal portion of a mandible, so the only diagnostic characters at the family level are the steeply-sloped anterior surface and completely fused and robust mandibular symphysis. At the generic level, *D. shoshonensis* is recognized as having large, robust canines and incisors that incrementally increase in size from i1

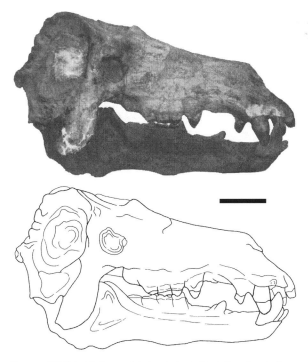

Fig. 9.7. UCMP 953, holotype of *Archaeotherium calkinsi*. Complete skull and jaws from the upper Turtle Cove Member of the John Day Formation, Oregon. Scale bar equals 10 cm.

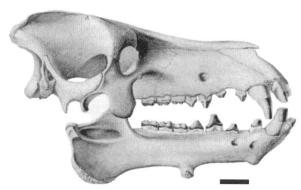

Fig. 9.8. CM 1594, *Daeodon hollandi*. Complete skull and jaws from the Agate Springs Fossil Quarry in the lower Harrison Formation, Nebraska. Scale bar equals 10 cm. (Reproduced from Peterson, 1909, plate 55.)

to i3, only a small recognizable i3–c1 diastema, and symphyseal tubercles absent (though often considered diagnostic of the family, they are small in *Daeodon hollandi* [Peterson, 1905]). Lucas et al. (1998) believed the difference between small and absent tubercles in giant entelodonts was well within the variation for a single genus and synonymized *Dinohyus* with *Daeodon*. They even considered that *Daeodon* may be comprised of only a single species, *D. shoshonensis*.

In 1868, Marsh named a new species of giant entelodont (*Elotherium leidyanum*) from the Squankum Marl deposits of Monmouth County, New Jersey. He later placed it in a new genus (*Ammodon leidyanus* Marsh, 1893). *Ammodon* is known from a single right p4 (YPM 12040, holotype) and a left m3 (YPM 12041, paratype); these two teeth are probably from the same specimen (Peterson, 1909:67). Although it comprises only two isolated elements, the material is in good condition and shares characters with the other giant entelodonts, such as an independently derived fifth posterior cone or heel (hypoconulid?) on m3. Peterson (1909) chose to maintain the generic status of *Ammodon* in spite of its similarities to *Dinohyus* (see Lucas et al., 1998, for a more complete discussion). Lucas et al. (1998) synonymized *Ammodon* with *Daeodon*. This would not have been possible without the holotype specimen of *Dinohyus hollandi* (CMNH 1594, complete skeleton), because *Daeodon shoshonensis* (AMNH 7387, mandibular symphasis) and *Ammodon* leidyanus (YPM 12040, rp4 and YPM 12041, lm3) do not have any skeletal elements in common.

Foss and Fremd (1998) added *Boochoerus humerosus* Cope, 1879a, (AMNH 7380, complete manus, pes, and forelimb) to this synonymy of *Daeodon*, citing similar foot bone arrangement and limb proportions to CMNH 1594 (complete skeleton). They continued to recognize *D. humerosum* as a valid species (Foss and Fremd, 1998: 65).

The priority of *Daeodon* (in spite of the fragmentary nature of AMNH 7387, mandibular symphasis) over *Dinohyus* (CMNH 1594, complete skeleton) is one that bears explanation. All of the type specimens of *Daeodon*, *Ammodon*, and *Boochoerus* are similar to *Dinohyus* at the generic level (Lucas et al., 1998; Foss and Fremd, 1998) and therefore

should be recognized as a single genus. The holotype (CMNH 1594, complete skeleton) and referred specimen (UNSM 1150, complete skeleton) of *"Dinohyus hollandi"* are clearly the best preserved and most complete specimens of "giant" entelodonts. Unfortunately, by taxonomic convention, the geno-holotype name must bear tribute to the original named in literature, in this case *Daeodon*.

If one were to argue that the type specimen of *Daeodon shoshonensis* Cope, 1878, were insufficient as a holotype of the genus, *Boochoerus* Cope, 1879a, would assume holotype status. This may not be any more desirable than *Daeodon* and perhaps less so because the name *Boochoerus* (aside from Peterson, 1909, and Loomis, 1932) has never been in common usage. If one were to conclude that both *Daeodon* and *Boochoerus* were insufficient holotypes, then the name *Ammodon* Marsh, 1893, would default as the holotype of the group. This specimen (rp4 and lm3) is less appropriate as a holotype than *Daeodon* (mandibular symphysis). Only by assigning the three genera: *Daeodon*, *Boochoerus*, and *Ammodon*, as *nomina dubia* or through intervention by the ICZN can *Dinohyus* (CMNH 1594, complete skeleton) be legitimately designated the holotype of "giant" entelodonts, and at this time neither step is warranted.

By virtue of its early reference in literature (Cope, 1878d) and by its history of continued popular usage (Cope, 1881: 397, 1884b: 713, 1887: 1063; Zittel, 1892: 337; Rogers, 1896: 194; Trouessart, 1898: 740; Palmer, 1904: 214; Matthew, 1909a: 108; Peterson, 1909: 63; Loomis, 1932: 361; Simpson, 1945; Brunet, 1979: 88; Gallagher and Parris, 1996: 11; McKenna and Bell, 1997: 412; Lucas et al., 1998; Foss and Fremd, 1998: 64), the single genus of "giant" entelodonts from North America is *Daeodon*.

Included Species.   *D. shoshonensis* Cope, 1878a; *D. bathrodon* (Marsh, 1874); *D. humerosum* (Cope, 1879a, 1879b); *D. leidyanus* (Marsh, 1893); *D. potens* (Marsh, 1893); *D. hollandi* (Peterson, 1905a, 1905b); *D. mento* Allen, 1926; *D. minor* Loomis, 1932; *D. angustus* (Loomis, 1932); *D. minimus* Schlaijker, 1935.

Genus *Paraentelodon* Gabunia, 1964

Type Species.   *Paraentelodon intermedium* Gabunia, 1964.

Diagnosis.   Large entelodont with incisors that increase in size from I1/i1 to I3/i3; large p2 that exceeds p3 and p4 in overall size; and large square molars. The upper molars have six blunt cusps, including well-developed paraconules and metaconules, and a heavy precingulum (Lucas and Emry, 1999, in part). There are "flangelike" tubercles on the inferior ramus of the dentary below p2 (Aubekerova, 1969; Lucas and Emry, 1999).

Discussion.   *Paraentelodon* has been compared, mostly by overall size and tooth morphology, to *Daeodon* (Lucas et al., 1998; Lucas and Emry, 1999). However, the "flangelike" tubercle on the dentary is reminiscent of certain large Whitneyan archaeotheres of North America, specifically *"Pelonax"* (see *Archaeotherium* Subgenus A, above).

Paraentelodon is widely distributed across Asia, including the Georgian Republic, Kazakhstan, and China. Lucas and Emry (1996) synonymized all of the species of *Paraentelodon* and *Neoentelodon* into a single species. They went on to hypothesize that *Paraentelodon* may have given rise to the giant entelodonts of North America (*Daeodon*) via a late Oligocene Beringian immigration (Lucas et al., 1998), thus supporting a biogeographic hypothesis originally put forward by Brunet (1979).

Included Species. *P. intermedium* Gabunia, 1964.

INDETERMINATE TAXA

Helohyus lentus (Marsh, 1871: 39)

Diagnosis. Represented by a single left dentary with m3; the tooth has four blunt cusps and a hypoconulid in the center of the posterior cingulum, a twinned parametaconid, elevated trigonid, and small anterior cingulum. "There is a prominent rugose tubercle on the inner superior margin of the lower jaw, a short distance behind and above the last molar" (Marsh, 1871a: 40).

Discussion. If the specimen were an entelodont, it would be the oldest known member of the family. The size, "about one half the size of *Elotherium* [*Archaeotherium*] *mortoni*," is within the size range of *Eoentelodon*, *Brachyhyops*, and *Parahyus*, and the description of m3 is most consistent with the latter genus.

Marsh (in King, 1878) reassigned the specimen to *Helohyus*, but similar isolated dentary and maxillary fragments from the middle to late Eocene continue to be assigned to the Entelodontidae (see Peterson, 1934: 375–376; Gazin, 1956: 25–26).

Referred specimens (including AMNH 89612 and 89614) show a strong sharp cristid obliqua, trigonid and talonid of similar heights, relatively sharp cusps, separate para- and metaconids, and a fully developed hypoconulid shelf with a large, relatively sharp hypoconulid; all uncharacteristic of the Entelodontidae. It seems probable that this species is a member of the Helohyidae, consistent with Marsh's 1878 reassignment.

BIOGEOGRAPHY

Lucas et al. (1998) suggested that the giant entelodonts of North America are a result of a late Oligocene immigration event of large entelodonts (i.e. *Paraentelodon*) to North America via Beringia. This hypothesis is supported by the similar morphology of the two species and the apparent paucity of entelodont taxa during the North American Arikareean land-mammal "age." However, a taxon of large entelodont in the John Day region of Oregon (*Archaeotherium*

calkinsi) has features of both *Archaeotherium* and *Daeodon* as well as a unique suite of characters of its own. It is temporally contemporaneous with *Paraentelodon* of Asia. A late Oligocene Beringian association between *Daeodon* and *Paraentelodon* is morphologically supported. However, converse to Lucas's hypothesis of an Asian origin of giant entelodonts, the possibility of *Paraentelodon* and *Daeodon* sharing a common ancestor in North America must also be entertained (Foss and Fremd, 2001).

EVOLUTIONARY HISTORY

The skeleton of the earliest entelodonts (*Eoentelodon* and *Brachyhyops*) is unknown. Therefore, the phylogenetic placement of entelodonts with other early artiodactyl groups can be only provisionally inferred. Entelodonts first appeared in Asia during the late-middle Eocene (*Eoentelodon*) and appeared in North America shortly after that (*Brachyhyops*, approximately 38 Ma). The lineage is widely distributed throughout Asia and Europe (*Eoentelodon*, *Entelodon*, and *Paraentelodon*), but discoveries are isolated, widely dispersed, and usually fragmentary. Virtually any complete skull or articulated skeletal elements found in Europe or Asia will contribute significantly to the knowledge of the Eurasian taxa. In North America, the lineage appears temporally continuous (*Brachyhyops*, *Cypretherium*, *Archaeotherium*, *Daeodon*). However, interchange between North America and Eurasia is likely to have taken place. The last appearance of the Entelodontidae is correlated with the Agate Springs quarry in northwestern Nebraska, which is earliest Hemingfordian (approximately 19 Ma) in age (Hunt, 1990).

ACKNOWLEDGMENTS

I thank my colleagues for many years of patience and support while I investigated the entelodont family, including Jessica Theodor, Spencer Lucas, Christine Janis, Don Prothero, Robert Hunt, Jr., R. Matthew Joeckel, Theodore Fremd, J. Michael Parrish, and many others. Numerous curators, collections managers, and registrars in both North America and Europe have spent many hours assisting me with observing, photographing, measuring, and casting entelodont specimens in their care. To the many research librarians who have successfully assisted me with tracking down obscure references and the museum registrars and archivists who have offered their expertise and patience while tracking down original field notes and correspondence, I am indebted. Thank you to Jude Higgins for producing the original illustrations on very short notice. As always, I thank Jessica Theodor for her continued technical assistance and collaboration, and Lia Vella for her thorough proofreading, comments, and never-ending friendship.

JOHN M. HARRIS
AND LIU LI-PING

10

Superfamily Suoidea

STUDY OF THE PIGLIKE SUPERFAMILY Suoidea has been contentious throughout the past half century, with much disagreement as to the number of families and subfamilies represented in the fossil record and the content and relationships of those entities. McKenna and Bell (1997) attributed Suidae (pigs), Tayassuidae (peccaries), and the extinct Sanitheriidae (sanitheres) to the Suoidea. They also included the Hippopotamidae, but hippos are now more properly regarded as close relatives of anthracotheriids (Colbert, 1935b; Viret, 1961; Gentry and Hooker, 1988; Boisserie et al., 2005a, 2005b). As the systematics of the fossil suoids has yet to be unambiguously resolved, we here summarize the apparent status quo pending future detailed reappraisal of this group.

Compared to the highly diverse ruminants, the suoids today form a small and cohesive group of artiodactyls, with only seven genera and 12 species worldwide (Table 10.1). The sister relationship between Suidae and Tayassuidae is well documented on both morphological (Gentry and Hooker, 1988) and molecular grounds (Irwin and Arnason, 1994; Randi et al., 1996). Pigs and peccaries occupy similar adaptive zones (*sensu* Simpson, 1944) in the Old and New World, respectively. Both are primarily omnivorous and, of the living artiodactyls, display the most primitive traits, retaining low-crowned cheek teeth with simple bunodont cusps, four distinct digits, separated foot bones, absence of frontal appendages, and a simple, nonruminating stomach.

The early history of the superfamily has remained obscure. Until the recent discoveries of Eocene suoid fossils from South Asia, the geological range of suoids was considered to be early Oligocene to Recent (Simpson, 1984). The oldest suoid fossils now known are from the late Eocene of China (Tong and Zhao, 1986; Liu, 2001) and Thailand (Ducrocq, 1994a; Ducrocq et al., 1998). Although the evidence is still sparse

Table 10.1. The living suids and their occurrences by Nowak (1991) and McDonald (1999)

Family	Genus	Species	Geographic distribution
Suidae			
	Sus	*S. scrofa*	Most of Eurasia, North Africa
		S. barbatus	Malay Peninsula, Rhio Archipelago etc.
		S. celebensis	Sulawesi and nearby small islands
		S. verrucosus	Java, Madura, and Bawean
		S. salvanus	Himalayan foothills of Assam
	Potamochoerus	*P. porcus*	Sub-Saharan Africa and Madagascar
	Hylochoerus	*H. meinertzhageni*	Central African Congo Basin
	Phacochoerus	*P. aethiopicus*	Sub-Saharan Africa
	Babyrousa	*B. babyrussa*	Sulawesi, Togian, Sula, and Bura Islands
Tayassuidae			
	Catagonus	*C. wagneri*	Southern Bolivia, Paraguay, Northern Argentina
	Tayussu	*T. tajacu*	Arizonas and Texas to Northern Argentina
		T. pecari	SE Veracruz State, Mexico to Northern Argentina

and incomplete, it appears that suoids originated in eastern Asia during the Eocene, subsequently dispersing into the New World (Tayassuidae) and elsewhere in the Old World (Suidae) during the Oligocene (Ducrocq, 1994a; Ducrocq et al., 1998; Liu, 2001).

A group of early suoids known as palaeochoeres or "Old World peccaries" was interpreted by Pickford (1993) as a subset of the Tayassuidae and by van der Made (1997a) as a separate family, but McKenna and Bell (1997) divided constituent species between the Palaeochoerinae (Suidae) and Doliochoerinae (Tayassuidae). Liu (2003) confirmed that the palaeochoeres were not a monophyletic group. Furthermore, she demonstrated that none of the "Old World suoids" is a true peccary and redistributed these taxa to either the Suidae or an unresolved stem group of Suoidea (Table 10.2).

Opinions on the taxonomic position of sanitheres range from independent family status (Sanitheriidae: Pickford, 1993) to a genus of the Palaeochoerinae (van der Made, 1997a).

Superfamily SUOIDEA Gray, 1821

The Suoidea is united by the uniquely derived features (autapomorphies) of absence of a lingual cingulum on the upper molars, i2 similar to i1, rootless lower male canine, paraconid fused to metaconid, tympanic process of squamosal dorsoventrally elongate, and external auditory meatus opening dorsally, and the presence of ossified tympanic bullae (Liu, 2003). Eocene species recovered from eastern Asia during the past 15 years were originally assigned to the Tayassuidae (*Odoichoerus unicornis* Tong and Zhao, 1986; *Egatochoerus jaegeri* Ducrocq, 1994) or Suidae (*Siamochoerus banmarkensis* Ducrocq et al., 1998; *Eocenchoerus savagei* Liu, 2001) or Palaeochoeridae (*Siamochoerus viriosus* and *Huaxiachoerus guangiensis* Liu, 2001). Subsequent cladistic analysis by Liu (2003) indicated that the Palaeochoeridae were not a monophyletic group, and she placed *Huaxichoerus, Egatochoerus* (Fig. 10.2), *Siamochoerus* (Fig. 10.3), and *Odiochoerus* as an unresolved group within

the Suoidea while identifying *Eocenchoerus* (Fig. 10.4) and *Palaeochoerus* as "true" suids (Table 10.2).

Family SUIDAE Gray, 1821

The family Suidae is separated from the other suoids by the absence of the angular process of the mandible (Liu, 2003). Suid molars are characterized by the presence of three furrows or "furchen" on each of the four main cusps that are numbered 1–12 (Hünermann, 1968; Pickford, 1986) (see Fig. 10.1). The Suidae has been divided into various subfamilies, of which seven were recognized by McKenna and

Table 10.2. Primary classification of the Suoidea by Liu (2003)

Superfamily	Family	Subfamily	Genus
Suoidea	Indeterminate		*Huaxiachoerus*
			Egatochoerus
			Siamochoerus
			Odiochoerus
	Suidae	Indeterminate	*Eocenchoerus*
			Palaeochoerus
			Sinapriculus
			Hyotherium
			Miochoerus
		Schizochoerinae	*Taucanamo*
			Schizochoerus
			Yunnanochoerus
		Listriodontinae	*Listriodon*
			Bunolistriodon
		Kubanochoerinae	*Kubanochoerus*
		Tetraconodontinae	*Conohyus*
			Parachleuastochoerus
		Suinae	*Chleuastochoerus*
			Microstonyx
			Propotamochoerus
			Hippopotamodon
			Sus

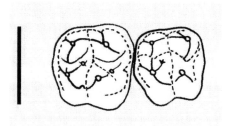

Fig. 10.1. *Egatochoerus jaegeri* Ducrocq, 1994; TF 2672, right M1–M2; scale bar equals 10 mm (after Ducrocq, 1994a: Fig. 1).

Fig. 10.2. *Siamochoerus banmarkensis* Ducrocq et al., 1998; TF 2905; lower left dental row (p4–m3); scale bar equals 10 mm (after Ducrocq et al., 1998: Fig. 3).

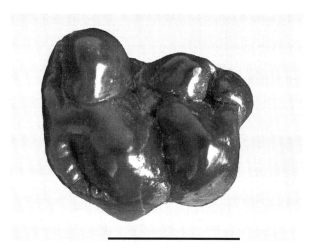

Fig. 10.3. *Eocenchoerus savagei* Liu, 2001; V 7881; holotype right M3; scale bar equals 10 mm.

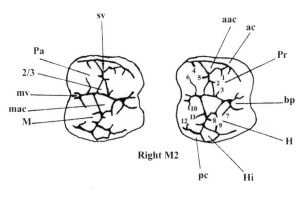

Right M2

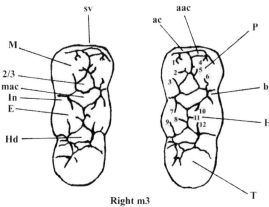

Right m3

Fig. 10.4. Upper and lower molars of *Hippopotamodon sivalense* with anatomical features labeled (after Pickford, 1986: Fig. 2). 1–12 = groove (furchen) numbers; 2/3 = 2/3 cusplet; aac = accessory cusp; ac = anterior cingulum; bp = basal pillar; E = entoconid; H = hypocone; Hd = hypoconid; Hl = hypoconule; Hld = hypoconulid; l = lingual notch; M = metacone; mac = median accessory cusp; Md = metaconid; mv = median valley; P = protoconid; Pa = paracone; pc = posterior cingulum; Pr = protocone; sv = sagittal valley.

Bell (1997): Palaeochoerinae, Tetraconodontinae, Hyotheriinae, Kubanochoerinae, Listriodontinae, Namachoerinae, and Suinae. Liu (2003) subsequently determined that palaeochoeres were not a monophyletic group. She moved Schizochoerinae from Tayassuidae to Suidae and grouped its component genera on the basis of the deep transverse valley floor of their molars. The Eurasian hyotheriines were interpreted as the root stock of the true suids by Pickford (1993), although, again, Liu (2003) found no support for this subfamily as a monophyletic entity. The Kubanochoerinae, Listriodontinae, Tetraconodontinae, and Suinae are well supported (Table 10.2). The Namachoerinae and Cainochoerinae were not reviewed in Liu's (2003) analysis. The earliest recognized suid is *Eocenchoerus savagei* (Liu, 2001), although, because of the sparse record of Paleogene suoids, this cannot be assigned to an existing subfamily (Liu, 2003).

The subfamily Palaeochoerinae is not recognized here. The taxonomic status of palaeochoeres or the "Old World peccaries" has been highly controversial. The dental characters used to define them are plesiomorphic and equally well support their assignment to either Tayassuidae (Van der Made, 1989–1990a; Van der Made and Han, 1994; Pickford, 1993) or Suidae (Pickford, 1988a) or even a separate family (Pickford, 2004). Liu (2003) reassigned most of the genera previously identified as palaeochoeres elsewhere, leaving *Palaeochoerus* as an unresolved genus of the family Suidae.

Palaeochoerus is characterized by an unreduced eutherian dentition and by a basicranium in which the glenoid cavity is located low down, more or less at the level of the occlusal surface of the cheek teeth (Pickford, 1993; Pearson, 1928). It is known in Europe from the early and middle Oligocene and from the early Miocene (= *Xenohyus*). Ducrocq (1994a) suggests that direct ancestry of the European taxa should be sought among conservative forms resembling the Chinese *Odoichoerus*, which Liu (2003) grouped among the unresolved suoid genera. African palaeochoere teeth reported from Fort Ternan (12.5 to 13.5 Ma) by Pickford (1993) were close in shape and size to those of *Albanohyus* (which van der Made [1989–90a] compares to the similarly small *Cain-*

ochoerus from South Africa). In Asia, the palaeochoeres that appeared as early as the late Eocene (Ducrocq, 1994a; Ducrocq et al., 1998; Liu, 2001) are now interpreted to represent the early radiation of suoids (Liu, 2003).

The Eurasian hyotheriines were interpreted as the rootstock of the true suids by Pickford (1993), but Liu (2003) found no support for this subfamily as a monophyletic entity. Genera previously attributed to the Hyotheriinae were primitive, relatively short-snouted forms that retained complete eutherian dentitions, bunodont cheek teeth, and other plesiomorphic features. They ranged from MN 1 to MN 9 (van der Made, 1989–90b) and were also documented from China MN 5 (Liu et al., 2002). They were formerly considered to be an ancestral group that gave rise to all other suid subfamilies by periodically producing longer-snouted forms with dental specializations that could be attributed to separate subfamilies on account of their derived morphology (Pickford, 1993). Liu et al. (2002), for example, suggested that *Hyotherium shangwangense* (Fig. 10.5) lay close to the ancestry of the tetraconodonts.

As with the palaeochoeres, there has been little agreement on what constitutes a hyotheriine. Pickford (1993) placed *Hyotherium* and *Aureliachoerus* in this subfamily. Hellmund's (1992) revision of the palaeochoeres relocated *Dubiotherium* (= *Palaeochoerus*) to the Hyotheriinae. McKenna and Bell (1997) list *Dubiotherium, Hyotherium,* and *Chleuastochoerus* in the Hyotheriinae but place *Aureliachoerus* in the Palaeochoerinae. Van der Made and Morales (1999) interpreted *Aureliachoerus* as a suid, not a palaeochoerid. Liu et al. (2002) placed the early Miocene *Sinapriculus linquensis* in the Hyotheriinae although they recognized that it may represent the survival of an earlier suid radiation. Liu later (2003) interpreted *Chleuastochoerus* as a suine and provisionally grouped *Hyotherium* and *Sinapriculus* together with *Eocenchoerus, Miochoerus,* and *Palaeochoerus* as unresolved genera within the family Suidae.

Subfamily LISTRIODONTINAE Lydekker, 1884

Listriodontinae appear in the late middle Miocene of Africa, India, and Europe (MN 4). Listriodonts are charac-

terized by a specialization of the incisor area, the anterior part of the snout being enlarged with a concurrent widening of the incisors, and the crowns of the central upper incisors form a transverse ridge (Orliac, 2006). An early form from Spain, identified as *Eurolistriodon adelli* by Pickford and Moyà-Solà (1995), is very bunodont, and its teeth differ from those of *Hyotherium* only by their greater size. Its skull, however, displays features seen in the more lophodont *Listriodon,* including retired distal margin of palatines, elongated diastemata, spatulate premaxilla, laterally flaring canine flanges, and broad but short neurocranium. The canine flanges are sexually dimorphic, male flanges often doubling the width of the anterior part of the snout. In the bunodont listriodonts the P4 cusps are discrete, whereas in the more lophodont forms they are closely applied to each other. P1 is rudimentary in the listriodonts; where present it is located just behind the canine and is widely separated from the rest of the premolars. The basic suid furchen are distinguishable in unworn listriodont molars, but they soon become obliterated by wear. The talon is simple, and the enamel is moderately thick. Progressive listriodonts are immediately identifiable by their lophodont molars (Figs. 10.6 and 10.7), which represent an adaptation to folivory (Hunter and Fortelius, 1994). The four main cusps unite to form two transverse lophs, and the anterior, median, and posterior accessory cusps are reduced.

Listriodonts have a very long mandible, in which the symphysis is elongated and the ascending ramus is retired. The symphysis is splayed outward, so that the lower canines emerge almost horizontally.

Listridodontinae are represented by four genera—*Bunolistriodon, Eurolistriodon, Lopholistriodon,* and *Listriodon*—the latter descending directly from *Bunolistriodon* but differing by the more perfectly developed lophodonty and by extremely spatulate upper incisors that occlude with the first two lower incisors. Pickford and Moyà-Solà (1995) erected *Eurolistriodon* as a replacement for *Bunolistriodon* Arambourg, 1963, but van der Made (1997b) contended that *Bunolistriodon* is a valid name, and *Eurolistriodon* is a junior synonym, and McKenna and Bell (1997) agreed. Orliac (2006), however, pointed out that the orientation and modality of growth of the upper male canine differed markedly in *Bunolistriodon* and *Eurolistriodon* and retained the latter genus (for *E. adelli* and *E. tenarezensis* Orliac, 2006) even though it was a *nomen nudum.*

Fig. 10.5. *Hyotherium shanwangense* Liu et al., 2002; V 11942.1; holotype skull, ventral view; scale bar equals 10 mm.

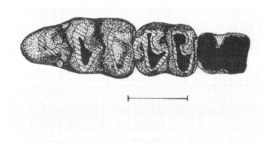

Fig. 10.6. *Listriodon pentapotamiae* Falconer, 1868. GSP 4527; right m1–m3, occlusal view; scale bar equals 20 mm (after Pickford, 1988a: Fig. 49).

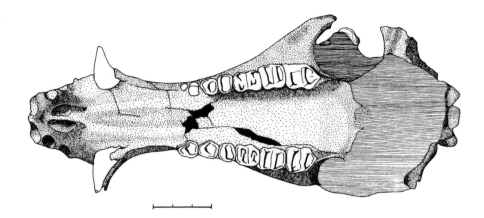

Fig. 10.7. *Lopholistriodon kidogosana*
Pickford and Wilkinson, 1975; KNM-BN
992; holotype cranium, palatal view;
scale bar equals 30 mm (after Pickford,
1986: Fig. 63).

Characteristically lophodont *Listriodon* occurs in both Africa and Eurasia, but there are also some bunolophodont species (*Bunolistriodon intermedius*: Ye et al., 1992; Liu and Lee, 1963). Ginsburg (1977) and van der Made (1999a) suggested that an initial radiation of bunodont listriodontines was followed by radiation of lophodont forms; Pickford and Morales (2003) suggested that a single radiation of bunodonts was followed by independent development of lophodonty in the different continents. During much of the middle Miocene South and Central Europe, South Asia and India, and Africa formed a single biogeographic realm with unrestricted faunal interchange between the different regions throughout the time that listriodonts were present. *Listriodon* was recognized from Mbagathi, Maboko, and Fort Ternan by Pickford (1986), although van der Made (1992) thought these specimens belonged in *Lopholistriodon*. Pickford (1995) relocated the diminutive *Lopholistriodon moruoroti* to the Namachoerinae.

Subfamily KUBANOCHOERINAE Gabunia, 1958

Although Pickford (1986, 1993) recognized Kubanochoerinae as a distinct subfamily, van der Made (1997b) noted that three groups of derived characters are shared by kubanochoeres and listriodonts: mesiodistal diameter of the incisors and related characters such as anterior mandible shape, decrease in the height of the crown of the incisors, and morphology and orientation of the first upper incisor. Accordingly, van der Made viewed the kubanochoeres as a tribe within the Listriodontinae rather than a separate subfamily. McKenna and Bell (1997) follow Pickford (1993) in attributing *Nguruwe* (Fig. 10.8), *Kenyasus* (Fig. 10.9), *Libycochoerus* (Fig. 10.10), *Kubanochoerus,* and *Megalochoerus* to the Kubanochoerinae. Liu (2003) further supported this by documenting that this group shared some unique autapomorphies such as a frontal bone horn in males and strong orbital protuberances, neither of which are found in the Listriodontinae.

Kubanochoerinae basicrania have elevated glenoid cavities, widely separated glenoid fossae, bullae, and paroccipi-

Fig. 10.8. *Nguruwe kijivium* (Wilkinson, 1976); Nap I′64; holotype left M1–M3, occlusal view; scale bar equals 25 mm (after Pickford, 1986: Fig. 6).

Fig. 10.9. *Kenyasus rusingensis* Pickford, 1986; KNM-RU 2810, right M3, occlusal view; scale bar equals 25 mm (after Pickford, 1986: Fig. 30).

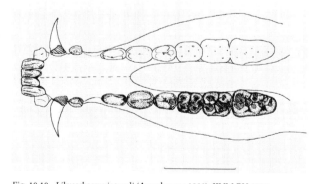

Fig. 10.10. *Libycochoerus jeanneli* (Arambourg, 1933); KNM-RU 2785; reconstructed mandible, occlusal view; scale bar equals 50 mm (after Pickford, 1986: Fig. 46).

tal processes. The posterior choanae are V-shaped and open immediately behind M3 except in some advanced forms. The snout is elongated, and diastemata are present. Canine flanges are not developed. Some later kubanochoeres develop a frontal horn and paired postorbital swellings. The premaxillae are spatulate and do not meet anteriorly. The zygoma is not widely flared, and the snout is not particularly squared in section.

Kubanochoerinae upper premolars are similar to those of listriodonts but differ by the greater separation of the two labial cusps of P4 and the presence of an internal cingulum. In addition, the anterior loph is less complete than that of listriodonts. P1 is a prominent tooth in kubanochoeres and is not widely separated from its neighboring premolars. A smaller version of P2, P1 is fully functional but does become reduced in size in progressive species such as *Kubanochoerus robustus.*

Kubanochoerinae possess rather simple molars in which the uppers have labial cingula, a feature that in suids is otherwise common only in tetraconodonts. There are traces of cingula in the lower molars too. Another unique feature of kubanochoeres is that furchen 6 of the upper molars migrates lingually to such an extent that it is visible from the lingual aspect (Pickford and Ertürk, 1979).

Kubanochoere mandibles are characterized by a shallow horizontal ramus, by the ascending ramus hiding the rear portion of m3 in side view, by the presence of short diastemata, by a robust p1, and by the elongated though not particularly flared symphysis. The incisive border is not retired. In the lower molars, the main cusps are more evenly sized than in other suids, and there is similar flare on labial and lingual cusps, although the main cusps are almost all the same height above the cervix.

The earliest Kubanochoerinae occur in Africa. *Nguruwe* and *Kenyasus* are superficially similar to *Hyotherium,* leading Wilkinson (1976) to identify them as hyotheriines. Later genera increased greatly in size and emigrated from Africa to Turkey, Georgia, Pakistan, and China where they are represented by *Libycochoerus, Kubanochoerus,* and *Megalochoerus.* At least two lineages have horns on the frontal part of the skull (Gabunia, 1960; Arambourg, 1963; Qiu et al., 1988). *Libycochoerus* has small bosses above each orbit, and there may have been a central frontal horn in males. *Kubanochoerus* developed a long centrally placed frontal horn that extended forward over the snout and smaller horns above the orbits. The posterior part of the palate extended farther posteriorly in *Kubanochoerus* than in *Libycochoerus,* and other important differences in the dentitions and skull support separation of the African and Asian kubanochoeres at the generic level (Pickford, 1986; Qiu et al., 1988). The later kubanochoeres sometimes attained gigantic size.

Subfamily TETRACONODONTINAE Lydekker, 1876

Tetraconodonts depart widely from the primitive suid condition by the hypertrophy of their premolars and heteromorphy between the first two premolars and last two premolars (Liu, 2003); P3 and P4 become very large, while P1–2 become relatively smaller and more bunodont. The disparity in size between the anterior pair of premolars and the posterior pair is characteristic of tetraconodonts and not found in any other group of suids. The tetraconodonts are

moderately long-snouted but do not generally develop long diastemata. The sagittal crest is retained or enlarged, and the zygoma is moderately flared. The lower border of the zygoma sweeps downward and backward from its root so that, in side view, M2–3 are hidden from view. The posterior choanae open up behind M3 and are U-shaped.

The main cusps of P3–4 are greatly inflated, and the enamel is thick and wrinkled. The posterolabial and anterolabial cusps of P4 become very closely applied and, with wear, coalesce and function as a single large cusp. The lingual cusp is also enlarged, but the sagittal valley remains unobscured. The anterior and posterior cingula are relatively small. P3 is also greatly inflated, with thick and wrinkled enamel. The simple posterolingual accessory cusplet is prominent, but the cingula are not well developed. The lower P2 is so reduced that it does not make contact with P3 until the latter is heavily worn. Pl–2 are rather bunodont but are mesiodistally elongated in comparison with anterior premolars of *Palaeochoerus.*

Tetraconodonts have rather simple molars, but the enamel is thicker than in other suids, and the upper molars have labial cingula. The thick enamel of tetraconodont molars has inflated the cusps, and the furrows are not very clearly defined on the surface of the tooth. The talon is simple.

Tetraconodont mandibles have a short symphysis, sharply curved incisive border, minimal flare of the symphysis, and marginally flared canines. The ascending ramus originates moderately far forward, thereby hiding the rear portion of m3 in side view. In occlusal view there is a strong waisting of the jaw level with p1–2. As in the upper premolars, the large p3–4 contrast sharply with the small p1–2.

Tetraconodontinae appeared a little later than the listriodonts (about MN 5), and early records from India may indicate an eastern origin for the subfamily. Somewhat surprisingly, Pickford (1993), McKenna and Bell (1997), and van der Made (1999b) all agree on the genera constituting the Tetraconodontinae. *Conohyus* (Fig. 10.11) quickly radiated into Europe, Africa, and Asia but remained relatively conservative for much of the middle Miocene. At the beginning of the late Miocene, coincident with the arrival of *Hipparion* into the Old World, tetraconodonts in the different continents began to diverge. In Europe, the diminutive *Parachleuastochoerus* is unknown before MN 7/8 (Fortelius et al., 2004), and its last known record is from MN 10.

Sivachoeres (Fig. 10.12) appeared in Africa by 11 Ma, giving rise to seven or eight species. *Sivachoerus* and *Nyanzachoerus* are conceivably synonymous, although van der Made

Fig. 10.11. *Conohyus sindiensis* (Lydekker, 1884); GSP 1375; right upper cheek teeth (P1–M3), occlusal view; scale bar equals 50 mm (after Pickford, 1988a: Fig. 72).

Fig. 10.12. *Sivachoerus syrticus* (Leonardi, 1952); KNM-LT 316; right upper cheek teeth (P3–M3), occlusal view; scale bar equals 50 mm (after Harris and White, 1979: Fig. 4).

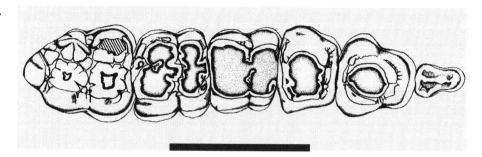

Fig. 10.13. *Notochoerus euilus* (Hopwood, 1926); KNM-ER 2773; right lower cheek teeth (p3–m3), occlusal view; scale bar equals 50 mm (after Harris and White, 1979: Fig. 43).

(1999b) retains the latter name for the narrow-toothed siva-choeres from the Western Rift Valley. Sivachoeres became extinct about 3.2 Ma but gave rise to *Notochoerus* (Fig. 10.13), which survived until about 2 Ma. The demise of the tetra-conodonts in Africa coincides with the influx of the Suinae.

The late Miocene of India was the acme of tetracon-odonts that were represented by the diminutive *Lopho-choerus,* perhaps independently derived from *Parachleuasto-choerus,* and also giant tetraconodonts such as *Tetraconodon* with immensely enlarged third and fourth premolars. *Siva-choerus* appeared in India at the end of Miocene, probably an immigrant from Africa. Tetraconodonts became extinct in India by the beginning of the Pliocene but persisted in Africa until the end of the Pliocene.

Subfamily NAMACHOERINAE Pickford, 1995

On the basis of material recovered from the middle Miocene site of Arrisdrift, Namibia, Pickford (1995) created the subfamily Namachoerinae for a species that had been described from East Africa as *Lopholistriodon moruoroti* but that the Namibian material proved to be a precociously lophodont short-snouted suid that was close to *Hyotherium* and *Palaeochoerus* yet differed significantly from even the most primitive listriodontines. The lophodonty that char-acterizes *Namachoerus* is a parallelism shared with *Listriodon* and *Lopholistriodon* among the listriodonts and with *Schizo-choerus* and *Yunnanochoerus* among the schizochoerines.

Namachoerines are short-snouted lophodont suids that differ from the listriodonts by the shortness of the snout and from the schizochoerines by laterally splayed lower canines and ungrooved laterally splayed upper canines. The palate of *Namachoerus* is narrower than that of *Lopholistriodon,* and the upper and lower molars differ from those of *Lopholistri-*

odon by the possession of well-developed anteroposterior crests that run centrally along the crown.

Subfamily CAINOCHERINAE Pickford, 1988

Diminutive *Cainochoerus africanus* is known from the late Miocene of South and East Africa (Hendey, 1976; Pickford, 1988b; Harris and Leakey, 2003). Hendey (1976) originally identified it as a tayassuid on the basis of the canine orien-tation and molar morphology. Pickford (1993) thought it probably derived from the Hyotheriinae, but van der Made (1997b) also interpreted *Kenyasus* (kubanochoere) and *Al-banohyus* (unresolved suid) as cainochoerines and postulated that *Cainochoerus* was a descendant of *Albanohyus.*

The basicranium of *Cainochoerus* is typically suiform with a high craniomandibular joint and short external auditory meatus. It has a full eutherian cheek teeth complement but lacks the third lower incisor. The metapodial articular mor-phology is suine, but the third and fourth metacarpals are of-ten fused. The lower incisors are ever-growing and have lost the enamel on the lingual surface. Pickford (1988a) noted that the molars recall those of cercopithecoids, whereas the postcranial articulations converge on bovids: reasons he thought were sufficient to accord it its own subfamily.

Subfamily SCHIZOCHOERINAE Thenius, 1979

This subfamily has had a checkered history, but Liu (2003) recognized its constituent genera—*Taucanamo, Schizo-choerus, Yunnanochoerus*—as forming a monophyletic group nested within the Suidae. Although their lophodonty re-sulted in their previous assignment to the Tayassuidae or Pa-leochoeridae, Liu (2003) included them in the Suidae be-cause of the absence of an angular process on their mandibles and grouped them together in the Schizochoeri-nae because of the deep transverse valley in their molars. Thus, they are no longer the youngest representatives of the "Old World peccaries." Unfortunately, the name *Schizo-choerus* as used by Crusafont-Pairó and Lavocat (1954) seems to be preoccupied by the tapeworm *Schizochoerus* Poche, 1922, but this issue will need to be addressed elsewhere.

The genus *Taucanamo* has narrow premolars (a feature that distinguishes it from *Schizochoerus;* van der Made, 1997) and first appeared in Europe in MN 4. *T. pygmaeum* was widespread over the continent, *T. sansaniensis* occurred in western Europe, and the large species *T. inonuensis* is known from Turkey. Medium-sized schizochoeres are known from isolated teeth and a partial skull and mandible from the middle Miocene of Kenya (Muruyur, Kirimun, Ngorora: 15 to 12 Ma). More complete material was retrieved from Moroto, Uganda and assigned to *Morotochoerus ugandensis* (Pickford, 1998). Pickford (1993, 1998) speculated these might have given rise to the genus *Schizochoerus* known from several sites in Europe (Zone MN 9–10, 11 to 9 Ma) and Pakistan (Nagri, 9 Ma). *Yunnanochoerus* has been recorded from the late Miocene of India (Pickford, 1987) and China (van der Made and Han, 1994).

Subfamily SUINAE Gray, 1821

The Suinae appeared toward the end of the middle Miocene (MN 8 in Europe, Chinji Fm of Pakistan). Suinae crania are distinguished by an elongated flange—the prezygomatic shelf—that projects anteriorly from the zygomatic root. This flange separates the chewing musculature from the muscles that operate the snout. It was not present in the hyotheriines, kubanochoeres, listriodonts, and tetraconodonts, where the anterior tendinal guides for the snout musculature in the vicinity of the canines were generally poorly developed. This suggested to Pickford (1993) that the characteristic rooting habit of extant pigs is restricted to the Suinae.

Apart from the prezygomatic shelf, suines differ little from *Hyotherium* in the sculpture of their skull except in some very derived forms such as *Phacochoerus* and *Babyrousa* that possess long gaps between canines and the anterior premolar. The posterior choanae are U-shaped and open up behind M3; the zygomatic arches are well developed and flare laterally from the facial surface at a high angle. The parietal crests are often widely separated, and the dorsal surface of the skull is wide and flat.

The suines are more progressive than *Hyotherium* in their premolar structure by virtue of the development of sagittal cusplets in P4. In the hyotheres, the labial pair of cusps extend as ridges into the sagittal valley, but in suines, definite cusps develop at the ends of the ridges and help to fill up the sagittal valley. For Pickford (1993), the suine stage is reached when the cusplets are mesiodistally longer than they are labiolingually wide. These cusps are not present in the other subfamilies and constitute an apomorphic character.

P3 to P1 are slightly more complex in the suines than in *Hyotherium* because of the increased hypsodonty of the anterior and posterior cingula and the development of a more or less complete lingual cingulum. In some genera, the anterior premolars are lost, but the basic structure of the remaining teeth is constant. Thus, even progressive forms such as *Phacochoerus* retain the three main cusps in P4 plus the sagittal pair, making a total of five. In phacochoeres, the enamel is frequently covered by cement, which represents a further apomorphy. Even in very hypsodont genera such as *Sivahyus* and *Hippohyus,* five cusps are present in P4, a feature shared with all Suinae genera. *Babyrousa* can also be shown to be a suine as it, too, has five cusplets in P4.

The Suinae radiated rapidly and are represented by some 15 genera during and after the late Miocene. The heterogeneous nature of suine molars permits this subfamily to be subdivided into several component tribes but, not unexpectedly, there isn't much consensus of opinion as to the tribal compositions. Pickford (1988a) recognized three tribes: Hippohyini (hypsodont suines from Asia), Phacochoerini (hypsodont suines from Africa), and Suini (all the other suines). He later (Pickford, 1993) recognized four tribes, limiting the Suini to *Sus* plus African and eastern Asian species and relocating most of the suins from Eurasia into the Dicoryphochoerini. Van de Made (1997a) also recognized four tribes, whereas McKenna and Bell (1997), recognized five. Not unexpectedly, that *Sus* belonged in a different tribe from *Propotamochoerus* and that the hypsodont African suids formed a natural grouping (Table 10.3) were the only items all authors agreed upon.

In Europe, suines were represented by *Propotamochoerus* (= *Korynochoerus* Schmidt-Kittler, 1971) (first appearing in MN 8), *Microstonyx* (MN 10), and *Sus* (MN 14), whereas *Eumaiochoerus* (MN 12) was an island endemic of Tuscany (Agusti et al., 2001; van der Made and Moyà-Solà, 1989; van der Made, 1989–1990b).

The late Miocene Suinae of the Indian subcontinent were represented by *Propotamochoerus* (Fig. 10.14) and *Hippopotamodon* (Fig. 10.1). Van der Made (1992) report *Microstonyx* as common on the Indian subcontinent, but Pickford (1988a, 1993) and Liu et al. (2004) believe van der Made misidentified *Hippopotamodon* as *Microstonyx.* However, *Microstonyx* (Fig. 10.15) is present in eastern Asia (e.g., Liu et al., 2004), along with *Propotamochoerus* and *Chleuastochoerus.*

Latest Miocene to Pliocene assemblages of northern India were characterized by the appearance of *Hippohyus* and its smaller relative *Sivahyus.* These have typical suine skulls but hypsodont cheek teeth. The skull of *Hippohyus* resembles that of *Propotamochoerus* behind the snout, but the snout is shortened, and the upper incisors are vertically implanted.

Sus appears in the Indian subcontinent, eastern Asia, and Europe in the earliest Pliocene. Pickford (1993) interpreted the diminutive *Sus salvinius* (= *Porcula salvinia*) as a direct derivative of *Sus,* as were *Babyrousa* and its possible ancestor *Celebochoerus. Sus* got to North Africa in the Pleistocene but did not enter Sub-Saharan Africa until introduced there by humans.

The Suinae first appeared in Africa at about 3.4 Ma. *Kolpochoerus* (Fig. 10.16), *Potamochoerus* (Fig. 10.17), and *Metridiochoerus* (Fig. 10.18) represent the ancestral stocks of the extant forest hogs (*Hylochoerus*), bush pigs (*Potamochoerus*), and wart hogs (*Phacochoerus*), respectively (Harris and White, 1979; Cooke, 1978). *Potamochoerus* remained virtually unchanged during the ensuing 3+ million years, but both *Kolpochoerus* and *Metridiochoerus* species displayed progressive increase in length, height, and complexity of the third molars that made these species useful for biostratigraphic correlation.

Table 10.3. Contrasting Suinae classifications of Pickford (1993), McKenna and Bell (1997), and van der Made (1997)

Pickford (1993)	McKenna and Bell (1997)	van der Made (1997)
Dicoryphochoerini	**Potamochoerini**	**Dicoryphochoerini**
Propotamochoerus	*Propotamochoerus*	*Propotamochoerus*
Eumaiochoerus		*Eumioaochoerus*
Hippopotamodon		
Microstonyx		*Microstonyx=Hippopotamodon*
	Hylochoerus	*Hylochoerus*
	Kolpochoerus	*Kolpochoerus*
Chleuastochoerus		
	Potamochoerus	
	Celebochoerus	
Hippohyini	**Hippohyini**	
Hippohyus	*Hippohyus*	
Sivahyus	*Sivahyus*	
	Sinohyus	
Suini	**Suini**	**Suini**
Sus	*Sus*	*Sus*
Babyroussa		
	Korynochoerus	
Celebochoerus		
	Hippopotamodon	
Kolpochoerus		
	Eumaiochoerus	
Hylochoerus		
	Microstonyx	
Potamochoerus		
Phacochoerini	**Phacochoerini**	**Phacochoerini**
Potamochoeroides	*Potamochoeroides*	*Potamochoeroides*
Metridiochoerus	*Metridiochoerus*	*Metridiochoerus*
Phacochoerus	*Phacochoerus*	*Phacochoerus*
Stylochoerus	*Stylochoerus*	
Afrochoerus		
		Sivahyus
		Hippohyus
	Babirusini	**Babirusini**
	Babyrousa	*Babyrousa*
		Celebochoerus
		Potamochoerus

Family TAYASSUIDAE Palmer, 1897

The similarity of New World peccaries to Old World pigs has been recognized for a long time (Tyson, 1683; Wright, 1998). Synapomorphies include external auditory meatus enclosed by tympanic and posttympanic process of squamosal, ethmoid exposed on orbital wall, neural arch pedicels of cervical vertebrae 3–6 perforated by foramina, and a disklike rhinarium (Wright, 1998). Gentry and Hooker (1988) listed additional dental and postcranial traits uniting peccaries and suids. There has been a long history of attribution of Old World primitive species to the Tayassuidae (Stehlin, 1899, 1900; Pearson, 1927; Colbert, 1933; Simpson, 1945; Pickford, 1993; etc.), and some New World species were originally attributed to Old World genera (Leidy, 1856b; Cope, 1869; etc.). However, according to Wright (1998), none of the New World tayassuid species are represented in the Old World. Wright (1998) pointed out that the alliance between the Old World "tayassuids" and North American tayassuids is not based on synapomorphy, and at least some of the Old World species in question share apparent apomorphies with suids. His argument is fully supported by Liu (2003).

Woodburne (1969) recognized two major evolutionary radiations for the New World tayassuids (Fig. 10.19) with *Perchoerus* and *Cynorca* being considered as the root genera. *Perchoerus* and *Cynorca* are very similar to the Old World genera *Egatochoerus* and *Huaxiachoerus,* so the ancestors of the New World tayassuids conceivably lie among the Old World suoids that have dentitions similar to those of *Egatochoerus* and *Huaxiachoerus.*

New World Tayassuidae (*Thinohyus, Perchoerus*) exploited North America during the Chadronian Mammal Age (early Oligocene), probably as part of the Eocene/Oligocene faunal turnover described in Europe as the *Grande Coupure* and in Asia as "Mongolian Remodeling" (Meng and McKenna,

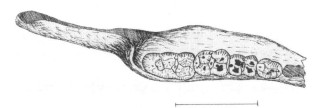

Fig. 10.14. *Propotamochoerus hysudricus* Falconer and Cautley, 1876; GSP 10998; left mandible (p4–m3), occlusal view; scale bar equals 50 mm (after Pickford, 1988a: Fig. 115).

Fig. 10.15. *Microstonyx major* (Gervais, 1848); HMV 0976; partial cranium, ventral view; scale bar equals 50 mm.

Fig. 10.16. *Kolpochoerus limnetes* (Hopwood, 1926); KNM-ER 3381; lower right cheek teeth (p2–m3), occlusal view; scale bar equals 50 mm (after Harris and White, 1979: Fig. 74).

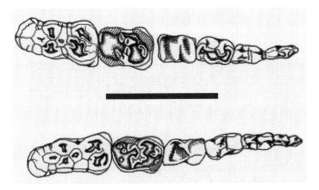

Fig. 10.17. *Potamochoerus porcus* (Linnaeus, 1758); left and right lower cheek teeth (p2–m3) from Laetoli, Tanzania (mid-Pliocene), occlusal view; scale bar equals 50 mm (after Harris and White, 1979, Fig. 56).

1998). This occurred at more or less the same time that *Palaeochoerus* exploited Europe, presumably migrating from Asia where Eocene suoids have been reported. Having settled in North America, tayassuids underwent their own autochthonous evolution on that continent, where they are known from an almost continuous fossil record from early Oligocene to the present (*Dyseohyus, Prosthennops, Cynorca, Macrogenis, Desmathyus, Platygonus, Mylohyus*). Late in their

history, New World tayassuids migrated to South America, perhaps twice. The genus *Argyrohyus* is known from the Chapadmalalan (lower Pliocene), whereas the North American genus *Platygonus* appeared in South America during the Uquian (Pleistocene). Later in the Pleistocene (Ensenadan and Lujanian), the South American tayassuids diversified, and three genera are known, *Tayassu, Platygonus,* and *Catagonus.* The modern representatives comprise the genera *Dicotyles, Tayassu,* and *Catagonus* (Woodburne, 1969; Wright, 1998).

Family SANITHERIIDAE Simpson, 1945

Sanitheres were widely distributed during the Miocene, having been found in Africa, Europe (Fig. 10.20), and the Indian Subcontinent (Pickford, 1984), but their fossil record is very fragmentary. Originally considered to be bunoselenodont pigs, this group of suoids was accorded family rank by Pickford (1984) because of its peculiar cranial and dental morphology. His interpretation is followed here, although van der Made (1997b) placed *Sanitherium* and *Diamantohyus* in the Palaeochoeridae. The basicranium (known from fragmentary fossils) is basically suid-like, with a raised craniomandibular joint, short external auditory meatus, and short paramastoid process. The face and dentition are, however, widely divergent from those of all known suids, some features recalling peccaries and others even primitive ruminants.

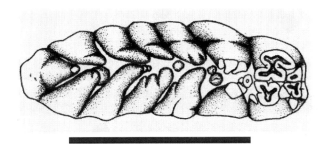

Fig. 10.18. *Metridiochoerus andrewsi* Hopwood, 1926; KNM-ER 3305; upper left M3, occlusal view; scale bar equals 50 mm (after Harris and White, 1979: Fig. 91).

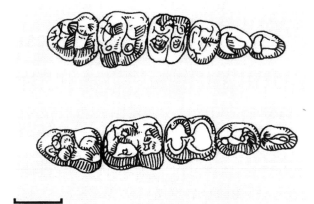

Fig. 10.19. Tayassuid cheek teeth: uppers = "*Cynorca*" *sociale* (Marsh, 1875); lowers = "*Cynorca*" *occidentale* Woodburne, 1969; scale bar equals 10 mm (after Wright, 1998: Fig. 26.3).

Sanithere cheek teeth are bunoselenodont, with a tendency to complicate the premolars by polycuspy and polycristy accompanied by heavy wrinkling of occlusal enamel. In the upper molars the buccal cusps are markedly narrower than lingual ones, the anterior crest of the protocone reaches the buccal edge of the crown, and a beaded buccal cingulum is present. The buccal cusps are compressed conically without "furchen" and oriented obliquely to the tooth row (the long axis of the cusp is oriented from anterobuccal to distolingual). The lingual cusps are selenoid with weak or absent "furchen" and heavily wrinkled enamel. The curvature of the occlusal surface of the cheek dentition is convex ventrally, whereas in tayassuids and suids this surface is generally flat or convex dorsally, the snout appearing to bend ventrally below the level of the cheek teeth. In this respect, sanitheres are convergent on selenodont artiodactyls such as ruminants. This similarity is enhanced by the lower molars possessing metastylids and a "*Palaeomeryx* fold," the upper premolars being stepped labially, the lower border of the mandible being markedly curved, and the tooth enamel being heavily wrinkled. The astragalus, however, is typically suoid (Pickford, 1984).

Sanitheres were confined to Africa during the early Miocene and spread to Europe and Asia only during the middle Miocene (MN 5, 15–16 Ma; Pickford, 2004). The latest Eurasian record is from the Chinji Area of Pakistan (MN 6), but the family persisted in Africa until about 14 Ma (Pickford, 2004). Pickford (1984) demonstrated that sanitheres occur preferentially in floodplain and swamp deposits, suggesting that the group was dependent on humid habitats. However, it is so rare in the fossil record that further speculation would be unwise. Pickford (2004) suggested that sanitheres may have been both more carnivorous and more cursorially adapted than the other suoids.

SUOID PALEOECOLOGY

Suoids may be thought of as archetypal omnivores, and most living species have strikingly broad dietary and environmental adaptations. Fortelius et al. (1996a: Table 28.2)

Fig. 10.20. *Sanitherium schlagintweiti* (von Meyer, 1866); THB-17; lower dentition (left p2–m3; right p4–m3), occlusal view; scale bar equals 10 mm (after de Bonis et al., 1997, Fig. 3b).

reviewed the body masses of European middle and late Miocene suids in terms of four classes: 1–20 kg (forest forms), 21–80 kg (closed habitat forms), 81–200 kg (open habitat forms), and >201 kg (woodland-grassland mosaic and open woodland). The simultaneous presence of all four class sizes implies maximum ecological as well as taxonomic diversity. This occurred in Southwest Asia only in MN 6 and in western Europe only in MN 9.

Living Class 1 suoids comprise *Sus salvanius* (6–10 kg, tall grasslands of Himalayan foothills) and *Tayassu tajacu* (17–25 kg; wet and dry tropical forests to open arid habitats). All fossil class 1 taxa molars have high pointed cusps and variable loph development; their functional similarity to the teeth of cercopithecoids and tragulids suggested to Fortelius et al. (1996a) that these were adaptations for processing forest vegetation.

Class 2 spans the body range 35–80 kg, where foraging in open habitats becomes increasingly feasible physiologically (Wheeler, 1992) and class 2 taxa are more dentally diverse. Extant Class 2 representatives comprise the two larger peccaries and the bush pig; these exploit a wide range of habitats, but all are at least partially forest animals. Combination of thick enamel and conical premolars with hyena-like macrowear suggests that cracking hard items such as seeds was important to tetraconodonts. *Schizochoerus* sp. from Sinap represents a lophodont Class 2 suid that exploited open habitat (Fortelius et al., 1996a).

Extant class 3 suoids include babirusa and the warthog and also medium-sized *Sus* species. The proliferation of minor molar cusps concomitant with loss of individual cusp identity was interpreted by Fortelius et al. (1996a) to indicate an omnivorous diet. *Bunolistriodon* retains bunodont browsing together with derived listriodont incisive cropping. *Listriodon* and *Schizochoerus* represent lophodont class 3 suids.

The only extant class 4 suid is the forest hog *Hylochoerus. Kubanochoerus,* whose sexually dimorphic horns suggest a social lifestyle, belongs in class 4 and is associated with ungulates that would be consistent with forest to grassland-woodland mosaics. The *Hippopotamodon–Microstonyx* lineage similarly indicates open country environments.

Changes in the diversity of Eurasian suids seem to follow overall climatic changes. The high diversity at the beginning of the middle Miocene was attributed to immigration from Africa and western Asia (Fortelius et al., 1996a). A subsequent trend of declining diversity equates with the general reduction of warm subtropical woodlands between 15.5 and 9 Ma succeeded by an explosive expansion of open country habitats between 9 and 7 Ma. Fortelius et al. (1996a) documented an overall increase in mean size accompanied by a decrease in diversity proceeding time-transgressive from east to west. Liu (2003) noted that, in the middle Miocene, Europe and western Asia show only a minor decline in diversity and are still dominated by small to medium-sized suids. This contrasts with northern China, where a lack of small to medium-sized suids in the middle Miocene indicates expansion of open-country habitats some 5 million years earlier than in Europe. However, southern China was the only region dominated by small suoids during the early late Miocene (Liu, 2003), reflecting a relatively stable humid environment that contrasted with the harsher and more arid conditions prevailing in most of Eurasia at that time (Fortelius et al., 1996a).

In Sub-Saharan Africa a major faunal turnover occurred toward the end of the Miocene, reflecting the opening up of forested habitats and the spread of C4 grasses (Leakey at al., 1996; Cerling et al., 1997a). The listriodonts, kubanochoeres, and sanitheres characteristic of the African early and middle Miocene were replaced by tetraconodonts in the late Miocene, and these were in turn replaced by suines that arrived during the late Pliocene. The contrast in size between the relatively small early and middle Miocene forms and their larger late Miocene and Pliocene counterparts reflects the more open habitats available to the later forms.

Hypsodonty is an adaptive response to the development of more fibrous or abrasive plants in a progressively more open and arid adapted vegetation, and Fortelius et al. (2002) used large mammal herbivore hypsodonty as a proxy for Neogene paleoprecipitation patterns in Eurasia. There were no hypsodont grazing suoids in western Eurasia, thereby contrasting with the Indian Subcontinent and Sub-Saharan Africa. In Africa, three lineages of suids—*Sivachoerus-Notochoerus, Metridiochoerus,* and *Kolpochoerus*—displayed increase in size, height, and complexity of the third molars through time, making them useful for biostratigraphic correlation (Cooke and Maglio, 1972; White and Harris, 1977; Cooke, 1978; Harris and White, 1979). Kullmer (1999) compared the molar morphology of Plio-Pleistocene African suids with that of their extant counterparts and correlated increases in hypsodonty, length, and complexity with adaptation to more open habitat and acquisition of a grazing diet. Harris and Cerling (2002) found, on the basis of carbon isotopes, that such a transition occurred in the *Sivachoerus–Notochoerus* lineage but that the earliest representatives of the suines *Metridiochoerus* and *Kolpochoerus* were already well-established grazers.

APPENDIX 10.1: ANNOTATED LIST OF SUOID GENERA

This appendix is concisely summarized in Table 10.4.

Primitive Suoids

Genus *Odoichoerus* Tong and Zhao, 1986 (Late Eocene: Asia)

Diagnosis.   p4 compressed laterally, consisting of a trenchant main cusp and a longitudinal talonid crest. Lower molars with four main cusps erected and well separated from each other, and crests running from the labial cusps weak but distinct. The ends of the anterior wing of protoconid and oblique crest inflated into small tubercles, especially on m3. Hypoconulid on m1–2 large, the third lobe of

m3 is only half the width of the tooth. Labial and lingual cingula absent.

Type Species.   *Odoichoerus unicornis* Tong and Zhao, 1986.

Genus *Egatochoerus* Ducrocq, 1994 (Late Eocene: Asia)

Diagnosis.   Suoid close to *Perchoerus* in its dental morphology. Size similar to that of *Odoichoerus unicornis.* Mandible deep, becoming shallower anteriorly, with a well-developed angular process and unfused mandibular symphysis. Vertical lower canine with triangular cross section, premolars increasing in size distally, p1 lost, diastema between c and p2, p4 with two main cusps (protoconid and metaconid), lower molars with a deep waist between the two lobes, and m3 with a strong hypoconulid. Upper molars with four main cusps and two (mesial and distal) lingually situated accessory cusps, fused internal roots, system of grooves (Furchenplan of von Hünermann) poorly expressed. Astragalus morphologically close to that of *Tayassu.*

Type Species.   *Egatochoerus jaegeri* Ducrocq, 1994.

Genus *Siamochoerus* Ducrocq et al., 1998 (Late Eocene: Asia)

Diagnosis.   Primitive suid, close to *Propalaeochoerus pusillus* in size. p3 simple and lacking a metaconid. p4 with small metaconid and hypoconid but lacking a paraconid. Lower molars show a marked increase in size from front to back, with the mesial lobe wider than the distal lobe, small hypoconulids, and extremely weak accessory cusps. m3 elongated with two-cusped hypoconulid. Upper molars simple, lacking accessory cusps. m3 without distally salient talon. Enamel very finely wrinkled.

Type Species.   *Siamochoerus banmarkensis* Ducrocq, 1998.

Other Species.   *Siamochoerus viriosus* Liu, 2001.

Genus *Huaxiachoerus* Liu, 2001 (Late Eocene: Asia)

Diagnosis.   Primitive suoid differing from other suoids in upper molar morphology: elongated teeth, metaconid obviously smaller than other main cusps, main cusps pointed, and a more elongated third molar. Differing from tayassuids in their longitudinal groove of upper molars.

Type Species.   *Huaxiachoerus guangxiensis* Liu, 2001.

Primitive Genera of Suidae

Genus *Eocenchoerus* Liu, 2001 (Late Eocene: Asia)

Diagnosis.   Medium-sized suid with four bunodont cusps and a talon on M3. Differing from other suids in having primitive dentition: P4 has only one labial cusp; M3 lacking all accessory cusps and furchen, lingual cusps larger than

Table 10.4. Chronological distribution of suoid genera discussed in this chapter

	Late Eocene	Early Oligocene	Late Oligocene	Early Miocene	Mid Miocene	Late Miocene	Early Pliocene	Late Pliocene	Early Pleistocene	Late Pleistocene	Recent
SUOIDEA											
Odoichoerus	AS										
Egatochoerus	AS										
Siamochoerus	AS										
Huaxiachoerus	AS										
SUIDAE Indet.											
Eocenchoerus	AS										
Hyotherium				EURAS	EURAS	EURAS					
Sinapriculus				AS							
Palaeochoerus		EUR	EUR	EUR							
Pecarichoerus				AS	AS						
Albanohyus					EUR						
Aureliachoerus				EUR	EUR						
Miochoerus					AS						
Listriodontinae											
Bunolistriodon				EURAS AF	EURAS AF						
Listriodon				EURAS AF	EURAS						
Lopholistriodon					AF						
Kubanochoerinae											
Kubanochoerus					EURAS AF						
Nguruwe				AF							
Kenyasus				AF							
Megalochoerus				AS AF							
Tetraconodontinae											
Tetraconodon						AS					
Sivachoerus					AS	AS? AF					
Conohyus			AS	AS	EURAS AF	AS					
Nyanzachoerus						AF	AF				
Notochoerus						AF	AF	AF	AF		
Lophochoerus						AS					
Parachleuastochoerus					EUR	EUR AS					
Namachoerinae											
Namachoerus				AF	AF						
Cainochoerinae											
Cainochoerus						AF	AF				
Schizochoerinae											
Taucanamo				EUR	EURAS AF?	AS					
Yunnanochoerus						AS					
Schizochoerus					AS AF	EURAS					
Morotochoerus					AF						
Suinae											
Propotamochoerus					AS	AS AF	AS	AS			
Microstonyx					EUR	EURAS					
Eumaiochoerus						EUR					
Hippopotamodon						AS	AS	AS	AS		
Chleuastochoerus						AS	AS	AS			
Hippohyus						AS	AS	AS	AS		
Sivahyus						AS	AS	AS			
Sus					AS?	EURAS	EURAS	EURAS	EURAS	EURAS	EURAS AF
Babyrousa											AS
Celebochoerus										AS	

Table 10.4. Continued

	Late Eocene	Early Oligocene	Late Oligocene	Early Miocene	Mid Miocene	Late Miocene	Early Pliocene	Late Pliocene	Early Pleistocene	Late Pleistocene	Recent
Potamochoerus					AS	EURAS	AS	AS	AS	AS	
							AF	AF	AF	AF	AF
Kolpochoerus							AF	AF	AF		
Hylochoerus											AF
Metridiochoerus							AF	AF	AF		
Phacochoerus										AF	AF
Sanitheriidae											
Diamantohyus					EURAS						
				AF	AF						
Sanitherium					EURAS						
					AF						
TAYASSUIDAE											
Thinohyus		N AM	N AM	N AM							
Perchoerus	N AM	N AM	N AM	N AM							
Hesperhys				N AM	N AM	N AM	N AM				
Chaenohyus				N AM							
Cynorca				N AM	N AM						
Dyseohyus				N AM	N AM	N AM?					
Prosthennops					N AM	N AM	N AM				
						C AM					
Platygonus							N AM	N AM	N AM	N AM	
								S AM	SAM	SAM	
Mylohyus								N AM	N AM	N AM	
Tayassu								S AM	S AM	S AM	S AM
									NAM	N AM	
Catagonus									S AM	S AM	S AM

Note: AS = Asia; Eur = Europe; EURAS = Eurasia; AF = Africa; N AM = North America; S AM = South America.

labials, and with a labial talon. Differing from *Palaeochoerus* in having a developed talon on M3 and lacking a posterior cingulum.

Type Species.   *Eocenchoerus savagei* Liu, 2001.

Genus *Hyotherium* Meyer, 1834 (Early to Late Miocene: Asia and Europe)

Diagnosis.   Suidae in which the male canines do not flare out much, bullae ovoid with an anteroposteriorly directed long axis, diastemata small or absent, premolars elongated, p4 with two large cusps on the trigonid (van der Made, 1994).

Type Species.   *Hyotherium soemmeringi* von Meyer, 1834.

Other Species.   *Hyotherium meissneri* von Meyer, 1829; *H. major* (Pomel, 1847); *H, shanwangense* Liu et al., 2002.

Genus *Sinapriculus* Liu, Fortelius, and Pickford, 2002 (Early Miocene: Asia)

Diagnosis.   Medium-sized suid differing from *Hyotherium* in the following special character states: symphysis very long, diastemata well developed between i3–c, c–p1, p1–p2; mandible shallow and thick; canine vertically upward, p3 and p4 slightly longer than m1.

Type Species.   *Sinapriculus linquensis* Liu, Fortelius, and Pickford, 2002.

Genus *Palaeochoerus* Pomel, 1847 (Early Oligocene–Early Miocene: Europe)

Diagnosis.   Small size, the snout is short, and no diastema between the canine and premolar, sagittal crest strong, the paroccipital processes are short, the canines are vertical, no canine alveolar crest, zygoma unflared. Premolars simple and with one main cusp, molars are simple bunodont, no "furchen" on the cusp, accessory cusplet undeveloped, with a real talon on M3.

Type Species.   *Palaeochoerus typus* Pomel, 1847.

Other Species.   *Palaeochoerus gergovianus* (de Blainville, 1846); *P. leptodon* (Pomel, 1848); *P. waterhousi* (Pomel, 1853); *P. paronae* (Dal Piaz, 1930); *P. venitor* (Ginsburg, 1980).

Genus *Pecarichoerus* Colbert, 1933 (Early–Mid-Miocene: Asia)

Diagnosis.   Small size, molar bunodont, with a large talonid on m3, weak anterior cingulum on the lower molar, no deep "furchen" on the cusps of molars.

Type Species.   *Pecarichoerus orientalis* Colbert, 1933.

Other Species.   *Pecarichoerus sminthos* (Forster Cooper, 1913).

Genus *Albanohyus* Ginsburg, 1974 (Mid-Miocene: Europe)

Diagnosis.   A small suid with four roots in the lower first and second molars and two roots in the anterior lobe of m3, the protoconule is fused to the cingulum, cusps of the molars are rounded, and the canine are sexually dimorphic.

Type Species.   *Albanohyus pygmaeus* (Depéret, 1892).

Other Species.   *Albanohyus castellensis* Golpe-Posse, 1977.

Genus *Aureliachoerus* Ginsberg, 1974 (Early–Mid-Miocene: Europe)

Diagnosis.   Small size. Upper canine short and rounded as in *Hyotherium major*. External wall of P4 less often divided than in *Hyotherium* and always more asymmetrical. Lower male canine larger than in *Hyotherium*. Lower premolars narrower than in *H. major*.

Type Species.   *Aureliachoerus aurelianesis* (Stehlin, 1899).

Other Species.   *Aureliachoerus minor* (Golpe, 1972).

Genus *Miochoerus* Chen, 1997 (Mid-Miocene: Asia)

Diagnosis.   Small-sized suid, with short diastema between p1 and p2, almost vertical "verrucose" lower canine, and two cusplets on the main cusp of p4.

Type Species.   *Miochoerus youngi* Chen, 1997

Subfamily LISTRIODONTINAE

Genus *Bunolistriodon* Arambourg, 1933 (Early–Mid-Miocene: Africa, Eurasia)

Diagnosis.   Bunolophodont Listriodontinae; post-cranial bones elongated and gracile; snout adorned with large laterally oriented hornlike processes above the canine.

Type Species.   *Bunolistriodon lockharti* (Pomel, 1848).

Other Species.   *Bunolistriodon latidens* (Biedermann, 1873); *B. meidamon* Fortelius et al., 1996; *B. intermedius* Liu and Li, 1963; *B. affinis* (Pilgrim, 1908); *B. anchidens* Van der Made, 1996; *B. jeanneli* (Arambourg, 1933); *B. akatikubas* (Wilkinson, 1976); *B. guptai* (Pilgrim, 1926).

Genus *Eurolistriodon* Orliac, 2006 (Early–Mid-Miocene: Europe)

Diagnosis.   I1 with long concave postcrista and reduced basal fossa compared to *Bunolistriodon;* male upper canine small with wide radius of curvature, curving upward and backward; compared to *Listriodon* the posterior extension of the palatine fissure is reduced, the posttympanic

process is weak, and the condyloid process of the mandible is globular and much wider.

Type Species.   *Eurolistriodon adelli* Pickford and Moyà-Solà, 1995.

Other Species.   *Eurolistrodon tenarezensis* Orliac, 2006.

Genus *Listriodon* von Meyer, 1846 (Mid-Miocene: Africa; Mid–?Late Miocene: Eurasia)

Diagnosis.   Lophodont Listriodontinae with high crowned upper canine that curves outward and upward (van der Made, 1996b); postcranial bones short and robust; canine flange more marked than in *Eurolistriodon* (Pickford and Morales, 2003).

Type Species.   *Listriodon splendens* von Meyer, 1846.

Other Species.   *Listriodon pentapotamiae* (Falconer, 1868); *L. bartulensis* Pickford, 2001; *L. retamaensis* Pickford and Morales, 2003.

Genus *Lopholistriodon* Pickford and Wilkinson, 1975 (Mid-Miocene: Africa)

Diagnosis.   A genus of Listriodontinae distinguished by its small size and the extreme development of transverse crests in the molars and fourth premolars with suppression of accessory cusps. The upper premolars possess enlarged cingula and wide cingular platforms. The nasal ridge is narrow.

Type Species.   *Lopholistriodon kidogosana* Pickford and Wilkinson, 1975.

Other Species.   *Lopholistriodon pickfordi* Van der Made, 1996; *L. akatidogus* (Wilkinson, 1976).

Subfamily KUBANOCHOERINAE

Genus *Kubanochoerus* Gabunia, 1955 (Mid-Miocene: Eurasia, Africa)

Diagnosis.   Giant bunodont suid with large cranial appendages and fully bunodont cheek teeth.

Type Species.   *Kubanochoerus robustus* Gabunia, 1955.

Other Species.   *Kubanochoerus marymunguae* Van der Made, 1996; *K. khinzikebirus* (Wilkinson, 1976); *K. mancharensis* Van der Made, 1996; *K. gigas* (Pearson, 1928); *K. minheensis* (Qiu et al., 1981).

Genus *Nguruwe* Pickford, 1986 (Early Miocene: Africa)

Diagnosis.   A small genus of Kubanochoerinae in which the I1 is labiolingually compressed, not meeting interproximally; P4 with two main cusps and complete cingulum; molars with thick enamel, inflated main cusps, closed lingual notches; simple talon/id on the third molar; occlusal outline of m3 symmetrical; p3 with wide distal lingual cingular cusp or platform; p4 with innenhugel almost completely suppressed; dm4 labiolingually inflated; lower canine scrofic; upper canine with dorsal cement cover.

Type Species. *Nguruwe kijivium* (Wilkinson, 1976).

Other Species. *Nguruwe namibiensis* Pickford, 1986.

Genus *Kenyasus* Pickford, 1986
(Early Miocene: Africa)

Diagnosis. Small to medium-sized suids with I1 not spatulate, i1–2 lightly built, p4 with very small innenhugel, closely applied to main cusps; P4 with reduced posterior labial cusp closely applied to anterior labial cusp; lingual cusp of P4 located near anterior border of tooth opposite anterior labial cusp; P3 with prominent lingual cingulum but no internal; posterior cusp; short diastemata between c–p1–p2; upper molars with labial cingulum of variable strength, talon of M3 very simple in most individuals, being little more than a posterior cingulum; hypocone of M3 small; main cusps of lower molars of subequal height, minimal buccal or lingual flare; upper molars with strong lingual flare and zygodont crests (six furchen) visible in labial view, cusps moderately inflated, enamel moderately thick, furchen shallow and lingual notches wide, V-shaped; lingual cingulum of P4 generally complete, but labial cingulum only partial.

Type Species. *Kenyasus rusingensis* Pickford, 1986.

Genus *Libycochoerus* Arambourg, 1961

Diagnosis. Medium to large-sized kubanochoeres in which the posterior margin of the palatines is not retired far behind M3; first premolars elongate oblong robust teeth (length breadth index is 220 for P1 in comparison with *Kubanochoerus* in which it is 138).

Type Species. *Libycochoerus massai* Arambourg, 1961.

Other Species. *Libycochoerus jeanelli* (Arambourg, 1933); *L. khinzekebirus* (Wilkinson, 1976); *L. fategadensis* (Prasad, 1967); *L. affinis* (Pilgrim, 1908); *L. juba* Ginsburg, 1977.

Genus *Megalochoerus* Pickford, 1993
(Mid-Miocene: Africa and Asia)

Diagnosis. Gigantic kubanochoerine with talonid on the lower third molars doubled.

Type Species. *Megalochoerus homungous* Pickford, 1993.

Subfamily TETRACONODONTINAE
Genus *Tetraconodon* Falconer, 1868 (Late Miocene: Asia; Mid?-Miocene–?Pliocene)

Diagnosis. Tetraconodontine with extremely enlarged third and fourth premolars but relatively small third molars. DAP × DT of the p3 and p4 about 160 × 150 and 140 × 150, respectively.

Type Species. *Tetraconodon magnum* Falconer, 1868.

Other Species. *Tetraconodon minor* Pilgrim, 1926; *T. intermedius* van der Made, 1999.

Genus *Sivachoerus* Pilgrim, 1926
(Late Miocene: Africa; Late Miocene–Late Pliocene: Asia)

Diagnosis. Tetraconodontine with wide but not extremely wide premolars (both DT and I) and with a tendency to enlarge but not so much elongate (I) the posterior molars. DT' of the p3 and p4 about 110–140 and 110–140 respectively. DAP' m3 about 160–320.

Type Species. *Sivachoerus prior* Pilgrim, 1926.

Other Species. *Sivachoerus sindiensis* (Lydekker, 1878); *Sivachoerus indicus* (Lydekker, 1884); *Sivachoerus syrticus* Leonardi, 1952 (*S. s. syrticus* Leonardi, 1952; *S. s. pattersoni* [Cooke and Ewer, 1972]); *Sivachoerus australis* (Cooke and Hendey, 1992) (*S. a. australis* [Cooke and Hendey, 1992]; *S. a. megadens* van der Made, 1999).

Genus *Conohyus* Pilgrim, 1926
(Late Oligocene–Late Miocene: Asia; Mid-Miocene: Africa and Europe]

Diagnosis. Nyanzachoerin with relatively small, not very elongate, and simple M3. Third lobe of m3 with large pentaconid and hexaconid small or absent. m3 DAP about 140–200; index I about 160–190.

Type Species. *Conohyus simorrensis* (Lartet, 1851).

Other Species. *Conohyus giganteus* (Falconer and Cautley, 1847); *C. steinheimensis* (Fraas, 1870), *C. huenermanni* (Hessig, 1989), *C. sindiensis* (Lydekker, 1884); *C. indicus* (Lydekker, 1884); *C. ebroensis* Azanza, 1986; *C. olujici* Bernor, Bi, and Radovèiê, 2004.

Genus *Nyanzachoerus* Leakey, 1958
(Late Miocene–Early Pliocene: Africa; Late Miocene: Asia)

Diagnosis. Nyanzachoerin with elongated but not very hypsodont M3. DAP m3 about 200–360.

Type Species. *Nyanzachoerus kanamensis* Leakey, 1958 (*N. k. kanamensis* Leakey, 1958; *N. k. waylandi* [Cooke and Coryndon, 1970]).

Other Species. *Nyanzachoerus cookei* Kotsakis and Inigo, 1980.

Genus *Notochoerus* Broom, 1925
(Early Pliocene–Early Pleistocene: Africa)

Diagnosis. Nyanzachoerin with small premolars, elongate molars, very long and hypsodont M3. DAPs: p3 about 80–90; p4 about 70–80; m3 over 300. Index I m3 over 250.

Type Species. *Notochoerus capensis* Broom, 1925.

Other Species.   *Notochoerus euilus* (Hopwood, 1926), *N. scotti* (Leakey, 1943), *N. jaegeri* (Coppens, 1971), *N. harrisi* van der Made, 1999; *N. clarki* White and Suwa, 2004.

Genus *Lophochoerus* Pilgrim, 1926 (Late Miocene: Asia)

Diagnosis.   Very small nyanzachoerin with small and simple m3. M1 length about 8–9 mm. m3 DAP about 19–20 mm; index I about 180–190.

Type Species.   *Lophochoerus nagri* Pilgrim, 1926.

Genus *Parachleuastochoerus* Golpe-Posse, 1972 (Mid–Late Miocene: Europe, Asia)

Diagnosis.   Tetraconodontine with narrow and relatively small premolars. DAP x DT of the p3 about 110 x 70, of the p4 about 100 x 90; indices I over approximately 180 and 120, respectively.

Type Species.   *Parachleuasterochoerus crusafonti* Golpe-Posse, 1972.

Other Species.   *Parachleuasterochoerus sinensis* Pickford and Liu, 2001; *P. kretzoii* Fortelius, Armour Chelu, Bernor, and Fessaha, 2004.

Subfamily NAMACHOERINAE

Genus *Namachoerus* Pickford, 1995 (Early–Mid-Miocene: Africa)

Diagnosis.   Small suids with complete eutherian dentition; upper dentition closed except for diastema between I3 and C; molars and posterior premolars highly lophodont; molars with crests running anteroposteriorly between the centers of cusp pairs; central upper incisors not divided by deep sulci; incisive foramen opposite the root of I1; palatine bone ends distally opposite front of M3; palate narrow; zygomatic root located above P4, sweeping backward without prezygomatic flange; orbit above M2–3; central upper incisor alveoli well separated; upper canines short with no distal groove, emerging anteroventrally and slightly laterally, bordered dorsally by a canine flange; lower canines scrofic and splayed slightly laterally, fitting into niche in maxilla which has sharp dorsal border; no diastemata in lower dentition.

Type Species.   *Namachoerus moruoroti* (Wilkinson, 1976).

Subfamily CAINOCHOERINAE

Genus *Cainochoerus* Pickford, 1988 (Late Miocene–Early Pliocene: Africa)

Diagnosis.   Small suid (ca. 7 kg) lacking the third lower incisor. Teeth with smooth enamel; no postcanine diastemata; I1 appreciably larger than I2 and I3; i2 and i3 approximately equal in size; C short, nearly straight and pointed downward; c short, curved, and pointed upward and slightly

outward; first to third premolars double rooted, single cusped, and relatively high crowned; P4 triple rooted with paired principal cusps flanked posteriorly by a prominent cingulum; p4 double rooted with paired principal cusps flanked posteriorly by a smaller accessory cusp; M1 to M3 quadricuspid, each tooth with a transverse enamel ridge anteriorly and a smaller oblique one posteriorly (talon); m1 to m3 quadricuspid, each tooth with a centrally situated posterior accessory cusp (talonid). Metatarsals II and IV completely fused proximally.

Type Species.   *Cainochoerus africanus* (Hendey, 1976).

Subfamily SCHIZOCHOERINAE

Genus *Taucanamo* Simpson, 1945 (Early–Mid-Miocene: Europe; Mid-Miocene?: Africa; Mid–Late Miocene: Asia)

Diagnosis.   Small suid with elongated premolars. The molars show a slight tendency toward lophodonty; the paraconule is fused to the protocone.

Type Species.   *Taucanamo sansaniense* (Lartet, 1851).

Other Species.   *Taucanamo grandaevum* (Fraas, 1870); *T. inonuensis* Pickford and Etürk, 1979; *T. primum* van der Made, 1997.

Genus *Yunnanochoerus* van der Made and Han, 1994 (Late Miocene: Asia)

Diagnosis.   Suid with lophodont molars, enlarged elongate premolars and p4 with one main cusp.

Type Species.   *Yunnanochoerus lufengensis* (Han, 1983).

Other Species.   *Yunnanochoerus gandakasensis* (Pickford, 1977).

Genus *Schizochoerus* Crusafont and Lavocat, 1954 (Mid-Miocene: Africa; Mid–Late Miocene: Asia; Late Miocene: Europe)

Diagnosis.   Large-size suid in which the molars are lophodont, upper incisors vertically emplaced in the premaxillae and of circular to oval section. Large deep canine flanges for lower canines. Great posterior extension of the palate to the rear of M3.

Type Species.   *Schizochoerus vallesiensis* Crusafont and Lavocat, 1954.

Other Species.   *Schizochoerus anatoliensis* van der Made, 1997; *S. sinapensis* van der Made, 1997.

Genus *Morotochoerus* Pickford, 1998 (Mid-Miocene: Africa)

Diagnosis.   Small schizochoerine with sublophodont molars, in which the lophs are scored vertically by a sagittal groove; presence of zygodont crests at the end of the median valley in upper molars; median accessory cusplet reduced in stature, forming a ridge running forward from the midline of the posterior loph; talon small, cingulum-like; upper canine ovoid in section with a deep posterior groove; lower canine hypsodont and scrofic in section; cheek tooth row closed from C to M3; marked differential wear gradient (m1 deeply worn before m3 is fully erupted); orbit located far forward, the anterior margin sited over the M1 and the postorbital process of the malar sited on a level with M3; zygomatic arches leave the face at a small angle; lacrimal foramen large and protected by a lacrimal flange; deep precanine niches in front of the upper canines; maxillary recess extends anteriorly forming a platform separating the upper orbital cavity from the palate; palate convex from P2 to M3; palatine foramina located opposite the front of M2.

Type Species. *Morotochoerus ugandensis* Pickford, 1998.

Subfamily SUINAE

Genus *Propotamochoerus* Pilgrim, 1925 (Mid-Miocene–Pliocene: Asia; Late Miocene: Africa)

Diagnosis. Medium-sized Suidae with facial part of skull longer than the cranial part; zygomatic arches depart from maxilla at right angles opposite M1; shallow but extensive fossae present on dorsal and ventral surfaces of prezygomatic shelf; snout nearly square in cross section; posterior choanae U-shaped, open immediately behind M3; large canine flanges reach backward from I3 to P2; basicranium advanced; parietal lines not widely separated but no sagittal crest; 2 cusplets in sagittal valley of P4; no labial or lingual cingula on P4; p4 with variable innenhugel and low to medium hypsodont anterior cusp and cingulum; upper canine almost surrounded by cementum; molar enamel thickness moderate; furchen shallow; no cingula on molars, except for occasional basal pillars in ends of median valleys; talonid of third molar relatively simple, no extra cusp pairs; mandible deep and robust; symphysis reaches back to rear of p2; canines not at corners of anterior ends of symphysis; anterior and posterior cusp of premolars variable but never reach same height as main cusps.

Alternate Diagnosis. Small Dicoryphochoerini with canines that are not much reduced (van der Made and Han, 1994).

Type Species. *Propotamochoerus hysudricus* (Stehlin, 1899).

Other Species. *Propotamochoerus palaeochoerus* (Kaup, 1833); *P. hyotheroides* (Schlosser, 1903); *P. provincialis* (Gervais, 1859); *P. wui* van der Made and Han, 1994; *P. parvulus* (Chang, 1974).

Genus *Microstonyx* Pilgrim, 1926 (Mid–Late Miocene: Europe; Late Miocene: Asia)

Diagnosis. Giant Suinae with small and weak splaying male canine. Braincase broad and flat, lachrymal notch very deep, zygomatic arch strongly inflated, P1 present or absent, diastema between the canine and first premolar long, alveolar crest above the upper canine strong, snout elongated.

Type Species. *Microstonyx major* (Gervais, 1848).

Other Species. *Microstonyx antiquus* (Kaup, 1833).

Genus *Eumaiochoerus* Hürzeler, 1982 (Late Miocene: Europe)

Diagnosis. Elevated orbit with anterior rim above the anterior edge of M3, elongated incisors and molars; canines reduced in size; mandible shallow.

Type Species. *Eumaiochoerus etruscus* (Michelotti, 1861).

Genus *Hippopotamodon* Lydekker, 1877 (Late Miocene–Early Pleistocene: Asia)

Diagnosis. Giant Suinae in which the males have large flaring canines; molar enamel relatively thin; molars relatively simple with well-developed furchenplan; labial cusps in lower molars lower crowned than lingual ones; premolars strong and stout, P4 with posterior accessory cusp almost as large as two main labial cusps; posterior choanae U-shaped, open immediately behind M3; p4 with prominent innenhugel and 2–3–4 cusp, anterior cingulum and ac–1 cusp moderately high; short diastemata between c–p1, p1–p2; braincase with broad flat dorsal surface.

Type Species. *Hippopotamodon sivalense* Lydekker, 1877.

Genus *Chleuastochoerus* Pearson, 1928 (Late Miocene–Pliocene: Asia)

Diagnosis. Skull, otherwise primitive in structure, is at once marked off from any other known type of pig by the great bony arch over the canine teeth of the male and by the curious shelf-like expansion of the anterior end of the zygomatic arch (Pearson, 1928).

Type Species. *Chleuastochoerus stehlini* (Schlosser, 1903).

Genus *Hippohyus* Falconer and Cautley in Owen, 1840–1845 (Late Miocene–Early Pleistocene: Asia)

Diagnosis. Suinae with hypsodont cheek teeth without cement cover, short snout with relatively vertically implanted incisors. Parietal crests close together but not joined to form sagittal crest. Molar and premolar enamel thin; furchen deep, forming complex infolding of enamel surfaces. No labial pillar in lower molars. Orbits and zygomatic arches sited farther forward than in *Propotamochoerus*. P3 with two labial main cusps and two ridges leading from labial cusp tips down lingual surface onto lingual cingulum.

Type Species. *Hippohyus sivalensis* Falconer and Cautley in Owen, 1840–1845.

Other Species. *Hippohyus lydekkeri* Pilgrim, 1910.

Genus *Sivahyus* Pilgrim, 1926 (Late Miocene–Pliocene: Asia)

Diagnosis. Small Suinae with hypsodont molars and premolars. Furchen deep. Molar enamel relatively thin.

Type Species. *Sivahyus punjabiensis* (Lydekker, 1878).

Genus *Sus* Linnaeus, 1758 (Mid-Miocene?–Recent: Asia; Late Miocene–Recent: Elsewhere)

Type Species. *Sus scrofa* Linnaeus, 1758.

Other Species. *Sus arvanensis* Croizet and Jobert, 1828; *S. nanus* van der Made, 1988; *S. strozzi* Forsyth Major, 1881; *S. sangirensis* von Koenigswald, 1963.

Genus *Babyrousa* Perry, 1811 (Recent: Asia)

Diagnosis. Similar to *Sus* but the upper tusks protrude through the snout and arc backward to a maximum length of 31 cm.

Type Species. *Babyrousa babyrussa* (Linnaeus, 1758).

Genus *Celebochoerus* Hooijer, 1948 (Late Pleistocene: Asia)

Diagnosis. Giant suid with incisors as in *Sus* but less hypsodont; enamel of I1 only 1.5 times higher than wide labially; labial height of enamel of i1–2 only two times the width. Lower canines as in *Sus verrucosus* in cross section. Upper canines much wider than lowers, subtriangular in cross section, slightly constricted at pulp cavity; anterior surface at right angles to upper surface and slightly narrower than the latter; upper canines projecting more sideways and placed more backward than in *Sus,* the ends extending beyond the lower canines; enamel restricted to one band of variable width at anterior edge below, developed in fewer than half of the specimens. Female canines similar to male but about three-fourths as large. Two premolars in the mandible, three in the maxilla, shaped as in *Potamochoerus* but less hypsodont; p3 with strong protoconid slightly compressed transversely, and talonid half as high as protoconid, anterior cingular point small or absent; p4 with massive and conical protoconid, uncleft at the apex, talonid two-thirds as high as protoconid, anterior point of cingulum one-third as high at most; P2 small, degenerate; P3 sectorial with conical paracone, well-developed inner talon and deep but narrow posterointernal fossa; P4 with paracone and metacone rounded and separated by a distinct cleft, anteroposterior ridge lingually of paracone connected with the latter up to the apex, central fossa narrow, and protocone rounded. Premolars not enlarged relative to molars: p4 as wide as m1 and P3 decidedly less wide than M1; premolar series slightly diverging to the front. Molars with thick and very weakly folded enamel and no lobe formation, cusps distinct, transverse valleys wide open; talon/id of third molar short. Skull with the zygomatic process of the maxilla springing out

rather abruptly from the surface of the cheek, as in *Potamochoerus,* and placed more forward relative to the molars than in *Sus* or *Potamochoerus:* the anterior margin of the orbit above the middle of M3 , and facial crest anterior to M3 as in *Propotamochoerus.* Lateral angulation of anterior frontal and nasal regions absent, as in *Phacochoerus* and *Hylochoerus.* Strong tubular alveolus for upper canine without jugum caninum above. Mandible rather thick, and medial concavity for pterygoid muscle less extended than in *Sus.* Skeleton large and massive; distal articular surface of radius as in *Sus verrucosus;* unciform with obtuse angle between magnum and third metacarpal facets; third metacarpal robust; fifth metacarpal smallish proximally as in recent Suidae.

Type Species. *Celebochoerus heekereni* Hooijer, 1948.

Genus *Potamochoerus* Gray, 1854 (Mid-Miocene–Late Pleistocene: Asia; Late Miocene: Europe; Late Pliocene–Recent: Africa)

Diagnosis. A small sexually dimorphic genus of Suinae. Similar in size and morphology to *Sus,* from which it differs by its more brachyodont and less trenchant premolars, simpler third molars, facial tuberosities (nasal and supracanine region in male skulls), and more divergent zygomatic region.

Type Species. *Potamochoerus porcus* (Linnaeus, 1758).

Other Species. *Potamochoerus larvatus* (Cuvier, 1822).

Genus *Kolpochoerus* van Hoepen and van Hoepen, 1932 (Late Pliocene–Mid-Pleistocene: Africa)

Diagnosis. A genus of Suinae of small to large size, sexually dimorphic, especially in canine size and development of canine flange and zygoma. Upper premolars with well-developed protocone. Molars well rooted, low crowned, and similar in morphology to *Sus* and *Potamochoerus* but slightly taller. Mandible inflated laterally beneath cheek teeth.

Type Species. *Kolpochoerus heseloni* (Leakey, 1943).

Other Species. *Kolpochoerus olduvaiensis* (Leakey, 1942); *K. paiceae* (Broom, 1931); *K. majus* (Hopwood, 1934); *K. afarensis* Cooke, 1978; *K. maroccanus* (Ennouchi, 1953); *K. deheinzelini* Brunet and White, 2001; *K. cookei* Brunet and White, 2001; *K. phacochoeroides* (Thomas, 1884).

Genus *Hylochoerus* Thomas, 1904 (Late Pleistocene–Recent: Africa)

Diagnosis. Sexually dimorphic genus of Suinae of moderate size. Only P3 and P4 and p4 retained in adults, and these tend to be shed in aged individuals. Molars tall (not hypsodont) with thick cementum cover; major lateral cusps modified to form tall and widely spaced loph(id)s.

Type Species. *Hylochoerus meinerzhageni* Thomas, 1904.

Genus *Metridiochoerus* Hopwood, 1926 (Late Pliocene–Mid-Pleistocene: Africa)

Diagnosis. Asexually dimorphic genus of Suinae of small to large size. Progressive species exhibit reduction of the premolar row but increase in size and hypsodonty and delay in root fusion of the molars. Except in the least progressive examples, fusion of the crown elements of the teeth begins shortly after wear commences. In the most progressive species the remaining anterior cheek teeth are shed after maturity.

Type Species. *Metridiochoerus andrewsi* Hopwood, 1926.

Other Species. *Metridiochoerus modestus* (van Hoepen and van Hoepen, 1932); *M. hopwoodi* (Leakey, 1958); *M. compactus* (van Hoepen and van Hoepen, 1932).

Genus *Phacochoerus* Cuvier, 1817 (Late Pleistocene–Recent: Africa)

Diagnosis. A sexually dimorphic species of Suinae of small to moderate size. Cranium with broad zygomatic arches lacking distinct knobs, elevated orbits, and short cranial region. Upper incisors reduced to one pair. Upper canines lacking enamel except at tips. Premolars reduced and commonly shed in adults. Molars hypsodont, formed of closely packed columnar elements and well cemented; lateral pillars flattened externally, elongate and oval to subtriangular in shape.

Type Species. *Phacochoerus aethiopicus* (Pallas, 1767).

Other Species. *Phacochoerus antiquus* Broom, 1948.

Family SANITHERIIDAE Simpson, 1945

Genus *Diamantohyus* Stromer, 1926 (Early–Mid-Miocene: Africa; Mid-Miocene: Asia)

Diagnosis. A genus of the family Sanitheriidae in which the molarization of the premolars is not as complete as in *Sanitherium*; P4 with three main cusps and two subsidiary ones; anterior and posterior lingual cusps less developed than in *Sanitherium*; metastylid prominent in barely worn molars; upper and lower incisors low crowned with short roots; lower canine not hypsodont.

Type Species. *Diamantohyus africanus* Stromer, 1926.

Other Species. *Diamantohyus jeffreysi* (Forster-Cooper, 1913); *D. nadirus* (Wilkinson, 1976).

Genus *Sanitherium* von Meyer, 1866 (Mid-Miocene: Africa and Eurasia)

Diagnosis. A genus of the family Sanitheriidae in which P4 has six cusps of subequal size. Lower molars wider than in *Diamantohyus*.

Type Species. *Sanitherium schlaginweiti* von Meyer, 1866.

Other Species. *Sanitherium leobense* (Zdarsky, 1909).

Family TAYASSUIDAE Palmer, 1897

Genus *Thinohyus* Marsh, 1875 (Early Oligocene–Early Miocene: North America)

Type and Only Species. *Thinohyus lentus* Marsh, 1875.

Genus *Perchoerus* Leidy, 1869 (Late Eocene–Early Miocene: North America)

Type and Only Species. *Perchoerus probus* Leidy, 1869.

Genus *Hesperhys* Douglass, 1903 (Early Miocene–Early Pliocene: North America)

Type Species. *Hesperhys vagrans* Douglass, 1903.

Included Species. *Hesperhys pinensis* (Matthew, 1907), *H. olseni* (White, 1941).

Genus *Chaenohyus* Cope, 1879 (Early Miocene: North America)

Type Species. *Chaenohyus decedens* Cope, 1879.

Genus *Cynorca* Cope, 1867 (Early–Mid-Miocene: North America)

Type Species. *Cynorca proterva* Cope, 1867.

Genus *Dyseohyus* Stock, 1937 (Early to ?Late Miocene: North America)

Type Species. *Dyseohyus fricki* Stock, 1937.

Genus *Prosthennops* Gidley, in Matthew and Gidley, 1904 (Mid-Miocene–Early Pliocene: North America; Late Miocene: Central America)

Type Species. *Prosthennops serus* (Cope, 1878).

Genus *Platygonus* Le Conte, 1848 (Late Miocene–Late Pleistocene: North America; Late Pliocene–Late Pleistocene: South America)

Type Species. *Platygonus compressus* Le Conte, 1848.

Genus *Mylohyus* Cope, 1889 (Early Pliocene–Late Pleistocene: North America)

Type Species. *Mylohyus fossilis* (Leidy, 1860).

Genus *Tayassu* Fischer de Waldheim, 1814 (?Late Pliocene–Recent: South America; ?Late Pleistocene–Recent: North and Central America)

Type Species. *Tayassu pecari* Fischer, 1814.

Genus *Catagonus* Ameghino, 1904 (Mid-Pleistocene–Recent: South America)

Type Species. *Catagonus metropolitanus* Ameghino, 1904.

ACKNOWLEDGMENTS

Images from Pickford (1986) (Figs. 2a–2b, 6, 30, 46, 63) are reproduced by kind permission from Backhuys Publishers B.V., Leiden, Netherlands. Images from Pickford (1988a) (Figs. 49, 72, 115) are reproduced by kind permission of Dr. Friedrich Pfeil, Munich, Germany. Figure 3b from de Bonis et al. (1997) is reproduced by kind permission of the Muséum d'Histoire Naturelle de Genève. Images from Harris and White (1979) (Figs. 4, 9, 43, 56, 74) are reproduced by kind permission of the American Philosophical Society. Figure 1 from Ducrocq (1994a) and Figure 3 from Ducrocq et al. (1998) are reproduced by kind permission from Blackwell Publishers, Oxford, UK. Figure 26.3 from Wright (1998) is reproduced by kind permission of Cambridge University Press. We acknowledge the considerable volume of pioneering work by Martin Pickford and Jan van der Made that helped our current understanding of suid systematics. Liu thanks Mikael Fortelius for support and advice. Harris thanks the Natural History Museum of Los Angeles County and the California Institute of Technology for logistic support. We thank Terry Harrison, Scott Foss, and Ray Bernor for helpful reviews of this chapter.

JOSHUA A. LUDTKE

11

Family Agriochoeridae

MEMBERS OF AGRIOCHOERIDAE (Leidy, 1869), or agriochoerid-grade oreodonts, represent the earlier and more basal members of an extinct, taxonomically diverse group of artiodactyls. The abundance of oreodont remains in the fossil record has contributed to our knowledge of the wide ranges of individual variation in their cranial and dental morphologies. A lack of recognition of these ranges of variation often resulted in creating spectacular numbers of oreodont taxa. For agriochoerids, 10 genera that now are considered junior synonyms have been erected, and upward of 35 species have been described.

The two named agriochoerid genera now recognized, *Protoreodon* and *Agriochoerus,* were united in Agriochoeridae based on the retention of character states that are basal for oreodonts: an incomplete postorbital bar and a lack of lacrimal fossae. Most, if not all, members of Oreodontidae (Leidy, 1869) develop a complete postorbital bar, lacrimal fossae, and completely selenodont molars. Virtually all agriochoerids, unlike oreodontids, develop molariform P3s with a partially split parametacone and P4s with a split parametacone and at least a trace of a hypocone. Later diverging species in Agriochoeridae possess ungual phalanges that are better described as claws than hooves, making them unique among Cetartiodactyla. The lack of such features in oreodontids suggests that the origin of Oreodontidae lies within early members of Agriochoeridae. Derived features that unite agriochoerids with oreodontids in Oreodontoidea (Gill, 1872) are their development of enlarged caniniform teeth (C and p1), which are apparently slightly larger in males than in females. Diastemata in both the upper and lower anterior dentitions developed to accommodate these enlarged teeth.

Oreodonts are selenodont artiodactyls whose exact phylogenetic position in Cetartiodactyla remains unknown, although recent morphological analyses have recognized several possibilities that will be discussed. Agrio-choerid-grade oreodonts appeared in North America in the middle Eocene (early Uintan NALMA; Robinson et al., 2004), at approximately the same time as, or just before, other selenodont artiodactyls such as oromerycids, proto-ceratids, and cameloids also appeared. Agriochoeridae remained endemic to North America throughout its existence and last occurred in the late Oligocene (late early Arika-reean NALMA; Tedford et al., 2004).

SYSTEMATICS OF AGRIOCHOERIDAE

Oreodontidae (more commonly called Merycoidodonti-dae, following Thorpe, 1937) and Agriochoeridae were understood by Leidy (1869) to be closely related. The first clade named to contain these two groups was Oreodontoidea (Gill, 1872). In his synoptic table, Gill demoted both families to subfamily rank but still retained Oreodontidae to include both agriochoerids and oreodontids. Scott (1890) later introduced Protoreodontinae as a new subfamily of Oreodontidae. Hay (1902) subsequently placed Protoreodontinae and Agriochoerinae in a resurrected Agriochoeridae.

With Hay's (1902) inclusion of *Protoreodon,* Agriochoeridae was changed from a monophyletic taxon to a para-phyletic grade of oreodonts. *Protoreodon* is a paraphyletic group that includes some species that show agriochoerid attributes, some species that exhibit oreodontid attributes, and some species that, in displaying neither of these derived states, seem to reveal basal oreodont character states. Thorpe (1937), Scott (1940, 1945), and Simpson (1945) were perhaps the last authors to place *Protoreodon* in Oreodontidae. Subsequent authors (Gazin, 1955; Romer, 1966; Wilson, 1971a; McKenna and Bell, 1997; Lander, 1998; among others) have continued Hay's (1902) usage of a *Protoreodon*-inclusive Agriochoeridae, despite the observation by many of these authors that such a group was paraphyletic.

This chapter uses the conventional (Hay, 1902) definition of Agriochoeridae, although it recommends not using the name until it has been redefined as a true clade; it also comments on some other phylogenetic scenarios, but recommendations are intended for inclusion in a later, more comprehensive, contribution. Therefore, familiar names are employed to identify clades when names do exist, but such usage should not be interpreted as support for the monophyly of these groups.

Understanding of Agriochoeridae suffers from a long-standing absence of a systematic review. Some of this absence is a result of possible avoidance. Schultz and Falkenbach, for example, published eight monographs on oreodontids but did not publish on agriochoerids. Another setback has been works that have refined individual genera (such as Thorpe, 1937; Scott, 1945; Gazin, 1955; Wilson, 1971a; Golz, 1976) but done little to address the relationship of different genera within Agriochoeridae. Lander's (1998) article was the first attempt to resolve this issue, but space constraints limited the depth and breadth of that review.

Because only two reports on agriochoerids has appeared since 1998 (Theodor, 1999; Lander and Hanson, in press), this chapter does not provide any substantial revision of the family. Moreover, E. B. Lander (pers. comm.) plans to publish additional studies on agriochoerid systematics. Therefore, this chapter does not describe any new taxa and mainly relies on Lander's (1998) systematics. The chapter's use of Lander's (1998) unnamed taxa is preferred until these taxa have been defined, diagnosed, and named by Lander or another worker. Many of that researcher's revisions to the genus *Protoreodon* still await formal explanation, and the summary of *Protoreodon* systematics in Theodor (1999) presents a historically more accepted view. Lander's (in press) and Lander and Hanson's (in press) works on *Agriochoerus* contain alternate views on some of the taxa discussed below, and subsequent research may resolve a majority of the taxonomic issues regarding Agriochoeridae.

PROTOREODON SCOTT AND OSBORN, 1887

Scott and Osborn (1887) described *Protoreodon parvus* (Fig. 11.1) from the middle Uintan Uinta B horizon of the Uinta Formation in Utah. Since then, a considerable number of species and genera of similar morphology have been described. A taxonomic history of *Protoreodon* was given by Gazin (1955), and more recently Wilson (1971a), Golz (1976), and Theodor (1999) have added new or existing species to this genus. Lander (1998) removed several species from the genus, transferring one to *Agriochoerus,* and four to new, unnamed genera. In doing so, he restricted *Protoreodon* to one evolutionary lineage that included the type species. This chapter follows that work's systematics while recognizing that formal support of these systematics is still needed.

As *Protoreodon* now stands, the genus contains six identified species. *P. parvus* is attributed only to the middle Uintan of Utah. *P. annectens* (Scott, 1899, although often considered *Agriochoerus pumilus* Marsh, 1875) has a much more cosmopolitan distribution and is reported from the late Uintan to early Duchesnean of Utah, California (Golz, 1976), Texas (Wilson, 1971a), Wyoming (Emry, 1975), and Montana (Tabrum et al., 1996). *P. tardus* (Scott, 1945) is known from middle to late Uintan California (Golz, 1976) and Wyoming (Eaton, 1985). *P. pacificus* is attributed only to late Uintan and early Duchesnean records from California by Golz (1976), but Lander (1998) also lists its possible occurrence in South Dakota. *P. paradoxicus* (Scott, 1898, 1899), lumped with *P. parvus* by Lander (1998), may represent a separate genus and species (E. B. Lander, pers. comm.). *P. walshi* (Theodor, 1999) is known from late Uintan records of California, and its P4, which lacks a split parametacone, is suggestive of a basal position in Oreodontoidea (E. B. Lander, pers. comm.). In addition to these six named species, Lander (1998) also lists at least one unnamed species similar to *P. annectens.* Undescribed specimens, possibly attributable to *Protoreodon,* are known from California (Walsh, 1996), Saskatchewan (Storer,

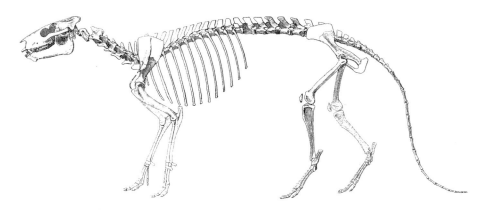

Fig. 11.1. Skeleton of *Protoreodon annectens*, based on specimens identified as *Protoreodon medius*. From Peterson (1919: 157, Plate XLI).

1996), Montana (Tabrum et al., 1996; A. R. Tabrum, pers. comm.), New Mexico (Lucas, 1983b), Colorado (Stucky et al., 1996), and Arkansas (Westgate and Emry, 1985).

Unnamed Genus A

This genus is based on specimens from the late Uintan Candelaria local fauna of Texas that were assigned to *Protoreodon petersoni* by Wilson (1971a). Wilson's suggestions for taxonomic relations within Agriochoeridae (1971, Fig. 3) made no distinction between specimens with a P4 with a split parametacone and those in which this cusp is unsplit. This decision was reflected in his choice to expand the definition of *"P." petersoni* rather than "complicate the nomenclature" by recognizing the Candelaria specimens as a separate species. Similar specimens found in other late Uintan and early Duchesnean localities in Texas (Wilson, 1986; Westgate, 1990) may also be assignable to unnamed genus A. Lander (1998) separated this genus from *Protoreodon* on the basis of its unusually conical, tall, and erect M1–3 paracone and metacone, less inflated or bulbous parastyle and mesostyle, and prominent paraconule. These characters distinguish this genus from all other contemporary and later agriochoerid species.

Unnamed Genus B

Gazin first (1955) reported *Protoreodon petersoni* from the late Uintan Myton Member of the Uinta Formation in Utah and then (1956) reported this species from late Uintan localities of the Badwater fauna of the Wind River Basin in Wyoming. Wilson (1971a) included this species, like *P. paradoxicus* and *P. minor*, part of the *"Protoreodon parvus* species group," as a possible synonym of *P. parvus*. Lander (1998) supports the recognition of *Protoreodon petersoni*, along with two currently undescribed species, as a separate genus, based on their small size and the appearance of a small enamel-lined pit at the labial end of the transverse valley of M1–3 postprotocrista in later forms. Elsewhere, this pit has been observed only in one Orellan individual of *Agriochoerus* (E. B. Lander, pers. comm.).

Protoreodon petersoni has appeared to several workers to be a basal oreodontid. Gazin (1955) believed it to be near the ancestry of all oreodontids and supported a monophyletic Oreodontidae. Wilson (1971) considered it ancestral to *Bathygenys* and *Leptauchenia* and supported a polyphyletic Oreodontidae. Lander (1998) weighed the evidence for both hypotheses and also proposed a third scenario in which *Bathygenys* and *Protoreodon petersoni* could be derived from an earlier member of *Protoreodon*. Although further work is necessary to resolve which of these hypotheses, if any, is most accurate, the monophyly of Oreodontidae would perhaps best be served if this genus were included within it.

Unnamed Genus C

Douglass (1902) reported *Agriochoerus minimus* from the earliest Chadronian White River Formation of Montana, and Lander (1998) lists it from other earliest Chadronian localities in Montana and Wyoming. Wilson (1971a) defined five characters that change in the evolution of *Agriochoerus* from *Protoreodon* and supported assigning a species to *Agriochoerus* only if it possessed all of these characters. Therefore, he transferred *A. minimus* to *Protoreodon*, citing its retention of a paraconule on the upper third molar. Lander (1998) placed this species into its own separate taxon, based on a size that is intermediate between contemporary records of unnamed genus B and *Protoreodon*.

Unnamed Genus D

Stock (1949) identified *Agriochoerus transmontanus* from specimens collected in the late Duchesnean or early Chadronian Titus Canyon Formation of California. The species was transferred to *Protoreodon* by Golz (1976), based on its primitive characters: a lack of a hypocone on the fourth upper premolar and the retention of a paraconule on the upper molars, both of which are variable characters in earlier agriochoerids. In addition, Golz thought it was a descendent of the late Uintan to early Duchesnean *Protoreodon pacificus*. Lander (1998) supported the addition of at least one other undescribed species to this genus. Lander (1998) reported a middle Duchesnean record from the Lapoint local fauna of Utah, late Duchesnean records of this genus from the Porvenir local fauna and Montgomery Bonebed of Texas, and late Duchesnean or early Chadronian records from the Rancho Gaitan local fauna of Chihuahua and the Coffee Cup local fauna of Texas. The characters that currently define Unnamed genus D are also evident in Unnamed genera A and C, and, thus, these three genera may be closely related.

AGRIOCHOERUS LEIDY, 1851

The systematics of *Agriochoerus* (Fig. 11.2) has recently been revised by Lander (1998). Lander (in press) and Lander and Hanson (in press) further refine and substantiate the systematics of the genus. To summarize, the characters that were used to distinguish *Diplobunops,* such as a P1–P2 diastema, enlarged caniniform teeth, and the correspondingly expanded rostrum, are also found in Chadronian and even Orellan examples of *Agriochoerus.* Therefore, these characters may reflect sexual dimorphism rather than a phylogenetic signal (Peterson, 1919; Black, 1978b; Lander and Hanson, in press). The clawed ungual phalanges of *Diplobunops* and *Agriochoerus,* unique to Cetartiodactyla, suggest that these two taxa are closely related, contrary to the phylogenies set forth by other workers (Gazin, 1955; Wilson, 1971a) in which this character would have had to evolve twice. This chapter employs the systematics of Lander (1998, in press) and Lander and Hanson (in press), while noting that descriptions of several of the taxa discussed below have not yet been published.

Agriochoerus pumilus (Marsh, 1875 and 1894), the earliest and most basal member, is found in the middle Uintan of Utah and has priority over Gazin's (1955) *Diplobunops vanhouteni. A. matthewi* (Peterson, 1919) is known from the late Uintan of Utah. Scott (1945) recognized a larger species, *A. crassus,* from the overlying late Uintan Randlett fauna of the Duchesne River Formation. This species also is known from late Uintan and early Duchesnean localities in Texas and Oregon (Wilson and Stevens, 1986; Hanson, 1996; Lander, 1998; Foss et al., 2004; Lander and Hanson, in press). *Agriochoerus* new species A is recognized from late Duchesnean Texas (Lander, 1998; Lander and Hanson, in press). *Agriochoerus* new species B (divided by Lander, 1998, and Lander and Hanson, in press, into several subspecies) is a Chadronian-aged representative of the genus, with a widespread distribution. The original genus and species of the family, *Agriochoerus antiquus* (Figure 11.2), is based on a palate and lower jaw that were collected in the Nebraska Territory, in modern-day South Dakota. *A. antiquus* also is known from Nebraska and Wyoming and spans the late Chadronian to the Orellan. *A. major* (Leidy, 1856) is restricted to the late Orellan of Montana and South Dakota, appearing to be a slightly larger form of similar morphology to *A. antiquus.* Both of these taxa seem to have been replaced in the Whitneyan and early Arikareean by *A. guyotianus,* most prevalent in, and first named from, the John Day Formation of Oregon (Cope, 1879c) but also occurring in South Dakota, Wyoming, and Nebraska (Lander, 1998). *A. gaudryi* is based on a hind foot described by Osborn and Wortman (1893). The diagnostic dental characters for this species are nonexistent, and thus, it might properly be considered a *nomen dubium,* probably equivalent to *A. guyotianus.*

Fig. 11.2. *Agriochoerus antiquus.* From Scott (1940: 727, Fig. 136).

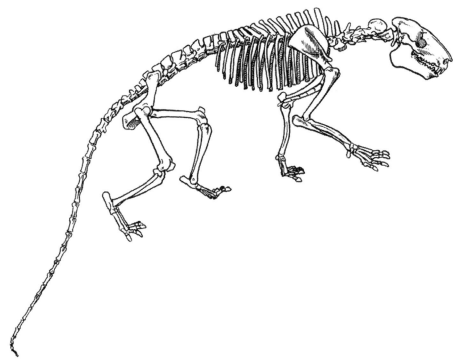

Unnamed Genus E

Agriochoerus maximus was first described by Douglass (1902) from the middle Chadronian of Montana. In addition, it has been recognized from the middle Duchesnean of Utah (Emry, 1981; Lander, 1998) and the late Duchesnean or early Chadronian of Chihuahua (Ferrusquìa, 1967). Additional material belonging to this species is also present in the earliest and early Chadronian of Wyoming (Emry, 1981, 1992; Lander, 1998). Douglass (1902) suggested that it might be a separate taxon, and Emry (1981) supported this idea. Lander (1998) also endorsed this suggestion and placed it in a distinct unnamed genus. *Agriochoerus maximus* is distinguished from *Agriochoerus* by its distinctively larger size and its retention of a premolariform p4, an oreodont plesiomorphic character shared with earlier species such as *A. matthewi* of *Agriochoerus,* but not contemporary species, in which the p4 is incipiently molariform (Lander, 1998). In fact, Lander and Hanson (in press) suggest a derivation of unnamed genus E from *Agriochoerus matthewi,* making unnamed genus E the sister taxon to later species of *Agriochoerus.*

PHYLOGENETIC RELATIONSHIPS OF AGRIOCHOERID-GRADE OREODONTS

Agriochoeridae belongs in the clade Artiodactyla, or Cetartiodactyla as the group has more recently been termed. This relationship is demonstrated by the now-familiar morphological hallmarks of artiodactyls: the hexacuspid deciduous fourth lower premolar, the paraxonic arrangement of the manus and the pes, and the double-trochleated astragalus. Leidy (1869) considered oreodonts to be "ruminating hogs" but placed them nearer ruminants, based on their shared selenodont dentitions. However, a selenodont dentition might have evolved several times in the course of cetartiodactylan evolution, and, thus, this character's presence alone might not be phylogenetically informative.

Similarly, the astragalus of oreodonts has been used to determine the place of oreodonts within Artiodactyla. When compared to extant artiodactyls, the oreodont astragalus most resembles that of Suina. This morphological character, along with others, was used by Simpson (1945) and McKenna and Bell (1997) to place Oreodontoidea in Suiformes. However, a suid-like astragalus is a primitive character also found in other middle Eocene selenodont artiodactyls (Lander, 1998). Therefore, only more derived astragalar morphologies are phylogenetically informative (Lander, 1998).

Other workers considered Oreodontoidea to be a basal member or close relative of Tylopoda, as perhaps best elaborated by Scott (1940), Romer (1966), and Carroll (1988). This view was supported by the cladistic analysis of Gentry and Hooker (1988). Janis et al. (1998a: 340, Fig. 22.1) reviewed the proposed members of Tylopoda and concluded that oreodonts were members of, or closely related to, Tylopoda.

The other possible conclusion is that the morphology of oreodonts is so unusual that they cannot confidently be placed in any of the modern subdivisions of Artiodactyla. Osborn and Wortman (1893) created Artionychia, a clade established for artiodactyl-like relatives of chalicotheres, including their *Artionyx gaudryi.* Artionychia was abandoned when it was demonstrated that *Artionyx* was an agriochoerid, although Lander and Hanson (in press) recently resurrected the term. Osborn (1910) created Oreodonta to contain Oreodontidae, but its narrow definition ("American primitive ruminants") makes this taxon equivalent to Gill's (1872) Oreodontoidea. Matthew (1929a), seeking to create subdivisions of Artiodactyla based on dental and foot characters, placed oreodonts in Ancodonta, artiodactyls with bunoselenodont to selenodont teeth and tetradactyl feet. This placement situated oreodonts near anthracotheres, an enigmatic group containing some species whose upper molar dentition is highly similar to that of *Agriochoerus* (E. B. Lander, pers. comm.). This hypothesis, however, seems to have had little support among other workers; Scott (1940) and Simpson (1945) used Ancodonta but did not place oreodonts in it.

Recent phylogenies have lent support to several hypotheses on the placement of Oreodontoidea in Cetartiodactyla. Norris (1999) supported Gentry and Hooker (1988) in their recognition of Oreodontoidea as basal members of Tylopoda. Thewissen et al. (2001b), Geisler (2001b), and Geisler and Uhen (2003) regarded Oreodontoidea as the sister group to Ruminantia plus Tylopoda, a clade termed Neoselenodontia by Webb and Taylor (1980). Theodor and Foss (2005) also supported this placement, although their consensus tree presented a paraphyletic Oreodontoidea that included *Leptotragulus,* a taxon commonly incorporated in Protoceratidae. Geisler and Uhen (2005) considered Oreodontoidea to be members of the basal-most branch of the clade containing Ruminantia. This placement also showed an unnamed clade consisting of Oreodontoidea and *Amphimeryx,* a European taxon included by Gill (1872) in his unnamed clade linking oreodonts with Eocene Old World taxa.

The geographic location of the origin of Oreodontoidea is also unknown at present. The most basal grades of Cetartiodactyla possess bunodont and bunoselenodont dentitions. These bunodont and bunoselenodont artiodactyls are believed to have given rise to all the later diverging taxa in Cetartiodactyla. Stucky (1998) for instance, placed the homacodontid genera *Pentacemylus* and *Mesomeryx* near the origin of Ruminantia. Lander (1998) supported the derivation of oreodonts and oromerycids from a homacodontid similar in morphology, but older in age, than *Pentacemylus.* However, the middle Eocene is marked by several immigration events among Europe, Asia, and North America. The recent description of *Asiohomacodon myanmarensis* (Tsubamoto et al., 2003), a taxon similar in geological age, dental morphology, and size to *Protoreodon parvus,* underscores the possibility that oreodonts may have arisen outside of North America.

Recent studies such as Métais (2006) show that phylogenetic analyses of Eocene Cetartiodactyla need to include both New and Old World taxa to best understand evolutionary relationships with this group.

FUNCTIONAL MORPHOLOGY

Once *Agriochoerus,* a selenodont artiodactyl, was shown to have clawed unguals, several researchers tried to determine their function. Matthew (1911) was the first to suggest that *Agriochoerus* might have been scansorial, as exhibited by the mounted skeleton at the American Museum of Natural History. Scott (1940) entertained this idea and also raised the possibility that agriochoerids were fossorial, digging for food much as *Vombatus,* the modern wombat, is known to do, or ground sloths, such as *Nothrotherium,* are supposed to have done. In a review of the possible use of claws by large mammalian herbivores, Coombs (1983) looked for skeletal adaptations that supported a digging or bipedal browsing lifestyle. Although *Agriochoerus* has some characters that are suggestive of either lifestyle, such as a long tail, they are plesiomorphic for Oreodontoidea and, thus, not an evolutionary innovation. Coombs agreed with Matthew (1911) that the scansorial hypothesis for *Agriochoerus* is the most likely, although the animal probably was not an agile climber, based on its limited limb flexibility and lack of opposable digits.

If *Agriochoerus* were a medium-sized scansorial mammalian herbivore, then the tree kangaroo would be its closest modern analogue. Indeed, Wilhelm (1993), in her analysis of oreodont limb morphology, compared the large manus-to-forelimb length ratio of *Agriochoerus* to that of tree kangaroos. The largest extant tree kangaroo, *Dendrolagus bennettianus,* has a maximum weight of 14 kg and a total body length of 1.6 m. The modern leopard, *Panthera pardus,* can reach a length of 2 m and a weight of 90 kg, and is able to carry dead prey into trees. *Agriochoerus,* at an estimated weight of between 38 (Radinsky, 1978) and 50 kg (Janis, 1982), would have been intermediate in weight between these two modern analogues, and, thus, would not have been limited from a scansorial lifestyle by its weight.

ACKNOWLEDGMENTS

I thank D. R. Prothero for inviting me to contribute to this volume. J. M. Theodor, M. S. Stevens, S. G. Lucas, J. D. Archibald, and three anonymous reviewers helped greatly by editing an earlier version of this chapter. E. B. Lander and J. M. Theodor edited its second iteration. E. B. Lander also provided prepublication copies of Lander (in press) and Lander and Hanson (in press) for use in this study. A. R. Tabrum and R. J. Emry provided helpful commentary on aspects of the agriochoerids with which they are most familiar. The Doris and Samuel P. Welles Fund (UCMP) and a Harry M. Hamberg Scholarship (SDSU) supported the author's research. Also, Julie, you are exceptional in your willingness to listen to me ramble on about Paleogene cetartiodactyls whom you have never met.

MARGARET SKEELS STEVENS
AND JAMES BOWIE STEVENS

12

Family Merycoidodontidae

OREODONTS, FAMILY MERYCOIDODONTIDAE, were artiodactyls of the late Paleogene and the Neogene that were mostly confined to North America and that were "successful" for about 30 million years. Piglike in general format, from jackrabbit to domestic pig in size, slender to stout, they were primitive in detail and displayed an unreduced selenodont dentition, an incisiform lower canine, a C/p1 caniniform pair, P4 with an undivided parametacone, a short diastema between C and P1, a deep jaw with a much-expanded gonial angle, and primitive digitgrade tetradactyl feet in which digits 2–5 contacted the ground. Oreodonts in general are regarded as browsers (Janis et al., 2000) (presumably low level), but some probably were mixed feeders or grazers (Wall and Shikany, 1993) and may have stripped bark seasonally. Their niche precludes direct competition with taller, speedier herbivores but is restrictive. The reoccupation of similar habitats by different oreodonts during the Tertiary commonly led to the development of convergent traits. By "convergent traits," we intend functionally similar apomorphies that are in a sense homoplasic conditions of homologues. Identification of a lineage must associate a suite of characteristics with stratigraphy.

The initial adaptive radiation of oreodonts began in the late Duchesnean and earliest Chadronian (later Eocene), and a second radiation began in the Whitneyan (early Oligocene; see Fig. 12.6). No Merycoidodontidae were present in the early Duchesnean or before that time (Black and Dawson, 1966), although the Agriochoeridae make their first appearance in Uinta A, early Uintan. None of the earliest oreodonts seem to be direct descendants of any known agriochoerid. *Aclistomycter, Bathygenys, Oreonetes, Limnenetes, Merycoidodon,* and *Miniochoerus* were already distinct when they first entered the fossil record, and they must have had long earlier histories. The immediate ancestors of these genera may have been among the "modernized" mammals that came from Asia into the New World in the late Eocene (Simpson, 1947; Webb, 1977; Prothero, 1985; Graham, 1999). *Miniochoerus* continued on to the earlier Whitneyan, *Limnenetes* gave rise to the leptaucheniin oreodonts that survived through the early Arikareean, and *Merycoidodon* produced the second and more extensive adaptive radiation that characterizes the Neogene (see Fig. 12.6).

FAMILY MERYCOIDODONTIDAE

Merycoidodontidae Hay, 1902: 665.
Merycoidodontidae Thorpe, 1937: 31, in part.
Merycoidodontidae (Oreodontidae) Schultz and Falkenbach, 1940, 1941, 1947, 1949, 1950, 1954, 1956.
Merycoidodontidae Schultz and Falkenbach, 1968.
Merycoidodontidae Stevens and Stevens, 1996: 510.
Oreodontidae Leidy McKenna and Bell, 1997: 408.
Merycoidodontidae Lander, 1998: 409.

Type Species.   *Merycoidodon culbertsoni* Leidy, 1848: 48.

Included Subfamilies.   Aclistomycterinae; Miniochoerinae and Oreonetinae; Leptaucheniinae; Merycoidodontinae; Promerycochoerinae; Merycochoerinae; Eporeodontinae; Ticholeptinae; Merychyinae; Ustatochoerinae; and Brachycrurinae.

Distribution and Age.   Rare occurrences in southern Saskatchewan, eastern Delaware, and Florida; common throughout the western United States; and known from Mexico and Central America: late Duchesnean to earliest Hemphillian, late Eocene to late Miocene.

Discussion.   We maintain oreodonts in the family Merycoidodontidae because the maxillary and ramal fragments of a ?single individual on which *Merycoidodon culbertsoni* from the *Poebrotherium*-bearing White River beds, Nebraska Territory, the first oreodont described and illustrated, can be tied to more intact specimens from the same rocks. *Merycoidodon* is a properly published name (International Code of Zoological Nomenclature, 1999, Article 8). Leidy (1851c) coined the term *"Oreodon"* when he described *"O. priscum"* and the much smaller *"O." gracile*. He (Leidy, 1852) then concluded that *"O. priscum"* is a synonym of *M. culbertsoni*. Leidy (1869) recharacterized *"O." culbertsoni* and placed this and related animals in an *"Oreodontidae."* Subsequent authors have vacillated between the use of *Merycoidodon* and *"Oreodon,"* hence the existence of a Merycoidodontidae or an *"Oreodontidae."* Schultz and Falkenbach (1940, 1941, 1947, 1950, 1954, 1956, 1968) placed oreodonts in a Merycoidodontidae (Oreodontidae) to indicate a preference and a synonym, but in their (1968) revision of *Merycoidodon* and their concluding summary, they placed all oreodonts in a Merycoidodontidae. Paragraph 23.3.7 (Article 23) and Recommendation 40A and its example (Article 40, particularly paragraph 40.2, International Code on Zoological Nomenclature [1999]), stress seniority and the importance of prevailing usage. We believe that Thorpe (1937) as the first reviewer, Schultz and Falkenbach as second reviewers, and the revision of *Merycoidodon* by Stevens and Stevens (1996) and of *Sespia* by Hoffman and Prothero (2004) establish that oreodonts, a vernacular name, are members of the Family Merycoidodontidae.

SUBFAMILY ACLISTOMYCTERINAE

Aclistomycterinae Lander, 1998: 409.

Type Species.   *Aclistomycter middletoni* Wilson, 1971a: 40.

Included Species.   Type species and ?*Aclistomycter dunagani*.

Distribution and Age.   Chambers Tuff Formation, the lowermost part of the Bandera Mesa Member, Devil's Graveyard Formation, Texas; and the Upper Tuff member, Prietos Formation, Rancho Gaitan, Chihuahua, Mexico: late Duchesnean to late Duchesnean/earliest Chadronian; late Eocene.

Discussion.   The oldest known oreodont-like animal is *Aclistomycter middletoni* of the Porvenir local fauna, Trans-Pecos Texas (Fig. 12.1A). ?*A. dunagani*. Little Egypt local fauna, Texas, is a presumed descendant. The placement of these animals in Aclistomycterinae Lander, 1998, seems appropriate on the basis of the open postorbital bar of *A. middletoni* and the documented retention of protoconules in ?*A. dunagani*.

SUBFAMILY OREONETINAE

Oreonetinae Schultz and Falkenbach, 1956: 453, in part.
Oreonetinae Schultz and Falkenbach, 1968: 382.
Oreonetinae McKenna and Bell, 1997: 408.
Miniochoerinae Lander, 1998: 410, in part.

Type Species.   *Limnenetes* (?) *anceps* Douglass, 1901b: 262.

Included Genera.   *Bathygenys* Douglass and *Oreonetes* Loomis.

Distribution and Age.   Chambers Tuff and Capote Mountain Tuff formations, Bandera Mesa Member, Devil's Graveyard Formation, Texas; Climbing Arrow Member, Renova Formation and Renova Formation undifferentiated, Montana; White River Formation, Cameron Springs, Flagstaff Rim, and Alcova areas and Yoder Member, Chadron Formation, Wyoming; and the Prietos Formation, Rancho Gaitan, Mexico: late Duchesnean through the medial Chadronian.

Discussion.   The Oreonetinae contains *Bathygenys* and *Oreonetes*. Two sequential species are recognized for *Bathygenys, B. reevesi* (Fig. 12.1B), latest Duchesnean/earliest Chadronian, Little Egypt local fauna, Texas, and *B. alpha*, early medial Chadronian, Pipestone Springs, Montana, the Yoder, Flagstaff Rim, and other sites in Wyoming and Nebraska, the Airstrip and Red Hill local faunas of Trans-Pecos

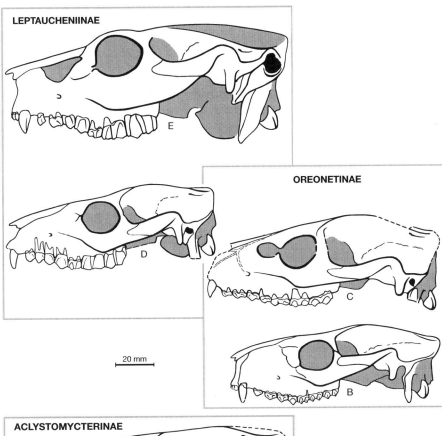

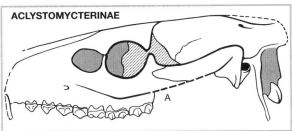

Fig. 12.1. Selected examples of Aclistomycterinae, Oreonetinae, and Leptaucheniinae. (A) Aclistomycterinae, *Aclistomycter middletoni*, late Duchesnean, TMM 41213, type, reversed, Porvenir local fauna, Chambers Tuff Formation, Presidio County, Texas. (B, C) Oreonetinae: (B) *Bathygenys reevesi*, latest Duchesnean–earliest Chadronian, composite skull based mostly on TMM 40209–198, type; missing parts supplied by –202, and –203, Little Egypt local fauna, Chambers Tuff Formation; (C) *Oreonetes anceps*, earlier Chadronian, CMNH 1052, McCarty's Mountain local fauna, Beaverhead County, Montana. (D, E) Leptaucheniinae: (D) *Limnenetes platyceps*, early Chadronian, composite skull based mainly on TMM 40504–63, missing parts supplied by –148, snout, –241, occiput, Airstrip local fauna, Chambers Tuff Formation, Presidio County, Texas; (E) *Leptauchenia decora*, upper Brule, late Whitneyan, USNM 2074 (redrawn from Scott, 1940), Fall River County, South Dakota. CMNH, Carnegie Museum of Natural History, Pittsburgh; TMM, Texas Memorial Museum of History and Science, The University of Texas at Austin, Austin; USNM, United States National Museum, Washington, D.C. Scale bar applies to all.

Texas, and the Calf Creek local fauna, Saskatchewan, Canada. We abandon a "Bathygeniinae" (Lander, 1998) because the defining features, atavistic, are not based on *B. alpha,* the type species, but on *B. reevesi*. Although the skulls of *B. reevesi* and *B. alpha* are generally similar, *B. reevesi* is larger, has a less firmly closed postorbital bar, and retains various primitive dental traits. *Bathygenys* is absent in the approximately contemporary assemblages that yield *Oreonetes* and *Limnenetes*. *Bathygenys* lacks descendants but probably shares a common ancestry with *Oreonetes* (see Fig. 12.6).

We do not accept the synonymy of *Miniochoerus* and its synonyms with *Oreonetes* or the placement of *Oreonetes* within a "Miniochoerinae" as Lander (1998) suggested. *Oreonetes* contains *O. anceps*, the early Chadronian type species (Fig. 12.1C), and *O. douglassi,* a medial Chadronian, larger and presumed descendant. *Oreonetes* does not occur in the Pipestone Springs faunas, Montana, the earlier Chadronian faunas from the north-central High Plains, nor in the earlier Chadronian Vieja faunas of Trans-Pecos Texas that contain *Bathygenys*. *O. anceps* is sympatric with *Limnenetes platyceps* at Thompson Creek, late early Chadronian, and *O. douglassi* oc-

curs with *Limnenetes* sp., middle Chadronian, in the highest part of the Renova Formation undifferentiated at McCarty's Mountain, Montana. *Oreonetes* appears to lack descendants.

Early authors were unable to relate the teeth of *Oreonetes anceps,* well preserved, with an intact auditory bulla and lacked the association of a much-inflated bulla with identifiable teeth in *Limnenetes platyceps;* specimens of either tended to be misidentified. These similar-sized contemporary genera were regarded as closely related and appeared to differ only by minor cranial features and the presence or absence of a preorbital fossa. Study of the leptauchenin-like teeth of *Limnenetes* from the Vieja of Texas, the largest sample known, shows that *Oreonetes* and *Limnenetes* are unrelated.

SUBFAMILY LEPTAUCHENIINAE

Leptaucheniinae Schultz and Falkenbach, 1940: 215, name only.
Leptaucheniinae Schultz and Falkenbach, 1968: 227, in part.
Leptaucheniinae CoBabe, 1996: 574, in part.

Leptaucheniinae McKenna and Bell, 1997: 409, in part.
Leptaucheniinae Lander, 1998: 410.
Leptaucheniinae Hoffman and Prothero, 2004: 156.
Leptaucheniinae Prothero and Sanchez, in press.

Type Species.   *Leptauchenia decora* Leidy, 1856a: 88.

Included Genera.   *Leptauchenia, Limnenetes,* and *Sespia.*

Distribution and Age.   Capote Mountain Tuff Formation, Texas; Thompson Creek, Climbing Arrow Member, Renova Formation, and McCarty's Mountain area, Renova Formation undifferentiated, Montana; Orella/Scenic, and Whitney/Poleslide members, Brule Formation, Wyoming, Colorado, South Dakota; and Gering and Monroe Creek formations, or equivalents, Colorado, Wyoming, and Nebraska; and upper Sespe and Otay formations, California: early Chadronian to late early Arikareean, late Eocene to medial Oligocene.

Discussion.   The Leptaucheniinae, adapted to dry environments, contain *Limnenetes* (Fig. 12.1D), *Leptauchenia,* and the more specialized, *Sespia. L. platyceps* and *Limnenetes* sp. represent an increasing earlier to medial Chadronian size cline. *L. platyceps* is absent in the Pipestone Springs faunas and from approximately contemporary faunas from the north-central High Plains. Most importantly, only the development of a nasal-facial vacuity and the late Chadronian separate *Limnenetes* sp. from the earliest *Leptauchenia* (Figs. 12.1E and 12.6), *L. decora. L. decora,* early Orellan to earlier Arikareean, *L. major,* late Whitneyan to early Arikareean, and *L. lullianus,* early Arikareean, are recognized (Prothero and Sanchez, in press). *Sespia californica,* earliest Arikareean, *S. nitida,* and *S. ultima,* earlier Arikareean, and, tentatively, *S. heterodon,* ?early middle Arikareean, are recognized (Hoffman and Prothero, 2004).

SUBFAMILY MINIOCHOERINAE

Miniochoerinae Schultz and Falkenbach, 1956: 391.
Miniochoerinae Schultz and Falkenbach, 1968: 382.
Miniochoerinae McKenna and Bell, 1997: 410.
Miniochoerinae Lander, 1998: 410, in part.

Type Species.   *Miniochoerus gracilis,* Leidy, 1851b: 239.

Included Genera.   *Miniochoerus* only.

Distribution and Age.   White River Group undifferentiated, Flagstaff Rim, Ahearn Member, Chadron, and Chadron Formation, Wyoming; Orella and Whitney members, Brule Formation, Nebraska; Chadron Formation, Pennington County, Scenic and Poleslide members, Brule Formation, South Dakota; and the Fitterer Badlands, White River Group, North Dakota: medial Chadronian to earlier Whitneyan, latest Eocene to earliest Oligocene.

Discussion.   *?Miniochoerus forsythae* Chadron Formation or equivalents (Fig. 12.2F), Wyoming and South Dakota, a generally *Merycoidodon*-like species except for size, is the earliest of the Miniochoerinae. *Miniochoerus* were relatively small and cranially stereotyped oreodonts and retained a minute auditory bulla, but their teeth became distinctive with thin-enameled, quickly abraded fossettes and fossettids (traits convergent with *Limnenetes*), and foreshortened premolars. *?M. forsythae,* medial Chadronian, *M. chadronensis,* upper part, Chadron Formation, late Chadronian, *M. affinis,* "Lower nodules" and equivalents, lower part, Orella Member, Brule Formation, early Orellan, and *M. starkensis,* "Upper nodules," upper part, Orella Member, and lower part, Whitney Member, and equivalents, Brule Formation, form the central lineage (Stevens and Stevens, 1996). A minor episode of speciation produced the dwarfed *M. gracilis* in the early Orellan (Fig. 12.2I).

SUBFAMILY MERYCOIDODONTINAE

Merycoidodontinae Hay, 1902: 665.
Merycoidodontinae Schultz and Falkenbach, 1968: 24.
Promerycochoerinae Schultz and Falkenbach, 1949: 140, in part.
Desmatochoerinae Schultz and Falkenbach, 1954: 163, in part.
Merycoidodontinae Stevens and Stevens, 1996: 512.
Oreodontinae McKenna and Bell, 1997: 409, in part.
Miniochoerinae Lander, 1998: 410, in part.

Type Species.   *Merycoidodon culbertsoni* Leidy, 1848: 24.

Included Genera.   *Merycoidodon* and *Mesoreodon.*

Distribution and Age.   Undifferentiated Vieja Group, Texas; Chadron and Brule formations, White River Group, Montana, Colorado, Nebraska, Wyoming, North and South Dakota; Fort Logan, Cabbage Patch, and Toston formations, Montana; lower part, Turtle Cove Member, John Day Formation, Oregon; and Gering Formation, Wyoming and Nebraska: medial Chadronian to earliest Arikareean, latest Eocene to early Oligocene.

Discussion.   The earliest *Merycoidodon* appears to be *M. presidioensis,* early Chadronian, Capote Mountain Tuff Formation, Texas. Dental similarity and size connect *M. presidioensis* with *M. culbertsoni* latest Chadronian–early Orellan, the type species for the Family Merycoidodontidae, but *M. presidioensis* had a deeper, narrower, and more arched rostrum. *M. culbertsoni* (Fig. 12.2A) occurs in the ?late Chadronian Ash Spring local fauna, Texas. *M. bullatus,* a late Orellan presumed descendant (Fig. 12.2B), initiated the trend in *Merycoidodon* for progressive inflation of the auditory bulla. *M. bullatus* or an immediate early Whitneyan descendant produced two branches (see Fig. 12.6), one leading to *M. major* (Fig. 12.2C) and the other to the smaller *Eporeodon occidentalis* (see Fig. 12.4A). *M. major* (Fig. 12.2C) is the presumed ancestor of *Mesoreodon,* founder of all of the "gigantic" Neogene oreodonts. *Mesoreodon* traditionally has been viewed as "brachycephalic" because of the biased referral of only dorsoventrally crushed specimens to a referred species, "*M.*" *megalodon, M. chelonyx,* the type species (Fig. 12.2D), is as mesocephalic as *Merycoidodon culbertsoni.*

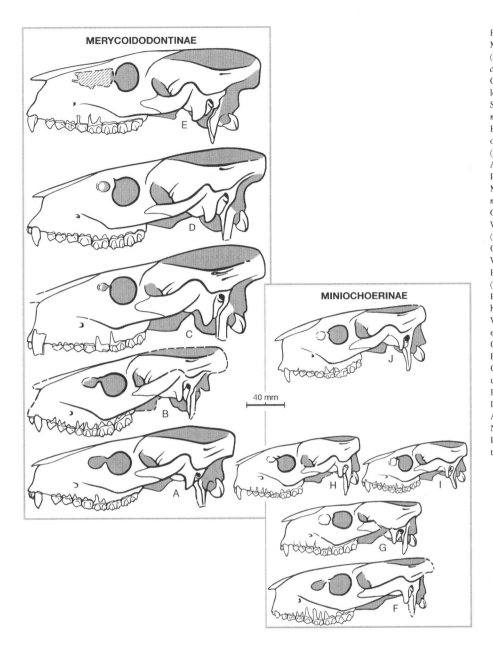

Fig. 12.2. Selected examples of Merycoidodontinae and Miniochoerinae. (A–E) Merycoidodontinae: (A) *Merycoidodon culbertsoni*, lower Brule, early Orellan, F:AM 45155; (B) *M. bullatus*, lower Brule, late Orellan, F:AM 45267, Shannon County, South Dakota; (C) *M. major*, upper Brule, late Whitneyan, F:AM 45298, "Washabaugh" (now a part of Jackson) County, South Dakota; (D) *Mesoreodon chelonyx*, earliest Arikareean, PU 10418, dentition from PU 11796, Cabbage Patch Formation, Meagher County, Montana; (E) ?*M. minor*, early Arikareean, F:AM 45430, Gering Formation, Goshen County, Wyoming. (F–J) Miniochoerinae: (F) ?*Miniochoerus forsythae*, medial Chadronian, F:AM 72303, above Ash D, White River Group undifferentiated, Flagstaff Rim, Natrona County, Wyoming; (G) *M. chadronensis*, late Chadronian, F:AM 45489, type, Chadron Formation, high, Seaman Hills, Niobrara County, Wyoming; (H) *M. affinis*, lower Brule, early Orellan, F:AM 44977, Niobrara County, Wyoming; (I) *M. gracilis*, lower Brule, early Orellan, F:AM45363, Niobrara County, Wyoming; (J) *M. starkensis*, upper Brule, early Whitneyan, type, F:AM 49585, Harding County, South Dakota. F:AM, Frick Collection, American Museum of Natural History, New York; PU, Princeton University, Princeton, New Jersey. Scale bar applies to all.

M. chelonyx, earliest Arikareean, is the presumed ancestor for ?*M. minor,* early Arikareean. ?*M. minor* (Fig. 12.2E) began the trend to exaggerate the upward "hook" and the increased lateral spread of the zygomatic arch and is believed to be ancestral to both the Promerycochoerinae and Merycochoerinae. *M. floridensis* (MacFadden and Morgan, 2003) is an endemic *Mesoreodon* from Florida.

SUBFAMILY PROMERYCOCHOERINAE

Promerycochoerinae Schultz and Falkenbach, 1949: 84, in part.
Desmatochoerinae Schultz and Falkenbach, 1954: 163, in part.
Desmatochoerinae Schultz and Falkenbach, 1968: 380, in part.
Promerycochoerinae McKenna and Bell, 1997, in part.

Merycochoerinae Lander, 1998: 412, in part.

Type Species. *Promerycochoerus superbus* Douglass, 1901b: 82.

Included Genera. *Desmatochoerus, Promerycochoerus,* and *Megoreodon.*

Distribution and Age. Upper Sespe Formation, California; Turtle Cove Member, John Day Formation, Oregon; Lemhi County, Idaho; Fort Logan and Toston formations, Montana; Colter Formation, Wyoming; Gering, Monroe Creek, and Harrison formations, Wyoming, Nebraska; and "Harrison" formation, South Dakota: early to late Arikareean.

Discussion. The Promerycochoerinae originated from within *Mesoreodon.* "*M.*" *megalodon,* long regarded as "brachycephalic," is conspecific with *Desmatochoerus "curvidens"* on

Fig. 12.3. Selected examples of Promerycochoerinae and Merycochoerinae, and *Mesoreodon*, Merycoidodontinae, for reference. (A) *?Mesoreodon minor*, early Arikareean, F:AM 45430, Gering Formation, Goshen County, Wyoming. (B–D) Promerycochoerinae: (B) *Desmatochoerus megalodon*, Monroe Creek Formation, early Arikareean, F:AM 33344 (type of *D. "hatcheri niobrarensis"*); C, *Promerycochoerus superbus*, ?early medial Arikareean, AMNH 7431, Turtle Cove Member, above Deep Creek Tuff, John Day Formation, central Oregon; (D) *P. carrikeri*, late Arikareean, F:AM 33352, Harrison Formation, Niobrara County, Wyoming. (E–H) Merycochoerinae: (E) *?Hypsiops latidens* (Douglass), ?late early Arikareean, F:AM 42986, reversed, type Rosebud Formation, Todd County, South Dakota; (F) *Hypsiops breviceps*, late Arikareean, AMNH 9731 (type of *H. "brachymelis,"* reversed), Six Mile Creek Formation, Jefferson County, Montana; (G) *Submerycochoerus bannackensis*, latest Arikareean, rostrum, F:AM 34317, Grasshopper Creek, Beaverhead County, occiput, F:AM 34481, type of *"Pseudomesoreodon rolli,"* North Boulder Valley, Jefferson County, Montana; (H) *Merycochoerus proprius*, late early Hemingfordian, F:AM 42469B, ?female, upper part, Runningwater Formation, Dawes County, Nebraska. Scale bar applies to all.

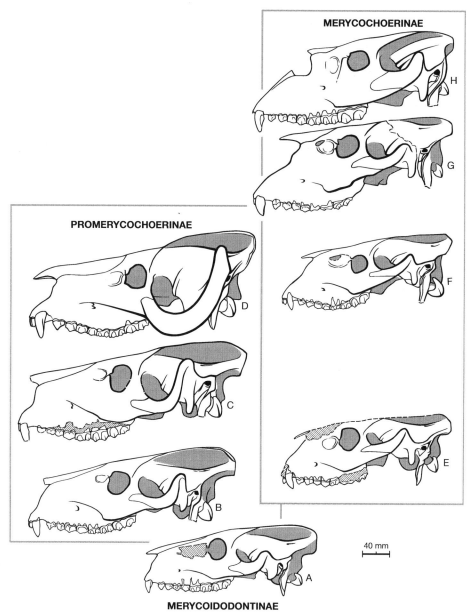

which the presumably dolichocephalic "Desmatochoerinae" was based. The "brachycephaly" of *"M."* *megalodon* and the "dolichocephaly" of *"D."* *curvidens* result from apposing directions of postmortem deformation of the type and referred specimens. The equivalence of *"M."* *megalodon* and *"D."* *curvidens,* and the placement of *"M."* *megalodon* in *Desmatochoerus* cannot alter the priority and validity of *Desmatochoerus* (Article 40, 40.1; International Commission on Zoological Nomenclature, 1999). *D. megalodon* (Fig. 12.3B), early Arikareean, is a presumed descendant of *?Mesoreodon minor* (Figs. 12.3A and 12.6) but is larger and with a more upwardly "hooked" zygomatic arch, the hallmark of the Promerycochoerinae. *D. megalodon* is independently ancestral to *Megoreodon grandis,* late early Arikareean, an early venture in gigantism that produced the largest oreodont that ever lived, and to *Promerycochoerus superbus,* medial Arikareean, a secondary experiment in gigantism. *M. gran-*

dis has the same dental morphologies and mesocephally as *D. megalodon* but is a fifth to a quarter larger, and we do not consider *M. grandis* to have been males of *D. megalodon,* when contemporary. *M. grandis* is not ancestral to *P. superbus,* a sister group, because the skull of *M. grandis* is disharmonic (longer yet relatively narrower) relative to *P. superbus* with its larger size but more gracile zygomatic arch. *P. superbus* (Fig. 12.3C) continues the trend begun in *D. megalodon* for a size increase, a broadening of the bizygomatic diameter that makes the skull "brachycephalic," a deepening of the malar, and for increased "hooking" of the zygomatic arch, with the teeth remaining the same except for size.

Promerycochoerus superbus, best known from the John Day Formation, Oregon, between the Deep Creek and Tin Roof tuffs (see Fremd et al., 1994), Turtle Cove Member, lived during the time of the development of a 3- to 4-million-year medial Arikareean hiatus in the Wyoming and Nebraska area (MacFadden and Hunt, 1998) that separates the earlier Arikareean Monroe Creek Formation, rocks that preserve *Desmatochoerus megalodon* from the late Arikareean Harrison Formation rocks that contain *P. carrikeri. P. superbus,* however, occurs in the "Harrison formation" of southwestern South Dakota, rocks that are younger than the Monroe Creek but older than the Harrison Formation. *P. superbus* is the probable ancestor for *P. chelydra,* a John Day taxon that is morphologically close to *P. carrikeri. P. carrikeri* (Fig. 12.3D), early late Arikareean, is notable for the great bizygomatic width and the grossly upwardly "hooked" zygomatic arch. *P. carrikeri* lacks descendants.

MERYCOCHOERINAE

Merycochoerinae Schultz and Falkenbach, 1940: 216, in part.
Phenacocoelinae Schultz and Falkenbach, 1950: 100, in part.
Merycochoerinae McKenna and Bell, 1997: 411, in part.
Merycochoerinae Lander, 1998: 412, in part.

Type Species. *Merycochoerus proprius* Leidy, 1858b: 24.

Included Genera. *Hypsiops, Submerycochoerus,* and *Merycochoerus.*

Distribution and Age. Haystack Valley, and Rose Creek members, John Day Formation, Oregon; North Boulder Valley fauna, Renova Formation, Montana; the Lower channel, Martin Canyon beds, Colorado; type Rosebud Formation, South Dakota; Monroe Creek and Harrison formations, Wyoming; and the Anderson Ranch and Runningwater formations, Nebraska: late early Arikareean to early Hemingfordian, early late Oligocene to medial early Miocene.

Discussion. The Merycochoerinae are related to the Promerycochoerinae, but neither one is ancestral to the other. Both share a common ancestry, presumably from ?*Mesoreodon minor* (Fig. 12.3A). ?*Hypsiops latidens,* early medial Arikareean, *H. breviceps,* late Arikareean, *Submerycochoerus bannackensis,* latest Arikareean–earliest Hemingfordian, genera once placed in a paraphyletic "Phenacocoelinae," and *Merycochoerus matthewi,* early early Hemingfordian, and *M. proprius,* late early Hemingfordian, have an ancestor–descendant relationship (Figs. 12.3E–H and 12.6), where the trend with time was for a size increase and nasal "retraction." The retraction of the nasal notch and the shortening of the nasal are associated with the progressive adaptation of the lineage to a riparian paleoenvironment, and probably to a tapir-like ecology and feeding strategy. *Merycochoerus* is not ancestral to *Brachycrus.*

SUBFAMILY EPOREODONTINAE

Eporeodontinae Schultz and Falkenbach, 1940: 215, name only.
Eporeodontinae Schultz and Falkenbach, 1968: 193.
Promerycochoerinae Schultz and Falkenbach, 1949: 84, in part.
Phenacocoelinae Schultz and Falkenbach, 1950: 100, in part.
Desmatochoerinae Schultz and Falkenbach, 1954: 162, in part.
Merycoidodontinae Schultz and Falkenbach, 1968: 24, in part.
Eporeodontinae McKenna and Bell, 1997, in part.
Eucrotaphinae Lander, 1998: 411, in part.
Merycochoerinae Lander, 1998: 412, in part.

Type Species. *Eporeodon occidentalis* Marsh, 1873: 409.

Included Genera. *Eporeodon* and *Merycoides.*

Distribution and Age. Upper Sespe Formation, South Mountain Fauna, California; Turtle Cove Member, John Day Formation, Oregon; Gering and Monroe Creek formations, Wyoming and Nebraska; and the lower part of the Harrison and the Anderson Ranch formations, Nebraska: latest Whitneyan to latest Arikareean.

Discussion. The Eporeodontinae represent the base of the second branch that diverged directly or indirectly from *Merycoidodon bullatus* (Figs. 12.2C and 12.6) and contain *Eporeodon* and *Merycoides.* All of the dwarfed or partially dwarfed oreodonts that characterize the Neogene appear to have come from *Eporeodon.* The White River oreodonts, Merycoidodontinae, with an inflated auditory bulla are excluded from *Eporeodon,* as is *"Eucrotaphus,"* all taxa of which, except the type species, are regarded as *nomina dubia. E. pacificus,* as extrapolated from the faunal lists provided by Fremd et al. (1994), is a slightly larger, early Arikareean descendant of *E. occidentalis.* We are not convinced about the validity of ?*E. trigonocephalus,* similar to *E. occidentalis* in size and age, because the skull width may be exaggerated by the advanced ontogeny of the virtually edentulous type. *E. thurstoni,* southern California, is believed to be separable from *E. occidentalis,* Pacific Northwest, on the basis of subtle morphological differences and because *E. thurstoni* belonged to a different botanical realm (Axelrod, 1958; Graham, 1993). *Merycoides pariogonus,* early Arikareean, is a slightly modified *E. pacificus*–like species (Fig. 12.4B). *Merycoides* underwent a size increase, but unlike many lineages that became larger, crania of *Merycoides* remained "refined" with their rounded contours and low crests and ridges. *M. pariogonus* seems ancestral to ?*M. longiceps,* late Arikareean (Fig. 12.4C), which in turn is the "best" ancestor for ?*M. stouti,* originally placed in *Phenacocoelus* Peterson. Absence or presence of a small facial vacuity, premolar foreshortening, taller teeth, and a broadening of the occiput of ?*M. longiceps* and ?*M. stouti* relative to *M. pariogonus* are derived traits that progressively removed *Merycoides* from *Eporeodon.* ?*M. stouti* lacks descendants.

SUBFAMILY TICHOLEPTINAE

Ticholeptinae Schultz and Falkenbach, 1940: 215, name only.

Fig. 12.4. Selected examples of Eporeodontinae, Ticholeptinae, and Merychyinae. (A–C) Eporeodontinae: (A) *Eporeodon occidentalis* (Marsh), latest Whitneyan, YPM 10142, type (reversed), lowermost part, Turtle Cove Member, John Day Formation, central Oregon; (B) *Merycoides pariogonus,* late early Arikareean, CMNH 1222 (type of *M. "cursor,"* reversed), Canyon Ferry area, Toston Formation, Lewis and Clark County, Montana; (C) *?Merycoides longiceps,* late Arikareean, AMNH 9732, type, Six Mile Creek Formation, Jefferson County, Montana. (D–F) Ticholeptinae: (D) *Paroreodon parvus,* late medial or early late Arikareean, YPM 12420, reversed, Haystack Valley Member, John Day Formation, Oregon; (E) *Phenacocoelus typus* Peterson, late Arikareean, F:AM 33660A (type of *P. "kayi"*), Harrison Formation, Niobrara County, Wyoming; (F) *Ticholeptus zygomaticus,* early to medial Barstovian, F:AM 43073 (reversed), Sinclair Draw, Olcott Formation, Sioux County, Nebraska (occiput of F:AM 24492, Pawnee Creek Formation, Colorado). (G, H) Merychyinae: (G) *Oreodontoides curtus,* late early Arikareean, AMNH 13817, Monroe Creek Formation, "Washabaugh" (now Jackson County), South Dakota; (H) *Merychyus arenarum,* latest Arikareean, F:AM 34419, Anderson Ranch Formation, Sioux County, Nebraska. YPM, Yale Peabody Museum, Yale University, New Haven. Scale bar applies to all.

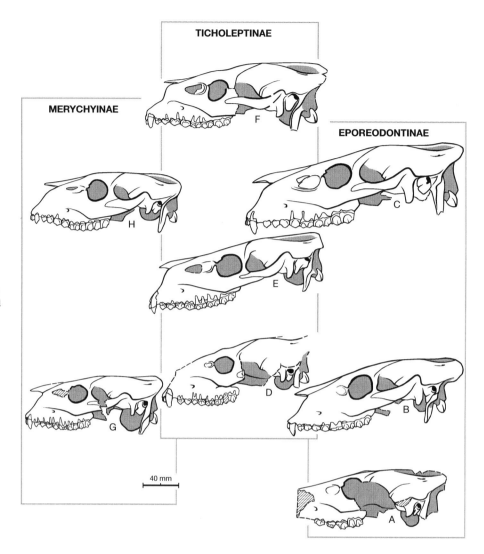

TICHOLEPTINAE

MERYCHYINAE

EPOREODONTINAE

40 mm

Ticholeptinae Schultz and Falkenbach, 1941: 6, in part.
Ticholeptinae Schultz and Falkenbach, 1968: 371, in part.
Merycoidodontinae Schultz and Falkenbach, 1968: 164, in part.
Ticholeptinae McKenna and Bell, 1997: 411, in part.
Ticholeptinae Lander, 1998: 414, in part.

Type Species. *Ticholeptus zygomaticus* Cope, 1878a: 129; 1878b: 380.

Included Genera. *Paroreodon, Phenacocoelus,* and *Ticholeptus.*

Distribution and Age. Turtle Cove, Kimberly, and Haystack Valley members, John Day Formation, Oregon; Harrison Formation, Wyoming and Nebraska; Runningwater, Sheep Creek, Box Butte, and Olcott formations, Nebraska; Martin-Anthony road cut locality, Marion County, Florida: late early Arikareean to medial Barstovian, early late Oligocene to middle Miocene.

Discussion. We remove *Ustatochoerus* and *Mediochoerus* from the Ticholeptinae of Schultz and Falkenbach (1941) (Fig. 12.5), and *Merychyus* (including *Ustatochoerus*),

Mediochoerus, and *Brachycrus* (Fig. 12.5) from Lander's (1998) proposed Ticholeptinae. The presumed relationship between these taxa and *Ticholeptus zygomaticus* (Fig. 12.4E) rests on cranial "refinement" and nasal "retraction," homoplasic, convergent traits. Teeth of *T. zygomaticus,* earlier Barstovian, and the early to late Hemingfordian ancestors; *T. bluei* and *T. calimontanus,* however, are more primitive with their anteroposteriorly elongated premolars and low dental crowns relative to *Merychyus, Mediochoerus,* and surely *Ustatochoerus,* of similar age, now excluded genera. The origin of the Ticholeptinae is more remote than *Merychyus calaminthus,* early late Arikareean, the earliest *Merychyus,* because *M. calaminthus* is dentally more specialized than *T. bluei,* the earliest probable *Ticholeptus. Paroreodon parvus* (Fig. 12.4D) of the presumed Northwest center of origin for the Ticholeptinae is the most adequate ancestor known for *Ticholeptus. P. parvus* also is a suitable ancestor for *Phenacocoelus typus* Peterson (Fig. 12.4F), late Arikareean, another cranially "derived" but dentally "primitive" species with relatively elongated but exceptionally molarized premolars and a larger facial vacuity. MacFadden (1980) reports *Phenacocoelus* from Marion County, Florida. The re-

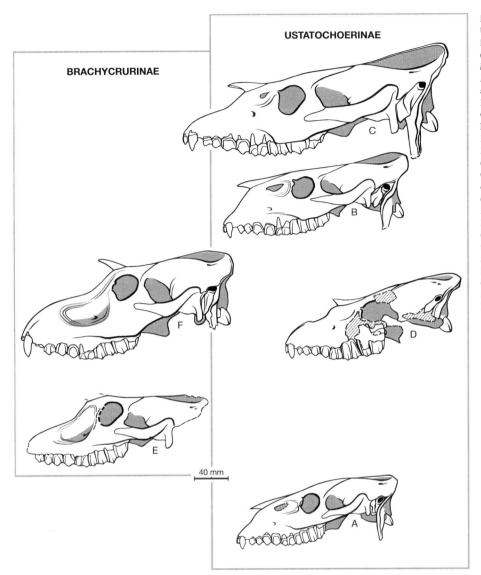

Fig. 12.5. Selected examples of Ustatochoerinae informal, and Brachycrurinae informal. (A–D) Ustatochoerinae: (A) *Ustatochoerus leptoscelos* (Stevens), earliest late Arikareean, TMM 40635-26 (snout augmented slightly by –6), Lower Member, Delaho Formation, Brewster County, Texas; (B) *U. medius* (Leidy), late Barstovian–earliest Clarendonian, F:AM 43030B, lower Valentine Formation, Brown County, Nebraska; (C) *U. major,* early and medial Clarendonian, F:AM 34220, Kat Quarry channel, Cherry County, Nebraska; (D) *Mediochoerus blicki,* early Barstovian, F:AM 43172, Sinclair Draw, Olcott Formation, Sioux County, Nebraska. (E, F) Brachycrurinae: (E) *Brachycrus rusticus,* latest Hemingfordian, F:AM 36105 (reversed), upper part, Split Rock Formation, Fremont County, Wyoming; (F) *B. siouense,* earlier Barstovian, F:AM 36113, Olcott Formation, Sioux County, Nebraska. Scale bar applies to all.

moval of *P. typus* (and *Hypsiops* and *Submerycochoerus*) from the "Phenacocoelinae" vacates that subfamily.

SUBFAMILY MERYCHYINAE

Merychyinae Simpson, 1945: 149.
Merychyinae Schultz and Falkenbach, 1947: 168.
Merychyinae Schultz and Falkenbach, 1968: 377, in part.
Merychyinae McKenna and Bell, 1997: 410, in part.
Ticholeptinae Lander, 1998: 414, in part.

Type Species.   *Merychyus elegans* Leidy, 1858b: 25.

Included Genera.   *Merychyus, Oreodontoides,* and *Paramerychyus.*

Distribution and Age.   Turtle Cove Member, John Day Formation, Oregon; Monroe Creek Formation, South Dakota; Tick Canyon, Hector, Kinnick formations, Tropico Group, and 4th. Division, Barstow Formation, California; Zia Formation, and Skull Ridge Member, Tesuque For-

mation, New Mexico; Lower member, Delaho Formation, and Closed Canyon "formation" ("Member 9," Rawls Formation), Texas; Aguas Calientes area, Mexico; Harrison and Anderson Ranch formations, Wyoming and Nebraska; Runningwater, Sheep Creek, Box Butte, and Olcott formations, Nebraska; "Upper channel," Martin Canyon local fauna, Colorado; upper part, Colter Formation, Wyoming; and equivalents elsewhere: late early Arikareean to medial Barstovian, early late Oligocene to middle Miocene.

Discussion.   The abandonment of the Merychyinae, the synonymy of *Ustatochoerus* with *Merychyus,* the placement of *Merychyus, Ticholeptus, Mediochoerus,* and *Brachycrus* within a "Ticholeptinae," and the placement of *Oreodontoides* within a "Phenacocoelinae" as proposed by Lander (1998) lack merit. *Oreodontoides,* Picture Gorge Ignimbrite to approximately Tin Roof Tuff (see Fremd et al., 1994), Turtle Cove Member, John Day Formation, is a notably dwarfed oreodont with a relatively very large auditory bulla. Two closely related species, *O. oregonensis,* Turtle Cove

Member, John Day Formation, early Arikareean, and *O. curtus* (Fig. 12.4G), Monroe Creek Formation, an early Arikareean immigrant to the north-central High Plains, are recognized. *O. curtus* is the presumed direct ancestor for *Merychyus* and initiated the hallmark trends that characterize *Merychyus,* such as premolar foreshortening, loss of premolar labial cingula, and a marked increase in dental crown height. *Merychyus* remained relatively small, probably as a result of niche segregation and predator escape strategy. The main lineage consists of *M. calaminthus,* early late Arikareean, *M. arenarum* (Fig. 12.4H), latest Arikareean, *M. elegans,* earlier Hemingfordian, and *M. relictus,* late Hemingfordian–earlier Barstovian. Minor speciation in the early Hemingfordian produced *M. verrucomalus,* the only oreodont to have developed an extremely deep, almost horizontally laterally flared, and rugose malar. The Barstovian demise of *Merychyus* (and *Ticholeptus*) may have resulted from competition with *Ustatochoerus,* an immigrant from the Southwest, from the rise of grazing camelids and equids, and/or from the rise of aelurodont canids (Wang et al., 1999). *O. curtus* is the "best," remote ancestor for *Paramerychyus,* a short-lived late Arikareean ?riparian browser with low-crowned teeth that probably was separated ecologically from *Merychyus.*

USTATOCHOERINAE, INFORMAL

Ticholeptinae Schultz and Falkenbach, 1941: 6, in part.
Ticholeptinae McKenna and Bell, 1997, in part.
Ticholeptinae Lander, 1998: 414, in part.

Defining Species. *Ustatochoerus major* (Leidy, 1858b: 26).

Included Genera. *Ustatochoerus* and *Mediochoerus.*

Distribution and Age. Lower member, Delaho, Closed Canyon *sensu lato* ("Member 9, Rawls Formation"), Goliad formations, and Ogallala Group of Texas; Runningwater, Olcott, Snake Creek, Valentine, and Ash Hollow formations of Nebraska; Pojoaque, Chama-el Rito, Ojo Caliente members, Tesuque Formation, Chamita Formation, and the Cerro Conejo Formation, New Mexico; Dove Spring, Hector, and Barstow formations of California; Pawnee Creek Formation, Colorado: latest middle–earliest late Arikareean to latest Clarendonian–earliest Hemphillian; earliest to late Miocene.

Discussion. *Ustatochoerus* can neither be placed in the Ticholeptinae as believed by Schultz and Falkenbach (1941) nor regarded as a synonym of *Merychyus* as suggested by Lander (1998). Teeth of *Ustatochoerus* are higher crowned than those in *Ticholeptus* and have more prominent, shelf-like labial premolar cingula. Premolar cingula are essentially lacking in *Merychyus.* Further, the earliest *Ustatochoerus, U. leptoscelos,* early late Arikareean, and the much smaller *M. calaminthus* were contemporaries. A single lineage is recognized, *U. leptoscelos* (Figs. 12.5A and 12.6), *Ustatochoerus* sp., early Hemingfordian, *U. medius,* later Barstovian–earliest

Clarendonian, *U. major,* early to late Clarendonian, and *U. californicus,* latest Clarendonian–earliest Hemphillian. *Ustatochoerus* originated in the southwestern United States or northern Mexico within the evolving Madro-Tertiary Geoflora (Axelrod, 1958; Graham, 1993) and persisted there as the last known member of the Family Merycoidodontidae. *Ustatochoerus* is probably a remote descendant of *Eporeodon thurstoni,* earliest Arikareean of southern California. Early *Ustatochoerus* were morphologically stereotyped, but the following trends—increased crown height of the molars, especially M3/m3, diminished size of the facial vacuity, broadening of the occiput, and an increase in skull size—accelerated from *U. medius* (Fig. 12.5B) over time to the time of extinction. *Ustatochoerus* apparently followed the spread of brushy grasslands (Graham, 1999) northward up along the High Plains corridor to become common on the midcontinent during the later Barstovian and may have replaced *Merychyus* and *Ticholeptus.* The extinction of *Ustatochoerus* may coincide with the rise of C_4 photosynthesis (Thomasson et al., 1990) and pressure imposed by borophagine canids (Wang et al., 1999). The status of *U. "profectus"* remains unsettled because without knowledge of the snout of *U. "profectus"* from the type area, the taxon cannot be shown to differ from *U. major* (Fig. 12.5C). The rare *Mediochoerus* probably arose from early *Ustatochoerus* and contains *M. johnsoni, M. blicki* (Fig. 12.5D), and *M. mohavensis* with a late early Hemingfordian to ?medial Barstovian range. The cranial and dental specializations seen for *Mediochoerus* are elaborations on the morphology of *Ustatochoerus. Mediochoerus* lacks descendants.

BRACHYCRURINAE, INFORMAL

Merycochoerinae Schultz and Falkenbach, 1940: 216, in part.
Merycochoerinae McKenna and Bell, 1997: 411, in part.
Ticholeptinae Lander, 1998: 415, in part.

Defining Species. *Brachycrus rusticus* (Leidy, 1870b: 109).

Included Genera. *Brachycrus.*

Distribution and Age. Flint Creek, Deep River, and Madison Valley formations, Montana; Upper Porous Sandstone sequence, Split Rock Formation, Wyoming; Olcott Formation, Nebraska, Green Hills, "Third Division," Barstow Formation, California; Skull Ridge Member, Tesuque Formation, New Mexico; and the Cucaracha Formation, Panama, Central America: latest Hemingfordian to earlier Barstovian, middle Miocene.

Discussion. *Brachycrus* is a bizarre, uniquely specialized group that suddenly appears in the latest Hemingfordian North American fossil record. Specializations (not unlike those for the modern *Saiga* antelope) include a strongly arched rostrum, upwardly vaulted, triangular, and very short nasal, nasal and nasal notch retracted to above the orbit, a narrow and elongated external nasal aperture that is

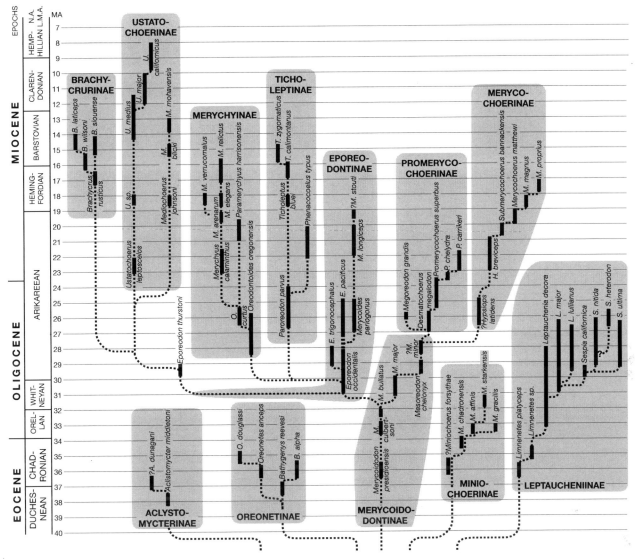

Fig. 12.6. A phylogenetic history of the Family Merycoidodontidae on the species level.

bordered by a ridge, and a teardrop-shaped facial depression (Fig. 12.5E and F). *Brachycrus* is not descended from the tapir-like *Merycochoerus,* nor is it a member of the Ticholeptinae. The bizarre morphology, lack of suitable earlier Hemingfordian or Arikareean ancestors, and the general stereotypy through the about 3-million-year known history (except for increase in skull size) suggest that *Brachycrus* evolved slowly and did not originate in the United States. The occurrence of *Brachycrus* in Panama suggests that *Brachycrus* may have originated in Central America or southern Mexico. It is possible that *Brachycrus* shares a remote common ancestry with the Ustatochoerinae, and the teeth of these two are similar with their tall crowns and shelf-like premolar labial cingula. *B. rusticus* the smallest, earliest, and ancestral species known (Fig. 12.5E) produced two lineages. One comprises *B. wilsoni* and *B. laticeps,* sequential earlier Barstovian taxa that became relatively large

and were confined to the Wyoming-Montana area. The second lineage contains *B. siouense* (Fig. 12.5F), a contemporaneous group of smaller animals that lived at lower latitudes and ?elevations, that are known from southern California, New Mexico, and the north-central High Plains.

CONCLUSION

High levels of scholarship common in paleontology and embodied in the bibliographies of the Society of Vertebrate Paleontology, and guided by the International Code of Zoological Nomenclature, prevent or solve many nomenclatural situations. Nevertheless, oreodont nomenclature has long been in disarray. Most of the taxonomic oversplitting that obscures an understanding of oreodont history arises not from poor scholarship but from diverse perceptions of the fossil remains and from varying understandings of paleobiological and stratigraphic contexts. Our phylogeny (Fig. 12.6) rests on the synonymy of many taxa, unlisted, that we regard as junior to those acknowledged. The taxa that are recognized are based on multiple lines of evidence, such as a reevaluation of the effects of postmortem deformation, acceptance of a broadened geographic distribution, and a reasonable range of intraspecific variation, on statistical analyses, and on a reassessment of current radiometric and stratigraphic determinations. New information and additional fossils will show that this synopsis is imperfect, but we believe that the synthetic approach shows that oreodonts can serve as useful biostratigraphic indicators.

DONALD R. PROTHERO
AND JOSHUA A. LUDTKE

13

Family Protoceratidae

THE PROTOCERATIDAE WERE a relatively rare family of artiodactyls endemic to North America from the middle Eocene (early Uintan) until the earliest Pliocene (latest Hemphillian). Their most distinctive feature is the large bony horns with peculiar shapes in the male individuals of later species. These horns were found above the orbits, on the rostrum, and sometimes even on the parietals or occiput. For example, *Protoceras* had broad flange-like horns over the maxillae and short knob-like horns over the orbits and parietals (Figs. 13.1 and 13.2C). *Paratoceras* had inwardly curved horns over the orbits and a forked "propeller"-like horn on the occiput (Figs. 13.1 and 13.2D). The synthetoceratines were characterized by Y-shaped "slingshot" horns on their rostrum and long curved horns over the orbits. In addition, most protoceratids show nasals that are retracted at least to the level of P3 or M1, suggesting the presence of a short proboscis or prehensile lip. Protoceratid cheek teeth tend to be rather low crowned and bunoselenodont, and the upper molars are much wider than they are long, with a strong lingual cingulum and parastyles and mesostyles. Males displayed an enlarged upper canine and a caniniform P1/p1 isolated by diastemata. The lower jaw shows a distinctively short coronoid process, and the symphysis curves ventrally. Compared to most other selenodont artiodactyls, it appears the limbs tended to remain relatively short and unspecialized, with unfused radius and ulna (except in the most derived synthetoceratines), unfused cuboid and navicular, almost no loss of digits II and V, and no fusion of the metapodials into a cannon bone. These features suggest that protoceratids were not as cursorial as most camels and ruminants and may have lived in marshy terrain and/or dense brush, where speed was not a critical factor.

Fig. 13.1. Family tree of the Protoceratidae (drawing by C. R. Prothero).

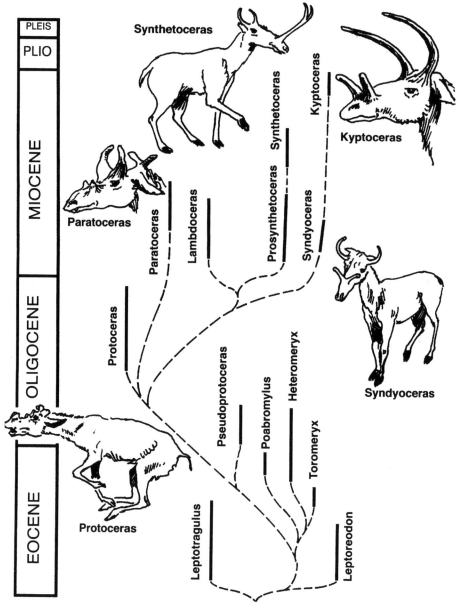

SYSTEMATICS

Basal Protoceratids ("Leptotragulinae")

Protoceratids were never very diverse or abundant, but in some faunas they were important and distinctive in their presence (e.g., the late Whitneyan deposits of South Dakota have been named the *"Protoceras* beds"). In the late Uintan and early Duchesnean, they were at their peak in diversity and in some cases were common elements of the fauna (for example, *Leptoreodon* from the late Uintan of California; *Poabromylus* in nearly every Duchesnean locality). Prothero (1998a) recognized 13 valid genera of protoceratids, and no new genera (but two new species) have been described since 1998. In this chapter we do not repeat all the information summarized by Prothero (1998a) but, instead, provide a broad overview with updates on the developments since publication of that chapter.

The earliest protoceratids (the paraphyletic "leptotragulines") were small hornless forms, known from mostly incomplete material, that are hard to distinguish from Uintan oromerycids, oreodonts, and ruminants. In fact, most of them were placed in groups such as the Camelidae, Hypertragulidae, and Leptomerycidae until formally placed in the protoceratids by Wilson (1974). Gazin (1955: 14), Stirton (1967: 35), and Patton and Taylor (1973: 407–408) positioned them in the Leptomerycidae but considered them ancestral to the protoceratids. The paraphyletic "leptotrag-

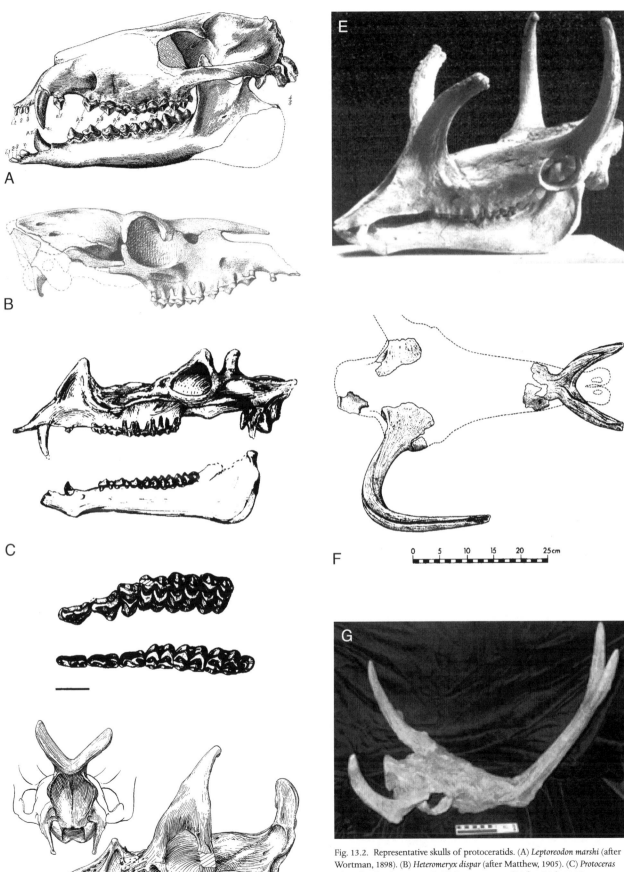

Fig. 13.2. Representative skulls of protoceratids. (A) *Leptoreodon marshi* (after Wortman, 1898). (B) *Heteromeryx dispar* (after Matthew, 1905). (C) *Protoceras celer* (after Prothero, 1998a). (D) *Paratoceras wardi* (after Frick, 1937). (E) *Syndyoceras cooki* (after Osborn, 1910). (F) *Kyptoceras amatorum* (modified from Webb, 1981). (G) *Synthetoceras tricornatus* (photo courtesy P. Holroyd and E. Davis, UCMP).

ulines" include the small primitive forms *Leptotragulus* and *Leptoreodon* plus more advanced forms *Poabromylus*, *Toromeryx*, and *Heteromeryx*. However, Prothero (1998a: Fig. 29.3) showed that all of these genera except *Leptotragulus* form a monophyletic group, so it would be appropriate to name this clade something other than "Leptotragulinae."

In addition to *Leptoreodon* and *Leptotragulus*, the late Uintan and Duchesnean beds of the western United States yield rare specimens of other primitive hornless protoceratids, including *Poabromylus*, *Toromeryx*, and *Heteromeryx*. Thus, the late Uintan and Duchesnean represent the high point in protoceratid diversity, with at least five genera and over a dozen species recognized during this interval.

Leptotragulus *Scott and Osborn, 1887*

Leptotragulus was a small, very primitive protoceratid, distinguished from *Leptoreodon* in having a smaller p4 metaconid, with the p4 anterior crest sharply curved and bearing a large parastylid. Similar features are also found in the p3. The type species *L. proavus* is known only from the Uinta Formation, Uinta Basin, Northeast Utah. *L. medius* occurs in the Uinta Formation and in the Badwater Formation of Wyoming. *L. clarki* occurs in the Uinta Formation and in the Big Sandstone Draw Lentil in Beaver Divide, Wyoming. "*L.*" *profectus* (probably referable to *Trigenicus* Douglass, 1903, according to J. E. Storer, pers. comm.) occurs in the late Duchesnean Porvenir l.f. of Trans-Pecos Texas and in a variety of Chadronian localities, especially in Wyoming, Nebraska, South Dakota, Saskatchewan, and Montana.

Leptoreodon *Wortman, 1898*

Leptoreodon (Fig. 13.2A) can be distinguished from *Leptotragulus* by its large bulbous metaconid and broadly curved anterior crest on p4; the talonid basin also tends to be closed posteriorly. Ludtke and Prothero (2004) reviewed all the known material of *Leptoreodon*, including previously undescribed material from the San Diego Eocene. They upheld six previously named species of *Leptoreodon* and named a new species, *Leptoreodon golzi*, restricted to the late Uintan of San Diego County. Most of the material of *Leptoreodon* is known from the California Uintan (Sespe Formation in Ventura-Los Angeles Counties; Friars, Mission Valley, and Santiago formations in San Diego County) and from Trans-Pecos Texas, but the type species *L. marshi* occurs in Uinta "C" (late Uintan) in Utah, and other specimens are known from Saskatchewan and Montana.

Poabromylus *Peterson, 1931*

Poabromylus is very similar to *Pseudoprotoceras* in many features, except that it is more primitive with slender cheek teeth, p4 nearly as broad as m1, and smaller lingual stylids on the lower molars (Emry and Storer, 1981). The upper molar cingula are smaller than those of *Leptoreodon*. The type species *P. kayi* occurs in the late Duchesnean Lapoint

l.f. of Utah and in the Chadronian of Wyoming and South Dakota. *P. minor* occurs in the Duchesnean Galisteo Formation of New Mexico, the Porvenir l.f. of Trans-Pecos Texas, and the Chadronian of Wyoming and South Dakota. *P. golzi* occurs in the Duchesnean of Wyoming and Montana. *P. robustus* is restricted to the ?Duchesnean or ?Chadronian Titus Canyon l.f. in Death Valley, California.

Toromeryx *Wilson, 1974*

Wilson (1974) named *Toromeryx* for a more massive primitive protoceratid with large, thick labial cingula and low-crowned, bulbous, thick-enameled cusps. It is known only from fragmentary material from late Uintan of Texas (Candelaria l.f. of Trans-Pecos Texas; Casa Blanca l.f. of the Texas Gulf Coast).

Heteromeryx *Matthew, 1905*

In contrast to the fragmentary material of most Eocene protoceratids, *Heteromeryx* is known from a complete skull (Fig. 13.2B) and associated skeleton (AMNH 12326). It was a large taxon comparable in size to *Pseudoprotoceras*, except that it has a more massive, more complex P2–3, a large protocone on P2, and less retracted nasals. The type skull was hornless, although it might have come from a female individual (except that the enlarged upper canine suggests that it was from a male). Only one species, *H. dispar*, is known, and it occurs in the late Uintan (Candelaria l.f.) and late Duchesnan (Porvenir l.f.) of Trans-Pecos Texas, and the Chadron Formation of South Dakota and Nebraska.

Pseudoprotoceras *Cook, 1934*

Pseudoprotoceras was originally known only from a badly crushed skull from the Chadronian of Nebraska, and Wilson (1974) placed the genus in synonymy with *Poabromylus*. Emry and Storer (1981) revived the name *Pseudoprotoceras* and showed how it was distinct from *Poabromylus*. All of the known skulls were hornless with the nasals retracted to the level of the first molar. The cheek teeth appear more hypsodont and broader than those of *Poabromylus* or *Heteromeryx* and lack the thick enamel of *Heteromeryx*. A number of other diagnostic features are discussed by Emry and Storer (1981) and Prothero (1998a). The type species, *P. longinaris*, occurs in the early-middle Chadronian of Wyoming and Nebraska. *P. semicinctus* comes from the early Chadronian of Saskatchewan, and *P. taylori* was named by Emry and Storer (1981) from the middle Chadronian of Flagstaff Rim, Wyoming.

Subfamily Protoceratinae

Despite being one of the most fossiliferous intervals of the entire Cenozoic (the incredibly rich "lower *Oreodon* beds" of the White River Group), the Orellan has yet to yield any protoceratid. The next two genera are closely related to each other in an unnamed taxon (Prothero, 1998a: Fig. 13.29.3, node 8), and that taxon is the sister group of the monophyletic Synthetoceratinae. Prothero (1998a: Fig. 13.29.3, node 6) placed *Pseudoprotoceras,* the *Protoceras-Paratoceras* clade, and the Synthetoceratinae in a monophyletic clade Protoceratinae (broader than the paraphyletic usage of that name by Patton and Taylor, 1973).

Protoceras Marsh, 1891

When protoceratids reappeared in the Whitneyan, they were represented by a single genus, *Protoceras,* with only one species in the Whitneyan (*P. celer*), one in the early Arikareean (*P. skinneri*), and one in the late Arikareean (*P. neatodelpha*). As mentioned above, male specimens of *Protoceras* are distinctive in having horns over the maxillae, orbits, and parietal regions, as well as large canines (Fig. 13.2C). Although the river channel sandstones of the Poleslide Member in the Big Badlands are known as the *"Protoceras* beds," *Protoceras celer* is actually quite rare (Patton and Taylor, 1973, list only about 50 specimens) and known only from two areas in the Big Badlands of South Dakota. The Arikareean species are even more rare and are known from different localities in Wyoming, Nebraska, and South Dakota.

Paratoceras Frick, 1937

Paratoceras is characterized by large triangular medially curved horns over the orbits, small triangular maxillary protuberances, and a peculiar forked occipital horn that looks vaguely like a propeller (Fig. 13.2D). Originally named and diagnosed on the basis of a fragmentary lower jaw with large unreduced premolars and low-crowned molars, the material described by Patton and Taylor (1973) revealed many other diagnostic features (listed by Patton and Taylor, 1973, and Prothero, 1998a). *Paratoceras* is known only from the south-central United States and Central America: the Clarendonian of the Texas Panhandle (the type species, *P. macadamsi*); the newly described early Miocene *P. tedfordi* from the amber mines of Chiapas, Mexico, and several other similar localities in Mexico (Webb et al., 2003): and *P. wardi* from the early Barstovian of the Texas Gulf Coast (Fleming Formation, Trinity River Pit 1) and the early Hemingfordian Cucaracha l.f. of Panama (MacFadden, 2006). Webb et al. (2003: 363) remark that protoceratines are about as common as horses in Central America, so clearly, there was some sort of ecological difference in the Central America–Gulf Coast region that favored them and prevented them from being widely distributed to the North.

Subfamily Synthetoceratinae

By the late Arikareean, a new radiation of protoceratids emerged, living alongside rare *Paratoceras,* in both the Gulf Coast and Texas Panhandle and occasionally in the High Plains, where they persisted in small numbers until the end of the Hemphillian. They are all characterized by some form of Y-branched rostral horn and long orbital horn and a small frontal protuberance including the lacrimal bone in males. Both males and females had longer muzzles and nasal bones, reduced premolars, and more hypsodont molars. They were extensively monographed by Patton and Taylor (1971) with only a few new discoveries since then.

Tribe Kyptoceratini

One tribe of sythetoceratines is the Kyptoceratini, with two genera, *Syndyoceras* and *Kyptoceras*. Their rostral horn is fused only at the base, with no "stalk" beneath the branching point. The horns are circular in cross section, and the orbital horns are upright or anteriorly directed. Although their spectacular horns have garnered much attention, they are also extremely rare, being known from only a few specimens.

Syndyoceras Barbour, 1905

Syndyoceras is distinctive in having a large rostral horn that diverges posterolaterally from the maxillaries and long frontal horns that extend over the orbit and curve dorsomedially (Fig. 13.2E). The dentition is relatively low crowned, and the premolars are large compared to the rest of the more derived Synthetoceratinae. Additional diagnostic characters are given by Patton and Taylor (1971: 130). The genus is monotypic, with its type species *S. cooki*. In addition to the spectacular type skull and skeleton (UNSM 1153) from the late Arikareean of Nebraska, Patton and Taylor (1971: 131) list a few additional specimens from the early Hemingfordian of Nebraska and Wyoming.

Kyptoceras Webb, 1981

Kyptoceras was a protoceratid with a rostral horn that diverges immediately from the fused base in the maxillaries and tall upright frontal horns; all of the horns curve inward and very strongly anteriorly (Fig. 13.2F). Unlike any other protoceratid, the cheek teeth are very hypsodont, and there is no exposure of the narial passage dorsally behind the rostral horn. *Kyptoceras* was described by Webb (1981) on the basis of a fragmentary skull from the late Hemphillian Bone Valley Formation of Florida, and its type species is *K. amatorum*. Webb et al. (2003: Fig. 14.9) show in a map (but do not discuss) other occurrences of *K. amatorum* from coastal North Carolina and from southern Mexico as well; these occurrences will be documented shortly (B. Beatty, pers. comm., 2006). *Kyptoceras* was the very last of the protoceratids, culminating the early Miocene lineage of *Syndyoceras* with a long hiatus (13.5 million years in duration) between the genera, whereas the Synthetoceratini are known from much of the Miocene.

Tribe Synthetoceratini

The more familiar lineage of the Synthetoceratinae includes the tribe Synthetoceratini, famous for the commonly illustrated skulls of *Synthetoceras* with their spectacular horns. They are diagnosed by having a distinct shaft beneath the "Y" branch of the rostral horn formed by fusion of the maxillaries above the nasal cavity (in male skulls), horns that are triangular in cross section, and by long diastemata and reduced anterior premolars. Most of the known material was described by Patton and Taylor (1971). The taxa are known primarily from the Gulf Coastal Plain and Texas Panhandle except for *Lambdoceras*, which is primarily known from the Great Plains.

Lambdoceras Stirton, 1967

Lambdoceras was described by Stirton (1967) based on *L. hessei* from the early Hemingfordian Batesland Formation of South Dakota. Stirton (1967), and Patton and Taylor (1971) considered *Lambdoceras* to be a subgenus of *Prosynthetoceras*, but Webb (1981) and Prothero (1998a) treated it as a distinct genus, and there seems to be no real justification for the subgenus. *Lambdoceras* shows the characteristic fusion of the maxillaries into a shaft or base for the rostral horn, but that shaft is very long, and the forked part is short. The frontal horns extend out directly over the orbit and curve posteromedially, as do the rostral horns. The basal part of the horns is less triangular in cross section than in *Prosynthetoceras*. The lower premolars are wider, thicker, and more wedge-shaped than in *Prosynthetoceras*, and the molars are lower crowned. *L. hessei* also occurs in the late Hemingfordian Sheep Creek Formation of Nebraska. Two other species are known: *Lambdoceras trinitiensis* from the early Barstovian Fleming Formation of Texas and *L. siouxensis* from the early Barstovian Olcott Formation of Nebraska.

Prosynthetoceras Frick, 1937

Prosynthetoceras has a more anteriorly directed shaft of the rostral horn than in *Lambdoceras*, and the frontal horns sweep outward and posteriorly, not anteriorly. All the horns have more sharply angled triangular cross sections. The lower premolars are more laterally compressed than in *Lambdoceras*. The lower molar stylids are reduced, and there is a small medial pillar at the base of the crown between the protoconid and hypoconid on m1–2. The type species, *P. francisi*, is known from the early Barstovian Cold Spring fauna of the Texas Gulf Coastal Plain. *P. texanus*, the more primitive species, is known from numerous middle to late Hemingfordian localities in Texas and Florida (see Patton and Taylor, 1971: 140–141).

Synthetoceras Stirton, 1932

This spectacular genus has the longest rostral horns of any protoceratid, with an anteriorly inclined rostral horn, and knobs on the tips of the frontal horns (unlike *Prosynthetoceras*). The p1–4 diastema is very long with p1–2 lost, and the p3 and p4 are very reduced and have a high median protoconid. The molars are larger and higher crowned than

those of *Prosynthetoceras*. *Synthetoceras* is known only from its type species, *S. tricornatus*, which occurs in the Clarendonian of the Texas Panhandle and Gulf Coastal Plain, and the Hemphillian of Florida and Alabama.

RELATIONSHIPS OF THE PROTOCERATIDAE

Ever since its recognition as a separate order within the Mammalia by Owen in 1848, Artiodactyla has had three main divisions: Suiformes, Tylopoda, and Ruminantia. With the rise of cladistics in the late twentieth century, these three divisions were understood to be stem- or branch-based monophyletic groupings, with Suiformes diverging off first, followed by Tylopoda. The sister taxa relationship between Ruminantia and Tylopoda formed a node-based monophyly, which was termed Neoselenodontia by Webb and Taylor (1980).

In the past decade, systematists have had to revise their evolutionary hypotheses in the face of molecular data. Recent analyses of mitochondrial and nuclear DNA from Mammalia have radically changed not only supraordinal relations but also the internal relationships within many orders. Artiodactyla is one of the most reorganized, with not only the inclusion of Cetacea but also the dissolution of Suiformes and the possible migration of Tylopoda to a basal position among extant artiodactyls. In this basal position, Bayesian estimates of divergence time between camels and all other cetartiodactyls arrive at somewhere between 64.6 and 59.9 Ma (Springer et al., 2005).

However, what is the position of Tylopoda among all artiodactyls, both extant and extinct? And, more importantly, what is a tylopod? These two questions have yet to be deciphered. This situation makes the question of where protoceratids fit in artiodactyl systematics more complicated than just trying to decipher if they are tylopods or ruminants.

Marsh, when describing *Protoceras* in 1891, considered it to be a ruminant, although a primitive one. Other authors of the late nineteenth or first half of the twentieth centuries (Osborn and Wortman, 1892; Scott, 1895b, 1899; Wortman, 1898; Matthew, 1905; Colbert, 1941; Stirton, 1944; Simpson, 1945) continued this trend, although the older, more ancestral members of Protoceratidae, the "Leptotragulinae," were often placed as tylopods (Wortman, 1898; Scott, 1899; Matthew 1905). This raises the idea that the morphologies of derived Oligocene and Miocene protoceratids were convergent on a ruminant or pecoran body plan, but basal, hornless protoceratids still retained significant features of their tylopod ancestry.

The dominant idea shifted to protoceratids as part of Tylopoda in the latter half of the twentieth century. Scott (1940) and Stirton (1967) started the shift, Patton and Taylor (1971, 1973) placed the Protoceratinae in the Tylopoda, and several authors (Wilson, 1974; Golz, 1976; Black, 1978b) placed the "Leptotragulinae" as basal members of the Protoceratidae instead of into any one of a range of hornless ruminants, most often, the Leptomerycidae. Webb and Taylor (1980) regarded the Tylopoda as composed of the oromerycids, protoceratids, xiphodontids, amphimerycids, and camelids, although they left the relationships within the Tylopoda unresolved. Gentry and Hooker (1988) suggested that protoceratids were the sister group to oromerycids and camelids, a hypothesis also adopted in the Janis volume (Janis et al., 1998b; Prothero, 1998a). This idea, however, was not found through computer algorithms but arose from simple manual cladistic analyses.

Without a stable cladogram, the position of Protoceratidae has remained under debate. The first argument was contained in the consensus cladogram of Gentry and Hooker (1988). Webb and Taylor's (1980) ideas on Tylopoda were contested, and protoceratids were placed as basal ruminants. Further support came from Joeckel and Stavas (1996), who compared the basicrania of *Syndyoceras, Protoceras, Lambdoceras, Synthetoceras,* and *Lama* with the aid of CT. This report found very little morphological similarity between protoceratids and camelids, but, as the authors admitted, this could partially be a result of using derived, instead of basal, taxa.

More recent discussions on the placement of Protoceratidae have occurred as a result of authors researching basal and derived members of different artiodactyl clades. Norris (1999) first supported the exclusion of protoceratids from a clade containing camelids, anoplotheroids (containing anoplotherids and xiphodonts), oreodonts, and "bunomerycids." Norris (2000) further fleshed out this idea of protoceratids as basal tylopods or, more likely in his analysis, basal ruminants, although Norris found that both hypotheses were supported by a single skeletal synapomorphy. The three scenarios he found possible were protoceratids as the sister group to Ruminants, or Tylopoda, or both; however, this third scenario had no known characters supporting it.

The most recent morphological analyses of Artiodactyla have lent evidence to more than one of these scenarios. Geisler's (2001b) analysis placed protoceratids in a clade with oreodonts as a sister group to a clade of amphimerycids and xiphodontids, which itself is a sister group to Ruminantia plus Cameloidea. Geisler and Uhen (2003) placed them in a similar position, but now in a more derived location just "up" the tree from Oreodontoidea. Geisler and Uhen (2005) provided a total evidence analysis that used a molecular phylogeny of extant Cetartiodactyla as a constraint for building a morphologically based tree including fossil taxa. Although being placed as the sister taxon to Camelidae was the most common result, no strict consensus for Protoceratidae existed, and many trees placed them as the sister group to Ruminantia. The employment of stratigraphic data to produce a stratocladistic tree added support for protoceratid placement as basal tylopods. Theodor and Foss's (2005) recent usage of deciduous dental characters showed that these too support the placement of Protoceratidae as basal members of Selenodontia, although they are less clear as to position within the clade.

Where does this leave protoceratids? Most morphological analyses see a close relationship between camels and ruminants, a view that conflicts with phylogenies produced by molecular workers. Until either morphological or molecular studies are shown to be in error, it is possible that no progress can be made in correctly assigning protoceratids to either Tylopoda or Ruminantia.

PALEOECOLOGY

The ecology of Eocene and Oligocene protoceratids is not well established, since they have few modern analogues. Presumably they were forest browsers along with most of the other similar-looking selenodont artiodactyls known from the Uintan and Duchesnean, such as the protoreodonts, agriochoeres, oromerycids, and hypertragulids. Certainly these protoceratids were a successful and diverse group in the Uintan, as evidenced by the large number of valid species of *Leptotragulus* and *Leptoreodon* and *Poabromylus,* and by their widespread distribution. As the climate became drier in the Chadronian, protoceratids apparently became more hypsodont in response to the vegetational change, and show evidence of nasal retraction (especially in *Heteromeryx* and *Pseudoprotoceras*) suggesting that they had a prehensile lip or snout for browsing.

After their hiatus in the Orellan, Whitneyan *Protoceras* is restricted to just two areas of the Big Badlands of South Dakota. The further retraction of the nasals and their exclusive preservation in river channel sandstones suggest that *Protoceras* is even more oriented toward browsing in the limited riparian forests of the Whitneyan (when most Whitneyan rocks represent eolian dune deposits, with an entirely different fauna dominated by hypsodont leptauchenine oreodonts). *Protoceras* continues to be scarce and low in diversity during the early Arikareean (restricted to Nebraska and Wyoming) before the appearance in the late Arikareean of *Syndyoceras* in the High Plains and *Lambdoceras, Paratoceras,* and *Prosynthetoceras* in the High Plains and Gulf Coast during the early Hemingfordian.

A number of authors (Patton and Taylor, 1971, 1973; Janis, 1982; Scott and Janis, 1987a; Webb et al., 2003) have looked at Miocene protoceratids, and pointed to the relatively low-crowned teeth, the retracted nasals suggesting a mooselike proboscis, the relatively wide rostrum (like that of a moose), and their relatively short limbs with unfused metapodials and a persistent four-toed manus and have suggested that protoceratids were adapted to more marshy forested habitat, rather than the open plains. Janis (1982) suggested a moose analogue, whereas Webb et al. (2003) thought they were more similar to the African bushbuck. Isotopic analyses of their bones (analyzed by Feranec; reported in Webb et al., 2003: 359) strongly suggest that they ate deep forest browse. In addition, the fact that the Miocene taxa are largely restricted to the more marshy forested regions of the Gulf Coastal Plain of Texas and Florida, as well as the Central American jungles, and extremely rare in most Miocene Plains and western localities, lends support to the hypothesis that the horned forms were deep forest/marsh browsers much like moose or bushbuck.

The sole exception is *Lambdoceras,* which is found both in the Texas Gulf Coast and in Nebraska and South Dakota. However, it has a much narrower snout, suggesting that it was not strictly an obligate folivore. Microwear analysis of *Lambdoceras* molars (*L. siouxensis* from the Olcott Formation and *L. trinitensis* from the Fleming Formation) suggests that *Lambdoceras* consumed both leaves and fruit (G. Semprebon, pers. comm., 2006).

ACKNOWLEDGMENTS

We thank Christine Janis, Chris Norris, Brian Beatty, and Gina Semprebon for helpful reviews of this chapter. Prothero was supported by a grant from the Donors of the Petroleum Research Fund, administered by the American Chemical Society, and by NSF grant EAR03-09538.

JAMES G. HONEY

14

Family Camelidae

PROTOLABIS AND MICHENIA SHARE skeletal and dental similarities that have resulted in their placement in the subfamily Protolabidinae. They show similar paleogeographic distributions, often co-occurring in Arikareean through earliest Hemphillian faunas (Honey et al., 1998). Moreover, they show similar morphological trends during the course of their evolution (Honey and Taylor, 1978): fusion and shortening of metapodials with the metacarpus length significantly less than the metatarsus length, development of hypsodonty, tooth reduction (I1–2 ultimately lost, P2–4 and p2–4 reduced in size and simplified with the p2 frequently lost), and increased to extreme rostral constriction.

In large samples from the early Hemingfordian Dunlap Camel Quarry (DCQ) of Nebraska and from the Clarendonian of Milk Creek (MC), Arizona, the co-occurring *Protolabis* and *Michenia* share similar, unique derived dental and skeletal features. If *Protolabis* and *Michenia* are two separate lineages, these similarities imply parallelism over a roughly 6-million-year time period; alternatively, these similarities may indicate that they are sexual dimorphs of the same species. Honey and Taylor (1978) considered the size disparity between the larger *Protolabis* and the smaller *Michenia* too great to result from sexual dimorphism. Nevertheless, the significance of similarity between *Protolabis* and *Michenia* warrants further consideration. This chapter examines in more detail the questions of whether these two taxa are distinct at the generic or only specific level or if they are really sexual dimorphs.

Editors' note: There has been relatively little systematic work on the camelids since Honey et al. (1998), so the present chapter deals with an important systematic issue within the Camelidae.

DESCRIPTION

Dunlap Camel Quarry

Protolabis and *Michenia* are the overwhelmingly predominant camelids at DCQ (Fig. 14.1). Other than larger size in *Protolabis*, there are a number of differences between the two taxa. Most important is a slightly more rounded braincase with a smaller sagittal crest in *Michenia*, its generally much narrower muzzle at the P1, and slenderer premaxillaries. The dentary is deeper in *Protolabis*, especially anterior to p2. The longitudinal depressions on the ventromedial surface of the horizontal ramus are deeper, the mandibular angle is relatively larger with slightly greater lateral flare, and the small mesial inflection and ventromesial shelf on the ramus are stronger in *Protolabis*. The I3 and C1/1 are large and robust in *Protolabis* but small in *Michenia*. The P1–M3/p1–m3 usually are larger in *Protolabis*, the p3 paraconid and hypoconid are more prominent, and the posterior lobe is relatively broader on larger *Protolabis* specimens. The M3 parastyle is more robust, and the m3 is enlarged relative to m1–2, in *Protolabis*. Limb elements of *Protolabis* are longer and more robust.

Especially significant, however, are identical morphologies in the P1–2/p1–2 of *Protolabis* and *Michenia* at DCQ. Typically, these teeth differ only in size between *Protolabis*

and *Michenia*. The P1 is tiny with the roots partly to completely fused. The upper part of the root and the crown are recurved posteriorly and, at the top of the tooth, slightly medially. The small crown perched atop the long root is unicuspid, with steep anterior and posterior ridges, is convex labially, and has an inflated base. The P2 has a high, buccally swollen paracone with a small parastyle and metastyle usually present. A very weak lingual cingulum is sometimes visible. The p1 is short, small, and weakly recurved posteriorly, with roots closely appressed or fused. The tall protoconid is lingually swollen, with short, steep anterior and posterior ridges. A low, tiny paraconid is present on only some specimens. The p2 protoconid is also high and lingually swollen. Anterior is a lingually inflected paraconid, which ranges in size from minute to nearly half the size of the protoconid. A small hypoconid is present in a few specimens. Compared to *Protolabis* and *Michenia* from Sheep Creek, the p1–2 of *Protolabis* and *Michenia* at DCQ are smaller and less elongate, and the p2 has a transversely unexpanded, lower posterior lobe; consequently, the p2 middle lobe is more prominent, with a swollen appearance in both DCQ *Protolabis* and *Michenia*. Also, the p1 roots of Sheep Creek *Protolabis* are more robust and sometimes widely flared compared to DCQ *Protolabis*. For P1–2, only those of *Protolabis* could be compared, but as for the lower premolars, the P1–2 at DCQ are smaller, and the P2 is reduced relative to P3–4.

Milk Creek

The MC sample (Honey and Taylor, 1978) is also dominated by two protolabines: *Protolabis coartatus* and *Michenia yavapaiensis*. Complete skulls and jaws are known for *P. coartatus*, but *Michenia* skull material consists only of palates, some with attached facial maxillary bone, and jaws without the angle. *M. yavapaiensis* differs from *P. coartatus* in smaller size, much smaller caniniform teeth (I3, C1/1), smaller cheek teeth with M3/3 relatively unexpanded, shallower horizontal ramus and symphysis, and shorter, more slender metapodials. The much more robust mandible of *P. coartatus* shows pronounced lateral flaring of the angle and a strongly developed mesial inflection on the posteroventral border. Although the mandibular angle on *M. yavapaiensis* is not preserved, the above features are lacking on specimens of other derived *Michenia* in the AMNH.[1] As at DCQ, there are some striking similarities between *Protolabis* and *Michenia* at MC (Table 14.1).

P. coartatus was reported (Honey and Taylor, 1978: 406, 413) to have a more extremely constricted rostrum (5 to 8.7 mm wide) than *M. yavapaiensis* (12.4 mm). This *Michenia* value, however, was extrapolated from the width (6.2 mm) of one preserved half of a rostrum (F:AM 73287) with the assumption that the maxillaries lie flat, which they do not. The rostral width depends in part on the angle from the horizontal at which the maxillaries lie; the steeper the angle, the

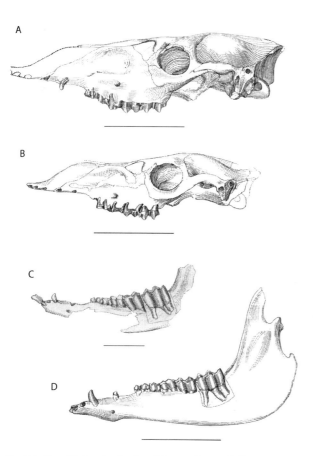

Fig. 14.1. Camelids from Dunlap Camel Quarry. (A) Large skull, F:AM 39777, curated as *Protolabis* sp. (B) Small skull, F:AM 42162, curated as *Michenia* sp. (C) Small jaw, F:AM 42069, curated as *Michenia* sp. (D) Large jaw, F:AM 39740, curated as *Protolabis* sp. Scale bar for A, B, and D equals 10 cm; scale bar for C equals 5 cm.

[1] Abbreviations: AMNH, American Museum of Natural History; F:AM, Frick American Mammals.

Table 14.1. Derived features found in *P. coartatus* and *M. yavapaiensis* from Milk Creek versus *P. heterodontus* and *M. agatensis*

P. coartatus vs. *P. heterodontus*	*M. yavapaiensis* vs. *M. agatensis*
I1–2 usually lost	I1–2 frequently lost in adults
P2 and P3 without parastyle	P2 relatively smaller, single prominent cusp, no parastyle; P3–4 reduced with smaller parastyles
p1 short, peglike; suppressed or absent in some cases; roots closely appressed to fused.	p1 nearly always suppressed, with fused roots
p2 very small; absent in some specimens	p2 relatively small or absent
p3–4 relatively smaller; p3 simpler	p3–4 smaller; p3 simpler

narrower the width. Reexamination and comparison with other *Michenia* suggest that the narrowest rostral width on F:AM 73287 could be 9.2 mm or less, considerably closer to the value for *P. coartatus.*

Barghoorn (1985: 231) found another derived feature of *P. coartatus,* namely that the premaxillaries at their anterior

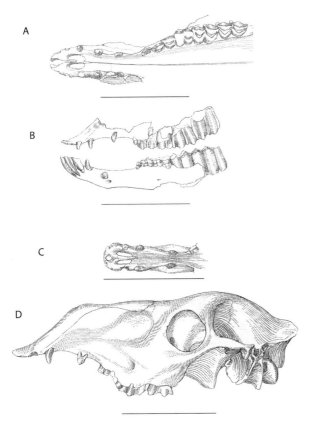

Fig. 14.2. (A) Palatal view *Protolabis* sp. F:AM 40932 from the Sheep Creek Formation, showing the primitive condition of less robust, nonupturned premaxillaries that do not meet anterior to the median fenestra. (B) Side view of *Michenia* F:AM 68323 from the Santa Fe Group, New Mexico, showing swollen, upturned anterior premaxillary. (C) *"Metalabis gracilis"* F:AM 36747 from the Ash Hollow Formation, Nebraska. *"Metalabis"* was an old Frick manuscript name given to some specimens that were later included in *Michenia.* Palatal view of rostrum is shown, illustrating the rugose anterior premaxillaries, which meet in front of the anterior palatine fenestra, and the pronounced lateral constriction of the maxillaries immediately behind the P1. (D) Side view of F:AM 36747 showing short, rounded braincase and caniniform C1. Scale bars equal 10 cm.

ends are robust and upturned and, in palatal view, meet medially to reduce or eliminate the median fenestra (Honey and Taylor, 1978: Fig. 14.8C). *Protolabis* cf. *P. barstowensis* and *Protolabis inaequidens* examined by Barghoorn, and *Protolabis* from DCQ and Sheep Creek, show a more primitive morphology of less robust and less anteriorly upturned premaxillaries that do not meet medially to eliminate the median fenestra (Fig. 14.2A).

The only complete anterior rostrum of *Michenia yavapaiensis* at MC (AMNH field number Bx7-112) shows striking similarities to that of *P. coartatus* (rostral fragment F:AM 99615, originally assigned to *Michenia yavapaiensis* [Honey and Taylor, 1978: Fig. 12C], is an early lamine, probably *Hemiauchenia*). On Bx7-112 the tip of each premaxillary anterior to I3 strongly curves lingually and is rugose, swollen, and flared upward, similar to *P. coartatus.* This derived premaxillary morphology is absent in Dunlap *Michenia.* The anterior-most portion of the premaxillaries is broken in the type of *Michenia, M. agatensis* AMNH 14255 (Frick and Taylor, 1971), but shows evidence of a deep median fenestra that was not bridged over.

Other Examples

This derived premaxillary morphology is also found on some late Barstovian and Clarendonian specimens curated as *Michenia* from the Santa Fe Group of New Mexico, such as ramus and palate F:AM 68323 (Fig. 14.2B) and F:AM 68327, a short, slender metacarpus and associated skull. Ramus F:AM 68323 is slender, has a suppressed p1 and a very reduced p2, and is nearly matched by some MC *M. yavapaiensis* (Honey and Taylor, 1978: Figs. 11E, 12E, 12F). The premaxillaries on F:AM 68323 and F:AM 68327 curve strongly medially anterior to I3, nearly contact each other, and are rugose, thickened, and slightly upturned. The narrowest rostral width for F:AM 68323 is 11.2 mm, but is only 6.0 mm for F:AM 68327. Specimen F:AM 68323 demonstrates that a *M. yavapaiensis*-like lower jaw goes with a rostrum morphologically similar to Bx7-112, and both specimens show that the anterior rostrum of derived *Michenia* is similar to that of *P. coartatus.*

From the Clarendonian Cap Rock Member of the Ash Hollow Formation in Nebraska is F:AM 37647 (Fig. 14.2C,D), a skull curated as *Michenia*. F:AM 37647 is small and has a rounded braincase and possible healed alveoli for I1–2. The cheek teeth are similar in size to those of *P. coartatus,* but the skull is shorter. The laterally constricted rostrum (narrowest width being 10.3 mm) shows weak upturning at the tip, is extremely rugose, and the two premaxillaries meet in front of the anterior palatine fenestra, although the fenestra is not filled by bone. F:AM 37647 differs from New Mexico *Michenia* skull F:AM 68327 in having a more rugose, bridged-over premaxilla and larger C1. Based on the anterior rostrum, the Nebraska specimen is more similar to *P. coartatus* but is *Michenia*-like in having a short skull and rounded braincase.

The morphological overlap described above between contemporaneous early and late species of *Protolabis* and *Michenia* raises again the question of whether they are really sexual dimorphs of the same species rather than two lineages that paralleled each other closely through millions of years. Let us look at the size distribution at Dunlap and evidence for any skeletal dimorphism in living camelids.

COMPARISONS

Kolmogorov-Smirnov Tests

Is the size spread on the teeth of the combined sample of *Protolabis* and *Michenia* more than what would be expected if the sample came from a normally distributed population? A one-sample Kolmogorov-Smirnov test for normality was run on the combined *Protolabis* and *Michenia* sample from Dunlap. For most dental measurements and all jaw depths, the null hypothesis, that the sample at hand could have come from a population exhibiting a normal distribution, is rejected.

Nonpelvic Sexual Dimorphism in Camelids

Camelines

The literature is scant concerning skeletal sexual dimorphism in living camelids. For the dromedary, Smuts and Bezuidenhout (1987) report that (1) the I3 is tusklike in the male and smaller or absent in the female; (2) the female C1/1 and P1/1 are smaller, with the P1/1 sometimes absent; (3) the longitudinal axis of the female sacrum is more strongly curved, and in lateral view the sacrum wing is more slender and pointed than in the male; (4) the pelvis shows several differences between the sexes (discussed below); and (5) the forefoot proximal phalanx is slightly longer and the distal articular surface slightly wider in the male. Metapodial lengths from a small dataset taken on sexed bactrians and dromedaries (mainly zoo and circus animals) show nearly complete overlap between males and females (M. Skinner and B. Taylor, unpub. data). Measurements taken at the AMNH showed only weak dimorphism in bactrian m3 length and jaw depth and in dromedary posterior jaw depth, but nothing to the degree seen in Dunlap *Protolabis* and *Michenia*.

Lamines

Little sexual size dimorphism has been reported for lamines. In his osteometric analysis of postcranial bones of New World camelids, Kent (1982) tested for sexual differences and found them nonsignificant; canine size and form, however, were useful in distinguishing lamine males and females. In some females, the C1 does not "develop" (presumably meaning "erupt"), although it is apparently always present in males; when present, the female C1 is "slender and peglike" versus "robust, curved, and sharply pointed" (Kent, 1982: 61) in the male.

Wheeler (1995) reported canine size dimorphism in guanacos and a small difference in shoulder height for *Vicugna vicugna mensalis*. Wheeler and Reitz (1987) reported sexual dimorphism in alpaca mandible and canine size, with females smaller. Dental measurements of a small sample of AMNH alpacas and guanacos confirmed that males tend to have deeper jaws between c1 and p4 but showed no dimorphism in molar size. Comparing the alpaca sample against the Dunlap combined *Protolabis* and *Michenia* sample shows that at Dunlap the range in narrowest jaw depth anterior to the closed premolar/molar row (14.5 mm) is nearly twice that for the alpacas (7.7 mm), as are the ranges for m3 length (11.5 versus 5.7 mm) and anterior width (4.6 mm versus 2.3 mm). If the range in measurements for *Protolabis* and *Michenia* reflects sexual dimorphism, then protolabine dimorphism was much greater than that in living alpacas.

Measurements collected on 18 llamas (M. Skinner, unpub. data) showed no significant sexual dimorphism in m3 length; the range was much less than in Dunlap protolabines (6.5 mm versus 11.5 mm) and overlapped only the smaller (*Michenia*) part of the Dunlap range.

Size Range of Postcrania for Lamines and the Dunlap Sample

A data set of postcranial length and width measurements for near-adult and adult male and female alpacas of the Huacaya breed was compiled by J. Wheeler (see Wing, 1988, for a bivariate plot derived in part from some of Wheeler's data). Although the bones are unsexed, these measurements do establish size ranges for bone in the living alpaca and are compared with those of the Dunlap protolabines. For metapodial length versus proximal width (Fig. 14.3A,B), there is complete separation of the *Protolabis* and *Michenia* distributions versus a nearly continuous alpaca distribution that approximates the *Michenia* distribution in size. The range in log lengths and widths for the alpaca metacarpals is only about half of that for the combined *Protolabis* and *Michenia* sample. For the alpaca metapodials, the null hypothesis, that the sample could have come from a population exhibiting a normal distribution, is not rejected for any of the variates. For the combined *Protolabis* and *Michenia* sample, however, the null hypothesis is rejected, with most probabilities less than 1%. Similar results are found for other limb bones (Fig. 14.3C); *Protolabis* and *Michenia* form two distinct clouds, but

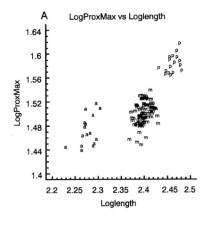

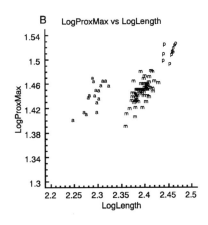

Fig. 14.3. (A) Log metacarpal length versus log maximum proximal width. (B) Log metatarsal length versus log maximum proximal width. (C) Log radius greatest length versus log distal maximum width. (D) Bivariate plot of metacarpal proximal anteroposterior (ProxAP) and transverse breadths (ProxTR). (E) Histograms of antero-posterior breadth of metacarpal distal trochlea. (F) Histograms showing distal anteroposterior length for the femur. Abbreviations: a, alpaca; g, guanaco; l, llama; m, *Michenia*; p, *Protolabis*; v, vicuna. Shading in histograms: diagonal lines used for *Michenia*, alpaca, and vicuna; solid fill used for *Protolabis*, guanaco, and llama.

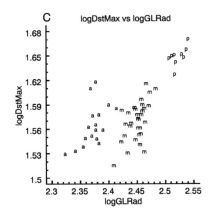

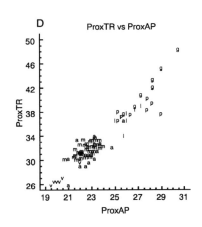

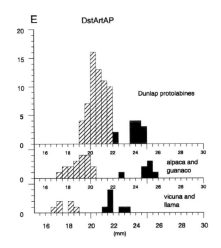

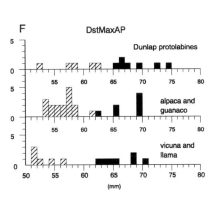

the alpaca distribution is continuous or shows only weak separation. The alpaca distribution approximates that of *Michenia* in size and is much less than that of the combined *Protolabis* and *Michenia* distribution, as for dental data of living lamines. For the combined Dunlap sample, most probabilities are quite low and lead to rejection of the null hypothesis.

J. D. Kent (1982) reported summary statistics for 166 different measurements on the postcrania of 56 specimens of the living lamines, divided as 8 vicunas, 26 alpacas, 10 llamas, 8 guanacos, 2 paco-vicunas, and 2 huarizo (alpaca–llama hybrid). Kent presented bivariate plots for selected combinations of two variables useful in partially separating the lamines. I estimated Kent's raw data by projecting points to the *X* and *Y* axes; summary statistics calculated using these estimated values were nearly the same as Kent's. Composite bivariate plots were generated using these estimated raw measurements and the values for the Dunlap protolabines.

For maximum proximal anteroposterior and transverse breadths for the metacarpals (Fig. 14.3D), the *Michenia* cloud lies directly atop the alpaca cloud, and the *Protolabis* cloud overlies the llama and small guanaco distributions. For the anteroposterior breadth of the metacarpal distal trochlea (Fig. 14.3E), the *Michenia* distribution mainly overlaps the smaller lamines, whereas the *Protolabis* distribution overlaps the larger lamines. Similar distributions are found using the metatarsals, radius-ulna, femur (Fig. 14.3F), and tibia. Thus, a pattern emerges whereby the spread on the Dunlap protolabine data usually occupies the spread of two species of living lamines; the *Michenia* cloud is similar to that of the alpaca, whereas the *Protolabis* cloud is similar to that of the llamas. The previous results for the dental data, and those for Wheeler's postcranial data, are entirely consistent with these results.

Nevertheless, the question remains open whether the size differences between *Protolabis* and *Michenia* reflect extreme sexual dimorphism or species differences. Another approach is to examine associated pelves to see if male pelves always occur with larger specimens (*Protolabis*) and if female pelves always occur with smaller specimens (*Michenia*).

Pelvic Sexual Dimorphism in Camelids

Living Camelids

Smuts and Bezuidenhout (1987: Figs. 1.91–1.100) discussed several features distinguishing the pelves of male and female *Camelus dromedarius,* some of which are present (Torres et al., 1986), but a little less pronounced, in the lamine pelvis. Examination of camelid pelves at the AMNH confirms the following characters as useful:

1. In male camelids, the sacral tuberosity is divided into a rugose cranial dorsal iliac spine and a caudal dorsal iliac spine, the latter a result of the greater expansion in males of the facies auricularis. In lateral view the dorsomedial margin of the iliac blade in males resembles the head of a battle-ax. In females, the caudal dorsal iliac spine is absent or only extremely weakly developed.

2. In male camels the ischiadic spine usually forms a prominent, slightly dorsomedially leaning ribbed sail in posterior view, whereas in females the spine is much weaker and slightly laterally leaning near the dorsal edge. In lamines the ischiadic spine is more elongate and usually taller in the male, giving the dorsal edge of the ischium a more convex appearance when viewed laterally. In the female lamine the ischiadic spine is vertical; in the male the ischiadic spine usually leans noticeably mesially, and is rarely vertical.

3. The major ischiadic incisure is deep in male camels and lamines, whereas in females it is shallower and elongate.

4. In male camelids the shaft of the ilium is more robust.

5. The pelvic symphysis is usually flattened in the female camelid but is thicker and more rounded in the male.

6. Suspensory tuberosities are absent, and the posterior pubic tubercle is small or absent, in the female camelid.

If *Protolabis* and *Michenia* are sexual dimorphs, then male pelves, and perhaps large caniniform teeth, should be confined to the larger taxon, *Protolabis,* and female pelves, and perhaps semiincisiform I3 and C1/1, should be confined to the smaller taxon, *Michenia.* If *Protolabis* and *Michenia* are different taxa, then male and female pelves, and possibly large and small I3 and C1/1, should be present in each taxon.

Pelvic material of *Protolabis* and *Michenia* was examined at the AMNH. Where associated cranial material was lacking, identification as *Protolabis* or *Michenia* rested on the size

Table 14.2. Pelvic features of Barstow protolabines

Specimen number	Taxon	Caudal dorsal iliac spine				Greater ischiadic notch		Ischiadic spine									
								Slant			Height		Curve			Length	
		st	w	ab	ax	sc	bc	m	v	l	t	sh	cx	fl	cv	sh	ln
FAM 24405	*Michenia*[1]			x			?										
FAM 62257	*Michenia*[2]			x			x										
FAM 62251A	*Michenia*[2]			x			x		?			x		x		x	
FAM 24440	Protolabis			x			x				i	i	?				x
FAM 24462	Protolabis	x			?	x					x		x				x
FAM 24494	Protolabis	x			x	x			?		x		x				x
FAM 62286A	*Michenia*[3]			x			x			x	i	i					
FAM 62250	*Michenia*[3]		x				x		?		i	i	x				x
FAM 62252	*Michenia*[2]		?	?			?		?		x		x				x
FAM 62253B	*Michenia*[3]		x				x										
FAM 62253A	*Michenia*[3]		?	?		x			?			?	?				
FAM 24452	Protolabis	i	i		?	i	i		?		i	i	x				x

Abbreviations: ab, absent; ax, battle-ax shape; bc, broad, open concavity; cv, weakly concave; cx, convex; fl, flat; i, intermediate between two categories listed; l, lateral; lg, large; ln, long; m, medial; sc, strongly concave; sh, short; sm, small; st, strong; t, tall; v, vertical; w, weak; wo, without horizontal ridge; wr, with horizontal ridge; x, present.
[1]Frick manuscript name is *Sublabis.*
[2]*Michenia mudhillsensis.*
[3]Frick manuscript name is *Metalabis.*

of the postcrania. Relevant samples included specimens from (1) the Barstow Formation, (2) the Marsland ("upper Harrison") and Runningwater Formations or their equivalents (none from DCQ), and (3) the Santa Fe Group.

Barstow Protolabines

From the Barstow Formation, relevant skeletons are from the latest Hemingfordian Rak Division Fauna, and the early Barstovian Green Hills and Second Division Faunas (Woodburne et al., 1990: Fig. 14.5; Tedford et al., 2004; Pagnac, 2005). Although only partially prepared, several Barstow pelves are separable into male and female specimens (Table 14.2).

Michenia from these faunas is smaller with more slender lower jaws and limbs than *Protolabis*. Rak Division *Protolabis* skulls show extremes in C1 size (Fig. 14.4A), from large and caniniform (F:AM 24442) to small and semiincisiform (F:AM 24377). Skull F:AM 24440 is associated with a left innominate of female type: lacking a caudal dorsal iliac spine and having a broadly concave greater ischiadic incisure (Fig. 14.5A). Thus, the expected size of female teeth in *Protolabis* cf. *P. barstowensis* is approximately the size of F:AM 24440 and smaller (Fig. 14.4A–C). Although no pelvis is associated with F:AM 24442, the large C1 of F:AM 24442 suggests it belonged to a male. This is supported by the close match of F:AM 24442 to *Protolabis* cf. *P. barstowensis* F:AM 24462, a skull, lower jaw, and associated pelvis from the Second Division Fauna. The pelvis F:AM 24462 shows the male characteristics of strong caudal dorsal iliac spine and strongly concave greater ischiadic notch (Fig. 14.5B). The I3 and C1/c1 are large and spikelike, with the C1/c1 slightly larger than on F:AM 24442 (Fig. 14.4A,C).

Comparison of female morphs (F:AM 24377 [canine; no pelvis] and 24440 [pelvis and canine present]) against male morphs (F:AM 24442 [canine; no pelvis], 24462 [pelvis and canine present]) and questionable male morphs (F:AM 24452 [canine present; pelvis suggestive of male]) shows that aside from larger I3 and C1/1 in males, the two sexes form a continuum in cheek tooth length and jaw depth (Fig. 14.4). Thus, female *Protolabis* at Barstow does not show a narrow, *Michenia*-like jaw but rather has a jaw depth similar to that of male *Protolabis*; likewise for molar size and tooth row length. Although the canines of female *Protolabis* are smaller than those of male *Protolabis*, they are not as small as in *Michenia* from Barstow.

Where associations occur and the pelvis is adequately preserved, *Michenia* pelves always show female features, best seen on specimens from Saucer Butte Quarry of the Second Division Fauna (F:AM 62286A, associated postcrania only; F:AM 62257 [holotype of *M. mudhillsensis*] and F:AM 62251A [referred *M. mudhillsensis*], both including crania): caudal dorsal iliac spine absent or very weak; greater ischiadic incisure broadly concave, merging smoothly with the iliac wing without the notch present on male *Protolabis*; ischiadic spine low and with a discernable lateral lean (Fig. 14.5C). *Michenia* skulls and jaws associated with pelves from Saucer Butte Quarry are smaller than those of most male and female *Protolabis* from the Second Division, show a diminutive C1/1, and have associated limbs that are less robust than *Protolabis* from the Second Division.

No *Michenia* skulls are associated with male-type pelves in the AMNH Barstow collection. However, not all *Michenia* skulls have associated pelves, such as F:AM 62254 (paratype of *M. mudhillsensis*) from Saucer Butte Quarry, a skull and mandible with a large C1/1 (Fig. 14.4), possibly indicating

Suspensory tuberosities			Craniodorsal surface of pubic ramus			Posterior pubic tubercle				Sex	Skull and/or jaws?	
lg	sm	ab	cx	fl	cv	lg	sm	wr	wo	Sex	Y	N
										F	x	
										F	x	
			x							F	x	
										F	x	
	?		x			i	i			M	x	
										M		x
										F		x
										F	x	
										F?	x	
										F	x	
								?	?	F?	x	
										M?	x	

Fig. 14.4. Bivariate plots of *Protolabis* and *Michenia* from the Barstow Formation, CA. Abbreviations: DJwBeP12, narrowest ramus height between p1 and p2; DJwP4, ramus height below p4; LC1L, length of lower canine; UC1L, length of upper canine; UM1t3L, combined length of upper M1 to M3. F:AM numbers for selected specimens indicated.

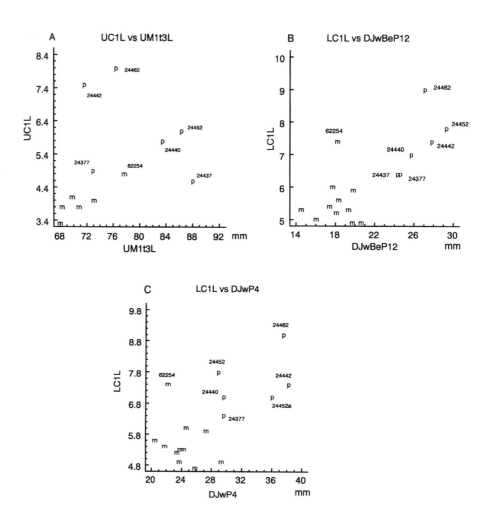

that it is a male; the I3 and C1 overlap in size with those of smaller female *Protolabis,* and the c1 overlaps in length with that of male *Protolabis.* The jaw depth of F:AM 62254 is similar to that of other *Michenia* and much less than that of female *Protolabis* (Fig. 14.4B).

Only one skull and jaws of *Michenia* (F:AM 24405) from the Rak Division Fauna had a fragmentary pelvis and metacarpus associated with it; the pelvis belongs to a female. Compared to female *Protolabis* specimen F:AM 24440 from the Rak Division, F:AM 24405 has a much more slender jaw with weaker ventral tuberosities, smaller anterior dentition with weaker, semiincisiform I3 and C1 and much smaller c1, and shorter, slender metapodials. The significant size difference in metapodial and jaw robustness between female *Michenia* F:AM 24405 and female *Protolabis* F:AM 24440 argues for their separation at the species level.

The occurrence of male and female *Protolabis* cf. *P. barstowensis* reveals two characters that appear to consistently distinguish female *Protolabis* from *Michenia.* Overall size does not always reliably distinguish the two: skulls of *Michenia* from Saucer Butte Quarry are nearly as long as the smaller skulls of female *Protolabis* from the Rak and Second Division Faunas. First, female *Protolabis* jaws are similar in depth to male *Protolabis* jaws and are less slender than those of *Michenia,* especially anterior to the p2 (Fig. 14.4B). Also, male and female *Protolabis* jaws have large, prominent tu-

bercles, forming a "chin" at the anteroventral margin of each ramus where they join at the symphysis. On *Michenia,* these tubercles are much less prominent; in the Saucer Butte Quarry sample, they are scarcely more than bony ridges.

Michenia agatensis

The type of *Michenia agatensis,* AMNH 14255 from the Marsland Formation, lacks the pelvis. However, some other specimens very similar to type *M. agatensis* have variably complete pelves. *Michenia* cf. *M. agatensis* AMNH (field number) 367-1809, a nearly complete skeleton with the lower jaw but lacking the skull, is only slightly larger than the type specimen. The left innominate is of female type (Fig. 14.5D): it lacks the caudal dorsal iliac spine, the greater ischiadic notch is a broad concavity, and the ischiadic spine is low and weakly canted laterally and has a smooth lateral surface without vertical ribbing. Similar features are present on skull and partial skeleton AMNH (field number) 402-3000, also referred to *Michenia* cf. *M. agatensis.*

New Mexico Protolabines

The American Museum contains a large collection of *Protolabis* and *Michenia* from the Santa Fe Group, New Mexico, some with pelvic fragments complete enough for gen-

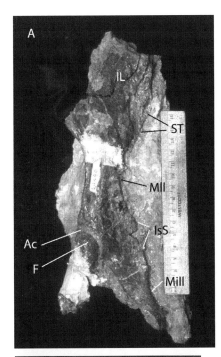

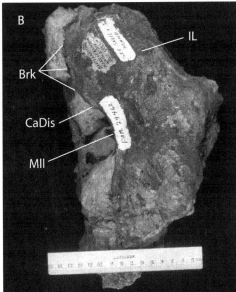

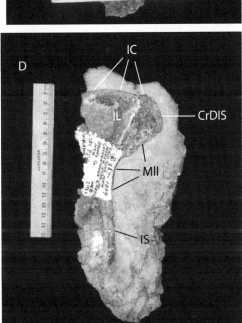

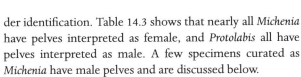

Fig. 14.5. (A) Partial pelvis of *Protolabis* F:AM 24440 from the Barstow Formation, California. (B) Partial pelvis of *Protolabis* F:AM 24462 from the Barstow Formation, California. (C) Partial pelvis of *Michenia mudhillsensis* (type) F:AM 62257 from the Barstow Formation, California. (D) Partial pelvis of *Michenia* cf. *M. agatensis,* AMNH field number 367-1809, from the Runningwater or Box Butte Formations, Nebraska. Abbreviations: Ac, acetabulum; Brk, broken edge of ilium; F, head of femur; CaDIS, caudal dorsal iliac spine; CrDIS, cranial dorsal iliac spine; IL, ilium; IsS, ischiadic spine; MII, major ischiadic incisure (greater ischiadic notch); MiII, minor ischiadic incisure on ischium; ST, sacral tuberosity.

der identification. Table 14.3 shows that nearly all *Michenia* have pelves interpreted as female, and *Protolabis* all have pelves interpreted as male. A few specimens curated as *Michenia* have male pelves and are discussed below.

Michenia pelvis F:AM 24233, without associated skull and jaws, belongs to a male. Pelvis F:AM 24233 is slightly smaller than some *Protolabis* pelves, notably F:AM 28165 and F:AM 24244, but is morphologically very similar to the latter. Associated with F:AM 24233 are slender limb bones that are the same length and slightly narrower than those of F:AM 28165, which Barghoorn (1985) indicated was the smallest *Protolabis* cf. *P. barstowensis;* if they had been found in the same quarry there would be little reason to separate the two sets of limbs into different taxa. It is thus problematic whether pelvis F:AM 24233 is from a (large) male *Michenia* or is a small *Protolabis.*

Another problematic *Michenia* is F:AM 24277, an associated broken pelvis and skull. The pelvis is questionably male, based on indications of a distinct caudal dorsal iliac spine, a large, convex, and vertically oriented ischiadic spine, a broad ramus of the ilium, and a ventral tubercle present. The associated skull is slightly shorter, and the rostrum is slightly narrower, than that of *Protolabis* F:AM 28165, but the P1–M3 are the same size as on F:AM 28165. Unlike *Protolabis* F:AM 28165, *Michenia* F:AM 24277 is not fully grown; the M3 is just beginning to erupt, and the skull may have lengthened. If so, F:AM 24277 would probably be the same size as small male *Protolabis* F:AM 28165, suggesting that F:AM 24277 is possibly a *Protolabis.*

In summary, only at Barstow is there an association of a pelvis and skull that shows the presence of both sexes in one taxon, in that case *Protolabis.* The small number of Run-

Table 14.3. Pelvic features of New Mexico protolabines

Specimen number	Taxon	Caudal dorsal iliac spine				Greater ischiadic notch		Ischiadic spine									
		st	w	ab	ax	sc	bc	Slant			Height		Curve			Length	
								m	v	l	t	s	cx	fl	cv	sh	ln
FAM 68320	*Michenia*[1]		x				x	i	i			x	i	i			x
FAM 68695	*Michenia*[1]	.	?				x										
FAM 28145	*Michenia*[1]			x			x										
FAM 38545	*Michenia*[1]		?				x	x	x			x		x			x
FAM 68771	*Michenia*[1]			x			x	x			i	i	x				x
FAM 38564	*Michenia*[1]			x			x	x				x		x			
FAM 38542	*Michenia*[2]			x			x	?				?					x
FAM 23951	*Michenia*[2]		?				x										
FAM 24233	*Michenia*[1]	x			x	i	i	x			x						x
FAM 23948	*Michenia*[1]								x			x		x		x	
FAM 23955	*Michenia*[1]			x			x	i	i			x	?				x
FAM 38711	*Protolabis*	x			x	x		?			x		x				x
FAM 28165	*Protolabis*	?			?		x	x			i	i	i	i			x
FAM 38613	*Protolabis*	?			?	x		x			x		?				x
FAM 24244	*Protolabis*	x			x			x				x		x			x
FAM 23960	*Protolabis*	x			?	x											x
FAM 24202	*Protolabis*	x			x	x		?			i	i	x				x
FAM 24078	*Michenia*[1]			x			x	i	i		i	i	x				x
FAM 24277	*Michenia*[1]	?					x	x			x		x				x
FAM 23933	*Michenia*[1]		?	?			x		x			x		x			x
FAM 68758	*Protolabis*	x			x	?											
FAM 24265	*Protolabis*	x				x		?			x		?				
FAM 38678	*Protolabis*	x				x											
FAM 38692	*Protolabis*	x			x	x		x			i	i	i	i			x
FAM 38616	*Michenia*		x			x	x	x				x		x	cv		

Abbreviations: ab, absent; ax, battle-ax shape; bc, broad, open concavity; cv, weakly concave; cx, convex; fl, flat; i, intermediate between two categories listed; l, lateral; lg, large; ln, long; m, medial; sc, strongly concave; sh, short; sm, small; st, strong; t, tall; v, vertical; w, weak; wo, without horizontal ridge; wr, with horizontal ridge; x, present.
[1] Frick manuscript name is *Metalabis*.
[2] Frick manuscript name is *Sublabis*.

ningwater specimens of *M. agatensis* and *M.* cf. *M. agatensis* with pelves are apparently all females. All New Mexico pelves with associated cranial material that are clearly male belong to *Protolabis,* and vice versa for *Michenia.* Two specimens curated as *Michenia* have male pelves but are problematic in their identifications.

DISCUSSION

Remarkable morphological similarities shared between *Protolabis* and *Michenia* in some samples warrant an explanation. Two explanations readily come to mind: sexual dimorphism and evolutionary parallelism. Previously (Honey and Taylor, 1978), the latter explanation was accepted. As at MC, size differences argue against DCQ protolabines being sexual dimorphs. Size ranges in teeth and postcrania of *Protolabis* and *Michenia* from DCQ are greater than those in a single species and overlap the ranges for two species of living lamines (alpaca and llama); although much larger than the DCQ protolabines, bactrian and dromedary camels also do not show the relative size differences seen at DCQ. One might suggest that the low degree of sexual dimorphism found in living camelids does not necessarily apply to all fos-

sil camelids and that the size difference between *Protolabis* and *Michenia* is really an example of extreme sexual dimorphism. However, I interpret the sparse fossil pelvic and canine evidence to indicate the presence of males and females in both taxa. Admittedly, in the rare specimens with pelves, most *Protolabis* have (interpreted) male pelves and all *Michenia* have (interpreted) female pelves, making the case for both sexes present in each taxon less than overwhelming; I attribute these differences to small sample size and the vagaries of preservation.

If not evidence for sexual dimorphism, are the morphological similarities between *Protolabis* and *Michenia* the result of parallel evolution? Evolutionary trends including loss of anterior teeth, fusion of metapodials, and increased hypsodonty, among others, are not unique to protolabines but occur independently in disparate camelids (Honey and Taylor, 1978; Pagnac, 2005). As discussed by Pagnac (2005), derived *Paramiolabis* shows similarities to *Michenia* in rostral constriction, tooth reduction and loss, and diastema elongation; furthermore, Pagnac noted that evolutionary changes from a more primitive, *Miolabis*-like ancestor to *Paramiolabis* mimic changes in a presumed *Michenia exilis* to *Michenia yavapaiensis* lineage. For *Protolabis* and *Michenia*, however, we

Suspensory tuberosities			Craniodorsal surface of pubic ramus			Posterior pubic tubercle					Skull and/or jaws?	
lg	sm	ab	cx	fl	cv	lg	sm	wr	wo	Sex	Y	N
										F?	x	
										F?	x	
	x			x			x		x	F		x
				?			x		x	F	x	
	x				?		x		x	F		x
				?			x		x	F	x	
										F	x	
										F	x	
	x		i	i		x			x	M		x
	x				x		x		x	F?	x	
	?						x		x	F		x
										M	x	
				x						M?	x	
										M	x	
				x						M	x	
				x						M	x	
	?					x			?	M		x
				x						F	x	
							?			M?	x	
			?							F	x	
										M	x	
				x		x			x	M	x	
				x						M	x	
				x		x			x	M	x	
				x						F	x	

are concerned with roughly contemporaneous changes in two supposedly distinct protolabines. The similar morphologies, particularly dental, seen in *Protolabis* and *Michenia* at the early Hemingfordian DCQ, and the acquisition in both taxa of similar derived features about 6 million years later in the Clarendonian MC locality and other sites, require a remarkable degree of parallelism, with changes occurring in both lineages at about the same time. In addition to the dental features listed in Table 14.1, the youngest samples of *Protolabis* and *Michenia* show shortened metapodials, upturned and rugose premaxillaries that tend to meet medially, and extremely constricted rostra. Such close parallelism is not seen, to my knowledge, in other camelid lineages.

The close similarities between *Protolabis* and *Michenia* at both DCQ and MC are not likely the result of evolutionary parallelism over great spans of time, nor of extreme dimorphism in a single taxon, but of retained similarities between two closely related and recently diverged species that are separated mainly by size. This hypothesis requires that the smaller of the two species (*Michenia* sp. at DCQ; *Michenia yavapaiensis* at MC) be removed from the genus *Michenia* because they are phylogenetically closer to the earlier named genus *Protolabis* than to the type species of *Michenia*,

Michenia agatensis. Within *Protolabis* at various times, speciation events occurred that gave rise to closely related species pairs, the smaller of which superficially resembled *M. agatensis*. The small size, slightly rounded braincase, and small sagittal crest of the *Michenia*-like species may in fact be a case of pedomorphosis, as suggested by Pagnac (2005) for *Paramiolabis minutus*, possibly indicating the recurrence of neoteny in speciation events in *Protolabis*.

In their original description of the new genus *Michenia*, Frick and Taylor (1971) compared the type species, *M. agatensis* of late Arikareean age (there being no other species described in that article), with only the type species of *Protolabis*, *P. heterodontus* of late Hemingfordian to early Hemphillian age. There are many significant differences between these two taxa, listed in Frick and Taylor's (1971) Table 1, showing unequivocally that they are not the same. However, it is important to realize that the comparisons were made between camelids of different geologic ages, no doubt in part because, to my knowledge, there are no good specimens of late Arikareean *Protolabis* at the AMNH that can be compared with *M. agatensis*. When comparisons are made between the two differently sized protolabines co-occurring in both DCQ and MC, the differences are less pronounced than between the type species of the two genera. Furthermore, at DCQ and MC, the co-occurring proto-labines share some unusual derived features not found in the genotypic species; these similarities resulted in modified diagnoses (Honey and Taylor, 1978; Honey et al., 1998) for both genera that contained fewer distinctions than in the original report on *Michenia* (Frick and Taylor, 1971).

Because of profound morphological differences with currently described species of *Protolabis*, *Michenia agatensis* is considered to be a valid taxon. However, the species *yavapaiensis* from MC (and morphologically similar highly derived small protolabines from New Mexico and Nebraska), and *"Michenia"* sp. from DCQ I now consider more closely related to *Protolabis* than to *M. agatensis*. Taxonomic review of other post-Arikareean small protolabines currently assigned to *Michenia* requires study beyond the scope of this chapter. The primitive *Michenia exilis* and *M. deschutensis* are still considered valid.

SYSTEMATIC PALEONTOLOGY

ARTIODACTYLA Owen, 1848

CAMELIDAE Gray, 1821

PROTOLABINAE Zittel, 1893

Tanymykter Honey and Taylor, 1978

Type Species.   *Tanymykter brachyodontus* (Peterson, 1904).

Included Species.   *T. longirostris* (Peterson, 1911).

Comments.   As I have indicated (Honey, 2004), the species *longirostris* is closely related to, and possibly directly descended from, late Arikareean *Tanymykter brachyodontus*. Originally named *Oxydactylus longirostris* by Peterson (1911), this camelid is morphologically very similar to *T. brachyodontus* but is significantly larger: it is more similar to *T. brachyodontus* than to *Oxydactylus longipes* Peterson, 1904,

the type species of *Oxydactylus*. The species *brachyodontus* was originally included in *Oxydactylus* as the second-named species of that genus. *Brachyodontus* was later made the type species of the newly named genus *Tanymykter*. Recognition of the close phylogenetic relation between *"Oxydactylus"* *longirostris* and *Tanymykter brachyodontus* necessitates the reassignment of *longirostris* to *Tanymykter*.

Protolabis Cope, 1876

Type Species.   *Protolabis heterodontus* (Cope, 1874).

Included Species.   *P. gracilis* (Leidy, 1858); *P. inaequidens* (Matthew, in Cope and Matthew, 1915); *P. coartatus* (Stirton, 1929); *P. barstowensis* Lewis, 1968; *P. yavapaiensis* (Honey and Taylor, 1978).

Michenia Frick and Taylor, 1971

Type Species.   *Michenia agatensis* Frick and Taylor, 1971.

Included Species.   *M. exilis* (Matthew, in Matthew and Macdonald, 1960); *M. deschutensis* Dingus, 1990; *M. mudhillsensis* Pagnac, 2005.

ACKNOWLEDGMENTS

I thank R. Tedford, M. McKenna, M. Norell, M. Novacek, and R. Hunt for access to relevant specimens and J. Wheeler and J. Kent for permission to use unpublished raw data. I thank R. Tedford for access to unpublished line drawings made by artists of the Frick Laboratory for Childs Frick. For darkroom help, I thank L. Meeker and C. Tarka, and for work on Figures 14.3 and 14.4, I thank J. M. Honey. For discussion I thank R. Tedford, R. Hunt, S. Barghoorn, K. Scott, and P. Robinson. Reviews and comments on this chapter were kindly provided by C. Janis, T. Kelly, P. Robinson, and D. Whistler.

GRÉGOIRE MÉTAIS
AND INESSA VISLOBOKOVA

15

Basal Ruminants

EARLY RUMINANTS ARE OFTEN CALLED "hornless ruminants" because they lack the cranial appendages that characterize most extant ruminants. A reliable modern analogue of these earliest ruminants is provided by tragulids, the most primitive extant ruminants that are confined to forested tropical habitats of Southeast Asia (Asiatic chevrotains or mouse deer, *Tragulus*) and Africa (water chevrotains, *Hyaemoschus*). Although the habitat of extant tragulids is likely to be close to that of Eocene Ruminantia, the living forms are more derived in many regards. Despite their supposed antiquity, the fossil record of tragulids remains very poor, thus introducing a bias in any attempt at phylogenetic reconstruction. The taxonomic diversity and geographic distribution of modern ruminants make them classically cited as the most successful group of extant ungulate mammals. Yet, although many studies have been devoted to the Neogene evolutionary history of extant groups of horned ruminants (Pecora), the first Eocene stages of ruminant evolution are comparatively poorly known. The fossil record of Ruminantia covers roughly 45 million years, but much fieldwork is needed, especially in Asia, to improve our knowledge of the emergence of bunoselenodont artiodactyls in general and Ruminantia in particular.

The clade Ruminantia rests basically on three morphological features: the fusion of the cuboid and navicular in the tarsus, the absence of upper incisors, and incisiform lower canines. Early ruminants were long included in the broad concept of Tylopoda on the basis of dental features (Scott, 1940). If the living sister group of ruminants is Camelidae, identifying their immediate sister group within the early-middle Eocene bunoselenodont dichobunoids remains a tricky business. Ruminants emerged as part of the bushy radiation of selenodont artiodactyls that diversified during the second half of the middle Eocene in North America and Asia (although it remains very poorly documented in that region). As stressed by Webb (1998), the main problem is to resolve phylogenetic relations within these early Selenodontia (in the sense of Gentry and Hooker, 1988: Fig. 9.8), especially relations between relatively well-documented European and North American forms that have often been considered separately in the literature. Webb and Taylor (1980) proposed the term Neoselenodontia, which includes the Ruminantia, and a limited concept of Tylopoda excluding Oreodontoidea, and comprising the North American Camelidae, Protoceratidae, and Oromerycidae and the European Amphimerycidae and Xiphodontidae. Beyond the origins of ruminants, understanding the basal phylogeny of Selenodontia is crucial because it is a critical clue for structuring the sequences of basal dichotomies in the phylogenetic tree of Cetartiodactyla as a whole (Geisler and Uhen, 2003).

We propose here to investigate the earliest stages of this evolutionary success story through an overview of the current state of knowledge on the diversity, systematics, phylogeny, and paleobiology of those stem Ruminantia usually referred to the infraorder Tragulina. Most living ruminants belong to the infraorder Pecora, which comprises the four extant families of horned ruminants (Bovidae, Cervidae, Giraffidae, Antilocapridae) plus the Moschidae (represented by the single genus *Moschus* [musk deer] living in highland forests of Asia).

Extant Neoselenodontia are mainly distinctive from other ungulate mammals in their capacity to ruminate ("chew the cud") because of both their complex digestive system including a multichambered stomach and the symbiotic association with bacteria in the gut for cellulose fermentation. However, tragulids do not possess a fully compartmentalized stomach, as they lack a distinct omasum and reticular groove, making their physiology intermediate between a nonruminant ungulate and a true ruminant (Langer, 1974, 2002). Numerous studies have been published to estimate the appearance of that important physiological feature that is likely responsible for current ruminant diversity (Jermann et al., 1995). These soft tissues are usually not preserved in the fossil record, but we can infer with a reasonable degree of confidence that the early ruminants were physiologically closer to extant tragulids than to any other living ruminants. The evolutionary changes of early ruminants from a nonruminant to pecoran type can also be traced through the morphology of their skulls, masticatory apparatus, and dentition.

The evolutionary history of Ruminantia (Fig. 15.1) as reflected by the fossil record indicates two major evolutionary radiations that initiated in Asia. The earlier radiation occurred during the Eocene and was probably conjoined to that of Tylopoda, which were fairly diversified in the middle-late Eocene of North America. The earliest remains of ruminants are known from the middle Eocene of Asia (Irdinmanhan ALMA), but the first representatives of the Asian selenodonts may have appeared earlier. The quality of the fossil record in Asia is certainly insufficient to evaluate reliably the timing and modalities of this radiation. The second radiation

occurred in the early Oligocene of Central Asia and corresponds to the emergence of Pecora (bovoids and cervoids). This second radiation produced the bulk of the total diversity of extant ruminants, tragulids being the only surviving family of the basal radiation. We focus here on this basal radiation of ruminants by reviewing the diversity of these earliest forms reported from Eurasia and North America.

Ruminants are known in North America and Asia as early as the middle Eocene. They appeared in Europe during the late Eocene but became more diversified after the *Grande Coupure* (beginning of the Oligocene) as a result of the invasion of Asiatic forms. Ruminantia probably migrated into Africa at the same time, although the ruminant fossil record becomes adequately documented on that continent only from the early Miocene (Barry et al., 2005). A limited number of ruminant taxa reached South America during the Pliocene Great American Interchange, but they remained moderately diversified taxonomically on that continent.

Faunal exchanges between Asia and North America occurred intermittently during the Eocene, as attested by the appearance in North America of groups that otherwise originated in Asia (rodents, lagomorphs, tillodonts, and perissodactyls). The North American leptomerycids and hypertragulids are also thought to have emigrated from Asia during the middle Eocene. This hypothesis would partly explain their sudden appearance in the late Unitan, but Stucky (1998) has recently reactivated the notion of a local evolution from North American homacodonts. This hypothesis may hold for the origin of tylopods, but it is unlikely to be true for ruminants, which are now significantly documented in the middle Eocene of Asia. Fossil specimens referable to *Archaeomeryx* sp. are known in Asia as early as the Irdinmanhan ALMA (Vislobokova, 2001) and thus predate the earliest occurrence of ruminants in North America in the late Uintan, with the appearance of the leptomerycids *Hendryomeryx* (Black, 1978b) and *Leptomeryx* (Webb, 1998). The systematic status of *Simimeryx* from the Sespe Formation is still unclear; should this genus prove to be an early ruminant, it would also move the earliest appearance of hypertragulids in North America back to the late Uintan. The first appearance of ruminants in Europe is controversial because it depends on whether or not amphimerycids are regarded as belonging to the ruminant clade. Several workers (e.g., Geraads et al., 1987; Gentry and Hooker, 1988; McKenna and Bell, 1997; Hooker and Weidmann, 2000) have considered this strictly European family of selenodont artiodactyls as the first offshoot within the Ruminantia. We do not endorse this view here, and amphimerycids are treated and discussed in detail in the chapter devoted to endemic artiodactyls from Europe (Erfurt and Métais, this volume).

PHYLOGENY OF EARLY RUMINANTS: BRIEF HISTORIC REVIEW AND PROSPECTS

The ruminant artiodactyls have long been recognized as a monophyletic group, and the Tylopoda are traditionally

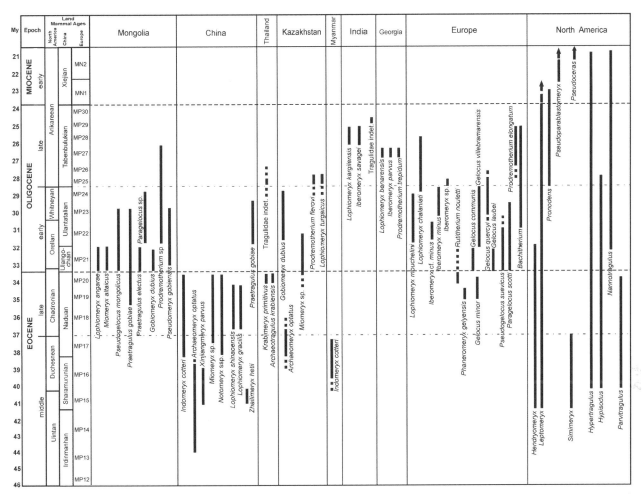

Fig. 15.1.   Stratigraphic ranges of early ruminant species on the three Holarctic landmasses; the Land Mammal Ages are based on Berggren et al. (1995), Russell and Zhai (1987), Schmidt-Kittler (1987), and Woodburne (2004b).

cited as the sister group of the Ruminantia in phylogenies based on morphological characters. By contrast, the clade Neoselenodontia (Ruminantia plus Tylopoda) is not supported by most phylogenies produced from molecular data, ruminants forming a clade with the Cetacea and the Hippopotamidae (Cetruminantia), excluding Tylopoda and Suoidea, which are alternatively placed as the first offshoot in the tree of Cetartiodactyla (e.g., Gatesy, 1998). Despite some tentative attempts to resolve this discrepancy (Geisler and Uhen, 2003), no real consensus reconciling the different signals has yet been reached (see Marcot, this volume; Geisler et al., this volume). The phylogenetic relationships of the basal ruminants are still ambiguous despite the efforts of generations of mammalian systematics. Cladistic models have succeeded gradistic models without making clearer the sequence of dichotomies among basal ruminants. Moreover, the fossil sampling is still insufficient, especially in Asia, and it remains an important limiting factor for resolving the phylogeny of early ruminants.

The primitive features (including morphology, physiology, behavior) exhibited by tragulids have led most authors to separate this small family from other living members of the Ruminantia and even sometimes to question their ru-

minant affinities (Duwe, 1969). Flower (1883) proposed the suborder Tragulina to differentiate this primitive grade of ruminants from both the Pecora (horned ruminants) and the Tylopoda (camels). The discoveries of fossil hornless ruminants have transformed the Tragulina into a heterogeneous group of primitive Holarctic ruminants of variously proposed relationships to the Tragulidae. The hornless ruminants were long gathered in three families (Tragulidae, Hypertragulidae, Geolocidae) whose relationships with each other and to the pecoran families were variably interpreted in the literature (see Webb and Taylor, 1980). The tendency to split off these three heterogeneous families into several familial entities was initiated by Gazin (1955) and encouraged by Webb and Taylor (1980). After a review of the Paleogene ruminants of Europe, Janis (1987) introduced the Lophiomerycidae and the Bachitheriidae and partially resolved the wastebasket status of gelocids. Recently, Vislobokova (2001) distinguished the Praetragulidae as a new family of Asian hypertraguloids and raised the Archaeomerycinae of Simpson (1945) and Hypisodontinae Cope, 1887 to familial rank.

The eight families reviewed herein probably originated in the Eocene of Asia. They were distributed in Asia and

North America during the middle-late Eocene and appeared in Europe only by the late Eocene, and mostly after the *Grande Coupure* (early Oligocene). The seminal paper of Webb and Taylor (1980), which includes a detailed description of the skull of *Archaeomeryx* as well as a rational phylogenetic scheme, evidenced the importance of taking into account characters from the basicranial, dental, and postcranial anatomy to undo the Gordian knot of early ruminant phylogeny. Recently, significant progress came with the description of new skeletons (Vislobokova, 1998) and analyses of more complete fossil material from the Eocene of Mongolia (Vislobokova, 2001) and China (Vislobokova and Trofimov, 2002). However, most of the Eocene Asian ruminants are still based on sparse dental material, making difficult any assessment of their phylogenetic relationships.

A lingering issue related to the radiation of selenodont artiodactyls is the apparent emergence of ruminant features independently at least in Europe and North America during the middle Eocene. Hypertragulids and amphimerycids have been suspected to be derived from a stock of North American and Europe dichobunoids, respectively, by several authors. Similarly, the parallel evolution of ruminant features acquired independently in the European amphimerycids on the one hand and in the North American hypertragulids on the other is implicit in the phylogeny proposed by Gentry and Hooker (1988). The phylogenetic position of hypertragulids (and by extension their ruminant status) has been diversely interpreted in the literature. Webb and Taylor (1980) considered this family to be the most plesiomorphic ruminants, whereas Sudre (1984) and Moyà-Solà (1988) believed them to be already too specialized and favored a North American origin from a local group of dichobunoids. Here, we consider Hypertragulidae as true Ruminantia because they possess the three characters that define the clade. However, these apomorphic characters are not demonstrated in Amphimerycidae, and their unusual dental morphology makes their inclusion within the Ruminantia very unlikely.

The Ruminantia were first defined on the basis of living taxa, and their classification was mainly based on both horns and limb structure. The Tragulidae are clearly the most primitive extant family, and they usually have been considered the first offshoot of the ruminant phylogenetic tree, although their fossil record before the Miocene is poor. The North American fossil record of early selenodonts is, in contrast, well documented, and most of our knowledge about early ruminant evolution and adaptation was formerly based on these rich fossil collections (Scott, 1899, 1940). The report (Matthew and Granger, 1925a) and the formal description (Colbert, 1941; Webb and Taylor, 1980) of *Archaeomeryx* from the middle Eocene of Mongolia actually focused attention on Asia as a probable center of origination and early evolution of the Ruminantia. Although *Archaeomeryx* is known by several complete individuals and was extensively studied, its phylogenetic position within the Tragulina is still variously interpreted. Many important issues such as the Eocene evolution of tragulids or the origin of hypertragulids remain unresolved.

Flower (1883) first introduced the term Tragulina as a poorly defined systematic unit including hornless ruminants showing a primitive grade within the Ruminantia. If extant Pecora (horned ruminants) are clearly distinguished from extant Tragulina (tragulids) by a series of anatomical features, the distinction when Paleogene fossil taxa are considered is much more problematic. Tragulina include both North American and Eurasian forms and are considered as a paraphyletic group of nonpecoran hornless ruminants by some researchers (Scott and Janis, 1993).

SYSTEMATIC REVIEW OF EARLY EURASIAN AND NORTH AMERICAN RUMINANTS

Paleogene ruminant artiodactyls reviewed below are placed into eight distinct families, which are divided into two superfamilies: Traguloidea and Hypertraguloidea. The state of knowledge concerning early ruminant genera is variable, and a significant number of them are based on inadequate and disputable fossil material. These basal ruminants recognized by ambiguous features often have a provisional taxonomic assignment that may change when additional material becomes available. We have tried to retrace the systematic history, sometimes complex, of each genus and in turn to retain the most likely (at least in our view) familial allocation.

Asian Land Mammal Ages (ALMA) are inspired by North American Land Mammal Ages (NALMA), which rest on a remarkable record of fossil mammals throughout the Paleogene. ALMA are mostly based on fossil faunas located in Mongolia or Northern China. Correlations with mammal faunas from Central Asia or South Asia have proven particularly difficult because of the "provincialism effect" that occurred during the Paleogene. We have tried to provide the most up-to-date data concerning the stratigraphic origin and age of the fossils discussed below. However, the age of some Asian fossiliferous localities is disputed and/or currently under investigation, and question marks generally indicate nondefinitive age assignments.

Capital letters indicate elements of the upper dentition, the lower-case letters indicate elements of the lower dentition. The Neogene and Recent Tragulidae are treated in detail in another chapter (Rössner, this volume).

Superfamily TRAGULOIDEA
Family ARCHAEOMERYCIDAE

Characteristics.   (after Vislobokova and Trofimov, 2002, based on *Archaeomeryx*). Skull with low and narrow cranial portion and relatively short facial portion. Sagittal, temporal, and occipital crests strongly developed. Sagittal crest long. Temporal crests fused close to coronal suture and arched anteriorly. Occipital crest strongly projecting posteriorly. Orbits small, located low, in central position, and closed posteriorly. Parietal foramina small and located near sagittal crest. Mastoid exposure mainly lateral. Auditory bullae small, external acoustic meatus extremely short. Vagina of styloid process weakly developed and widely open posteriorly. Petrosal small, short, broad, and closely adjoining basioccipital; ventral surface positioned almost horizontally. Anteroventral edge of petrosal located far from postglenoid process. Promontorium low, simple, occupying large part of ventral surface, and almost completely corresponding to main whorl of cochlea. Fenestra vestibuli (fenestra ovalis) small. Fossa for musculus tensor tympani small and located opposite posterior part of promontorium. Fossa for musculus stapedius narrow and placed behind fenestra vestibuli. Recessus epitympanicus located on petrosal. Foramen ovale small, oval, and located close to posterior edge of alisphenoid. Facial and orbital surfaces of lacrimal small. Nasolacrimal fissure probably undeveloped. Nasals long, not projecting anteriorly between premaxillae. Zygomatic bone almost approaching the tooth row ventrally and possessing a short lacrimal process and long temporal process. Palate flat. Infraorbital canal short and low. Posterior opening of infraorbital canal located on orbital surface of maxilla. Premaxilla low and short, with short nasal process extending dorsally, and almost overlapping the anterior opening of nasal cavity. Anterior opening of nasal cavity low and short, oval in dorsal view, and slightly narrowed posteriorly. Incisive foramina small. Mandible with strongly curved body; coronoid process low, well developed, and inclined posteriorly; angular process narrow and strongly projecting posteriorly. Upper incisors present. Upper canines small and procumbent. Lower canines incisiform but larger than first incisors. P1 present. Radius and ulna, fibula and tibia, and central metapodials separate. Trapezium and metacarpal I present. Astragalus elongated, with nonparallel trochleae. Manus pentadactyl. Pes probably tetradactyl.

Included Genera. *Archaeomeryx, Indomeryx, Miomeryx, Notomeryx, Xinjiangmeryx.*

Comments. This family, first proposed by Simpson (as a subfamily), was recently resuscitated by Vislobokova and Trofimov (2000, 2002) on the basis of a comprehensive examination of the complete and fragmentary skeletons of several *Archaeomeryx* individuals from Ula Usu (Shara Murun Formation, China) collected by the Soviet–Chinese Expedition in 1959.

Genus *Archaeomeryx* Matthew and Granger, 1925

Type Species and Type Specimen. *A. optatus* Matthew and Granger, 1925, from Ula Usu (Shara Murun Formation, Inner Mongolia, China); AMNH 20311, palate with P2–M3 and mandible (Matthew and Granger, 1925a).

Included Asian Species. *A.* sp. (cited in several Irdinmanhan localities [Russell and Zhai, 1987]).

Geographic Distribution. Mongolia and China.

Stratigraphic Range. Middle-late Eocene (Irdinmanhan-Sharamurunian).

Definition (after Vislobokova and Trofimov, 2002). Small sized. Mandible body shallow, ventral edge strongly curved under m3. Incisors procumbent, with almost symmetrical crowns; upper incisors small, reduced. Upper canines medium-sized, lower canines small, incisiform, but larger than incisors. P1 absent. C–P2 diastema short. p1 small, caniniform, located approximately in middle between canine and p2. Premolar row relatively long. Lower premolars narrow with cutting edges, p4 with well developed metaconid and relatively smaller paraconid, entoconid, and hypoconid. Molars brachydont and weakly crescentic with strongly developed parastyle and mesostyle and strong cingulum. Labial rib of metacone very weak. Valleys on m3 not deepened, talonid heel short. Dental formula: 3/3, 1/1, 3/4, 3/3.

Dimensions. M2 (*A. optatus*): L = 5.5 mm; W = 7.0 mm (Colbert, 1941).

Discussion. The familial status of *Archaeomeryx* has long been problematic. Matthew and Granger (1925a) and Colbert (1941) included *Archaeomeryx* within the Hypertragulidae; Simpson (1945) proposed a separate subfamily Archaeomerycinae (within the Hypertragulidae); and Webb and Taylor (1980) advocated its inclusion within the Leptomerycidae, an opinion followed by most subsequent workers (Sudre, 1984; Janis and Scott, 1988; McKenna and Bell, 1997). The numerous anatomical peculiarities of *Archaeomeryx* recently led Vislobokova (2001) to remove the genus from the Leptomerycidae and to propose the Archaeomerycidae (Vislobokova and Trofimov, 2000).

Genus *Indomeryx* Pilgrim, 1928

Type Species and Type Specimen. *I. cotteri* Pilgrim, 1928, from the environs of Sinzwe (Pondaung Formation, central Myanmar); G.S.I. no B768 fragmentary right dentary with p4–m3.

Included Asian Species. *I. arenae* Pilgrim, 1928; *I. pilgrimi* Métais et al., 2000; *I. minus* Métais et al., 2000; *I.* sp. (cited in Qiu (1978: 9]).

Geographic Distribution. Myanmar and China.

Stratigraphic Range. Middle-late Eocene (Sharamurunian-Naduan).

Definition. Small bunoselenodont ruminant, short diastema between p2 and p3, p4 with a paraconid, metaconid tends to merge into the bifurcated postprotocristid, lower molars with a groove on the anterior side of the entoconid (*Zhailimeryx* fold), trigonid closed anteriorly, metaconid and paraconid twinned on some specimens; upper molars increasing in size from M1 to M3 and having a weakly crested metaconule; astragalus with nonparallel trochleae.

Dimensions. *I. cotteri*: m2 of the type, L = 6.9 mm; W = 4.2 mm (Pilgrim, 1928: 10).

Discussion. Tsubamoto et al. (2003) doubted the validity of the species *I. pilgrimi* and *I. minus*. Even if these two species are based on incomplete material, the homogeneity of the type species *I. cotteri* may be discussed in regard to the wide range of variation of both size and morphological structure. This genus is known only by premolar and molar material, and assessment of its affinities is still poor. It is thus provisionally included within the Archaeomerycidae because of its primitive dental (premolar) morphology.

Genus *Miomeryx* Matthew and Granger, 1925

Type Species and Type Specimen.   *M. altaicus* Matthew and Granger, 1925, from Ergilin Dzo (=Ardyn Obo) (Ergilin Dzo Fm., Mongolia); AMNH 20383 left and right P2–M3 (Matthew and Granger, 1925b).

Included Asian Species.   *M.* sp. (cited in several Sharamurunian-Ergilian localities [Russell and Zhai, 1987]).

Geographic Distribution.   Asia.

Stratigraphic Range.   Middle Eocene–early Oligocene (Sharamurunian–Ergilian).

Definition.   Small sized. Mandible body almost not narrowed anteriorly, lower edge weakly curved under m2 and m3. Diastemata between C and P2 and between c and p1 weakly elongated. Lower p1 small, caniniform, and separated from p2 by short diastema. Premolar row slightly shortened. Lower premolars very narrow with cutting edges. p4 with well-developed metaconid and relatively smaller paraconid, entoconid, and hypoconid. Molars brachydont and weakly crescentic, with strongly developed parastyle and mesostyle, distinct external rib of the paracone, and stout cingulum. Crowns of M2 and M3 slightly extended lingualabially. Valleys on m3 deepened, talonid slightly elongated. Cingulum well developed. Enamel rugose.

Dimensions.   *M. altaicus*: m2 of the type, L = 6.9 mm; W = 4.2 mm (Matthew and Granger, 1925b).

Discussion.   This genus has been poorly known since it was first recovered from Ergilin-Dzo. *Miomeryx* sp. is cited from several other localities in Asia, in particular from the late Eocene Caijiachong Fm. (southern China; Wang and Zhang, 1983), early Oligocene of Tatal Gol and Khoer Dzan (Mongolia; Vislobokova and Daxner-Höck, 2002), and the Buran Svita in the Zaysan Depression (Kazakhstan; Emry et al., 1998). Formerly included in the Leptomerycidae, it is here removed to the Archaeomerycidae, but additional evidence is needed to determine its affinities.

Genus *Notomeryx* Qiu, 1978

Type Species and Type Specimen.   *N. besensis* Qiu, 1978 from the Bose Basin (Naduo Formation, Guangxi, South China); IVPP V4957.1, fragmentary maxilla with M1–M3, and IVPP V4957.2, fragmentary dentary with m2–m3.

Included Asian Species.   *N. major* Guo et al., 1999, from the Naduo Fm., Bose Basin; *N.* sp. (from the Naduo Fm. cited in Russell and Zhai, 1987).

Geographic Distribution.   China.

Stratigraphic Range.   Middle-late Eocene (Naduan).

Definition.   Small but larger than *Archaeomeryx*. Horizontal ramus of mandible thickened. Molars brachydont with relatively well-developed crescentic pattern and deepened valleys. Upper molars with strong external rib of paracone and well-developed external rib of metacone. Crowns of M2 and M3 only slightly extended linguolabially: length almost equal to width. Lower m3 with elongated talonid. Cingulum well developed. Enamel strongly rugose.

Dimensions.   *N. besensis*: M2, L = 8.5 mm; W = 9.6 mm; m2, L = 8.5 mm; W = 5.7 mm (Qiu, 1978: 10).

Discussion.   This genus was first considered to be a hypertragulid by Qiu (1978), who distinguished *Notomeryx* from *Xinjiangmeryx* and *Archaeomeryx* by its larger size and more crescentic molars. This author suggested close ties with *Miomeryx*, *Gelocus*, and *Lophiomeryx* on the basis of cingula development. The most recent revision of the material collected from the Naduo Formation in the Bose Basin led Guo et al. (1999) to distinguish a large species, *N. major*. Because of the scarcity of the material, additional insights into the affinities of the genus *Notomeryx* appear premature.

Genus *Xinjiangmeryx* Zheng, 1978

Type (and Unique) Species and Type Specimen.   *X. parvus* Zheng, 1978, from the Turfan Basin (Liankan Formation, Sinkiang, China); fragmentary skull with lower jaw, IVPP-V4054.

Geographic Distribution.   Type locality only.

Stratigraphic Range.   Late middle Eocene (Sharamurunian).

Definition.   Small-sized. Upper incisors possibly present; p1 absent. Molars brachydont and weakly crescentic with weakly deepened valleys.

Dimensions.   *X. parvus*: m1–m3: L = 18.3 mm (Zheng, 1978).

Discussion.   This genus is based on a single specimen on which observation of characters is very difficult. It was also reported (but never described in detail) in the Irdinmanhan levels of Inner Mongolia (Li and Ting, 1983). The Irdinmanhan age of *Xinjiangmeryx* makes it one of the earliest occurrences of Ruminantia. It is geographically restricted to the Turfan Basin, and both its dental morphology and its size make it closer to *Archaeomeryx* than to any other early ruminant from the middle Eocene of Asia.

Family LOPHIOMERYCIDAE

Characteristics.   Brachycephalic skull with most likely a short muzzle, orbit centrally placed and probably posteriorly opened, lateral exposure of the mastoid. P2 and P3 elongated and lacking a broad lingual lobe (it is weakly developed and posteriorly situated on P3); P4 fully crescent-shaped with salient styles, subquadrate upper molars with rounded, salient styles, retention of a mesiolingual cingulum around the protocone, and tendency to the reduction of the metaconule on M3. Symphysis of the mandible extending backward up to p1, angular process posteriorly projected, i1 larger than other incisors, c incisiform, p1 reduced and separated from c and p2 by diastemata, p2 and p3 elongated and narrow, p4 with a developed metaconid that tends to extend posteriorly into a spur closing part of the posterior heel, which otherwise remains generally open posterolingually; lower molars with trigonid open anterolingually, *Dorcatherium* fold variable, postentocristid absent, groove on the anterior side of the entoconid variable. Radius and ulna unfused, trapezium and Mt 1 absent in the manus, reduced fibula and trochleas of the astragalus nonparallel.

Included Genera. *Lophiomeryx, Iberomeryx, Krabimeryx,* and *Zhailimeryx.*

Comments. This family was created by Janis (1987) with the intention of finding a solution to the great morphological heterogeneity of the Gelocidae. *Lophiomeryx* and *Iberomeryx* (including *Cryptomeryx*, see Bouvrain et al., 1986) were removed to the new familial entity essentially on the basis of the typical lingually open trigonid (mesial extension of the preprotocristid and lack of a premetacristid) on lower molars, which is the defining dental character of the family. Vislobokova (2001) confirmed the validity of the family after an extensive review of cranial and postcranial material presently available. Guo et al. (2000) and Métais et al. (2001) reported new lophiomerycids from the late middle Eocene of China and from the late Eocene of Thailand, respectively, thus confirming the Asian origin of the family. However, the relations of lophiomerycids with other families of stem ruminants are still unclear, and many interpretations have been proposed (Janis, 1987; Geraads et al., 1987; Moyà-Solà, 1988; Guo et al., 2000; Métais et al., 2001; Vislobokova, 2001).

Genus *Lophiomeryx* Pomel, 1854

Type Species and Type Specimen. *L. chalaniati* Pomel, 1854, from Cournon (MP 22, Central France).

Included Species. *L. minor* Lydekker, 1885; *L. gaudryi* Filhol, 1877; *L. mouchelini* Brunet and Sudre, 1987; *L. benaraensis* Gabunia, 1964; *L. turgaicus* Flerov, 1938; *L. angarae* Matthew and Granger, 1925b; *L. gracilis* Miao, 1982 from the Shinao Fm. (Guizhou, China); *L. shinaoensis* Miao, 1982 from the Shinao Fm. (Guizhou, China); *L. kargilensis* Nanda and Sahni, 1990 from the Kargil Fm. (Indian Kahsmir); *L.* sp. Vislobokova and Daxner-Höck, 2002.

Geographic Distribution. Eurasia.

Stratigraphic Range. Late Eocene to early Oligocene.

Definition (adapted from Brunet and Sudre, 1987). C reduced, P2 and P3 elongated with a lingual shelf variously developed, P4 with a strongly concave labial wall and salient

styles, upper molars tetraselenodont and brachydont, M3 metaconule variously developed, labial rib on the paracone, labial wall of metacone strongly concave, parastyle and mesostyle salient, faint metastyle present on M3; c incisiform, p1 reduced and leaflike, separated from c and p2 by diastemata, p2 and p3 elongated and narrow, p4 with a well-developed metaconid and an elongated posterior heel open lingually, lower molars with trigonid open mesiolingually, no fold or groove on the metaconid, third lobe of m3 transversely compressed; Mc and Mt 3 and 4 unfused, Mt 2 and 5 strongly reduced but still present.

Dimensions. m2 (holotype of *L. chalaniati*): L = 13.5 mm; W = 8.2 mm (Brunet and Sudre, 1987).

Discussion. *Lophiomeryx* is fairly frequent in the early Oligocene assemblages from Europe, and the last revision of the genus by Brunet and Sudre (1987) gives detailed data on its stratigraphic range, specific diversity and variation, and evolution of the genus in Europe. *Lophiomeryx* immigrated into Europe shortly after the famous *Grande Coupure*, although it is unknown before MP 22 in Europe. The genus is represented in the late Eocene of Asia by several species, which are generally smaller than the European ones.

Genus *Iberomeryx* Gabunia, 1964

Type Species and Type Specimen. *I. gaudryi* (Filhol, 1877) from Raynal (Phosphorites du Quercy, Southwest France), probably early Oligocene in age; fragmentary lower jaw with left p4–m3 from the older collections of Quercy (figured in Filhol [1877: 227]).

Included Species. *I. parvus* Gabunia, 1964, from Benara (Georgia); *I. minus* (Filhol, 1882) from the Phosphorites du Quercy (old and undated collections); *I. savagei* Nanda and Sahni, 1990, from the Kargil Fm. (Indian Kahsmir); *I.* sp. (cited in Sudre, 1984).

Geographic Distribution. Europe and Caucasus.

Stratigraphic Range. Oligocene.

Definition. Undifferentiated incisors, c incisiform and adjacent to i3, one-rooted, leaflike, and reduced p1 separated from c and p2 by short diastemata, p2 and p3 elongated with small paraconid, p4 showing triangular groove-like heel as a result of posterior extension of the metaconid, molars with trigonid transversely narrower than talonid, cuspids mesially bent, trigonid largely open mesiolingually, *Dorcatherium* fold, ectostylid present; P2 and P3 shortened and weakly developed lingually, P4 triangular and crescent shaped, cingulum lingual to the protocone, styles and paraconal rib well developed.

Dimensions. M2 (*I. parvus*): L = 5.7 mm; W = 6.4 mm (Gabunia, 1964a).

Discussion. Sudre (1984) did an extensive review of the material from western Europe formerly referred to *Cryptomeryx* Schlosser, 1886, which was then thought to belong to the Tragulidae. This assumption was mainly based on the presence of a *Dorcatherium* fold on the lower molars, and it would have filled the supposed gap in the fossil record of tragulids stressed by Webb and Taylor (1980). Sudre (1984: 20) suspected that *Cryptomeryx* and *Iberomeryx* (then restricted to the Oligocene of the Caucasus) might be congeneric or at least closely related. *Iberomeryx* is essentially documented by dental remains, and its chronological and geographic distribution extends from early Oligocene (MP 22 to MP 25) of western Europe to the late Oligocene of India (Nanda and Sahni, 1990). The affinities of *Iberomeryx* within lophiomerycids are unclear, although it may lie closer to the basal Eocene lophiomerycids than to *Lophiomeryx*. Additional material (especially postcranial elements) will certainly provide important data on the polarity of characters in the family.

Genus *Krabimeryx* Métais et al., 2001

Type (and Unique) Species and Type Specimen.   *K. primitivus* Métais et al., 2001, from the Wai Lek pit (Krabi Basin, southern Thailand); fragmentary dentary with left p4–m3 (TF 2676).

Geographic Distribution.   Known only from the type locality.

Stratigraphic Range.   Late Eocene.

Definition.   Small ruminant slightly larger than *Iberomeryx parvus;* lower molars morphologically close to those of *Zhailimeryx* with lingual cuspids transversely compressed, entoconid displaced forward with respect to the hypoconid, lack of both rudimentary paraconid and hypoconulid on m1–2, p4 with a metaconid more distally situated and without distinct entoconid, faint remains of a *Dorcatherium* fold, distinct groove on the anterior side of the entoconid; upper molars with weakly developed crests, labial rib of the paracone strongly developed, styles salient labially, and presence of lingual cingulum around the protocone.

Dimensions.   m2 (holotype): L = 7.7 mm; W = 4.3 mm (Métais et al., 2001).

Discussion.   Métais et al. (2001) assigned *Krabimeryx* to the Lophiomerycidae on the basis of the typical morphology of the trigonid on lower molars and the presence of both an entoconid groove and a *Dorcatherium* fold. The morphology of p4 is unusual among lophiomerycids because of the absence of cusps lingual to the protoconid, and the molars are more bunodont than those of other representatives of the family.

Genus *Zhailimeryx* Guo et al., 2000

Type (and Only) Species and Type Specimen.   *L. jingweni,* Guo et al., 2000 from Locality 1 of Zdansky (1930), Heti Formation, Zhaili Member (Yuanqu Basin, Shanxi, China); IVPP V11385-1, fragmentary left dentary with p3–m3, and IVPP V11385-2, associated right dentary bearing dp1–m3.

Geographic Distribution.   Known only from the type locality.

Stratigraphic Range.   Late middle Eocene (Sharamurunian).

Definition   (adapted from Guo et al., 2000). Small, brachydont, and bunoselenodont lophiomerycid; DP1 semicaniniform, simple, P3 with protocone posterolingual in position, small metaconid on P4, M3 metaconule slightly reduced; the postprotocrista of P4 is short and does not intersect the postparacrista, whereas the postprotocrista of P4 in *Lophiomeryx* is longer and united medially to the postparacrista. Upper molars of *Zhailimeryx* show a damlike ridge along the protocone–metaconule gully slightly buccal to the end of the postprotocrista, which does not contact the anterior surface of the metaconule (in *Lophiomeryx* the ridge is absent, and the postprotocrista contacts the anterior surface of the metaconule or joins the premetaconule crista directly). Lower molar metaconid and entoconid cuspate and with shallow *Dorcatherium* fold on the former and *Zhailimeryx* fold, a shallow notch on the anterior surface, on the latter. Moreover, lower molars of *Zhailimeryx* show a reduced protolophid; the preentocristid extends toward the middle of the protolophid without contacting it. The cheek teeth of *Zhailimeryx* are slightly more brachydont than those of *Lophiomeryx,* and the metaconid of p3 is much weaker in *Zhailimeryx. Iberomeryx* differs from *Zhailimeryx* in having the metaconid of p4 extending posteriorly as a long lingual lophid and having lower molars with a large ectostylid, transversely compressed trigonid, and a more deeply incised *Dorcatherium* fold.

Dimensions.   m2 (holotype): L = 7.0 mm; W = 4.6 mm (Guo et al., 2000).

Discussion.   *Zhailimeryx* is the most primitive and the earliest representative of the family known so far. The *Dorcatherium* fold and the distinct groove on the anterior face of the entoconid (*Zhailimeryx* fold) are probably primitive dental features in the family that are absent in the more selenodont genus *Lophiomeryx. Iberomeryx* displays a very peculiar dental morphology and probably represents a separate lineage within the Lophiomerycidae.

Family TRAGULIDAE

Characteristics.   Braincase more enlarged and expanded than in other traguloids. Sagittal and temporal crests very weak. Sagittal crest short. Temporal crests curved posteriorly. Orbits large and closed posteriorly. Postorbital bar formed mostly of jugal. Postglenoid process absent. Auditory bulla large, inflated, with medium-long external acoustic meatus. Stylohyoid vagina deep, narrow, encroached on bulla, and enclosed posteriorly. Mastoid exposure mainly on lateral surface. Mastoid foramen large and lateral. Foramen ovale small and placed posteriorly. Pterygoid canal sometimes present. Promontorium well developed and elongated, with two posterior whorls, almost equal in height.

Fenestra vestibuli large. Tensor tympani fossa broadened, encroached on promontorium, and pocketed in medial wall. Stapedial muscle fossa placed opposite fenestra vestibuli. Subarcuate fossa deep and pocketed anteromedially. Lateral wall of epitympanic recess formed by squamosal. Lacrimal with enlarged facial and orbital parts and single lacrimal foramen within orbit. Posterior opening of infraorbital canal placed between lacrimal and maxilla. Nasofrontal suture anterior to orbits. Ethmoidal fissure absent or very small and mainly triangular. Nasals relatively short, gradually narrowed anteriorly, and strongly projecting above anterior opening of nasal cavity. Upper incisors absent. Lower i1 enlarged. Upper canine in males enlarged, not procumbent, and almost vertical in lateral view. P1 and p1 absent in *Tragulus* and *Hyemoschus* but present in Miocene taxa; posterior longitudinal groove and lack of distinct metaconid on p4. Presence of M structure (*Dorcatherium* fold plus *Tragulus* fold) at the distal side of the trigonid. Radius and ulna separate. Fibula sometimes fused distally with tibia. Metacarpals III and IV separated. Astragalus with almost parallel trochleae. Calcaneum with concave fibular facet. Metatarsals III and IV fused.

Included Genera. *Tragulus* (extant), *Hyemoschus* (extant), *Archaeotragulus*, *Dorcatherium*, *Dorcabune*, *Siamotragulus*, and *Yunnanotherium*.

Comments. The Tragulidae has long been considered the most primitive extant ruminant family in regard to primitive morphological (Milne-Edwards, 1864; Carlsson, 1926; Webb and Taylor, 1980), physiological (Duwe, 1969; Todd, 1975), and ecological (Dubost, 1975, 1978; Pérez-Barbería and Gordon, 2000) characters. Their primitiveness within the Ruminantia is also unambiguously supported by molecular data that make them the first offshoot of the extant ruminant radiation and the sister group of Pecora (Hassanin and Douzery, 2003). However, the fossil record of tragulids has long been restricted to the Neogene. This unexpected situation led several authors to cast doubt on their antiquity and suggests that recent tragulids would have secondarily acquired their primitive features from a pecoran ancestor. Recent discoveries of tragulid remains in the late Eocene of Thailand thus fill a part of the gap of the evolutionary history of tragulids (Métais et al., 2001). However, the Paleogene record of this key group remains very sparse, and additional fossil material from the Eocene of Asia would certainly resolve several issues related to ruminant origins.

Genus *Archaeotragulus* Métais et al., 2001

Type (and Only) Species and Type Specimen. *A. krabiensis* Métais et al., 2001 from the Wai Lek pit (Krabi Basin, southern Thailand); fragmentary dentary with left p2–m2 (TF 2997).

Geographic Distribution. Known only from the type locality.

Stratigraphic Range. Late Eocene.

Definition (adapted from Métais et al., 2001). Small primitive ruminant with lower molars resembling those of *Dorcatherium* in the presence of a typical tragulid M structure at the rear of the trigonid (see Fig. 15.4) and the derived pattern of its lower premolars, including the lack of a metaconid and the presence of a longitudinal groove on the posterior half of p4. Differs from *Dorcatherium* in its smaller size, its cusps more labiolingually compressed, the lack of an ectostylid, the presence of a well-marked "entoconidian groove" opening forward, the presence of a rudimentary hypoconulid on m2–3 and in the transversely compressed hypoconulid on m3. Differs from *Iberomeryx* in its larger size, its cristid obliqua more lingually oriented, a well-marked M structure on the posterior side of the trigonid, the absence of an ectostylid, and in its transversely compressed and pinched hypoconulid on m3. *Archaeotragulus* further differs from *Zhailimeryx* in its M structure, the lack of rudimentary paraconid on the lower molar, and in its p4 lacking both distinct metaconid and entoconid.

Dimensions. m2 (holotype): L = 7.3 mm; W = 4.6 mm (Métais et al., 2001).

Discussion. The earliest member of the Tragulidae is based on poorly preserved dental material that shows a typical M structure (combination of a *Dorcatherium* fold on the posterior side of the metaconid, and a *Tragulus* fold on the posterior side of the protoconid) at the rear of the trigonid. Métais et al. (2001) suggested that this character might be the most reliable dental feature for recognizing the early representatives of the family.

Further discussion of the later tragulids can be found in Rössner (this volume).

Family LEPTOMERYCIDAE

Characteristics (adapted from Scott, 1940; Webb and Taylor, 1980; Webb, 1998). Skull roof lower, with greater prolongation of an elongated rostrum than in hypertragulids; orbits centrally located and closed posteriorly by a complete postorbital bar, no or limited posterior exposure of the mastoid; enlarged and procumbent i1 (this character is also present in *Archaeomeryx*); lower canine adjacent to the third incisor; p1 caniniform, generally reduced in size, and isolated by anterior and posterior diastemata; p2 consists of three anteroposteriorly aligned cuspids and is directly adjacent to p3; the latter possesses a distinct paraconid; p4 with a well developed paraconid and metaconid and a strong posterolingual cuspid; brachydont lower molars showing crescent-shaped cuspids. Upper incisor absent; C vestigial (or present); P1 absent; P2 and P3 with lingual protocone; P4 with a crescent-shaped protocone; upper molars with mesostyle. The lateral digits (2 and 5) of the manus tend to be reduced; moderately spoutlike odontoid process of the axis; forelimb shorter than in hypertragulids, fused magnum and trapezoid in the carpus, tetradactyl manus with a marked trend toward paraxony; fibula reduced to a proximal remnant associated with a disjoint distal malleolar bone, trochleae of the astragalus not parallel.

Included Genera.   *Leptomeryx, Hendryomeryx, Pronodens,* and *Pseudoparablastomeryx.*

Comments.   *Leptomeryx* was the first Paleogene ruminant to be described in North America (Leidy, 1853), and the time range of leptomerycids in North America extends from the late middle Eocene to the middle Miocene, making them the longest-lived family of ruminants on that continent. Leptomerycids were long considered as a subfamily of the Hypertragulidae until Gazin (1955) recognized the noticeable morphological differences between the two main families of North American early ruminants. The geographic range of the leptomerycids originally extended to Asia with the inclusion of *Archaeomeryx* (Webb and Taylor, 1980), which was recently removed into its own family (the Archaeomerycidae) together with *Miomeryx, Xinjiangmeryx, Notomeryx,* and possibly *Indomeryx.* By their present definition, leptomerycids are thus restricted to North America and constitute a homogeneous assemblage of early ruminants that may have had an Asian ancestor close to the current Archaeomerycidae.

Genus *Leptomeryx* Leidy, 1853

Type Species and Type Specimen.   *L. evansi* Leidy, 1853 from South Dakota (also reported from Nebraska; Cedar Creek, Colorado; Pipestone, Montana); USNM 157, partial skull and jaw fragment.

Included Species.   *L. blacki* Stock, 1949; *L. mammifer* Cope, 1885; *L. obliquidens* Lull, 1922; *L. speciosus* Lambe, 1908; *L. yoderi* Schlaikjer, 1935; *L.* sp. (cited from several localities by Webb [1998]).

Geographic Distribution.   North America.

Stratigraphic Range.   Duchesnean (late middle Eocene) to Arikareean (early Miocene).

Definition   (adapted from Scott, 1940; Webb, 1998). Dental formula is 0/3-1/1-3/4-3/3, skull with elongated muzzle, upper incisors absent, vestigial upper canine, P1 absent, upper premolars triangular with weak development of cingula, subquadrate upper molars with well-developed labial ribs and styles; the first incisor in *Leptomeryx* is much larger than the others; incisiform lower canine, which is coalescent to i3; p1 vestigial, caniniform, and isolated from the other teeth by a diastema in front and behind; p3 with paraconid and incipient heel; p4 with a prominent posterolingual hypoconid.

Dimensions.   m2 (*L. evansi*): L = 7.1 mm; W = 4.3 mm (Scott, 1940).

Discussion.   *Leptomeryx* is common in the White River Group (northwest Great Plains) where it is known by abundant dental material and significant postcranial material

(Scott, 1940). This genus first appeared in the late Duchesnean and became extinct by the late Arikareean (or early Hemingfordian, Webb, 1998). The definition of the different species is uncertain because of the high dental variation in size and morphology within a stratigraphically and geographically local sample (Heaton and Emry, 1996). However, a reevaluation of Oligocene species of *Leptomeryx* and dental characters used in systematics led Korth and Diamond (2002) to differentiate different species.

Genus *Pseudoparablastomeryx* Frick, 1937

Type Species and Type Specimen.   *P. scotti* Frick, 1937 from Observation Quarry, Dawes County, Nebraska; F:AM 33763, left lower jaw with diastema, p2 alveolus, and p3–m3.

Included Species.   *P. francescita* (Frick, 1937); *P.* sp. (cited from several localities by Webb [1998]).

Geographic Distribution.   North America.

Stratigraphic Range.   Hemingfordian (early Miocene) to Barstovian (middle Miocene).

Definition   (adapted from Frick, 1937; Taylor and Webb, 1976; Webb, 1998). Medium-sized leptomerycid; brachycephalic cranium; lingual cingula of upper molars more developed than in *Leptomeryx;* p1 absent, premolars coalescent and reduced in length; lower molars are high crowned and transversely compressed.

Dimensions.   m2 (*P. scotti*): L = 7.0 mm; W = 4.1 mm (Taylor and Webb, 1976).

Discussion.   Frick (1937: 219) proposed a new subgenus to differentiate "a diminutive mandibular ramus with large, stubby-proportioned premolars and an abbreviation of the diastema exceeding that of *Parablastomeryx.*" It is worth noting that *Pseudoblastomeryx* displays the most derived postcranial features because the metatarsals 3 and 4 tend to fuse in a cannon bone as in more advanced pecoras.

Genus *Pronodens* Koerner, 1940

Type (and Unique) Species and Type Specimen.   *P. silberlingi* Koerner, 1940, from the Fort Logan Formation, Meagher County, Montana; YPM 13952, left lower jaw with i1, i2, alveoli of p3 and c, p2 to m2, and the anterior part of m3.

Included Species.   *P.* sp. (cited from several localities by Webb [1998]).

Geographic Distribution.   North America.

Stratigraphic Range.   Arikareean (late Oligocene–early Miocene).

Definition   (adapted from Koerner, 1940). Medium-sized ruminant with large, strongly procumbent tusklike incisors; diastema between c and p2 is usually short; the mandibular foramen and the posterior border of the mandibular symphysis lie below the anterior portion of p2. A high, narrow ridge extends from c to p2, adding considerable depth to that part of the jaw.

Dimensions. m2 (holotype): L = 10.3 mm; W = 7.2 mm (Koerner, 1940).

Discussion. Koerner (1940) compared the incisors of *Pronodens* to those of diprotodont marsupials to stress their procumbent and spatula-like morphology. The advanced morphology of its premolars and the transverse enlargement of its lower molars also suggest its derived condition within the Leptomerycidae.

Genus *Hendryomeryx* Black, 1978

Type Species and Type Specimen. *H. wilsoni* Black, 1978, from the Tepee Trail Formation, Bad Water Creek, Natrona County (Wind River Basin, Wyoming); CM 29102, left maxilla with P3–M3.

Included Species. *H. defordi* (Wilson, 1974); *H. esulcatus* (Cope, 1889); *H.* sp. (cited from several localities by Webb [1998]).

Geographic Distribution. North America.

Stratigraphic Range. Duchesnean (late middle Eocene) to Whitneyan (early Oligocene).

Definition (adapted from Black, 1978b). Paraconule absent on M1–3; all styles on M1–3 present but narrow; paracone rib present on M1–3 but no metacone rib; posterior protocone crest very short, directed toward metaconule; p4 talonid forming a rectangular posterior groove, transverse crest connects the protoconid higher than the metaconid; low-crowned lower molars with stylids reduced or absent; entoconid and metaconid connected through a straight lingual crest; selenes moderately developed on m1–3.

Dimensions. M2 (*H. wilsoni*): L = 5.3 mm; W = 5.8 mm (Black, 1978b).

Discussion. *Hendryomeryx* is one of the earliest ruminants, appearing in North America as early as the earliest Duchesnean. Only *Simimeryx* might be slightly older. The relative bunodonty of the molars of *Hendryomeryx* is often proposed to make it the most primitive representative of the family. Postcranial features of this genus are still unknown.

Family GELOCIDAE

Characteristics (after Viret, 1961). Fourth lower premolar slender, with small metaconid. Lower molars with cusps brachydont; bunoselenodont protoconid and hypoconid; metaconid and entoconid conical and slightly compressed laterally; rudiment of paraconid present; metaconid and entoconid crowded; metastylid absent; trace of "*Dorcatherium* fold" present. Prehypocristid joins posterior

of protoconid, postmetacristid joins postprotocristid and preentocristid at center of tooth.

Included Genera. *Gelocus, Paragelocus, Pseudogelocus, Phaneromeryx, Prodremotherium, Gobiomeryx, Pseudoceras,* and *Pseudomeryx.*

Comments. Gelocids are the only ruminants to be reported in the late Eocene of Europe with *Phaneromeryx gelyense* and *Gelocus minor;* both species are unknown in the Oligocene. The family Gelocidae forms a heterogeneous assemblage of small to medium-sized Eurasian hornless ruminants that are thought to share some "protopecoran" features. Unfortunately, most of those forms are poorly known genera showing morphological characters more or less intermediate between the early traguloids and true pecorans. Accordingly, gelocids have played the role of a *faute de mieux* family for these ambiguous and poorly defined taxa. Janis (1987) questioned the validity of this family defined with disputable dental characters, but in the absence of an adequate fossil record and real consensus concerning these poorly documented forms, gelocids remain the temporary familial assignment of the several genera described below. Additional material allied with a systematic review of the available material would certainly be profitable.

Genus *Gelocus* Aymard, 1855

Type Species and Type Specimen. *G. communis* (Aymard, 1846) from the lacustrine limestone of Ronzon (MP 21, southwest France).

Included Species. *G. minor* Pavlov, 1900, from Mormont-Entreroches (MP 19); *G. laubei* Schlosser, 1901, from the Jura Franco-Souabe (MP 21–MP 22); *G. whitworthi* Hamilton, 1973, from the early Miocene of Gebel Zelten (Libya); *G. villebramarensis* Brunet and Jehenne, 1976, from Villebramar (MP 22); *G. quercyi* Jehenne, 1987, from the old and undated collections of Quercy.

Geographic Distribution. Europe, and possibly North and East Africa.

Stratigraphic Range. Late Eocene–earliest Oligocene (Ronzon).

Definition (adapted from Viret, 1961). Skull with reduced rostrum, orbit closed by a postorbital bar, anterior border of the orbit situated at the level of M1; brachydont molars, upper molars with large external style, paracone supported by a median labial rib, parastyle and metastyle are low but well developed. The internal selenes on upper molars are large; the protocone is surrounded by a strong cingulum and has a reduced postprotocrista that is not directed toward the central portion of the molar (resembling the ancestral condition as exemplified by *Archaeomeryx*). P1 is lost, upper incisors absent, C is elongated (resembling the condition in *Hyemoschus*); lower molars have conical lingual cuspids and crescentic labial cuspids, rudiment of paraconid at the extremity of the preprotocristid, slight labial fold in the enamel on the posterior side of the metaconid in addition to the postmetacristid, which extends toward the center of the tooth to join the preentocristid and the postprotocristid; the talonid of p4 has three transverse crests but has a small metaconid. Mc III and IV unfused, Mt III and IV fused, lateral digits reduced to thin, but complete, splints; astragalus with parallel trochleae.

Dimensions.   m2 (*G. communis*): L = 9.6 mm; W = 5.9 mm.

Discussion.   *Gelocus* illustrates the ambiguous status of this family, which includes hornless ruminants of a more or less advanced prepecoran grade. As mentioned by Janis (1987), the petrosal of *Gelocus* is more reminiscent of that of Pecora than that of traguloids. However, postcranial and dental features would rather favor traguloid affinities. *Gelocus* is reported in the late Eocene (MP 19) of Europe, although "*G.*" *minor* from Mormont-Entreroches might eventually prove to be a lophiomerycid close to *Iberomeryx* (Sudre and Blondel, 1996). If the late Eocene occurrence of the genus is sustained, *Gelocus* may have survived through the *Grande Coupure*, as it occurred in early Oligocene until MP 25.

Genus *Pseudoceras* Frick, 1937

Type Species and Type Specimen.   *P. skinneri* Frick, 1937 from Cherry County, Nebraska; F:AM 33723, left lower jaw with i3-m3 (figured in Frick, 1937: 648).

Included Species.   *P. potteri* Frick, 1937, from Cherry County, Nebraska; *P. wilsoni* Frick, 1937, from Brown County, Nebraska; *P.* sp. (cited from several localities by Webb [1998]).

Geographic Distribution.   North America.

Stratigraphic Range.   Late Miocene (Barstovian and Hemphillian).

Definition   (adapted from Frick, 1937, and Webb, 1998). Small hornless ruminant. Skull with orbits at midlength; inflated hollow bullae; narrow triangular occiput. First incisor large and procumbent; others smaller and somewhat erect; lower canine large and recurved. Cheek teeth transversely compressed; p2-3 with posteriorly directed metaconid; p4 with posteriorly directed metaconid and hypoconid enclosing long narrow fossettid.

Dimensions.   m2 (*P. skinneri*): L = 9.8 mm; W = 6.3 mm.

Discussion.   First doubtfully referred to camelids by Frick (1937), *Pseudoceras* was removed to the Gelocidae by Tedford et al. (1987) and thus constitutes the only gelocid known in the New World. It is worth noting that this fairly enigmatic genus is known in the late Miocene of Central America (Webb and Perrigo, 1984) and is most abundant in the southern United States, although originally defined from specimens from the Miocene of Nebraska (Frick, 1937). According to Webb (1998), the skeletal proportions and the dentition of *Pseudoceras* are closer to those of *Moschus*

and thus clearly distinctive from the postcranial adaptations exhibited by *Gelocus*. The morphology of *Pseudoceras* and its doubtful inclusion into gelocids illustrate the extreme heterogeneity of this "wastebasket" group that puts together derived hornless ruminants from Eurasia, North America, and Africa.

Genus *Paragelocus* Schlosser, 1902

Type Species and Type Specimen.   *P. scotti* Schlosser, 1902, from Hochberg and Veringenstadt (MP 21, Germany); maxilla with M1–M3, Inst. Paläont. Munich 1083.

Geographic Distribution.   Europe.

Stratigraphic Range.   Early Oligocene (MP 21–MP 23).

Definition.   Differs from *Gelocus* in having smaller and shorter snout, more brachydont teeth, very strong cingulum in the transversely elongated upper molars with well-developed para- and mesostyles, very narrow p4 with only four cusps (largest conical protoconid, posteriorly situated metaconid, and lower para- and hypoconids). According to Schlosser (1902), the upper molars retain a small paraconule.

Dimensions.   *P. scotti*: LM1–M3 = 17 mm; Lp4 = 5.4 mm.

Discussion.   The validity of this genus has been questioned by several authors who successively considered it as a junior synonym of *Gelocus* (Ginsburg and Hugueney, 1987) or *Pseudogelocus* (Sudre and Blondel, 1996). The M3 of *P. scotti* displays a strongly reduced metaconule and a thick lingual cingulum; the mesostyle is strong. The p3 referred to this genus is reminiscent of that of *Rutitherium nouletti*. Vislobokova and Daxner-Höck (2002) resurrected *Paragelocus* by tentatively adding material from the middle-late Oligocene of Mongolia to the type species. This noticeably extends the geographic distribution of this genus, which was previously restricted to Western Europe.

Genus *Pseudogelocus* Schlosser, 1902

Type (and Unique) Species and Type Specimen.   *P. suevicus* Schlosser, 1902, from Oerlingerthal (MP 21, South Germany); lower jaw preserving right p4–m1 (No 1881-IX-537, Inst. Palaont. Munich).

Included Species.   *P. mongolicus* Vislobokova and Daxner-Höck, 2002, from Tatal Gol (Mongolia); *P.* cf. *scotti* (including *I. matsoui* [Sudre, 1984] from Mas de Got and La Plante 2 [MP 22, Phosphorites du Quercy]).

Geographic Distribution.   Eurasia.

Stratigraphic Range.   Early Oligocene (MP 21–MP 23).

Definition   (adapted from Jehenne, 1985). Fourth lower premolar with a strong paraconid, an isolated and well developed metaconid, and a talonid with distinct entocristid and hypocristid extending lingually as in the Pecora.

Dimensions.   m2 (*P. mongolicus*): L = 5.6 mm; W = 3.5 mm (Vislobokova and Daxner-Höck, 2002).

Discussion. The complex systematic history of this genus is summarized in Sudre and Blondel (1996). *Pseudogelocus* was long believed to be restricted to the Bavarian localities of South Germany, where it is sometimes referred to *Gelocus* (Ginsburg and Hugueney, 1987). However, *P. suevicus* has been recognized in the MP 22 Quercy localities, but its exact stratigraphic range in the Quercy remains poorly known (Blondel, 1997). The genus has recently been noted in the early Oligocene of Mongolia, considerably extending its geographic distribution (Vislobokova and Daxner-Höck, 2002) and confirming Asia as a major center of ruminant evolution before these creatures invaded western Europe immediately following the *Grande Coupure*.

Genus *Prodremotherium* Filhol, 1877

Type Species and Type Specimen. *P. elongatum* Filhol, 1877, from the undated old collections from Quercy, complete skull and part of limbs (see Jehenne, 1977, 1985).

Included Species. *P. flerowi* Trofimov, 1957, from the early Oligocene of Tchelkar-Teniz (Kazakhstan); *P. trepidum* Gabunia, 1964, from the late Oligocene of Benara (Georgia); *P.* sp. Vislobokova and Daxner-Höck, 2002, from Tatal Gol (Mongolia).

Geographic Distribution. Eurasia.

Stratigraphic Range. Oligocene.

Definition. P1 lost, P2 and P3 elongated, very weak or no cingulum, but a distinct entostyle on upper molars, p1 absent, long diastema between c and p2, p4 with a strong metaconid and four crests lingually orientated, no *Palaeomeryx* or *Dorcatherium* folds on lower molars but a distinct metastylid, entoconid transversely compressed and crested anteriorly and posteriorly, preprotocristid closing trigonid lingually, ectostylid present; Mc III and IV fused proximally, Mc II and V reduced and not functional, Mt III and IV fused proximally, and metatarsal gulley closed distally, astragalus with a transversely extended sustentacular facet and with parallel trochleae.

Dimensions. m2 (*P. elongatum*): L = 11.1 mm; W = 7.2 mm (Blondel, 1997).

Discussion. The single European species, *P. elongatum*, is particularly well documented in the old collections from Quercy, and its biochronological range extends from MP 25 to MP 28. Blondel (1997) claimed that *P. elongatum* is more primitive than the Asian species of the genus, but given the paucity of the fossil material presently referred to the latter, we consider this conclusion as premature.

Genus *Gobiomeryx* Trofimov, 1957

Type (and Unique) Species and Type Specimen. *G. dubius* Trofimov, 1957, from Ergilin-Dzo (Ergilin-Dzo Svita, Mongolia); PIN 473/42, fragmentary right lower jaw with m1–m3.

Included Species. *G.* sp. Vislobokova and Daxner-Höck, 2002.

Geographic Distribution. Central Asia.

Stratigraphic Range. Late Eocene (Ergilian)–early Oligocene (Shandgolian).

Definition. Adapted from Trofimov, 1957; Musakulova, 1963. Lower molars transversely narrow; cuspids relatively acute and mesially inclined; lingual cuspids transversely compressed, metastylid increasing in size from m1 to m3, no particular fold or groove is visible on the lower molars; presence of an elongated tubercle (entoconulid) on the mesial border of the third lobe of m3.

Dimensions. m2 (holotype): L = 10.3 mm; W = 5.2 mm (Trofimov, 1957).

Discussion. *G. dubius* is also known from the early Oligocene Kiin-Kerish locality, Zaysan Depression, Kazakhstan (Musakulova, 1963, 1971), and the early Oligocene Tatal-Gol, Mongolia. Vislobokova and Daxner-Höck (2002) reported additional specimens referred to as *Gobiomeryx* sp. showing the lower p3 and p2, with p2 separated from a large alveolus (probably for the canine) by a significant diastema. The familial status and the relationships of *Gobiomeryx* are still enigmatic, and improvements in this regard would benefit from knowledge of the morphology of the upper dentition. This genus is temporarily included in the paraphyletic gelocids, principally based on the presence of a metastylid on lower molars.

Genus *Pseudomeryx* Trofimov, 1957

Type (and Unique) Species and Type Specimen. *P. gobiensis* Trofimov, 1957, from Tatal-Gol (Shand-Gol Svita, Mongolia); PIN 475/3142, left lower jaw with m1–m3.

Included Species. *P.* sp. Vislobokova and Daxner-Höck, 2002.

Geographic Distribution. Asia.

Stratigraphic Range. Early Oligocene (Shandgolian).

Definition (adapted from Trofimov, 1957, with addition from Vislobokova and Daxner-Höck, 2002). Small ruminant with bunoselenodont lower molars; metastylid and ectostylid present; lower p4 is without entoconid and possesses posteriorly elongated metaconid, forming an almost closed fossetid with the hypoconid.

Dimensions. m2: L = 6.5 mm; W = 4.5 mm (Vislobokova and Daxner-Höck, 2002).

Discussion. The lower molars of the type specimen are fairly worn, limiting any precise observation. There is no trace of any folds or grooves. The hypoconulid is duplicated on the third lobe of m3. *P. gobiensis* is also reported from Hsanda-Gol and Tatal-Gol (early Oligocene, Mongolia), and Vislobokova and Daxner-Höck (2002) described additional material housed in the PIN collections of Moscow. These authors assumed an isolated p4 they consider as belonging to *P. gobiensis;* this specimen displays a strong para-conid, a conical metaconid, and a V-shaped groove opened posterolingually. This genus remains poorly documented, and its familial status should be considered as provisory. There are a number of undescribed specimens of the genus from Mongolia in the PIN collection in Moscow that could provide additional morphological information.

Poorly Defined and Doubtful Gelocids
Genus *Phaneromeryx* Schlosser, 1886

Type (and Unique) Species and Type Specimen. *P. geliensis* (Gervais, 1848) from Saint-Gely-du-Fesc (southern France); fragmentary lower jaw reportedly with p4–m3.

Geographic Distribution.   Southern France.

Stratigraphic Range.   Late Eocene (MP 18).

Discussion.   The type and unique specimen of *Phanero-meryx* is a fragmentary lower jaw with p4–m3, and this taxon is rarely mentioned in recent works on early ruminants. *Phaneromeryx* is associated with a fauna of upper Ludian age close to that of La Debruge (MP 18), which was collected in the nineteenth century (Hartenberger et al., 1969). The specimen undoubtedly belongs to a primitive ruminant (J. Sudre, pers. comm., August 2005), but Jehenne (1985) casts some doubt on its gelocid affinities because of the derived characters of the dentition. If the stratigraphic origin of *Phanero-meryx* is correct, this still-enigmatic form brings further evidence for a pre–*Grande Coupure* occurrence of ruminants in Europe.

Genus *Rutitherium* Filhol, 1876

Type (and Unique) Species and Type Specimen.   *R. nouletti* Filhol, 1876, from an unknown locus of the Quercy fissure fillings (southwestern France).

Geographic Distribution.   France.

Stratigraphic Range.   Questionably late Eocene to late Oligocene (Sudre, 1984).

Definition.   p1 and p2 separated by a very short diastema; three crests lingually oriented on p3, four on p4; lower molars relatively bunoselenodont, bearing an ectostylid, faint *Palaeomeryx* fold, and slight groove on the posterolingual side of the metaconid.

Dimensions.   m2 (*R. nouletti*): L = 6.5 mm; W = 4.7 mm.

Discussion.   *Rutitherium* is rarely discussed in the systematic reviews of gelocids, and it is particularly poorly defined. Schlosser (1886) and Zittel (1893) concluded that this genus was close to *Gelocus* and *Prodremotherium,* and Bouvrain et al. (1986) cast some doubt on the validity of the genus. Created from the undated old collections of Quercy, the genus was subsequently mentioned in the late Oligocene of France under various generic names (Sudre, 1984; Janis, 1987). This genus remains poorly documented, and

additional material would be needed to test its validity and to refine its definition.

Family BACHITHERIIDAE

Characteristics.   See the definition of *Bachitherium.*

Included Genera.   *Bachitherium.*

Comments.   The origin and phylogenetic relationships of this monogeneric family, proposed by Janis (1987), with other families of hornless ruminants are unclear. *Bachith-erium* possesses a mixture of primitive and derived features that have made its systematic history particularly confusing. Janis (1987) stressed the difficulties in assigning this genus to an existing family and failed to find apomorphic features for the bachitheriids.

Genus *Bachitherium* Filhol, 1882

Included Species.   *B. curtum* (Filhol, 1877); *B. insigne* (Filhol, 1882); *B. vireti* Sudre, 1986; *B. lavocati* Sudre, 1986; *B. guirounetensis* Sudre, 1995.

Geographic Distribution.   Europe.

Stratigraphic Range.   Oligocene (MP 22 to MP 28).

Definition   (Adapted from Geraads et al., 1987). Elongated muzzle with small orbits placed relatively posteriorly (the anterior border of the orbit is situated above M2), postorbital bar partly formed by the jugal, sagittal crest present and extending posteriorly into temporal crests, small auditory bullae, lacrimal fossa absent, strong paraoccipital apophysis, limited posterolateral exposure of the mastoid, ethmoidal fissure moderately developed, postglenoid process well developed; angular region of mandible extended posteriorly and upward, and the coronoid process is much higher than the articular joint, suggesting strong abductor muscles. Dental formula: 0/3, 1/1, 3/4, 3/3, upper canine tusklike, slightly curved posteriorly, and occludes against anterior side of caniniform p1, P1 lost, very long diastema between C and P2, lower incisors and incisiform canine small, small diastema between c and p1, and long diastema between p1 and p2. Radius and ulna separate, tibia and fibula partially fused, distal extremity reduced to malleolar bone, Mc III and IV unfused, Mt III and IV fused proximally, lateral metatarsals absent, astragalus with nonparallel trochleae, crural index close to that of the extant genus *Moschus.*

Dimensions.   m2 (*R. curtum*): L = 7.6 mm; W = 5.2 mm (Sudre, 1995).

Discussion.   *Bachitherium* is one of the rare Paleogene ruminants known by a nearly complete skeleton (Geraads et al., 1987). Lavocat (1951) and Viret (1961) advocated close relationships with the Hypertragulidae; Hamilton (1973) placed *Bachitherium* within the Gelocidae; Webb and Taylor (1980) favored affinities with the Leptomerycidae; and Janis (1987) finally created a new family for the genus. The main apomorphic feature of the family is the development of diastemata and the partial to complete fusion of the central metatarsals. The lower molars retain a faint and oblique fold on the posterolingual side of the metaconid, but this fold is not homologous to that seen in tragulids and in some lophiomerycids. *Bachitherium* appeared in Europe in MP 22 (as did *Lophiomeryx* and *Iberomeryx*), and it became extinct in the late Oligocene (MP 28). The specific diversity of *Bachitherium* is maximal in MP 25, which corresponds to the greatest generic diversity of hornless ruminants in Europe with the occurrence of *Lophiomeryx, Iberomeryx, Bachitherium, Gelocus,* and *Prodremotherium.* Mammal level MP 25 roughly correlates with the latest early Oligocene and records the last occurrences of *Iberomeryx* and *Gelocus* and the first occurrence of *Prodremotherium* (Blondel, 1997).

Superfamily HYPERTRAGULOIDEA
Family HYPERTRAGULIDAE

Characteristics (adapted from Scott, 1899; Webb and Taylor, 1980; Webb, 1998). Skull roof high, short rostrum; orbits centrally located and opened posteriorly, large lateral exposure of the mastoid; c incisiform and adjacent to i3; tusklike p1 isolated by anterior and posterior diastemata; p2 secant, triangular in lateral view, and separated from p3 by a short diastema, p4 with a paraconid, a metaconid, and a reduced subcircular heel; brachydont lower molars showing crescent-shaped cuspids. Upper incisor absent; tusklike C (when present); conical P1 separated from C and P2 by diastemata; P3 and P4 bearing a conical protocone lingually; mesostyle absent on upper molars. Short peglike odontoid process of axis; pentadactyl manus and four digits in pes, the lateral digits (2 and 5) tend to be reduced; unfused magnum and trapezoid; complete fibula fused distally with tibia, trochleae of astragalus not parallel.

Included Genera. *Hypertragulus, Nanotragulus,* and *Hypisodus.*

Comments. Hypertragulids are often considered as the most basal ruminants, and the issue of whether or not they should be included within the Ruminantia is recurrent (Webb and Taylor, 1980). *Hypertragulus* is the most completely known genus of the family. It displays a mixture of primitive and derived features that are not observed in the Asian taxa known so far. Hypertragulids became extinct in the Miocene. Vislobokova (2001) retained only *Hypertragulus* and *Nanotragulus* in the Hypertragulidae and placed *Hypisodus* in a separate family. *Simimeryx* was combined with *Praetragulus,* the first hypertraguloid outside of North America described from the early Oligocene Mongolia, in a new family Praetragulidae (Vislobokova, 1998, 2001).

Subfamily HYPERTRAGULINAE
Genus *Hypertragulus* Cope, 1873

Type Species and Type Specimen. *H. calcaratus* Cope, 1873, from Cedar Creek, White River Formation, Logan County, Colorado; AMNH 6815, left maxilla with P3–M3.

Included Species. *H. heikeni* Ferrusquia-Villafranca, 1969; *H. hesperius* Hay, 1902; and *H. minutus* Lull, 1922.

Geographic Distribution. North America.

Stratigraphic Range. Late middle Eocene (Duchesnean) to early Miocene (late Arikareean).

Definition (adapted from Scott, 1940; Webb, 1998). Dental formula is 0/3-1/1-4/4-3/3, large hypertragulid characterized by its brachydont dentition; prominent anterior cingula and accessory cusps and cuspids on upper and lower molars, respectively; P1/1 and P2/2 are simple, secant, and separated from adjacent teeth by diastemata; protocone very weak on P3, more developed and conical on P4; p4 with well-developed paraconid and metaconid; subcircular posterior heel enclosing a faint hypoconid.

Dimensions. m2 (*H. calcaratus*): L = 6.0 mm; W = 3.2 mm (Scott, 1940).

Discussion. *Hypertragulus* is the best-documented hypertragulid known so far, but paradoxically, that does not make clear its relationships with other early ruminants. Its skull appears primitive (short muzzle, mastoid exposed, incomplete postorbital bar, petrosal features), but some of its postcranial features are clearly derived (fibula reduced and fused with the tibia, ulna fused with the radius). Its dentition is also derived in many regards (tusklike p1, diastemata). *Hypertragulus* probably lived in forested habitats relying on soft food such as fruits or leaves, and possibly insects.

Genus *Nanotragulus* Lull, 1922

Type Species and Type Specimen. *N. loomisi* Lull, 1922, from Big Muddy River, "?Monroe Creek" (Arikareean), Wyoming; YPM 10330, palate and partial mandible.

Included Species. *N. ordinatus* (Matthew, 1907); and *N. planiceps* Sinclair, 1905.

Geographic Distribution. North America.

Stratigraphic Range. Early Oligocene (Whitneyan) to early Miocene (late Arikareean).

Definition (adapted from Frick, 1937; Webb, 1998). Larger and more hypsodont than *Hypertragulus;* as a consequence, accessory cusps and cingula are reduced; diastema between p2 and p3 is reduced in some derived species; auditory bullae are enlarged.

Dimensions. m2 (*N. loomisi*): L = 7.0 mm; W = 3.7 mm (Lull, 1922).

Discussion. The characters of the dentition are more derived than those of *Hypertragulus,* especially in the reduction of the premolars and the trend toward more hypsodont molars. *Nanotragulus* is the only hypertragulid to be unknown from the Eocene.

Subfamily HYPISODONTINAE

Genus *Hypisodus* Cope, 1873

Type (and Unique) Species and Type Specimen. *H. minimus* Cope, 1873, from Cedar Creek, White River Formation, Logan County, Colorado; AMNH 6543, left maxilla with M1–3.

Geographic Distribution.   North America.

Stratigraphic Range.   Late middle Eocene (Duchesnean) to late Oligocene (early Arikareean).

Definition   (adapted from Frick, 1937; Webb, 1998; Vislobokova, 2001). Much smaller than *Hypertragulus;* transversely compressed muzzle, large orbits closed posteriorly, very large auditory bullae meeting medioventrally, small anterior opening of the nasal cavity; P1 and P2 absent, molars showing a clear hypsodont trend; lower p2 tends to be absent in adult individuals, long diastemata before P3/3, upper molars with almost flat labial wall; metatarsals 3 and 4 nearly fused, and lateral digits strongly reduced.

Dimensions.   m2 (*H. minimus*): L = 4.0 mm; W = 2.2 mm (Scott, 1940).

Discussion.   In size and morphology, *Hypisodus* is so distinctive from other hypertragulids that Vislobokova (2001) removed it to its own family, the Hypisodontidae. This small form experimented with a bovid-like morphology and is the most derived among early ruminants known so far. Webb (1998) suggested that *Hypisodus* might have been ecologically similar to the extant dik-dik (*Madoqua*). It appears likely that *Hypisodus* lived in more open habitats that became widespread after the Eocene–Oligocene transition in North America.

Family PRAETRAGULIDAE

Characteristics.   Skull brachycephalic, with shorter snout than in hypertragulines and hypisodontines. Orbits not very large. Postorbital bar apparently open. Foramen ovale small, ovate, and placed as in hypertragulids. Petrosal short and broad. Promontorium weak, short, and corresponding to main whorl of cochlea. Tensor tympani fossa pocketed in both lateral and medial walls. Stapedial muscle fossa narrow, sinuous, and located posteriorly. Upper canines large. Lower canines not completely included in incisor row, small or enlarged. P1 and p1 lost. Diastemata behind p2 absent. Parastyle not compressed. Mesostyle present in unworn teeth. Radius and ulna usually separate. Fibula distally detached, not fused to tibia. Central metapodials unfused.

Included Genera.   *Praetragulus, Simimeryx,* and *Parvitragulus.*

Comments.   *Simimeryx* and *Parvitragulus* were combined with *Praetragulus,* the first hypertraguloid outside of North America described from the early Oligocene Mongolia, in a new family Praetragulidae (Vislobokova, 1998, 2001).

Genus *Praetragulus* Vislobokova, 1998

Type Species and Type Specimen.   *P. electus* Vislobokova, 1998, from Khoer-Dzan (Ergilin-Dzo Fm, Mongolia); PIN 3110/731, incomplete skull with P2–M3 and the associated lower jaw.

Included Species.   *P. gobiae* (Matthew and Granger, 1925a) Vislobokova, 1998, from Ergilin-Dzo–Tatal-Gol, Mongolia.

Geographic Distribution.   Mongolia.

Stratigraphic Range.   Late Eocene (Ergilian)–early Oligocene.

Definition.   Possesses mesostyle in unworn upper molars. Differs from *Simimeryx* in less-developed lower canines, loss of protoconule, compressed parastyle in upper molars, and better-developed selenodonty. Differs from *Parvitragulus* in shorter snout, more anterior position of posterior mental foramen, lower-crowned cheek teeth, and less complex upper and lower premolars.

Dimensions.   M2 (holotype) average: L = 10.6 mm; W = 11.4 mm; m2 (holotype) average: L = 9.9 mm; W = 6.4 mm (Vislobokova, 1998).

Discussion.   This genus now includes the material from Ardyn-Obo (=Ergilin-Dzo) formerly referred to *Lophiomeryx gobiae* by Matthew and Granger (1925a). *L. gobiae* is otherwise known from the early Oligocene of Tatal-Gol, Mongolia (Vislobokova and Daxner-Höck, 2002). *Praetragulus* is the only hypertraguloid known in Asia so far, and its forelimb morphology is more primitive than that of North American hypertragulids. Very little is known about the early evolution of hypertraguloids in Asia, but Vislobokova (1998) suggested that they originated in Asia, and probably invaded North America by the late middle Eocene. This hypothesis would explain the sudden appearance of the Hypertragulidae in the late Uintan of North America. Hypertraguloids would have evolved separately in Asia during the late Eocene, but their fossil record is still to be documented for this time interval. *Praetragulus* is suggested to be an Asian survivor of the basal radiation of hypertraguloids in Asia and North America, which appears morphologically derived (e.g., loss of p1, presence of mesostyle on upper molars) with respect to its North American counterparts. However, *Praetragulus* remains clearly identifiable as a hypertraguloid because of its cranial (short symphysis, brachycephalic cranium, orbits open posteriorly, deep subarcuate fossa, lateral exposure of the mastoid, premaxilla short) and postcranial characters (radius and ulna separated, magnum and trapezoid unfused, metapodials unfused, tetradactyl pes).

Genus *Simimeryx* Stock, 1934

Type Species and Type Specimen.   *S. hudsoni* Stock, 1934 from the Sespe Formation (late Uintan or early Duchesnean Ventura County, California); LACM (CIT) 1764, fragment of palate with P2–M3.

Included Species. *S. minutus* (Peterson, 1934).

Geographic Distribution. North America.

Stratigraphic Range. Late middle Eocene (early Duchesnean).

Definition (adapted from Golz, 1976; Webb, 1998). Slightly smaller and molars lower crowned than in *Hypertragulus;* molars more bunodont, and cingula well developed; upper molars tend to be transversely compressed and to retain tiny paraconule, mesostyle absent, premolar diastemata reduced.

Dimensions. m2: L = 5.0 mm; W = 2.6 mm (Stock, 1934a).

Discussion. This genus is poorly known, and its systematic status remains uncertain (Emry, 1978: 1005). Golz (1976) suggested that *Simimeryx* might be close to early agriochoerids or to the Uintan homacodont *Mesomeryx*. The relative bunodonty of the molars and the retention of a paraconule are plesiomorphic dental features unknown in hypertragulids. Additional material is needed to enlighten the affinities of this bunoselenodont artiodactyl and to confirm whether or not it should be considered as belonging to the Hypertragulidae. Only in *Simimeryx*, is there a small anterior intermediate cusp in the upper molars, which is regarded as the protoconule (paraconule) by Stock (1934a). However, it is not improbable that the cusp is an additional fold on the anterior wing of the protocone because it looks quite different from a true paraconule of other selenodont artiodactyls and ancient ungulates (Vislobokova, 2001).

Genus *Parvitragulus* Emry, 1978

Type (and Unique) Species and Type Specimen. *P. priscus* Emry, 1978, from Flagstaff Rim, Natrona County, Wyoming (Chadronian); USNM 243971, right maxillary with P3–M3.

Geographic Distribution. North America.

Stratigraphic Range. Late middle Eocene (Duchesnean) to late Eocene (Chadronian).

Definition (adapted from Emry, 1978). Smaller than all genera except *Hypisodus;* no diastemata between P2/2 and the other teeth, upper premolars bearing a lingual protocone, p3 and p4 with a paraconid, p4 with a distinct metaconid and a small hypoconid, cingula less developed than in *Hypertragulus* and *Nanotragulus*.

Dimensions. m2: L = 4.6 mm; W = 2.6 mm (Emry, 1978).

Discussion. Because of the relative complexity of its premolars, *Parvitragulus* is clearly distinctive from *Hypertragulus* and *Nanotragulus*. Webb (1998) tentatively placed *Parvitragulus* and *Simimeryx* as successive sister taxa of the other hypertragulids, whereas Vislobokova (2001) proposed to remove these two genera together with *Praetragulus* into

the Praetragulidae. The characteristics of this family are more extensively discussed by Vislobokova (2001); this new taxonomic configuration implies the middle Eocene dispersal of early ruminants from Asia to North America.

PHYLOGENETIC RELATIONSHIPS

Ruminantia, first defined on the basis of living taxa, have been widely recognized as a monophyletic group. The basic subdivision of the Ruminantia into the infraorders Tragulina and Pecora (Flower, 1883) is usually accepted and is supported by molecular data (Hassanin and Douzery, 2003). Tragulina have long been considered as a paraphyletic assemblage of stem ruminants that includes ancestry of Pecora (Janis, 1987; Janis and Scott, 1988; Scott and Janis, 1993). Except for the Tragulidae, most of the traguline taxa occupying basal position in ruminant phylogeny are now extinct and diversely documented in the fossil record. We here provide a manual cladogram (Fig. 15.2) and a directly testable hypothesis about phylogenetic relationships within extant and fossil tragulines based on morphological characters (Figs. 15.3 and 15.4).

A range of cranial, dental, and postcranial features was chosen to be as objective as possible, resulting from both direct observation and the literature. A taxon-character matrix including 70 characters regarded as a priori homologous was compiled; character descriptions are provided in Appendix 15.1, and the matrix is provided in Appendix 15.2.

The ingroup taxa cover the eight different families treated in the systematic part of the chapter, with tragulids representing the only extant taxon. Rooting and outgroup comparisons were based on two families, the Diacodexeidae and the Leptictidae. All characters were equally weighted. The characters 37, 38, 42, 46, 51, 55, 58, 61, and 70 were ordered, and no other constraints were applied on the analysis. Data were managed using NEXUS Data Editor version 0.5.0 (Page, 1999), and phylogenetic reconstruction was performed with PAUP*4.0β10 (Swofford, 2002). Although nonpreserved and inapplicable character states are distinct, both were treated as "missing" by PAUP. A heuristic search for the most parsimonious trees using a randomized stepwise addition (1,000 replications with randomized input order of taxa) was made, and tree-bisection reconnection (TBR) branch-swapping options were applied. The analysis generated four equally most-parsimonious trees with a total of 120 steps (consistency index [CI] = 0.716, retention index [RI] = 0.693). The strict and majority rule consensus trees are represented in Fig. 15.3. The topology depicted in the strict consensus tree (Fig. 15.3A) supports the clade Hypertraguloidea (Hypertragulidae plus Praetragulidae), and Lophiomerycidae and Archaeomerycidae consistently emerge as very basal groups of ruminants. This basal position of the two families within ruminants is also concordant with their relatively early appearance during the middle Eocene. Tragulidae are not closely related to Lophiomerycidae as suggested in several studies (Sudre, 1984; Métais et al., 2001) and supported in our manual

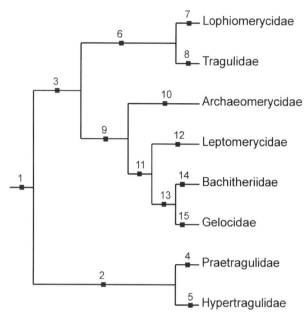

Fig. 15.2. Manual cladogram showing hypothetized relationships within traguline ruminants. Main characters (defining each taxa) at nodes.
(1) Ruminantia: reduction and loss of upper incisors, inclusion of lower canines in incisor row, fusion of cuboid and navicular. (2) Hypertraguloidea: strong medial concavity of posterior edge of palate, lack of complete postorbital bar, orbital part of lacrimal enlarged, mastoid exposure mainly lateral, premolar row shortened, reduction of mesostyle on upper molars, retention of trapezium. (3) Traguloidea: weaker medial concavity of posterior edge of palate, P1 lost, premolar row rather long, dorsal curvature of basipodium, fused magnum and trapezoid. (4) Praetragulidae: lower canines not completely included in incisor row, enlarged upper canines, loss of P1 and p1, radius and ulna separate, tibia and fibula separate. (5) Hypertragulidae: longer snout, nasals projecting anteriorly between premaxillae, caniniform P1 and p1, loss of mesostyle on upper molars, radius and ulna partly fused, tibia and fibula fused distally, pentadactyl manus. (6) Nasals projecting anteriorly between premaxillae, mastoid exposure mainly lateral, upper incisors lost, enlarged upper canines. (7) Lophiomerycidae: orbits apparently open posteriorly, small and noninflated auditory bullae, fenestrae vestibuli small, promontory sulcus absent, trigonid of lower molars open mesiolingually, metatarsals separate. (8) Tragulidae: orbits closed, optic foramina fused, fenestrae vestibuli larger, promontory sulci present, auditory bullae enlarged and strongly inflated; malleolar bone tends to be fused with tibia, astragalus with parallel proximal and distal trochleae, metatarsals III and IV fused.
(9) Postorbital bar closed posteriorly, small and noninflated auditory bullae (10) Archaeomerycidae: very short snout, nasals not projecting anteriorly between premaxillae, narrow braincase, mastoid exposure mainly lateral, retention of upper incisors, incomplete inclusion of lower canine in incisor row, very weak selenodonty, premalleolar bone condition of tibia (complete tibia with thin shaft and enlarged distal end), pentadactyl manus. (11) More elongated snout, nasals projecting anteriorly between premaxillae.
(12) Leptomerycidae: almost fused optic foramina, strongly procumbent i1, small caniniform p1, postorbital bar formed half of frontal and half of jugal, metatarsals III and IV fused. (13) Postglenoid foramen enclosed. (14) Bachitheriidae: postorbital bar formed mostly from jugal, bifurcated paracone of P4, large caniniform p1, talonid of p3 and p4 consist of two parallel walls, metatarsals III and IV incompletely fused. (15) Gelocidae: postorbital bar mostly from frontal, fenestra vestibule large, metatarsals III and IV fused.

cladogram (Fig. 15.2), but they appeared as a stem group of a clade including the North American Leptomerycidae and the European Bachitheriidae and Gelocidae. Moreover, despite the numerous basicranial features considered, this analysis does not support the inclusion of *Archaeomeryx* in the Leptomerycidae as convincingly demonstrated by Webb and Taylor (1980), mostly on the basis of basicranial characters. On the majority rule tree (Fig. 15.3B), the Lophio-

merycidae become the sister group of Archaeomerycidae, but their basal branching remains unresolved, and the Tragulidae shift as the sister group of a clade comprising the most protopecoran families.

PALEOBIOLOGY OF EARLY RUMINANTS

Body Size and Locomotion Patterns

The body size of an animal is important in determining its metabolic requirements. Early ruminants are among the smallest ungulates ever recorded, and this has broader implications in their paleobiology. Their size varies from that of a small rabbit (*Hypisodus*) to that of a musk deer (*Prodremotherium*). Most of them were probably not true ruminants physiologically, as the minimum weight for an animal to develop a viable rumination process is around 5 kg (Van Soest, 1982). The small size and relatively restricted locomotor function of the earliest ruminants may indicate a relative ecological specialization within the herbivorous guild during the middle Eocene. The best modern analogue from an ecological point of view is certainly represented by tragulids living confined to moist tropical forests of Southeast Asia and Central Africa. The postcranial morphology of early ruminants primarily shows various degrees of cursorial adaptations in the different families.

Hypertraguloids and traguloids were ecomorphologically distinct from each other with a primitive and fairly conservative limb morphology in a number of the earliest representatives, and a clear trend toward cursoriality, which is marked in leptomerycids and gelocids. Important transformations occurred in the postcranial morphology of traguloids throughout the Eocene–Oligocene transition.

The postcranial morphology of *Archaeomeryx* shows an example of a transition from a primitive rebounding jump of early eutherians to the cursorial locomotion typical of the majority of ruminants except hypertraguloids (Vislobokova and Trofimov, 2002). Postcranially, and considering hypertraguloids apart, *Archaeomeryx* is undoubtedly the most primitive ruminant because of the retention of the ulna and fibula, complete metapodals 2 and 5 (and probable retention of the metacarpal 1).

The Hypertragulidae possessed an original association of primitive and derived postcranial features (Webb and Taylor, 1980). The reduction of the stylopodium is accompanied by elongation of the zeugopodium with a tendency to fusion of the two elements in the forelimb (an autapomorphic feature of *Hypertragulus*). The reduction of the fibula diaphysis and the fusion of tarsal elements (mostly the cuboid and navicular) strengthened the basipodium and limited lateral motions, whereas the parallel astragalar trochleae along with the tranverse extension of the sustentacular facet on the astragalus favored the parasagittal movement of the hind limb. Similarly, the fusion of magnum and trapezoid in the carpus and the loss of the trapezium (except in *Hypertragulus*) contributed to making the osteomuscular system of the basipodium more rigid. The

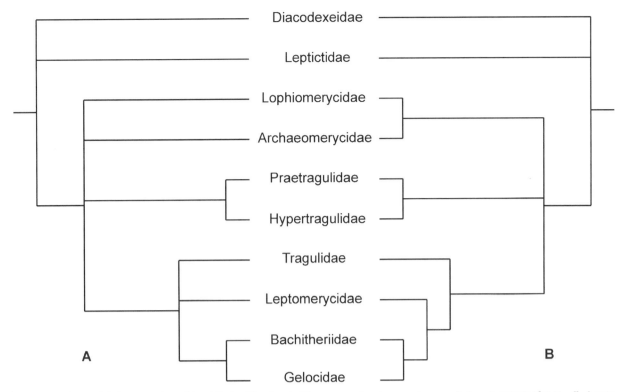

Fig. 15.3. Hypotheses of phylogenetic relationships within basal Ruminantia. These trees are based on a parsimony analysis run in PAUP*4.0β10 (Swofford, 2002) using 70 cranial, dental, and postcranial characters (Appendix 15.1). (A) Strict consensus of four equally most parsimonious trees (TL = 120; CI = 0.716; RI = 0.693); B, Majority rule consensus.

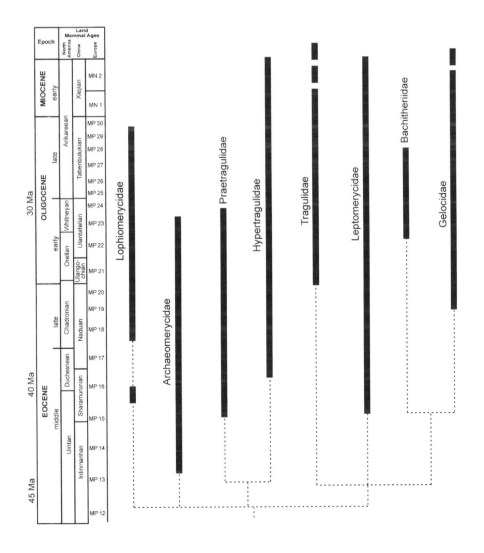

Fig. 15.4. Phylogenetic relationships of early ruminants based on the character-taxon matrix of Appendix 15.1, combined with the stratigraphic ranges of each family. Solid lines are temporal distributions of taxa documented by fossils; dashed lines are inferred presence given the presumed phylogeny. This diagram shows the importance of ghost lineages that certainly bias our phylogenetic inferences (e.g., Tragulidae).

elongation of the autopodium and the reduction of lateral digits are also observed (except in *Hypertragulus*, which keeps five-toed and [secondarily?] shortened autopodial pattern).

Metatarsals 3 and 4 are not completely fused, as seen in the Pecora, but they are closely associated along their entire length in advanced gelocids such as *Prodremotherium*. All the tragulina possessed five or four toes in the manus, unfused metapodials, and relatively unspecialized postcranial features.

Diet and Habitats

Structure of the masticatory apparatus, temporomandibular joint, and tooth morphology of early ruminants suggest a herbivorous diet. Early ruminants were most likely browsing herbivores, and they certainly had an important frugivore component in their feeding habits because of their small size and their probable inability to digest a large quantity of the structural part of plants, which requires fermentation. Similar to extant tragulids, early ruminants could be forest dwellers and probably mixed feeders; most of them were likely selective browsers and also ate berries, small fruits, and seeds (Nowak, 1999; Janis, 1984). Most extant ruminants are browsers (ca. 80%), but tragulids remain relatively omnivorous and have essentially an opportunistic feeding behavior. Ruminants became the dominant large-sized elements of the herbivorous guild after the replacement of perissodactyls that were dominant during the early Paleogene (Cifelli, 1981).

There are, however, significant differences in the skull and dentition of these forms that seem to indicate differences in their dietary preferences. The elongated first incisor of leptomerycids may have played a role in grooming by analogy with extant ruminants. The lower canine or the first premolar can be tusklike, but the role of teeth specialized in this fashion is not known; the canines do show a significant separation into two size classes, but sexual dimorphism in early ruminants is, at this time, not clearly demonstrable. However, the large upper canines of hypertraguloids were possibly used as sexual display as in extant tragulids and moschids.

Archaeomeryx possesses the most primitive enamel microstructure among ruminants, which could be inherited from primitive eutherians (Vislobokova and Trofimov, 2002). In most tragulines the dental enamel is not structured in Hunter-Schreger bands as in most herbivorous mammals, suggesting a low level of specialization to herbivory (Vislobokova and Dmitrieva, 2000), though several trends of increasing enamel durability have been traced (Vislobokova and Dmitrieva, 2000; Vislobokova and Trofimov, 2002).

The dental morphology indicates browsing habits in early ruminants, whereas most of the earlier or contemporaneous dichobunoids were large-spectrum herbivores. It is interesting to note that Janis (2000b) hypothesized that the diversification of traguloid by the middle Eocene may have been linked to their granivorous diet (occupied today by bunolophodont granivorous rodents). When the climate

changed during the late Eocene to a drier mix of forest and grassland (Berggren and Prothero, 1992; Prothero, 1994) in temperate latitudes, such as North America and Europe, ruminants responded by becoming larger and by developing more crests on their molars. Oligocene forms display a clear trend toward mesodonty and more crested premolars and molars.

Evolutionary Patterns and Climate

Both paleobotanical and fauna diversity data indicate an increase in seasonality during the later Eocene. In North America, concomitantly with the climatic deterioration, tropical forests became more restricted toward the lower latitudes, whereas more decidous-type forests appeared in the higher latitudes (Wing, 1998). The late middle Eocene is also marked by the extinction of many groups of condylarths and the appearance of several groups of modern mammals, including several families of rodents, carnivorans, and artiodactyls (Janis, 2000b). The same climatic pattern can be observed in Europe during the late middle Eocene (Collinson and Hooker, 1987), with similar effects on the biodiversity of mammals, although faunal exchanges with the Asian mainland were fairly limited. These climatic and diversity patterns are still poorly known in Asia, but a severe cooling is observed throughout the late Eocene and early Oligocene epochs (Meng and McKenna, 1998). Even if these climatic events that characterized the later Eocene were favorable to the diversification of early ruminants, the first appearance of basal ruminants (earlier in the Eocene) in Asia may not be directly related to these abiotic events and may be linked to the peak in mean annual temperatures in northern latitude of the late early Eocene.

It is important to note that the emergence of selenodont artiodactyls in North America during the middle late Uintan corresponds to both an immigration event from Asia and the start of the global cooling that affected the terrestrial ecosystems during the late Eocene and early Oligocene (Berggren and Prothero, 1992). According to Janis (2000a), the late middle Eocene would have mostly been marked by dryer but not cooler climate, thus explaining the radiation of early ruminants. However, the sparse fossil record of dichobunoids and stem ruminants in Asia does not allow testing of this hypothesis. The small size and relatively restricted locomotor function of the earliest ruminants may indicate a relative ecological specialization within the herbivorous guild during the middle Eocene.

From the middle Eocene into the Oligocene, mammalian communities of the three Holarctic landmasses were severely restructured (e.g., Prothero, 1994). The resulting increase of diversity of browsing and cursorial herbivores seems to be correlated with the emergence of ruminants. Both biotic and abiotic factors have strongly influenced the emergence of Ruminantia, although the earliest stages of their evolutionary history still remain obscure.

The turnover and increase of ruminant diversity are well traced at the Eocene–Oligocene transition, with cooler and dryer climatic conditions. This increase in diversity is particularly well marked in Europe because of the invasion of Asian taxa during the *Grande Coupure*. The environmental changes that occurred around the Eocene–Oligocene transition forced changes in herbivorous and cursorial adaptations of early ruminants. Archaeomerycids are replaced by more derived gelocids, bachitheriids, and protopecorans in the Old World. In the New World, the changes during that time were less expressive, and only praetragulids (*Parvitragulus*) did not pass through the Eocene–Oligocene boundary.

CONCLUSION

The early stages of the evolutionary history of the most diversified group of living ungulates remain obscure, and their phylogenetic relationships with other Eocene groups of Selenodontia, are still enigmatic. Vislobokova (2001) suggested an ancestry close to earliest ungulates such as *Protungulatum* in regard to the plesiomorphic basicranial features of *Archaeomeryx*. However, the fossil record of early Selenodontia remains poorly documented, notably in Asia where cetartiodactyls are thought to have originated. The relatively sudden appearance of Ruminantia in the late middle Eocene of Asia and North America has led most authors to correlate the emergence of this group with the important climatic change that affected the northern continent at the end of the Eocene. The occurrence of archaeomerycids in various Irdinmanhan faunas of Asia indicates that the first appearance of ruminants or protoruminants in Asia predates the first record of leptomerycids and hypertragulids in North America (Fig. 15.1). Much work is needed, especially in Asia, to understand the interrelationships within the Tragulina as well as their branching pattern within the Cetartiodactyla. Likewise, the phylogenetic relationships within early Pecora and the emergence of the living families remain confusing prior to the middle Miocene. The questions of the phylogenetic origin and sister taxon of ruminants are difficult to address in regard to the relative morphological homogeneity of middle Eocene artiodactyls and the lack of adequate documentation in the early and middle Eocene of Asia.

APPENDIX 15.1:
MORPHOLOGICAL CHARACTERS

The appendix defines the 70 morphological characters used for cladistic analyses. Characters in bold have been ordered. Lower-case letters designate elements of the lower dentition, upper case letters elements of the upper dentition.

Cranial Characters (36)

1. Postorbital bar:
   Open (0)
   Close (1)
2. Postorbital formed by:
   Mostly jugal (0)
   Mostly frontal (1)
   Both (half–half) (2)
3. Lacrimal:
   Small facial and orbital parts (0)
   Expanded facial exposure (1)
4. Jugal:
   Lowly situated, weakly extended anteriorly, with a long posterior spine (0)
   Higher situated, strongly extended anteriorly, with a short posterior spine (1)
5. Contact premaxilla–nasal:
   Present (0)
   Absent (1)
6. Mastoid exposure:
   Lateral (0)
   Occipital (1)
7. Position of the mastoid foramen:
   Lateral (0)
   Posterodorsal (1)
8. Size of the mastoid foramen:
   Small (0)
   Moderate (1)
   Large (2)
9. Tympanohyal vagina:
   Small, posterolateral (0)
   Moderate, subcentral (1)
10. Fenestra vestibuli:
    Small (0)
    Large (1)
11. Promontory sulcus:
    Present (0)
    Absent (1)
12. Subarcuate fossa:
    Deep (0)
    Moderately deep (1)
    Shallow (2)
13. Fossa for stapedial muscle:
    Narrow, and posteriorly situated (behind the fenestra vestibuli) (0)
    Displaced anteriorly to the level of the fenestra vestibuli (1)
14. Lateral wall of epitympanic recess:
    Mostly formed by the petrosal (0)
    Mostly formed by the squamosal (1)
15. Stylohyoid vagina:
    Shallow, broadly open posteriorly, and situated between the auditory bulla and tube (0)
    Bulla deeper, encroached on, with sharp lateral border (1)
    Bulla deeper, narrower, encroached on and enclosed posteriorly (2)
16. Optic foramen:
    Fused (0)
    Separate (1)
17. Postglenoid process:
    Present (0)
    Absent (1)

18. Postglenoid foramen:
    Enclosed in the auditory bulla (0)
    Laterally open (1)
19. Foramen ovale:
    Small, ovate (0)
    Slitlike (1)
20. Ethmoidal fissure:
    Absent or small (0)
    Well developed (1)
21. Infraorbital canal:
    Lowly situated and small (0)
    Higher situated and larger (1)
22. Medial concavity of posterior edge of palate:
    Accentuated (0)
    Reduced (1)
23. Position of the posterior opening of the infraorbital
    canal:
    Higher than the sphenopalatine foramen (0)
    Opposite or lower than the sphenopalatine foramen (1)
24. Infraorbital canal opens in:
    Maxilla (0)
    On the suture separating the lacrimal and maxilla (1)
    On the triple point where lacrimal, palatine, and
    maxialla connect (2)
25. Palatine foramina situated:
    Anteriorly (0)
    Medially (1)
    Both locations on the palatine (2)
26. Jugular foramen:
    Not confluent with the posterior lacerate foramen (0)
    Confluent with the posterior lacerate foramen (1)
27. Basioccipital:
    Elongated, poorly expanded posteriorly (0)
    Shorter, expanded posteriorly (1)
28. Basisphenoid:
    Elongated, poorly expanded posteriorly, and strongly
    convex ventrally (0)
    Shorter, expanded posteriorly, and weakly convex or
    flat ventrally (1)
29. Alisphenoid:
    Poorly expanded laterally (0)
    Expanded laterally (1)
30. Alisphenoid canal:
    Present (0)
    Absent (1)
31. Pterygoid canal:
    Present (0)
    Absent (1)
32. Pterygoid process:
    Low, with strongly oblique posterior edge (0)
    Higher, with weakly oblique posterior edge (1)
33. Ventral border of the mandible:
    Convex (0)
    Flattened (1)
    Convex anteriorly and concave posteriorly (2)
34. Angular process of the mandible:
    Strongly convex posteriorly (0)

Moderately convex posteriorly (1)
    Weakly convex posteriorly (2)
35. Coronoid process of the mandible:
    High with oblique anterior border (0)
    High with subvertical anterior border (1)
    High with vertical and slightly convex anterior
    border (2)
36. Articular process of the mandible:
    Low (0)
    High (1)

Dental Characters (18)

37. Upper incisors:
    Present (0)
    Vestigial (1)
    Absent (2)
38. P1:
    Present (0)
    Absent (1)
39. P2:
    Subconical (0)
    Three labial cups, no lingual cusp (1)
    Three labial cusps, and lingual extension (2)
40. Premolar row:
    Shortened (0)
    Elongated (1)
41. Mesostyle on upper molars:
    Present (0)
    Absent (1)
42. Upper molar paraconule:
    Well developed (0)
    Vestigial (1)
    Absent (2)
43. i1–i3:
    Procumbent (0)
    Nonprocumbent (1)
44. i1:
    Spatulate (0)
    Fan shaped (1)
    Enlarged, procumbent (2)
45. Lower canine:
    Incisiform (0)
    Caniniform (1)
46. p1:
    Present (0)
    Absent (1)
47. p1 shape:
    Tusklike (0)
    Leaflike (1)
    Small, conical (2)
48. p1 position:
    Coalescent to p2 (0)
    Distant from p2 by a short diastema (1)
    Distant from p2 by a long diastema (2)
49. p4 metalophid:
    Absent (0)

Present (1)

50. Trigonid on lower molars:
    Open mesiolingually (0)
    Closed (1)

51. Lower molar paraconid:
    Present (0)
    Vestigial (1)
    Absent (2)

52. Lower molar *Dorcatherium* fold:
    Absent (0)
    Present (1)

53. Lower molar metastylid:
    Absent (0)
    Present (1)

54. Lower molar postentocristid:
    Absent (0)
    Present (1)

Postcranial Characters (16)

55. Radius and ulna:
    Separate (0)
    Partly fused (1)

56. Facet for the triquetrum on the distal articular surface
    of the radius:
    Present (0)
    Absent (1)

57. Trapezoid and magnum:
    Separate (0)
    Fused (1)

58. Trapezium:
    Present (0)
    Absent (1)

59. Metacarpals III and IV:
    Separate (0)
    Partly fused (1)

60. Metacarpals II and V:
    Nonreduced (0)
    Reduced (1)

61. Metacarpal I:
    Present (0)
    Absent (1)

62. Tibia and fibula:
    Separate (0)
    Partly fused (1)
    Reduced to a malleolar bone (2)

63. Fibular facet on the calcaneum:
    Present (0)
    Absent (1)

64. Metatarsals III and IV:
    Separate (0)
    Partly fused (1)

65. Metatarsals II and V:
    Slightly reduced (0)
    Strongly reduced (1)
    Lost (2)

66. Shape of the fibular facet on calcaneum:
    Large and convex (0)
    Concave (1)
    Large proximal convexity and small distal concavity (2)

67. Cuneiforms II and III:
    Separate (0)
    Fused (1)

68. Ectomesocuneiform and cubonavicular:
    Separate (0)
    Fused (1)

69. Trochlea of astraglus:
    Nonaligned (0)
    Aligned (1)

70. Cuboid and navicular:
    Fused (0)
    Separated (1)

APPENDIX 15.2: MATRIX OF MORPHOLOGICAL CHARACTERS

Matrix of the 70 morphological characters coded for each family of early ruminants and relevant outgroups (Diacodexeidae, Leptictidae). Characters were coded 0-1-2, polymorph character states are (A) either state 0 or state 1; (B) the character has both states (0 and 1); (C) either state 1 or state 2; or (D) the character has both states (1 and 2). "?" and "–" indicate missing data and nonapplicable character, respectively (see text for further details).

	1	2	3	4	5	6	7	8	9	10	11	12	13	14	15	16	17	18	19	20	21	22	23	24	25	26	27	28	29	30	31	32	33	34	35	36	37
Praetragulidae	0	–	1	1	0	0	0	?	?	1	0	0	0	0	?	?	0	?	?	?	0	0	?	?	?	0	?	?	0	0	?	?	1	1	2	0	2
Hypertragulidae	0	–	1	1	0	0	0	1	0	1	0	0	1	0	2	1	0	0	0	1	0	0	0	1	1	0	0	0	0	0	?	0	2	1	1	0	2
Lophiomerycidae	0	–	0	?	?	0	?	?	?	0	1	?	0	1	0	1	0	1	?	0	0	1	0	0	2	1	0	1	0	1	1	?	2	0	0	0	?
Tragulidae	1	0	1	1	B	0	0	1	0	1	0	1	1	1	2	0	1	1	0	0	1	1	1	2	2	1	1	1	1	1	B	1	1	2	1	1	2
Archaeomerycidae	1	1	0	0	0	1	0	2	1	0	1	1	0	0	0	1	0	1	1	0	0	1	0	0	1	0	0	0	1	1	0	2	0	0	0	0	0
Leptomerycidae	1	2	1	0	1	0	2	1	0	0	0	0	1	?	1	0	0	1	1	1	?	1	1	2	1	1	1	1	0	1	0	0	0	1	?	1	2
Bachitheriidae	1	1	1	1	0	1	1	0	?	?	?	?	?	?	?	?	?	?	0	?	1	0	1	?	?	1	1	1	?	?	?	?	?	1	2	1	2
Gelocidae	1	1	1	1	0	1	1	0	?	1	0	2	?	?	?	?	0	0	?	?	1	1	?	?	1	1	?	?	?	?	?	?	2	2	?	?	2
Diacodexeidae	0	–	?	?	0	0	0	0	?	?	0	–	?	?	?	?	0	1	0	0	0	?	0	?	?	0	0	0	0	1	?	0	?	?	?	0	?
Leptictidae	0	–	?	?	0	0	0	0	?	?	0	–	?	0	2	1	0	1	0	0	0	?	0	0	2	0	0	0	0	0	1	0	2	0	0	0	0

continued

APPENDIX 15.2: *continued*

	38	39	40	41	42	43	44	45	46	47	48	49	50	51	52	53	54	55	56	57	58	59	60	61	62	63	64	65	66	67	68	69	70
Praetragulidae	1	?	0	B	2	1	?	0	1	–	–	?	1	2	0	0	1	1	1	0	?	0	0	?	?	?	0	?	0	1	0	0	0
Hypertragulidae	0	0	0	1	2	1	0	0	0	0	1	0	1	2	0	0	1	1	0	0	0	0	0	0	1	1	0	0	0	1	0	0	0
Lophiomerycidae	1	1	1	0	2	1	1	0	0	1	B	1	0	C	A	0	0	0	?	1	0	0	0	?	0	1	0	1	1	0	–	1	0
Tragulidae	1	1	1	0	2	1	1	0	B	–	–	–	1	2	1	0	A	0	0	1	1	B	1	1	D	1	1	1	1	1	1	1	0
Archaeomerycidae	1	1	1	0	2	0	2	0	0	2	1	1	1	2	0	0	0	0	1	1	0	0	0	0	0	1	0	0	2	0	–	0	0
Leptomerycidae	1	2	1	0	2	0	2	0	0	2	1	1	1	2	0	A	1	0	1	1	1	1	0	1	1	2	0	1	2	2	1	0	1	0
Bachitheriidae	1	2	1	0	2	1	1	0	0	0	2	–	1	2	A	A	1	0	?	1	1	0	?	1	2	0	1	2	2	1	0	1	0
Gelocidae	1	2	1	0	2	1	1	0	0	?	B	1	1	2	0	1	A	0	0	1	1	B	1	1	2	0	1	2	2	1	0	1	0
Diacodexeidae	0	0	0	1	0	1	?	1	0	1	?	0	–	0	0	0	0	0	?	0	0	0	0	0	0	–	0	0	–	0	0	0	1
Leptictidae	0	0	0	1	0	1	–	1	0	–	1	0	–	0	0	0	0	0	?	?	0	0	0	0	0	–	0	0	–	0	0	0	1

ACKNOWLEDGMENTS

We thank J. Sudre (Murviel-lès-Montpellier) for his insightful comments, J. J. Hooker (NHM, London), P. Tassy (MNHN, Paris), and M. R. Dawson, K. C. Beard, and A. R. Tabrum (CMNH, Pittsburgh) for access to collections in their care. Thanks to Mary Dawson for her comments on the chapter. We are grateful to Don Prothero and Scott Foss for inviting us to contribute to this volume. I.V.'s research was supported by the boards of the Russian Federation (SSch 1840.2003.4). G.M. thanks the Fondation Singer-Polignac (Paris) and the Carnegie Museum of Natural History (Pittsburgh) for their financial support. An anonymous reviewer and John Barry provided helpful comments on this chapter.

16

Family Tragulidae

MODERN TRAGULIDS ARE A GROUP of small-sized artiodactyls including the smallest living hoofed mammal (Figs. 16.1, 16.2). They are nearly unknown to the public, although they are often captured, kept as pets or laboratory animals, and even bred in zoos. These enigmatic little creatures with enlarged upper canines in males are fascinating for study: on the one hand they are similar to all the other living ruminants (and therefore included in the suborder Ruminantia), but on the other hand they differ strikingly, which led us to grade them as primitive. Arising from their shy, nocturnal lifestyle in the dense undergrowth of tropical rainforests, they are hard to observe, and only fragmentary knowledge exists about their behavior. Their tiny, graceful appearance somewhat resembles a deer or antelope without antlers or horns, which results in the names mouse deer or chevrotains (French = juvenile goat) in common speech. Two genera live in two relict areas in Southeast Asia: Myanmar, Thailand, Vietnam, Malaysia, Sumatra, Indonesia, and the Philippines (*Tragulus,* the Lesser Malay mouse deer, and the Greater Malay mouse deer); and India and Sri Lanka (*Moschiola,* the Indian chevrotain). A third genus lives in the West and Central African rainforest from Sierra Leone to Uganda (*Hyemoschus,* the water chevrotain). With this small number of genera and species, tragulids are one of the least-diversified groups among existing ruminants with the exception of the Antilocapridae and Moschidae. All species are endangered in terms of loss of habitat by human destruction. *Hyemoschus* of Ghana is even included in the CITES (Convention on International Trade in Endangered Species of Wild Fauna and Flora) list of species because its survival is currently threatened.

The ancestors of these creatures were much more widely distributed in Eurasia and Africa: they were more diversely adapted, often with several sympatric species,

Fig. 16.1. A male lesser mouse deer, *Tragulus javanicus,* with the distinctive enlarged upper canines (photo by D. R. Prothero).

and were also more common. Because their present distribution is disjunct and their physique has remained primitive, they fulfill the criteria for being designated as "living fossils" (Janis, 1984; Thenius, 2000).

BIOLOGY OF TRAGULIDS

In general, studies on the biology of tragulids are rare. Although *Tragulus* has been examined in several physiological and ethological aspects, there are few studies on its behavior in the wild. *Hyemoschus* has been investigated both in the wild and in captivity, but its physiological processes have only been deduced. No physiological or behavioral studies on *Moschiola* exist at all.

Tragulids are nocturnal or crepuscular, as indicated by their comparatively large eyes. With their arched back and short limbs, they are well adapted to slip through the dense undergrowth in various forests. During the day, they rest in the vegetation, *Tragulus* sometimes in the branches of low trees. Their dark-colored coats, plain brown in *Tragulus* and with white stripes and spots in *Moschiola* and *Hyemoschus,* greatly affect camouflage. With their small body size, they are easy and important prey for snakes, crocodiles, birds of prey, and forest-dwelling cats.

Hyemoschus is always near water, and the animal takes to the water when pursued, hiding there under waterweeds or overhanging banks and roots of trees (Dubost, 1978). In this respect *Hyemoschus* occupies a particular ecological niche (forest-dwelling, fruit-eating, terrestrial, aquaphilious, small-sized, and nocturnal), unique in the African continent (Dubost, 1978). It is the ecological Old World equivalent of the Paca, a caviomorph in tropical forests of South America (Dubost, 1968; Eisenberg and McKay, 1974). Within *Tragulus* and *Moschiola,* some species prefer wetter and some drier areas in densely covered habitats (Pickford et al., 2004; Groves and Meijaard, 2005), but none is revealed to be as water-loving as *Hyemoschus.*

Hyemoschus is known to live alone (Dubost, 1978), and individuals of *Tragulus* also lead a largely solitary life (Eisenberg and McKay, 1974). They do not possess marking glands as do most other ruminants, but only preputial and anal ones. Urine and feces presumably serve to mark home ranges, particularly in males. *Tragulus* occasionally mark twigs with a chin gland. Ralls et al. (1975) recorded stamping as communication between widely separated individu-

Fig. 16.2. Skeleton of a male *Tragulus javanicus* stored at the Naturhistoriska Riksmuseet Stockholm. Phalanges are partly missing. Length 35 cm (photo by Mikael Axelsson).

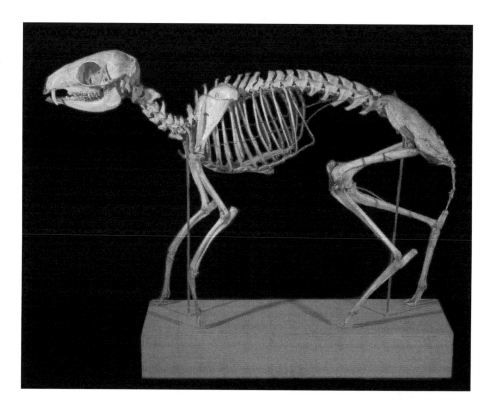

als of *Tragulus* and found that various vocalizations during encounters form an important social interaction. The same is true for the barking alarm of *Hyemoschus* (Dubost, 1978). Territorial and ritualized fighting behavior is primitive. The males stand in an antiparallel and not in a head-to-head stance (Dubost, 1965; Ralls et al., 1975). With their sabre-like elongated upper canines, the opponents try to slash each other's neck, sides, and bellies, causing long gashes on head and body (Fig. 16.3). Often death is a consequence, as the decreasing ratio of males to females with increasing age in the wild leads us to assume (Dubost, 1978). A toughened skin forms an effective shield on the back to protect the animal from antagonistic strokes (Dubost and Terrade, 1970). It ensures protection of the viscera both during intraspecific attacks and as the animal slips through the dense undergrowth. The same functions can be assumed for the bony dorsal shield in the pelvic region of the male *Tragulus javanicus* and *T. napu* (Milne-Edwards, 1864: pl. IV, Fig. 2, 2a; Dubost and Terrade, 1970: Fig. 3).

Tragulids become sexually mature at 9–26 months (Dubost, 1975). The courting and mating behaviors are primitive in many respects, e.g., absence of "Flehmen" in males (curling of lips in response to urine licking) (Dubost, 1965; Ralls et al., 1975). In the female there is a so-called diffuse placenta similar to that of camels and pigs (Milne-Edwards, 1864: pl. VII, Fig. 1-2; Turner, 1876; Strahl, 1905) and unlike the cotyledonous placenta of ruminants. One infant is produced at a time, which does not follow its mother but hides in dense vegetation (Eisenberg and McKay, 1974). Mothers nurse in a unique position: they stand and raise the hind leg on the side toward the infant (Ralls et al., 1975). Tragulids live for 8–12 years in the wild; in captivity, a lifespan of 16 years and more has been known (Jones, 1993).

Tragulids search for food on the ground. They are unable to stand on their hind legs to browse on vegetation as higher ruminants may do (Dubost, 1978). Their food consists of easily digestible forage providing an extremely high percentage of protein (fallen fruit and seeds, flowers, leaves, shoots, petioles and stems, and mushrooms) (Nordin, 1978; Dubost, 1984). This feeding strategy indicates that tragulids are so-called "concentrate-selectors." *Hyemoschus* is also known to eat invertebrates, fish, small mammals, and carrion occasionally (Dubost, 1964).

Metabolic processes of tragulids differ markedly from those of higher ruminants (e.g., Nolan et al., 1995). Investigations have been made into body temperature (Whittow et al., 1977), food intake (Nordin, 1978), digestibility characteristics (Paden and Nordin, 1978), and body-water turnover (Kamis, 1981). The gross morphology of their stomach is similar to that of pecoran concentrate-selectors (Agungpriyono et al., 1992), but tragulids lack or have only a small third of the four chambers of pecorans, the omasum, a condition evaluated as primitive (Langer, 1974). The name-giving digestion strategy in Ruminantia, the chewing of the cud (Latin = ruminare), is perfomed by tragulids also. However, the rumen, the first chamber enlarged from a forestomach performing cellulose fermentation in symbio-

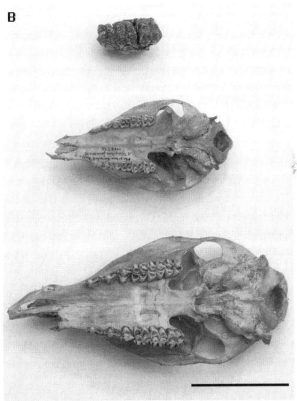

Fig. 16.3. (A) Lateral/labial view of skulls and upper dentition of fossil and extant tragulids. From above to below: *Dorcatherium crassum*, left upper molar row from the fossil site of Sandelzhausen (middle Miocene, Germany), Bayerische Staatssammlung für Paläontologie und Geologie in München 1959 II 4144; female *Tragulus javanicus*, provenance unknown, Bayerische Staatssammlung für Paläontologie und Geologie in München 1952 I 22; male *Hyemoschus aquaticus*, canines are missing, Makokou/Gabun, property of Gérard Dubost (Paris), H 30? 15-3-63 (photo by Georg Janssen). Scale bar equals 5 cm. (B) Ventral/occlusal view of the specimens from A (photo by Georg Janssen). Scale bar equals 5 cm.

sis with bacteria, is comparatively small (Milne-Edwards, 1864: pl. V, Fig. 1-5; Carlsson, 1926). The selection of a very digestible diet allowing rapid fermentation and swift passage through the gut (Kay, 1987) guarantees rapid and efficient digestion (Agungpriyono et al., 1992), necessary in tragulids because of their small body size and thus their requirement for high energy per kilogram body weight. In contrast, their ability to digest higher-fiber diets is more limited than that of larger, more derived ruminants (Bernard et al., 1994).

Tragulids exhibit the lowest daily water intake known for ruminants of the humid tropics (Macfarlane et al., 1974), which must be associated with their relatively dry and well-formed fecal pellets (Nolan et al., 1995). Nevertheless, they maintain a higher water content in their bodies than in all other ruminants (Panaretto, 1968). Tragulids are known to have a gallbladder and appendix. Their body contains very little fat and an extremely high percentage of muscle as well as specialized anatomical proportions (e.g., gut more than 25% of body mass) (Vidyadaran et al., 1983).

MORPHOLOGICAL TRAITS AND EVALUATION OF SYSTEMATIC POSITION WITHIN ARTIODACTYLS

Modern tragulids stand 20–40 cm high at the shoulder with males being smaller than females. *Hyemoschus aquaticus* is the largest, with a body length of 80 cm and a body mass of roughly 8–15 kg, and *Tragulus kanchil* is the smallest with a length of 45–55 cm and a mass of 1.5–2.5 kg. Their body is rabbit-like with an arched back and a short neck. The limbs are short, as well, with four-toed feet. Radius and ulna are unfused. The fibula is reduced to the malleolar bone, a small compact bone fused with the distolateral end of the tibia of *Tragulus*. Further fusions of bones are present in all tragulids with a cubonavicular formed of cuboid and navicular in the tarsus and a magnum of trapezoid and capitate in the carpus. In addition, the cubonavicular is fused with the ectomesocuneiform. The trapezium has been lost. The calcaneum has a nearly flat, slightly concave fibular facet. The metapodials comprise four complete elements after metacarpal and metatarsal I were lost. The median metatarsals (III and IV) are fused into a cannon bone. In *Tragulus* even metacarpals II and IV are fused as in pecorans. The relative proportions of metapodials and toes differ. In living tragulids they are longer and more slender in *Tragulus* and *Moschiola* compared with *Hyemoschus* (Smit-van Dort, 1989; Groves and Meijaard, 2005) and even show differences from species to species, indicating diverse adaptations to the habitat and locomotion. The side metapodials are completely present, slender in *Tragulus* and *Moschiola* and strong in *Hyemoschus*, but not functional except perhaps on swampy ground. A peculiarity is the bony dorsal shield of the male *Tragulus napu* and *T. javanicus* (Milne-Edwards, 1864, pl. IV, Fig. 2, 2a; Gray, 1869) above the pelvis, forming an internal turtle-like carapace.

The skull has nearly equal preorbital and postorbital proportions. The very large orbit has only one lacrimal foramen in its rostral rim. The postorbital bar is complete and primarily composed of the jugal. The snout is tapered. The auditory bulla is expanded and cancellous (with the exception of the hollow bulla of *Moschiola*). The closed plan of the facial skull lacks a lacrimal fossa in general. An antorbital vacuity is present in *Moschiola* and *Hyemoschus*. The premaxilla and nasal are in contact in *Tragulus* and *Moschiola* but separated in *Hyemoschus*. Cranial appendages are absent. The corpus of the mandible is low and without the shallow incision of the pecorans rostral to the angle.

Although the upper incisors are lost, the upper canine is saber-like and even enlarged in males, used for intraspecific combats (see above). Two-thirds of the canine is housed in a long maxillary alveolus, which curves back above the cheek tooth row, close to the nasomaxillar contact, to the level of P2 or P3. The lower canine is incisiform closely positioned to the third incisor. The cheek teeth are generally low crowned, lowest in *Hyemoschus* and a little higher in *Moschiola* and *Tragulus*. In the lower and upper premolar rows, the first tooth (P1 and p1) is totally lacking in extant species, but p1 is occasionally found in fossil species in which it has a premolariform crown, variable in length, with one or two conids and a bifurcated root (Gentry et al., 1999, Fig. 23.2). It is sometimes separated from the p2 by a small diastema. The other premolars are long and slender with one longitudinal main crest on the buccal side and a shorter lingual transverse crest that quickly turns to run toward the posterior end of the tooth. P4 is an exception, displaying a triangular shape. The crown of the molars is formed by four rounded to sickle-shaped cusps showing the so-called bunoselenodont morphology. The lower molars show a special crest complex called the "*Dorcatherium* fold." It is formed by the bifurcation of the posterior slopes of the protoconid and the metaconid resulting in a "Σ" shape and named after a fossil tragulid. The upper molars are characterized by labially projecting styles and cones.

Many of the described tragulid features (short limbs with unfused full-length lateral digits II and V, and with unfused or partly fused central metapodials; incomplete distal keel on metapodials; elongate, relatively narrow astragalus with the distal articulation pulley medially deflected; short and peglike odontoid process of the axis; lower premolars without lingually projecting cusps) are evaluated as plesiomorphic (primitive) because they are found in the primitive ruminant families Amphimerycidae (Europe), Archaeomerycidae (Central Asia), Leptomerycidae, and Hypertragulidae (North America), which are all documented from middle to late Eocene deposits (Webb and Taylor, 1980; Janis, 1987; Gentry and Hooker, 1988). Because tragulids lack derived pecoran characters, they are placed at the most basal systematic position within living Ruminantia. Based on these plesiomorphic features, the appearance of living tragulids is considered to resemble that of early ruminants. The cubonavicular fusion and the incisiform lower canines are shared derived features of all members of the suborder Ruminantia. Within this the infraorder Tragulina, including the family Tragulidae, is a paraphyletic assemblage of primitive ruminants (e.g., Webb and Taylor, 1980; Janis, 1987; Gentry and Hooker, 1988; Gentry, 2000a). The Tragulina is considered a sister group to the "higher" ruminants, mainly taxa bearing cranial appendages (deer, antelopes, cattle, giraffes, and pronghorns), which are today a highly diverse radiation grouped as the infraorder Pecora.

Unique (autapomorphic) features that separate the tragulids from all other ruminants are a malleolar bone fused with the distal end of the tibia, an ectocuneiform fused with the cubonavicular in the tarsus, no postglenoid process, a very small external exposure of the mastoid, as well as a closed postorbital bar. Exceptions are known in fossil tragulids, however (Gentry, 1978; pers. obs.).

A primitive status of tragulids within Ruminantia can be assumed not only from morphological features but also from aspects of the physiology, behavior, and ecology of the living representatives. Because similarities with peccaries are obvious (e.g., Milne-Edwards, 1864: 117; Janis, 1984), they comply with conditions for a transitional position between Suoidea and Ruminantia. This positioning is also supported by the observation that from peccaries through tragulids to ruminants, the forestomach common to all these groups increases in complexity (Langer, 1988). From an analysis of their muscle antigens, Duwe (1969) even concluded that the tragulids are most closely related to the tayassuids. If this is so, the loss of the upper incisors, the incisiform lower canine, and the cubonavicular would be traits developed in parallel with the pecorans, but much earlier.

Extensive descriptions of the tragulid skeleton and dentition can be found in Milne-Edwards (1864), Pilgrim (1910, 1915), Carlsson (1926), Mottl (1961), and Fahlbusch (1985).

ORIGIN, PHYLOGENY, INTRAFAMILIAR SYSTEMATICS, BIOGEOGRAPHY, AND PALEOECOLOGY

In 1864 Milne-Edwards established the family Tragulidae within the Ruminantia based on the structure of the stomach, dentition, and feet. He separated the Asian *Tragulus* and African *Hyemoschus* from the Moschidae, another living ruminant group without cranial appendages.

Investigations of extant Tragulidae since the first half of the nineteenth century have served to establish the generic distinction between the African *Hyemoschus* and the Asian *Tragulus* (e.g., Gray, 1843; Milne-Edwards, 1864), based on clear morphological differences in the skeleton and dentition, still in common use. Although the monotypic status of *Hyemoschus* has not been challenged (however, three subspecies have been named), a constant disagreement about the taxonomic composition of the Asian representatives has persisted. Since the 1940s, the opinion has prevailed that the genus *Tragulus* comprises two species in Southeast Asia (*T. napu* and *T. javanicus*) and one species (*T. meminna*) in India.

Since then, a generic distinction between a Southeast Asian genus and an Indian genus has been recommended, *Tragulus* and *Moschiola* respectively (Groves and Grubb, 1987), based on differences in the skeleton (Flerov, 1931; Smit-van Dort, 1989), and coat color (Pocock, 1919). Recently, Meijaard and Groves (2004a) have recognized six species within *Tragulus* (*T. javanicus, T. williamsoni, T. kanchil, T. nigricans, T. napu, T. versicolor*) with 24 subspecies, and Groves and Meijaard (2005) followed on with three species in *Moschiola* (*M. indica, M. meminna, M. kathygre*) and a potential fourth species (*?Moschiola* sp.).

Within the extant tragulids, *Hyemoschus* is regarded as the most primitive (e.g., Carlsson, 1926; Smit-van Dort, 1989). The more slender limbs, higher crowned teeth, and less intense water dependence of the Asian genera seem to resemble the more pronounced adaptation of advanced artiodactyls to drier and more open landscapes. In addition, body, especially limb, proportions of *Tragulus* and *Moschiola* are different and correlate with different habitat conditions (Smit-van Dort, 1989; Groves and Meijaard, 2005). In any case, the relict aspect of tragulid disjunct geographic distribution reflects their complex history.

The restriction of extant tragulids to the tropical climate zone can be explained by the availability of fruit for at least 9 months of the year. It is also known that *Hyemoschus* prefers very low seasonality in areas with rainfall equal to or greater than 1,500 mm per year, and it does not occur in areas that are even moderately seasonally arid (Dubost, 1978).

The origin of the Tragulidae is still a mystery. For a long time the geologically oldest fossils known were specimens of the genus *Dorcatherium* (Fig. 16.4) from lower Miocene deposits of Africa and Europe (Gentry et al., 1999; Pickford, 2001a). Thus, the evolutionary history of tragulids before the early Miocene remained unknown, although an Eocene appearance has been anticipated by morphological and molecular data (e.g., Webb and Taylor, 1980; Miyamoto et al., 1993). The genus *Cryptomeryx* (assessed as synonymous with *Iberomeryx* by Bouvrain et al., 1986) from the Oligocene of France and India has been considered to be a direct ancestor of the Tragulidae (Schlosser, 1902) or even the oldest tragulid (Sudre, 1984). Some years later it was suggested that *Cryptomeryx* should be included in a new family named Lophiomerycidae (Janis, 1987), which is grouped in the infraorder Tragulina with the Tragulidae and Hypertragulidae (Webb and Taylor, 1980) and whose roots can be followed to the late middle Eocene (Guo et al., 2000). Recently described early ruminant material from the late Eocene of Thailand (Métais et al., 2001) includes two mandibles with brachyodont dentition of what has been claimed to be the oldest known member of the Tragulidae, *Archaeotragulus krabiensis*. It suggests that the origin of the family should be searched for in the middle Eocene of Southeast Asia (see Métais and Vislobokova, this volume). Moreover, together with early ruminants from the middle Eocene of Myanmar (Métais et al., 2000), *Archaeotragulus* testifies to a contempo-

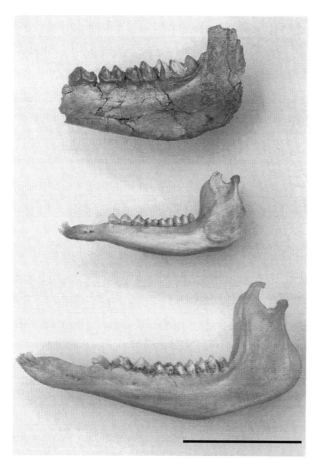

Fig. 16.4. Left jaws and lower dentition of fossil and extant tragulids. From above to below: *Dorcatherium crassum*, jaw fragment with fourth premolar to third molar from the fossil site of Sandelzhausen (middle Miocene, Germany), Bayerische Staatssammlung für Paläontologie und Geologie in München 1959 II 4146; female *Tragulus javanicus*, provenance unknown, Bayerische Staatssammlung für Paläontologie und Geologie in München 1952 I 22; male *Hyemoschus aquaticus*, Makokou/Gabun, property of Gérard Dubost (Paris), H 30? 15-3-63 (photo by Georg Janssen). Scale bar equals 5 cm.

raneous occurrence of a possible tragulid representative alongside lophiomerycids and opens the possibility of a common origin for both groups. This direction of thought is supported by a diverse, mainly undescribed fauna of early ruminants with still unknown familial affiliation from the middle Eocene of Myanmar and China, reflecting an early ruminant radiation (Métais et al., 2000; Tsubamoto et al., 2003; Métais et al., 2005). The same age of origin, or even earlier, has been proposed by recent molecular data (44.3–46.3 Ma; Hassanin and Douzery, 2003).

Apart from the Eocene *Archaeotragulus*, no tragulid fossils are known from the Paleogene. This fact still leaves a gap in the known evolution of tragulids until their sudden and widespread appearance in the lower Miocene of Southeast Asia (Mein and Ginsburg, 1997), Africa (Witworth, 1958; Hamilton, 1973), and Europe (Mein, 1989). The majority of fossil tragulids, including all of the European and African ones as well as some Asian forms, are referred to the genus *Dorcatherium,* established in the very first description of a fossil or living member of this family (Kaup, 1833) (Table

16.1). *Dorcatherium* comprises bunoselenodont to selenodont species of various body sizes. Its strong similarity with the extant *Hyemoschus* has often been mentioned (e.g., Gentry, 1978b), and the two genera have sometimes been combined. In fact, comparison of dentition and limb structure of *Dorcatherium* with those of extant tragulids shows a greater resemblance to *Hyemoschus* than to *Tragulus* or *Moschiola,* but it is not congruent in all details. Additionally, only a single *Dorcatherium* skull ever has been recorded (Kaup, 1933), and the taphonomic damage to this specimen obscures its original morphology and proportions, which makes it nearly impossible to compare cranial details pertinent to a synonymization with *Hyemoschus.* In its more selenodont cheek teeth, lacking p1 and a cingulum, more robust jaws, and lack of contact between premaxilla and nasals, *Hyemoschus* contrasts with *Dorcatherium.*

As yet the genus *Dorcabune* Pilgrim, 1910 is restricted to several Miocene Asian members of the family plus one Pleistocene species and a European probable occurrence in the late Miocene of Crete (Van der Made, 1997a) (Table 16.1). Blunt cusps of the cheek teeth were originally given as the diagnostic feature. The current morphological range in the genus resembles that of *Dorcatherium* in that there are bunoselenodont as well as selenodont species of different sizes. A revision of both genera is urgently needed. It would most probably result in two main lineages of Miocene tragulids with an obvious ecomorphological differentiation into bunoselenodont omnivores and selenodont herbivores (Mottl, 1961; Fahlbusch, 1985; Qiu and Gu, 1991), adaptations which might have appeared several times in parallel in different radiations producing a much more complex phylogenetic pattern. In fact, the interspecific relationships are difficult to reconstruct because usually we are faced with sudden appearances of tragulid assemblages of several sympatric species (Pilgrim, 1915; Mottl, 1961; Fahlbusch, 1985).

Two additional fossil genera, more recently established, show more similarities with the extant Asian genera. *Siamotragulus* is an early to middle Miocene genus from South and Southeast Asia (Thomas et al., 1990) with simpler lower premolars than *Dorcatherium* or *Dorcabune* and more advanced limb structures with completely fused median metapodials and more reduced side metapodials. Thus, it strongly resembles *Tragulus,* although it clearly differs by its still-separate isolated malleolar bone. These traits indicate a further Miocene but exclusively Asian lineage. Its relationship with the extant *Tragulus* has not yet been discussed. The Chinese *Yunnanotherium* (Han, 1986) of the late Miocene is characterized by small body size, upper cheek teeth with a weak cingulum, and an incompletely developed *Dorcatherium* fold; it thus also shares more features with the extant Asian tragulids and might belong to a lineage leading directly to them.

Unfortunately, Pliocene and Pleistocene tragulid specimens are nearly unknown and are more seldom described (e.g., Bakalov and Nikolov, 1962), which keeps the later history of the family in obscurity. An upper molar from the early Pliocene in Kenya recently referred to *Hyemoschus aquaticus* (Pickford et al., 2004) would be better placed in

Table 16.1. Tragulid species and their spatiotemporal distributions

Species	Distribution
Archaeotragulus krabiensis Métais, Chaimanee, Jaeger and Ducrocq 2001	Late Eocene, Thailand
Dorcatherium crassum (Lartet, 1851)	Latest early to late Middle Miocene, Europe
Dorcatherium guntianum von Meyer, 1846	Latest early to middle middle Miocene, Europe
Dorcatherium vindobonense von Meyer, 1846 (= *D. rogeri* Hofmann, 1909)	Latest early to middle middle Miocene, ?early late Miocene, Europe
Dorcatherium peneckei (Hofmann, 1893)	?Latest early Miocene, middle Miocene, Europe
Dorcatherium naui Kaup and Scholl, 1834	Late Miocene, Europe
Dorcatherium jourdani (Déperet, 1887)	Late Miocene, Europe
Dorcatherium puyhauberti Arambourg and Piveteau (1929)	Late Miocene, Europe
Dorcatherium bulgaricum Bakalov and Nikolov, 1962	?Pliocene, Bulgaria
Dorcatherium pigotti Whitworth, 1958	Early to middle Miocene, Africa
Dorcatherium chappuisi Arambourg, 1933	Early to middle Miocene, Africa
Dorcatherium parvum Whitworth, 1958	Early Miocene, East Africa
Dorcatherium songhorensis Whitworth, 1958	Early Miocene, East Africa
Dorcatherium iririensis Pickford, 2002	Early Miocene, East Africa
Dorcatherium moruorotensis Pickford, 2001	Latest early Miocene, East Africa
Dorcatherium libiensis Hamilton, 1973	Latest early to earliest middle Miocene, North Africa, Arabia
Dorcatherium sp. in Pickford, Senut and Mourer-Chauviré, 2004	Early Pliocene, East Africa
Dorcatherium orientale Qiu Zhanxiang and Gu Yumin, 1991	Middle Miocene, China
Dorcatherium minimus West, 1980	Earliest late Miocene, Pakistan
Dorcatherium nagrii Prasad, 1970	Late Miocene, Pakistan, India
Dorcatherium minus Lydekker, 1876	Late Miocene to early Pliocene, Pakistan, India, China
Dorcatherium birmanicus (Noetling, 1901)	Late Miocene to early Pliocene, Myanmar
Dorcatherium majus Lydekker, 1876	Late Miocene, Pakistan
Dorcabune welcommi Ginsburg, Morales, and Soria, 2001	Early Miocene, Pakistan
Dorcabune anthracotheroides Pilgrim, 1910	Late Miocene, Pakistan
Dorcabune nagrii Pilgrim, 1915	Late Miocene, Pakistan
Dorcabune sindiense Pilgrim, 1915	Late Miocene, Pakistan
Dorcabune progressus (Yan, 1978)	Late Miocene, China
Dorcabune liuchengense Han, 1974	Early Pleistocene, China
Siamotragulus sanyathanai Thomas, Ginsburg, Hintong, and Suteethorn, 1990	Middle Miocene, Thailand
Siamotragulus bugtiensis Ginsburg, Morales, and Soria, 2001	Early Miocene, Pakistan
?*Siamotragulus indicus* (Forster Cooper, 1915)	Early Miocene, Pakistan
Yunannotherium simplex Han, 1986	Late Miocene, China
Tragulus javanicus (Osbeck, 1765)	Extant, Southeast Asia
Tragulus napu (G. Cuvier, 1822)	Extant, Southeast Asia
Tragulus williamsoni Kloss, 1916	Extant, Southeast Asia
Tragulus kanchil (Raffles, 1821)	Extant, Southeast Asia
Tragulus nigricans Thomas, 1892	Extant, Southeast Asia
Tragulus versicolor Thomas, 1910	Extant, Southeast ASia
Tragulus ?*sivalensis* Lydekker, 1882	Late Miocene, India
Moschiola indica Gray, 1852	Extant, India
Moschiola meminna Erxleben, 1777	Extant, Sri Lanka, Dry Zone
Moschiola kathygre Groves and Meijaard, 2005	Extant, Sri Lanka, Wet Zone
?*Moschiola* sp. in Groves and Meijaard, 2005	Extant, Sri Lanka
Hyemoschus aquaticus (Ogilby, 1841)	Extant, West Africa

Note: Modified from Pickford (2001a).

Dorcatherium because of its strong cingulum, which is lacking in *Hyemoschus*.

The Tragulidae have always been an exclusively Old World family. Miocene occurrences are reported from Southeast and South Asia (e.g., Pilgrim, 1915; Forster-Cooper, 1915; Prasad and Satsangi, 1968; West, 1980a; Han, 1986; Thomas et al., 1990; Zhanxiang and Yumin, 1991; Gaur, 1992; Guo et al., 2000; Ginsburg et al., 2001), Southeast, West, and North Africa (e.g., Hopwood, 1929; Walker, 1969; Whitworth, 1958; Hamilton, 1973; Pickford, 2001a, 2002) as well as Europe (e.g., Kaup, 1833; von Meyer, 1846; Thenius, 1952; Rinnert, 1956; Mottl, 1961; Bakalov and Nikolov, 1962; Hünermann, 1983; Fahlbusch, 1985; Ginsburg, 1989; Ginsburg et al., 1994; Rössner, 2004) and testify to an early geographic division between Eurasian and African family branches. Their disjunct distribution in the modern world is thereby revealed to be a breakup of an earlier, much wider distribution extending over vast regions of Eurasia and Africa. Additionally, they were more common and more diverse in paleocommunities than they are today, and recently an adaptive success of tragulids, as even serious competitors to higher ruminants, in occupying a newly present wetland area in the middle Miocene of Central Europe has been documented (Rössner, 2004). There the tragulids used the drying Eurasian relict basins of the Paratethys, side arm of the Tethys, with limnofluvial environments to immigrate to Central Europe and were quantitatively much better represented than the Pecora.

ACKNOWLEDGMENTS

I thank Donald Prothero for inviting me to contribute to this volume. I am grateful to Gérard Dubost (Paris), who put a *Hyemoschus* skeleton (extremely rare in public collections) in his possession at my disposal and who reviewed the paragraphs on tragulid biology. Alan Gentry (London), Colin Groves (Canberra), and Christine Janis (Providence) gave helpful comments on the scientific content and language on an earlier draft of this chapter. Donald Prothero took the photo for Figure 16.1; Mikael Axelsson (Stockholm) took the photo for Figure 16.2; and Georg Janssen (Munich) took the photos for Figure 16.3.

17

Family Moschidae

THE FAMILY MOSCHIDAE, or the musk deer (Flerov, 1952; Groves, 1976; Nowak, 1999; Pyrchodko, 2003), today consists of only one living genus, *Moschus* (Fig. 17.1) with five or six living species (Groves et al., 1995). However, the moschids have an extensive fossil record in the Oligocene and Miocene of Eurasia and an endemic Miocene radiation in North America, known as the blastomerycines. Living *Moschus* are relatively small deerlike creatures, with body lengths around a meter in adults, and weighing only 7–17 kg. They lack any kind of cranial appendages, but instead, the males have prominent dagger-like upper canines. Today they inhabit the mountainous forests of eastern Asia, especially China, Siberia, Korea, Manchuria, Mongolia, Tibet, Kashmir, and Nepal. Because they are nocturnal and secretive and live in high-altitude forests, very little is known about their population biology. Most of their systematics (Groves et al., 1995) must necessarily be based on skulls and skins because few species have been extensively studied in the wild. For decades, only Linnaeus' type species, *Moschus moschiferus*, was recognized, but Groves et al. (1995) now recognize the following additional species: *M. berezovskii* from the central Chinese mountains and plateaus and Vietnam; *M. fuscus* from the coniferous forests of Tibet and Yunnan Province; *M. chrysogaster* from the eastern Tibetan plateau; *M. leucogaster* from Bhutan, Sikkim, southern Tibet, Nepal, and northern India; and the problematic form ?*M. cupreus* from high altitudes in Kashmir. Su et al. (2001) have shown that *Moschus anhuiensis* is a valid species in addition and that the Siberian musk deer, *M. moschiferus*, is the sister species to the several Chinese and Himalayan forms.

The natural history of the musk deer has been described on the basis of field research by Green (1987) and by Yang et al. (1990). Living *Moschus* tend to be solitary

Fig. 17.1. The living species *Moschus fuscus* (photo courtesy Colin Groves).

(except during rutting season) and extremely territorial. Males mark their turf by rubbing scent glands against stones, trees, and vegetation. They are very timid creatures and usually flee when humans are present. They feed on a variety of plant matter, especially young shoots, leaves, flowers, grasses, buds, mosses, and lichens. Their name derives from the musk gland in the male animal that is used to mark its territory. This musk has become highly prized in many Asian societies because it can be used for soaps and perfumes; its biochemical structure is described by Bi et al. (1985) and Sokolov et al. (1987). In Chinese medicine, moschid musk is used for the treatment of fever, sore throats, and rheumatism as well as many other maladies. Gram for gram, the musk is more valuable than gold or cocaine or rhino horn (currently worth more that US$45,000 a kilogram), one of the most expensive natural substances in existence. Consequently, the living moschids are now endangered because of extensive poaching for their musk (even though the females and juveniles have no musk glands, and are worthless to poachers). Experimental programs in Sichuan, China, have shown that captive musk deer can be bred, and the adult males can have their musk extracted without harming them. However, this farming of musk deer has not yet been widely adopted because captives do not breed well or live long, and the poaching pressure is still intense. According to several estimates, wild populations have declined by more than 50% in the past few decades as a result of poaching (Groves, 1976; Nowak, 1999; Prychodko, 2003).

SYSTEMATICS

The relationships of the Moschidae to other ruminant artiodactyls are still highly controversial. Traditionally, most authors (e.g., Gray, 1821; Brooke, 1878; Matthew, 1934; Stirton, 1944; Simpson, 1945; Flerov, 1952; Viret, 1961; Romer, 1966; Hamilton, 1978b; Eisenberg, 1981) have regarded moschids as the sister group of cervids or the link between cervids and tragulids because they possess some cervid synapomorphies (such as a closed metatarsal gully) but lack others (such as the double lacrimal orifice). Webb and Taylor (1980) argued that they were the sister group of all higher ruminants (bovids, giraffids, antilocaprids, and cervids), whereas Leinders and Heintz (1980), Moyà-Solà (1986, 1987), Groves and Grubb (1987), Janis and Scott (1987), Gentry (2000a), and Janis (2000a, 2000b) placed moschids as sister group to the cervids plus antilocaprids. Su et al. (1999) studied the cytochrome *b* gene sequence and suggested that cervids and moschids were sister groups. Hassanin and Douzery (2003) argued that the morphological evidence allied the moschids with the cervids plus bovids. This view was also supported by their mitochondrial and nuclear DNA sequence analysis, which actually placed moschids closer to bovids than to cervids. However, Fernández and Vrba (2005) combined a larger data set of molecular, morphological, and ethological information and found that the moschids were monophyletic and the sister group of cervids.

Because the family has only one living genus, the family definition might coincide with the generic definition of *Moschus*. Adding the fossil genera means that the definition of the family must be modified. Janis and Scott (1987: 65) defined moschids as cervoids with a closed metatarsal gully, a *Palaeomeryx* fold on the brachydont members of the family, a raised lip on the cubonavicular facet, an entostyle and a metastyle on the upper cheek teeth, and a P3 with a lingually directed protocone, but moschids have only a single lacrimal orifice and no cranial appendages. The fossil forms have the gelocid condition of a small metaconule on M3, but they are more derived than many ruminants in having a fused metatarsal V.

The cladistic relationships within the moschids are given by Janis and Scott (1987: Fig. 19) and Webb (1998) and so are not reprinted here.

Dremotherium *Geoffroy Saint-Hilaire, 1833*

Dremotherium is the oldest fossil taxon recognized as a moschid by most paleontologists (Sigogneau, 1968; Janis and Scott, 1987: 62; Ginsburg et al., 1994; Gentry et al., 1999). First appearing in the late Oligocene (MP 28), it is represented by the type species, *D. feignouxi,* in MN 2 and by *D. quercyi* in MP 28–30, *D. guthi* in MP 28–30, and *D. cetinensis* in MN 2. *Dremotherium* apparently vanished before the late early Miocene. It also occurs in the early Miocene of Asia (Vislobokova, 1997). *Dremotherium* was a relatively small to medium-sized moschid with no cranial appendages but a very long facial region, an orbital cavity positioned centrally between the braincase and face, a single lacrimal orifice, lacrimal fossae and antorbital vacuities anterior to the orbits, and very long canines with open roots running deep in the skull and maxilla along the nasal border almost to the orbit. Its teeth have a fairly advanced tooth crown pattern with p1 absent. It has elongated cervical vertebrae and a long basioccipital and high occipital region, suggesting that these creatures fed by browsing high vegetation, like the modern gerenuk (Janis and Scott, 1987: 64).

As discussed by Janis and Scott (1987: 62–64) *Dremotherium* and *Amphitragulus* have often been confused with one another. In some collections, all the skull material was labeled *Amphitragulus,* and all the postcranials were labeled *Dremotherium.* Janis and Scott (1987) argue that *Dremotherium* is more derived than *Amphitragulus* in certain features and that it is a moschid. On the other hand, *Amphitragulus* has features that ally it with the Palaeomerycidae.

Pomelomeryx *Ginsburg and Morales, 1989*

This Eurasian taxon is dainty and rabbit-like in size and proportions. Its teeth, however, are relatively primitive with low crowns, poorly developed selenodonty with conical cusps, fewer and weaker internal crests on the cheek teeth, less-rounded lingual premolar wall, as well as slender P3 and p3 (Gentry et al., 1999). *Pomelomeryx* ranges through the early Miocene (MN 1 to MN 3), longer than any other early Miocene European genus. Two species are recognized, *P. gracilis,* and *P. boulangeri* (Rössner and Rummel, 2001; Fahlbusch et al., 2003). Ginsburg et al. (1994) argued that *Pomelomeryx* was actually a gelocid, but Blondel (1997) gave evidence that it was indeed a moschid descended from *Dremotherium,* and recent workers have concurred with this allocation (Rössner and Rummel, 2001; Fahlbusch et al., 2003).

Hispanomeryx *Morales et al., 1981*

When they described *Hispanomeryx duriensis,* Morales et al. (1981) regarded it as a moschid. It is known from the early late Miocene (MN 9 to MN 10) levels in Spain. Janis and

Scott (1987: 66) stated that although the upper canines were then unknown, most of the described features are clearly moschid, and the molars resemble those of *Parablastomeryx.* However, Moyà-Solà (1986: 269) described upper canines that confirm its moschid affinities. Gentry et al. (1999: 241) placed *Hispanomeryx* in the "Bovoidea" without fully justifying this decision. Without detailed evidence as to its bovoid affinities, I follow Morales et al. (1981), Moyà-Solà (1986), Janis and Scott (1987), and McKenna and Bell (1997) and place it with the moschids.

Bedenomeryx *Jehenne, 1988*

Bedenomeryx occurs from the late Oligocene (MP 29) until the early Miocene (MN 2). It is significantly larger and more heavily built than *Dremotherium,* with less molarized premolars and p1 present in some specimens (Jehenne, 1988). It also had two lacrimal orifices, which would place it with *Amphitragulus* in the Palaeomerycidae, not the Moschidae, according to Janis and Scott (1987). Gentry et al. (1999: 232) include it with the Moschidae, along with *Oriomeryx* (another taxon that Janis and Scott, 1987, regard as a palaeomerycid). Three species are recognized: the type species, *B. paulhiacensis,* from the Paulhiac locality in France (MN 1); *B. milloquensis* from the late Oligocene (MP 29–30); and *B. truyolsi* from the Cetina de Aragon locality in Spain (MN 2).

Friburgomeryx *Becker et al., 2001*

This early Miocene (MN 2) taxon is known from a few specimens from the Wallenreied section in Switzerland. *F. wallenriedensis,* the type and only species, is a small moschid with selenodont–brachydont teeth with cusps that are more conical than crescentic in shape (Becker et al., 2001: 556). The upper molars bear a neocrista, a strong parastyle, paracone, and mesostyle as well as a strongly connected cingulum. The lower molars have a *Palaeomeryx* fold and rounded lingual wall of the metaconid. The p1 is present, and the rest of the premolars are wide and bulky and poorly molarized. The p3 has a short postprotocristid and postprotoconulidcristid, and p2 has a short postprotocristid and entocristid. Becker et al. (2001: 556–557) gave other characters that distinguish *Friburgomeryx* from other primitive moschids.

Oriomeryx *Ginsburg, 1985*

Oriomeryx is a poorly known and very primitive hornless form from the earliest Miocene (MN 2) with dentition very similar to that of the most primitive palaeomerycids. Ginsburg (1985) and McKenna and Bell (1997) regard it as a very primitive palaeomerycid, but other authors (e.g., Gentry et al., 1999; Becker et al., 2001) place it in the moschids. I follow the latter taxonomic allocation here, although clearly this designation needs to be further assessed.

Hydropotopsis *Jehenne, 1991*

This poorly known taxon from the late Oligocene of France has been questionably allocated to the Moschidae

(Becker et al., 2001), although McKenna and Bell (1997: 223) synonymized it with *Amphitragulus* (which is considered a moschid by some but a palaeomerycid by others, including Prothero and Liter, this volume). According to Gentry et al. (1999: 233), *Hydropotopsis* has the cervid synapomorphy of two lacrimal orifices.

Subfamily MOSCHINAE Gray, 1821

Micromeryx Lartet, 1851

After the extinction of most European moschids, only *Micromeryx* remained and survived in small numbers in faunas ranging in age from middle to late Miocene (MN 5 to MN 11) in Europe and from the middle Miocene of Asia. *Micromeryx* was a tiny but very long-legged moschid. It had many derived cervoid features, with greater hypsodonty, long premetacristid on p4 closing the anterior valley, and a third lingual cusp on the m3. Two species are recognized: *M. flourensianus,* the type species, primarily from France and Germany and ranging from MN 5 to MN 11; and *M. styriacus* from the Goriach locality (MN 5 in age) in Austria.

Moschus Linnaeus, 1758

As noted above, *Moschus* is the only living genus of the family, with six known species. In addition to its modern distribution in eastern Asia, it is also recorded from the late Miocene and early Pliocene and Pleistocene of Asia. This fossil history connects it with all the Oligocene–Miocene species recorded from Europe as well as the Miocene blastomerycines of North America.

Subfamily BLASTOMERYCINAE Frick, 1937

As discussed by Webb (1998) and Prothero (in press), the blastomerycines were an endemic North American radiation of moschids that have been poorly understood

until recently. Frick (1937) named numerous genera, subgenera, and species of blastomerycines, usually without adequate diagnoses. According to the International Code of Zoological Nomenclature (1999), these names are often *nomina nuda* and should be abandoned, but their use by Webb (1998) and other authors has validated many of them, and they are now revised and rediagnosed (Prothero, in press). Much of Frick's (1937) taxonomy was typological oversplitting, with no statistical analysis, populational thinking, or adequate comparisons between and among species. In many cases, the only difference between Frick's (1937) species is that they come from different localities in the same formation. Over 22 species were recognized by Frick (1937); McKenna and Bell (1997) allocated them to three genera, but my recent revision (Prothero, in press) recognized eight valid species in six genera.

Blastomeryx Cope, 1877

This is the most common and oversplit of the blastomerycines, with eight species recognized by Frick (1937), of which only five (*B. gemmifer, B. elegans, B. medius, B. mefferdi, B. francesca*) pertain to *Blastomeryx.* Prothero (in press) synonymized all these species with the type species *B. gemmifer.* This genus was a medium-sized, fairly advanced blastomerycine (Fig. 17.2) first occurring in the early Miocene (late Hemingfordian) of Nebraska and then abundantly represented in the Barstovian of Nebraska, Colorado, Montana, and New Mexico. *B. gemmifer* continued until the early Clarendonian, where it is found in both Nebraska and Texas. When Cope (1874c) first described the type specimen of *Blastomeryx gemmifer* based on a single tooth, he referred it to the antilocaprid genus *Merycodus.* Cope (1877) later erected the genus *Blastomeryx* for this specimen but was still not sure of its affinities. Not until much more complete material had been described by Matthew (1908) was it clear

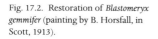

Fig. 17.2. Restoration of *Blastomeryx gemmifer* (painting by B. Horsfall, in Scott, 1913).

that *Blastomeryx* was a cervoid, but Webb and Taylor (1980) were the first to formally place it in the Moschidae.

Problastomeryx Frick, 1937

This is a very large, primitive blastomerycine with unreduced premolars (compared to the molar row length) and longer, more robust limb elements. It is also one of the first to appear in the fossil record, occurring in a number of late Arikareean localities in Nebraska, South Dakota, and Texas as well as early Hemingfordian localities in Nebraska, South Dakota, and Idaho. The type species was originally described as *Blastomeryx primus* by Matthew (1908), along with a second species, *B. olcotti*. Frick (1937) retained both species but placed them in the genus *Problastomeryx*. Prothero (in press) could not find any justification for both species, so only *P. primus* is valid.

Pseudoblastomeryx Frick, 1937

This genus is also very early (first appearing in the late Arikareean) and primitive, with unreduced premolars compared to the molars. Unlike *Problastomeryx,* however, it is distinctly smaller and does not have the primitive robust limb elements (Fig. 17.3). Although it was originally described by Matthew (1907) as *Blastomeryx advena,* Frick (1937) placed it in the new genus *Pseudoblastomeryx,* and recognized four additional species: *P. tantillus, P. falkenbachi, P. schultzi,* and *P. marsa*. Prothero (in press) could find no justification for these species and synonymized them with the type species, *P. advena*. In addition to late Arikareean specimens in South Dakota, Wyoming, and Nebraska, *P. advena* occurs in the early Hemingfordian of South Dakota, Nebraska, and possibly California as well.

Parablastomeryx Frick, 1937

This was the largest genus of blastomerycine, with many primitive features (brachydont molars, *Palaeomeryx* fold,

large premolars, short diastemata, and well-defined protocones on P2–3), yet it is not the earliest species (first occurrence in the early Hemingfordian, but still very primitive in morphology even by the late Clarendonian). Two species of *Parablastomeryx* are still considered valid: the type species, *P. gregorii,* from the late Clarendonian of Nebraska; and the fragmentary *P. floridanus* from the early Hemingfordian Thomas Farm l.f. in Florida.

Machaeromeryx Matthew, 1926

This genus was by far the smallest blastomerycine, with many highly derived features (reduced premolars, more hypsodont, and cementum on molars). *M. tragulus* (Matthew, 1926) was based on a beautiful complete skeleton (Fig. 17.4) from the late Arikareean of Nebraska, although Matthew (1926) described only the teeth, and the skeleton has still not been fully described. The only other species is *M. gilchristensis,* known from fragmentary material from the early Hemingfordian Thomas Farm l.f. in Florida.

Longirostromeryx Frick, 1937

The last and most distinctive of the blastomerycines was *Longirostromeryx*. As the name implies, the skull has an elongated rostrum and symphysis on the lower jaw, with long diastemata and great reduction of the premolar row (resulting in the loss of p2 in many specimens). A beautiful complete articulated skeleton of *L. wellsi* from the Clarendonian Ashfall Fossil Beds State Park in northeastern Nebraska is in the University of Nebraska State Museum, but it has yet to be described. It shows not only the distinctive long snout but also a very long neck and long legs, suggesting that this animal was more gazelle-like or even gerenuk-like than most of the blastomerycines. The type species, *L. wellsi,* was originally named *Blastomeryx wellsi* by Matthew (1904) because the type material was a fragmentary ramus that did not show the characteristic elongate symphysis and long diastemata. On the basis of better material, Frick

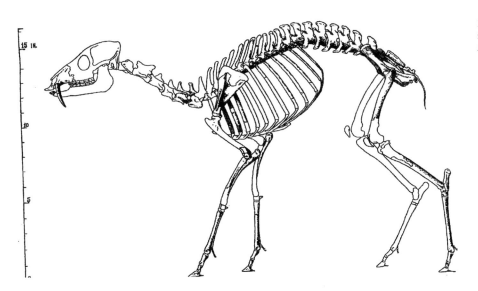

Fig. 17.3. Matthew's (1908: Fig. 1) illustration of the type specimen of *Pseudoblastomeryx advena* (AMNH 13014).

15 IN.

10

5

Fig. 17.4. Articulated type skeleton of *Machaeromeryx tragulus*. (AMNH 20548) (photo by D. R. Prothero).

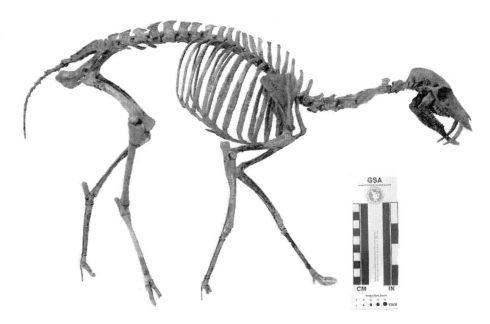

(1937) recognized this feature and erected the genus *Longirostromeryx*. Frick (1937) erected three more species (*L. merriami, L. serpentis, L. blicki*), but Prothero (in press) synonymized all of them with *L. wellsi*. *L. wellsi* occurs from the late Barstovian of Texas and New Mexico, the Clarendonian of Nebraska and New Mexico, and the late Hemphillian of Nebraska and New Mexico (the last known blastomerycine specimens). Another species, *L. clarendonensis*, is valid for much more derived smaller specimens with greater hypsodonty, more elongate diastemata, more reduced premolars, and p2 lost. It is known only from the early Clarendonian of Texas.

EVOLUTIONARY HISTORY

The earliest moschids include forms such as *Dremotherium* and *Bedenomeryx* from the Oligocene of Eurasia. By the early Miocene, they had a modest radiation in Eurasia, but most species vanished before the middle Miocene except for the tiny long-legged *Micromeryx*. This genus lasted until MN 11, after which moschids vanished from the European record and survived only in Asia. Gentry et al. (1999) regard *Walangania* from the early Miocene of North Africa as a moschid, which would indicate that the family had spread to that continent as well. But Janis and Scott (1987) gave evidence that *Walangania* was a primitive sister group to the higher cervoid radiation and not a member of the Moschidae.

During the early Miocene, moschids also spread to North America, where they diversified into the blastomerycines, a group that was widespread throughout the continent although never very abundant. In the late Arikareean and early Hemingfordian, there were the primitive (with unreduced premolars) *Problastomeryx primus* and *Pseudoblastomeryx advena* as well as the tiny but derived *Machaeromeryx tragulus*. By the late Hemingfordian, these taxa were replaced by the more derived *Blastomeryx gemmifer*, and the large, very primitive *Parablastomeryx floridanus*. In the Clarendonian, relict *B. gemmifer* and *Parablastomeryx gregorii* lived alongside the highly specialized long-snouted gazelle-like *Longirostromeryx*. The last blastomerycines (*Longirostromeryx*) vanished from North America in the late Hemphillian, very close to the end of the Miocene.

ACKNOWLEDGMENTS

I thank C. Groves, G. Rössner, and C. Janis for helpful reviews of this chapter. I thank J. J. Flynn, C. Norris, and J. Meng for access to specimens in the AMNH. Funding for museum travel for this research was provided by the Donors of the Petroleum Research Fund, administered by the American Chemical Society, and by NSF grant EAR03-09538.

Family Antilocapridae

IN A PATTERN SIMILAR TO THAT OF THE EQUIDAE, the Antilocapridae have an evolutionary history that begins with a grade of early Miocene stem taxa, historically placed in a paraphyletic "family," that leads to a diverse, monophyletic group of high-crowned taxa in the middle to late Miocene. That late Miocene diversity has dwindled to a single species, *Antilocapra americana,* today. The origin of the antilocaprids lies in the unresolved late Oligocene Asian mammalian fossil record, and a good understanding will have to await both more extensive phylogenetic research into the origins of the pecoran radiation and the discovery of new fossil species to break up the long branches at the base of the Pecora. Antilocaprids spring fully formed into the North American fossil record in the late Hemingfordian with the first appearance of fossils that already possess the apomorphy that unites the family: unshed paired, branching frontal horns located supraorbitally (Janis and Manning, 1998a). By the Hemphillian, the family was represented by 15 species in nine genera, with a diverse range of body sizes and horn morphologies that suggest a complex radiation related to changes in feeding and reproductive behavior. Today the family is represented by *Antilocapra americana,* found in the open plains and deserts of western North America (O'Gara, 1978, 1990).

Antilocapra has horns covered with a keratinous sheath that appear at first glance to be bovid-like. The "horns" are forked, however, a developmental impossibility under standard bovid horn growth. This fork in the horn sheath is what gives *Antilocapra* its common name, the pronghorn. The horn sheaths are grown and cast annually in a fashion reminiscent of antler growth and loss in cervids (and unlike the permanent sheaths of bovid horns), and this unusual horn growth has been the subject of a number of studies by O'Gara (1990), who has established the cellular mechanisms by which the horn sheaths are grown and shed. Extending the antilocaprid horn ontogeny into the fossil record and understanding how and when it began to evolve have been, and continue to be, important parts of antilocaprid systematics. Interestingly, the horn core that underlies the pronged horn sheath of *Antilocapra* is a simple blade of bone with only an anterior bump to indicate a prong, unlike the horn cores of most genera of fossil antilocaprids, which have two or more distinctively arranged horn parts.

In fact, the horn-core morphology of antilocaprids has been the main source of taxonomic characters in the group. Dental or postcranial apomorphies are not often used because of clear intergradation between characters, particularly dental characters (Stirton, 1932a; Frick, 1937; Janis and Manning, 1998a). With the current state of antilocaprid systematics, positive taxonomic identification often cannot be made without a horn core, and most of our understanding of evolutionary relationships within the clade hinges on changes in horn-core growth or dimensions. Future work, with a greater focus on the postcrania and dentition of these animals than has already been given, may be able to cast additional light on antilocaprid evolutionary relationships.

Based on the impression of blood vessels, the antler-like horns of the primitive, merycodont-grade Antilocapridae were covered only with skin, which sloughed off at full growth (Furlong, 1927; Janis and Manning, 1998a). These merycodont-grade horns possess burrs that appear similar to those of cervids but have a different growth structure (deposited on the surface of the horn, not integral to it), and no specimens are found cast off at the burr (Matthew, 1924; Furlong, 1927). In the systematic paleontology, I refer to the bony cranial appendages of the merycodont-grade antilocaprids as horns because these appendages were exposed in the life of the animal.

Antilocaprine horns were covered with a keratinous sheath, possibly evolved from a heterochronic retention of the "velvet" of the earlier "merycodont" grade (Janis and Manning, 1998a). This keratinous sheath is the synapomorphy that unites the Antilocaprinae within the Antilocapridae. The annual regrowth of horn sheaths, as in *Antilocapra*, may not have been possible for some of the more elaborately branched antilocaprines, such as *Ilingoceros* or *Hexobelomeryx*. It remains unclear whether the annual shedding seen in *Antilocapra* was widespread through antilocaprines or an autapomorphy of the extant species, but it is more likely the latter. Because decay has removed the horn covering of antilocaprines, the bony horn cores are now all we have to work with in the fossil record. In the systematic paleontology, I refer to the bony cranial appendages of the antilocaprine antilocaprids as horn cores because these appendages were covered in the life of the animal.

To emphasize homology among the diverse horn and horn-core forms within the antilocapridae, I refer to the base of the horn (core) leading to the first bifurcation as the "shaft" and the two branches of the horn (core) above the bifurcation as the anterior and posterior "tines." In that way,

modification of the ancestral antilocaprid horn form (Fig. 18.1) can be understood for all of the genera diagnosed here.

The systematics presented here is based on the work of Janis and Manning (1998a) and Kurtén and Anderson (1980). I have updated the biochronology from these authors following Tedford et al. (2004) and Bell et al. (2004) where possible. The stratigraphic information included here arises from four main sources: Kurtén and Anderson (1980), Janis and Manning (1998a), the Paleobiology Database (www.pbdb.org), and the MIOMAP database (miomap.berkeley.edu). I have tried to reconcile differences among these sources through reference to the primary literature and accept responsibility for any mistakes.

The last major revision of antilocaprids appeared in Frick's (1937) monograph on horned ruminants. Although Frick organized the state of knowledge up to that time and published several previously unknown taxa, he had an idiosyncratic method of presenting his information. Fossils were presented in several sections, each of which exclusively discusses horns, dental material, or postcrania, and there is no simple index to connect specimens to his taxonomic framework. Additionally, descriptions of his new taxa were not put forth as a unified systematic biology but, rather, spread over several sections, with incomplete cross references between them. Some taxa lack any sort of differential diagnosis, with descriptions limited to lists of specimens. Additionally, Frick (1937: 322) made explicit that his species-level taxonomy was often based on geography, without reference to morphological differentiation.

The antilocaprids need a species-level revision within a phylogenetic framework, much like the ones undertaken for canids (Wang, 1994; Wang et al., 1999). In this chapter, I

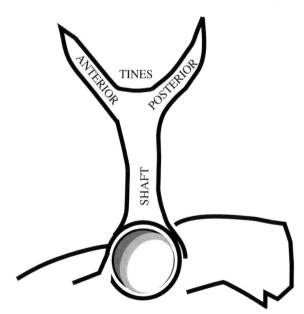

Fig. 18.1. Schematic diagram of idealized primitive antilocaprid horn structure, illustrating the shaft and anterior and posterior tines referred to in systematic descriptions.

have simply gathered a thorough literature review of the antilocaprids and have not made any attempt to revise the clade. Nor have I made an attempt to revise the genera or species discussed, and all are presented simply in their state in the current literature, taken at face value.

Antilocaprids have been considered to be related to the cervoid radiation on the basis of skeletal morphology (Janis and Scott, 1988; Janis and Manning, 1998a), but in the past they had been placed with the bovids on the basis of convergently derived dental and horn developmental characters (Matthew, 1904; O'Gara, 1978, 1990). New molecular evidence places them with the Giraffidae, near the base of the pecoran ruminant radiation (Hassanin and Douzery, 2003; Hernández Fernández and Vrba, 2005), which could mean that the evolution of keratinous horns in Antilocapridae is a true convergence and not a simple parallelism with the evolution of horns in Bovidae. If this molecular hypothesis withstands additional tests, Antilocapridae would be a key lineage for understanding the origin of cranial appendages in ruminants.

LIST OF ABBREVIATIONS

Stratigraphic Names. Cr., Creek; Fm., Formation; Loc., Local; Mbr., Member; Mt., Mountain; Spr., Spring.

Biostratigraphic Designations (Miocene designations from Tedford et al., 2004). He1, early Hemingfordian; He2, late Hemingfordian; Ba1, early Barstovian; Ba2, late Barstovian; Cl1, early Clarendonian; Cl2, middle Clarendonian; Cl3, late Clarendonian; Hh1, early early Hemphillian; Hh2, late early Hemphillian; Hh3, early late Hemphillian; Hh4, late late Hemphillian; Bl, Blancan; Ir, Irvingtonian; RLB, Rancholabrean.

SYSTEMATIC PALEONTOLOGY
Order ARTIODACTYLA
Family ANTILOCAPRIDAE Gray, 1866

The grade of antilocaprids with exposed bony horns, found at the base of the family, was classically called the subfamily "Merycodontinae" (Matthew, 1909a), which was "diagnosed" by a combination of the apomorphies of the whole antilocaprid clade and the plesiomorphic trait of exposed bony horns (Fig. 18.2). The apomorphies that diagnose the Family Antilocapridae are branched, permanent supraorbital horns present, antorbital vacuity present, lacrimal fossa absent, auditory bulla hollow and somewhat inflated, and body long-limbed and gracile (Janis and Manning, 1998a). Antilocaprids, and therefore "merycodontines," have inherited many plesiomorphic characters from their pecoran common ancestor, including large orbits with complete postorbital bar, upper incisors absent, lower canine incisiform and grouped with incisors, metapodial fusion, complete distal metapodial keels, parallel-sided astragali, and reduction of fibula and ulna (Janis and Manning,

1998a). Other plesiomorphic characters that mark the "merycodont" grade are body size smaller, horn shaft round in cross section and possibly marked by one or more "burrs" at or near the base, and vestigial side toes possibly present in forelimb (Janis and Manning, 1998a). Family Antilocapridae is known from western North America from the early Hemingfordian to the present.

Genus *Paracosoryx* Frick, 1937

Diagnosis. Burr located high on shaft relative to other "merycodontines," horn shaft tends toward extreme elongation, shaft projects posteriodorsally, tips of tines reduced and recurved, anterior tine equal to or smaller than posterior (Fig. 18.3); diastema relatively short, premolars not reduced, upper canine may be preserved, cheek teeth marked by primitive characters: lower crowns, small entostyle, ectostylid, and metastylid all present; metacarpal splints present (Frick, 1937; Janis and Manning, 1998a).

Discussion. *Paracosoryx* was originally designated as a subgenus of *Cosoryx* by Frick (1937) but was elevated to genus level by Tedford et al. (2004). Janis and Manning (1998a) support the elevation but suggest that *Paracosoryx* may actually represent a basal paraphyletic grade, with some species more closely related to other genera than they are to each other. If a species-level phylogenetic analysis breaks this genus apart, the taxonomic problem will be similar to that faced by Froehlich (2002) in his work with basal equids. A rank-free approach could simplify this sort of taxonomic reorganization.

Genus Range. California (Ba1–Cl3), Colorado (He2–Ba1), Idaho (Ba2), Montana (Ba1), Nebraska (He1–Ba1), Nevada (Ba1–Cl2), Oregon (Ba1–Ba2), and Wyoming (He2–Ba2).

Included Species. *Paracosoryx wilsoni* Frick, 1937 (type of genus): Sheep Cr. Fm. (type) (He2); Olcott Fm. (Ba1). *P.* cf. *P. wilsoni:* Massacre Lake Fauna (He2); Runningwater Fm. (He1).

P. alticornis Frick, 1937: Barstow Fm., middle Mbr. (type) (He2–Ba1); Tesuque Fm., Pojoaque Mbr. (Ba2); Sheep Cr. Fm. (He2); Valentine Fm., Burge Mbr. (Cl1). *P.* cf. *P. alticornis:* "Esmeralda Fm.," Fish Lake Valley Fauna (Cl2); Wood Mt. Fm. (Ba2).

P. dawesensis Frick, 1937: Hay Sprs. Area, Sheep Cr. Fm. equivalent (type) (He2); Box Butte Fm. (He2).

P. furlongi Frick, 1937: synonymies: *Merycodus* near *M. necatus* Leidy: Merriam, 1919: 582, Fig. 234; *Merycodus* cf. *M. furcatus* Leidy: Furlong, 1927: 147, Pl. XXIV, Fig. 3; Pl. XXVI, Fig. 1; Pl. XXVII, Fig. 4. Dove Spr. Fm., Iron Canyon Fauna (type) (Cl1–Cl2); Carlin Fm. (?Ba1). *P.* cf. *P. furlongi:* Bana Fm., South Tejon Hills Loc. Fauna (Cl1); Chanac Fm., North Tejon Hills Loc. Fauna (Cl3).

P. loxoceros (Furlong, 1934): synonymy: *Merycodus loxoceros* Furlong, 1934: 4, plates 1–4. Siebert Fm., Tonopah Fauna (type) (Ba1–Ba2). *P.* cf. *P. loxoceros:* "Esmeralda Fm.,"

Fig. 18.2. Phylogeny of Antilocapridae based on North American Land Mammal Age divisions from Tedford et al. (2004) and Bell et al. (2004). Structure of phylogeny based on cladogram of Janis and Manning (1998a). *Ottoceros* probably belongs to the Ilingocerotini, but there is insufficient evidence to support a particular position within the clade. See text for potential positions of *?Antilocapra garciae* and *Ceratomeryx*.

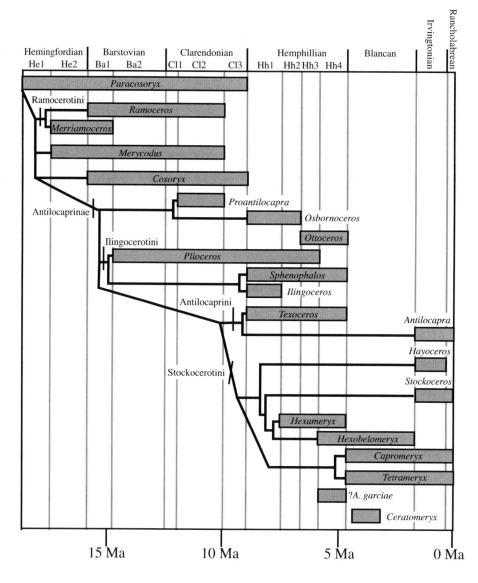

Stewart Sp. Fauna (Ba1); Butte Cr. Volcanic Sandstone, Beatty Buttes and Corral Buttes Faunas (Ba1).

?P. nevadensis (Merriam, 1911): synonymies: *Merycodus nevadensis* Merriam, 1911: 284, Figs. 64, 65. Washoe Fm., High Rock Fauna (type) (Ba2); Virgin Valley Fm., (Ba1). *?P.* cf. *?P. nevadensis*: Sucker Cr. Fm., Quartz Basin (Ba1).

P. sp.: Pawnee Cr. Fm (Ba2); Marsland Fm. (He1); Valentine Fm., Devil's Gulch Mbr. (Ba2); Six Mile Cr. Fm., McKanna Spr. Fauna (Ba1); Butte Cr. Volcanic Sandstone, Red Basin Fauna (Ba2), Sucker Cr. Fm., Sucker Cr. Fauna (He2-Ba1).

Tribe RAMOCEROTINI Frick, 1937

This tribe includes only *Ramoceros* and *Merriamoceros*, the two genera of primitive antilocaprids that have more than two tines in their horns (Fig. 18.2). That trait combined with the posterior tilt of the main horn shaft and a general similarity of preserved crania and postcrania serve to unite the taxon.

Genus *Ramoceros* Frick, 1937

Diagnosis. Depressed, postorbital three- to four-tined cervid-antler-like horns (Fig. 18.3). The main shaft of the horn is directed outwardly and posteriorly, and the tips of the tines flare widely and may express asymmetry (Frick, 1937).

Discussion. The included subgenera are diagnosed based on the length of the "secondary" shaft, the posterior branch that splits to form the novel third tine. *R. (Ramoceros)* has a long secondary shaft that ends in two or three tines (Fig. 18.3). *R. (Paramoceros)* has a reduced second shaft, resulting in an apparent trichotomy in the horns (Frick, 1937). Frick (1937: 323), in a curious phrasing, seemed to name both *R. ramosus* and *R. osborni* as type species: "[*Ramoceros*] is typified by *R. ramosus* (Cope) from New Mexico and *R. osborni* (Matthew) from Colorado, selected as the genotypic species." Janis and Manning (1998a) have taken this to mean "species" singular, modifying *R. osborni*, as I do here. Frick (1937) also described a subspecies of *R. ramosus*, *R. r. quad-*

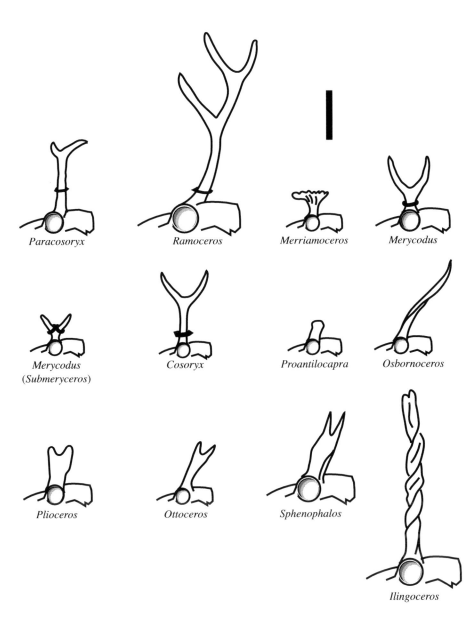

Fig. 18.3. Schematic illustrations of horn forms of primitive "merycodont" grade and some antilocaprine taxa, indicating relative sizes. Forms based on type species when possible but do not represent individual specimens. Scale bar equals 10 cm.

Paracosoryx   *Ramoceros*   *Merriamoceros*   *Merycodus*

Merycodus (Submeryceros)   *Cosoryx*   *Proantilocapra*   *Osbornoceros*

Plioceros   *Ottoceros*   *Sphenophalos*

Ilingoceros

ratus, typified by forking or incipient forking of the anterior ("brow") tine, producing a four-pointed "antler."

Genus Range. California (Ba1–Cl2), Colorado (Ba2), Kansas (Ba2–Cl1), Nebraska (Ba2–Cl1), New Mexico (Ba1–Cl2), Texas (Cl1–2), and Wyoming (Ba2–Cl1).

Included Species. *R. osborni* (Matthew, 1904) (type of genus): synonymies (Voorhies, 1990): *Merycodus osborni* Matthew, 1904: 107, Figs. 1–4, 7–16, Plate III; *Ramoceros hitchcockensis* Frick, 1937: 329; *Ramoceros howardae* Frick, 1937: 330; *Ramoceros kansanus* Frick, 1937: 331. Ogallala Group, Cedar Cr. (type) and Vim-Peets Faunas (Ba2), Sappa Cr. (Cl1); Valentine Fm.: Crookston Bridge, Devil's Gulch, and Burge Mbrs. (Ba2–Cl1). *R.* cf. *R. osborni:* Valentine Fm., Cornell Dam Mbr. (Ba2).

R. ramosus (Cope, 1874): synonymy: *Cosoryx ramosus* Cope, 1874: 148. Tesque Fm.: Pojoaque (type), Skull Ridge,

Ojo Caliente Mbrs. (Ba1–Cl1); Chamita Fm. (Cl2); Caliente Fm., Quarry Point (Cl1–2); Goliad Fm., Lapara Mbr. (Cl1–2).

R. (*Paramoceros*) *brevicornis* Frick, 1937: Barstow Fm., Barstow Fauna (type) (Ba2).

R. (*P.*) *marthae* Frick, 1937: Tesque Fm., Pojoaque Mbr. (type) (Ba2). *R.* (*P.*) ?*marthae:* Chamita Fm. (Cl2).

R. (*P.*) *palmatus* Frick, 1937: Chamita Fm. (type) (Cl2).

R. sp.: Caliente Fm., Doe Spr. Fauna (Ba2); Ash Hollow Fm., Escarpment Quarry (Ba2–Cl1); Olcott Fm. (Ba1).

Genus *Merriamoceros* Frick, 1937

Diagnosis. Small-bodied with small, short-shafted horns that end in an elongate horizontal platform with many small tines extending dorsolaterally (Fig. 18.3). Janis and Manning (1998a) describe these horns as mooselike, and point out that the mandibular dentition is marked by plesiomorphic states of the diastema and premolars. The horn morphology is the main apomorphy that diagnoses this monotypic genus.

Discussion.   *Merriamoceros* was originally named as a subspecies of *Ramoceros*, but Gregory (1942) elevated it to generic level, arguing that its palmate horn was a generically distinct character. The developmental series discussed by Frick (1937) supports the hypothesis of homology between the multifurcate horns of *Ramoceros* and *Merriamoceros*, with young individuals of *Merriamoceros* presenting three horns, reminiscent of those of *Paramoceros*. The horn ontogeny of these two genera definitely marks an evolutionarily distinct lineage.

Genus Range.   California (Ba1), Montana (Ba1), Nebraska (He2–Ba1), Nevada (He2), and Utah (He2).

Type and Only Species.   *Merriamoceros coronatus* (Merriam, 1913): synonymy: *Merycodus coronatus* Merriam, 1913: 335, Figs. 1–3. Barstow Fm., Middle Mbr. (type) (Ba1). *M.* cf. *M. coronatus:* Washoe Fm., High Rock Fauna (He2).

M. sp.: Carlin Fm. (?Ba1); Upper Sheep Cr. Fm. (He2); Olcott Fm. (Ba1); Six Mile Cr. Fm., McKanna Spr. Fauna (Ba1); Salt Lake Fm. (He2) .

Genus *Merycodus* Leidy, 1854

Diagnosis.   Horns with tines of equal to subequal size, with tines approaching or exceeding length of basal shaft (Fig. 18.3) (Janis and Manning, 1998a). *M.* (*Meryceros*) horns larger and narrower in cross section than *M.* (*Merycodus*). *M.* (*Submeryceros*) horns as in *M.* (*Meryceros*), but smaller and with individual burrs at the base of each tine instead of a single burr on the main shaft of the horn.

Discussion.   The taxonomic problems surrounding the genus *Merycodus* began with the original type of Leidy (1854), which was never figured and disappeared after Matthew (1918a). This original type was a partial lower jaw, and the absence of both this type and a diagnostic horn core inspired Frick (1937) to erect a new genus, *Meryceros*, to contain the non-*Cosoryx* species of advanced "merycodontines" in his taxonomic revision. Skinner and Taylor (1967) secured neotype material from the Bijou Hills and redescribed *M. necatus* but preserved Frick's (1937) *Meryceros* on the basis of size and narrowness of horn cross sections.

Gregory (1942) suggested that *M. warreni*, the type of Frick's (1937) *Meryceros*, was a junior synonym of *M. necatus*. He was also one of the first to suggest that Frick (1937) had oversplit the "merycodontines" at the generic level, reading too much phylogenetic importance into minor differences in horn morphology. Many of the characters of *Merycodus, Meryceros,* and *Cosoryx* discussed by Frick (1937), including the flattening and "wedging" of horn cross sections through time, suggest that this series of organisms represents a grade evolving toward the Antilocaprinae, so taxonomic revision will be futile until a species-level phylogenetic analysis clarifies our understanding of the relationships in this grade. Should *Meryceros* turn out to be a mono-

phyletic clade and not intergrade with *Merycodus*, then it is probably more closely related to the later antilocaprids than to *Merycodus*. Here I have treated *Meryceros* as a subgenus of *Merycodus*, as the relative narrowness and robustness of the horns of *Meryceros* (Skinner and Taylor, 1967) do not seem enough to warrant a generic distinction, but I wish to retain the systematic information, if any, that this distinction exists. Janis and Manning (1998a) suggest that *"Merycodus" sabulonis* should be placed in a different or distinct genus.

Originally described as a subgenus of *Meryceros* by Frick (1937), *Submeryceros* was elevated to generic level by Storer (1975) and Voorhies (1990) on the basis of its distinctive burr and horn core morphology. Janis and Manning (1998a) consider *Submeryceros* a likely junior synonym of *Merycodus*, and I am inclined to agree. It may turn out, however, that *Submeryceros* creates a distinct clade within *Merycodus*, in which case the name could be retained at its original subgeneric rank or as a clade name in a rank-free taxonomy. As with *Meryceros*, I have included it as a subgenus here in order to preserve systematic information, if any exists. *M.* (*?Submeryceros*) *minimus* and *M.* (*?Submeryceros*) *minor* were based on dentitions by Frick (1937), who only provisionally attributed them to *Submeryceros* on the basis of geography and size. The lack of diagnostic features in antilocaprid dentitions renders these species questionable in their identities and affinities.

Genus Range.   Arizona (Cl1–2), California (He2–Cl3), Colorado (He2–Ba2), Kansas (Cl2), Montana (Ba1–Hh4), Nebraska (He1–Cl3), Nevada (Ba1–Cl2), New Mexico (He1–Cl3), Oregon (Ba1–2), South Dakota (Ba1–Cl3), and Wyoming (He2-Cl1).

Included Species.   *M. necatus* Leidy, 1854 (type of genus): Fort Randall Fm. (type) (Ba2); Pawnee Cr. Fm. (Ba2); Valentine Fm., Cornell Dam Mbr. (Ba2); Wood Mt. Fm. (Ba2); Tesuque Fm., Skull Ridge Mbr. (Ba1). *M.* cf. *M. necatus:* Madison Valley Fm., Anceney Fauna (Ba1).

"M." sabulonis (Matthew and Cook, 1909): synonymies (Skinner and Taylor, 1967): *M. necatus sabulonis* Matthew and Cook, 1909: 411, Fig. 24. *M. necatus* Leidy: Matthew, 1918: 219, Fig. 18. *Cosoryx* (*Paracosoryx*) *sabulonis* (Matthew and Cook): Frick, 1937: 353, Fig. 44. *Cosoryx* (*Subparacosoryx*) *savaronis* Frick, 1937: 353, Figs. 28, 28A. Sheep Cr. Fm. (type); Split Rock Fm. (He2); Olcott Fm. (Ba1); Valentine Fm., Devil's Gulch Mbr. (Ba2); Wood Mt. Fm. (Ba2); Tesuque Fm., Ojo-Caliente Mbr. (Ba2–Cl1). *"M."* cf. *"M." sabulonis:* Pawnee Cr. Fm. (Ba2).

M. (*Meryceros*) *warreni* (Leidy, 1858) (type of subgenus): synonymies: *Cervus warreni* Leidy, 1858: 23. *Meryceros warreni* (Leidy): Frick, 1937: 361, Fig. 39. Valentine Fm., Crookston Bridge Mbr. (type) and Devil's Gulch Mbr. (Ba2); Brown's Park Fm. (Ba2); Pawnee Cr. Fm. (Ba2). *M.* ?*M. warreni:* Ogallala Gp., Kennesaw Fauna (Ba2).

M. (*Meryceros*) *crucensis* (Frick, 1937): synonymy: *Meryceros crucensis* Frick, 1937: 356, Fig. 39, 39A. Tesuque Fm., Pojoaque Mbr. (type) (Ba2), Skull Ridge Mbr. (Ba1); Chamita Fm. (Cl2).

M. (Meryceros) hookwayi Furlong, 1934: synonymy: *Meryceros hookwayi* Frick, 1937: 370. Siebert Fm. (type) (Ba1).

M. (Meryceros) joraki (Frick, 1937): synonymies: *Merycodus necatus* Leidy: Merriam, 1919: 517, Figs. 113–117, 133–134. *Merycodus necatus* Leidy: Furlong, 1927: 147, Plates 24, 25. *Meryceros joraki* Frick, 1937: 367, Fig. 39. Barstow Fm., Barstow Fauna (type) (Ba2), Second Division Fauna (Ba1). *M.* cf. *M. joraki:* Barstow Fm., Red Division Fauna (He2), Rak Division Fauna (He2), Green Hills Fauna (Ba1). *M.* cf. *M. joraki:* Caliente Fm. (Cl2).

M. (Meryceros) major (Frick, 1937): synonymy: *Meryceros major* Frick, 1937: 355, Fig. 39A. Tesuque Fm., Skull Ridge Mbr. (type) (Ba1), Pojoaque Mbr. (Ba2).

M. (Meryceros) nenzelensis (Frick, 1937): synonymy: *Meryceros nenzelensis* Frick, 1937: 359, Fig. 40. Valentine Fm., Crookston Bridge Mbr. (type) (Ba2), Devil's Gulch Mbr. (Ba2).

M. (Submeryceros) crucianus Frick, 1937 (type of subgenus): synonymies: *Dicrocerus necatus* Leidy: Cope, 1877: 353, Pl. 57, Fig. 4. *Meryceros (Submeryceros) crucianus* Frick, 1937: 370, Fig. 35. Tesuque Fm., Pojoaque Mbr. (type) (Ba2), Nambé Mbr. (He2).

M. (?Submeryceros) minimus Frick, 1937: synonymy: *Meryceros (Submeryceros) minimus* Frick, 1937: 404, Fig. 41. Tesuque Fm., Pojoaque Mbr. (type) (Ba2); Chamita Fm. (Cl2).

M. (?Submeryceros) minor Frick, 1937: synonymy: *Meryceros (Submeryceros) minor* Frick, 1937: 403, Fig. 41. Tesuque Fm., Pojoaque Mbr. (type) (Ba2); Pawnee Cr. Fm. (Ba2); Sheep Cr. (He2); Hay Sprs. (He2); Olcott Fm. (Ba1); Valentine Fm., Cornell Dam Mbr., Crookston Bridge Mbr. (Ba2); Wood Mt. Fm. (Ba2).

M. sp.: Bidahochi Fm. (Cl1–2); Green Valley Fm. (Cl3); Bopesta Fm. (He2–Ba1); Mehrten Fm. (Cl1–2); Hector Fm. (He2); North Park Fm. (He2–Ba1); Wagontongue (He1–2); Bert Creek Fauna (Ba2); Hepburn's Mesa Fm. (Ba1–2); Flint Creek Beds (Ba1); Raine Ranch Fm. (Ba2); Esmeralda Fm. (Cl2); Zia Fm. (He1); Deer Butte Fm. (Ba2); Butte Creek Fm. (Ba2); Mascall Fm. (Ba1); Oak Creek Fm. (Cl1–2); Colter Fm. (Ba2); Joe's Quarry, WY (He2–Ba1).

Genus *Cosoryx* Leidy, 1869

Diagnosis. Tall, slender horns with elongated tines, rotated so that the anterior tine is pointed somewhat medially (Fig. 18.3). Main shafts circular in cross section, with an apparent distinction between shaft and pedicle at burr; shafts directed slightly anterolaterally (Frick, 1937). Horns range in size from small to large, and though size may simply reflect a sample of multiple ontogenetic stages, there does seem to be some phylogenetic or biogeographic information in this character, with specimens from New Mexico larger than those from the Great Plains (Frick, 1937). *C. (Subcosoryx) cerroensis* has reduced premolars, indicating it may be closer to the Antilocaprini than to the rest of *Cosoryx* proper.

Discussion. The reason for the question in generic assignment of *?C. agilis* is that the hornless type is either a ju-

venile individual that has not grown horns or a female (Douglass, 1899; Frick, 1937). *?C. agilis* was originally described in Douglass' (1899) master's thesis, and this description was accepted as indeterminate by Frick (1937) because of the lack of diagnostic horns. Additional morphometric work with cranial and dental remains may be able to discern both whether this is a unique species of antilocaprid and where it properly fits into pronghorn systematics, but the current systematics based on horns leaves the status of this taxon uncertain.

Webb (1969) lists *C. burgensis* Frick, 1937 as a junior synonym of *C. furcatus* from the Burge Quarries of Cherry Co., Nebraska. I have been unable to locate Frick's (1937) description of this species; Frick (1937), like Webb (1969), seems to attribute all of the Burge *Cosoryx* to *C. furcatus*.

Genus Range. California (Ba1–Cl3), Colorado (He2–Ba2), Montana (Ba1–Cl3), Nebraska (He2–Cl3), New Mexico (Ba1–Cl3), Nevada (He2–Cl3), Oklahoma (Cl2), Oregon (Ba1–2), South Dakota (Ba1–Cl3), and Wyoming (He2–Ba2).

Included Species. *C. furcatus* Leidy, 1869 (type of genus): synonymy? (Webb, 1969): *Cosoryx (Paracosoryx) burgensis* Frick, 1937. Valentine Fm., Burge Mbr. (type) (Cl1), Devil's Gulch Mbr. (Ba2); Pawnee Cr. Fm. (Ba2); Ash Hollow Fm. (Cl1–3); Fort Logan Fm., Deep River Fauna (Ba1); Roglove Fauna (Cl1–3). *C.* cf. *furcatus:* Ash Hollow Fm. (Cl1–3). *C. ?C. furcatus:* Olcott Fm. (Ba1).

?C. agilis Douglass, 1899: synonymy: *Merycodus ? agilis* (Douglass): Douglass, 1903: 155. Madison Valley Fm. (type) (Ba1).

C. ilfonsensis Frick, 1937: Chamita Fm. (type) (Cl2); Tesuque Fm., Skull Ridge Mbr. (Ba1), Pojoaque Mbr. (Ba2).

C. (Subcosoryx) cerroensis Frick, 1937: Chamita Fm. (type) (Cl2). *C.* cf. *C. cerroensis:* Caliente Fm., Matthews Ranch Fauna (Cl1–2), Nettle Spr. Fauna (Cl2); Avawatz Fm. (Cl2); Esmeralda Fm. (Cl2); Laverne Fm. (Cl2–3).

C. sp.: Dove Spr. Fm. (Cl1–3); Ogallala Group, Uhl Pit (Ba2); Ash Hollow Fm. (Cl1–3).

Subfamily ANTILOCAPRINAE Gray, 1866

This apparently monophyletic subfamily is marked by the loss of "burrs" from the horn core shafts and the addition of an apparent keratinous horn sheath, revealed through an absence of wear at the tips of tines and a texture to the horn cores similar to that of *Antilocapra* (Frick, 1937; Janis and Manning, 1998a). Additionally, all have horn core shafts that are mediolaterally flattened in cross section, highly hypsodont dentitions, enlarged auditory bullae, and distal limb elements that independently converge on those of other clades of derived pecorans, with loss of side toes, elongated metapodials, and a cervid-like closed metatarsal gully (Janis and Manning, 1998a). Although the situation may not be as dire as it seemed to Frick (1937) when he suggested that isolated teeth of any member of the Antilocaprinae might be indistinguishable from *Antilocapra* itself, the differences in dentition within this group are subtle.

Genus *Proantilocapra* Barbour and Schultz, 1934

Diagnosis. Horn core bladelike, superficially similar to *Antilocapra* with a reduction of anterior tine, but with tuberosity at distal end (Fig. 18.3) (Barbour and Schultz, 1934; Frick, 1937). Horn core cross section wedgelike at midshaft, with anterior width 25% of posterior width (Webb, 1973). Cheek teeth similar to *Merycodus* but higher crowned, with a smaller p3; angle of mandible more pronounced than *Merycodus* (Barbour and Schultz, 1934).

Discussion. Stirton (1938) suggested that *Proantilocapra* might be on the line to *Antilocapra,* expanding on the argument of Barbour and Schultz (1934). On the basis of its horn core morphology, Stirton (1938) argued that *Proantilocapra* could not be on the line from *Sphenophalos* as Barbour and Schultz (1934) had originally suggested. Webb (1973) suggested that *Proantilocapra* presents convergent features with *Antilocapra.* I follow Janis and Manning (1998a) in grouping *Proantilocapra* with *Osbornoceros* on the basis of the posterior tilt and wedge-shaped cross section of the horn core shaft.

Genus Range.   Nebraska (Cl2).

Type and Only Species.   *P. platycornea* Barbour and Schultz, 1934: Ash Hollow Fm., Cap Rock Mbr. (type) (Cl2).

Genus *Osbornoceros* Frick, 1937

Diagnosis (Frick, 1937). Unforked horn cores with distinctive slender, twisted shafts having a dorsoventrally compressed and laterally extended base below a dorsoposteriorly sweeping tip (Fig. 18.3). Horn cores make one-half of a twist from base to tip. Cross section of horn cores near base is like an airfoil in shape, with narrow anterior, wide posterior, and lateral side concave. Horn cores superficially similar in shape to the modern nyala antelope *Tragelaphus angasii.* Dentition with primitive short diastema and small m3 third lobe.

Genus Range.   New Mexico (Hh1–4) and Texas (Hh2).

Type and Only Species.   *O. osborni* Frick, 1937: Chamita Fm., *Osbornoceros* Quarry (type) (Hh1–2). *O.* cf. *O. osborni:* Popotosa Fm. (Hh1).

?O. sp.: Ogallala Group, Hemphill beds: Box T Fauna (Hh2).

Tribe ILINGOCEROTINI Frick, 1937

This tribe is marked by a reduction in the tines of the horn cores, with a greater emphasis on the shaft. Additionally, there is a lateral twist to the shaft that is only slightly pronounced in most of the genera but is extremely evident in *Ilingoceros,* which has a long, twisting horn core (Fig. 18.3) that caused Merriam (1909) to mistakenly attribute it to the twist-horned tragelaphine (bovid) antelope such as a kudu. Frick (1937) named this taxon as a subfamily and included *Osbornoceros,* which now is not considered part of the natural monophyletic group with *Ilingoceros* and *Sphenophalos.* Frick (1937) also had *Plioceros* in its own monotypic subfamily, but I follow Janis and Manning (1998a) and include it in the Ilingocerotini. Finally, *Ottoceros* Miller and Downs (1974) is included in this subfamily because, as Janis and Manning (1998a) suggest, its horn core characteristics share apparent synapomorphies with *Ilingoceros* and *Sphenophalos.* Because my own work examining the postcrania of *Ilingoceros* and *Sphenophalos* indicates little morphological difference between the two besides the horn cores, and because there is little in the horn core morphology of *Plioceros, Ottoceros,* and *Sphenophalos* to differentiate among them, a conservative taxonomic approach might synonymize all four of the genera in this subfamily. However, this reductionist approach would generate more confusion than it would resolve, so I have left these genera distinct, awaiting a revision that includes morphometric analyses of dental, cranial, and postcranial remains in addition to the traditional horn core characters.

Genus *Plioceros* Frick, 1937

Diagnosis.   Two-tined horn cores with shaft tall and laterally compressed, with dumbbell-shaped cross section at midpoint. Anterior tine shorter and narrower than posterior tine, depressed posterolaterally. Horn cores on upper posterior of orbit, angled poserolaterally, with a tendency to widen distally (Fig. 18.3). Metaconid and entoconid of p3–p4 not connected. Diastema relatively long. Talonid of m3 moderate in size (Frick, 1937; Janis and Manning, 1998a).

Discussion.   *Plioceros* was erected by Frick (1937) as a replacement for *Sphenophalos,* which he considered to have an indeterminate type. As Webb (1969) notes, subsequent collections of *Sphenophalos* in the type Thousand Cr. Fm. have confirmed Merriam's (1909) diagnoses; this led Webb (1969) to place *Plioceros* as a subgenus of *Sphenophalos.* Janis and Manning (1998a) suggest that some species of *Plioceros* may need to be reassigned to *Sphenophalos* after species-level phylogenetic work.

Genus Range.   California (Cl2–Hh4), Kansas (Hh1), Montana (Hh1–2), Nebraska (Ba2–Hh1), New Mexico (Ba1–Hh2), Oregon (Hh1–4).

Included Species.   *P. blicki* Frick, 1937 (type of genus): Tesuque Fm., Pojoaque Mbr. (type) (Ba2); Chamita Fm. (Cl2, Hh1). *P.* cf. *P. blicki:* Chamita Fm. (Hh2–3); Deer Lodge Valley Loc. Fauna (Hh1–2).

P. dehlini Frick, 1937: Ash Hollow Fm., Merritt Dam Mbr. (type) (Cl2–3).

P. floblairi Frick, 1937: Valentine Fm., Devil's Gulch Mbr. (type) (Ba2); Ash Hollow Fm., Merrit Dam Mbr. (Cl2–3).

P. sp.: Dove Spr. Fm (Hh1); Popotosa Fm. (Hh1); Ash Hollow Fm., Merrit Dam Mbr. (Cl2–3). Cf. *P.* sp.: Ash Hollow Fm. (Cl1–3). *?P.* sp.: Ogallala Gp. (Hh1).

Genus *Ottoceros* Miller and Downs, 1974

Diagnosis. Tall two-tined horn cores, transversely flat-tened, with lateral flange above orbit that terminates well below the bifurcation. Horn core on posterior region of or-bit, slopes posteriorly (Fig. 18.3). Horn core less massive than that of *Sphenophalos* and longer relative to skull than that of *Antilocapra*. Slight twist to horn cores (anterior tine twists medially), with dumbbell-shaped cross section near the bifurcation. Indication of slight sexual dimorphism in horn cores, but not as pronounced as in *Antilocapra*. Meta-conid and entoconid of p4 not connected. Lower first and second molars highly hypsodont but not as hypsodont as *Tetrameryx* or *Antilocapra* (Miller and Downs, 1974; Janis and Manning, 1998a).

Discussion. The only known specimens of *Ottoceros* are of the five individuals from the Peace Valley Formation described by Miller and Downs (1974). These specimens were laterally crushed in diagenesis, concealing or disrupt-ing some cranial characters, but the shape of the horn cores suggests a placement within the Ilingocerotini (Janis and Manning, 1998a).

Genus Range. California (Hh3–4).

Type and Only Included Species. *O. peacevalleyensis* Miller and Downs, 1974: Peace Valley Fm. (type) (Hh3–4).

?*O.* sp.: Hungry Valley Fm. (Hh3–4).

Genus *Sphenophalos* Merriam, 1909

Diagnosis. Tall two-tined horn cores, transversely flat-tened, with anterior edge narrower than posterior edge. Slight twist to tines, with anterior tine twisting laterally. Horn cores positioned above posterior edge of orbit, with slight posterolateral slant (Fig. 18.3). Flange present at an-terior edge of shaft. Cheek teeth almost as hypsodont as *An-tilocapra* (Merriam, 1909, 1911; Stirton, 1932a; Janis and Manning, 1998a).

Discussion. There are many similarities between *Otto-ceros* and *Sphenophalos*, but the important difference is that the horn cores of the first twist medially, whereas those of the second twist laterally. Also, *Sphenophalos* is much larger and more robust than *Ottoceros*. As noted above, Frick (1937) considered the type of *Sphenophalos* indeterminate and erected *Plioceros* as a replacement. Many workers have con-sidered them synonymous, but there is some evidence that *Plioceros blicki* represents a more primitive form in the Ilin-gocerotine clade (Janis and Manning, 1998a).

Genus Range. Arizona (Hh3), California (Hh1–4), Ne-braska (Hh2), Nevada (Hh1), and Oregon (Hh2–3).

Included Species. *S. nevadanus* Merriam, 1909 (type of genus): Thousand Cr. Fm. (type) (Hh1); Big Sandy Fm. (Hh3); Drewsey Fm. (Hh3).
S. middleswarti Barbour and Schultz, 1941: Ash Hollow Fm. (type) (Hh2).

S. sp.: Mehrten Fm. (Hh3–4); Rattlesnake Fm. (Hh2); Ju-niper Cr. Canyon Sites (Hh2). Cf. *S.* sp.: Pinole Fm. (Hh3–4).

Genus *Ilingoceros* Merriam, 1909

Diagnosis. Horn cores with exceptionally tall shafts and very reduced, possibly absent tines. Horn cores have an extreme spiral, with anterior side twisting laterally; may have as many as two complete twists over length of horn core in larger adult form (Fig. 18.3). Horn cores >30 cm in length and have largest diameter near midpoint of shaft; cross section nearly circular. Horn cores on posterior side of orbit with slight posterolateral tilt (Merriam, 1909, 1911; Frick, 1937; Janis and Manning, 1998a).

Discussion. *Ilingoceros* is one of the most striking of the late Miocene antilocaprines, with its spiraling, elongate horn cores (Fig. 18.3). *I. schizoceras* was named for a small horn core discovered subsequent to the description of *I. alexandrae*. Frick (1934) and Stirton (1932a) both suggested that *I. schizoceras* was actually a juvenile or female form of *I. alexandrae*, a posi-tion I support. I have included *I. schizoceras* as a separate species here because no one has formally synonymized it with *I. alexandrae*, and this chapter is only a literature review. De-spite promising initial results (Davis, 2004), my own work with material from the type Thousand Creek Formation has sug-gested that *Ilingoceros* and *Sphenophalos* postcrania are indis-tinguishable, grading one into another in size (unpub. data). The presence of *Sphenophalos* at other sites without *Ilingoceros* (Furlong, 1941) indicates the two are not male and female of the same species, as originally suggested by Stirton (1932a); however, the indeterminacy of the postcrania of these taxa leaves me unable to support reported occurrences of *Ilingo-ceros* that have not had associated identifiable horn cores. The known occurrence of *Ilingoceros* is limited to Thousand Creek; the other occurrences might better be classified as "Ilingo-cerotini indet.," as they could as easily be *Sphenophalos*.

Genus Range. Nevada (Hh1).

Included Species. *I. alexandrae* Merriam, 1909 (type of genus): Thousand Cr. Fm. (type) (Hh1).
I. schizoceras Merriam, 1911: Thousand Cr. Fm. (type) (Hh1).
?*I.* sp.: Dove Spr. Fm. (Hh1); Horned Toad Fm (Hh4); Carlin Fm. (Hh1–2); Hay Ranch Fm. (Hh3–4); Chamita Fm. (Hh2–3); Big Sandy Fm. (Hh3–4); Rattlesnake Fm. (Hh2).

Tribe ANTILOCAPRINI Gray, 1866

The Antilocaprini contains the only extant member of the Antilocapridae, *Antilocapra americana*, and is marked by several synapomorphies: reduction of anterior tine, denti-tion extremely hypsodont, diastema very long, and expan-sion of the fourth lobe on the m3.

Frick (1937) included *Proantilocapra* in this subdivision of his systematic hierarchy and put *Texoceros* in his "Stocko-cerotinae." Following the phylogeny presented by Janis and Manning (1998a), *Texoceros* and *Antilocapra* form a monophyletic group to the exclusion of all other antilocaprines, supported by the apomorphies listed above.

Genus *Texoceros* Frick, 1937

Diagnosis. Horn cores with two tines and relatively short shaft, but shaft not completely reduced as in Stockocerotini, so there is not a four-horned appearance. Anterior tine laterally flattened; posterior tine larger, with a rounder cross section. Anterior tine twisted somewhat laterally, posterior tine somewhat medially (Fig. 18.4). Diastema twice the length of premolar row; jaw not shortened as in *Capromeryx;* m3 lobe expanded; p4 does not have paraconid–metaconid closure (Frick, 1937; Janis and Manning, 1998a).

Discussion. Some species of *Capromeryx* were included in *Texoceros* by Frick (1937). Morgan and Morgan (1995) describe the taxonomic history of *Capromeryx* and explain this synonymy. Dalquest (1983) synonymized *T. texanus* with *T. altidens* and suggested that *T. altidens,* based on an isolated m3, might be synonymous with *T. guymonensis.* Janis and Manning (1998a) mistakenly list *T. altidens* as the type of this genus, but Frick (1937: 483) explicitly makes a horn core from *T. guymonensis* the type specimen of the type species. If Dalquest's (1983) suggestion is upheld by a detailed revision, *T. altidens* would then become the type species.

Genus Range. Arizona (Hh3), California (Hh3–4), Colorado (Hh2), Kansas (Hh3), Nebraska (Hh2), Nevada (Hh3–4), Oklahoma (Hh3), and Texas (Hh1–4).

Included Species. *T. guymonensis* Frick, 1937 (type of genus): Ogallala Group, Optima Loc. Fauna (type) (Hh3). *T. cf. T. guymonensis:* Ash Hollow Fm. (Hh2); Ogallala Group (Hh3).

T. altidens (Matthew, 1924): synonymies: *Merycodus altidens* Matthew, 1924: 200, Fig. 60. *Capromeryx altidens* (Matthew): Hesse, 1935: 308, Figs. 1–3. *Capromeryx texanus* Hesse, 1935: 311, Figs. 4–5. *Dorcameryx optimae* Reed and Longnecker, 1932: 66. Snake Cr. Fm., Johnson Mbr. (type) (Hh2); Ogallala Group, Higgins Fauna (Hh1), Optima Loc. Fauna (Hh3). *T. cf. T. altidens:* Caliente Fm. (Hh3–4); Hemphill Beds, Coffee Ranch Loc. Fauna (Hh3). *T. ?T. altidens:* Snake Cr. Fm., Johnson Mbr. (Hh2).

T. edensis Frick, 1937: Eden Beds (type) (Hh4).

?*T. minorei* Frick, 1937: Ogallala Group, Optima Loc. Fauna (type) (Hh3), Higgins Fauna (Hh1); Big Sandy Fm. (Hh3); Bidahochi Fm. (Hh3–4). ?*T. ?T. minorei* (questionably attributed): Goliad Fm., Lahabia Mbr. (Hh1–2); Ogallala Group (Hh1).

?*T. vaughani* Frick, 1937: Ogallala Group, Wray Fauna (type) (Hh2).

Fig. 18.4. As for Figure 18.3, illustrating the remainder of the antilocaprine taxa. Scale bar equals 10 cm.

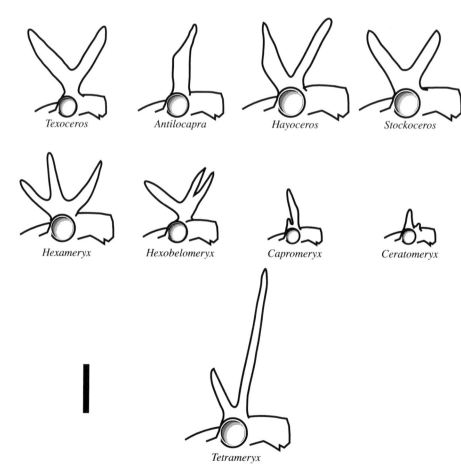

Texoceros *Antilocapra* *Hayoceros* *Stockoceros*

Hexameryx *Hexobelomeryx* *Capromeryx* *Ceratomeryx*

Tetrameryx

T. sp.: Panaca Fm. (Hh4); Quibris Fm. (Hh3–4); Ogallala Fm (Hh3–4). ?*T.* sp.: St. David Fm., Benson Loc. Fauna (Bl).

Genus *Antilocapra* Ord, 1818

Diagnosis.   (O'Gara, 1978). Horn cores bladelike, with reduction of anterior bony tine to a lump on the posterior tine, posterior tine continues bladelike shape of horn core; cross section somewhat triangular (Fig. 18.4). Sexually dimorphic for horn cores, with females showing a range of smaller horn cores from 3 to 12 mm, male horn cores 12 to 15 cm in length. External, keratinous horn sheaths of males are two-pronged, with the anterior prong smaller and pointing slightly laterally; posterior prong inclines posteromedially; horn sheaths dark colored with whitish tips; keratinous sheath is shed annually. Females typically have unbranched horn sheaths if any are present; they are shed variably. Lateral digits completely lost (Kurtén and Anderson, 1980). Connection between paraconid and metaconid closed on p4; fourth lobe developed on m3 (Janis and Manning, 1998a). For a review of the biology of modern *Antilocapra*, see O'Gara (1978, 1990) and references therein.

Discussion.   Richards and McCrossin (1991) and Janis and Manning (1998a) suggest that ?*A. garciae* would be better placed in *Sphenophalos*. Although Webb (1973) originally suggested that this species represented a transitional form between *Sphenophalos* and *Antilocapra americana,* the phylogeny presented by Janis and Manning (1998a) nests *Sphenophalos* in a clade distinct from the *Antilocapra–Texoceros* clade, which would make such a transition impossible. A decision on whether ?*A. garciae* occupies a transitional position between the ilingocerotine and antilocaprine clades or belongs within *Sphenophalos* will have to await a species-level phylogenetic analysis.

Genus Range.   California (Ir-RLB), Florida (Hh4), and RLB to recent as for *A. americana.*

Included Species.   *A. americana* (Ord, 1815) (type of genus): synonymies: *Antilope americanus* Ord, 1815: 292. *Antilocapra americana* (Ord): Ord, 1818: 149. *Antilope (Dicranocerus) furcifer* Hamilton-Smith, 1827: 170. *Antilocapra anteflexa* Gray, 1855: 10. Known from extensive Pleistocene localities throughout and slightly beyond its Recent distribution, mostly from RLB (see Kurtén and Anderson, 1980; FAUNMAP working group, 1994). Recent distribution: Arizona, California, Colorado, Idaho, Iowa, Kansas, Minnesota, Missouri, Montana, Nebraska, Nevada, New Mexico, North Dakota, Oklahoma, Oregon, South Dakota, Texas, Utah, Washington, and Wyoming in the United States; southern Alberta, Saskatchewan, and Manitoba in Canada; and Baja California Norte, Baja California Sur, Sonora, Chihuahua, Coahuila, and Durango in Mexico (O'Gara, 1978).

A. pacifica Richards and McCrossin, 1991: Big Break (Ir-RLB).

?*A. garciae* Webb, 1973: Upper Bone Valley Fm. (type) (Hh4).

Tribe STOCKOCERATINI Frick, 1937

Marked by the reduction of the central shaft, this tribe is composed of lineages that appear to have four or six horn core elements above their orbits, with each horn core element homologous to a tine from the horn cores of other antilocaprids.

Genus *Hayoceros* Frick, 1937

Diagnosis.   Each horn core consists of two major tines with an almost complete reduction in the shaft, giving a four-horned appearance (Fig. 18.4). The anterior tine is wedge-shaped in cross section and has a texture similar to the horn core of *Antilocapra,* which caused Frick (1937) to suggest that the animal expressed six horns in life, with the anterior tine pronged as in *Antilocapra.* The posterior tine is either cylindrical (*H. falkenbachi*) or bladelike (*H. barbouri*) and is much smaller than the anterior (Frick, 1937; Skinner, 1942).

Discussion.   When Frick (1937) named this taxon, he ranked it as a subgenus but did not make clear which existing genus he considered to contain *Hayoceros.* In fact, all of the taxonomic references to *Hayoceros* in his 1937 volume are formatted as though it were a full genus. Its context in Frick (1937) suggests that *Hayoceros* was intended to subdivide *Tetrameryx,* but subsequent workers, following Skinner (1942), have treated it as an evolutionarily distinct lineage from *Tetrameryx.*

Genus Range.   Nebraska (Ir), and Texas (Ir).

Included Species.   *H. falkenbachi* Frick, 1937 (type of genus): Hay Sprs. (type) (Ir);

H. barbouri Skinner, 1942: Gordon Quarry (type) (Ir); Rock Cr. (Ir).

Genus *Stockoceros* Frick, 1937

Diagnosis.   Short, symmetrically forked horn core with loss of main horn core shaft, so that each tine emerges from the skull as an apparently distinct horn core (Fig. 18.4). Nutrient foramina around base of tines indicates that each was individually sheathed, as in *Tetrameryx* (Janis and Manning, 1998a). Small, variable fourth lobe on m3 (Skinner, 1942).

Discussion.   As with *Hayoceros,* Frick (1937) named this taxon as a subgenus without clearly indicating the parent genus. The context has been taken to indicate that he intended *Stockoceros* to fit within *Tetrameryx,* but there is some reason to think that Frick (1937) meant for it to fit within *Texoceros.* Subsequent workers have, in general, treated this taxon as an evolutionarily distinct lineage. Furlong (1943) described a growth series of *S. conklingi* from Mexico, with particular reference to the changing angle between the two tines with increasing ontogenetic age, and Skinner (1942) made a similar analysis of *S. onusrosagris.* There, he suggested that *S. onusrosagris* might be synonymous with *S. conklingi,* but resolution must wait for statistical analyses of the large samples available (Kurtén and Anderson, 1980).

Genus Range.   Arizona (RLB), Nebraska (Ir), New Mexico (RLB); Mexico: Aguascalientes (RLB), Nuevo León (RLB).

Included Species.   *S. conklingi* (Stock, 1930) (type of genus): synonymies: ?*Tetrameryx conklingi* Stock, 1930: 6, Figs. 1–3. *Stockoceros conklingi* (Stock): Frick, 1937: 522, Fig. 53. Shelter Cave (type) (RLB); San Josecito Cave (RLB); Ventana (RLB); Cedazo (RLB).

S. onusrosagris (Roosevelt and Burden, 1934): synonymies: *Tetrameryx onusrosagris* Roosevelt and Burden, 1934: 4, Fig. 1. *Stockoceros onusrosagris* (Roosevelt and Burden): Frick, 1937: 522, Fig. 53. Papago Sprs. (type) (RLB); Burnet (RLB); Mullen II (Ir); U-Bar Cave (RLB).

Genus *Hexameryx* White, 1941

Diagnosis (Webb, 1973).   Horn cores with three tines each and loss of main shaft, resulting in a six-horned appearance (Fig. 18.4). Each tine circular in cross section, with a separate sheath. Anterior and medial tines directed dorsolaterally, posterior tine directed dorsomedially. Each tine larger than next anterior. Upper P4 with strong labial rib, like *Hexobelomeryx* and unlike *Antilocapra.* Upper molars less hypsodont, with weaker styles than *Antilocapra,* resemble *Stockoceros.* Variable, small fourth lobe present on m3, as in *Stockoceros, Hexobelomeryx,* and *Tetrameryx,* but not as developed as in *Antilocapra.*

Discussion.   *H. elmorei* was shown by Webb to be the (larger) male of the sexually dimorphic *H. simpsoni,* based on a larger sample than White (1941, 1942) originally had available. Simpson (1945) listed *Hexameryx* as a junior synonym of *Hexobelomeryx* in his mammal classification, but Webb (1973) made a good case for separating the genera on the basis of differences in horn core growth and morphology. At the same time, Webb (1973) suggested that these two taxa share a unique evolutionary history, a hypothesis supported by Janis and Manning (1998a) and reflected in the topology of the phylogeny presented here (Fig. 18.2), based on their cladogram.

Genus Range.   Florida (Hh2–4).

Type and Only Species.   *H. simpsoni* White, 1941 (type of genus): synonym: *Hexameryx elmorei* White, 1942: 88, Plate XVII, Figs. A, B. Bone Valley Mbr., Peace River Fm. (type) (Hh4); "Alachua Clays," Withlacoochee Fauna (Hh2).

Genus *Hexobelomeryx* Furlong, 1941

Diagnosis.   Each horn core has three tines, as in *Hexameryx,* but two of the three tines are closely aligned and share a single horn sheath. Anterior tine largest in cross-sectional area, with the posterior two tines subequal in size.

Middle tine projects more medially than either anterior or posterior tines (Fig. 18.4). Main horn core shaft not as reduced as in *Hexameryx,* so that tines diverge about 2.5 cm above cranium; main shaft emerges from posterior roof of orbit, so that horn cores are more posteriorly located than in other antilocaprines. Sexual dimorphism (Webb, 1973) expressed in twinning of tines: males have medial tine tied to posterior tine (Fig. 18.4), whereas females have medial tine tied to anterior tine. Because of these connections between tines, the horn cores appear to have two tines, one of which has bifurcated, unlike the three-tined state apparent in *Hexameryx.* Upper P4 with strong labial rib, like *Hexameryx* and unlike *Antilocapra.* Variable, small fourth lobe present on m3, as in *Stockoceros, Hexameryx,* and *Tetrameryx* but not as developed as in *Antilocapra* (Furlong, 1941; Webb, 1973; Janis and Manning, 1998a).

Discussion.   The dimorphism in horn core form was noted by Furlong (1941) and ascribed to sexual dimorphism by Webb (1973). Again, the three-tined horn cores and similarities in dentition suggest a close relationship between *Hexameryx* and *Hexobelomeryx.*

Genus Range.   Nevada (Hh4), Texas (Hh4); Mexico: Chihuahua (Hh4–Bl), Guanajuato (Hh4), Hidalgo (Hh4).

Included Species.   *H. fricki* Furlong, 1941 (type of genus): Rincón Fauna (type) (Hh4); Matachic Fauna (Bl).

H. sp.: Tehuichila Fauna (Hh4); Ocote Loc. Fauna (Bl); Coecillo Fauna (Hh4). Cf. *H.* sp.: Goodnight Beds (Hh4). ?*H.* sp. Panaca Fm. (Hh4).

Genus *Capromeryx* Matthew, 1902

Diagnosis (Janis and Manning, 1998a).   Two-tined horn cores with a relatively reduced base give a four-horned appearance (Fig. 18.4). Anterior tine has a smaller diameter than posterior tine and may be much shorter. Anterior tine cylindrical, posterior tine somewhat laterally flattened. Horn cores supraorbital with tines parallel, extending straight dorsally. Deep groove in posterior side of posterior tine. Auditory bullae enlarged. Relatively short diastema and premolar row. Lower m3 lacks fourth lobe. Apparent trend toward metaconid–entoconid closure on p4.

Discussion.   Hesse (1935) suggested that *Capromeryx* was the ancestor of *Antilocapra,* but his argument was based on the morphology of dental and postcranial specimens. Differences in dentition and horn core growth, as well as trends through time, now suggest *Capromeryx* is not directly related to *Antilocapra* (Morgan and Morgan, 1995; Janis and Manning, 1998a).

C. arizonensis has tines of subequal length on each horn core, unlike the other species of *Capromeryx,* which have a much smaller anterior than posterior tine.

Morgan and Morgan (1995) suggest that *Capromeryx* underwent progressive reduction in body size through the Pliocene and Pleistocene, with *C. tauntonensis* slightly smaller than *Antilocapra*, progressing through *C. arizonensis, C. furcifer, C. minor,* and *C. mexicanus,* with the last two 66% of the size of the first in linear dimensions, based on m3 length. Additionally, there is a trend toward gracility in the horn cores, with the diameters of *C. mexicanus* and *C. minor* horn cores 50% of those of *C. tauntonensis.* There also seems to be a trend toward a more posterior location of the horn cores above the orbits (Morgan and Morgan, 1995).

Genus Range. Arizona (Bl), California (RLB), Florida (Bl-Ir), Kansas (RLB), Nebraska (Bl-Ir), New Mexico (RLB), Texas (Ir-RLB), Washington (Bl); Mexico: Aguascalientes (RLB), Guanajuato (Bl), México (RLB), Sonora (RLB).

Included Species. *C. furcifer* Matthew, 1902 (type of genus): synonymies: *Capromeryx minimus* Meade, 1942: 89, Figs. 1–2. *Breameryx minimus* (Meade): Furlong, 1946: 137, Pl. 1, 2 (Figs. 1–3). Hay Sprs. (type) (Ir); Angus (Ir); Slaton Quarry (Ir); Cragin Quarry (RLB); Gordon Quarry (Ir).
C. minor Taylor, 1911: synonymy: *Breameryx minor* (Taylor): Furlong, 1946: 138, Pl. 3 (Fig. 6). Rancho La Brea (type) (RLB); Blackwater Draw (RLB); Ingleside (RLB); McKittrick (RLB); Schuilling (RLB); Lone Tree Point (RLB). *C.* cf. *C. minor:* Rancho La Brisca (RLB).
C. mexicanus Furlong, 1925: synonymy: *Breameryx mexicana* (Furlong): Furlong, 1946: 138, Pl. 3 (Fig. 5). Tequixquiac (type) (RLB); Cedazo (RLB).
C. gidleyi Frick, 1937: synonymy: *Breameryx gidleyi* (Frick): Furlong, 1946: 138. St. David Fm. (type) (Bl).
C. arizonensis Skinner, 1942: synonymy: *Breameryx arizonensis* (Skinner): Furlong, 1946: 138. Dry Mountain (type) (late Bl); Broadwater (Bl); Santa Fe River (Bl); Inglis 1A (Ir); Kissimmee River (Bl); De Soto Shell Pit (Ir).
C. tauntonensis Morgan and Morgan, 1995: Ringold Fm., Taunton (type) (early Bl); Rancho Viejo Beds, San Miguel de Allende Area (Bl). *C.* cf. *C. tauntonensis:* Sand Draw Fauna, Kiem Fm (Bl).

Genus *Ceratomeryx* Gazin, 1935

Diagnosis (Gazin, 1935). Two-tined horn cores with extreme reduction of main shaft, resulting in a four-horned appearance (Fig. 18.4). Tines transversely flattened, anterior tine much larger than posterior tine. Anterior tine bladelike and located supraorbitally, centered on the midline of the orbit, with a slight posterior tilt. Posterior tine blunt and reduced, located postorbitally. Smaller than *Antilocapra.*

Discussion. *Ceratomeryx* is known from two skulls, one adult and one juvenile, and associated postcrania from the Hagerman Fossil Beds. The lack of dental material makes the phylogenetic affinities of this genus hard to diagnose. Gazin (1935) suggested that *Ceratomeryx* might be related to *Tetrameryx,* but the work of Frick (1937) suggested more of a relationship with *Texoceros* or *Sphenophalos* (Colbert and Chaffee, 1939; Kurtén and Anderson, 1980).

Genus Range. Idaho (Bl).

Type and Only Species. *C. prenticei* Gazin, 1935: Glenns Ferry Fm., Hagerman Fauna (type) (Bl).

Genus *Tetrameryx* Lull, 1921

Diagnosis. Asymmetrical horn cores with main shaft completely reduced; posterior tine much taller than anterior (Fig. 18.4). Each tine was surrounded by an individual sheath, as in *Stockoceros* (Janis and Manning, 1998a). Anterior tine is similar to horn core of *Antilocapra,* but with a slightly more rounded cross section near the base (Lull, 1921). Upper molars larger than in *Antilocapra,* with upper premolars relatively much smaller (Frick, 1937). Upper M3 with more pronounced mesostyle than *Antilocapra,* but lacking a distinct hypocone (Lull, 1921).

Discussion. *Ceratomeryx* and *Tetrameryx* are united in the phylogeny (Fig. 18.2) because of the apparently homologous cylindrical shape of their anterior tines (Janis and Manning, 1998a: 500). Lull (1921) suggested that the anterior tines of *Tetrameryx* were homologous with the horn cores of *Antilocapra,* indicating that the anterior tines would be covered in a similarly shaped, annually shed keratinous sheath. This is similar to the suggestion of Frick (1937) for *Hayoceros,* which he seems to have originally named as a subgenus of *Tetrameryx.* The homology I suggest, which follows that of Janis and Manning (1998a), links the anterior tine of *Tetrameryx* and, for that matter, all of the Stockocerotini to the coreless anterior tine of the horn sheath of *Antilocapra.* The posterior tine is consequently suggested to be homologous to the upper portion of the horn core of *Antilocapra.* Careful character analysis will be needed to finally sort out this homology.

Genus Range. Arizona (Bl), California (Hh3–Bl), Texas (Bl–RLB); Canada, Saskatchewan (Ir); Mexico, Aguascalientes (RLB), Sonora (Bl-RLB).

Included Species. *T. shuleri* Lull, 1921 (type species): Dallas Sand Pits (type) (RLB); Moore Pit (RLB); Slaton Quarry (Ir); Hill-Schuler (RLB); Iron Bridge (RLB).
T. irvingtonensis Stirton, 1939: Irvington Gravels (type) (Ir); Turlock Lake Fm., Fairmead Landfill (Ir).
T. knoxensis Hibbard and Dalquest, 1960: Seymour Fm., Gilliland (type) (Ir), Burnett Quarry (Ir).
T. mooseri Dalquest, 1974: Cedazo (type) (RLB).
T. tacubayensis Mooser and Dalquest, 1975: Cedazo (type) (RLB).
T. sp.: Mehrten Fm. (Hh3–4); Palm Spr. Fm. (Ir-Bl); Ogallala Group, Cita Canyon Fauna (Bl); Radec (Bl); Ocotillo Fm. (Ir); El Golfo de Santa Clara (Bl); Rancho La Brisca (RLB). Cf. *T.* sp.: Duncan (Bl). ?*T.* sp.: California Oaks (Ir); Wellsch Valley (Ir).

ACKNOWLEDGMENTS

I thank D. Prothero and S. Foss for asking me to contribute this chapter. Reviews by Earl Manning (Tulane University),

Christine Janis (Brown University), and Richard White (International Wildlife Museum) greatly contributed to the quality of this work. My wife, Samantha Hopkins, has been very forgiving of my antilocaprid obsession. A. Barnosky (UC Berkeley) provided support through MIOMAP (NSF grant EAR-0310221)–funded equipment and space. I thank J. Alroy (NCEAS) and the other workers at the Paleobiology Database for making their data available online; it was an invaluable resource for checking occurrences. Similarly, I thank M. Carrasco and B. Kraatz (UC Berkeley) for their work with the MIOMAP database. R. Irmis (UC Berkeley) provided me with a useful reference on Papago Springs Cave. This work was completed under a postdoctoral fellowship at the UC Museum of Vertebrate Zoology, funded by California State Parks and the Resources Law Group. This is UCMP publication number 1926.

DONALD R. PROTHERO
AND MATTHEW R. LITER

19

Family Palaeomerycidae

THE PALAEOMERYCIDAE COMPRISED A GROUP of cervoid artiodactyls both diverse and widespread in the northern continents during the early to late Miocene. Although they were deerlike in bearing cranial appendages, these appendages were not deciduous as are cervid antlers, nor apparently did they have deciduous tips, like those found in antilocaprids, or keratinous sheaths, like those found in bovids. These appendages may have been covered in skin, as giraffid ossicones are, but because the Palaeomerycidae are extinct, it is impossible to determine this beyond doubt. Most presumed male palaeomerycid skulls had unbranched paired supraorbital appendages, and male skulls of several genera (primarily in the North American tribe Cranioceratini and in several skulls found in Europe) had a medial occipital appendage as well. In addition to a diversity of cranial appendages (paralleling the diversity in cervids and bovids), palaeomerycids also exhibited brachydont to mesodont dentitions and apparently had deerlike antorbital glands (based on the antorbital vacuity in the skull) for scent marking. In several primitive taxa, the presumed male skulls possessed large saber-like upper canines as well.

The family is defined primarily by a suite of dental features: reduced metastyles in the upper molars, attenuated metacone on P4, bifurcated posterior wing of the metaconule, and double posterior lobe on m3 that is closed posteriorly (Janis and Scott, 1987; Janis and Manning, 1998b). Many of the taxa possess the primitive character of the *Palaeomeryx* fold (a short enamel ridge running posteromedially from the protoconid to the labial face of the lower molars), although in more derived and hypsodont taxa, this fold is lost.

Most palaeomerycids had deerlike limb proportions, and they were comparable in size to small to medium-sized living deer. Compared to many running ruminants, however, their limbs are relatively short and heavily proportioned, although a few taxa had relatively long slender limbs. Metapodials III and IV are fully fused, and several taxa still retain lateral digits in the carpus. The distal articular surfaces of the metapodials bear complete phalangeal keels. The metatarsal gully is closed (a derived cervoid feature), and there is a tuberosity on the lateral side of the posterior proximal metatarsals in most genera. There is a tubercle on the medioproximal end of the metatarsus for the fused proximal end of metatarsal II, but metatarsal V is unfused. The cubonavicular facet is somewhat raised, typical of cervids (Janis and Scott, 1987; Janis and Manning, 1998b).

Based on these cranial and postcranial features, the Palaeomerycidae are thought to have been browsers that lived in subtropical to temperate habitats ranging from dense woodlands to open brushy habitats and grasslands (Janis and Manning, 1998b). They reached their peak in diversity in the early and middle Miocene, both in North America and in Eurasia, and then declined to just a few taxa by the late Miocene and earliest Pliocene, presumably because their subtropical woodland habitat was vanishing in the expansion of the temperate grasslands.

SYSTEMATICS

The genus *Palaeomeryx* (from the early to middle Miocene of Eurasia) was first recognized by von Meyer in 1834, and Lydekker (1883c) erected the family Palaeomerycidae to include *Palaeomeryx* and his genus *Propalaeomeryx*. Roger (1904) added the genus *Lagomeryx* for smaller specimens that had been referred to *Palaeomeryx*. A number of other genera (such as *Blastomeryx, Dremotherium, Procervulus, Climacocerus, Amphitragulus, Dicroceras,* and *Micromeryx*) have also been referred to this family, although most of these taxa are no longer considered palaeomerycids. Janis and Scott (1987: 52–53) reviewed the complex taxonomic history of the Eurasian palaeomerycids and lagomerycids, so there is no need to repeat this analysis here. Janis and Scott (1987: 71) also redefined the Palaeomerycidae to include *Palaeomeryx, Amphitragulus, Prolibytherium,* and possibly *Triceromeryx*. In addition, several new taxa, including *Tauromeryx* and *Ampelomeryx,* have been added to this list of Eurasian palaeomerycids (Astibia and Morales, 1987; Duranthon et al., 1995).

The North American dromomerycines were originally confused with blastomerycids (musk deer) or antilocaprids (pronghorns), but Scott (1895a) first recognized their affinities with the palaeomerycids. When the first nearly complete material of *Dromomeryx borealis* (Fig. 19.1) was described by Douglass (1909), he recognized that it was neither a blastomerycid nor an antilocaprid. Even when the distinctiveness of *Dromomeryx* was recognized, palaeomerycids were still placed within the cervids, and many taxa that were subsequently described were mistaken for other kinds of ruminants. Sinclair (1915) described *Drepanomeryx* as an antelope; Matthew (1918a) detailed *Cranioceras* and also thought it was an antelope. Lull (1920) described *Aletomeryx* but con-

cluded it was a pronghorn. Frick (1937) was the first to unite all these previously described genera, along with all the new taxa he named, as a unified group, the Dromomerycinae, within the family Cervidae. Stirton (1944) placed the dromomerycines in the Palaeomerycidae, which he allied with Giraffidae rather than Cervidae. This assignment was followed by a number of authors (Crusafont, 1961; Viret, 1961; Hamilton, 1978b), although others (Romer, 1966; Leinders, 1983) placed them with cervids.

Hamilton (1978a) suggested that the family Palaeomerycidae be abandoned because it had no unifying characters, and this recommendation was followed by later authors (e.g., Leinders, 1983). The discovery of important new specimens of *Palaeomeryx* and *Lagomeryx* (Chow and Shih, 1978; Qiu et al., 1985) allowed Janis and Scott (1987) to redefine the Palaeomerycidae as a monophyletic group and clearly show that they are a sister group to the Cervidae. However, Duranthon et al. (1995) argued that the cranial homologies of the palaeomerycid *Ampelomeryx* (which apparently had giraffe-like ossicones) are different from the dromomerycid cranial appendages (which are apparently frontal outgrowths). Janis and Manning (1998b) and Janis (2000b) did not find this argument compelling, and their position is followed here because cranial appendages are notoriously labile and variable (Bubenik, 1982; Janis and Scott, 1987), and other characters (Janis and Manning, 1998b; Janis, 2000b) support the monophyly of the Palaeomerycidae (including the dromomerycines).

Dromomerycines have been raised to the family-rank name Dromomerycidae by many recent authors (Hamilton, 1978b; Janis, 1982; Leinders, 1983; Webb, 1983; Stucky and McKenna, 1993; Janis and Manning, 1998b), but Janis and Scott (1987) recommended that they be referred to the European family Palaeomerycidae because *Palaeomeryx* is clearly the sister taxon of the dromomerycines; this was also followed by McKenna and Bell (1997). Even though the family-rank name Dromomerycidae is widely used in the current literature, we agree with Janis and Scott (1987) and McKenna and Bell (1997) that the more natural grouping should include the North American dromomerycines with their Eurasian sister taxa in a single monophyletic family.

As mentioned above, Janis and Scott (1987) placed the genera *Palaeomeryx, Amphitragulus, Prolibytherium,* and possibly *Triceromeryx* within the Eurasian Palaeomerycidae, and Duranthon et al. (1995) described *Ampelomeryx* as a new genus of palaeomerycid. McKenna and Bell (1997: 423) also included *Oriomeryx* Ginsburg, 1985, which Gentry et al. (1999: 234) considered to be a moschid ancestor of the palaeomerycids, along with *Bedenomeryx* Jehenne, 1988. McKenna and Bell (1997) regarded *Bedenomeryx* and *Sinomeryx* Duranthon et al., 1995 as junior synonyms of *Palaeomeryx*; they synonymized *Hydropotopsis* Jehenne, 1985 and *Pomelomeryx* Ginsburg and Morales, 1989 with *Amphitragulus* Pomel, 1846. Thus, the generic content of the Eurasian Palaeomerycidae has fluctuated in recent years, but at least five and possibly nine valid genera are known.

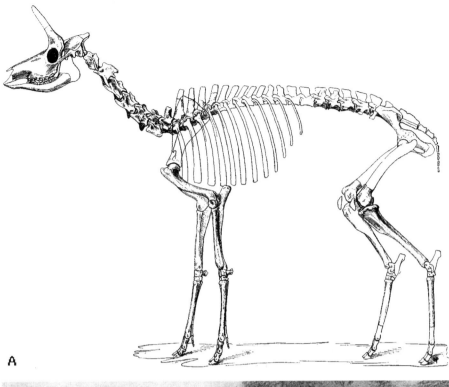

Fig. 19.1. (A) Skeleton (after Douglass, 1909) and (B) restoration (from Scott, 1913) of *Dromomeryx borealis*.

A

B

The North American radiation of dromomerycines was first fully described by Frick (1937), who recognized over 50 species and dozens of genera and subgenera. Unfortunately, most of Frick's taxa were inadequately diagnosed, and many were based on slight differences in specimens from different localities without regard to intrapopulational variability and without any statistical techniques to separate species objectively (Prothero and Liter, in press). Janis and Manning (1998b) rediagnosed Frick's (1937) genera and performed the first cladistic analysis of the group but did not attempt to deal with the systematics of all of Frick's in-

valid species. Prothero and Liter (in press), using modern statistical techniques and concepts of populational variability, reduced Frick's (1937) 50 (mostly invalid) species to just 17 species in 11 genera (most of which are now monotypic) with no subgenera. Thus, the apparent high diversity of palaeomerycids that appears in the older literature has now been reduced to just five to nine genera in Eurasia and 11 genera in North America. Nonetheless, this smaller number of valid taxa still shows a distinct pattern of diversification and extinction, already discussed by Janis and Manning (1998b) and Janis (2000b).

Order ARTIODACTYLA Owen, 1848
Family PALAEOMERYCIDAE Lydekker, 1883
Amphitragulus Pomel, 1847

Amphitragulus is a small hornless ruminant known from the early Miocene of Eurasia. As discussed in detail by Janis and Scott (1987: 61–64), there is a longstanding controversy over this taxon and its distinction from *Dremotherium*. However, Janis and Scott (1987: 64) argue that it has derived characters of a more advanced cervoid, closer to the palaeomerycids and cervids than to the moschids (such as *Dremotherium*). Janis and Scott (1987: 78, Fig. 19) showed that the derived characters place *Amphitragulus* as a sister taxon to the dromomerycines plus *Palaeomeryx*, with *Prolibytherium* as a more remote sister group, in an enlarged conception of the Palaeomerycidae. This assignment was also followed by McKenna and Bell (1997).

Prolibytherium Arambourg, 1961

Prolibytherium (Fig. 19.2) is a peculiar taxon from the early Miocene Gebel Zelten locality of Libya. It had "flattened horizontal wing-like cranial appendages divided into anterior and posterior lobes in a butterfly-like pattern" (Janis and Scott, 1987: 71). Most authors have considered it to be a primitive giraffid, but Janis and Scott (1987: 78) considered it to be a primitive sister taxon to the Palaeomerycidae. McKenna and Bell (1997: 423) placed it in the Palaeomerycidae, as a primitive relative of the Palaeomerycinae and Dromomerycinae. Until better fossils are known, however, this hypothesis needs further testing. If *Prolibytherium* is indeed the sister group of all the Palaeomerycidae, it might suggest an African origin for the group before it spread across Eurasia and North America.

Fig. 19.2. Restoration of *Prolibytherium* (drawing by S. E. Foss and J. Higgins).

Subfamily PALAEOMERYCINAE Lydekker, 1883
Palaeomeryx von Meyer, 1834

Although *Palaeomeryx* was the first named taxon of the family, the type material consists of about 20 isolated teeth and fragments of the skeleton (Duranthon et al., 1995), and no complete skull is known (even though many other specimens have been referred to this genus). Thus, we have no knowledge of its cranial appendages (or lack thereof). Because it is so fragmentary and primitive, the genus is still poorly diagnosed, and it may well constitute a paraphyletic wastebasket of primitive palaeomerycids that have no clear diagnostic features of other genera. The earliest specimens are known from late MN 3 in Europe (17.5 Ma) and occur until MN 6 in Europe (15 Ma). Most of the examples show a trend toward increased body size during the Miocene, except for *Palaeomeryx kaupi* (De Bruijn et al., 1992) from MN 6, which is considerably smaller. Duranthon et al. (1995) placed *Palaeomeryx tricornis* Qiu et al., 1985 from MN 6 in China in a new genus, *Sinomeryx*, but McKenna and Bell (1997: 423) regarded this genus as an invalid junior synonym of *Palaeomeryx*. McKenna and Bell (1997: 423) also placed *Bedenomeryx* Jehenne, 1988 in synonymy with *Palaeomeryx*.

Ampelomeryx Duranthon et al., 1995

Ampelomeryx was based on 15 complete skulls and associated skeletal material from MN 4 age localities in France and Spain. They show a bifurcated occipital appendage in adult males (unbifurcated in juvenile males, absent in presumed females), and triangular laterally flattened supraorbital appendages. The presumed males have a large upper canine and pneumatic frontal bone as well. Both males and females display deep lacrimal fossae and a single lacrimal foramen on the orbital edge of the fossa. Duranthon et al. (1995) argued that these specimens reveal the cranial appendages are dermal ossicones that only fuse to the frontal in old individuals. By contrast, the cranial appendages of the dromomerycines are outgrowths of the frontal bones, and thus Duranthon et al. (1995) consider them to be nonhomologous. As discussed above, Janis and Manning (1998b) and Janis (2000b) did not find this argument convincing and gave many other derived characters that supported the monophyly of Palaeomerycidae (both Old World and New World forms).

Triceromeryx Villalta et al., 1946

Triceromeryx is known from the MN 5 level in Spain and differs from *Ampelomeryx* in having a less strongly bifurcated occipital appendage with a more rounded cross section. Two species are recognized from Spain: *T. pachecoi*, the type species, and *T. turiasonensis*, from the MN 5 level in Ebro, Spain (Astibia and Morales, 1987). Janis and Scott (1987) pointed out that their dentitions are much more primitive than those of other palaeomerycids and placed *Triceromeryx* in Cervoidea *incertae sedis*. McKenna and Bell (1997: 423) considered *Hispanocervus* Viret, 1946 to be a junior synonym of *Triceromeryx*.

Ta<mark></mark>uromeryx Astibia et al., 1998

Tauromeryx is also known from the MN 5 level in Spain but differs from *Ampelomeryx* and *Triceromeryx* in having thinner, more rounded, and sharper supraorbital ossicones (Gentry et al., 1999). Its dentition is less brachydont, and its limb bones are more robust than those in other contemporary palaeomerycines.

NORTH AMERICAN PALAEOMERYCIDS

The current systematics of all the palaeomerycids from North America was reviewed by Prothero and Liter (in press). The section below is a summary of the key taxa and their status.

Subfamily ALETOMERYCINAE Frick, 1937

Tribe ALETOMERYCINI

Aletomeryx Lull, 1920

Aletomeryx (Fig. 19.3) was the first immigrant palaeomerycid, arriving in North America in the latest Arikareean of Nebraska and found in the early Hemingfordian of Nebraska, Colorado, and California. Both males and females had short upright cylindrical cranial appendages. It was also a relatively small palaeomerycid but nonetheless had very long and slender metapodials. Hundreds of specimens of the type species, *A. gracilis*, are known from *Aletomeryx* Quarry and from other quarries in the early Hemingfordian Runningwater Formation of Nebraska. *A. marshi* Frick, 1937, *A. scotti* Frick, 1937, and *A. lugni* Frick, 1937 are all considered to be junior synonyms of *A. gracilis*. Prothero and Liter (in press) recognized only two other valid species: *A. marslandensis* from the early Hemingfordian of Nebraska, which is considerably larger than material referred to *A. gracilis*; and *A. occidentalis* Whistler, 1984, based on fragmentary material from the early Hemingfordian Boron l.f., in the Mojave Desert of California. This species is disjunctly larger than either of the Nebraska species.

Sinclairomeryx Frick, 1937

Closely related to *Aletomeryx* was *Sinclairomeryx*, known from a variety of late Hemingfordian localities in Nebraska

Fig. 19.4. Restoration of *Sinclairomeryx riparius* (modified from Scheele, 1955, after Frick, 1937).

and also Saskatchewan. Male skulls have spectacular cranial appendages that curve forward and downward in a broad arc over the orbits and are terminated by small knobs (Fig. 19.4). A large sample is known from the late Hemingfordian Box Butte and Sheep Creek formations, which Frick (1937) split into three species: *S. riparius*, *S. sinclairi*, and *S. tedi*. The latter two species are junior synonyms of the type species, *S. riparius*, according to Prothero and Liter (in press).

Subfamily DROMOMERYCINAE Frick, 1937

Tribe DROMOMERYCINI Frick, 1937

Drepanomeryx Sinclair, 1915

Drepanomeryx falciformis was originally based on a fragmentary horn pedicle that had a distally widened tip, flattened ovoid profile, and anterodorsal curvature, somewhat like a very thick heavy blunt saber (Fig. 19.5). The type specimen and nearly all other material came from the early Barstovian of Nebraska, except for a few specimens from the early Barstovian of the Texas Gulf Coastal Plain. Frick (1937) named another slightly different set of curved horns *D. (Matthomeryx) matthewi*. These horns narrowed distally rather than expanding as in *D. (D.) falciformis*. Prothero and Liter (in press) examined all the known material and regard the type of *D. falciformis* as simply an aberrant specimen of *D. matthewi*. They synonymized the two species (*D. falciformis* is the senior synonym) and discarded the redundant and undiagnosed subgenera of Frick (1937).

Rakomeryx Frick, 1937

Rakomeryx was based on a distinctive dromomerycine skull with thick bowed horns (Fig. 19.5) from the Barstow Formation (named for Joe Rak, Frick's main collector in the Barstow area). This name is the first valid one for material

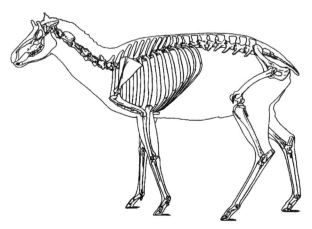

Fig. 19.3. Skeleton of *Aletomeryx gracilis* (after Lull, 1920).

Fig. 19.5. Restoration of (left) *Rakomeryx sinclairi* and (right) *Drepanomeryx falciformis* (modified from Scheele, 1955, after Frick, 1937).

named by Matthew (1918a) as *"Cervavus" sinclairi,* so the senior species of the genus is now *R. sinclairi.* Frick named a number of additional species of *Rakomeryx,* mostly without diagnoses and none with adequate comparisons or statistical analysis. They included *R. jorakius, R. yermonensis,* and *R. gazini.* Prothero and Liter (in press) showed that all three of these species can not objectively be distinguished from the type material of *R. sinclairi,* so they are all junior synonyms, and this genus is also monotypic.

Dromomeryx Douglass, 1909

Dromomeryx was the first taxon of the family Palaeomerycidae to be known from a nearly complete skull and skeleton (Fig. 19.1), found in the early Barstovian of Montana. The original material was mistaken by Cope (1878b) for the moschid *Blastomeryx* but assigned to *Palaeomeryx* by Douglass in 1903 before he erected his new genus *Dromomeryx* in 1909 based on better material. *Dromomeryx* is distinctive in having large, broad forwardly inclined supraorbital cranial appendages. Although Frick (1937) recognized *D. whitfordi* and *D. pawniensis* in addition to the type species, *D. borealis,* Prothero and Liter (in press) found no justification for these species, so they are both classified junior synonyms of *D. borealis,* making it monotypic. *Dromomeryx borealis* is first found in the late Hemingfordian of Nebraska, and by early Barstovian, it is widespread in Montana, Oregon, Nevada, Nebraska, New Mexico, and Colorado. It also occurs in the late Barstovian of Nebraska, Colorado, Wyoming, Nevada, and California. A few specimens are known from the early Clarendonian of California and South Dakota.

Subdromomeryx Frick, 1937

Subdromomeryx bears the same cranial appendages as *Dromomeryx,* so Frick (1937) placed it as a subgenus of *Dromomeryx,* with four named species. The primary distinctions are much smaller body size and a few other diagnostic features (cited in Janis and Manning, 1998b; Prothero and Liter, in press). In their analysis of the material, Prothero

and Liter (in press) found that there was no justification for the subgenus and so raised *Subdromomeryx* to generic rank. They also argued that the type species, *S. antilopinus,* is the senior synonym of all the other named species (*S. scotti, S. wilsoni,* and *S. kinseyi*), so the genus is now monotypic.

Tribe CRANIOCERATINI Frick, 1937

A second major lineage (Fig. 19.6) of North American palaeomerycids is the cranioceratins, which appear in the late Arikareean (*Barbouromeryx*) simultaneously with the appearance of the more primitive Aletomerycini. As Janis and Manning (1998b) point out, because both lineages are widely separated on the cladogram, this suggests at least two separate immigration events to North America. Unlike the other groups, cranioceratins are distinguished by supraorbital cranial appendages that lack the basal flange seen in dromomerycins and, in most male skulls of most taxa, a medial occipital horn. Their limbs tend to be shorter and more heavily proportioned than those in the dromomerycins, especially compared to *Aletomeryx.* Unlike the dromomerycins, which vanished after the early Clarendonian, the cranioceratins persisted until the late Hemphillian or earli-

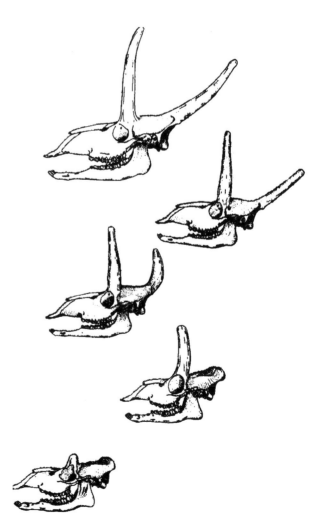

Fig. 19.6. Lateral views of cranioceratin skulls of (bottom to top) *Barbouromeryx trigonocorneus, Bouromeryx americanus, Procranioceras skinneri, Cranioceras unicornis,* and *Pediomeryx hemphillensis* (after Webb, 1983:Fig. 1).

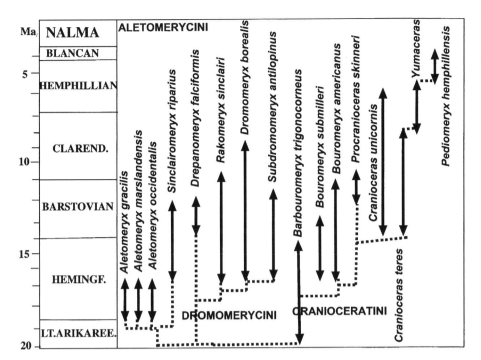

Fig. 19.7. Temporal ranges of valid dromomerycine taxa, plotted against the time scale of the North American land mammal ages (NALMA), after Tedford et al. (2004) and Janis and Manning (1998b). Solid line terminated by arrows shows known temporal ranges; dashed line shows inferred temporal ranges (based on cladogram in Janis and Manning, 1998b) and ghost lineages. Lt. Arikaree. = Late Arikareean; Hemingf. = Hemingfordian; Clarend. = Clarendonian.

est Pliocene (*Pediomeryx*). This was long after they had vanished from Eurasia, so they were the very last members of their family (Fig. 19.7).

Barbouromeryx Frick, 1937

The earliest known cranioceratin is *Barbouromeryx* (Fig. 19.6), which first appears in the latest Arikareean of Nebraska and also occurs in small numbers in the early Hemingfordian of Nebraska, South Dakota, Colorado, and Texas and in the middle Hemingfordian of Nebraska and Wyoming. The genus is distinctive in having short upright supraorbital cranial appendages with bulbous tips and a medial occipital appendage that is short and bulbous as well. The type skull was originally named *Dromomeryx trigonocorneus* by Barbour and Schultz (1934) but then renamed *Barbouromeryx* by Frick (1937). Frick also named two other subgenera: *Probarbouromeryx*, with one species, *B. (P.) sweeti, and Protobarbouromeryx*, with one species, *B. (P.) marslandensis*. Prothero and Liter (in press) found no morphological or statistical justification for Frick's largely undiagnosed subgenera and species and reduced all of them to *B. trigonocorneus*.

Bouromeryx Frick, 1937

Bouromeryx is a slightly larger cranioceratin with medium-sized supraorbital appendages and a short occipital cranial appendage. In addition to larger body size and longer horns, it also has a number of other derived features (listed by Janis and Manning, 1998b, and Prothero and Liter, in press). Although there are still no good complete adult male skulls of the genus, there are enough partial skulls and broken horn cores to show what it looked like (Fig. 19.6). *Bouromeryx* was very widespread in the late Hemingfordian of Montana, Oregon, Nevada, California, Nebraska, and Saskatchewan, and

even more broadly distributed in the early Barstovian of Montana, Oregon, Nevada, California, Colorado, Nebraska, Florida, and Texas. By the late Barstovian it was restricted to Nebraska, Montana, and Texas, after which it vanished. As befitting a widespread and usually fragmentary taxon dealt with by numerous authors (including hypersplitters such as Frick), no fewer than eight junior synonyms of *B. americanus* (Douglass, 1899) are now known, all of which were synonymized by Prothero and Liter (in press). Only the distinctly smaller late Hemingfordian taxon *B. submilleri* Frick, 1937, was recognized as valid by Prothero and Liter (in press).

Procranioceras Frick, 1937

Procranioceras was originally erected as a subgenus of *Cranioceras* by Frick (1937). It is based on a spectacular complete type skull from the late Barstovian of Nebraska (Fig. 19.6). It is much larger than *Bouromeryx*, with much longer supraorbital and median occipital cranial appendages. Unlike *Cranioceras*, it has a shorter more anteriorly curved medial occipital appendage. Even Frick (1937) recognized only one species, *P. skinneri*, and no new material has changed this conclusion. Janis and Manning (1998b) and Prothero and Liter (in press) found no justification for including it as a subgenus of *Cranioceras* (making the latter genus paraphyletic), so they raised *Procranioceras* to generic rank.

Cranioceras Matthew, 1918

The first recognized specimen of *Cranioceras* was a fragmentary horn pedicle from the early Clarendonian of New Mexico described by Cope (1874) as *Cosoryx teres* (but it does not pertain to the pronghorn *Cosoryx*). Later, Matthew (1918) described *Cranioceras unicornis,* based on a fragmental occipital cranial appendage that clearly showed that it was straight and pointed posteriorly. Frick (1937) transferred Cope's species "*Cosoryx*" *teres* to *Cranioceras* but added four additional species to the genus without adequate diagnoses or descriptions. Prothero and Liter (in press) recognized only two species in the genus *Cranioceras*: the larger species, *C. unicornis*, and the distinctly smaller species, *C. teres*. *Cranioceras* (Fig. 19.6) is distinct from *Procranioceras* in its long, upright supraorbital appendages that taper distally and curve slightly inward and long posteriorly directed occipital cranial appendage. *Cranioceras* actually occurs earlier (early Barstovian of California, New Mexico, and Nebraska) than the more primitive late Barstovian taxon *Procranioceras*. *Cranioceras* also occurs in the late Barstovian of Nebraska and New Mexico, the early Clarendonian of California, New Mexico, Texas, South Dakota, and Nebraska and persists until the late Clarendonian of California and Nebraska.

Yumaceras Frick, 1937, and *Pediomeryx* Stirton, 1936

Yumaceras figginsi was named by Frick (1937) based on very incomplete material from the middle Hemphillian Wray localities in Colorado. Just months earlier, Stirton (1936) had named *Pediomeryx hemphillensis* (Fig. 19.6) based on slightly better material from the late Hemphillian of Texas. Both genera were later shown by Webb (1983) to have long occipital cranial appendages, making them cranioceratins. Unlike *Cranioceras*, the occipital appendage is much longer and rises vertically, not posteriorly, and has an anteroposteriorly flattened tip. Stirton (1944) suggested that Frick's *Yumaceras* might be a junior synonym of his 1936 genus *Pediomeryx*, and Webb (1983) placed *Yumaceras* as a subgenus of *Pediomeryx*. This decision was also followed by Janis and Manning (1998b), although they also noted that the two genera were as distinct as any of the subgenera recognized in the Palaeomerycidae. Prothero and Liter (in press) argued that these genera deserve to be distinct because none of the subgenera in the family have proven justifiable. *Pediomeryx* can be distinguished from *Yumaceras* by its much more hypsodont cheek teeth and much smaller size. According to Webb (1983) and Janis and Manning (1998b), *Yumaceras* has three valid species, *Y. figginsi*, *Y. hamiltoni*, and *Y. ruminalis*. It occurs in the late Clarendonian of Florida and Texas, the early Hemphillian of Florida, Texas, Oklahoma, Oregon, and California, and the middle Hemphillian of California, Colorado, Nebraska, and Texas. *Pediomeryx* has only one valid species, *P. hemphillensis* (senior synonym of two Frick species). It is mostly known from the middle to late Hemphillian of Florida, California, Oklahoma, Texas, and Nebraska. As reported by Webb (1983: Fig. 7), the *Yumaceras-Pediomeryx* lineage shows increasing body size (*Yumaceras*) through the early and middle Hemphillian, then decreasing body size (*Pediomeryx*) until the lineage (and the family) vanished in the late Hemphillian.

PALEOECOLOGY

Janis (1982), Janis et al. (1994), and Janis and Manning (1998b) compared the diverse range of body sizes and cranial appendages in palaeomerycids to the variation seen in modern antelopes of Africa. According to Janis and Manning (1998b), the cranial appendages were skin covered (as evidenced by a healed broken horn in *Sinclairomeryx*) and probably used both for lateral display and for butting and neck wrestling. Most palaeomerycids were apparently dimorphic in their cranial appendages, and some had a deep lacrimal vacuity, which may have housed a scent gland seen in most living cervids. This combination of features suggested to Janis and Manning (1998b) that the more primitive palaeomerycids were territorial and lived in dense brush and woodlands, as do many modern antelopes. By contrast, the lack of dimorphism in *Aletomeryx*, combined with the proportionally longest limbs of any palaeomerycid, suggests that these animals may have preferred a more open habitat, as do modern gazelles.

Semprebon et al. (2004a) studied the mesowear, microwear, and the skull proportions of the full range of North American palaeomerycids. They found that most had features indicative of normal browsing habitat, consistent with their relatively low-crowned teeth. Only the more derived late Clarendonian and Hemphillian cranioceratins (*Cranioceras*, *Yumaceras*, and *Pediomeryx*) show signs of mixed feeding or some grazing, and some of these taxa are also more hypsodont as well.

As already mentioned, Webb (1983) noticed a trend of increasing size followed by size decrease in the *Yumaceras–Pediomeryx* lineage, which is comparable to the size trend found in many other North American lineages during the Hemphillian. According to Webb (1983) and Janis and Manning (1998b), this may have represented an attempt by the last of the palaeomerycids to survive in the face of reduced woodland habitats and expanding grasslands. The last of the palaeomerycids occurs in the earliest Pliocene (latest Hemphillian) in Nebraska, where it coexists with the earliest true cervid in North America. This is proof that the two groups briefly overlapped in time. The appearance of cervids may have driven the palaeomerycids to extinction. Alternatively, Janis and Manning (1998b) argue that the cervids may have been better adapted to the more open arid grassy habitats of the Pliocene, and dromomerycids could not find a tropical forest refuge to ensure their survival.

ACKNOWLEDGMENTS

We thank G. Rössner and C. Janis for helpful reviews of this chapter. We thank J. J. Flynn, C. Norris, and J. Meng for access to specimens in the AMNH. Funding for museum travel for this research was provided by the Donors of the Petroleum Research Fund, administered by the American Chemical Society, and by NSF grant EAR03-09538.

20

Family Cervidae

T HE DEER FAMILY CERVIDAE is the most diverse of the Artiodactyla/
Cetartiodactyla after the Bovidae. The exact number of species is uncertain but
has certainly been underestimated in the past, both because the important
"catch-all" species *Cervus elaphus* has turned out to be nonmonophyletic (Pitra et al.,
2004) and because an overzealous application of the biological species concept has
led most authorities, up to the 1990s, to shoehorn allopatric taxa into as few species
as possible. Proposals to rectify this as far as East Asian deer are concerned have been
made by Groves (2006), and for Venezuelan *Odocoileus* by Molina and Molinari (1999).

In this chapter I attempt to combine disparate sources of evidence, including fossils and molecular genetics, into an overall picture of the Cervidae. Essentially, I attempt to fit confirmatory data from published sources into the jigsaw rather than
bring forward new sources of evidence.

RELATIONSHIPS OF THE CERVIDAE

Among the living fauna there has long been an assumption that the sister group
of Cervidae is the family Moschidae, the musk deer (see, for example, Gentry, 1994).
Probably this is a legacy of old arrangements, lasting up to about the middle of the
twentieth century (Ellerman and Morrison-Scott, 1951), whereby the musk deer
were regarded as constituting a subfamily, Moschinae, of the Cervidae. A recent molecular study has cast doubt on this. Using three mitochondrial and four nuclear sequences, Hassanin and Douzery (2003) found that the Moschidae constitute a sister
group not to the Cervidae but rather to the Bovidae, with 95% bootstrap support.
Applying a molecular clock, these authors estimated that the Cervidae would have

separated from the Bovidae/Moschidae clade about 27–28 million years ago; the combined group separated from a Giraffidae/Antilocapridae clade somewhat earlier.

The fossil relationships of the Cervidae were extensively reviewed by Janis and Scott (1987). These authors regarded *Lagomeryx* as the earliest member of the Cervidae; this early Miocene genus has frequently been assigned to its own family (to which two other Miocene genera, *Stephanocemas* and *Dicrocerus,* have often been assigned), but Janis and Scott (1987) drew attention to some apparently derived conditions that it shared with at least some of the Cervidae, including a lacrimal fossa (which is not, however, confined to the Cervidae) and a *Palaeomeryx* fold in the lower molars. It also had appendages that they identified as antlers—long pedicels crowned by an arrangement of palmated tines— although there is no sign that they were or could ever have been deciduous.

Certainly, there should be some objective criteria for combining taxa into a single family or separating them into different families (and, for that matter, similar criteria for designating genera as well). Goodman et al. (1997), working on primates, suggested a time criterion: in order to be considered distinct families, monophyletic clades should have separated at least by the Oligocene–Miocene boundary (and, for genera, by the Miocene–Pliocene boundary). Groves (2001: especially pp. 17–20) brought fossil data to bear on ungulates and carnivores and modified the proposals slightly. The time–depth criterion would work well for the living fauna; in combination with the plesion concept, it might not be too problematic for fossils either, given adequate evidence. The earliest *Lagomeryx* dates from MN 3, as does the earliest undoubted cervid, *Procervulus;* this makes its position a little equivocal, but an earliest Miocene separation is perfectly feasible, and including *Lagomeryx* in the Cervidae seems justifiable at least provisionally.

The fossil sister group to the Cervidae, according to both Janis and Scott (1987) and Gentry (1994), is the family Palaeomerycidae. Janis and Scott (1987) recognized a further family, Hoplitomerycidae for the genus *Hoplitomeryx,* but Gentry (1994) argued that this genus should be included in the Palaeomerycidae. The early Miocene genus *Dremotherium* would be plesion to the cervid/palaeomerycid clade (Ginsburg et al., 1994; Gentry, 1994).

The origin of antlers has been discussed by Bubenik (1990, and elsewhere), Ginsburg and Azanza (1991), and Gentry (1994). Those of *Procervulus* apparently were never shed, in that they lacked a burr; they were present in males only (Ginsburg and Azanza, 1991), as in almost all modern cervids. Those of *Dicrocerus,* a genus known first from MN 5, were cast; specimens with small antlers were identified by Ginsburg and Azanza (1991) as females.

RELATIONSHIPS WITHIN THE CERVIDAE

Groves and Grubb (1987) divided the Cervidae into three subfamilies: Hydropotinae, for the antlerless Chinese water deer, genus *Hydropotes;* Odocoileinae for the New World deer ("neocervines") plus reindeer (*Rangifer*), moose (*Alces*), and roe deer (*Capreolus*); and Cervinae for the remaining Old World deer including the muntjacs. This was predicated on the fact that the reduction of lateral metacarpals has proceeded differently in the two antlered groups: the Odocoileinae have retained the distal ends, whereas the Cervinae have retained the proximal ends. These two types are known as telemetacarpal and plesiometacarpal, respectively, terms that are not available in a formal nomenclatural sense because they are not based on generic names but have nonetheless achieved wide currency. A cranial character separating the two antlered groups was described by Bouvrain et al. (1989). In all plesiometacarpal deer, the postglenoid foramen is entirely within the squamosal; in the telemetacarpal deer, on the contrary, the petrous forms its medial border. There appear, finally, to be behavioral features that also separate the two groups (Cap et al., 2002). These studies not only confirm that moose and roe deer truly are part of the neocervine clade but also render unnecessary the argument of Bubenik (1990) and Groves and Grubb (1990) that antlers evolved independently in muntjac and cervines.

Within Groves and Grubb's (1987) subfamily Odocoileinae, the neocervines and reindeer are characterized by a complete division of the choanae by the vomer, which forms a vertical partition. The common possession of this undoubted synapomorphy clearly unites them to the exclusion of moose and roe deer.

Does the only living antlerless genus, *Hydropotes,* fit into one of these two groups, or is it sister to them both? Implicit in the first view is that the lack of antlers of this genus is secondary; if *Hydropotes* occupies the second position, it might well have been antlerless from the start. *Hydropotes* is telemetacarpal, but a possibility exists that this condition may actually be primitive (it is also the condition in the Moschidae and Bovidae); its common occurrence in *Hydropotes* and the odocoileines would therefore be a symplesiomorph. Bouvrain et al. (1989) and Cap et al. (2002), however, both argued, using cranial and behavioral evidence, respectively, for the first view: that *Hydropotes* is related to the odocoileines.

Molecular studies have universally substantiated the telemetacarpal/plesiometacarpal split. Miyamoto et al. (1990) and Li and Sheng (1998), both using the mitochondrial cytochrome *b* gene, confirmed a close relationship between *Muntiacus* and *Cervus.* Randi et al. (1998a), using the same gene and the nuclear ê-casein gene, considered the monophyly of the telemetacarpal group probable but could not confirm it because of low support values and in the end opted for three equal clades: the telemetacarpalians with the choanal division, those without, and the plesiometacarpalians. Hassanin and Douzery (2003) and Pitra et al. (2004), however, found clear support for the two groups.

The hypothesis of Bouvrain et al. (1989) and Cap et al. (2002), that the affinities of *Hydropotes* lie with the other (antlered) telemetacarpalians, is abundantly confirmed by molecular studies. In addition, Randi et al. (1998a), Hassanin and Douzery (2003), and Pitra et al. (2004) have all positioned it as a clear sister group to *Capreolus*. Evidently, its ancestors had antlers, but these have been lost during evolution.

Finally, a reorganization of nomenclature has proved necessary (Grubb, 2000). The subfamilial name Odocoileinae, used by Groves and Grubb (1987) and many other authors, dates from 1923 and is antedated by more than a century by Capreolinae, Rangiferinae, and Alcinae. Accordingly, it is not the correct name for the telemetacarpalians.

In summary, the classification of genera of living deer accepted here is as follows (modified from Grubb, 2000):

Subfamily Lagomerycinae Pilgrim, 1941
Plesion *Procervulus*
Subfamily Pliocervinae Symeonidis, 1974
Subfamily Capreolinae Brookes, 1828
    Tribe Rangiferini Brookes, 1828
        Odocoileus
        Blastocerus
        Ozotoceros
        Hippocamelus
        Mazama
        Pudu
        Rangifer
    Tribe Capreolini Brookes, 1828
        Capreolus
        Hydropotes
    Tribe Alceini Brookes, 1828
        Alces
Subfamily Cervinae Goldfuss, 1820
    Tribe Muntiacini Knottnerus-Meyer, 1907
        Elaphodus
        Muntiacus
    Tribe Cervini Goldfuss, 1820
        Dama
        Axis
        Rucervus
        Panolia
        Elaphurus
        Cervus

MIOCENE CERVIDS

Pride of place among early cervids must go to *Procervulus dichotomus,* from the early Miocene (MN 3–5) of Europe (Gentry, 1994; Azanza and Menendez, 1990). The short antlers, confined to presumed males, sit on the end of long pedicels that are parallel to each other; most examples have two prongs, but a few have three. The surface sculpture of the antler suggests strong vascularization, implying the presence of velvet (Bubenik, 1990; Azanza, 1993a), but there is no burr, so it has been inferred that it was never shed. Ginsburg and Bulot (1987), however, found two antlers with sharp downwardly concave breaks, which they argued might after all represent the earliest form of natural shedding. Several species of *Procervulus* are known; they were re-

vised by Rössner (1995), who cautioned that the relationships of the genus are still very difficult to resolve.

Procervulus was succeeded by *Heteroprox,* from MN 5–7, and the two are commonly combined in a subfamily Procervulinae; and by *Euprox,* from MN 6–10, which is considered by Gentry (1994) to be developing toward modern deer. In *Euprox,* the posterior of its two tines is longer and stronger than the anterior, as if it were antecedent to a beam. The immediate precursors to the crown-group cervids in Gentry's (1994) scheme were the late Miocene *Pliocervus, Cervavitus,* and *Cervocerus:* telemetacarpal deer (where this condition is known), retaining long maxillary canines and the *Palaeomeryx* fold, but with short pedicels and three or more tines.

In the phylogenetic scheme of Gentry (1994), the *Lagomeryx* clade includes *Stephanocemas* and *Dicrocerus,* but Azanza (1993b) identified a "protocoronet" in these two genera and assigned them a position intermediate between *Procervulus* and the crown group of modern deer. For Azanza (1993b), *Euprox* is within the crown group and belongs to the muntjac clade; like Bubenik (1990), she would place the muntjacs outside the clade to which all other modern deer, with their shortened pedicels, belong; this would require that the plesiometacarpal condition was acquired independently, and in addition such an interpretation now appears incompatible with the consistent molecular findings.

CERVINAE: MUNTIACINI

The muntjac group are not primitive antlered deer, *pace* Azanza (1993b) and Bubenik (1990), the latter of whom was impressed by the normally nondeciduous nature of the antlers of the Bornean endemic *Muntiacus atherodes;* rather, muntjacs have apparently reverted to a form with small antlers and long pedicels as part of a forest (mainly rainforest) adaptive strategy ("duiker syndrome"). Like another small deer, the telemetacarpal genus *Pudu,* the Muntiacini have fused their cubonavicular and external and median cuneiforms into a single bone.

The reputed diversity of species in *Muntiacus,* with several new species described over the past 10 years or so (beginning with the giant muntjac, originally placed in a separate genus and named *Megamuntiacus vuquangensis*), is getting out of hand. It is surely time to gather the available specimens of little black muntjac in one place and compare them morphologically and morphometrically and run several mtDNA and nDNA sequences. That said, we may all now agree that the revision by Groves and Grubb (1990) was somewhat overlumped. There is also a good deal of geographic variability in the sister genus *Elaphodus* (the only other living genus of the Muntiacini), but with so little available material, Groves and Grubb (1990) were unable to make much of it.

The earliest muntiacin is the late Miocene (7–9 Ma) *Muntiacus leilaoensis,* from Yuanmou, Yunnan (Dong et al., 2004). Like so many deer, this is known only from antlers and pedicels, and we do not know how close it is to living *Muntiacus* or *Elaphodus* in other aspects of its morphology. It does, however, seem to confirm that antlers are reduced in such small-antlered taxa as *Muntiacus atherodes* and *Elaphodus* spp. and makes the Cervini/Muntiacini split earlier than depicted by Pitra et al. (2004).

CERVINAE: CERVINI

The phylogeny of the Cervini has been most recently explored by Pitra et al. (2004). It is possible that four genera (*Axis, Rucervus, Dama,* and *Cervus*) should be recognized in the living fauna, but the recent discovery of a 7- to 9-million-year-old muntjac (Dong et al., 2004) indicates that the calibration chosen by Pitra et al. (2004) for their molecular clock has evidently been set too late, and I here propose to recognize two further genera, *Panolia* and *Elaphurus.*

The genus *Axis* is now shown to have but a single living species, the chital or spotted deer *A. axis* from India, Nepal, and Sri Lanka. The hog deer (*Cervus porcinus*) and its relatives are not closely related to it; their common possession of slender antlers, built on a plesiomorphic three-point plan (Geist, 1998), would seem to have misled most commentators. Actually, the presumed primitive antler type may be derived even in *Axis axis* itself because the early Pleistocene Mediterranean *A. nesti* and *A. eurygonos* had antlers built on a four-point plan (Di Stefano and Petronio, 2002). The skull of *Axis* is distinctive. The anterior ends of the nasal bones are bifid, the lateral prong of each nasal being equal to or longer than the median, so that together they form a concave anterior margin. The posterior ends form a shallow, blunt wedge into the frontals. There are no upper canines. The lower central incisors are very wide, their width exceeding the combined width of the two lateral incisors plus the canine.

Rucervus contains the swamp deer (traditionally a single species, *R. duvauceli*) of Nepal and India and the presumed-extinct Schomburgk's deer (*R. schomburgki*) of Thailand. Pitra et al. (2004) extracted DNA from a Schombrugk's deer specimen and showed that it is indeed the sister species of the swamp deer and not a representative of a separate genus or subgenus *Thaocervus* as had sometimes been thought. Cranially, *Rucervus* resembles *Dama* and is readily distinguished from other Cervini by the form of the nasofrontal suture: the nasal bones together make a deep, acute-angled wedge back into the frontals. At their free ends, each has two prongs, a very long lateral one and a rudimentary median one, as in *Axis*. There are no upper canines.

The genus *Dama* (fallow deer) is closer to *Cervus* than are the previous two genera, according to Pitra et al. (2004). It contains two living species, the European (*D. dama*) and Persian (*D. mesopotamica*); these are quite distinct, despite the urge of many authors to make them subspecies of a single species. The fossil record of the genus does not begin until the middle Pleistocene, with *D. clactoniana,* of which the modern *D. mesopotamica* is essentially a size-reduced version, although Pfeiffer (1999) plausibly ascribed some late

Pliocene fossils to the *Dama* stem. The nasal bones resemble those of *Rucervus*. There are no upper canines (except as an anomaly). The lower central incisors are widened, as in *Axis.*

The giant deer, *Megaloceros,* has long been postulated to be closely related to *Dama* (Lister, 1994), a hypothesis that has recently been spectacularly confirmed by the sequencing of its DNA (Lister et al., 2005). The genus is first known from the middle Pleistocene of Europe and western Siberia; Lister (1994) documented the changes over time from the early *M. verticornis* and *M. savini* to the terminal Pleistocene/early Holocene *M. giganteus* and showed changes in body shape between open and woodland habitats. Lister et al. (2005) placed the split between *Megaloceros* and *Dama* at 4–5 Ma, which would be on the cusp of generic recognition in the scheme of Goodman et al. (1997), but they accepted the 7-Ma muntiacin–cervin split postulated by Pitra et al. (2004) as their calibration point, and, as suggested above, this might be too late. Azzaroli (1994) and Azzaroli and Mazza (1992b) noted resemblances between *Eucladoceros boulei* from Nihewan (China) and *Megaloceros* and suggested that they might represent an ancestor–descendant series.

The Eld's deer complex, here tentatively regarded as forming a separate genus *Panolia,* is cranially distinct from other Cervini, especially in that the posterior ends of the nasal bones form only a blunt, shallow wedge into the frontals, whereas anteriorly each nasal is single pointed, the points diverging from one another leaving a midline V-shaped gap—the form usually seen in *Cervus*. The antlers of this genus are highly distinctive, the brow tine and the beam forming a continuous, almost unbroken arc. Traditionally only a single species, *P.* (formerly *Cervus*) *eldi,* has been recognized. Balakrishnan et al. (2003) showed, however, that there is a deep split between a western clade, from Burma and Manipur, and an eastern clade, from the Indochinese region and Hainan. There are also considerable morphological differences between the two as well as between the dryland-living thamin of Burma and the critically endangered sangai, confined to the floating reed beds (phumdi) of Logtak Lake in Manipur. Groves (2006) suggests that Eld's deer be reexamined with a view to ascertaining whether they ought to be reclassified into two or even three distinct species.

The apparent affinities of the East Asian genus *Elaphurus,* Père David's deer, vary according to what system is being studied. As listed by Meijaard and Groves (2004b; Table 7), its morphological features (except for its unique antler conformation) generally recall those of *Cervus;* protein electrophoresis and a nDNA sequence (κ-casein) similarly align it *Cervus;* yet mtDNA puts it in a clade with *Panolia*. The best explanation of this at present is that it originates from an ancient hybridization between stem representatives of *Cervus* (male) and *Panolia* (female) (Meijaard and Groves, 2004b; Pitra et al., 2004). *Elaphurus* is known as far back as the late Pliocene (*E. bifurcatus*), and according to the modification of the molecular clock of Pitra et al. (2004) that I propose here, its separation from *Panolia* would date from about the Miocene–Pliocene boundary. Deer hybridize readily, at least within the same genus, and it may be that other species will also be shown to be of hybrid origin, but no case at present seems as plausible as that of Père David's deer. Y-chromosome DNA analysis in the Cervidae would be of enormous interest.

The genus *Cervus* is cranially most similar to *Panolia*, except that the posterior nasal bones form a distinct, but relatively short, point. The males possess small canines. Even shorn of some of its erstwhile components, the genus is large and unwieldy. As argued by Pitra et al. (2004), specializations of the display organs (antlers, mane, rump patch, voice) are indicators of habitat and reproductive seasonality rather than of phylogenetic affinity. The three well-separated clades are as follows:

1. Tropical clade, including *C. porcinus* (hog deer) and its relatives, *C. timorensis* (rusa) and *C. unicolor* (sambar) and its relatives. The so-called *Axis lydekkeri*, from 1.5-million-year-old deposits at Sangiran, Java, is related to *C. porcinus* and may reflect the plesiomorphic morphology of this clade, according to Meijaard and Groves (2004b). Di Stefano and Petronio (2002) suggest that the so-called *Rusa elegans* and *R. hilzheimeri* of the late Pliocene already possess derived conditions in common with *Cervus unicolor*; if so, then the diversification of the lineage must have begun at least by then. Nothing is known of the fossil antecedents of *C. timorensis*.

2. Western temperate clade, *C. elaphus* and its relatives. Pitra et al. (2004) suggested that the geographically isolated Central Asian *C. yarkandensis* (including *bactrianus?*) constitutes a distinct species; the spotted form *C. maral* of Turkey and Iran and the small, secondarily simplified North African deer (introduced to Corsica and Sardinia) are further candidates for species status. The earliest representative of what is presumably this clade is the early Pleistocene *C. magnus*, which has a relatively simple four-point antler plan resembling that of a sika. The European/West Siberian *Cervus acoronatus*, which resembles modern *C. yarkandensis* in lacking the multi-tine antler "crown" of *C. elaphus sensu stricto*, appears at the time of the Matuyama/Brunhes boundary, 0.8 Ma; it is not until about 0.5 Ma that the earliest "crowned" deer, *C. eostephanoceros*, appears.

3. Eastern temperate clade, recently treated by Groves (2006). This includes *C. albirostris* (white-lipped deer), and a subclade containing the *C. nippon* group (sika) and the large Sino-Rosso-American deer (wapiti and shou), which form a progressive cline running southwest-northeast from *C. wallichi* (Tibet) via *C. macneilli* (Sichuan) and *C. xanthopygus* (Primoriye, Manchuria, East Mongolia) to *C. canadensis* of the Central Asian mountains and North America. One of the most unexpected but most consistently corroborated findings of molecular studies on deer has been that the wapiti and shou are not eastern subspecies of *C. elaphus*, as had always been as-

sumed, but are from an entirely separate clade, to which sika also belong. The early Pleistocene Chinese *Cervus grayi* is very close to sika (Di Stefano and Petronio, 2002); the fossil history of wapiti and shou is unknown.

The origin of the tribe Cervini itself is uncertain, but a number of European Pliocene and Pleistocene genera are assigned to it: *Croizetoceros*, *Arvernoceros*, *Eucladoceros*, *Pseudodama* (probably a synonym of *Dama*), and the enigmatic *Candiacervus* from the Pleistocene of Crete (Heintz and Aguirre, 1976; Heintz and Dubar, 1981; Lister, 1994; Azzaroli, 1992; Azzaroli and Mazza, 1992a, 1992b). Perhaps the best known of these is *Eucladoceros*, characterized by its complex multibranched antlers; it is best known from the late Pliocene to early Pleistocene of Europe but extended well outside Europe to the late Pliocene site of Nihewan in China. As noted above, Azzaroli (1994) and Azzaroli and Mazza (1992b) hypothesized that the Chinese species, *E. boulei* would constitute a plausible ancestor for *Megaloceros*.

CAPREOLINAE: CAPREOLINI

It has been clear for some time that living roe deer (*Capreolus*) belong to two species, although the relative distributions of the two have yet to be completely delimited (they may have changed historically: the western *C. capreolus* seems to have extended its range at the expense of the eastern *C. pygargus*). Mitochondrial control region sequences separate them completely and indicate that the split between them goes back 2 to 3 million years; there are two clusters within each of the species, opening the possibility that the taxonomic diversity is still underestimated (Randi et al., 1998a).

The earliest known fossil of the tribe, *Procapreolus ucrainicus*, is late middle Miocene in age (about 10 Ma according to Lister et al. [1998], but 12 Ma according to Azanza and Menendez [1990]). Other species assigned to the same genus are known as late as the early Pliocene (Di Stefano and Petronio, 2002). Compared to *Capreolus*, the antlers are very similar but with weaker development of the "pearling" so characteristic of modern roe deer, and the two top tines are sometimes each bifurcated; there are primitive features in the teeth, and large canines are present in a specimen of *P. wenzensis* (Lister et al., 1998). The earliest species assigned to *Capreolus*, *C. constantini* from the middle Pliocene of Central Asia, more resembles modern *C. pygargus* according to Lister et al. (1998), who were uncertain whether this represents the beginning of geographic separation or indicates that *C. pygargus* is the more plesiomorphic of the two living species.

Because it now seems evident that *Hydropotes*, the Chinese water deer, is a second genus of this tribe, some of the evolutionary trends seen in the *Capreolus* lineage may make sense. The possession of more complex antlers in *Procapreolus ucrainicus* is evidently the primitive state, which became somewhat reduced in *Capreolus* and lost altogether in *Hy-*

dropotes, and the large canines of *P. wenzensis* may well be not primitive but derived in the direction of *Hydropotes.* These hypotheses will be tested by later discoveries.

CAPREOLINAE: ALCEINI

Most authors place all living *Alces* (elk [Europe] or moose [North America]) in a single species, but Boyesko-rov (1999) has convincingly argued that there are in fact two species: *A. alces* from Europe and western Siberia, and *A. americanus* from eastern Siberia and North America. The boundary seems to be the Enisei River. They differ in chromosome number, color and color pattern, body proportions, skull characters, and antler form. The small moose from the Manchuria/Primoriye region, which Boyeskorov refers to as *A. americanus cameloides,* has single-palm antlers like those of *A. alces* rather than the double-palm antlers of the larger *A. americanus* and may rank as a third species. The status of the recently extinct Caucasus moose has yet to be settled.

The fossil history of *Alces* is fairly well known. Azzaroli (1981, 1985) recognized a separate genus *Cervalces* for all the fossil species, reserving *Alces* for just the living forms. Lister (1993), on the other hand, incorporated them all in the same genus and traced the changes from early Pleistocene *Alces gallarum* via middle Pleistocene *A. latifrons* to the modern European *A. alces,* which appeared in the late Pleistocene. (An intermediate species, *A. carnutorum,* has sometimes been recognized at the early/middle Pleistocene boundary, between *A. gallarum* and *A. latifrons* [Heintz and Poplin, 1981].) The changes involved fluctuation in body size and directional modification of skull form and shortening of the antler beam, indicating the adoption of a more forested, less steppic environment (see also Breda and Marchetti, 2005, who retain the genus *Cervalces*).

In North America, modern moose appeared only at the end of the Pleistocene, where they replaced a late surviving population of *A. latifrons* (classed by Azzaroli as a distinct, endemic North American species, *A. scotti*).

CAPREOLINAE: RANGIFERINI

The living genera of Rangiferini (as understood here) are *Rangifer* (Holarctic reindeer and caribou), *Odocoileus* (white-tail, blacktail, and mule deer, from North America and the northwestern part of South America), *Mazama* (brockets, mainly South American but extending into southern Mexico), and the entirely South American *Hippocamelus* (huemul), *Blastocerus* (marsh deer), *Ozotoceros* (pampas deer), and *Pudu* (pudu). These genera form a well-defined clade, and, despite such unusual features as the possession of antlers in the female, *Rangifer* is an integral member, although it may well be sister taxon to the other genera (Pitra et al., 2004). The tribe is distinguished from Capreolini and Alceini by the de-

rived condition of the vomer (mentioned above), by the presence of a stylohyoid–paroccipital contact, and by the antler pedicels being well separated (Webb, 2000).

A cladogram has been proposed by Webb (2000), as follows:

((*Pudu* (*Rangifer* (*Hippocamelus*)))(*Mazama* (*Ozotoceros* (*Blastocerus, Odocoileus*))))

This was based on craniodental characters, but the molecular data of Pitra et al. (2004) indicate instead the following:

(*Rangifer* (*Blastocerus, Pudu*)(*Odocoileus, Mazama*))

The remaining two genera were not available.

Much work remains to be done on the distinctiveness and α-taxonomy of all these genera. At present, both *Hippocamelus* and *Pudu* are assigned two species, and there is no indication that this is not correct. *Blastocerus* and *Ozotoceros* are each considered monotypic, although the status of the supposed subspecies, especially of the latter, needs to be re-examined.

Odocoileus is a very difficult genus taxonomically. In North America there are two species groups, each currently being classed as a single species, but this is almost certainly overlumped. One problem is that white-tailed deer (*O. virginianus:* if this really is a single species, rather than a species complex) and mule deer (*O.* cf. *hemionus*) are known to interbreed in West Texas. It was discovered 20 years ago that in this region the two share a mtDNA restriction type characteristic of white tail, suggesting that interbreeding had occurred between male mule deer and female whitetails, and the whitetail phenotypic characteristics have been lost by generations of backcrossing with mule deer (Carr et al., 1986). A subsequent study (Cathey et al., 1998) confirmed that F_1 hybrids are rare and found that Y chromosome DNA, by contrast with mtDNA, assorted more clearly along species lines.

There is a decline in size down the east coast of North America, culminating in the diminutive deer of the Florida Keys (currently known as *Odocoileus virginianus clavium*). Study of mtDNA of deer in Florida, including Key deer, found three different haplotypes, but these corresponded only poorly with the described subspecies, including *clavium* (Ellsworth et al., 1994). This raises the question, frequently posed in recent years, of whether putative subspecies would better be defined as possessors of unique mitochondrial haplotypes rather than, as traditionally, on morphological characters. It has been argued that the function of the subspecies category is to delimit geographically restricted lineages; against this, I would maintain first that of course mtDNA is inherited only matrilineally, and the depicted lineages might be entirely different if Y chromosome DNA were studied (as in the West Texas hybridization study cited above). In any case, morphological characters are themselves (broadly speaking) heritable, and there is some value in continuing to recognize gene pools that are strongly divergent as a whole.

An analogous situation recurs in Venezuela, at the southern end of the distribution of whitetails. Molina and Molinari (1999) recognized three distinct species in Venezuela and suggested that there was evidence for a fourth. These are *O. margaritae* (Margarita I.), *O. lasiotis* (Mérida Andean Highlands), and *O. cariacou* (from the rest of the range). In contrast, a study of mtDNA by Moscarella et al. (2003) identified four clades, but these corresponded rather poorly with the proposed species. It is clearly time for a new look, utilizing both methods and using the same samples for both.

Mazama is also a taxonomically complex genus, with more species than the four that had been previously assigned to it (see, for example, Medellín et al., 1998, who identified a distinct species for Yucatan).

Rangifer (called reindeer in Europe, caribou in North America) are currently regarded as a single species, *R. tarandus,* which is broadly divided into woodland, tundra, and high-Arctic subspecies. The woodland forms (*R. t. fennicus* in the Old World, *R. t. caribou* in the New World) are very different in appearance—antler form, color, and build—from the tundra forms (*R. t. tarandus* in the Old World, *R. t. granti* and *groenlandicus* in the New), and there is said to be some seasonal overlap in range when tundra reindeer/caribou enter the northern fringes of the coniferous forest. Precisely how much interbreeding there may be, however, is not known. This situation suggests that speciation is well under way or complete, but before we can contemplate a new taxonomic arrangement, we need to know whether Old and New World woodland forms constitute a clade separate from the tundra forms or are separately derived from them (or even vice versa). The monophyly of the three high-Arctic forms, however, has already been investigated: the two New World forms, *R. t. pearyi* and *eogroenlandicus,* do indeed form a clade, related to the nearest tundra subspecies, *R. t. groenlandicus* (known as barren-ground caribou in Canada), whereas the Svalbard reindeer, *R. t. platyrhynchus,* although morphologically not dissimilar, is a clear derivative of Old World tundra reindeer (Gravlund et al., 1998). Tundra and high-Arctic reindeer are said to be seasonally sympatric on some of the Canadian high-Arctic islands and are strikingly different in appearance, but in this case one wonders how much the differences might be the result of phenotypic plasticity.

The earliest example of a living genus in the Americas is a species of *Odocoileus* from the early Pliocene; the earliest known of all American deer is *Eocoileus gentryorum* Webb, 2000, of the late Miocene (5 Ma) of Florida, which according to its describer is "most comparable to *Ozotoceros* and *Mazama.*" His placement of it on one of two major New World clades (see above), rather than at the base, implies that still earlier deer fossils may be expected in North America—as indeed would be predicted on molecular clock grounds (Pitra et al., 2004). Other genera described from the Americas are as follows:

Bretzia. Best known from the Rancholabrean (latest Pleistocene), the earliest possible occurrence of this genus, consisting of a small palmated antler, is also late Pliocene in age (Webb, 2000). Morejohn et al. (2005) recently added further material to the known sample of the genus, describing some unusual features of the carpals and tracing the evolution of its antler form from earliest through intermediate to latest occurrences.

Navahoceros. A somewhat stockily built mountain deer with small antlers. Best known from Rancholabrean deposits, it may go back to some 3 Ma (Webb, 2000).

Sangamona. Described by Kurtén (1979) as a stilt-legged deer from the late Pleistocene, is a phantom. Churcher (1984) performed a notable hatchet job on it, showing that the type specimen is indeterminable and that every other specimen that has been referred to it from time to time is also either indeterminable or else demonstrably something else (usually *Odocoileus*). No such deer ever existed.

Torontoceros. Described by Churcher and Peterson (1982), this genus is known only from a partial skull with heavy, horizontal but otherwise reindeer-like antlers, dated by ^{14}C at 11,315 ± 325 B.P.

Morenelaphus, Epieuryceros, Agalmoceros, Chaitocerus. These are poorly known deer from the middle (?) Pleistocene of South America. Their relationships are unclear.

CONCLUSION

The broad outlines of cervid phylogeny and taxonomy are now clear, but the details remain to be filled in. Further DNA work—nuclear, including Y-chromosome, as well as mitochondrial—is needed, and renewed morphological and behavioral studies are necessary to find where the species (*sensu* phylogenetic species concept) begin and end, and hence, what are the units of biodiversity in the genus.

EPILOGUE

During a visit in 1994 to the Institute of Ecology and Biological Resources, Hanoi, the late Shantini Dawson drew my attention to a very unusual frontlet and antlers labeled "Black Muntjac," collected on September 9, 1974, at Gia Lai, Kontum. It bears an uncanny resemblance to *Procervulus* in the symmetrical lack of a brow tine as well as in the remarkably straight, parallel antler beams. I sent the photos (see Fig. 20.1) to the late Tony Bubenik, who was extremely intrigued and excited by the resemblance.

Careful examination suggests, however, that this may be an unusual example of hog deer, *Cervus porcinus.* It does have a clear burr, unlike *Procervulus,* and the antlers are bent back just above the burr, as if to accommodate a brow tine whose development has been suppressed.

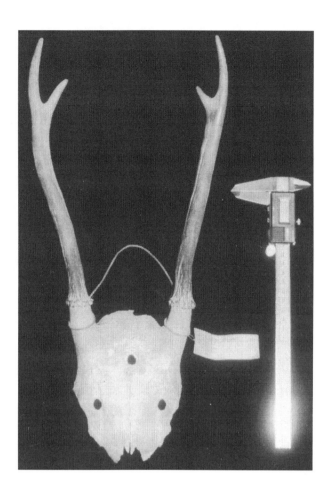

Fig. 20.1. "Black Muntjac" from Gia Lai, Kontum. Probably an unusual example of hog deer, *Cervus porcinus*.

The significance of the specimen is that it shows the degree to which occasional anomalies can mislead one's assessment of affinities but that the sort of crown cervid features elucidated by Janis and Scott (1987), Bubenik (1990), and Azanza (1993a) are sure guides to the true nature of a deer, however bizarre.

ACKNOWLEDGMENTS

I thank Nikos Solounias, Alan Gentry, and another referee for their useful comments. Erik Meijaard and the late Peter Grubb have been valued collaborators over the years, and without them, a chapter such as this could never have been written. My association with molecular workers, especially Christian Pitra and Joerns Fickel, has been fruitful, and I have learned from it. I learned more than I can say from my all too brief association with the late Tony Bubenik. Finally, I thank Don Prothero for inviting me to contribute to this volume.

21

Family Giraffidae

THE GIRAFFE AND THE OKAPI are the only living species of the Giraffidae. A celebrated species on its own merit, the giraffe also represents Africa. Every person knows of the giraffe, and Sherr (1997) has shown many of its popular expressions. Its trivial name, *Giraffa camelopardalis,* means spotted (pardali) camel (camela) in Greek. There are many extinct species of giraffids from Eurasia and Africa, and they were common in Eurasian and African faunas during the late Miocene (Bohlin, 1926; Hamilton, 1978a). The okapi (*Okapia johnstoni*) is not well known to the public but is fascinating to zoologists and to paleontologists because it resembles the many now extinct species in having a simple ruminant form with a short neck. Little is known about the okapi because it inhabits forbidding dense forests in Zaire, so it has not been well studied, and its relationship with the giraffe and the extinct giraffids is problematic. The okapi was discovered in 1901, and it was immediately envisioned as a modern representative of already known fossil forms such as *Palaeotragus, Helladotherium,* and *Samotherium,* taxa from the localities of Pikermi and or Samos in Greece. Its discovery was reminiscent to that of the coelacanth (*Latimeria*), which was found living long after similar fossils had been unearthed and studied. Even the Hollywood film industry's notion of unexplored deep rain forests containing still living taxa formerly presumed extinct, including dinosaurs, is in part inspired by the discovery of the okapi. Was the okapi, then, a survivor like a Miocene taxon of *Helladotherium* or *Palaeotragus?* Scientists have compared the okapi to other giraffids (Colbert, 1938c; Hamilton, 1978a; Lankester, 1907a, 1907b). A more recent interesting approach was that of Geraads (1986), who suggested a close affinity of the okapi to the giraffe on the basis of the presence of ossicones and strong pneumatization of the frontals. Fossil taxa did not have ossicones, but instead the horns were outgrowths of the frontal bone. It appears, however, that ossicones are more widespread than in *Giraffa* and *Okapia* and were probably present in all Giraffidae (Solounias, 1988a). Nevertheless, both *Giraffa* and *Okapia* inhabit central Africa, and a close affinity of only those two taxa may be possible.

Throughout this chapter, museums and collections are indicated by abbreviations.[1]

MORPHOLOGICAL DESCRIPTIONS AND EVALUATION

Palaeomerycidae as Sister Taxon of Giraffoidea and Origins of Giraffidae

Giraffoidea probably originated from Gelocidae before the early Miocene (Janis and Scott, 1987). Tragulidae are also good sister taxa of most ruminants including the Giraffoidea. Several phylogenetic analyses are available; for summaries see (Janis and Scott, 1987; Janis et al., 1998; Gentry et al., 1999; Gatesy and O'Leary, 2001; Fernández and Vrba, 2005). Clearly more research is needed to reveal the interrelationships of the ruminants. One could place the origin of Bovidae and Giraffidae around 19 Ma, but bovids may slightly predate giraffids. The Giraffoidea include the Giraffidae and Climacoceridae. The Palaeomerycidae (Prothero and Liter, this volume) could be closely related to Giraffoidea because they also have ossicones above the orbits. Janis and Scott (1987) emphasize that ossicones are rather unique in formation and, as such, may hint at close relationships of taxa with ossicones. One could say that Palaeomerycidae are the broad sister taxon to the Giraffidae. Palaeomerycidae may still be part of the Eucervoidea (Janis and Scott, 1987) and be close to the Giraffidae. Another potential sister taxon of the Giraffidae could be the Antilocapridae (Janis and Scott, 1987; Fernández and Vrba, 2005).

From the early and middle Miocene have come various species of Giraffoidea. Some of these giraffoids may be close to Giraffidae, but others are probably more distantly related. The origins and paleogeography of Giraffidae are not well known. There are potential sister taxa and archaic Giraffidae in both Eurasia and Africa (basal Siwaliks of Pakistan, Chios of Greece, Gebel Zelten in Libya as well as Muruarot Hill and Rusinga in Kenya). The fossil record is incomplete, and thus, detailed studies of these early species are difficult. Most of the late Miocene species are undisputed members of Giraffidae, but going back in time into the middle and early Miocene, it is difficult to recognize the

earlier species of giraffids (e.g., Morales et al., 1999). Hamilton (1978a) used the bilobed canine as the basic unifying character of the Giraffoidea and the large size of the second lobe on the canine as unifying the Giraffidae. Probably all giraffids possess a bilobed lower canine with an enlarged second lobe. Sivatheres also show the enlarged bilobed canine, indicating that they are indeed Giraffidae, although Janis and Scott (1987) were inclined to separate the sivatheres on the basis of their branched ossicones. In addition to the enlarged second lobe, it is presently proposed that the characteristic long slender limbs and the large body size of Giraffidae are also characters of the Giraffidae contrasting to the small-sized giraffoid relatives. In addition, the nasolacrimal canal appears to have been closed in the Okapiinae and perhaps even in forerunners of that clade. During the late Miocene and lingering afterward, there was a radiation of Giraffidae in Eurasia and Africa. Many late Miocene deposits in China, Southern Russia, Siwaliks of Pakistan, Greece, Spain, and North Africa had more than three species of Giraffidae per locality. The fossil record of Africa is poor, but the few giraffid fossils show that giraffids were present there as well (early and middle Miocene giraffoid fossils from Southwest Africa, Begila, Kalodirr, Muruarot Hill, and Maboko; during the Pleistocene, however, there are many localities with Giraffidae). The late Miocene Eurasian taxa *Samotherium, Palaeotragus,* and *Sivatherium* are commonly cited in studies and in textbooks. *Samotherium* and *Palaeotragus* are envisioned as similar to the okapi with short necks and simple ossicones. *Sivatherium* was also similar to the okapi, but it was a very large ruminant. With respect to diet, the giraffe is a browser, and the okapi may be either a browser or a fruit browser (Solounias and Semprebon, 2002). The extinct taxa were mostly browsers or mixed feeders but rarely grazers (Solounias et al., 2000).

Literature

Key descriptive papers are those of Bohlin (1926), Crusafont-Pairó (1952), Churcher (1970), Hamilton (1973, 1978a), and Godina (1979) on a variety of species from the late Miocene of China, Greece, Russia, Africa, and Spain. Studies that are more focused in resolving relationships include that of Colbert (1938c) on the relationships of the okapi, containing a key figure that displays a variety of skulls in profile with important proportional differences (Colbert 1938c: Fig. 1). Hamilton (1978a) and Geraads (1986) have presented cladograms on the systematics of the family that to a large extent still hold but also show that a more rigorous analysis is needed. These two studies provide a wealth of information on morphological characters and on various species. Fraipont (1907) is a classic on the zoology of the okapi with many nice plates of osteological comparisons to the giraffe. Dagg and Foster (1976) and Spinage (1968a) are textbooks mainly on the giraffe but with comments on the okapi. Lankester (1907a, 1907b) described the unusual ossicones of the okapi and suggested that these ossicones may include miniantlers. The okapi ossicones have bare apices

---

[1]Abbreviations: AMNH, American Museum of Natural History in New York; GSP, Geological Survey of Pakistan, currently at Peabody Museum at Harvard; NMK, National Museums of Kenya in Nairobi; NHM, Natural History Museum in London; NHMW, Natural History Museum in Vienna; PIW, Paleontological Institute at the University of Vienna; SMF, Senckenberg Museum in Frankfurt; SNHM, State Natural History Museum of Stuttgart; SPGM, Institute for Paleontology and Historical Geology in Munich; USNM, United States National Museum of Natural History, Smithsonian, in Washington.

with pits and constrictions. Bohlin (1926) provided a wealth of information on extinct taxa, especially those of the late Miocene of Shanxi in China. Bohlin pointed out the bare ossicones in *Palaeotragus microdon* and figured skulls of *Bohlinia*, *Samotherium*, and *Honanotherium*. Bohlin also included many diagrams of postcranial skeletal differences between the various species. Lankester (1908) described the proximal cervicals of the okapi and the giraffe and presented the trend of the first thoracic vertebra of the giraffe to be cervicalized. Solounias (1999, 2000) showed that the neck of the giraffe has an extra vertebra (a total of eight). Solounias has studied the premaxillary shapes and the paleodiet of various Miocene species (Solounias et al., 1988b; Solounias and Moelleken, 1993a; Solounias et al., 2000; Franz-Odendaal and Solounias, 2004).

A Working Cladogram, Descriptions, and Discussion

This presentation utilizes a working cladogram (Fig. 21.1) that has not yet been algorithmically tested, but it is offered as a framework for a revised presentation of the various morphologies and species. There is no space for separate descriptions that traditionally precede the cladogram. Thus, the nodes are also where a short description can be found. Each node lists derived characters and species and is followed by a brief discussion. Several new observations and original ideas are also introduced at various nodes. Expansion is not possible because of limitations of space. In addition, note that not all taxa are included, as Giraffidae are awaiting a major revision. For example, *Orangemeryx*, *Progiraffa*, and certain species from Russia, China, and North Africa need to be included. The selected group of taxa is provided as a starting point. A brief paleoecology and biostratigraphy of the group are also provided at the end.

Three proposed new genera need to be researched, and their detailed descriptions have not been completed. Quotes imply new systematic names are needed. Such taxa are included in this study under the names (1) *"Palaeotragus" tungurensis*—currently known as *Palaeotragus tungurensis* from Tung Gur, Mongolia, (2)*"Palaeotragus" primaevus*—currently known as *Palaeotragus primaevus* from Fort Ternan, (3)*"Palaeotragus primaevus"* currently known as *Palaeotragus primaevus* from Ngorora Formation (Baringo), Kenya, and (4) *"Samotherium" major*—currently known as *Samotherium boissieri* variety *major* from Samos, Greece. In addition, *"Palaeotragus primaevus"* from Ngorora, Kenya requires a new trivial name.

Node 1. Reduction of lateral digits in manus and pes and reduction of upper incisors; Pecora with secondary bone growth alongside the line of supraorbital foramina; Bovidae, Antilocapridae, and Cervidae are not part of this study but are at this node.

Node 2. Reduced front limbs; lower incisors specialized in forming a central cleft. Procession of rudimentary horn-

ossicone-type bony structures seen alongside their supraorbital foramina and fused to the frontal.

> Tragulidae Milne-Edwards, 1864
> > *Hyemoschus* Gray, 1845
> > > *Hyemoschus aquaticus* (Ogilby, 1841)
> > *Dorcatherium* Kaup, 1833
> > > *Dorcatherium naui* Kaup, 1833

In *Tragulus* and *Hyemoschus* the lower incisors are specialized in forming a central cleft where vegetation can be trapped and pulled. In *Tragulus* metapodials 3 and 4 fuse to form cannon bones. Inseparable third and fourth elements occur only in the hindlegs of *Hyemoschus*. Metapodials 3 and 4 support the two hooves. Digits 2 and 5 are reduced and are compressed onto 3 and 4. The upper incisors are lost in these Pecora. Tragulid morphology is not within the scope of this study. The tragulids are commonly accepted as the nearest or only living group related to Pecora (Gentry et al., 1999).

A major proposal of this study is that Tragulidae are basal Pecora because of their procession of small bone spurs (growths) along the supraorbital foramina, which can be interpreted as initial prehorn and preossicone structures (Fig. 21.2). Bone spurs are a type of secondary bone growth. Thus, this new character is presented for the tragulids that clearly needs to be investigated further. Most researchers consider tragulids as taxa with no horns-ossicones. It is proposed that *Hyemoschus* has some type of secondary bone growth in precisely the right place on the frontals that may be a precursor for the formation of ossicones.

Secondary bone growth is a term for accretions of bony spurs and bony lumps over the ossicones of the giraffe (Spinage, 1968b). These accretions are fused to the surface of the ossicone. It is not known if such secondary bone growth is generated by bony cells originating within the ossicones, by the overlying integument, or by an interaction of the two. It is also unknown if unfused accretions occur in the integument. Yet to be determined are how many types of secondary bone growth exist and if any of these is synonymous with "overgrowth." A preliminary distribution of secondary bone growth *sensu lato* is in certain Tragulids, *Palaeomeryx*, Bovidae (*Boselaphus*, *Tragoportax*, *Tetracerus*), and many Giraffidae. It occurs mostly in adults. In giraffes, the spurs result from clash combats of males, but they also have a genetic basis as they are found in captive individuals. Tragulidae (e.g., *Hyemoschus*) possess small oval surface ossifications that are fused on the frontals, which can also be termed growth spurs. These spurs are situated longitudinally along the lateral margin of multiple supraorbital foramina. These are definitely not horns of any description, but they are a form of secondary bone growth situated next to the supraorbital neurovascular bundle and may have a protective function (Fig. 21.2).

Node 3. Ossicones. The following description holds for giraffoid and giraffid ossicones with the exception of Cli-

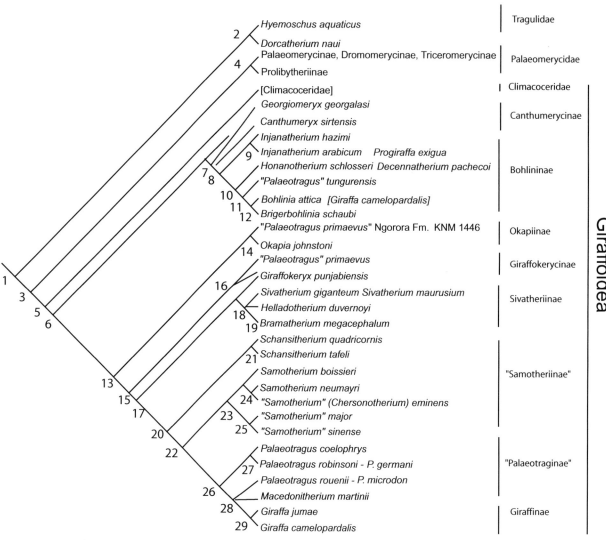

Fig. 21.1. A working cladogram of the Giraffidae and sister taxa. Thirteen key nodes are used to break the text discussion into sections. Additional characters are discussed with references to species or specimens but without assignment to the cladogram with numbers.

macoceridae. An ossicone is derived developmentally from a cluster of small oval ossifications that form a disk (Lankester, 1907a, 1907b). They are dermal ossifications that fuse later in development with the frontal bones. Ossicones are mostly skin covered. The ossicone is preformed in cartilage with subsequent endochondral ossification. In bovids the horn formation is induced by the connective tissue and dermis superficial to the frontal bone and its periosteum. Frontal bone and periosteum alone cannot form a horn. Complex bony lamellae form the horn, which is covered by keratin and is derived from the frontal bone. In bovids the ossification is intramembranous and similar to that of the flat bones of the skull. Janis and Scott (1987) summarize and categorize all types of horns and the relevant literature. The ossicone is more independent on the frontals as a cartilaginous disk, which can be moved by palpation. The bovid horn is intimately related to the connective tissue, dermis, and the frontal bone.

The ossicone grows in length and fuses to the frontal bone. In *Palaeomeryx* the shaft of the ossicone is bent medially, and there is disorganized secondary bone growth on the surface (Ginsburg and Heintz, 1966). The giraffe (*Giraffa*) ossicone displays an additional layer of growth later in adulthood, which has been termed secondary bone growth, and it resembles the original small oval ossifications of the developing ossicone. The secondary bone growth is extensive and covers the ossicone and the adjacent dorsal lateral and occipital part of the skull (Spinage, 1968b). The issues of the secondary bone growth of the giraffe and the prevalence of ossicones within the Giraffidae are complex and debated (Geraads, 1986). Solounias (1988a) has proposed that probably all giraffids have ossicones.

The secondary growth has appeared independently in *Okapia, Giraffokeryx,* Sivatheriinae, *Giraffa,* and a few other taxa. Spinage (1968b), Solounias (1988a), Solounias and Tang (1990), Solounias and Moelleken (1991), and Spinage (1993)

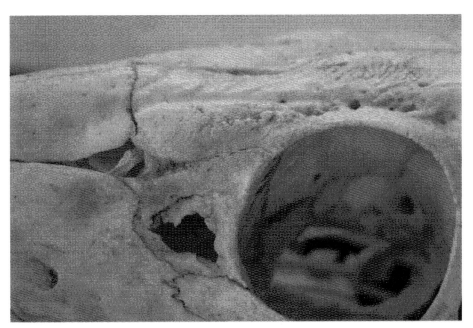

Fig. 21.2. Bony spurs in *Hyemoschus aquaticus* USNM 220395. (A) Lateral view; (B) close-up of supraorbital bony spurs.

A

supraorbital bony spurs

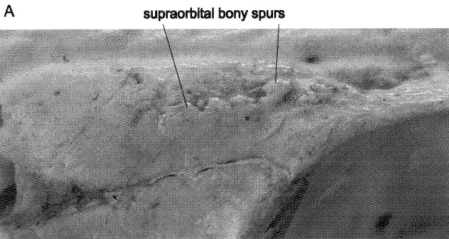

B

have addressed issues about ossicones. Solounias and Tang (1990) were wrong in suggesting that the median single ossicone of the giraffe is constructed only by secondary bone growth. Spinage (1993) has corrected this hypothesis and showed that the median ossicone is also based on an ossicone. It is probable that ossicones evolved from growth spurs that became detached and formed the small osseous disks observed in juvenile giraffes and okapis. The ossicones are larger and distinct at birth in the giraffe and the okapi and grow rapidly, but, unlike bovid horns, they do not fuse to the skull until later in life. The frontal grows upward into the growing ossicone, forming a substantial supporting boss. When growing, the ossicone mass is diffuse—it appears as clusters and subclusters of small oval disks of bone—and is full of vasculature and heavily invested integument. Later in life, it becomes more compact and often forms a smooth external surface. It appears that ossicones are not branched. In sivatheres, the apparent branches are

large bumps that occur, unlike those of the giraffe, in regular positions. The ossicone base more commonly has large bumps (excrescences) posteriorly or anteriorly (*Giraffokeryx* or *Sivatherium*, respectively). In *Bramatherium* (Lewis, 1939) the ossicones appear to be branched. However, the anterior ossicone pair of *Giraffokeryx* is so similar to that of *Bramatherium* that it is almost certain that they can be best interpreted as two fused ossicones rather than a single branched one. The same is true for the four ossicones of *Schansitherium tafeli* and *Schansitherium quadricornis;* these are species with four ossicones. Most ossicones grow long, retaining an oval cross section, and the frontal sinus invades the ossicone. Male combats of the giraffe are termed "necking." The long neck is used as a flexible club that wraps around the opponent's neck, and then the attacker's head administers lateral blows onto the head and neck of the opponent. Male giraffes also wrestle by pushing their chests and necks into each other. The okapi has short spikelike os-

Fig. 21.3. Apices of ossicones of *Okapia johnstoni*. (A) AMNH 51904; (B) AMNH 51902; (C) AMNH 51215. (D and E) Sectioned ossicone AMNH 51228: (D) parasagittal plane section; (E) median plane section.

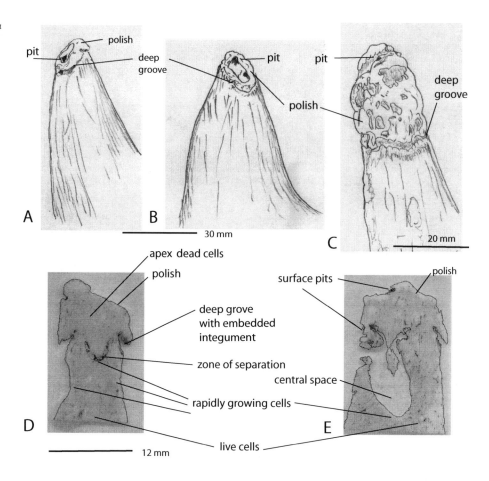

sicones, but its fighting is not well known. Almost all extinct species and the okapi have "hornless" females; the giraffe (*Giraffa camelopardalis*) does not. Ossicones are covered by integument in the giraffe (*Giraffa*), and it appears that the same situation was true in many of the extinct taxa. Vein impressions cover many fossil ossicones.

In the okapi, the ossicone apices are frequently bare (Fig. 21.3). Exposed bare bone is not common in nature. In deer antlers, when bone is exposed, the antler is dead and separated from the living pedicle below at the burr. In a dead antler, the blood circulation has ceased, and the periosteum and integument have peeled off. The okapi apex is a distinct piece of bone from the ossicone shaft. It has separated from the shaft during the maturation of the animal. The apex has a polished surface and often has small pits on its surface between the smooth polished regions. The ossicone shaft is covered by integument below the apex, and there is often a bony constriction at the termination of the integument, a deep groove separating the apex from the shaft. Often the integument is situated below the constriction, but presumably it always started at the constriction. No second constriction is formed by the integument in its final position of retreat. I have examined live okapis closely in zoos. Sections of a single ossicone from a wild specimen show a zone of separation of the apex from the shaft (Fig. 21.3). The separation is by an actual gap, which in this specimen (AMNH 55228) contains a median space. The gap is

peripheral and exits at the constriction, and the space is central to the ossicone. The space opens to the surface as a pit. The internal gap has normal cells above at the apex. The apical cells are separated by the gap from the cells below, which are in the ossicone shaft. The apical cells at the surface show a clean cut through them where the polish has occurred. This observation implies that cells or parts of cells are being stripped off. No remodeling or reorganization has been observed at the surface, and presumably the cells are dead. Below the gap an active cell growth zone is evident. This zone encapsulates the lower surface of the gap. It is a zone distinct from the cells below that comprise the shaft of the ossicone. It appears that the apical cells are separated from the live cells below, and therefore the apices are dead and hence do not bleed when an okapi scrapes tree bark with its ossicones. The okapi has several facets of polish on a single apex. So far, no other specimens have been studied.

In a few *Palaeotragus* and rarely in "*Samotherium*" *major*, the apex is also bare with wear facets (e.g., Bohlin, 1926: Text Fig. 2-5) (Fig. 21.4). These are also wear facets on bone. There are, however, differences from the okapi that are not presently studied. In the later taxa there are single large facets and no pits or constriction separating the shaft below from the worn apex. Similar in appearance are single large facets found on the horn sheaths of the nilgai (*Boselaphus tragocamelus*). These horn sheaths are keratin.

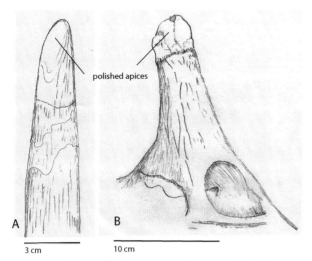

polished apices

A   B

3 cm   10 cm

Fig. 21.4. Wear facets on fossil ossicones. A, *Palaeotragus coelophrys* from Samos AMNH 20773 QX, lateral view; B, *"Samotherium" major* from Samos SMF 3600 posterolateral view. Note that SMF 3600 never grew the left ossicone.

Node 4. Occipital horns of frontal origin positioned posteriorly at the nuchal crest (at the median plane of the skull).

Palaeomerycidae Lydekker, 1883
  Palaeomerycinae Lydekker, 1883
    Palaeomeryx von Meyer, 1834
  Dromomerycinae Frick, 1937
  Triceromerycinae new rank
    Triceromeryx Villalta, Crusafont Pairó, and Lavocat, 1946
  Prolibytherinae new rank
    Progiraffa Pilgrim, 1908
    Prolibytherium Arambourg, 1961

Palaeomerycidae (including, as subfamilies, Palaeomerycinae, Dromomerycinae, Triceromerycinae, Prolibytherinae) are taxa with supraorbital ossicones and median occipital horns. Two supraorbital ossicones are present, and in addition, a single posterior median horn that is not an ossicone. The posterior horn forms at the occipital, and it is variable in orientation, shape, and branching according to subfamily. Palaeomerycidae clearly differ from other Pecora in possessing the combination of these two horn types. All subfamilies except the Prolibytherinae have median occipital horns of similar form. These occipital horns are not paired and appear to be derived primarily by backward extensions of the frontal bone. The occipital horn has small contributions at its base from the parietal and the occipital. Juvenile specimens are needed to clarify the frontal origin of the occipital horn. In the cow (Bovidae, *Bos taurus*), as an example, the frontal bones override the parietals to form a double cranial vault. Hence, the overriding moves the frontal cow horn posteriorly. An alternative description of this is an expansion of the frontals at the expense of the parietals (Hooijer, 1958a: 44–50, 57–59). The posterior median horn of Dromomerycidae may even be homologous to the bovid

horns or cervid pedicels. In *Triceromeryx pachecoi* the base of the median horn overlaps into the most posterior part of the fossa of the temporalis muscle. This back part of the temporalis fossa in other pecorans is formed by the parietal bone. Thus, the median horn most likely contains part of the parietal bone as well. The distal part of the horn is forked in *Triceromeryx*.

The paired ossicones of Palaeomerycidae, probably as in giraffids, form above the orbits as small supraorbital bony disks. Fully grown ossicones are simple shafts, oval in cross section, and often terminate in small blunt knobs. Typical ossicones have compact and smooth surfaces as in *Palaeomeryx, Triceromeryx, Cranioceras, Dromomeryx,* and *Barbouromeryx*. The ossicones of *Palaeomeryx* and *Triceromeryx* have regions of secondary bone growth. There are only two individuals figured (one of each), and there is no way to know if the secondary bone growth is ordered as in *Giraffokeryx* or more randomly distributed as in *Giraffa*.

The enamel of the teeth is rugose in these subfamilies. This may be an adaptation that produces higher enamel friction with the vegetation.

The Dromomerycinae, which are North American, may not be distinct from Palaeomerycinae at the subfamily level (see Prothero and Liter, this volume). Palaeomerycinae are distinguished by showing a *"Palaeomeryx* fold" in the lower molars. Most dromomerycinae do not have such a fold. Triceromerycinae do not seem to be clearly distinct from Dromomerycinae.

In the present view, Prolibytheriinae are specialized Palaeomerycidae and are distinguished by flat complex supraorbital ossicones that are fused to the occipital ones, forming a functional platelike structure as in a moose. It is not clearly known that their horns are ossicones, but examination of the margin and surface (texture and venation) suggests that they are. Within this plate there are four elongated cores (thickenings). Two cores are directed anteriorly, and two directed posterolaterally. The posterior cores are larger. These cores are suggestive that *Prolibytherium* evolved from a taxon with four ossicones by forming a web of bony connection among the four branches. A possible candidate could be a frontlet specimen figured by Hamilton (1973: pl. I, Fig. 6, NHM M 26690 from Gebel Zelten) that shows two posterior ossicones, but the anterior frontal surface is broken, and the anterior side is not known. The posterior ossicones are where the thickenings within the ossicones of *Prolibytherium* are located. The occipital of *Prolibytherium* is rather broad and rounded in posterior view, unlike that of Giraffidae.

All known Pecora have occipital condyles with a deep cleft separating the left and right condyle ventrally at the median plane. *Prolibytherium* has an unusual feature among Pecora. It has occipital condyles that are ventrally united, forming a complete ring and allowing an almost 360-degree rotation of the head on the atlas. Apparently the two occipital condyles are ventrally fused, and the narrow universal cleft is absent. Such a rotation is problematic because the typically longish pecoran neck would allow sufficient flexibility as in most taxa. Perhaps a type of specialized display or combat resulted in the unusual occipital condyles. It is also possible that the cervical vertebrae of *Prolibytherium* were atypical. A possible display could be where the dorsal surface of the ossicone plate is placed into the ground by a strong rotation of the head. Barry et al. (2005: Fig. 13) have identified a cranium from the Zind Pir of Pakistan as *Progiraffa*, which has these specialized occipital condyles and, in my opinion, is *Prolibytherium*. *Progiraffa* of the Siwaliks is known only from fragmentary material, but it may turn out to be congeneric with *Injanatherium* or *Prolibytherium*.

Node 5.   Bilobed canine.
Climacoceridae Hamilton, 1978
   Climacoceras afircanus McInnes, 1936
   Climacoceras gentryi Hamilton, 1978
   Orangemeryx Morales et al., 1999

Climacoceridae is probably a valid family. They are placed closer to Giraffidae than the Palaeomerycidae on the working cladogram (Fig. 21.1) only because of the acquisition of bilobed lower canines and possibly the loss of the fang canines and the loss of the occipital median horns. Climacoceridae and Giraffidae are the only Ruminantia known with bilobed lower canines. The "horns" of the Climacoceridae are distinct from those of many pecorans, and their placement close to the giraffids based on their horns is tentative. The arguments on the value of the bilobed canine are weak. Ruminants have the six lower incisors and the two canines compressed anteriorly to form a broad collecting feeding dental edge composed of eight teeth. The edentulous premaxilla occludes with this edge. The bilobed canine of the Climacoceridae and Giraffidae only slightly increases the overall breadth of the dental plate, acting as 10 teeth. The size of the extra lobe in the canine appears to me to be a character of minor adaptational significance when considered in isolation. Nevertheless it is a character that unites all Giraffidae and Climacoceridae (Hamilton, 1978a). Morales et al. (1999) have questioned the presence of the second lobe on the canine for certain Climacoceridae. A. W. Gentry (2006, pers. comm.) thinks the second lobe in the Fort Ternan species (*Climacoceras gentryi*) is delicate and could be convergent to that of the Giraffidae. Additional fossil material and a more detailed study are needed for these archaic taxa. In *Giraffa* and *Okapia*, it appears that the upper lateral lip may differ from those of other Pecora. Extra muscle fibers on the lateral upper lip (unpublished dissection of *Giraffa*) may play a special role in the collection of vegetation. In his perspective, the bilobed canine may gain more significance as only a part of a more complex adaptation.

Climacoceridae (*Climacoceras africanus* and *C. gentryi*) have branching horns that superficially resemble the antlers of cervids. In contrast, Giraffidae do not have branching ossicones. It cannot be certain that the horns of *Climacoceras* are ossicones or that they were permanently (continuously) covered with integument (velvet). These are assumptions.

The ossicones of *Climacoceras* are very different from those of Giraffidae (Hamilton, 1978a; Janis and Scott, 1987; Morales et al., 1999). The branching of Climacoceridae is similar to that of cervids and of the merycodontids, which are related to antilocaprids. In *Merycodus* there are wear facets on the apices as in fossil giraffids. In cervids, the antler tines are formed by localized accelerated growth and are covered by integument when the antler is alive. The horns of Climacoceridae could be similar to antlers in that they were covered with integument that enabled them to grow and branch. However, they could also be similar to giraffids in having permanent horns (not deciduous). One could envision Climacoceridae as not that different from Giraffidae and having true ossicones that, unlike those of Giraffidae, are branched. Climacoceridae may have had long necks and average metapodials (Morales et al., 1999). With such a body form they resembled llamas.

Node 6.   Giraffidae. The second lobe of the canine is enlarged; large-sized animals with slender limbs; ossicones at posterior edge of superior orbital rim margin; occipital narrow; occipital, at the superior horizontal edge of the nuchal crest, is wider than the descending supporting occipital ridges.

Brief Description of Giraffidae.   Giraffidae can be briefly described by the following characters (plesiomorphic and apomorphic): (1) The possession of ossicones, which is shared with Palaeomerycidae and possibly Climacoceridae. The skull of Giraffidae (Fig. 21.5) is simple, and the ossicones are hollowed by the frontal sinus. The frontals often form raised hollow bosses where the ossicones attach. In many forms the bosses and frontal sinuses are less developed (samotheres, *Palaeotragus*, *Giraffokeryx*, "*Palaeotragus*"). The frontal sinuses are expanded above the orbits in the giraffe, okapi, and the sivatheres. (2) The brain case is long, and the occipital is notably narrow and rectangular in a posterior view. The superior occipital edge extends backward in most taxa. However, in *Okapia* and *Giraffa* the occipital extends less and in *Schansitherium* not at all. (3) The brain case axis is not bent in relation to the palatal axis. (4) The lacrimal canal is closed in the bony orbit. (5) The ethmoidal fissure is open in most except in samotheres and *Palaeotragus*. In *Schansitherium*, a plesiomorphic member of the "Samotheriinae," it is open. (6) The mandibular symphysis and lower incisors are horizontal in relation to the basipalatal axis in most taxa but not in *Samotherium* and *Okapia*, where it curves upward to occlude with the premaxillae. (7) The diastema is long for many species. It is especially long in *Palaeotragus microdon*. (8) The masseteric morphology and its relation to diet are complex as in extant species. Overall, grazers have larger masseteric fossae than browsers (Solounias et al., 1995). Figure 21.5 shows two arbitrary angles designed to represent the size of the masseteric fossa. Angle *a* is defined by a line from the external auditory meatus passing at the inferior orbital margin. The second line is from the external auditory meatus to the preorbital foramen. Angle *b* is defined by an apex at the posterior root of M3. A line passes the orbital margin; the other line passes through the anterior root of P2. Most species have angles that are similar to those of a mixed feeder or browsing bovid. In *"Samotherium" sinense* and *"Samotherium" major,* the masseteric fossa is the largest. In these species the skull in lateral view resembles those of Alcelaphini, which also have large masseters and are grazers. *Canthumeryx* and *Okapia* have very small angles. *Giraffa* has a negative angle *a* in that the first line is below the second. (9) The nasals and premaxillae are horizontal in many forms because they terminate above the line from posterior root of M3 to anterior root of P2. This line approximates the direction of the hard palate. Only in *Giraffa* are the nasals and anterior snout strongly down-turned. (10) The position of the soft palate varies (adults only are consid-

ered). Most giraffids have a median palatine indentation for the soft palate that is situated either posterior to the M3 or at the level of the M3. Both of these conditions are typical in ruminants (Fig. 21.6). It is interesting that two taxa (*Giraffa* and *"Samotherium" major*) have an anteriorly positioned indentation of the soft palate at the level of the M2s. It is probable that this is related to the verticality of the distal neck near the head. In the giraffe the neck is vertical and pushes the pharynx–larynx complex anteriorly. The same was probably true for *"Samotherium" major.* In *Samotherium boissieri* the median palatine indentation is at the level of M3 and more plesiomorphic than that of *"Samotherium" major.* *Bohlinia* is similar to *Palaeotragus* (pers. obs.). It is possible that the neck of *Bohlinia,* although long, was not as vertical as that of *Giraffa. Samotherium boissieri* probably had a distal

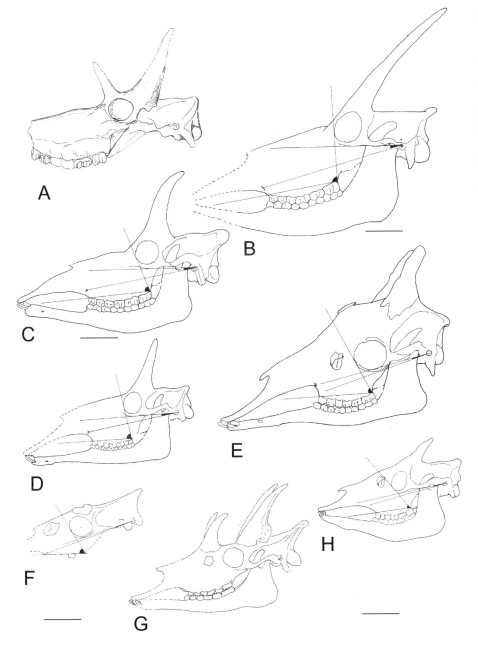

Fig. 21.5. A selection of skulls of Giraffidae. (A) *Schansitherium tafeli* type Shanxi China (pers. obs.). (B) *"Samotherium" major,* Samos Greece (a composite of unpublished data). (C) *Samotherium boissieri,* Samos Greece type NHM M 4215 (reconstruction). (D) *Palaeotragus microdon,* Shanxi China (based on Bohlin, 1926: figures and reconstruction). (E) *Giraffa camelopardalis,* AMNH 14135 male. (F) *Canthumeryx sirtensis,* Gebel Zelten NHM M 26670. (G) *Giraffokeryx punjabiensis,* Chinji Pakistan, AMNH 19475 (reconstruction). (H) *Okapia johstoni* AMNH 15228. Scale bar equals 100 mm.

Fig. 21.6. A selection of soft palate indentations of Giraffidae. (A) *Samotherium boissieri*, Samos Greece type NHM M 4219. (B) *"Samotherium" major*, Samos Greece NHM Ba 30. (C) *Palaeotragus rouenii*, Samos Greece SNHM 44240. (D) *Okapia johnstoni* AMNH 15228. (E) *Giraffa camelopardalis* AMNH 14136. Scale bar equals 40 mm.

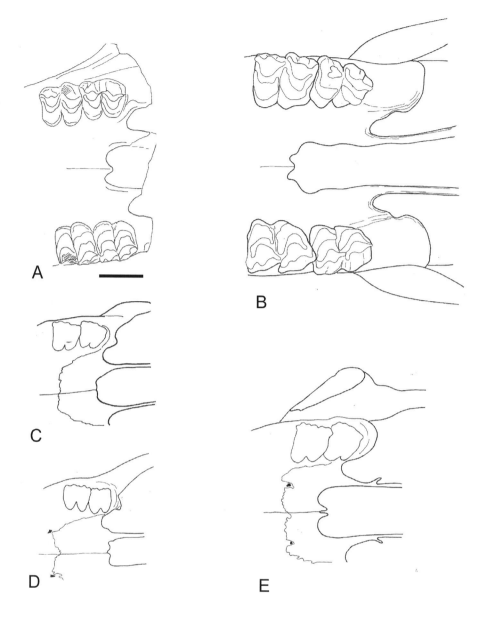

neck less vertical than *"Samotherium" major* (pers. assumption). (11) The second lobe of the canine is large (Hamilton, 1978a). (12) The dentition is brachydont in most species (including *Giraffa* and *Okapia*) or slightly mesodont in *Samotherium* and *"Samotherium" major*. (13) The surface of the enamel is commonly rugose.

(14) The neck so far shows two stages of elongation. Primitively cervicals are short as in *Okapia, Giraffokeyrx, Bramatherium, Helladotherium*, and *Sivatherium*. Medium-length cervicals are those of *Samotherium* and *Palaeotragus* (Bohlin,1926: Text Fig. 37, pl. VIII Figs. 3 and 4; Godina, 1979: Figs. 2 and 3; Solounias, 1999: Fig. 4; pers. obs.). *Giraffa* and *Bohlinia* have the longest necks (unpublished data). Long cervicals may have evolved twice depending on whether *Bohlinia* and *Giraffa* are closely related (cladogram, Fig. 21.1). *Giraffa camelopardalis* has an extra cervical vertebra, probably a duplicated C3. C6 has specializations and loses the characteristic laminae for the longus coli and longus

thoracis muscles. The first rib attaches on C7 as if it were a T1. In turn, T1 occupies the position of T2 and hence it carries the second rib (Solounias, 1999, 2000). C7 and T1 are not known for *Bohlinia*. However, C1 and C2 of *Bohlinia* are similar to those of *Giraffa*. (15) Almost all giraffid species are large animals relative to other ruminants. Early giraffid forms (e.g., *Canthumeryx* and *Injanatherium*) of about 16-14 Ma are as large as modern *Okapia* and much larger than their contemporaneous bovids. The following three characters are known for the okapi and the giraffe: (16) The orbicularis oculi muscle is incomplete on the dorsal side of the orbital rim. One of the explanations for this character may be that this rim is the original supporting boss of the ossicone. The ossicones have relocated medially in *Giraffa* and *Okapia*, but the position of the original boss remains at the orbital rim. (17) The tongue is very long. (18) Locomotion is by pacing as in camels. (19) The body in *Okapia* and *Giraffa* is anteroposteriorly compact because the thorax is deep. A new observation is that this character can be inferred from the proportions of the scapula, which is longer than that of a typical ruminant with a shorter thorax. When not known, the depth of the thorax can be approximated by the length of the scapula because it traditionally spans the entire lateral side of the thorax. *"Palaeotragus"* primaevus from Fort Ternan has a short scapula, so the thorax was presumably similar to the thorax of a gazelle. *Honanotherium, Samotherium, "Samotherium," Okapia,* and *Giraffa* have long scapulae and probably deep thoraxes. (20) Long limbs and slender proportions are common for many giraffid species. Length of limbs is not so much attributable to the humerus or femur. Limb length can be best appreciated by the radius, tibia, and the metapodials. This is especially true for the archaic taxa (e.g., *Canthumeryx* and *Injanatherium*). *Injanatherium* has metapodials very similar in length to those of modern *Giraffa*. Relative to the size of its skull, which is similar to that of *Okapia,* the radius and the metapodilas are even longer The longest *Injanatherium* would be more slender than *Ammodorcas clarkei* (the dibatag). Some derived species such as the Sivatheriinae and "Samotheriinae" have developed stouter limbs. *Sivatherium* is the extreme with short metapodials as in the Bovini (see figures in Geraads, 1996).

Eight primary clades of giraffid species have been discerned in the present study based on metacarpal morphology:

1. Canthumerycinae: *Georgiomeryx* and *Canthumeryx* (*Zarafa*)
2. Bohlininae: *Injanatherium* spp., *"Palaeotragus"tungurensis, Bohlinia,* and *Birgerbohlinia*
3. Okapiinae: *"Palaeotragus primaevus"* Ngorora KNM 1446 and *Okapia*
4. Giraffokerycinae: *"Palaeotragus" primaevus* Fort Ternan, *Giraffokeryx*
5. Sivatheriinae: *Sivatherium* spp., *Helladotherium,* and *Bramatherium* spp.
6. "Samotheriinae": *Schansitherium tafeli, S. quadricornis, Samotherium boissieri, Chersonotherium eminens, "Samotherium"* major
7. "Palaeotraginae": *Palaeotragus rouenii, P. coelophrys, P. microdon, Macedonitherium martinii*
8. Giraffinae: *Giraffa camelopardalis, G. jumae*

For clarification on the numerous species of *Palaeotragus* and *Samotherium* where quotations imply a new generic or trivial name is needed:*"Palaeotragus"* tungurensis from Tung Gur Mongolia, traditional name is *Palaeotragus tungurensis* (Colbert, 1936a); *Palaeotragus primaevus* from Ngorora KNM 1446 traditional name is *Palaeotragus primaevus,* and it is assumed to be conspecific with *Palaeotragus primaevus* from Fort Ternan (Hamilton, 1978a); *"Palaeotragus" primaevus* from Fort Ternan, traditional name is *Palaeotragus primaevus* (Churcher, 1970); *Samotherium boissieri* (type species) from Samos (Major, 1889). "Samotheriinae" as they lack derived characters to connect the taxa *Schansitherium tafeli* from Shanxi, *Samotherium boissieri, Chersonotherium eminens* from Chersona Russia, *"Samotherium"* major from Samos

(Major, 1889; Alexeyev, 1915; Killgus, 1922)."*Samotherium*" major from Samos, traditionally *"Samotherium" boissieri variety major* (Bohlin, 1926); "Palaeotraginae": *Palaeotragus rouenii* Pikermi (the type species) and Samos (Gaudry, 1861), *Palaeotragus microdon* Shanxi China (Bohlin, 1926), *Palaeotragus coelophrys* Maragheh and Samos, *Schansitherium quadricornis* Samos traditionally *Palaeotragus quadricornis* (Rodler and Weithoefer, 1890; Bohlin, 1926).

There is always a need for a novel set of characters that can eventually be tested in a more rigorous morphometric and cladistic analyses. Presently, it is proposed to study the body proportions in Giraffoidea expanding on the ratios that Bohlin (1926) and Colbert (1938c) have initiated. Observations on metacarpals are presented as an initial step to a more elaborate study. The limbs are slender in most species except in Sivatheriinae and "Samotheriinae." It is proposed that the length and the cross-sectional morphology of the metacarpals can be effectively used for the discrimination of the giraffid subfamilies (Figs. 21.7 and 21.8). Examination of giraffid metacarpals reveals four interesting patterns (pers. obs.): (1) Elongation. Sufficient elongation of the metapodials seems to have been achieved from the onset of Giraffidae. Specimens from the Kamlial Formation of Pakistan (16 Ma), which is temporally close to the beginning of the family, are already slender and long (Bohlininae). Proportionally to the width, some of the Kamlial specimens are relatively even longer and more slender than those of the modern *Giraffa*. Thus, temporal trends of additional metapodial lengthening and thinning within Giraffidae are minor. (2) Broadening. A preliminary study of the detailed morphology of various metacarpals of giraffids suggests that long and slender plesiomorphic metapodials evolved into stout robust ones by maintaining the original length and increasing primarily in width. For example, *Injanatherium* could evolve into *Birgerbohlinia* and *Giraffokeryx* into *Bramatherium* and *Helladotherium* by increasing primarily the width (unpublished data). *Bramatherium* and *Helladotherium* have broadened metacarpals, but the length remains approximately that of *Giraffokeryx*. Such metapodials could have been derived from those of *Giraffokeryx* by broadening. (3) Shortening. This is rare but is observed in *Sivatherium* species. (4) The depth of the posterior longitudinal hollowing of the metacarpals containing the palmar interosseous muscles also reveals patterns. A bony metacarpal trough contains the palmar interossei or specialized tendinous derivatives of these muscles that aid in the adduction of the phalanges (pers. obs.). The depth of the bony trough is visible on the posterior side of the metacarpals and can be appreciated by cross sections. Three general types of depth of trough are observed. (a) The most common ones are those with a medium-depth trough. They are prevalent throughout the Miocene and are the ones that resemble most bovid metapodials (Okapiinae, Giraffokeryxinae, "Samotheriinae," and "Palaeotraginae"). The other two types are much rarer. (b) One is with metapodials of extremely flat posterior side, where the interosseous muscles are reduced (Giraffinae). (c) The third metapodial type has

Fig. 21.7. A selection of metacarpals and cross sections, which are displayed at the location of the cross section. All are schematic, and some are composites. (A) *Canthumeryx sirtensis* KNM 41 B Moruarot Hill (reconstruction). (B) *Injanatherium arabicum* Al Jadidah S. Arabia cast AMNH 127351b (reconstruction). (C) Cf. *Injanatherium* from the Chinji Siwaliks GSP Y 20415. (D) *"Palaeotragus" tungurensis* AMNH 92306 distal metatarsal and 808 proximal metacarpal. (E) *Bohlinia attica* from specimens in Paris and in Athens, Pikermi Greece. (F) *Birgerbolinia schaubi* AMNH 20610 Q1 Samos Greece. (G) *Giraffa camelopardalis* AMNH 14136. (H) *Okapia johstoni* AMNH 15228 (its cross section is similar to that of *"S." major*, Fig. 21.8D). (I) *Giraffokeryx punjabiensis* Chinji Pakistan GSP Y 24306. (J) *Bramatherium megacephalum* AMNH 19460. (K) *Helladotherium duvernoyi* Pikermi Greece from specimens in London, Athens, and Paris. (L) *Sivatherium* Siwaliks NHM M 17102a. (M) *Sivatherium maurisium* Kenya NMK I 7242/620 (610). Scale bar equals 200 mm.

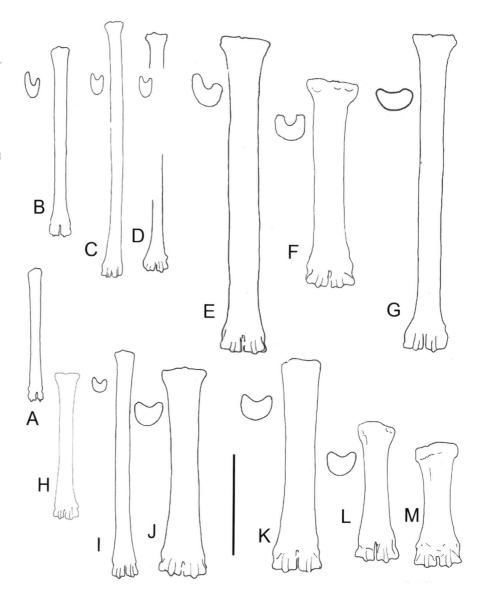

deep troughs for extremely well-developed interosseous muscles (Bohlininae). The use of such palmar interosseous muscles in ruminants has not been studied in detail. From what is known of humans and cows, one would expect that the palmar interossei of Giraffidae also adduct the digits. Enhanced adduction at the phalanges probably relates to the substrate, as it would hold the two hooves uniformly together. Dissections of *Rangifer tarandus* (reindeer) have shown well-developed interosseous muscles when compared to *Odocoileus* (pers. obs.). Reindeer traverse snowfields and perhaps are adducting the side hooves and the main hooves to avoid heterogeneous phalangeal sinking in the soft ice. It is hypothesized that Bohlininae with their characteristic deep troughs and presumably well-developed interossei inhabited soft substrates. During the Miocene, these habitations could be either muddy lake margins or sandy areas but not ice. The metatarsals are similar to the metacarpals in length and slenderness. However, additional study is needed because the metatarsals may have, in some instances, less pronounced troughs.

In summary the metacarpal patterns are:

1. Canthumerycinae: Medium-length metapodials. The metacarpal has a medium in depth trough posteriorly, which extends approximately one-fourth down the shaft.
2. Bohlininae: Long and slender metapodials. The metacarpal has a deep trough posteriorly, which extends approximately two-thirds down the shaft. In *Birgerbohlinia* the metapodials become wide, but the posterior trough remains and is well developed. The lateral metacarpals 2 and 5 are well developed in *Birgerbohlinia*.
3. Okapiinae: Medium-length metapodials. The metacarpal has a medium-depth trough.
4. Giraffokerycinae: Long metapodials. The metacarpal has a medium-depth trough, which extends approximately two-thirds down the shaft.
5. Sivatheriinae: Medium and short metapodials. The metacarpal has a shallow trough. In *Sivatherium* the metapodials are short.

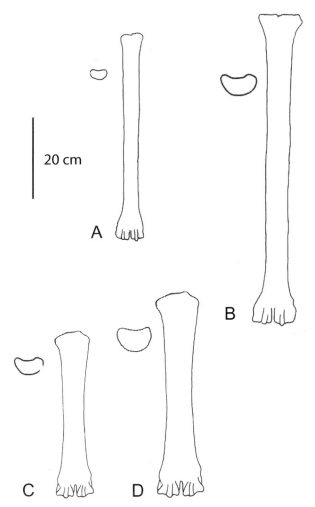

Fig. 21.8. A selection of metacarpals and cross sections, which are displayed at the location of the cross section. (A) *Palaeotragus rouenii* from Samos Greece (reconstructed from three specimens). (B) *Giraffa camelopardalis* AMNH 14136. (C) *Samotherium boissieri* NHM M 4289 Samos (lower beds of Vrysoula or Stefana) Greece. (D) "*Samotherium*" *major* AMNH 20595 Q1 Samos Greece. Scale bar equals 200 mm.

6. "Samotheriinae": Medium-length metapodials. The metacarpal has a medium-depth trough.
7. "Palaeotraginae": Long metapodials. The metacarpal has a shallow trough, which extends approximately two-thirds down the shaft.
8. Giraffinae: Long and slender metapodials. The metacarpal has a very shallow trough posteriorly.

Each clade of Giraffidae is presented, and possible interrelations of taxa within each of the clade are discussed (Fig. 21.1).

Node 7. Ossicones posterolaterally above orbital rims, at the very edge of the rims; ossicones reduced in size, taxa obtaining a large body size; notable shortening in length of the lower p2 and especially the p3.

Canthumerycinae new rank
  Georgiomeryx georgalasi Paraskevaidis, 1940
    Canthumeryx sirtensis Hamilton, 1973 (=Zarafa Hamilton, 1973)

These are the most plesiomorphic Giraffidae. The ossicones are situated posterolaterally above the orbits and are reduced in size to small cones. The occipital is very narrow, suggesting a strongly flexible neck. It is narrowest in *Georgiomeryx*. The length of the lower p2 and p3 is reduced in Giraffidae. *Canthumeryx sirtensis* has normal-shaped metapodials, which are long. The metapodials may have a deep trough posteriorly, but this trough is restricted to the most proximal part of the metapodial. More distally, the posterior side of the metapodial becomes flattened.

Two stages can be discerned from the occipital. *Georgiomeryx* has a narrower occipital than *Canthumeryx*. The narrow occipitals indicate a different adaptation in the mode of fighting from a typical bovid or *Prolibytherium*, where the occipitals are broad. *Tragulus*, *Hyemoschus*, and *Dorcatherium* also have strongly approximated temporal lines and a narrow occipital. Narrow occipitals, especially in *Georgiomeryx*, suggest a flexible neck in the lateral positions. The narrow occipital also implies an expanded fossa for the origin of the temporalis muscle and closely approximated temporal ridges. Thus, the temporal fossae are wide for longer (larger) temporalis muscles that occupied the lateral surface of the parietals. Other Giraffidae have broader occipitals and much more restricted temporal fossae. The expanded temporal fossae for the temporalis are also found in *Triceromeryx*.

Canthumeryx sirtensis and *Georgiomeryx georgalasi* have small ossicones with a central axis located posteriorly on the orbital rim. These ossicones are simple, short triangular cups. The axis of each ossicone is directed posterolaterally.

There is a notable increase in body size of early giraffids compared to Climacocerinae. The large body size is a characteristic of the entire family of Giraffidae, whereas palaeomerycids include many small-sized taxa. For example, Giraffidae from the lower Chinji Formation in Pakistan are as large as a modern elk (red deer) in contrast to the contemporaneous small Chinji bovids which average the size of a Thompson's gazelle. The larger size is probably related to metabolic and other adaptational differences in the Giraffidae such as feeding from vegetation growing higher up from the ground than the bovids.

The loss of median occipital horns and of the upper canine along with the retention of ossicones contrasts the Giraffidae from the Palaeomerycidae. Until other fossil taxa are studied, the definite direct sister taxon of Giraffidae is not known. The indirect sister taxon of the Giraffidae is the Climacocerycinae, indirect as these specimens are specialized in having branching ossicones. The lower p2 and especially the p3 of *Canthumeryx* are more molariform than in previous taxa.

Node 8. Share a deep posterior trough in the metapodials.

Node 9. Extremely sharp ends on the trough of the metapodials; four ossicones.

Node 10. Large size; two ossicones.

Node 11. Upper premolars with inward-curving styles; premolars round in occlusal view.

Node 12.   Large-sized species. *Birgerbohlinia* has short and robust metapodials. The metapodials have a deep trough.

    Bohlininae new rank
        Injanatherium arabicum Morales, Soria and Thomas, 1987
        Injanatherium hazimi Heintz, Brunet and Sen, 1981
        "Palaeotragus" tungurensis Colbert, 1936
        Bohlinia attica Matthew, 1929
        Honanotherium schlosseri Bohlin, 1926
        Decennatherium pachecoi Crusafont, 1952
        Birgerbolinia schaubi Crusafont-Pairo, 1952

Bohlininae share extremely deep troughed metapodials with a trough extending even distally. The edge of the trough is well defined, and the metapodials are long. The palmar interosseous muscles must have been well developed, facilitating adduction of the digits. The metapodials are very long, suggesting a slender and tall body construction. There is a second pair of ossicones in *Injanatherium* anteriorly, but they are known only from the broken base connection to the frontals. The posterior ossicones fully cover the orbital rim; including the most anterior aspect of the orbital rim and the frontal boss expands into the borer with the lacrimal. *Birgerbolinia schaubi* also had four ossicones, and probably *Decennatherium pachecoi*. Elongated cervicals (only C1 and C2 are known for *Bohlinia attica*) is a character that must go together with extremely elongated metapodials. It would be impossible to have long limbs and short necks because of the need to drink water. However, long neck can exist without long limbs (as in camelids and *Orangemeryx*). The humerus in *Bohlinia* is plesiomorphic and different from *Giraffa* in having a small lateral proximal medial tuberosity (for the origin of pectoralis and subscapularis) (Geraads, 1979: plate I, Fig. 2).

Injanatherium arabicum, I. hazimi, and *"Palaeotragus" tungurensis* are the most plesiomorphic species. The ossicones protrude laterally in *I. arabicum,* and they also terminate in a knob. *I. jazimi* has larger ossicones. *"Palaeotragus" tungurensis* is not known for ossicones, but the jaw and the dentition of this species are specialized in having a deep shaft below the molars and a strongly vertical ascending ramus.

"Palaeotragus" tungurensis and *Bohlinia attica* possess upper premolars with strongly curved anterior and posterior parastyles and postmetacristae. On an occlusal view, the buccal wall of the premolars reveals a characteristic curved-in appearance (Fig. 21.9). *Honanotherium schlosseri* is more plesiomorphic on the premolars, where the parastyles and postmetacristae are simple and slightly reduced as in *Samotherium.* The dentition of *Honanotheirum schlosseri* does not resemble that of *Bohlinia attica. Bohlinia attica* and *Honanotherium schlosseri* are much larger than *"Palaeotragus" tungurensis* and *Injanatherium.* This is a derived condition and is also in agreement with the stratigraphy. *Bohlinia attica* and *Honanotherium schlosseri* are later Miocene species (9–7 Ma), whereas *"Palaeotragus" tungurensis* is probably older than 12 Ma (found in Tung Gur dated by a pre-*Hippotherium* datum).

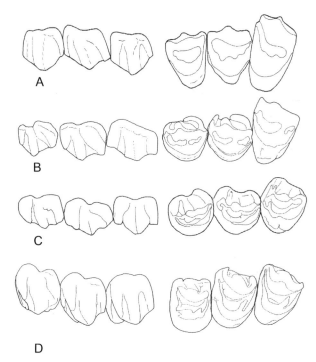

Fig. 21.9. Upper premolar sequences showing the similarity of *"Palaeotragus" tungurensis* and *Bohlinia attica* and their difference from *Palaetragus rouenii* and *Giraffa camelopardalis.* (A) *Palaeotragus rouenii* from Pikermi type in Paris. (B) *"Palaeotragus" tungurensis* AMNH 26 586 from Tung Gur Mongolia. (C) *Bohlinia attica* from Pikermi Greece specimen in SPGM AS II 640 Munich. (D) *Giraffa camelopardalis* AMNH 81824. Lateral and occlusal views of the same specimens. Not to scale.

Birgerbolinia schaubi is the most specialized taxon of the subfamily with secondarily broadened metapodials. The metacarpals retain the deep posterior trough that unites the Bohlininae. *Birgerbolinia* had well-developed metacarpals 2 and 5. *Birgerbolinia* is a big stout-shaped species in which several of the anatomical features render it similar to sivatheres. However, large size and graviportality can evolve independently.

Decennatherium pachecoi is a problem species. Crusafont-Pairó (1952) originally found dental similarities to *Giraffa.* Morales (1985) and Gentry (1994: 137) following Morales placed *Decennatherium* in the Sivatheriinae. The dentition is primitive, but a partial cranium makes it similar to *Honanotherium* (my opinion) in the position of the ossicones, the squared appearance of the nasals, and the morphology of the metapodials. *Decennatherium pachecoi* has brachydont dentition but very large incisors and canines in relation to the molars. The diastema is very long (see figure in Geraads, 1989: pl. II Fig. 1). The astragalus is similar to those of *Honanotherium* and *Samotherium,* but that is a plesiomorphic feature. It differs from the astragalus of *Giraffa,* which is apomorphic (a wide one).

If *Giraffa camelopardalis* has evolved from *Bohlinia,* the following changes must have taken place: The upper premolars reverted to a more plesiomorphic condition; there was total reduction of the posterior metapodial trough; the astragalus became squared in anterior view (length similar to width);

the ossicones moved inward on the frontals and became positioned slightly posteriorly. The width of the ossicones at base was reduced. The soft palate was anteriorly positioned.

Node 13. Closure of the lacrimal foramen (the bony canal) in the orbit for the nasolacrimal duct; the thorax is deep in lateral view (known only for *Okapia*, *Giraffa*, and *Samotherium*); the scapula is short and broad in *"Palaeotragus"* primaevus from Fort Ternan.

Node 14. Large bullae and secondary bone growth on the ossicone apices; short and long canals in the ossicones; the thorax is deep; the scapula is long and narrow; scapula with a reduced coracoid process. (This is in conflict with the same feature in *Giraffokeryx punjabiensis*.) In *Okapia*, lower premolars become widened; ossicones move medially and posteriorly; the medial epicondyle of the humerus is well developed.

Okapiinae Hamilton,1978

"Palaeotragus primaevus" Partial cranium and ossicones from Ngorora Formation KNM 1446 (Hamilton, 1978a)

Okapia johnstoni Lankester, 1901

Species with large bullae; ossicones with internal canals or distal pits. Nasolacrimal bony canal closed in the orbit (note that this feature is not known for the Injanatheriinae; in Canthumericinae the canal is open). The humerus of *Okapia* is apomorphic, having a large lateral proximal medial epicondyle (for the origin of pectoralis and subscapularis). The metapodials are known for the okapi and are simple with re-

duced palmar interosseous complex (reduced but not as much as in *Giraffa;* hence, they are classified as medium). The metapodial lengths are also slightly reduced in the okapi.

The Ngorora KNM 1446 cranium has large bullae (*"Palaeotragus primaevus"*). The ossicones (Fig. 21.10) are broad and flattened as in *Palaeomeryx*. In anterior view the ossicones have a straight long axis. In *Palaeomeryx* the axis is abruptly bent medially (Ginsburg and Heintz, 1979). The surface of the KNM 1446 ossicone is less compact than those in other taxa, suggesting a deep investment of the integument. In *Giraffokeryx* there are longitudinal ridges separated by grooves where the integument would penetrate. It can be stated that the grooves of secondary bone growth are ordered in *Giraffokeryx*. In *"Palaeotragus primaevus"* Ngorora KNM 1446 there are lumpy oval ridges separated by depressions where the integument would penetrate. The ossicone has a central canal that starts at the base of the ossicone and was probably confluent with the frontal sinus. Halfway into the ossicone the canal splits, forming a Y. The lateral branch opens as a wide pit on the distal part of the lateral surface. The canal strongly differs from a typical frontal sinus in having continuous well-defined inner walls as in a pipe. The canals of the two ossicones are asymmetric. Hamilton (1978a) did not observe these details about the KNM 1446 ossicones. His Figures 38 and 39 actually show the left ossicone mounted incorrectly with the anterior edge as posterior. In Figure 38 the base appears too wide, but careful examination shows that the base is actually the right ossicone overlapping pictorially with the left. In Figures 38 and 39 the internal opening of a canal is visi-

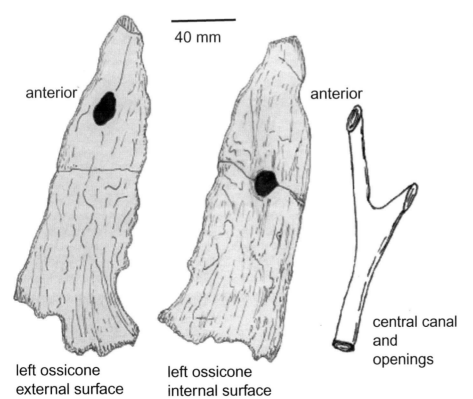

Fig. 21.10. Left ossicone of *"Palaeotragus primaevus"* from Ngorora Formation of Kenya, NMK 1446.

40 mm

anterior

anterior

central canal and openings

left ossicone external surface

left ossicone internal surface

ble. In Figure 40 the compact canals of both ossicones are visible, and it can be noted that they differ from traditional frontal sinuses invading the ossicone. These canals have a thick cortical rim.

In the okapi, the ossicones are apomorphic for the following reasons: They are positioned medially but retain a broad boss over the top of the orbits with the frontal sinus investing the boss and ossicone. The ossicone apices become exposed as bare bone. The sectioned ossicone AMNH 51215 has a 14-mm-long canal within the apex, but the canal does not penetrate through to the frontal sinus as in the Baringo specimen KNM 1446. Note that *Samotherium* and *Palaeotragus* ossicones may have wear facets. These facets are simple and beveled and thus differ from those of the okapi. No pits, canals, or constrictions have been observed in *Palaeotragus* and *Samotherium*. The okapi also has the second and third tarsal cuneiforms fused to the cubonavicular (autapomorphic). The neck is short in this clade but known only for the okapi. The thorax is deep, and the scapula is long.

Node 15. Four ossicones; irregular surface on the ossicones of secondary bone growth (secondary bone growth arranged in lumps or ridges); upper and lower premolars become more molariform; the thorax is deep; the scapula is long and narrow; scapula with a reduced coracoid process (in contrast with the same feature in *Okapia*).

Node 16. Slender metapodials with metacarpals where the trough is reduced to medium depth. *Giraffokeryx* has more molariform lower premolars than *"Palaeotragus" primaevus.*

Giraffokerycinae new rank
    "Palaeotragus" primaevus (Churcher, 1970)
    Giraffokeryx punjabiensis Pilgrim, 1910

Normal-shaped and -sized metapodials and four ossicones. The neck is short in *Giraffokeryx punjabiensis* but may be longer in *"Palaeotragus" primaevus* from Fort Ternan. The metapodials are the most typical in form. That is, they resemble gazelline bovids. There are four ossicones in most species: two anterior and two posterior to the orbit. The surface of the ossicones has irregular thick ridges of secondary bone growth in the adults. The ridges run along the length of the ossicone and are fine in juveniles. The ridges are less developed in *"Palaeotragus" primaevus.* The anterior ossicones protrude laterally. The posterior pair is superolateral, and each has a massive growth at its base like a flange. The cross section of the ossicones is triangular. Thus, there is a ridge within the ossicone presumably for strengthening. The ossicones terminate in small knobs, which are different from the characteristic knobs of *Giraffa* and from the pointed apices of *Palaeotragus rouenii, P. microdon,* and *P. colepophrys. "Palaeotragus" primaevus* from Fort Ternan has more plesiomorphic p4 than *G. punjabiensis,* which has a more advanced p4. In *Giraffokeryx punjabiensis* the occipital condyles are specialized. They have a depressed area where the tectorial membrane

attached. The intercondylar groove is narrow. These morphologies resemble those of *Prolibytherium* and suggest specialized head rotations during combats.

Node 17. Large size and metapodials thickened or robust.

Node 18. Posterior ossicones compressed in cross section (in *Helladotherium* inferred from female skull frontal bosses). In *Sivatherium,* metapodials very short.

Node 19. In *Bramatherium* anterior ossicones fused at their base forming a Y.

Sivatheriinae Bonanparte, 1850
    Sivatherium giganteum Cautley and Falconer, 1835
    Bramatherium megacephalum Falconer, 1845
    Helladotherium duvernoyi Gaudry, 1860

The sivatheres are large animals. Their ossicones are similar to those of *Giraffokeryx,* but the frontals are much more inflated by sinuses. *Bramatherium* is more plesiomorphic in the metapodials, which are not short, and it is stratigraphically older than *Sivatherium. Sivatherium giganteum* is more plesiomorphic in the ossicones, which are similar in shape and position to *Giraffokeryx punjabiensis. Bramatherium* has fused anterior ossicones and simple posterior ones. In *Sivatherium,* the anterior ossicones are directed anteriorly instead of laterally. The posterior ossicones are flattened with local growths on the lateral edge. In *Bramatherium megacephalum* the autapomorphy is the fusion of the anterior ossicones to form a single median shaft, but the apices remain separate and form a Y in anterior view. The posterior ossicones are simple and rounder than those in *Sivatherium. Helladotherium duvernoyi* is known from only one hornless skull from Pikermi. It has been portrayed as a hornless taxon on the argument that ossicones would surely have been found in the fossil-rich Pikermi bone beds. However, there are other taxa at Pikermi known from single specimens. The female skull of *Helladotherium* has two anteroposterior elongated frontal bosses, which clearly have supported long flat ossicones. This observation can be substantiated by a similar example in *Sivatherium,* where the female hornless specimen at The Natural History Museum (London) M 39523 also has long ridges (bosses) at the same region where the male *Sivatherium* has the flat long ossicones. The same is true from another *Sivatherium maurusium* specimen, where the anterior ossicones form long ridges as in the female *Helladotherium* skull (see Geraads, 1985: Fig. 1).

Bramatherium and *Helladotherium* have broadened metacarpals, but the length remains approximately that of *Giraffokeryx.* Such metapodials could have been derived from those of *Giraffokeryx* by broadening. *Bramatheirum* and *Helladotherium* metapodials are very similar. *Sivatherium* has much shorter metapodials than the other two. It is also interesting that in *S. maurusium* the metatarsal may be much taller than the metacarpal (Geraads, 1996: Fig. 1). This is different from *Bramatherium* and *Helladotherium* and from all other giraffids where the two metapodials are approximately the same length. This characteristic is reminiscent of a similar situation of cervids and many early artiodactyls. One autapomorphy is very short metacarpals similar to those of Bovini (e.g., *Bos*). *Bramatherium megacephalum* retains the long size of metacarpals, but they are stout (Fig. 21.7).

Node 20. Two ossicones with smooth surface loss of second pair; posterior cuspids of lower p4 reduced and anterior cuspids long.

Node 21. The ossicone is split at the base, forming a Y; nuchal crest extension reduced.

Node 22. Loss of split posterior ossicones.

Node 23. Ossicones curve back.

Node 24. Ossicones reduced in size; masseteric fossa reduced in size; premaxillae squared in dorsal view.

Node 25. Ossicones very long; masseteric fossa enlarged; soft palate anteriorly positioned; larger body size; cervical spinous processes more reduced and broader in dorsal view.

"Samotheriinae" Hamilton, 1978

Two simple spike ossicones at the posterior superior margin of the orbital rim. Secondary bone growth on ossicones reduced to absent. Taxa more plesiomorphic than *Palaeotragus* as the ossicones emanate from the posterior orbital rim.

Schansitherium tafeli Killgus, 1922 (includes *Palaeotragus decipiens* Bohlin, 1926: 30)
Schansitherium quadricornis (Bohlin, 1926: 42)

Schansitherium tafeli is dentally larger than *Schansitherium quadricornis*. Four simple spike-shaped ossicones; the second pair is anterior, smaller, and merging at the base to the main one. Presently they are interpreted as four ossicones. The pairs are anteroposterior, but they are similar to left and right seen in the anterior ossicones of *Giraffokeryx* and in *Baramatherium*. In *Giraffokeryx* a fusion line is observable between the left and the right ossicones, and the sinuses inside are also split. In *Baramatherium* and *Shanistherium* no anteroposterior fusion is observable. *Shanistherium tafeli* differs from *Samotherium boissieri* in having four ossicones and an occipital that does not extend back. Secondary bone growth is occasionally present in *Shanistherium*. Such growth has never been observed in *S. boissieri*. Pachyostotic limbs and mandibles are occasionally present in *S. tafeli*. The neck is probably elongated to medium length. The metacarpals show a medium-level trough. Metacarpals are of medium slenderness. *S. quadricornis* has been envisioned to be a *Palaeotragus* (e.g., Solounias, 1981). *S. quadricornis* is newly recognized as a smaller species of *Schansitherium*. Its dentition figured by Bohlin (1926: Text Figs. 55–56) shows many differences from *Palaeotragus rouenii*.

Chersonotherium eminens Alexejev, 1915 (also spelled as *Kersonotherium*)
Samotherium boissieri Major, 1889
Samotherium neumayri (*Alcicephalus neumayeri* Rodler and Weithoefer, 1890)

These taxa can be described as follows: Neck elongated to medium-length. Simple spike ossicones (one pair) with extra smooth surface. Limbs stout; Reduction of p4 posterior cuspids. Geraads (1977) selected a type specimen for *Samotherium boissieri* from Samos: a skull and jaw (NHM M 4215). In *Samotherium boissieri*, the ossicones are large and curved posteriorly. The masseteric origin at the anterior edge of the zygomatic is weak. The coracoid process of the scapula is underdeveloped. The distal radial articulation is thinner at the lateral side, where the ulna protrudes. In *Samotherium neumayri* the masseteric origin at the anterior edge of the zygomatic is strong. *S. neumayri* is the only giraffid known to have had a very flat (grazing) premaxilla. However, the skull has a very small masseteric fossa, which appears to be a contradiction. Distal neck was less vertical in *Samotherium boissieri* than in "*Samotherium*" *major*. Borissiak (1914–1915: Fig. 1a,b) figured *Achtiaria expectans*, which is probably a small *Samotherium*, tentatively *Samotherium expectans*. *Chersonotherium eminens* (Alexeyev, 1915: 138) may be a more plesiomorphic species than *Samotherium*. In this species, the ossicones and the frontal sinuses are small. The origin at the anterior edge of the zygomatic is strong. The coracoid process of the scapula is better developed. The distal radial articulation is fuller at the lateral side, where the ulna protrudes.

"*Samotherium*" *major* (*Samotherium boissieri* variety *major* Bohlin, 1926: 87; but not *S. major* Geraads from Turkey, which probably is *S. neumayri*)
"*Samotherium*" *sinense* (Bohlin, 1926)

"*Samotherium*" *major* differs from *Samotherium boissieri* in its larger size and straighter, more pointed, and longer ossicones and a distinctly larger fossa for the masseter superficialis and profundus. It is actually the only giraffid with a masseteric fossa as large as that of Alcelaphini. As a result, the orbit is positioned more posteriorly. Also, the angle of the jaw shows a stronger masseter insertion. In "*S.*" *major* the middle indentation of the hard palate is anteriorly positioned, between the M2s. In *S. boissieri* the indentation is more plesiomorphic and is between the M3s.

When the distal neck is vertical as in *Giraffa* and "*Samotherium*" *major*, the pharynx is pushed more anteriorly, resulting in an anteriorly positioned soft palate (a hypothesis). This reorganization provides sufficient space for the motion of the soft palate and for swallowing. In *Samotherium boissieri* the indentation is more posterior, and hence, the pharynx is also more posterior. This assumption is in agreement with observed cervical differences between "*Samotherium*" *major* and *Samotherium boissieri* (pers. obs.). The masseteric fossae of "*Samotherium*" *major* and "*Samotherium*" *sinense* are similar (large in both). "*Samotherium*" *major*, however, differs in that it has a broad indentation for the soft palate. "*Samo-*

therium" sinense differs from *"S." major* in having a very narrow indentation in the attachment of the soft palate. These differences are sufficient for a specific separation of *"S." sinense* from *"S." major.* In *S. boissieri* the palatine indentation is in a plesiomorphic position. In *"S." sinense,* the palatine indentation forms a narrow cleft but is still posterior as in *Samotherium boissieri.* In *"S" major* the palatine indentation is anteriorly positioned and broad. One could envision a sequence in which such a narrow cleft forms first (evolution from *S. boissieri* to *"S." sinense*), and subsequently, the palate broadens, and the soft palate moves anteriorly in *"S." major,* which would allow the pharynx to move forward and the neck to become more vertical. The neck is elongated more in these taxa but is still within the medium-length range. In *S. major* the distal neck is more vertical than that of *Samotherium boissieri.*

Node 26.   Reduction is size; ossicones move more medially from the orbital rims; the posterior cuspids of the lower p4 are enlarged and oriented anteroposteriorly.

Node 27.   *P. coelophrys* are a larger-sized species; ossicones curved; ossicones reduced in size in the North African varieties. *Palaeotragus germaini* and *Palaeotragus robinsoni* probably are conspecific with *P. coelophrys.*

Node 28.   *P. rouenii,* a smaller-sized species; ossicones slender, straight, and vertical; the apices are often exposed (not covered by integument); in *Macedonitherium* ossicones become thick and strongly curved; apices very pointed.

"Palaoetraginae" Pilgrim 1911
  Palaeotragus coelophrys (Rodler and Weithofer, 1890)
  Palaeotragus germani Arambourg, 1959
  Palaeotragus robinsoni Crusafont, 1979
  Palaeotragus rouenii, Gaudry, 1861
  Macedonitherium martinii Sickenberg, 1967

Palaeotragus rouenii is the type species from Pikermi. In "Palaeotraginae" the ossicones move medially from above the very edge of the orbital rim. Note that in *Samotherium boissieri* they are at the very edge of the rim. But the ossicones are still on the thin plate of the orbit. In *Okapia* they are scientifically more internally positioned. The texture of the *P. rouenii* ossicones is extra smooth. The lower p4 has posterior cuspids as large as the anterior cuspids. Also, the orientation of the cuspids is anteroposterior. Limbs are slender. The metacarpal trough is medium in depth in *Samotherium, Schansitherium,* and *Palaeotragus rouenii.* The *Samotherium* and *Schansitherium* share broad and flat metacarpals. In *Palaeotragus* the metacarpals are more slender. The metapodials of *"Samotherium" major* are approximately the same length as the metapodials of *Palaeotragus rouenii.* It is theorized that the slenderness of *Palaeotragus rouenii* results from a reduction of the width and retention of the original length. *Palaeotragus* may have evolved from *Samotherium boissieri* by reduction of body size, of width in metapodials, and the placement of ossicones medially. The more plesiomorphic

Palaoetragus species than *P. rouenii* were probably larger and with longer metapodials.

Palaeotragus coelophrys from Samos and Maragheh is just a larger version of *P. rouenii* but very similar to it. *P. coelophrys* is probably a more plesiomorphic taxon because it is closer to *Samotherium boissieri* in size and in the shape of the origin of the masseter on the maxilla and the medial indentation of the hard palate for the soft palate. It differs from *Samotherium boissieri* in having smaller dentition, ossicones more medially positioned, and having slenderer metapodials. *P. coelophrys* is a good intermediate between *S. boissieri* and *P. rouenii.* It could be derived from *Samotherium* by the placement of the ossicones more medially and reduction in size. Also the dentition is smaller, and the lower p4 is different in the orientation of the posterior cuspids as in all species of *Palaeotragus. Palaeotragus robinsoni* and *P. germaini* are similar to or slightly larger than *P. coelophrys* and may be revealed to be conspecific to *P. coelophrys. P. microdon* is more similar to *P. rouenii* in size and morphology, but the diastema is longer (pers. obs.).

Removal of *Palaeotragus quadricornis* from *Palaeotragus.* Originally it was assumed that *Palaeotragus quadricornis* was similar or conspecific to *P. coelophrys* (Solounias, 1981). Bohlin figured the specimens (1926: Figs. 53–56). Two partial frontlets from Samos were on display in Munich (SPGM) and were destroyed along with the crania of rhinoceroses from Samos during the latter part of the Second World War. It is currently proposed to move *Palaeotragus quadricornis* into *Schansitherium. Palaeotragus quadricornis* and *Schansitherium tafeli* have four ossicones that are positioned in identical fashion. *Palaeotragus quadricornis* differs from true *Palaeotragus rouenii* in having the ossicones on the supraorbital margin (Bohlin, 1926: Text Fig. 53). The four ossicones are the same in the two species of *Schansitherium* and are also positioned above the orbital margin as in *Samotherium boissieri.* The dentition of *Schansitherium quadricornis* is very different from all other *Palaeotragus* species and is more similar to *Schansitherium tafeli* (pers. obs.). One difference is that the premolars and molars are wider than in *Palaeotragus.* However, *Schansitherium quadricornis* is smaller than *S. tafeli.* There are more divergences than size between the two. It appears from unpublished specimens of *Schansitherium tafeli* that the later species is pachyostotic unlike *S. quadricornis,* which is normal. It is pachyostotic in the thickness of the mandible and possibly the limbs. There are also randomly positioned spurs of secondary bone growth on the ossicones of *S. tafeli.* Thus, *Schansitherium qudricorins* is probably the more plesiomorphic species of *Schansitherium.* More work is needed with these taxa as Godina (1979: plate IX) shows a pachyostotic radius for *Yuorlovia* a subgenus of *Palaeotragus.*

Palaoetragus rouenii and *Macedonitherium martinii. Palaeotragus rouenii* is the type species of *Palaeotragus.* It differs from *Schansitherium, Samotherium boissieri,* and *"S." major* in having ossicones positioned more medially on the frontals. *P. rouenii* is a smaller and a specialized species of *Palaeotragus.* The lower p4 of *P. rouenii* is specialized in having the two posterior cuspids anteroposteriorly oriented and as large as the anterior cuspids (Hamilton, 1978a). It appears that the more plesiomorphic taxon is *Palaeotragus coelophrys* rather than *P. rouenii* because it is larger and hence morphologically close to *Samotherium boissieri*. *Macedonitherium martinii* is similar to *Palaeotragus*, but its ossicones have become massive, and it is a larger species. The ossicones, however, are curved but retain the sharp apices of *Palaeotragus rouenii*.

Node 29. Ossicones with posterior blunt keels; skull top and ossicones covered with irregular secondary bone growth in *G. camelopardalis;* snout downturned; nuchal crest extension reduced; soft palate anteriorly positioned, third median (unpaired) ossicone in *G. camelopardalis;* cervical vertebrae elongated; C3 or C4 duplicated; spinous processes elongated in lateral view; C6 with modified lamina; C7 with an attachment for the first rib; T1 in position of T2 with attachment of the second rib. Limbs long; astragalus squared; metapodials with a reduced posterior trough.

Giraffinae Zittel, 1893
    Giraffa jumae Leakey, 1970
    Giraffa camelopardalis Brünnich, 1771

Two alternative hypotheses are proposed for the origin of *Giraffa*. (1) From *Bohlinia*, in the traditional view. This hypothesis would require reorganization of the metapodials in reducing their trough from deep to almost none. It would also require a reorganization of the strongly curved-in styles of the upper premolars. Both of these are possible. In this case, the extreme long neck and metapodials evolved once. (2) The second alterative is to derive *Giraffa* from *Palaeotragus rouenii*.

Based on the second alternative, *Giraffa* falls in the same clade as *Palaeotragus rouenii* and *P. coelophrys*. *Giraffa* shares similar p4 morphology to *P. rouenii*. They both have subequal anterior and posterior cuspids on p4. Also, the two posterior cuspids are directed anteroposteriorly. Both have ossicones that are situated medially from the orbital rim. In *Giraffa* they are more medial and more posterior. The ossicones of *Giraffa* are more massive and terminate in blunt ends as in young female giraffes (the knobs are the result of secondary bone growth). Thus, the ossicones are unlike those of *P. rouenii*, which are slender and terminate in pointed apices. The metacarpal of *P. rouenii* is slightly broader than that in *Giraffa*. The evolutionary change from *P. rouenii* to *Giraffa* is not isometric but would require metapodial elongation, narrowing in width, and flattening of the posterior surface (reduction of the interossei from medium to small). The overall change in metapodials is small between *Palaeotragus* and *Giraffa*. The neck had to elongate from medium to long.

Giraffa is anatomically better known because it is extant. Several major morphological changes are involved in the origin of *Giraffa*. One would be the reduction in the depth of the metapodial trough. The metapodials of *Giraffa* are flat on their posterior side. This is suggestive of a major ecological shift from *Palaeotragus* or from *Bohlinia*, and this shift may signify the habitation of savannas for *Giraffa*.

In *Giraffa*, as in *"Samotherium" major*, the middle indentation of the hard palate is anteriorly positioned between the M2s. It is possible that this is related to a vertically positioned neck in these two species (convergently). The verticality is evidenced by the short and very broad spinous cervical processes of C3 and C4. The laminar part of the nuchal ligament attaches on the spinous processes emanating from T1. When the distal neck is vertical, the pharynx is pushed anteriorly, resulting in a reconfiguration of the soft palate. Because the soft palate is anteriorly positioned, there is more space for swallowing. There may be other factors involved in the indentation and the soft palate because in camelids where the neck is also vertical, the soft palate is positioned posteriorly. Camelids have complex oral morphology involving a dulah sac.

The ossicones of modern *Giraffa* are covered with secondary bone growth (Spinage, 1968b), which is a layer of bone that results from combats of males but also has a genetic basis. Zoo giraffes have the secondary bone layer but do not fight. The shape of the ossicones of the giraffe cannot be easily observed because they are covered by secondary bone growth. *Giraffa camelopardalis* displays secondary bone growth that covers all ossicones and major parts of the frontal area and often spreads onto the occipitals. The best way to observe the shape of the ossicone is in young individuals and in females. The shape is oval with a small knob and a medioposterior blunt keel as the ossicone gently curves back. In adult males, the mass of the ossicone is composed of three parts: the frontal boss, the ossicone proper (a separate ossification), and the external encasing of the secondary bone growth (Solounias, 1988a; Solounias and Tang, 1990).

Giraffa camelopardalis also has a third ossicone in the median plane anterior to the paired ones. Solounias and Tang (1990) proposed the name giraffacone for this median single ossicone. The argument was that it is formed by a boss and secondary bone growth but without an ossicone (the separate ossification). Spinage (1993), however, showed that there is an ossicone component for the median single ossicone. The problem of the median single ossicone is probably resolved, but there are differences in how it grows. It may begin as a very small plate of bone, it may form later in life, or it may form more attached to the integument. I now question why there are so few detached juvenile specimens of the median ossicone in museum collections.

Giraffa has a long neck. The cervicals are notably longer than in *Palaeotragus* and *Samotherium*. The elongation has taken place on the posterior half of the vertebral body (between the posterior opening of the transverse foramen and the posterior edge of the vertebral body). The transverse process is inclined as in *S. boissieri*. The spinous processes are developed to span the entire dorsal side of the vertebral bodies. The atlas is compressed in lateral view. The giraffe also has a specialized C7 vertebra that caries the first rib, a highly unusual character. It also has a duplicated C3 or C4 vertebra and hence a total of eight cervicals. The vertebrae relate to muscles and nerves as in normal ruminants, but the cervical vertebral series has shifted down one to accommodate the additional duplicated vertebra. Vertebral duplication requires further study, but it may relate to more than one non-mutually exclusive possibilities: (1) balance of the long neck, (2) higher reaching for foliage to avoid competition from other shorter-necked giraffid species, or (3) for a different mode of fighting. The amount of new foliage available even if a neck is longer by a few centimeters is immense. Length is added by the duplicated vertebra, but length is also removed by incorporating C7 into the thorax. The net gain in height is approximately 20 cm. It is actually interesting that more species do not have such a specialization. There may be many unknown constraints preventing cervical duplications and cervical elongations in mammals.

PALEOECOLOGY AND PALEODIET

Giraffids are found in numerous Miocene faunas and co-existed with many bovids and other ungulates. Usually at least three giraffid taxa are found per locality. At Samos there are eight species: *Samotherium boissieri*, "*Samotherium*" *major*, *Schansitherium quadricornis*, *Helladotherium duvernoyi*, *Helladotherium* sp., *Bohlinia attica*, *Palaeotragus coelophrys*, and *P. rouenii* (sampling a small valley roughly from 8.2 to 7.2 Ma). Giraffids therefore were a dominant part of later Miocene faunas. They coexisted with many other ungulates. The species of Giraffidae are in general larger than contemporaneous bovids and cervids. This size difference

Fig. 21.11. Premaxillae all reconstructed except for A, B, F, K, and M. Lower incisors reconstructed series mostly from isolated teeth and fragments. (A) *Connochaetes taurinus*. (B) *Giraffa camelopardalis*. (C) *Samotherium neumayri* Maragheh Iran (based on Rodler and Withoefer, 1890: pl. 4, Figs. 2 and 3). (D) "*Samotherium*" *major* (reconstruction) NHMW V 35 and SMF M 2590 from Samos Greece. (E) *Honanotherium schlosseri* from Shanxi China, uncatalogued specimen at AMNH (field number 63 B 772). (F) *Samotherium boissieri* NHM M 4215 Samos Greece. (G) *Palaeotragus rouenii* from Godina (1979: pl. 4, Fig. 2) Sebastopol Russia. (H) *Palaeotragus coelophrys* from Maragheh, Iran (based on Rodler and Weithoefer, 1890: pl. 4, Fig. 4). It is reconstructed either with a groove in the center as in *Tragulus* or with a pointed center as a suiform. (I) *Bramatherium megacephalum* (Siwaliks) based on AMNH 19684. (J) *Sivatherium giganteum* based on NHM M 15288. (K) *Okapia johnstoni* from Fraipont (1907: pl. XX, Fig. 1 and pl. XXV, Fig. 3). (L) "*Palaeotragus*" *primaevus* from Fort Ternan Kenya NMK 3035, 3032, 3030, 3031, 3029 from isolated teeth. (M) *Tragulus* sp. (N) "*Palaoetragus*" "*primaevus*" from Ngorora Formation, Kenya NMK 950. Scale bar equals 70 mm.

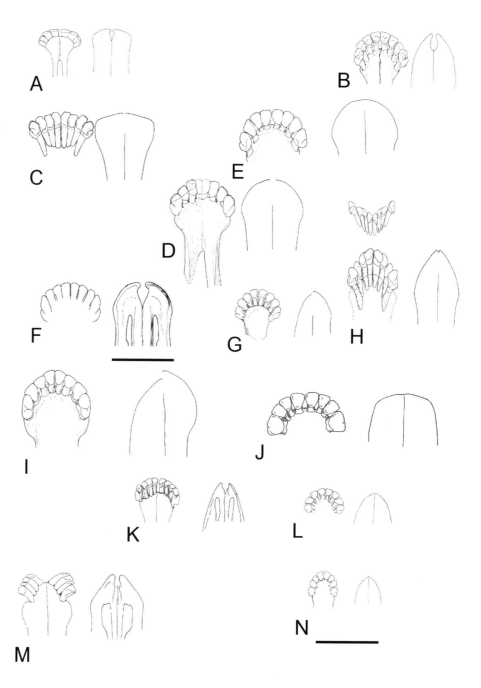

may have given them access to vegetation higher from the ground and thus to a different range of habitats.

The enamel is rugose, and the teeth remain brachydont in all species. Actually brachydonty and mixed feeding and grazing are found in many modern cervids. Thus, the hypsodonty of bovids is probably more related to metabolic differences. Giraffids (including the extinct species) and cervids are similar in diet preferences and dental anatomy.

Research on their paleodiet using tooth microwear analysis has shown that extinct species were either browsers or were mixed feeders (Solounias et al., 2000). The premaxillae can be reconstructed by the shape of the lower incisors (Fig. 21.11). Most species have premaxillae that are rounded as in mixed feeders or pointed as in browsers.

Problem Species. *Samotherium neumayri* has a small masseteric fossa as in browsers, but the premaxillae (Fig. 21.11) are flat as in extreme grazers. *Palaeotragus coelophrys* from Maragheh has a premaxilla that is either very pointed or with a median groove. If pointed, it is reminiscent of certain suids. If grooved, it resembles certain proboscidea and hippopotamuses. In *Tragulus* the incisors also form a median groove, but this species is much smaller than *P. coelophrys* (Fig. 21.11) (the restoration will depend on how one interprets the specimen in the NHMW Vienna).

Tooth microwear analysis confirms that several species were mixed feeders. Both the modern giraffe and the okapi are browsers. Many extinct species, however, were mixed feeders. The most primitive taxa were browsers. These include *"P." tungurensis* (mixed), *"P." primaevus* (mixed), *Giraffokeryx pujabiensis* (mixed), *Bramatherium megacephalum* (grazer), *Sivatherium giganteum* (mixed), *Helladotherium duvernoyi* (browser: pers. obs.), *Palaeotragus rouenii* (browser or mixed), *P. coelophrys* (grazer), *Samotherium boissieri* (mixed; possibly at times grazer), *"S." major* (grazer), *S. neumayri* (grazer), *Bohlinia attica* (browser), and *Honanotherium schlosseri* (grazer; Solounias et al., 2000).

STRATIGRAPHY

There are major gaps in the fossil record of Giraffidae. The majority of the studied taxa fall into three age clusters. (1) The oldest one is approximately 16 Ma and would include *Georgiomeryx, Canthumeryx, Injanatherium* (perhaps *=Progiraffa*), and *"Palaeotragus" tungurensis*. These taxa are found in Africa, Europe, and Asia, and their widespread distribution suggests that the early giraffids occurred throughout the Old World. (2) Ranging around 14 Ma are *"Palaeotragus" primaevus* from Fort Ternan and *Giraffokeryx punjabiensis* and other unstudied taxa from the Chinji Formation of Pakistan. The Ngorora species is about 12.5 Ma (A. Hill, pers. comm., 2006). (3) The majority of species (all taxa from clades 8–12) are of late Miocene, 7–9 Ma, and are primarily from Europe and Asia. The poor record from Africa is probably a result of depositional issues rather than of the absence of giraffids from Africa. The Plio-Pleistocene is rich in *Sivatherium* and related forms. Recent *Giraffa camelopardalis* inhabits the savannas of central, eastern, and southern Africa. Cave paintings suggest that it probably inhabited the Sahara a few thousand years ago. The okapi inhabits rainforests in Cameroon.

ACKNOWLEDGMENTS

I am grateful to Alan Gentry for his extensive help in editing this chapter. I also thank John Barry, John Harris, Andrew Hill, Robert Hill, Henry Galiano, Matthew Mihlbachler, and Renata Trister. I thank Guy Musser and Mammalogy at the AMNH for permission to section an okapi ossicone. I am grateful to the Brookfield Zoo, Chicago and the White Oak Conservation Center, Florida, for study of live okapis. Data from 1974 to the present were used, and all the curators and museums have been acknowledged in Solounias (1981).

22

Family Bovidae

BOVIDS ARE THE MOST DIVERSE GROUP of ruminants. They are naturally found in a variety of habitats and on all continents except South America and Australia. Bovids exist in most habitats ranging from deep rainforests to woodlands, savannas, deserts, mountains, and artic environments. They range in size from very large (Asian water buffalo and the East African Plio-Pleistocene bovine *Pelorovis*) to very small (dik-diks and the Siwalik Miocene *Elachistoceras*). There are approximately 158 extinct genera and 47 that are still extant (McKenna and Bell, 1998). With such a rich record and range of habitats and size, bovids are ideal for the study of herbivory and adaptation to various climatic conditions. Although they are suited perfectly for the study of evolution, their usefulness in biostratigraphy is poor.

Much has been written about Bovidae. Important summaries for the extant bovids include Hofmann and Stewart (1972) on the anatomy of the rumen and the key differences in feeding among browsers, grazers, and mixed feeders; Schaller (1977) on Caprinae from Eurasia; Eisenberg's (1981) general summary on the biology and classification; Chapman and Feldhamer (1982) on many details of the North American species, especially for behavior and diet; and Kingdon (1982a; 1982b) on details of the East African species focusing on behavior and diet. Kingdon also compiled the scattered African literature on these species; Nowak and Paradiso (1983) and Nowak (1999) revised the classic book *Walker's Mammals of the World* in which many nomenclaturial, behavioral, and dietary data can be found; Smithers (1983) covered the South African species; Macdonald (1984) is one of the best books for introducing many bovid species, providing systematic and adaptational summary data, and its colored pictures are useful; Vrba and Schaller (2000a) is a summary of conservation and taxonomy; Prothero and Schoch (2002) provides a general summary with an emphasis on Bovini.

A selection for the fossil bovids is provided by geographic region. The discussions of the following studies are not limited to their particular main geographic region: Moyà-Solà (1983) on species from Spain; Bonis et al. (1998), Bouvrain (1992, 1996, 1997), Gaudry (1862–1867), Gentry (1971), Kostopoulos (1996, 2004), Schlosser (1904), and Solounias (1981) on Miocene species from Greece; Azanza and Morales (1994), Korotkevich (1981), and Gentry and Heizmann (1996) on central Eurasian (Paratherys) taxa; Köhler (1987) and Gentry (2003) on the Turkish Miocene; Gentry (1999b) on Saudi Arabian taxa; Pilgrim (1937, 1939) and Thomas (1984a, 1984b) on taxa from the Siwaliks of Pakistan; Chen (1988), Bohlin (1935a), and Pilgrim (1934) on the Miocene of Mongolia and China; Gentry (1970, 1978a), Thomas (1981), Lehmann and Thomas (1987), and Vrba (1995, 1997) for African species. Significant are the contributions of Gentry on numerous bovid issues: Gentry (1990) on the evolution and dispersal of African bovids; Gentry (1992) on the subfamilies and tribes of Bovidae; Gentry (1994) on key Miocene genera and the differentiation of Pecora; Gentry (2000a) on the ruminant radiation; and Gentry (2000b) on the differences between Caprinae and Hippotragini; Gentry (2003) is seminal as a general review of ruminants. Gatesy et al. (1997) and Matthee and Davis (2001) describe cladistics based on DNA. Estes (1984) provides an evaluation of behavior. Vrba and Schaller (2000b) contribute on the classification based on behavior and glands.

The extant bovids have been classified into subfamilies and tribes by Simpson (1945), and some of these classifications are still considered valid (McKenna and Bell, 1997). In addition, other classifications of bovids have been constructed using molecular data (Matthee and Davis, 2001; Gatesy and O'Leary, 2001; Fernández and Vrba, 2005). The results of the two methods have a strong correspondence, but there are important differences to be considered in the future. For example, there are various systematic problems with certain extant genera such as *Oreotragus*, *Pelea*, *Aepyceros*, and *Budorcas* (Gentry, 1992). Furthermore, there are systematic problems with many fossil genera such as *Hypsodontus*, *Caprotragoides*, the *Criotherium* group, *Palaeoryx*, and so on (Gentry, 1992, 1994).

Bovids are closely related to the giraffids, the cervids, and the antilocaprids. They differ from these families, among other features, by the possession of horns. Horns are derived embryologically from the frontal bone, covered by a keratinous sheath, and are permanent (Bubenik, 1983; Janis and Scott, 1987). In contrast, the ossicone of giraffids is permanent and covered by integument; it begins as a separate ossification and fuses to the frontal later in time. Cervid antlers are derived from the periosteum, are covered by skin (which in this case is termed velvet), and are shed annually (Janis and Scott, 1987). Antilocaprids also have keratinous sheaths, but

these are shed completely; unlike those of the bovids, the sheaths are forked. The internal bony part of the bovid horn is termed the horn core. The shapes of horns and horn cores differ in bovid species and can serve as a key differentiating feature. Horn cores preserve well in the fossil record and are, therefore, the most essential part in paleontological studies. Other characters that differentiate bovid species are skull structure, dentition, premaxillae, and skeletons. This study summarizes some of these morphological characters. Figures and explanations of bovid characters can be found in Gentry (1970, 1990, 1992, 1994), Köhler (1993), Kingdon (1982a, 1982b), and Nowak and Paradiso (1983). Museums and collections are indicated by abbreviations.[1]

THE HORNS

Anatomy

Horns are frequently present and are the most characteristic feature of Bovidae for systematists. The morphology of the horns is consistent in diagnosis of extinct species because the extant species also have differently shaped horns. Horns consist of three parts: the horn core, the pedicle, and the horn sheath.

The horn core is a bony core that originates from the frontal but is induced by the integument (Janis and Scott, 1987). In the integument there is a tissue, a precursor, termed the os cornu. Originally the os cornu was thought to be a distinct bone. Now it is thought to be part of the integument that induces the development of the horn core. The rim of the horn core base is rough where the integument terminates and the sheath begins (Fig. 22.1A).

The pedicle is an upward extension of the frontal bone, and it is always thinner than the horn core (Fig. 22.1A). In contrast, the integument covering the pedicle is thick and is responsible for the growth of the horn sheath in a manner similar to that of hooves and nails. The horn core is totally fused to the pedicle.

The horn core is covered by a hollow keratinous sheath (Fig. 22.1B). The basal rim of the horn core is where the horn sheath begins. The horn core is covered by periosteum, fascia, and the keratinous horn sheath, which is shaped like a cone. The horn sheath is hollow inside where the horn core fits; hence, the old name Cavicornia for the Bovidae (hollow horns). The shapes of the horn and sheath are almost always congruent. Horn cores are shorter than the surrounding sheaths (Fig. 22.1C versus Fig. 22.1B). New sheaths naturally form inside old ones that sometimes may exfoliate. This is different from the shedding of the horn sheath of *Antilocapra*, where the entire sheath falls off. Horn sheaths often have keels. The keels are also expressed

---

[1]Abbreviations: AMNH, American Museum of Natural History, New York; MNHNP, Muséum National d'Histoire Naturelle de Paris, Paris; NS, Nikos Solounias, personal collection, New York; SMF, Senckenberg Museum Frankfurt, Frankfurt am Main.

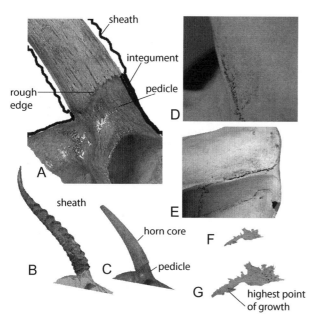

Fig. 22.1. (A) Close-up of the base of the right horn core of *Gazella dorcas* AMNH 54003 showing pedicle and the rough edge of the horn core base. (B) Horn sheath of same specimen. (C) Same specimen without the sheath showing the difference in size between sheath and horn core (B versus C). (D) *Bos taurus* (NS 14) adult domestic cow showing the beginning of growth of a horn core (posterolateral view). (E) Same specimen in lateral view. (F) The horn core isolated photographically. (G) The same specimen enlarged.

in the shape of the horn cores, but the transverse sheath ridges, as in *Capra, Aepyceros,* and *Gazella,* are not expressed on the horn cores. The new sheaths grow from inside out, and the older sheaths slowly exfoliate at the apices and the anterior region where combat impacts often occur for some species.

The horn core and pedicle comprise the basic unit that, unlike the sheath, commonly fossilizes. There are thousands of horn cores in the fossil record that have been used in the studies of fossil bovids. Accordingly, horn cores are more diagnostic than the dentition. If dentition had provided the only available fossils, the diversity of Bovidae would not have been known. Figure 22.1A–C shows a horn core, pedicle, and sheath of *Gazella dorcas.*

In addition, this chapter also includes Figure 22.1D–G, which shows the best obtainable example of a horn core at an initial stage of formation. Figure 22.1D and 22.1E illustrates two views of a growing horn core in an adult domesticated cow (same specimen). Apparently, the cow is too old to have such a small horn core; hence, it is a condition of a delayed juvenile character. The margins are reminiscent of the margins of a giraffid juvenile ossicone. This particular specimen is completely fused to the frontal but graphically isolated and magnified (Fig. 22.1F–G).

Another distinctive feature of some horns is the frontal sinus, which extends into the horn core, making it hollow (e.g., Caprini, Bovini). Thus, the horn cores are less heavy for these taxa. In contrast, many other species, such as the: *Boselaphus, Tetracerus,* Antilopini, Hippotragini, Reduncini, Cephalophini, and *Aepyceros,* have solid horn cores.

Types of Horns

This chapter proposes two subdivisions of bovid horn cores that have important ramifications for the evolution of Bovidae, although further developmental research is needed. The first subdivision is universal in Bovidae, and the second is restricted to most Boselaphini and Cephalophini.

The more common horn cores have a single consolidated horn core mass restricted above a well-defined pedicle (Fig. 22.1A–C). The pedicle is surrounded by integument in all directions, and the horn core base is above the pedicle. Examples of bovids that fall into this category include the very first bovids such as *Eotragus* as well as *Hypsodontus.*

Interestingly, new observations have been made regarding the less common horn cores. They lack a pedicle and are situated low on the frontals (Fig. 22.2A–D). This is observable in *Boselaphus, Tertacerus, Tragoportax, Sivaceros,* and *Sivoreas* of Boselaphini and in *Cephalophus* of Cepahlophini. Additionally, the horn core seems to emanate below the braincase in lateral view in Boselaphini and Cephalophini. Moreover, in certain Boselaphini (e.g., *Sivoreas* and *Miotagocerus monacensis*), a strong pedicle is visible medially but is engulfed by the horn core base both anteriorly and posteriorly. In these, the medial edge of the horn core base is arched. Such a pedicle may actually be a different type of

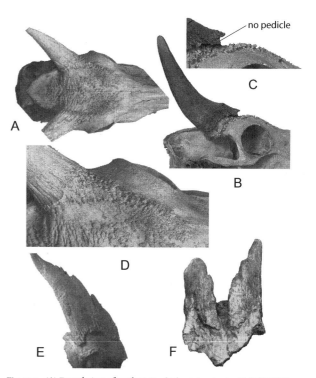

Fig. 22.2. (A) Dorsal view of a nilgai, *Boselaphus tragocamelus* AMNH 35520, showing the distribution of the field of bony growths. (B) Close-up of the same specimen in lateral view showing elevated bony pillars anterior to the base of the horn core. The horn is set low, and the absence of pedicle is clearly visible. (C) Magnified view of the same bony pillars. (D) Dorsal view showing the triangular distribution of bony growths anterior to the base of the horn core. This field is similar to the horn base of *Miotragocerus* and *Tragoportax.* (E) *Miotragocerus monacensis* from the Miocene of Samos, Greece SMF M 1968. Independent bony plates comprise the horn core. (F) Same taxon from the Miocene of Pikermi, Greece MNHNP no number.

frontal enlargement than a true pedicle that is surrounded with integument in all directions.

In *Boselaphus tragocamelus* and *Tragoportax salmontanus*, expanded fields of bony growths are often present in addition to the primary horn core. In adult *Boselaphus* the horns are surrounded by a bone growth field (Fig. 22.2A–D). It has been proposed that this field probably consists of additional rudimentary horn cores surrounding the main horn cores (Solounias, 1990). Figure 22.2C, which is a magnification of Figure 22.2B, shows the physical merging (continuity) of the bony growths into the anterior base of the horn core. These bony growths are found anterior to, between, and posterior to the horn cores with different morphological appearances. The field has four distinct regions: a triangular wedge of elevated pillars immediately in front of the horn core (Fig. 22.2B–D), compressed pillars between the horn cores anteriorly and above the orbits, elongated centrally interconnected pillars forming ridges between the horn cores, and irregular large pillars posterior to the horn cores between the two temporal lines (Fig. 22.2A). The posterior growth field is the only region observable in the extinct *Tragoportax salmontanus* (Boselaphini from the Siwaliks; A. W. Gentry, N. Solounias, J. C. Berry, and M. Raza, unpublished data). In addition to the bony growth fields, in certain extinct species of *Miotragocerus* (Boselaphini), the horn cores are composed of a series of distinct vertical plates, which appear as reminiscent of the triangular wedge of *Boselaphus* (Fig. 22.2D). The plates are located in the same region as the triangular wedge. The plates are not visible in all specimens, but a few probably young individuals clearly show these plates (Bohlin, 1935b) (Fig. 22.2E–F).

The horn cores of *Miotragocerus* are compressed with an anterior horn core demarcation and keel. These horn cores may form differently, perhaps from the conglomeration of multiple small bony growth fields one adjacent to the next. The combination results in a steplike pattern. Bohlin (1935b) described the mode of growth of the horns in *Miotragocerus amalthea*. The shapes of the plates seen in these horns are supportive of such growth.

Tetracerus quadricornis (the chusingha, Boselaphini) has four horns; two additional horn cores anteriorly (Fig. 22.3A). They are probably formed in the same manner as those of *Miotragocerus*. *Tetracerus* only has two bony growth fields that are situated between the anterior and posterior horn cores. In one specimen of *Tetracerus* (AMNH 35306), the anterior horn core shows a steplike pattern similar to *Miotragocerus* (Fig. 22.3B). Figure 22.3C is a magnification of Figure 22.3A and shows the exfoliation.

Horn Core Surface and Shapes

The horn core surfaces of bovids vary from deep grooves (*Prosinotragus kuhlmanii*, *Parurmiatherium*, *Oioceros rothii*, *Oioceros wegneri*) to fine grooves (*Gazella* and *Capricornis*) and to very smooth grooves (*Tragoportax* and *Oryx*). They also vary in shape. Horn cores can be long (*Tragelaphus*, *Pseudoryx*, and *Oryx*) or short (*Parurmiatherium* and *Ovibos*).

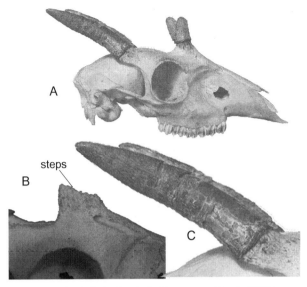

Fig. 22.3. (A) The skull of a chousingha, *Tetracerus quadricornis* AMNH 27732. The posterior horn is set low, and the absence of pedicle is clearly visible in the posterior horn. (B) Anterior horn core of *T. quadricornis* AMNH 35306 with a clearly defined step as in the Miocene genus *Miotragocerus*. (C) Exfoliation of the keratinous sheath (same specimen as A).

The majority are oval to round in cross section (*Rupicapra*, *Gazella*, and *Naemorhedus*). In some cases the horn core can be strongly compressed as in *Tragoportax*, *Kipsingiceros*, *Hemitragus*, and *Capra*. They are either almost straight (*Hypsodontus*, *Oreotragus*, *Ouribia*, and *Oryx*) or curved posteriorly (*Pachytragus*, *Capricornis*, *Gazella*, *Hemitragus*, and *Hippotragus*). The later is the most common type of horn core curvature. There are species in which the horn cores curve anteriorly (*Eotragus*, and *Ammodorcas*) or laterally (some *Hypsodontus*, *Bos*, and *Ovibos*). In some, the curvature is both posterior and anterior because the original backward curvature has become restricted to the very base (*Antidorcas*). In some rare cases there are also medially directed curvatures (*Syncerus*, *Kobus*, *Redunca*, and *Alcelaphus*). There are various spiral twists of the long axis (*Oioceros*, *Protragelaphus*, *Prostrepsiceros*, *Antilospira*, *Antilope*, and *Tragelaphus*) as well. The twist can be clockwise or counterclockwise as one examines the right horn core from the base up.

Horn cores can also have keels. These keels are either anterior (*Capra* and *Hemitragus*) or posterior (*Protragelaphus*) or both (*Prostrepsiceros*, *Oioceros*). Most horn cores are uniform in thickness, but there are species in which the proximal part is significantly thicker than the distal (*Ovibos*). Rarely the left and right horn cores fuse partially at their base (*Urmiatherium*) or fully to form a single shaft (*Tsaidamotherium*). The plesiomorphic position is only slightly posterior to the orbit (in the literature this position is termed above the orbit) (*Eotragus* and *Hypsodontus*). The horns often are located further posteriorly (*Boselaphus* and *Tetracerus*), even as far as the occiput (*Criotherium* and *Bos*). Combinations of all these characters and size differences are numerous and result in many species. Horn cores are useful for systematic studies, but one could not construct a satisfactory cladogram using the horns alone because there are many evolutionary convergences.

The Frontals

The frontals in bovids also vary in size. In some species the frontal sinuses enter the horn core and can extend all the way up to the apex. Horn core sinuses are primitively absent as in Reduncini, Hippotragini, and Antilopini. They are extensive in several groups as in Ovibovini, Caprini, and Bovini. In Bovini the frontals take the position of the parietals while the parietals move even more posteriorly and are situated vertically adjacent to the occipitals. This remarkable specialization of the frontals and parietals needs to be incorporated in cladograms.

The Insertion of Horns onto the Frontals

There are two classes of positions for the horn cores. The plesiomorphic position of the horn insertion on the frontals is slightly behind the orbits as in *Eotragus*. Many bovids retain this position.

The apomorphic position is for the horn cores to shift posteriorly. When the horns are located posteriorly as in Bovini, the frontal surrounding the horn is also repositioned. Thus, the actual horn core location is at the same place as before. In other words, in Bovini the entire frontal and the frontal sinus are plastically displaced posteriorly. Although the horn core appears to be in a new position, in actuality, its location does not move from its origin. Cross sections of the frontal sinuses show that they also extend back with the horn core and that these sinuses form a second layer over the parietals. The parietals form the braincase internally and are covered externally by the frontal bone and frontal sinuses.

A closer look at the cow, *Bos taurus*, can exemplify the apomorphic positioning of the horn cores. The frontal bones override the parietals to form a double cranial vault. In many specimens of *Bos*, one can see the sutures externally. The parietal is embedded in the temporal fossa, and the frontal is located dorsally, extending all the way to the back and forming the edge of the nuchal crest. So the occipital attaches to the frontal. In a cross section of *Bos* in our laboratory, the suture of the frontal can be seen in a coronal section (transverse across the auditory meatus). The parietal has a different and much more limited sinus in it. It is located below the frontal, and the frontal sinus extends along with the frontal bone above the parietal. The supraorbital foramen is also moved posteriorly along with the horn. Hence, the overriding moves the frontal horn posteriorly. Another way of describing this is as an expansion of the frontals at the expense of the parietals (Hooijer, 1958a: 44–50, 57–59).

The posterior positioning of the horn cores occurs in other genera besides Bovini. It would be interesting to investigate whether the movement of the frontals over the parietals occurs in the other genera similar to the Bovini or

whether it takes place in a different way. Some groups of interest may include *Criotherium*, *Urmiatherium*, *Tsaidamotherium*, *Oioceros wegneri*, Cephalophini, and Alcelaphini.

In an anterior view most of the frontals are relatively flat in relation to the orbits. Thus, the pedicles and horn cores are at the level of the orbital rims. However, in a few taxa such as Caprini and Alcelaphini the frontal sinuses form a thick buttress above the orbital rims, consequently raising the horn cores above the rims (Gentry, 2000b).

Sexual Dimorphism of Horns

All male bovids have horns. The plesiomorphic condition is for females to be hornless. Examples of hornless females are in *Cephalophus*, *Tragelaphus*, *Boselaphus*, *Tetracerus*, *Procapra*, *Litocranius*, and *Gazella dama*. On several occasions during evolution, females grow horns independently for reasons not clearly known (Janis, 1982; Estes, 1984; Roberts, 1996). One hypothesis is that they reduce the visual aggression in the herd. Another is for the defense of the young in open habitats. It may also make it more difficult for predators to single out females during a chase when the females look like males. Horned females are common in taxa that form herds. Multiple explanations are not necessarily mutually exclusive. The Miocene fossil record is poor, and it is difficult to distinguish between male and female horn cores. Hornless females are found in extinct *Sporadotragus*, Boselaphini, Antilopini, and Bovini. Horned females may be found in *Pachytragus*, *Oioceros*, Caprini, and *Criotherium*. The females grow horns more frequently after the end of the Miocene.

Adaptive Significance of Horn Core Shapes and Positions

Horns are used mostly for intraspecific fighting and in ritualized displays. One would assume that the shapes of the horns are functional. However, there may be major elements of chance in the details of the shapes of horns. Details could be local fixations involved with speciation in isolated clades of Bovidae. Some general patterns are related to the mode of male combat. For example, in *Capra* and *Ovis* the horns are straight and long with a slight curvature. This shape is in agreement with their mode of fighting in which the males first run and rise on their hind legs and then clash and fall onto their front legs making an arc in the air before landing. The horns themselves display the shape of that arc, facilitating combat during collision. On the other hand, in *Ovibos* the horns are short, broad, with laterally positioned apices. This shape allows the animal to clash head on. These clashes gouge the base of the keratinous sheath, but the apices are spared. In contrast, *Tragelaphus* and *Antilope* have twisted horns, so their combat is more like wresting, with ample lateral motion and twisting of the opponent's head.

Other functions of horns, in addition to combat, are to scrape off bark for feeding, as in *Boselaphus*, and to toss earth over the back of the body when sitting down, as in *Gazella* and *Aepyceros*. Horns may also be involved in thermoregulation.

OTHER CRANIAL MORPHOLOGIC FEATURES

The Ethmoidal Fissure

The ethmoidal fissure is a triangular or somewhat rectangular opening between the nasals and the maxilla. It is closed in *Tragulus* and *Moschus*. It is open in some bovid taxa such as in *Capra* and *Ovis*, but it is closed in *Bos*, *Connochaetes*, *Boselaphus*, and *Tetracerus*.

Bending of the Braincase

The bending of the braincase in relation to the palatal axis is an obvious feature in the skull of bovids (Gentry, 1970). Bovids have either a strongly bent or a less-bent braincase (strongly bent in *Capra* versus less-bent in *Bos* or *Tragelaphus*). In a strongly bent braincase, the hard palate, face, and maxilla are angled down in relation to the basicranial axis, which is defined as the median line formed by the basioccipital and basisphenoid bones in lateral view. For the face, a median line at the hard palate defines that plane. In a less-bent braincase, the two axes are relatively in the same orientation (subparallel). This feature can be appreciated by a lateral view of the skull. The term cyptocephaly (bent headedness) was given to this character by Osborn (1929: vol. 2: 823, 827; Figs. 740, 742).

In bovids, braincases are somewhat bent even in the more plesiomorphic taxa. *Eotragus*, one of the earliest bovids, has a bent braincase. In addition, the braincase is bent in tragulids and other archaic artiodactyls. In bovids, several independently derived lineages exist in which the braincase is strongly bent. The degree of bending needs to be accurately measured. For example, in *Kobus* the anterior basioccipital tuberosities are very large, and the basisphenoid in between these tuberosities gives a different angle than the angle obtained in a lateral view incorporating the tuberosities. A more reliable way to evaluate bending is by using the dorsal aspect of the face and braincase in a lateral view. The braincase is a rectangular box, and the basisphenoid is parallel to the parietals in lateral view.

In several taxa the degree of bending is small (*Eotragus*, *Cephalophus*, *Bos*, *Tetracerus*, *Boselaphus*, some extinct Boselaphini, and Miocene *Gazella*). In derived taxa, the braincase forms an acute angle (*Pachytragus*, Caprini, Alcelaphini, Ovibovini, Reduncini, Hippotragini).

A Novel Interpretation of the Bending

The bending of the braincase in relation to the cervical vertebrae has not been studied. Contemporary evaluation of the bending of the braincase utilizes an isolated skull, and the neck is not taken into consideration. However, the neck is an integral part of the posterior skull, with numerous muscles attaching on the occipital, mastoid, and basioccipital, and deserves consideration. The cervical vertebrae and the basicranium are related embryologically and functionally. The base of the braincase not only is covered with cervically emanating muscles but is also embryologically vertebral.

In lateral view, the vertebral bodies of C1–C3 and the basioccipital are aligned in some species but not in others. It is difficult to find correctly mounted or figured specimens, but preliminary examination shows that at least four possibilities exist. The vertebral bodies of C1–C3 and basioccipital can be aligned or not. In turn, the face is either bent or not bent.

It is also important to point out that commonly the bending is envisioned as taking place at the braincase. Another perspective is that the bending may actually be taking place in the face instead. The position of the face and the orbits needs to be investigated in relation to the standing posture of the animal. Moreover, the bent braincase may actually be interpreted as not bent when one considers the cervicals. For example, if the neck is taken into consideration, it appears that a strongly bent braincase of *Alcelaphus* is perfectly lined up with the vertebral bodies. From that view, it is the face that is bent. On the other hand, the braincase of *Bos* is considered not to be bent. However, if it is evaluated along with the neck, the braincase is strongly bent.

Based on these preliminary observations, it appears that more research needs to be done in order to investigate the bending of the braincase in relation to the face. The functional reasons for the bending require study as well. A cursory examination shows that most bovid grazers, such as *Alcelaphus* and *Hippotragus*, have more strongly bent braincases. However, bovid browsers such as *Capra* and *Cephalophus* also have bent braincases. Hence, other considerations to explain this are necessary. The masticatory and cervicocranial muscles may be oriented in different ways in species with bent braincases. In addition, a bent braincase would reduce the length and orientation of the temporalis muscle onto the coronoid process. Bending may also relate to the mode of fighting and head posture.

The Projection of the Orbital Rims

In a dorsal or anterior view of the skull, the orbital rims and orbits do not project in most species (*Miotragocerus*, *Boselaphus*, *Tragelaphus*, and *Gazella*). In a few species, however, the orbital rims do project. Projecting orbits is the derived condition. For example, *Ovis*, *Capra*, *Saiga*, and *Budorcas* have projecting orbits. This projection may have to do with vision or the size of the ethmoidal sinuses between the orbits. The ethmoid transmits olfactory nerves into the nasal cavity. Another explanation may be that this results from combat because the position of the orbits is interrelated to the position of the horns.

A projecting orbit also results in a projecting zygomatic, which in turn relates to muscles of mastication. The zygomatic is the location of the origin of masseter profundus and zygomandibularis muscles. Large muscles are needed in mastication, as in grazers, and these may result in projecting orbital rims as in *Ovis*. Many grazers, however, do not have particularly projecting orbits (e.g., *Damaliscus, Connochaetes,* and *Bos*).

The Temporal Fossa

The temporalis muscle attaches to the sides of the braincase at the temporal fossa. The term fossa leads one to envision a depression. This concept of depression (as in a temporal fossa) leads to confusion. The temporal fossa is centrally elevated and is mistakenly classified as a reduced fossa. That is, the parietals bulge centrally within the fossa. Perhaps the best way to define the temporal fossa is by its borders, which are the postorbital bar, the temporal line on the parietals and frontals where the deep fascia of the temporalis muscle attach, the lateral ridge of the occipital bone, and the dorsal margin of the zygomatic arch. Within the temporal fossae, some braincases are low as in *Bos,* and some are raised as in *Gazella, Turcocerus,* and other Hypsodontinae.

Plesiomorphically the fossa is extensive, and its temporal lines are small and are found close to the median plane (e.g., *Turcocerus grangeri* from Tung Gur and *Hypsodontus tanyceras* from Fort Ternan). In these taxa the fossa is so large that a single midsagittal ridge forms by the merging of the left and right temporal lines as in carnivores. This feature is uncommon in ungulates, but it is observed in tapirs. In contrast, the apomorphic bovids have the two parietals fused into a single unit, and the temporal lines are positioned laterally on the braincase, lacking a midsagittal ridge.

As mentioned, the temporal lines are repositioned laterally in most bovids. This results in a smaller temporal fossa for the temporalis. It is not currently known if this large group of bovids has smaller temporalis muscles or if the temporalis is reorganized in some way by becoming thicker. The temporal fossa becomes smaller in a few bovids (Bovini and *Connochaetes*), and in this case the temporalis is thicker.

The Morphology of the Masseter Muscles

The muscles of mastication in addition to the temporalis are the masseter superficialis, profundus, zygomandibularis, and the medial pterygoid. The masseteric fossa on the maxilla provides origin for the two masseter muscles of mastication (Solounias et al., 1995). The masseter superficialis and masseter profundus originate there. Both of these attach anteriorly on the maxilla, unlike in most mammals. In bovids, the muscles' position anterior to the zygomatic arch position is notable. This construction strongly differs from those of perissodactyls and camelids, where all masseters are confined below the zygomatic arch. In these taxa there are no masseter muscles anterior to the zygoma. In primitive artiodactyls (e.g., *Tragulus*), the masseter is also anterior as in bovids. However, in plesiomorphic artiodactyls, this condition has not been investigated. In cervids and in giraffids, the anterior distribution is also similar to that in the bovids.

Masseter superficialis has a weak origin in browsers and a strong one in most grazers (Fig. 22.4A versus 22.4B)

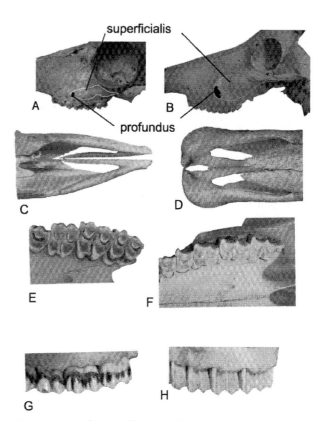

Fig. 22.4. Browsers have a small masseteric fossa, whereas grazers have a large one. (A) A gerenuk *Litocranius walleri* AMNH 16713 (browser). (B) The topi *Damaliscus korrigum* AMNH 82147 (grazer). Browsers have pointed premaxillae, whereas grazers have squared ones. (C) *Litocranius walleri* AMNH 16713 (browser). (D) *Connochaetes taurinus* AMNH 81850 (grazer). (E and G) Occlusal and lateral views of a boodont dentition *Cepalophus dorsalis* AMNH 52900. (F and H) Occlusal and lateral views of the aegodont dentition of a domestic goat *Capra hircus* NS 43.

(Solounias et al., 1995). In most grazers, the masseter superficialis has a restricted origin onto the maxilla, where it forms a bony protrusion. The insertion of the superficialis is on the very edge of the angle of the mandible, where a bony ridge forms and bony growths result from the pull of the muscle. In contrast, in browsers there is no protrusion of the superficialis at its origin.

The masseter profundus is larger in grazers (e.g., Bovini, Reduncini, Alcelaphini, and Hippotragini; Fig. 22.4A versus 22.4B). The masseter profundus covers the entire mandibular angle. In browsers, the profundus is smaller and covers a smaller area on the maxilla. The mandibular angle is slender.

Posteriorly, inferior to the zygomatic arch, the zygomandibularis is thin in browsers and thick in grazers. In grazers, as the zygomatic arch expands to accommodate a larger zygomandibularis, the orbit is also pulled laterally. Hence, the projection of the orbit laterally in conjunction with size of zygomandibularis deems further investigation.

The pterygoids are internal to the mandible. They attach on the margins of the palatine bone adjacent to the soft palate that is located at the median palatine indentation. They also attach on the inner side of the zygomatics. Details of their anatomy are not known. Preliminary dissections show that the pterygoid muscles are larger in taxa where the median indentation is positioned posteriorly. This is the plesiomorphic condition. Detailed dissection of the pterygoids in ruminants is not known. Veterinary literature usually provides a coronal cross section without detailed explanation.

The Soft Palate

The palatine bone of bovids has two lateral and one median palatine indentations. Two positions of the soft palate are dominant in bovids. In one, the plesiomorphic soft palate is situated posteriorly in relation to the lateral palatine indentations. The other position of the soft palate is at the level of M3 or the level of the lateral palatine indentations. Anterior positions of the soft palate are uncommon. Variation in the position of the soft palate can be found within the same genus (e.g., *Cephalophus dorsalis* versus *C. niger*; *Gazella gaudryi* versus *Gazella dorcadoies*).

The Premaxillae

Bovids have an edentulous fibrous pad covering the premaxillae. The shape of the premaxillae relates to the mode of food collection (Solounias and Moelleken, 1993a, 1993b). There are three predominant types of premaxillary shape: pointed, flat, or in between. The browsers have pointed premaxillae (Fig. 22.4C). This is an effective design because it acts as forceps in collecting and selecting browse vegetation. Browse often grows in isolated clumps, and a tweezers-like premaxilla is best designed for selectivity. In contrast, many grazing premaxillae are squared as in Bovini or some Alcelaphini (Fig. 22.4D). A flattened anterior edge allows an effective collection of grasses that grow in large clumps and have thin blades.

Other premaxillae are shaped in between. That is, they are slightly squared with rounded corners, as seen in some Reduncini, Hippotragini, and Caprini. Solounias and Moelleken (1993a) generalized that this shape represents a mixed feeder ruminant. However, examination by eye on large numbers of species suggests that this is not entirely true. Although most mixed feeders have this shape, there are grazers and browsers that also have the intermediate shape. Apparently the shape in between the two extremes can be utilized in grazing and in browsing. This intermediate shape is effective in all types of feeding but is most commonly used by the mixed feeders. A subcategory of the intermediate shape is that of a bony flange on the side of the premaxilla (e.g., *Kobus, Oryx, Ovis,* and *Capra*). The function of the flange is not presently known. Also, note that *Capra* is a browser and *Ovis* is a grazer, but they both have the same type of premaxilla (the later type with the flange). Other shapes are also present. In *Gazella thomsoni* the front is entirely flat but narrow, and the overall shape is pointed.

The Symphysis and Diastema

The disarticulated symphysis in bovids is usually an average oval in outline. The diastema is average in length.

These morphologies are notably uniform among bovids, unlike in Giraffidae where a large range of variation exists. Cervids have similar symphysis and diastema as bovids.

The Dentition

The second most important character in systematic determination after the horns is the dentition. Teeth are hard and fossilize readily. A basic division is that of boodontia and aegodontia (cow versus goat). Boodonts are the Bovinae, Cephalophinae, and Hippotraginae. Aegodonts are the Alcelaphinae, Caprinae, and Antilopinae. Boodonts tend to have less hypsodont teeth, more rugose enamel, basal pillars, a slower rate of fusion between the lobes of the molar teeth in ontogeny, and longer premolar rows (Gentry, 1970, 1978a: 564). Aegodonts display just the opposite features. Figure 22.4E–H shows the dentition of a cephalophine and a caprine.

Two trends dominate a multitude of patterns within both boodonts and aegodotns. The first and far more important is that of hypsodonty (tall teeth), a term applied especially for the molars (Janis, 1988). Plesiomorphic taxa are brachydont and have simple central cavities and usually thick styles at their base (e.g., *Cephalophus*, Figs. 22.4E and 22.4G). Hypsodonty evolved numerous times in bovids. For example, in *Capra* hypsodonty is well developed (e.g., Figs. 22.4 F and 22.4H), whereas *Cephalophus* is brachydont. A molar intermediate in height is called mesodont as seen in *Antidorcas marsupialis,* and mesodonty has been described in Fortelius and Solounias (2000).

The second trend is the reduction of the premolars. Plesiomorphic premolars are long as in *Cephalophus*. On the other hand, the reduction is notable in Reduncini, Caprini, and Antilopini (e.g., *Redunca, Capra,* and *Gazella).* In Hippotragini some premolars are reduced as in *Oryx,* but others are not as in *Hippotragus*.

Plesiomorphic teeth are brachydont, and the third molar is very similar in size to the first. Bovid hypsodonty is confined to the second and especially third molars, unlike those of perissodactyls. In cases of extreme hypsodonty, even the premolars become mesodont. Examples of the plesiomorphic dentition can be found in *Eotragus* and in *Cephalophus* (Fig. 22.4E and 22.4G). In these teeth the premolars are long, and the molars are brachydont. The ribs and styles are well developed and are thicker toward the roots.

The development of the infolding of the central cavities is a character that increases the overall surface of enamel. It can be seen in Bovini, Alcelaphini, Hippotragini, Reduncini, and others. A reduction or increase in the size of ribs and styles is also a common trend in bovids.

The roots adhere to the alveolar bone by the periodontal ligaments. The development of thick or thin cementum is coupled with hypsodonty. When a tooth is hypsodont, the additional enamel is embedded in the alveoli. The alveolar bone cannot attach to the enamel as it can to the root, which is made of dentine. The evolution of cementum occurs so that the alveolar bone can attach to the cementum via the periodontal ligaments. Many bovid hypsodont teeth have a veneer of thin but hard cementum, which can be iridescent (e.g., *Rupicapra, Capra*). Some Bovini, Reduncini, and Alcelaphini resemble *Equus* in having very thick and chalky cementum. Cementum also protects the walls of the enamel when a tooth is stressed during mastication.

Hypsodonty and Grazing in Ungulates

Hypsodonty occurs in equid evolution. Equids become more hypsodont as grassy plains replace forested environments. The Eocene forests are replaced by woodlands and eventually by grasslands. In parallel to the vegetation, Eocene browsing equids are replaced by mixed feeders and eventually by grassland grazers. This scenario has profoundly influenced the thinking about hypsodonty and grazing. Several ungulate groups evolved hypsodonty (equids, rhinocerotids, oreodonts, camelids, antilocaprids, and bovids). Actually, in many instances but not all, hypsodonty correlates well with grazing. Consequently, hypsodonty has been correlated with open habitats. Janis (1988), Janis and Fortelius (1988), Janis et al. (2002), and Mihlbachler and Solounias (2006) have addressed issues on hypsodonty for various selenodont ungulates. Other research found that the correlation is high. A review of the issues and pertinent literature is provided on the second page of Mihlbachler and Solounias (2006).

Attrition (tooth to tooth contact) and abrasion (food to tooth contact) are co-occurring forces on teeth (Fortelius, 1985; Fortelius and Solounias, 2000). It is impossible to have one without the other. However, the relative strength of each force may vary. Further research is needed in this category as well. It is impossible to have strong abrasion without strong attrition, although the opposite may occur. Typical grazers such as Alcelaphini or Hippotragini probably have both high abrasion and high attrition. High attrition, however, may not necessarily be coupled with high abrasion, as when an animal feeds on tough foods that are not so abrasive. Although no current data are available, one can envision a wide spectrum of possibilities where these two forces have different effects on tooth wear.

In conclusion, three extremes are envisioned: low forces for both attrition and abrasion, high attrition and low abrasion in cases of tough but not abrasive foods, and instances in which abrasive foods demand high attrition and high abrasion. The cases in which both forces are low or high are easy to understand. The difficult case to explain is that in which attrition is high and abrasion is low. Hence, it is possible to have mastication where high abrasion-attrition dominates and other instances where attrition dominates. Future experimental data will clarify the interplay of these forces.

Low Attrition and Abrasion.   Leaf-dominated bovid browsers such as *Tragelaphus (Boocercus) euryceros* and *Litocranius walleri* have overall brachydont teeth, and their vegetation probably results in low abrasion and low attrition.

High Attrition and Abrasion.   Grazers such as *Bison bison* and *Hippotragus niger* feed on grasses, which are rich in phytoliths, and hence these taxa are subjected to high abrasion and attrition. These taxa are hypsodont.

High Attrition and Low Abrasion.   It is proposed that *Capra, Rupicapra, Oreamnos, Saiga, Ovibos,* and *Budorcas* are browsers that fall in the category where attrition is high but abrasion is low. These taxa are hypsodont but are browsers. This category leads to the proposition that hypsodonty does not evolve only with abrasion but can also evolve from attrition. These genera are hypsodont browsers—a new category of feeding. Fruit browsers, such as the *Cephalophus* species, differ from the other three categories. These taxa appear to be subjected to higher attritional and perhaps abrasive forces from the fruit pits and tough covers of the fruits than the taxa of leaf browsers. Fruit browsers retain brachydont dentitions and may be the oldest type of bovid browsers. Further research is required to sort out the science behind these preliminary observations.

DIETARY INTERPRETATIONS

Tooth Microwear and Tooth Mesowear

Several studies have analyzed the diets of some extant and extinct bovids along with other ungulates. The method of tooth microwear analyzes the scratches and the pits vegetation makes on the molars. Research on microwear of extant bovids has shown distinctions among fruit-dominated browsers, leaf browsers, mixed feeders, and grazers. In addition, meal-by-meal mixed feeders were different from seasonal and regional mixed feeders (Solounias and Semprebon, 2002; Semprebon et al., 2004). Several extinct taxa have been studied by tooth microwear. *Eotragus sansaniensis* is one of the earliest bovids. It was found that *Dicroceros elegans* (an archaic cervid) was more of a brachydont species than *Eotragus sansaniensis* at the locality of Sansan, but *D. elegans* was more of a grazer (Solounias and Moelleken, 1994). Another study showed that *Kipsingicerus labidotus* (Boselaphini), although dentally more brachydont than *Hypsodontus tanyceras* (Hypsodontini), was more of a grazer at the locality of Fort Ternan (Solounias and Moelleken, 1992a, 1992b, 1993b, 1994; Cerling et al., 1997). Early Caprini from the Miocene of Samos were studied: *Pachytragus laticeps* was a grazer, and *Pachytragus crassicornis* was a mixed feeder (Solounias and Moelleken, 1992b). We studied late Miocene Boselaphni from Samos and Pikermi and found that *Tragoportax amalthea* was a mixed feeder at Pikermi and a grazer at Samos. *Tragoportax rugosifrons* from Samos was a grazer (Solounias and Hayek, 1993). Microwear analysis of the late Miocene gazelles from Pikermi and Samos found that *Gazella capricornis* was a mixed feeder (Solounias and Moelleken, 1999a).

Tooth mesowear analyzes the shape of molar apices. Sharp apices are common in browsers. Rounded apices are common in mixed feeders and grazers. Blunt apices are found in only a few grazers. There is a large overlap between the shapes of apices and diets. The trophic groups of extant species were defined as typical, conservative, and radical with different outcomes. The typical set of species classified 96% of the diets correctly. An analysis of the Serengeti grazing succession was replicated by mesowear. The succession from a grazing extreme to a browsing extreme is as follows: *Equus burchelli*, *Alcelaphus buselaphus*, *Connochaetes taurinus*, *Damaliscus lunatus*, and *Gazella thomsoni*. In two extinct Caprini, the study showed that *Pachytragus laticeps* from Quarry 1 of Samos and *Pachytragus crassicornis* from Quarry 5 of Samos differed. The mesowear results were the reverse of the data from these species based on microwear. Thus, *Pachytragus crassicornis* was more of a grazer (Fortelius and Solounias, 2000).

Tooth Wear Rates

Wear rates were estimated from longevity and tooth heights available in the literature (not all bovids). The wear rates were exaggerated by the body size distribution of the species examined. The browser wear was approximately 0.33 mm per day. The mixed feeders were 0.41–0.93 mm per day. The grazers were 2.03–3.65 mm per day (Solounias et al., 1994).

Foraminal Size Analysis

Browsers have a smaller foramen ovale than grazers because the corresponding motor cranial nerve innervating the muscles of mastication is probably smaller. A smaller nerve would correspond to smaller muscles of mastication for the browsers. Browsers have larger and more mobile lips than grazers. Hence, browsers have a large stylomastoid and infraorbital foramina. Cranial nerve 7 (motor) and the second division of the trigeminal nerve (sensory) will be larger to serve the larger lip and facial muscles. The analysis found these differences in extant bovids. Starting with these data, several extinct species were analyzed. *Pachytragus crassicornis* from Samos was more of a browser than *Pachytragus laticeps*. *Tragoportax amalthea* from Pikermi was more of a browser than *Tragoportax amalthea* from Samos. In addition, *Tragoportax rugosifrons* from Samos was a grazer (Solounias and Moelleken, 1999b).

Masseteric Muscle Estimation

Large muscles of mastication correlate to grazing. Outlines of muscles were analyzed, and differences were found between browsers and grazers. The mixed feeders overlapped significantly with the browsers. The late Miocene *Tragoportax amalthea*, *Tragoportax rugosifrons* (Boselaphini), and *Gazella capricornis* (Antilopini) were probably either browsers or mixed feeders. *Pachytragus crassicornis* and *Pachytragus laticeps* (Caprini) were probably mixed feeders (Solounias et al., 1995).

THE NECK AND THE SKELETON

The Occipital

The occipital of bovids is remarkably uniform. Most occipitals are an inverted U-shape, which is similar to that of cervids. They are low and broad. For example, it is U-shaped in *Tragoportax*, *Boselaphus*, *Gazella*, *Antidorcas*, *Oreamnos*, *Capra*, and *Ovis*. However, a few fossil taxa, namely *Hypsodontus*, *Sivaceros*, and the extant *Tetracerus quadricorins*, have narrow occipitals forming an inverted V. The V-shaped occipitals would have more slender necks attached to them.

The Neck

Most bovid necks are of an average length and appear to be deep in lateral view (Reduncini and Alcelaphini). There are a few short necks as in Bovini, and these are probably related to massive body built and graviportality. The necks are also short in *Connochaetes*, which has a slender body. In *Gazella dama*, *Ammodorcas*, and *Litocranius*, the neck is slender and long, probably indicating that they feed higher than the majority of other medium-sized bovids (*Gazella granti* and *Aepyceros*). The following section discusses the different types of neck length in different species along with the skeletal description.

Skeletal Proportions

The skeleton of bovids is complex. Gentry (1970) and Köhler (1993) have addressed various issues of paleoecology from the limbs. This chapter describes the general silhouette of the skeleton from observation of photographs. A key feature that can be appreciated in a lateral view is the relative length of the posterior limb in relation to the anterior. The position of the tibia in relation to the foot, which touches the ground in a gentle inclined slope, is notably horizontal. Tragulids and *Moschus* have very long posterior limbs, and the tibia is relatively horizontal in the stationary-standing position. The shape of the back is very visible in photographs.

Overall, Tragelaphini have tall limbs with a relatively horizontal or slightly arched back in lateral view. Their necks are longer than average. The two extant Boselaphini differ. *Boselaphus* has slightly higher front limbs and a somewhat inclined back. *Tetracerus* is small and has an arched back, and the posterior limbs are longer than the anterior. The Bovini are massive, with short necks and limbs. Their back is either arched (anoa) or flat (*Bubalus bubalis*), and in some taxa there is a massive development above the scapulae (*Bison*, *Bos gaurus*, and *B. javanicus*). The posterior limbs are longer in some breeds of cows. In Bovini the tibia is vertically positioned, perhaps because of graviportality. Cehalophini have strongly arched backs, longer posterior limbs, and notably short necks.

On the other hand, Reduncini have a concave back because of a protruding anterior iliac crest. The posterior limbs are slightly longer. The neck is longer than average. *Kobus megaceros* has a protruding iliac crest with a tuft of hair on it. Hippotragini have a massive muscular development around the scapulae. The neck and the limbs are average. The Alcelaphini also have a well-developed muscular complex around the scapula. The back can be slightly arched (*Damaliscus hunteri*), flatter (*Alcelaphus buselaphus),* or inclined (*Damaliscus pygmaeus*). The posterior limbs are probably slightly longer than the anterior. In *Connochaetes gnou* the iliac crest protrudes as in *Kobus megaceros.*

In addition, in *Oreotragus oreotragus* the posterior limbs are notably long. *Ourebia ourebi* also has long posterior limbs. Both of these have arched backs, which are common in small-sized taxa. *Raphicerus campestris* is similar to these species in limb proportions, but its back is horizontal. *Neotragus* and *Madoqua,* the smallest extant bovids, have long posterior limbs and arched backs. Their neck appears to be of average length. *Dorcatragus* is similar, but its back is also horizontal. *Aepyceros* and *Gazella* have the typical average neck length and a horizontal back. The posterior limbs are longer but not to the extent of the previous three. *Procapra* and *Pantholops* are similar to *Gazella* except that the thorax and abdomen are unusually bulky. *Procapra* also has deep neck with laryngeal specializations. Saiga has a proboscis and a short neck. The back is horizontal, and the thorax and abdomen are bulky.

Ammodorcas and *Litocranius* are the most gracile and have notably long necks and limbs. The posterior limbs are longer. Their back is flattened, and their shoulder musculature is slender. *Pseudoryx* resembles duikers in body silhouette. *Capricornis* and *Naemorhedus* and *Rupicapra* have short limbs and a horizontal back. The neck is deep. *Oreamnos* has a deeper thorax and abdomen and a well-developed scapular musculature.

The Ovobovini are similar in many respects. *Budorcas* and *Ovibos* have short horizontal necks, well-developed scapular musculature, and horizontal backs. The metapodials are short, and the limbs are positioned relatively vertically. Caprini appear to be homogeneous. Their metapodials are short, and the thorax and abdomen are deep. The neck is short and deep.

Most bovid hooves are of average length. In *Oreamnos* and *Oreotragus oreotragus* the hooves are very short. In *Tragelaphus spekeii* the hooves are long. These are adaptations for climbing rocks and walking on the edges of marshes. Little is known about most fossil taxa. Skeletons of *Miotragocerus* from Höweneg and *Tragoportax* from Pikermi seem similar to Reduncini and *Tragelaphus scriptus.*

Gentry (1970) indicated key characters on various bones of two bovids from Fort Ternan that can be related to closed versus open ecology (*Kipsingicerus labidotus* versus *Hypsodontus tanyceras*). Lengths of metapodials relate to ecology (Köhler, 1993). Köhler (1993) studied limbs and ecology of various ruminants and developed paleoecological diagrams into which some fossils can be placed. She found the following similarities: (1) In Boselaphini, similarities are drawn between *Austroportax* and *Taurotragus* and between *Trago-*

portax gaudryi and *Tragelaphus scriptus.* In addition, similarities are found between the limbs of *Miotragocerus* and those of Reduncini and *Tragelaphus spekei* and between *Tragoportax amalthea* and *Cervus.* (2) Similarities are also found between *Parabos* (Boselaphini) and *Anoa* (Bovini).

RELATIONSHIPS OF EXTANT BOVIDAE

It is necessary to summarize cladistic and phenetic results on the present bovids before considering the fossils. Gentry (1992) critically evaluated the subfamilies and tribes of Bovidae. Gentry's discussion is comprehensive, and the reader is referred to it for a more detailed presentation of the morphology, cladograms, and phenogram. His research found that the following are probably the most primitive extant genera: *Tetracerus, Tragelaphus,* and *Sylvicapra.* In addition, *Neotragus* is the most primitive neotragine (Gentry, 1992: 11). Gentry (1992: 28–29) found that bovids cluster around four foci (by foci I believe he means systematic concentrations of taxa): Boselaphini and allies; Antilopini and some Neotragini; Caprinae; and a fourth group centering on Alcelaphini. Boselaphini and allies constitute the most primitive group. Another critical finding was the Caprinae and Alcelaphini go together rather than either with Antilopini.

Boodontia (Gentry's Focus 1)

Boselaphini and allies (Gentry, 1992: 15). This is the most primitive group, and Bovini are part of this group. *Tetracerus* is problematic, but it can be an early separate lineage (primitive sister taxon) within the boselaphine–bovine group. Cephalophini may also be boodonts. Their relationships to Bovini are not clear. Tragelaphini are also boodonts.

Aegodontia (Gentry's Focus 2)

Antilopini with some Neotragini form another cluster (Gentry, 1992: 17). Neotragini are not monophyletic and appear as survivors from an early phase of bovid evolution. *Saiga* is placed within the Antilopini. *Ourebia* could be between Antilopini plus *Saiga.*

Gentry's Foci 3 and 4

Caprinae and Alcelaphini go together. *Oreotragus* is placed as a basal member of the *Pelea,* Caprinae, and the *Aepyceros-Damaliscus-Kobus-Hippotragus* cluster (Gentry's focus 3, p. 19). The Caprinae form a notably coherent group. *Rupicapra* is not clearly related but may be near the base of all the other Caprinae. *Ovibos* may not be close to *Budorcas* (Gentry, 1992: 20–21). Between these two genera *Budorcas* is the more problematic. *Capricornis* is close to *Nemorhaedus. Pantholops* is placed between Caprinae and the *Aepyceros-Damaliscus-Kobus-Hippotragus* cluster (Gentry's focus 4, p. 22). The Alcelaphini and allies are central to any concept of an African group such as the *Aepyceros-Damaliscus-Kobus-Hippotragus* cluster. In this cluster, *Aepyceros* is the most problematic genus because it resembles many different taxa (Gentry, 1992: 22). Hippotragini are a coherent group, but their ancestry is problematic. Hippotragini link best with Alcelaphini. Reduncini are the most difficult to place among all bovids (Gentry, 1992: 23).

In addition, in this study it is proposed that there are two groups of extinct Bovidae that probably need to be placed in new subfamilies: "Subfamily Hypsodontinae" (Hypsodontini) for *Hyposdontus* and "Subfamily Eotraginae" (Eotragini) for *Eotragus*.

PALEONTOLOGY

Origins

Archaeomeryx, which lived in Asia in the middle Eocene, is a representative ancestral taxon to many higher ruminants including the bovids. *Moschus* has recently been found to be close to bovids (Groves, this volume). During the Oligocene, the record is poor. Taxa that are near the origin of Bovidae are *Palaeohypsodontus* and *Hanhaicerus*. During the early Miocene, other taxa that are near the origin of bovids are *Andagameryx, Amphimoschus,* and *Hispanomeryx* (Wang, 1992; Gentry, 1994, 2000a).

In the early Miocene, *Walangania* and other archaic ruminants probably entered Africa from Asia. *Walangania* has very brachydont dentition and a first premolar. The horns have not been found but could have existed. *Walangania* is not definitely a bovid. Horn cores seem to be the defining character of Bovidae; in addition, however, the dentition is more hypsodont than that of the previous taxa, although it still is classified as brachydont. Approximately 18 Ma *Eotragus* and other taxa with horn cores represent what is known of the earliest bovids (Gentry, 2000a). *Eotragus* is widespread and is found in Central Europe, Spain, possibly East Africa, the Middle East, and in Pakistan. *Eotragus* is plesiomorphic for many characters including its long premolars and brachydonty. *Eotragus* has tall pedicles and thus may not be a boselaphine.

Hypsodontus and related taxa are early Miocene Asian animals that differ in having very hypsodont dentition, reduced premolar rows, narrow V-shaped occipitals, and large temporal fossae with approximated temporal lines (temporal lines at the median plane). They occur early and are temporarily disconnected from later aegodont hypsodont forms. It is possible that *Hypsodontus* evolved separately, in which case Bovidae may be diphyletic (Solounias, 1990). Later hypsodont taxa are numerous, and it is more likely that they evolved from *Eotragus,* while independently evolving hypsodonty from *Hypsodontus.* In this scenario the Hypsodontini may be a dead end. *Hypsodontus* and the possible origin of Antilocapridae in Central Asia suggest that ecological conditions for hypsodonty existed in Central Asia early. Synchronous *Eotragus* was less hypsodont and probably a browser (Solounias and Moelleken, 1992a). *Caprotragoides* and *Hypsodontus* are nonboodont and nonboselaphine Miocene taxa.

Numbers 1–4 in parentheses correspond to Gentry's four foci described above and to the last part of this chapter where a classification is provided.

Fossils from Africa

Eotragus is present in Africa only at Fort Ternan. *Caprotragoides* from Fort Ternan is a key taxon of problematic relationship to Eurasian taxa. *Hypsodontus* and *Nyanzameryx* are hypsodontines. At the end of the Miocene, the modern African tribes evolve in Africa. Gentry (1992) discussed how the African tribes may be related to Boselaphini and Caprini of the late Miocene.

Boodontia (1): Boselaphini

Fort Ternan in East Africa at 14 Ma samples a remarkable early group of Bovidae (Gentry, 1970; Thomas, 1984a, 1984b). *Kipsingicerus* from Fort Ternan is an advanced boselaphine in comparison to the younger European tribes. Bovini are *Pelorovis* and *Simatherium* in Olduvai Upper bed II and Shungura D. *Syncerus* is found at Shungura, and *Leptobos* at Sahabi, and *Bos* and *Ungadax* in the Kaiso. *Tragelaphus* is known from Lukeino and Shungura C-G and Langebaanweg, and *Taurotragus* in Olduvai. The Cephalophini record is poor and problematic. *Megalotragus* is from South Africa.

Aegodontia (2): Antilopini

Gazella are found in Maboko and many other localities; *Antidorcas* occurs in the Upper Ndolanya Beds near Laetoli. *Antilope* (a Pakistan–India taxon) at Shungura, C. *Prostrepsiceros* at Sahabi (also in Eurasia, at Samos and Maragheh). Neotragini are *Homiodorcas* from the Ngorora pre-*Hipparion* levels. *Homiodorcas* (Thomas, 1981) is a fossil neotragine that is very similar to Antilopini, and African antiliopines may have evolved from this form.

Caprinae and Alcelaphini (3 and 4)

Pachytragus from the Beglia (also in Eurasia—at Samos and Maragheh) and *Damalavus* from North Africa (3).

Ovibovini are *Makapania* from Makapansgat. Hippotragini are *Preadamalis* from Laetoli, *Wellsiana* and *Brabovus* from Makapansgat. *Hippotragus* and *Oryx* are found at Olduvai I, Shunguar upper G, and Makapansgat.

Alcelaphini (4)

Hypsodontus from Fort Ternan possibly ancestral to Alcelaphini (Gentry, 1992). *Damalacra* from Langebaanweg. *Parmularius* from Laetoli; *Damalops* from Hadar SH and DD (also in Asia, from Pakistan). *Betartagus* from Olduvai I and II and *Oreonagor* from N. Africa. *Megalotragus* from Olduvai II–IV. *Rhynotragus* and *Sigmoceros* are also early alcelaphines.

Other Taxa

The Reduncini are *Kobus* from Shungura E-G and Lange-baanweg, *Menelikia* from Shungura E-J and *Thaleroceros* from Olduvai IV.

Problem taxa are *Paraantidorcas* from Ain Boucherit, which is similar to *Oioceros* from Pikermi.

Fossils from Eurasia

Boodontia (1): Boselaphini

From 16 Ma onward there is a notable differentiation of early bovids. *Eotragus* and various boselaphines are found in faunas of Kamlial and the Chinji in Pakistan. Described taxa include *Selenoportax vexillarius, Helicoportax praecox, Strepsioprtax gluten, Tragoportax salmontanus, Tragoportax browni* and *T. punjabicus, Sivaceros gradiens, S. vedicus,* and *Sivoreas eremita* (Pilgrim, 1937, 1939). There are numerous new taxa in these faunas (A. W. Gentry, N. Solounias, J. C. Barry, and M. Raza, unpublished data).

Boselphines appear in Europe in the late of the middle Miocene and in China in the late Miocene. The group is diverse with several species in the genera: *Helicoportax, Selenoportax, Pachyportax, Strepsiportax, Miotragocerus, Tragoportax, Samokeros, Greacoryx, Portragocerus, Sivaceros, Sivoreas* (most from the Siwaliks but also Central Europe and Greece). The diversity is greatest in the Chinji and the U level and upper Siwaliks but also at Titov Veles, Thessaloniki (Vathylakkos various localities), Samos, Pikermi, Sinap, other Turkish localities, and Maragheh. Several localities from Ukraine are also notable, such as Sebastopol and Polgárdi. In contrast, Central Europe (Höweneg, Sansan, Epplesheim, and Dorn Dürkheim) sampled a few species. Bovini such as *Leptobos* and *Parabos* can be derived from *Pachyportax* but also from *Samokeros.* The middle Pliocene *Alephis* as well as *Pachyportax* are similar to bovines. *Parabos* is a boselaphine (Gentry, 1992). *Leptobos* and *Epileptobos* are bovines. *Samokeros* may be an early link between Boselaphini and Bovini (Solounias, 1981), but its bovine morphology could also be a result of parallelism.

Sivatragus is possibly a distant relative of *Hippotragus* from the Siwaliks, again suggesting an ancestry outside of Africa for Hippotragini.

Aegodontia (2): Antilopini

Gazella occurs throughout the latest part of the middle Miocene. It is abundant during the late Miocene of Eurasia. *Gazella* becomes widespread and possibly originated from *Homiodorcas* (Gentry, 2000a). Spiral-horned Antilopini appear in the late Miocene in Europe, North Africa, and the Siwaliks, which enter China only in the Pliocene. Some of these are *Lyroceros, Protragelaphus,* and *Prostrepsiceros.*

Tethytragus in Europe and *Homiodorcas* in Africa are not boseleaphines, but they are very similar dentally to boselaphines. Caprines may have evolved from *Tethytragus.* Early

caprines are *Pachytragus* (definitely an early Caprini), and *Protoryx* (Köhler, 1987) may also be a caprine. *Sporadotragus* and *Pseudotragus* are also Caprini. In addition *Palaeoryx* is a problem genus from many localities (Pikermi Samos, Sinap, and Maragheh). It may be close to *Tragoreas, Protoryx, Damalavus,* and *Pachytragus. Palaeoryx, Tragoreas,* and *Damalavus* can be intermediate between Caprini and Hippotragini.

In addition, there are taxa that resemble ovibovines, which were widespread from China to Greece. The ones most similar to ovibovines are *Criotherium, Tsaidamotherium, Parurmiatherium,* and *Urmiatherium; Plesiaddax* may be unique and not early Ovibovini. The extant *Ovibos* in several characters is more primitive than these fossils. *Budorcas* may not belong to Ovibovini (Gentry, 1992). In addition there are four genera that stand out as unique but appear to have affinities with the previous group. *Hispanodorcas* and *Oioceros* are also problematic taxa resembling caprines and ovibovines (Spain, Greece, Turkey, and Iran). In addition, the taxa *Sinotragus* and *Prosinotragus* resemble caprines but are also similar to ovibovines (Shanxi of China).

Sivacobus and other taxa are reduncines from the Siwaliks that perhaps migrated into Africa from the Siwaliks.

Fossils from North America

Bovids migrated into North America from Asia in the Pliocene. A few groups are represented: *Bison* and allies as part of Bovini, *Ovis* representing Caprini, and *Oreamnos* as part of Rupicaprini (Kurtén and Anderson, 1980; Prothero and Schoch, 2002).

CLASSIFICATION

Gentry (1992) provided a classification of Bovidae that best represents the subfamilies and tribes. Gatesy et al. (1997) also provided a similar molecular classification. *Eotragus* and *Hyposdontus* are added to the list as separate in this study. *Eotragus* has a tall pedicle unlike other Boselaphini. *Hypsodontus* is distinguished as an early hypsodont taxon with tall pedicles. The temporal lines approximate the median plane. It is also suggested to separate *Criotherium, Plesiaddax, Urimiatherium,* and *Parurmiatherium* from the Ovibovini (Gentry, 1992). Gatesy et al. (1997) provide data that support the following taxonomy and that an indeterminate subfamily and tribe include *Aepyceros, Pelea, Pantholops,* and *Saiga.*

Bovidae cluster around four foci according to Gentry (1992). In addition, certain fossil taxa appear to be separate. The pedicle / no-pedicle observation can also be used in the following summary:

Horn Cores with No Pedicle

Gentry's Focus 1

Subfamily Bovinae
    Boselaphini
    Cephalophini

Horn Cores with Pedicle

Subfamily Bovinae
    Tragelaphini
    Bovini (evolved probably polyphyletically from Bose-laphini, including *Pseudoryx*)

Gentry's Focus 2

Subfamily Antilopinae—some Neotragini (paraphyletic)
    and perhaps *Pelea, Saiga* and *Pantholops*—even
      Aepyceros
    Antilopini

Gentry's Focus 3

Pantholops and perhaps *Pelea,* and *Saiga*-even *Aepyceros*
Oreamnos
Capricornis
Nemorhaedus
Rupicapra
Caprinae
Ovibovini
Caprini
"Criotheriini" (suggested new tribe in this study)

Gentry's Focus 4

Subfamily Hippotraginae
    Reduncini
    Hippotragini
Subfamily Alcelaphinae
    Aepyceros and *Pelea* near each other and near Hippo-tragini, Caprinae, and *Pantholops*
    Alcelaphini

Additional Taxa, Probably to be Placed in Two New Subfamilies

"Subfamily Hypsodontinae"
    Hypsodontini, *Hyposdontus*
"Subfamily Eotaginae"
    Eotragini, *Eotragus*

ACKNOWLEDGMENTS

I thank Alan Gentry for extensive review of this chapter. I also thank Lotus Ahmed and Renata Trister for additional editorial help. For comments and reviews of this chapter, I thank John Barry, Alan Gentry, Henry Galiano, Christine Janis, and Matthew Mihlbachler. The American Museum of Natural History is acknowledged for permission to study fossil and modern material.

23

Artiodactyl Paleoecology and Evolutionary Trends

ARTIODACTYLS TODAY ARE BY FAR the predominant ungulates (i.e., hoofed mammals), comprising 10 families, 85 genera, and 217 species (excluding cetaceans). (An excellent, up-to-date definition of the order can be found in Theodor et al., 2005.) The other major order of ungulates, the perissodactyls, are today far less diverse, comprising three families, six genera, and 17 species (data from Nowak, 1999). Artiodactyls and perissodactyls are closely related to each other if not sister taxa as has long been assumed (see Springer et al., 2005). Some other mammal clades, such as the extinct South American litopterns and notoungulates, are often included as "ungulates" and have converged on artiodactyls and perissodactyls in many aspects of their morphology (e.g., longer legs, more hypsodont cheek teeth), but the relationships of these groups to extant ungulates are obscure. The other extant mammalian groups that are often termed "ungulates," such as hyraxes and proboscideans (i.e., paenungulates), were once considered to be closely related to the clade including artiodactyls and perissodactyls but are now placed in the more distantly related Afrotheria (Springer et al., 2005).

One way of thinking about the history of artiodactyls and their paleobiological patterns over the course of the Cenozoic is to compare and contrast them with perissodactyls. Both Artiodactyla and Perissodactyla contain taxa with craniodental specializations for herbivory (e.g., hypsodont cheek teeth, skulls accommodating large masseter muscles), with specialization of the digestive system for a diet containing cellulose (foregut or hindgut systems of fermentation) and for cursorial locomotion (e.g., elongation of distal limb elements, reduction or loss of digits, unguligrade stance). Similarities between artiodactyls and perissodactyls, in terms of either ecomorphological adaptations or the timing of evolutionary events (such as patterns of diversification and extinction), would highlight ways in which the external environment had influenced these two clades in a similar fashion. In contrast, differences between the two groups would draw attention to possible ways in which artiodactyls might have phylogenetic advantages or constraints in comparison with other ungulates.

Because the primary radiation of perissodactyls was in the Eocene, with the artiodactyls coming to prominence later, the common assumption some decades ago was that artiodactyls were somehow the superior group of ungulates and had outcompeted the less efficient perissodactyls (see discussion in Janis, 1976). However, detailed examination of patterns in the fossil record shows that competition was unlikely to be the reason for the mid-Cenozoic rise in diversity of artiodactyls and decline in diversity of perissodactyls (Cifelli, 1981; Janis, 1989). Another common assumption has been that the ruminant type of digestive physiology of many artiodactyls (camelids and ruminants [in the taxonomic sense]) is superior to the hindgut type of fermentation seen in perissodactyls (and also in proboscideans and hyracoids). Some years ago (Janis, 1976) I reviewed the existing evidence for this assumption and argued for these different types of digestive physiology being "separate but equal" strategies. Ruminants have the advantage where food is of limited quantity (but of relatively good quality), and hindgut fermenters have the advantage where food is of low quality (but of less limited quantity).

Ruminants also have an advantage in their possession of a complex digestive nitrogen cycle: their effective dietary protein is actually made by the foregut microbes, so they are less dependent on finding essential amino acids in their dietary intake and need less diversity of plants in their diet. In addition, waste urea is also fed into this nitrogen cycle, with the result that ruminants need to excrete less urea and so can conserve more urinary water. The ruminant foregut site of fermentation may also be advantageous in detoxifying plant secondary compounds. However, a disadvantage of rumination is that all free sugars, as well as cellulose, become fermented, meaning that eating fruit offers little nutritional advantage. It also seems apparent from the body size distribution of past and present ruminants that neither very small (under 5–10 kg) nor very large (over 1,000 kg) ruminants are viable (see later discussion for the possible reasons for this).

Although this chapter is primarily about artiodactyls, I consider a comparison with perissodactyls to be an interesting approach. In exploring the ways in which the two groups differ in their evolutionary patterns, one may gain insight into the ways in which artiodactyls are unique among ungulates and also ways in which their evolution has been constrained. I address the following issues:

1. The late Eocene displacement of perissodactyls by artiodactyls. Perissodactyls initially radiated into a diversity of families in the early Eocene, with a variety of body sizes, whereas until the late middle Eocene, artiodactyls were mainly small forms (under 5 kg in body mass). After around 45 Ma (late middle Eocene), the diversity of perissodactyls

started to decline, and that of artiodactyls increased. The later Eocene was a time of profound climatic changes in the higher latitudes; can these patterns of ungulate extinction and diversification be related to climatic events?

2. The radiation of omnivorous artiodactyls. A major portion of the artiodactyl radiation was contained in the radiation of the Suiformes, many of which (e.g., many extant pigs) had bunodont cheek teeth indicative of an omnivorous diet. However, no parallel radiation of omnivores was seen among the perissodactyls. Why were artiodactyls unique in this aspect of their evolutionary radiation?

3. The lack of artiodactyl "megaherbivores." Almost all ruminating artiodactyls, past and present, are under 1,000 kg in body mass, and certainly no ruminant was greater than 2,000 kg in body mass, in contrast to many proboscideans and perissodactyls. Is this absence of ruminants from the megaherbivore category somehow related to their type of digestive physiology?

4. The transition to the Neogene artiodactyl fauna. Present-day artiodactyls mainly comprise the pecorans (bovids, cervids, etc.), whose initial radiation was in the early Miocene, with the appearance of larger, specialized horned forms in the late early Miocene at around 19 Ma (Gentry et al., 1999). In contrast, many artiodactyls that were prominent in the late Paleogene, such as entelodonts, anthracotheres, oreodonts, and a diversity of traguloids, became extinct or greatly reduced in diversity about this time. What were the reasons for this faunal turnover, and what environmental changes favored the pecoran ecomorphological design?

5. The transition to the modern fauna, and the recent predominance of bovids. The majority of extant artiodactyls belong to the family Bovidae. Yet until relatively recently there was a great diversity of other types of ruminants. The later Miocene and Plio-Pleistocene was a time of severe climatic deterioration in the higher latitudes, with the savanna grasslands turning into prairie, and cold winters restricting the plant growing season in general. Were bovids somehow especially well adapted to cope with these climatic changes, or is their present-day predominance more of a matter of happenstance?

THE LATE EOCENE DISPLACEMENT OF PERISSODACTYLS BY ARTIODACTYLS

Both perissodactyls and artiodactyls are first known from the earliest Eocene of North America and Eurasia. Eocene perissodactyls ranged from small hyracotheriine equids (with body masses of around 5–10 kg) to very large lophiodontid tapiroids and brontotheres (with body masses up to several thousand kilograms). All of these forms had lophodont (including bilophodont or at least bunolophodont) cheek teeth indicative of varying degrees of folivory, whereas the contemporaneous artiodactyls were all of small body size with bunodont (or bunoselenodont) cheek teeth indicative of omnivory (or possibly folivory/frugivory at best) (Collinson and Hooker, 1991; Stucky, 1998; Janis,

2000a). These artiodactyls represented various basal groups (some or all of which may be paraphyletic) such as diacodexeids, dichobunids, homacodontids (North America only), leptochoerids (North America only), raoellids (Asia only), and helohyids (see Stucky, 1998; Theodor et al., 2005, this volume; Foss, this volume), plus members of the suiform families Cebochoeridae and Choeropotamidae in Eurasia (see Erfurt and Métais, this volume).

A profound change first commenced in the late middle Eocene, when larger-sized artiodactyls with more specialized selenodont cheek teeth made an appearance. Most of the late middle Eocene artiodactyls can be assigned to one of the three extant subfamilies (although some of the more primitive groups persisted into the Oligocene): the Suiformes (paraphyletic if not polyphyletic), the Tylopoda (probably paraphyletic), and the Ruminantia (see Theodor et al., 2005; Erfurt and Métais, this volume; Harris and Liu, this volume; Métais and Vislobokova, this volume). The later Eocene also was the time of the demise of many types of perissodactyls coincident with the increase in artiodactyls, including hyracotheriine equids, palaeotheres, eomoropine chalicotheres, and a diversity of "tapiroids" (see chapters in Janis et al., 1998; Meng and McKenna, 1998; Hooker, 2000; Tsubamoto et al., 2004). However, more modern types of perissodactyls also made their first appearance at this time: anchitheriine equids (North America only), tapirids, and rhinocerotids.

This pattern of faunal replacement can be understood in terms of the different digestive physiologies of artiodactyls and perissodactyls and how these influenced their ability to cope with vegetational changes. All living perissodactyls have a similar system of hindgut fermentation, both in terms of the anatomy involved (an enlarged cecum and an even more greatly enlarged colon) and the digestive physiology. It is thus parsimonious to assume that this type of fermentation system characterized the common ancestor of living perissodactyls, which can be dated to at least the earliest Eocene from the split in the equid and ceratomorph (rhino plus tapir) lineages. Because the earliest perissodactyls all had lophodont teeth indicative of folivory, and hence of food containing cellulose, it is likely that these animals would have required some system of cellulose fermentation.

The situation is quite different for artiodactyls. Different groups of extant artiodactyls have different systems of fermentation: most suiformes have no fermentation or have a type of foregut fermentation without complex stomach subdivisions and cud chewing; both camelids and ruminants have complex stomach subdivisions and chew the cud, but the three-chambered stomach of camelids and the four-chambered stomach of most ruminants suggest at least a considerable extent of convergent evolution. (Note that all artiodactyls may additionally have a small amount of hindgut fermentation; this appears to be a primitive mammalian condition.) The implication from this is that original radiation of artiodactyls ("dichobunids") had the primitive condition of no forestomach fermentation. The bunodont cheek teeth of early artiodactyls also suggest a relatively nonfibrous diet that would not require fermentation.

Thus, the acquisition of selenodont (or at least bunoselenodont) cheek teeth in ruminants, tylopods, and a few suiformes (e.g., anthracotheres), first seen in the late middle Eocene, was likely coincident with the acquisition of some sort of foregut fermentation in these same groups (possibly convergently) and the shift to a more folivorous (i.e., leaf-eating, whether browse or grass) diet from a more omnivorous one. The early Eocene high-latitude paratropical forests would have provided a variety of plant food year round for small herbivores and omnivores (Wing, 1998; Harrington, 2001). The demise of the smaller, bunodont artiodactyls at high latitudes in the later Eocene, as well as the bunodont archaic ungulates ("condylarths"), was likely related to the reduction in the availability of fruit as a food resource in the more seasonal climate that developed at this time (Collinson and Hooker, 1991; Hooker, 2000) as well as to a decrease in the size of fruits and seeds available (Eriksson et al., 2000).

How can these evolutionary events among artiodactyls be correlated with the contemporaneous environmental changes in the Northern Hemisphere higher latitudes, and why was the decline of perissodactyls broadly synchronous? As previously mentioned, ruminants (in the physiological sense of having complexly divided forestomach fermentation in combination with cud-chewing, this general definition thus including camelids as well as members of the Ruminantia) are at an advantage when food is of high quality but of limited quantity and indeed cannot cope with food that must be eaten in large quantities because of their lengthy digestive passage time (see also later discussion of ruminant digestive physiology in the section on megaherbivores). In contrast, perissodactyls have a shorter passage time, can consume greater quantities of food per day (for their size), and thus can survive on lower-quality food. Under conditions of low seasonality, which would have characterized the forests of the earlier Eocene, plants show little differentiation of fiber content between leaf and stem (see references in Janis, 1989). Thus, it is not possible for an ungulate to select better-quality (i.e., less fibrous) portions within the same plant. These vegetational conditions favor hindgut fermenters, which do not need to be selective in their feeding but do require relatively large quantities of food (as would be the case in the early Eocene nonseasonal environments).

The late middle Eocene marks the point at which the climate of the northern latitudes became more seasonal, in correlation with high latitude cooling at this time (Zachos et al., 2001). Higher-latitude vegetation became more seasonal in aspect, with evidence of winter frosts (e.g., Wolfe, 1985; Collinson and Hooker, 1991; Wing, 1998). In these more seasonal conditions, when plants are more likely to be deciduous and to invest less heavily in their leaves, the plant leaves contain less fiber than the stems, and selection of better-quality parts within a given plant (i.e., leaf versus stem) becomes possible. These vegetational conditions would favor ruminants, which need to feed selectively on high-quality food. Ruminants would also be favored over hindgut fermenters if the actual quantity of the vegetation decreased, a likely possibility in a more seasonal habitat, as a ruminant needs less food per day than a hindgut fermenter of the same size (see Janis, 1989, for a more extended discussion and references). In addition, other environmental changes at this time may also have affected the diversification of artiodactyls. During the Paleocene through mid-Eocene, levels of atmospheric CO_2 were much higher than today (as much as 4,000 ppm, as opposed to around 200 ppm in the recent preindustrial atmosphere) (Pearson and Palmer, 2000). How might higher levels of atmospheric CO_2, and a subsequent drop in these levels, affect the vegetation in ways that would influence the distribution of different types of ungulates?

Greenhouse studies of plants today (see Janis et al., 2004a, for references) show that, with higher levels of CO_2, plants tend to produce a greater amount of vegetative growth with a lower nitrogen-to-carbon ratio. That is, such plants would represent a resource of high quantity (more growth) but low quality (proportionally less protein). As discussed above, this type of vegetation would favor hindgut fermenters (perissodactyls) over ruminants. Additionally, plants with a C_3 type of photosynthesis (C_4 plants are not known from the early Cenozoic) require less water in the presence of higher levels of CO_2. Thus, under a regime of the high levels of CO_2 in the early Eocene, vegetation would be abundant, and the plants would tend to be more immune to the effects of rainfall seasonality and thus show less differentiation of fiber within the plant even under water-stress conditions. These vegetational conditions of high quantity and low quality would favor hindgut fermenters over ruminants.

Two episodes of plummeting levels of CO_2 were seen in the Eocene: an initial drop at 51 Ma from around 4,000 ppm to around 800 ppm, at the early/middle Eocene boundary, coincident with the Cenozoic thermal maximum, shortly followed by a rebound to around 2,500 ppm in the early middle Eocene, and a second drop at the start of the early late middle Eocene, around 46 Ma, to around 500 ppm, with levels increasing slightly thereafter but never returning to their earlier peak (Pearson and Palmer, 2000). Faunal changes were not as pronounced at 51 Ma, when global tropical conditions prevailed, as they were in the later Eocene when the second plunge in atmospheric CO_2 was accompanied by changes in temperature seasonality in the northern latitudes containing the artiodactyls and perissodactyls at this time. The late middle Eocene was also the time of the greatest episodes of volatility among the diversity of Cenozoic mammals, including the "primate/ungulate" switch—a great decline in the number of high-latitude primate taxa with a corresponding rise among the ungulates (Alroy et al., 2000). The decrease in atmospheric CO_2 at this point would have resulted in water stress on the plants, resulting in perceived aridity even if actual rainfall levels had remained constant, and vegetational abundance may have been lower because of slower growth rates under a regime of lowered CO_2. However, this same lower level of CO_2 would result in plants with a higher ratio of nitrogen to carbon. Thus, declining levels of CO_2 would create the converse of the vegetational quality/quantity conditions of the earlier Eocene, resulting in vegetation of lower quantity but higher quality, now favoring ruminant artiodactyls over perissodactyls.

Larger (greater than 10 kg) artiodactyls also became apparent at this time. A ruminating type of foregut fermentation is not possible at a smaller body size; smaller animals have relatively greater metabolic requirements and so must consume relatively more food, but the long passage time required for full rumination mitigates against this. Ruminants of less than 10 kg today (e.g., mouse deer and some duikers) consume a diet with very little fiber that is not delayed in its passage by retention in the rumen.

Of course, the actual pattern of artiodactyl diversification and perissodactyl decline was not a simple, single event, and the overall picture is much more complex than this. Nevertheless, the later Eocene changes in temperature and levels of atmospheric carbon dioxide provide an appropriate backdrop to the timing of the artiodactyl diversification, explaining not only why more larger, folivorous artiodactyls became common following this time but also why perissodactyls would not be so favored.

THE RADIATION OF OMNIVOROUS ARTIODACTYLS (WITH A NOTE ON INCISOR LOSS IN HERBIVORES)

A major portion of the artiodactyl radiation was contained in the radiation of the Suiformes, many of which (e.g., many extant pigs) had bunodont cheek teeth indicative of omnivory. However, no parallel radiation was seen among the perissodactyls. Why were artiodactyls unique in this aspect of their radiation? Today Suiformes traditionally comprise pigs (Suidae), peccaries (Tayassuidae), and hippos (Hippopotamidae): pigs and peccaries are sister taxa, and the relationship of hippos to other artiodactyls (including cetaceans) is currently a matter of much debate (see Theodor et al., 2005; Marcot, this volume). Extinct artiodactyl families included in this (likely paraphyletic) suborder are the entelodonts, anthracotheres, sanitheres, cebochoerids, and choeropotamids (these latter two families comprising small taxa limited to the Eocene or earliest Oligocene) (see Foss, this volume).

Suiformes are generally characterized by bunodont, brachydont cheek teeth indicative of an omnivorous diet. Bunodont cheek teeth are the primitive condition for ungulates, from which lophed teeth may be derived (Jernvall, 1995), even if the particular dental morphology of any bunodont group may be derived with respect to the generalized bunodont condition. Note that many suiformes are derived with respect to the primitive condition of bunodonty and omnivory. Hippos have complexly lophed cheek teeth that may be hypsodont, and extant hippos are strictly folivorous, subsisting almost entirely on grass. Anthracotheres and sanitheres had teeth that were somewhat selenodont, both pigs and peccaries have produced a diversity of bilophodont forms (e.g., the extinct listriodontid suids), and the warthog

(*Phacochoerus aethiopicus*) is a predominantly grazing taxon, with complexly lophed ("columnar"), hypsodont cheek teeth. Note, however, that current systematic evidence suggests that all of these suiformes with more complex teeth evolved from a bunodont form.

Suiformes have also been grouped together not only on the primitive dental condition of bunodonty but also on primitive characteristics of their postcranial skeleton, usually retaining four functional toes and unfused metapodials (although note that entelodonts were didactyl with fused metapodials). Even today people often consider pigs as "primitive artiodactyls," although their skulls and dentitions are highly derived from the primitive artiodactyl condition, albeit in a different direction from that of tylopods and ruminants. Note also that the tendency for extant suids to have litters of multiple young is also a derived condition, not a primitive one as often assumed (Gaucher et al., 2004).

However, a pertinent point to note here is that a radiation of specialized omnivorous forms (whether or not these groups constitute a clade) is a distinctly artiodactyl phenomenon: no such radiation ever occurred among the perissodactyls, nor among the endemic South American ungulates. Many lineages of archaic ungulates ("condylarths") had bunodont cheek teeth, but these were mainly small animals and went extinct before the radiation of omnivorous artiodactyls (see Janis, 2000a). There was also a significant radiation of large, piglike bunodont hyraxes in the later Paleogene of Africa, which apparently retained the primitive condition of bunodont cheek teeth (Rasmussen and Simons, 1988). Many types of proboscideans also had bunodont or bunolophodont ("zygodont") cheek teeth (e.g., gomphotheres, mastodonts). The primitive condition in proboscideans appears to be for a bilophed, bunolophodont tooth from which more specialized cheek teeth, either more lophed or more truly bunodont, were derived (Tassy, 1996).

It is possible that a constraint may exist in tooth formation, whereby a tooth that has become "lophed" in form (i.e., lophodont or selenodont) cannot be secondarily made bunodont. It seems that it is possible, in proboscideans at least, to derive a bunodont tooth secondarily from a bunolophodont one, but it might still be the case that a fully lophed tooth is not easily reversed. This constraint might be developmental or functional. Lophed teeth have relatively thin enamel and rely on initial wear to form the functional ("acquired") occlusal pattern of enamel ridges separated by dentine lakes (Fortelius, 1985); a viable functional intermediate from this form back to a thicker-enameled tooth with separate cusps might be difficult to evolve.

If a constraint against regaining a bunodont tooth form exists, this would explain why perissodactyls never produced a radiation of omnivorous forms, as the early Eocene forms were all more specialized, presumably folivorous forms, with lophodont or bilophodont cheek teeth (Collinson and Hooker, 1991; Janis, 2000a). In contrast, the initial Eocene radiation of artiodactyls into small, relatively unspecialized, largely bunodont forms meant that with the secondary radiation of both artiodactyls and perissodactyls

in the late Eocene (see above), the artiodactyls had the evolutionary flexibility (in dental terms) to evolve more larger, specialized omnivorous forms, retaining (or elaborating) the bunodont dentition. However, if it is indeed not possible to derive a bunodont tooth from a lophed form, a radiation into omnivorous habits would not have been an option for any perissodactyl lineage. This hypothesis of dental constraints is pure speculation at the present time, but it remains true that the evolutionary patterns of artiodactyls differ considerably from those of perissodactyls in their radiation of specialized omnivorous forms.

Another interesting craniodental difference between artiodactyls and perissodactyls is seen with the radiation of more folivorous forms. Ruminants (and, to a lesser extent, tylopods) tend to lose their upper incisors and replace them with a horny pad. In contrast, equids always keep both sets of incisors, and if rhinos lose incisors, they lose both uppers and lowers.

It is possible that these different evolutionary trends are related to differences in food selection between the two groups, which are in turn related to the differences in digestive physiology previously discussed. Folivorous artiodactyls, focusing on food quality rather than food quantity, are highly selective feeders and use their tongue to select food. (This is most obvious in specialized browsers, such as giraffe, but is also true of grazers; see comments below.) In contrast, perissodactyls are relatively unselective and use their lips for food prehension. Incisorless rhinos use their lips alone for food gathering, but horses also make extensive use of their lips despite a full set of front teeth. Note that if you approach domestic ungulates with a food treat, horses will attempt to prehend it with their lips, whereas cows investigate with their tongue. One consequence of this use of the lips in perissodactyls is the much greater elaboration of facial musculature than in ruminants: movies and TV shows featuring "talking" ungulates are limited to equids (a cow could never play Mr. Ed). The more extensive use of the lips in perissodactyls might also explain why this group has lost the primitive mammalian condition of a moist rhinarium (i.e., a wet nose), although this is retained to a greater or lesser extent in artiodactyls.

How might this use of tongue versus use of lips to select food relate to incisor loss? I offer the potentially plausible, although probably untestable, hypothesis that accidentally biting and damaging a protruding tongue might lead to selection pressure for the loss of the upper incisors. I do note that chalicotheres, although perissodactyls, also evolved a rather ruminant-like anterior dentition with the loss of the upper incisors. But perhaps, as derived chalicotheres seem to have been specialized tree browsers (Hessig, 1999), they also evolved a ruminant-like mode of using the tongue for food gathering: it is tree browsers such as giraffids among extant ruminants that have spectacularly long tongues for this purpose. Whatever the actual reason for incisor loss in ruminants, this difference in craniodental anatomy and mode of food selection between artiodactyls and perissodactyls remains a pertinent observation.

THE LACK OF ARTIODACTYL MEGAHERBIVORES

Megaherbivores are considered to be those weighing more than a metric ton (1,000 kg) (Owen-Smith, 1988). River hippos (*Hippopotamus amphibius*) are certainly megaherbivores, but although hippos are foregut fermenters, they are not true ruminants (see further discussion below). A few extant ruminants (e.g., giraffes) technically fall into this category (as would have some extinct camelids and sivatheriine giraffids), but none of these animals would have exceeded 2,000 kg (see discussion below). In general the megaherbivores have been proboscideans and perissodactyls such as rhinos, including the extinct indricotherine hyracodontids (with a body mass of up to 15 tons) and late Eocene brontotheres (some of which were larger than extant white rhinos and may have had a body mass of 5–6 tons).

Marcus Clauss (e.g., Clauss et al., 2003; Clauss and Hummel, 2005) provides an elegant explanation for why ruminants are absent from the megaherbivore size range (see these articles for references in the summary of this work below). Ruminants have longer passage times of the digesta than hindgut fermenters of a comparable body size: for example, a horse takes around 40 hours to process the food, whereas a grazing ruminant of similar size takes around 60 hours (although browsers have a more rapid passage). Larger herbivores have, in general, longer passage times than smaller ones: large rhinos have a passage time of around 60 hours (although elephants actually have a shorter passage time than rhinos; see Clauss et al., 2003, for extensive discussion, and comments below). The ruminant digestive tract is specifically adapted to delay the passage of food, in order to maximize cellulose digestion. Rumination likely evolved in a fairly small animal (around 20 kg, to judge from the size of early selenodont forms that would have been eating food requiring fermentation, see Janis, 1976). However, with increasing body size, the time for passage of the digesta increases, and here the ruminant digestive strategy may become problematic. After around 60–70 hours, digestion of plant material is more or less complete, and further retention holds no physiological advantage; thus, a ruminant would lose its advantage (in terms of completeness of cellulose digestion) over a similarly sized perissodactyl at a body mass approaching 1,000 kg.

The lack of advantage to large size in ruminants is well known (e.g., Demment and Van Soest, 1985). However, Clauss et al. (2003) suggest that there might actually be a disadvantage (rather than merely a lack of advantage) to being a larger-sized ruminant. This relates to the fact that with increasing passage time a problem develops with the growth of methogenetic bacteria. These bacteria convert acetic acid (the main volatile fatty acid product of cellulose fermentation) to methane and carbon dioxide, with resultant high energy losses, and this problem becomes acute with a passage time of more than 4 days. Elephants solve this problem by adopting relatively shorter and broader guts, thus speeding up the passage rate of the digesta, but this would not be a possible solution with the design of the ruminant digestive tract, with its built-in system of passage delays (such as retention in the rumenoreticulum region). Thus, ruminants are limited to body sizes where their digesta passage time would be less than 4 days.

Clauss et al. (2003) also note a difference here between browsing and grazing ruminants. In all ruminants, because of the delaying effect of the rumenoreticulum, the size of this structure needs to be relatively bigger (i.e., to have a disproportionally increased capacity) in larger animals (this can be observed empirically). Thus, there must be a built-in limitation to the maximum size of a ruminant in terms of the maximum size of the rumenoreticulum that it can accommodate (theoretically, at a body mass of 12 tons, a ruminant would consist entirely of rumen contents!). This body size threshold, above which a larger rumenoreticulum could not be accommodated, differs between browsers and grazers, as grazers can be observed to have a relatively larger rumenoreticulum than browsers. (This is perhaps because of the stratification of rumen contents that is induced with grass forage, which further delays passage time, thus necessitating a relatively larger rumenoreticular capacity.) Clauss et al. (2003) provide theoretical calculations to show that, at any given threshold for relative rumenoreticulum capacity, browsers would be able to be around 400 kg larger than the largest grazer: they note that this accords well with the modern fauna, where the largest browser (the giraffe) has a mean body mass of around 1,200 kg, and the largest grazer (the bison) has a mean body mass of around 800 kg. Larger ruminants have existed in the past: some Plio-Pleistocene sivatheres (giraffids) and North American Pliocene camelids (e.g., *Gigantocamelus, Titanocamelus*) probably had body masses approaching 2,000 kg, and, to judge from their dental remains (somewhat hypsodont), were probably mixed feeders rather than pure browsers. However, even the very large Pleistocene buffaloes (presumably grazers), such as *Pelorovis,* probably did not greatly exceed extant bison in body mass (see calculations in Clauss et al., 2003). It thus seems that, although the largest extant ruminant may not represent the maximum size possible, there is nevertheless a constraint on how big a ruminant can be, and ruminants over a body mass of 500 kg are certainly the exception rather than the rule.

Clauss et al. (2003) also note that, with a relatively larger rumenoreticulum, grazers would have more packing constraints within the abdomen than browsers. They point out that the larger rumenoreticulum is compensated for by a relative decrease in the size in the descending colon, the area of water resorption: grazers are more dependent on drinking water than browsers, and this lack of water retention explains why grazers defecate "pies" whereas browsers defecate pellets. I can only conclude that the latrines of giant camelids and sivatheres would have been particularly unpleasant!

An interesting corollary to this exists with considerations of the hippopotamus. Hippos are nonruminant foregut fermenters and can have a body mass of up to 4 tons, putting them assuredly in the megaherbivore category. However, hippos combine a huge stomach (up to 25% of their body mass), with an extremely short and undifferentiated hindgut, and a very high level of fecal water content. Clauss et al. (2003) hypothesize that hippos can only withstand such fecal water losses because of their amphibious lifestyle, and that they may be limited to this habitat by their gastrointestinal morphology and physiology. This provides a cautionary tale for the common assumption that extinct rhinos with an apparently hippo-like morphology (e.g., the North American rhino *Teleoceras*) would also have had an amphibious lifestyle, as a hindgut fermenter would not encounter such water-retention limitations.

THE TRANSITION TO THE NEOGENE ARTIODACTYL FAUNA

The Neogene was the time of the expansion and radiation of pecoran ruminants. Most pecoran families made their first appearance in the early Miocene, before 20 Ma, but it was not until somewhat later in the early Miocene, around 18–19 Ma, that larger, horned ruminants became prominent (Gentry et al., 1999). The later early Miocene also marked the appearance of modern types of camelids (Camelinae, see Honey et al., 1998), tayassuids (Tayassuinae, see Wright, 1998), and more derived types of suids (such as the now-extinct subfamilies Listriodontinae, Kubanochoerinae, Tetraconodontinae, and Schizotherinae), as well as the separate extinct suoid family Sanitheriidae (see Harris and Liu, this volume).

Almost all extant pecorans have horns or other types of cranial appendages (moschids are the main exception), and perhaps surprisingly these cranial appendages appear to have arisen convergently among the different groups (Janis, 1982). Many different types of pecorans also acquired more hypsodont cheek teeth at the same time that they evolved cranial appendages, around 18 Ma. Note that in North America some other forms of "tylopods" made an initial appearance and radiated at this time, such as synthetoceratine protoceratids (also with cranial appendages) (Prothero and Ludtke, this volume), and other types of more derived camelids (protolabines, camelines) (Honey et al., 1998); these forms were also in general more hypsodont than earlier members of their families.

The radiation of horned pecorans was broadly coincident with the decline of other artiodactyl groups that had been predominant in the Oligocene and earlier Miocene. This included all types of suiformes except for the extant families, a number of European "tylopods" (anoplotheriids, xiphodontids, cainotheriids, etc.; see Erfurt and Métais, this volume), some North American "tylopods" (oreodonts, oromerycids, protoceratine protoceratids, and more primitive camelids such as poebrotheriines and stenomylines), and the majority of the traguloid ruminants (including hypertragulids, leptomerycids, etc., the modern Tragulidae being an exception) (see chapters in Janis et al., 1998; Métais and Vislobokova, this volume). Some of these artiodactyls had a "second wind" of more derived forms that extended

at low diversity through the Miocene (e.g., merychyine oreodonts, see Stevens and Stevens, this volume; and anthracotheres, see Lihoreau and Ducrocq, this volume), but the late Paleogene artiodactyl radiations were essentially over by the mid-Miocene.

What made these earlier artiodactyls different from those of the later, Neogene radiations? In terms of the tylopods and ruminants, the Eocene–Oligocene forms were in general of small to medium body size (some oreodonts were larger, perhaps around the size of a tapir), and although they had derived, selenodont cheek teeth, in general they were not hypsodont. Some exceptions to this exist among the North American artiodactyls, where some precociously hypsodont forms may have been specialized for arid habitats (*Hypisodus* among the hypertragulids, leptaucheniine oreodonts, and stenomyline camelids); these taxa were likely highly selective mixed feeders (rather than grazers), as they all had narrow muzzles. Some later oreodonts were also somewhat hypsodont, but these were members of the Merychyinae, which were part of the Miocene artiodactyl radiation. With the exception of some protoceratine protoceratids, these forms did not evolve cranial appendages.

In addition Paleogene artiodactyls in general were not highly cursorially adapted. Some smaller forms, such as cainotheres and some of the traguloids, had relatively long, slender limbs, but such limb proportions (in combination with a flexed limb posture) are a feature of smaller mammals in general and are not necessarily equivalent to the cursorial specializations of larger forms (Bertram and Biewener, 1990). Some of the medium-sized camelids evolved somewhat longer limbs, especially by the late Oligocene (Janis et al., 2002), but they were not as specialized as Neogene camelids.

Among the suiformes, there was a significant radiation at this time of large forms such as entelodonts (likely omnivorous, with bunodont cheek teeth) and anthracotheres (likely more folivorous, with bunoselenodont cheek teeth). Entelodonts may have also been scavengers and/or somewhat carnivorous. Entelodonts actually show quite derived limb morphologies (didactyly with fused metapodials), convergent on the later pecorans and camelids (see Foss, this volume).

In considering the replacement of these artiodactyls with the more modern forms in the early Miocene, it is difficult to understand exactly what happened because the late Eocene–Oligocene artiodactyls seem to have few parallels among extant forms. Modern tragulids appear to be a Neogene radiation, and although they rather resemble the Paleogene traguloids, they are now found only in tropical forest habitats. This was not the type of habitat occupied by the Paleogene forms, at least in the higher latitudes, as tropical forests were confined to equatorial regions by the late Eocene. Thus, artiodactyls such as hypertragulids and leptomerycids could not have been the ecological equivalent of extant tragulids; they might have been more similar to extant moschids. Stenomyline camelids, also known as "gazelle-camels," were long-legged, hypsodont forms that possibly were ecological equivalents of arid-habitat modern gazelles, and the leptaucheniine oreodonts were rather hyrax-like and may have been rock dwellers, but other camelids (miolabines, floridatragulines) also have no modern counterparts. Oreodonts appear peculiarly stumpy-legged and short-faced in comparison with modern camelids and pecorans, and there is no present-day equivalent to the large suiforms. One possible exception, an extant form similar to the larger anthracotheres, might be the pygmy hippo (*Hexaprotodon* [*Choeropsis*] *liberiensis*), but this taxon is also confined to tropical forests today and so is subject to the same issues as discussed above for the tragulids.

Perhaps the closest modern equivalent to this late Paleogene radiation of artiodactyls is among the caviomorph rodents of South America: capybaras and pacas appear to be rather like oreodonts, agoutis are rather like traguloids, and maras are rather like cainotheres. It is not clear if these morphological similarities reflect an ecological similarity of the northern latitude late Paleogene environments with those of subtropical South America. Note that the amount of fruit and seed volume in angiosperm fossil faunas increased somewhat in the Oligocene (following the decline in the late Eocene) but then declined again in the Miocene (Eriksson et al., 2000). It is possible that fruits and seeds were of greater availability to the artiodactyls in the Oligocene faunas, which would favor those forms with little or no rumination.

In general, however, the Neogene radiation was comprised of artiodactyls that had craniodental morphologies indicative of a more fibrous diet (e.g., more hypsodont cheek teeth), and postcranial morphologies indicative of more cursorial locomotor behaviors (e.g., longer legs), than of early forms. Both of these morphological changes would signify environmental changes to more open habitats with a greater preponderance of grass, perhaps a shift from environments dominated by woodland to environments dominated by savanna (but see discussion below). Note, also, that regional aridity characterized many Oligocene environments (e.g., Singh, 1988; Legendre, 1989; Prothero, 1998b).

The hypothesis of environmental changes influencing the evolutionary changes among the artiodactyls is supported by the fact that the late early Miocene was also a time of systematic turnover and morphological change among the perissodactyls. The hypsodont equine equids first appeared at this time, and the majority of anchitheriines became extinct. (Note, however, that the specialized large hypohippine anchitheriines flourished into the later Miocene in both North America and in Eurasia, where they appeared as immigrants at 20 Ma.) As with the artiodactyls merychyine oreodonts and the Old World anthracotheres mentioned earlier, several Paleogene perissodactyl lineages produced a "second wind" of a surviving, more derived clade (such as the hypohippines) that can be considered as part of the Neogene radiation. Among the rhinos, the nonrhinocerotid families (Hyracodontidae and Amynodontidae) became extinct. Within the Rhinocerotidae smaller, more primitive forms such as diceratheriines disappeared, and

larger, more derived forms (also more hypsodont) such as aceratheriines, teleoceratines, elasmotherines, and rhinocerotines (the tribe containing the extant rhinos) made their first appearance. More primitive types of tapiroids became extinct, and more derived, true tapiriids (e.g., *Miotapirus*) made an appearance. Another group of perissodactyls to radiate at this time, now extinct, was the larger, clawed chalicotheriine chalicotheres, that appear to have been specialized browsers (see chapters in Janis et al., 1998; chapters in Rössner and Heissig, 1999; Agustí and Anton, 2002).

However, a paradox exists with this paleoenvironmental interpretation. The transition to hypsodonty in equids in the late early Miocene of North America (and also in a number of other ungulates, see Janis et al., 2004a) has long been interpreted as reflecting the initial spread of grassland habitats; yet paleobotanical evidence shows that grasslands were present from the start of the Miocene, some 5 million years earlier (Strömberg, 2006). Although craniodental morphologies apparently lagged behind environmental changes in North America, changes in postcranial morphologies to a more cursorial form were apparent by the early Miocene in both camelids (Janis et al., 2002) and equids (Janis et al., 2004b). Moreover, locomotor and dental changes in North American rodents and lagomorphs were more synchronous, both occurring at the start of the Miocene (Janis et al., in press).

The situation in the Old World provides no greater clarity. The transition to an ungulate fauna containing horned pecorans (including bovids, cervids, and giraffids) and more derived perissodactyls (including anchitheriine equids, schizotheriine chalicotheres, and aceratherine and teleoceratine rhinos) also occurred at around 18 Ma (Agustí and Anton, 2002). As in North America, these ungulates represent forms larger in body size, usually more cursorially adapted, and frequently more hypsodont (i.e., more "open-habitat adapted") that those of the Paleogene and earlier Miocene. Yet Old World habitats were not generally similar to those in North America at this time. Habitats in Western Europe, at least (Fortelius et al., 1996b; Solounias et al., 1999; Fortelius et al., 2003; Costeur et al., 2004), and also in Africa (Nesbit Evans et al., 1981) were woodland or forested for most of the Miocene, rather than savanna grassland, although those of eastern Europe and Asia were apparently more open in nature (Fortelius et al., 1996b, 2003).

This date of around 18 Ma was an interesting time in earth history for a variety of reasons. This was a warm time in earth history, just on the edge of the mid-Miocene climatic optimum (17 to 15 Ma) (Zachos et al., 2001). It marks the formation of the "*Gomphotherium* land bridge" allowing connection between Africa and Eurasia (Rögl, 1999); proboscideans now appeared in Eurasia, and shortly afterward in North America (Woodburne and Swisher, 1995), and both artiodactyls and perissodactyls gained entry to Africa at this time. Pecoran ruminants (blastomerycid moschids and dromomerycid palaeomerycids) arrived in North America a little earlier, at around 21 Ma (but again, the radiation of a diversity of horned forms was not until around 18 Ma, which also marks the first appearance of antilocaprids), and the early Miocene in general was a period of lowered sea level and intercontinental faunal exchange (Woodburne and Swisher, 1995). Certainly, increased faunal interchange would have promoted competition among ungulates and stimulated both extinctions and originations. Yet the similarity of the evolutionary changes in both artiodactyls and perissodactyls in the early Miocene to ecomorphological types apparently better adapted to more open habitats and more abrasive diets, bespeaks of some external forcing by the environmental change at this time, even if the traditional explanation of "spread of grasslands" no longer holds.

THE TRANSITION TO THE PRESENT-DAY FAUNA AND THE RECENT PREDOMINANCE OF BOVIDS

Today not only are the majority of artiodactyls members of the suborder Ruminantia (192 of 217 species), but the majority of ruminants belong to the family Bovidae (140 species) and, to a lesser extent, the family Cervidae (41 species) (data from Nowak, 1999). Yet during the later Miocene, and even into the Pliocene and Pleistocene, there was a great diversity of other types of ruminants (for example, among the extant families Giraffidae, Antilocapridae, and Moschidae and the extinct family Palaeomerycidae). In North America there was a great deal of diversity among the camelids in the late Neogene (Honey et al., 1998), and in the Old World several different subfamilies of suids were present (Harris and Liu, this volume).

The Neogene faunas discussed in the previous section had their acme, in terms of species richness and morphological diversity, in the middle Miocene, around 14 Ma, and declined throughout the late Miocene, especially in the numbers of brachydont (presumed browsing) taxa (e.g., Van Couvering, 1980; Fortelius et al., 1996b; Barry et al., 2002; Fortelius et al., 2003; Costeur et al., 2004; Janis et al., 2004a). However, the rise in predominance of bovids in the ungulate faunas, especially of the highly hypsodont grazing forms that we consider to be the "typical" bovids today, was primarily a Plio-Pleistocene phenomenon, as was the final reduction in diversity of the other previously mentioned artiodactyls. The present-day cervid radiation, especially of the New World forms, is also a relatively recent phenomenon.

Although bovids were fairly common by the mid-Miocene, especially in Africa and eastern Eurasia, bovid species richness and taxonomic composition did not start to resemble present-day conditions until the mid-Pliocene (Bobe and Eck, 2001). In addition, although many Miocene bovids were hypsodont, other indicators of diet (e.g., cranial morphology, dental wear) were not indicative of grazing, and most of these animals were likely mixed feeders (Solounias and Dawson-Saunders, 1988; Solounias, this volume). It was not until the Pliocene that the initial members of the modern grazing bovid tribes first appeared, such as members of the Alcelaphini, Hippotragini, and Bovini. It was only in the Pleistocene that bovids with specialized

craniodental adaptations for grazing made their first appearance (Spencer, 1997), and numbers of grazing bovid species approached the level seen in today's faunas (Reed, 1997). In terms of habitat changes, this was a time of increasing aridity (see Bobe and Eck, 2001), expansion of C_4 grasses in Africa (Cerling, 1992), and the establishment of extensive dry savannas comprised of secondary grasslands rather than woodland savannas or seasonally flooded edaphic grasslands.

A similar pattern of the diversification of larger, specialized hypsodont forms can be seen among the perissodactyls and proboscideans. By the Pliocene the diversity of equids was restricted to a handful of genera, and only *Equus* survived into the Pleistocene. Among the rhinos, only the Old World Rhinocerotini (the extant rhinos) and Elasmotheriini (extinct) survived past the earliest Pliocene, although note that the Pleistocene radiation of rhinos included some specialized high latitude hypsodont forms that are now extinct, such as *Coelodonta* (the woolly rhino, a rhinocerotine) and the long-legged *Elasmotherium*. Among the proboscideans, a few mastodonts, deinotheres (both brachydont), and gomphotheres (some forms hypsodont) survived into the Pleistocene, but the recent diversity is restricted to the highly hypsodont Elephantidae.

The great reduction of the diversity of North American ungulates and the extinction of almost all of the endemic South American ungulates by the start of the Pleistocene (i.e., not taking into account the further reduction resulting from the Pleistocene extinctions, which may have been human induced) predate the arrival of humans by over a million years and thus must be ascribed to climatic effects. High-latitude cooling on what were essentially island continents (despite a limited connection via the Isthmus of Panama by the middle Pliocene), turning productive savanna habitats into less productive prairie, combined with tropical habitats in South America that were likely forest rather than savanna, probably accounted for these faunal changes.

The question of why it was primarily the Bovidae among the pecoran artiodactyls that benefited from the Plio-Pleistocene climatic changes remains unanswered. Note that among the suoids, the radiation of the subfamily Suinae (including the specialized hypsodont warthog) and the extinction of other suoid subfamilies is also a Plio-Pleistocene phenomenon (Harris and Liu, this volume). Hippos, too, although first appearing in the late Miocene are primarily a Plio-Pleistocene radiation (Boisserie, this volume). But it is not clear why bovids alone dominated the late Cenozoic hypsodont, grazing niche, one that was not exploited by cervids (although note that many present-day bovids are browsers or frugivores, especially in Africa).

The greater species diversity of present-day bovids, in comparison with cervids, may be in part because of their long history of inhabiting Africa south of the Sahara (which cervids never colonized), and the retention in Africa today of a diversity of forest, woodland, and grassland habitats that support a diversity of bovid species ranging from diminutive frugivorous duikers to extremely large grazing bovines. Note that in tropical Asia the diversity of bovids and cervids is more equivalent. This biogeographic split in Eurasia between more northern cervids and more southern bovids dates back to the late Miocene (Costeur et al., 2004). Cervids also entered the New World much earlier than bovids (with a first appearance in North America at around 5 Ma, as opposed to 2 Ma for bovids), where they had an extensive radiation, especially in South America, which was never colonized by bovids (see Webb, 2000).

However, even in the absence of bovid competition, cervids never became true grazers. A few cervids are somewhat hypsodont, and either take fresh grass (such as the barashinga, *Cervus duvaucelli,* and Père David's deer, *Elaphurus davidianus,* in Asia, and the marsh deer, *Blastocerus dichotomus,* in South America), or are mixed feeders incorporating a significant amount of grass in the diet (such as the axis deer, *Axis axis,* in Asia, the wapiti, *Cervus canadensis,* in North America, and the huemul, *Hippocamelus bisulcus,* in South America). But it remains true that no cervid evolved into a grazing specialist, herd-forming species, to rival the North American bison or the East African wildebeest. Nor is it clear why the great diversity of Plio-Pleistocene sivatheriid giraffids, probably mainly mixed feeders (see Solounias, this volume), did not survive the Pleistocene (although see Colbert, 1936b). Is there something unique about bovid anatomy or physiology that enabled them to expand into the specialized grazing niches that opened up in the Pleistocene? The question remains unanswered.

CONCLUSIONS

The highly successful radiation of artiodactyls can be seen to be influenced in part by environmental changes and in part by some unique features of the order. Artiodactyls had four main evolutionary radiations. Their initial radiation, in the early to early middle Eocene, was of small, bunodont omnivorous forms. Their second radiation, commencing in the late middle Eocene, was into larger forms that were either more folivorous (with selenodont cheek teeth), such as the tylopods and ruminants, or more specialized for omnivory, such as suids and entelodonts. The rise of artiodactyls at this point in time was coincident with a decline in the diversity of perissodactyls. This was likely not a result of direct competition but can be related to the nature of the climatic changes in the Northern Hemisphere at this time (which was the site of the evolutionary radiations of both artiodactyls and perissodactyls during the Eocene). Declining temperatures, increasing levels of seasonality (of both temperature and rainfall), and declining levels of atmospheric CO_2 would have affected the nature of the vegetation (reduced quantity, but increased differentiation in quality between parts of the plant) that would have favored ruminating artiodactyls over perissodactyls.

The two later evolutionary radiations of artiodactyls were paralleled by the radiation of similar ecomorphological types among the perissodactyls, suggesting environmental changes that affected both groups in a similar fashion. A major change in artiodactyl faunas occurred in the early Neogene, around 18 Ma. This time period saw the rise of the horned pecorans and the decline or disappearance of a number of groups that radiated in the later Paleogene (e.g., entelodonts, anthracotheres, oreodonts, a variety of European tylopods, and traguloid ruminants). More derived types of camelids and suoids also appeared at this time. The pecorans (and the derived camelids) differed from the earlier artiodactyls in having craniodental modifications more suited to abrasive diets (including hypsodont cheek teeth, although not all pecorans were hypsodont), and postcranial modifications indicative of a more cursorial lifestyle (i.e., more open-habitat adapted). Similar ecomorphological changes were seen among the perissodactyls, with the evolution of the equine equids and the rhinocerotid rhinocerotoids. However, as discussed, the paleobotanical evidence does not support a universal change to more open habitats at this point in time, and the precise reasons for this evolutionary transition remain obscure. Certainly the morphological changes in both artiodactyls and perissodactyls imply a transition to a world where the diet was more abrasive, and long distance and/or fast locomotion was more of an issue, but a simple correlation with the rise of grasslands is clearly invalid.

The final artiodactyl radiation was the Plio-Pleistocene one, with a predominance of bovids, including specialized hypsodont grazing forms, and a decline of many other types of artiodactyls (camelids, giraffids, etc.). There was also a less pronounced radiation of cervids and of suine suids at this time. Among the perissodactyls, equids became restricted to the genus *Equus,* and rhinocerotids to a couple of tribes (including the extant Rhinocerotini). These changes likely reflect the extreme cooling and aridification of the higher latitudes in the past few million years. However, it is not clear why, out of all of the artiodactyls, it is only the bovids that diversified into the role of highly hypsodont grazing specialists.

There are a couple of ways in which the evolutionary radiation of artiodactyls differed significantly from that of most other ungulates, which can be related to some particular features of the order. In contrast to perissodactyls, artiodactyls have had a significant radiation of medium- to large-sized omnivorous forms. This may relate to the fact that early artiodactyls retained a bunodont dentition, later elaborated on in the more specialized larger omnivores. In contrast, the early radiation of perissodactyls consisted entirely of forms with more derived, lophed cheek teeth, adapted for a more folivorous diet. It may be the case that the morphology of a lophed tooth cannot easily be reversed into a bunodont form, making a radiation of omnivorous perissodactyls unlikely.

One way in which artiodactyls appear to have been limited, in contrast to other ungulates (not only perissodactyls, but also proboscideans and certain South American ungulates) is in the relative absence of megaherbivores (forms over a ton or so in body mass). This is likely a result of constraints imposed by the ruminant type of digestive system, which cannot easily be adapted to increase the relative passage time of the digesta important at a larger body size to avoid the growth of energy-sapping methanogenic bacteria. These constraints do not apply to hindgut fermenters such as perissodactyls and proboscideans. The one artiodactyl that has ever evolved a body mass greater than 2,000 kg, the hippo, may be restricted to a semiaquatic habit because of water conservation problems resulting from an extremely enlarged stomach (and correspondingly reduced colon).

ACKNOWLEDGMENTS

I thank Don Prothero for inviting me to contribute to this volume, Mikael Fortelius and Gertrud Rössner for comments on the chapter, and Marcus Clauss for checking that I summarized his work correctly.

DONALD R. PROTHERO
AND SCOTT E. FOSS

Summary

THE ARTIODACTYLS WERE RECOGNIZED as a distinctive group early in the history of systematics. Linnaeus (1758) first combined the camels and bovids into his Pecora, but the pigs and hippos were in other groups. Blumenbach (1779) placed *Sus* with the ruminants but not *Hippopotamus*. Cuvier (1822) recognized a group that included most of the modern artiodactyls except hippos, but it was de Blainville (1816, 1839) who first grouped all the living artiodactyls together, without, however, naming the group. Owen (1848a) formally named the Artiodactyla, based on their distinctive paraxonic foot symmetry. Since that time, many more authors have confirmed the monophyly of the artiodactyls, with additional characters such as the presence of a double-trochleated astragalus (Schaeffer, 1947) and many distinctive features of the teeth (Gentry and Hooker, 1988) and soft anatomy (Flower and Lydekker, 1891; Viret, 1961). Further discussion of the definition of the Artiodactyla and their early evolution was provided by Theodor et al. (2005).

For almost a century after the work of de Blainville and Owen, artiodactyls were informally divided into the ruminants (sometimes called Selenodontia) and the nonruminants (or Bunodontia), based largely on living taxa (Flower and Lydekker, 1891). The living families are all distinct and monophyletic, and most have been recognized since the days of Linnaeus (although as genera in Linnaeus' classification). However, the fossil forms were often difficult to allocate to these two categories. Osborn (1910) grouped them into "primitive artiodactyls" (including anoplotheres and anthracotheres), Suina (pigs, hippos, and peccaries), Oreodonta (oreodonts), Tylopoda (camels), Tragulina (including hypertragulids), and Pecora. Matthew (1929a) subdivided the artiodactyls into the Palaeodonta (dichobunids and enteolodonts), Hyodonta (suids, tayassuids, and hippopotamids), Ancodonta (anthracotheres, anoplotheres, cainotheres, and merycoidodonts), Tylopoda (xiphodonts and camels), and the Pecora (amphimerycids, tragulids, cervids, giraffids, antilocaprids, and bovids). Many authors followed Matthew's subdivisions (e.g., Viret, 1961; Romer, 1966), whereas Simpson (1945) recognized only three suborders: Suiformes, Tylopoda, and Ruminantia. Scott (1940) recognized the Palaeodonta, the Suina, the Ancodonta (excluding oreodonts), and the Ruminantia, consisting of Tylopoda (including not only camels, but also oreodonts and hypertragulids), the Tragulina (excluding hypertragulids), and the Pecora. However, Matthew (1929a) and Scott (1940: 365) expressed doubts about how natural some of the groups (e.g., Ancodonta and Palaeodonta) really were.

PHYLOGENY OF THE ARTIODACTYLS

Cladistic Analyses of Morphology

The first explicitly cladistic analysis of a large portion of the artiodactyls was the cladogram of selenodont artiodactyls of Webb and Taylor (1980), which clustered the Tylopoda and Ruminantia in the Neoselenodontia, excluding the Oreodonta, Cainotheriidae, and Anoplotheriidae. Janis and Scott (1987, 1988) proposed a cladogram of the ruminants, which placed the hypertragulids, tragulids, leptomerycids, and *Bachitherium* and *Lophiomeryx* as progressively more derived sister taxa of the Pecora (the horned ruminants). Within the Pecora, they placed the Giraffoidea as sister taxa of the Bovidae-Cervoidea group (the latter including moschids, antilocaprids, palaeomerycids, hoplitomerycids, and cervids).

Gentry and Hooker (1988) were the first to undertake a cladistic analysis of the entire Artiodactyla, using 36 taxa (mostly families plus some fossil genera) and over 116 characters, and subjecting the matrix to an analysis by PAUP software (Swofford, 1984) as well as performing a manual cladistic analysis. Another data matrix of the Ruminantia was analyzed separately. Their preferred manual cladogram (Gentry and Hooker, 1988: Fig. 9.8) produced a cluster they called Bunodontia. Within it, the entelodonts, suids, and tayassuids clustered, there was a cluster of cebochoerids, and another group included hippos, anthracotheres, helohyids, and raoellids. They found evidence to support the monophyly of Tylopoda and Ruminantia (the latter clustered with some of the "dichobunids" in the "Merycotheria"). Tylopoda was composed of camel-oromerycid-protoceratid and xiphodont-anoplothere-dacrythere clades clustered together plus clades of oreodonts-agriochoeres-mixtotheriids, cainotheres, and as remote sister taxa, bunomerycids and *Gujaratia* ("*Diacodexis*") *pakistanensis*. Their ruminant cladogram (Gentry and Hooker, 1988: Fig. 9.10) placed the tragulids, gelocids, leptomerycids, *Bachitherium*, palaeomerycids, cervids, moschids, giraffids, and antilocaprids as progressively more derived sister taxa of the bovids.

Molecular Phylogenies

As early as 1950, molecular data were available that suggested the close affinities of whales and artiodactyls (Boyden and Gemeroy, 1950). As summarized by Marcot (this volume), many additional molecular phylogenetic analyses have been performed since that time. Although there are many living taxa of artiodactyls, making them well suited to this sort of analysis, there are many more extinct groups whose phylogenetic relationships are inaccessible to this mode of analysis. This particularly hampers the analysis of the more basal groups, such as dichobunids, entelodonts, helohyids, oreodonts, anthracotheres, and the great diversity of endemic extinct European artiodactyls. Although they appear in cladograms such as that of Geisler et al. (this volume), their phylogenetic position is weakly resolved because it depends on relatively few morphological characters.

As reviewed by Marcot (this volume), different molecular studies in recent years have reached agreement on many points (such as the inclusion of Cetacea within the Cetartiodactyla). Marcot's (this volume, Fig. 2.1D) strict consensus of 21,100 most parsimonious trees places suids and tayassuids as sister groups in a Suiformes that excludes hippos (consistent with the definition of Geisler and Uhen, 2005). This breaks up the longstanding grouping of Suiformes as suids, tayassuids, and hippos plus many extinct groups but is demanded by the new molecular and morphological evidence placing hippos with cetaceans. The camels emerge as sister taxon to the "Cetruminantia," the ruminant plus hippo-cetacean clade. This placement is consistent with older notions that camels were closer to ruminants than they were to suiforms, except that the bunodont hippos now nest within the old "Selenodontia" (tylopods plus Ruminantia). Such an arrangement requires that many of the dental characters shared by tylopods and ruminants, plus the ruminating digestive tract with at least three chambers in the stomach (Webb and Taylor, 1980; Janis and Scott, 1987; Geisler, 2001b) must have either evolved in parallel or were secondarily lost in the hippos and cetaceans.

Within the living ruminants, the tragulids emerge as the most primitive sister taxon, which is consistent with nearly all previous analyses. However, unlike some phylogenies (e.g., Janis and Scott, 1987, 1988), the antilocaprids emerge as the next most derived group, followed by giraffids, as sister taxa to the cervoid plus moschid-bovid clade. Other molecular analyses (Hernández Fernández and Vrba, 2005) place the giraffes and antilocaprids as sister taxa to each other, a clade that is then sister taxon to the cervoid-bovid clade. As Marcot (this volume) points out, part of the problem may be the limited sampling, with only one living species of antilocaprid and two living giraffids, producing long-branch attraction.

The biggest controversy within the ruminants concerns the position of the moschids. They have traditionally been considered true deer (Simpson, 1945) or a sister group of the cervoids (Webb and Taylor, 1980; and many other studies, cited by Prothero, this volume). However, molecular data (Hassanin and Douzery, 2003) placed the moschids with the bovids, but combined morphological, molecular, and behavioral evidence still places them with the cervids (Hernández Fernández and Vrba, 2005). Marcot (this volume) argues that the multiple data sets and careful analyses of Hassanin and Douzery (2003) seem to place the moschids with the bovids, even though there is little morphological evidence to support this idea.

Within the cervids, four clades have long been recognized: muntjacs, cervines, odocoileines, and hydropotines. As reviewed by Marcot (this volume), a consensus now emerges (consistent with morphological data of Groves and Grubb, 1987) for the existence of two clusters: the muntjacs and cervines (Plesiometacarpalia) and the odocoileines and hydropotines (Telemetacarpalia). This differs from earlier molecular-morphological-ethological hypotheses (e.g., Hernández Fernández and Vrba, 2005) that place hydropotines as outgroup to all other cervoids, but is consistent with their clustering of muntjacs and cervines.

Within the Bovidae, the systematics are less well resolved. Morphological data suggested a split between the Boodontia (cattle, buffalo, kudus) and Aegodontia (antelopes, goats, sheep, etc.). Early molecular studies also supported these two groupings, but the results are now less conclusive as more and more taxa and molecular systems have been sampled (Marcot, this volume). As suggested by Rokas et al. (2005), the bovid radiation may have been so rapid that it will be difficult to resolve the phylogeny of this family despite thousands of base pairs of sequence data. Marcot (this volume, Figs. 2.1F and 2.1G) summarizes the latest molecular data on the living bovids but warns that this cladogram must be viewed with caution because it is poorly supported and more polytomous and unresolved than it appears.

Total Evidence Approaches

Since Gentry and Hooker (1988), no further attempt at a cladistic analysis of the morphological data for the entire Artiodactyla was performed until the issue of cetaceans as artiodactyls precipitated further analysis. As reviewed by Geisler et al. (this volume), the idea went from being considered outrageous to well supported as further molecular analyses were conducted and the discovery of the double-trochleated astragali in primitive whales was reported (Gingerich et al., 2001; Thewissen et al., 2001b). Several different analyses in response to these new data (Geisler, 2001b; Geisler and Uhen, 2003, 2005; Theodor and Foss, 2005) were performed, each integrating more and more morphological and molecular character sets. The latest analysis (Geisler et al., this volume) combines a large suite of over 200 morphological characters and an even larger suite of over 37,000 molecular characters, as well as a stratigraphic coding of characters, to find the most parsimonious trees. Geisler et al. (this volume) performed a total evidence analysis using several different software packages, including PAUP and MacClade, which yielded over 16,000 most parsimonious trees. However, strict consensus was resolved with 89% of nodes, so there is less ambiguity than there might seem. Mesonychids and then perissodactyls came out as the sister taxa to the Cetartiodactyla. Several primitive groups of dichobunids, choeropotamids, diacodexeids, helohyids, and leptochoerids were at the base of the cladogram, followed by an unresolved polytomy of four groups: *Antiacodon;* a

clade with camels, some oromerycids, cainotheres, and homacodonts; a clade of suids, tayassuids, and entelodonts; and all the rest of the artiodactyls (the Cetruminantia, node 2 in their cladogram). Within that group (node 6 of Geisler et al., this volume, Fig. 3.2) was a cluster of anthracotheres, cebochoerids, raoellids, then hippos plus cetaceans; another cluster with agriochoeres, xiphodonts, and oreodonts; and a third cluster with ruminants plus some protoceratids. Geisler et al. (this volume) note that the sampling of groups such as the oromerycids and protoceratids is very limited and does not include some of the characters traditionally used to unite these groups, so the fact that individual taxa are scattered around the cladogram may not be as significant as it appears.

Most of the features of this most recent phylogeny are not grossly inconsistent with previous hypotheses, suggesting that we are nearing resolution and agreement on many issues. There is now a clear consensus that hippos are related to cetaceans and that anthracotheres are closely related to hippopotamids (Boisserie et al., 2005a, 2005b; Boisserie, this volume), as originally suggested by Colbert (1935b). Suids and tayassuids still cluster, along with entelodonts (Suiformes *sensu* Geisler and Uhen, 2005), as many people have long thought. Camels and some oromerycids still cluster. This is not surprising because the oromerycids were long mistaken for primitive camels. Unlike many recent hypotheses of the Tylopoda (e.g., Webb and Taylor, 1980; Gentry and Hooker, 1988), the protoceratids again group with the ruminants, as most authors before 1950 thought (see review by Prothero and Ludtke, this volume). The oreodonts and xiphodonts are again in limbo, moving from close relationship to tylopods (Webb and Taylor, 1980; Gentry and Hooker, 1988) to closer relationships with protoceratids plus ruminants (Geisler et al., this volume). Because this phylogeny also recovers the Cetruminantia, it raises the same issues of the conflict with the characters once used to support the "Selenodontia." Either the dental and digestive similarities of camels and ruminants evolved in parallel, or they have been secondarily lost or modified in the hippos and cetaceans.

Although not every answer is yet assembled, nor every issue resolved, the evidence has been collected so fast that we are rapidly converging on consensus on many points since the efforts of Gentry and Hooker (1988) and Geisler (2001b). Clearly the monophyly of suids plus tayassuids is no longer in question, nor is the fact that cetaceans are a sister group to hippopotamids and possibly anthracotheres. The monophyly of the Ruminantia and the Pecora is well supported, with the tragulids being their most primitive sister taxon of the Pecora within the Ruminantia. Still to be resolved is the issue of the Cetruminantia and "Selenodontia" and whether the camels represent a parallel evolution of a three-chambered stomach and many dental similarities, or whether cetaceans and hippos secondarily lost these features. Also unresolved is the complex of relationships within the Ruminantia, particularly the position of antilocaprids, moschids, and giraffids. Although a huge number of morphological and molecular characters have already been analyzed, and their contributions are now well known, perhaps some additional understudied data set (such as embryology) might serve as the tiebreaker on many of these issues.

THE ORIGIN AND EARLIEST EVOLUTION OF ARTIODACTYLS

One of the great mysteries of mammalian paleontology is the origin of artiodactyls. Even though many more new specimens have been found, the earliest appearance of all members of the group is simultaneous in the early Eocene of North America and Eurasia. No fossils that can clearly be assigned to the Artiodactyla have yet been found in the Paleocene anywhere in the world. As reviewed by Theodor et al. (2005), various suggestions as to their place of origin have been proposed. North or Central America (Sloan, 1969; Gingerich, 1976; Schiebout, 1979) seems to be ruled out because the Paleocene record of this region is relatively good, and the abrupt appearance of Artiodactyla in the early Eocene strongly suggests immigration from elsewhere. Some (e.g., Gingerich, 1986, 1989; Estravis and Russell, 1989) have pointed to Africa for their origins, although the sparse fossil record of the Paleocene of northern Africa (Gheerbrant, 1987) still does not support this. Currently, the oldest known artiodactyls in Africa are anthracotheres from the upper Eocene Fayûm beds of Egypt. Recently, several authors have pointed to a possible Asian origin of artiodactyls (the "East of Eden" hypothesis of Beard, 1998), based on the primitive nature of the diacodexeids known there (Smith et al., 1996b; Thewissen et al., 1983), the great diversity of very primitive forms (Métais, 2006; Theodor et al., 2005), and the presence of extremely primitive suiforms from the early Eocene of China (Tong and Wang, 1998; Beard, 1998). Theodor et al. (2005) suggested that the co-occurrence of the earliest cetaceans and the primitive *Gujaratia* ("*Diacodexis* ") *pakistanensis* and the raoellids in Pakistan might point to an origin of the Cetartiodactyla in that region. Alternatively, Kumar and Jolly (1986) and Krause and Maas (1990) suggested that artiodactyls escaped from isolation when the Indian "Noah's Ark" docked with Asia in the early Eocene. Unfortunately, there are still no fossil mammals reported from the Paleocene of India (Russell and Zhai, 1987), and recent evidence of early Eocene mammals from Pakistan does not seem to support this theory (Clyde et al., 2003). Instead, the earliest Eocene fauna from the middle Ghazij beds consists of endemic anthracobunids and quettacyonids, and only in the upper part of the lower Eocene upper Ghazij beds do the dichobunids appear (along with adapids, perissodactyls, and hyaenodonts, apparently as late immigrants from elsewhere in Eurasia).

Related to this problem is the identification of the sister taxon of the Artiodactyla (or Cetartiodactyla) and how that group might dictate the topology of phylogenetic and biogeographic hypotheses. Many candidates have been proposed over the years, but more recent analyses have ruled most of them out. Simpson (1937) suggested that artiodactyls might be derived from hyopsodontids, based on dental similarities, but this has largely been contradicted by further evidence, especially from the postcranial material (Gazin, 1968; Theodor et al., 2005). Likewise, the mioclaenines has also been proposed based on their dental features but ruled out based on the rest of the skeleton (Theodor et al., 2005). Van Valen (1966) suggested the origin of both mesonychids and artiodactyls within the arctocyonids, and Rose (1987) focused on the Paleocene arctocyonid *Chriacus* as a possible candidate. As shown by Theodor et al. (2005), however, recent evidence from new early Paleocene specimens of *Chriacus* has weakened that conclusion.

Novacek (1982, 1986, 1992), Novacek and Wyss (1986), Prothero et al. (1988), and Thewissen and Domning (1992) saw the artiodactyls (then separated from the Cetacea) as the most primitive clade among the ungulates, branching much earlier than the clade that included cetaceans plus mesonychids, or the clade that included tethytheres plus perissodactyls. Most other cladistic hypotheses, however, have not focused on finding the closest sister taxon to artiodactyls but instead used convenient outgroups to determine character polarity and root the tree within the Artiodactyla. Only a few studies have addressed the question directly. Thewissen et al. (2001b) and Theodor and Foss (2005) placed the Mesonychia as closest sister taxon, with other archaic ungulates as the next most remote outgroup (Perissodactyla were not considered). However, the large molecular and morphological data matrices of Geisler (2001b), Geisler and Uhen (2003, 2005), and Geisler et al. (this volume) place Perissodactyla as the immediate sister group to Artiodactyla, followed by Mesonychia as the next most closely related group. Large-scale molecular studies of living taxa (so mesonychids were excluded) argue that Perissodactyla are the sister taxon of Carnivora plus Pholidota, and the Cetartiodactyla is their sister taxon, comprising the clade "Ferungulata" within the "Laurasiatheria" (Madsen et al., 2001; Murphy et al., 2001a, 2001b; Springer et al., 2003, 2005).

As indicated by Theodor et al. (2005), if either the Mesonychia or Perissodactyla are considered closest sister taxon to the Cetartiodactyla, this again points to an Asian origin of the group because both sister taxa have their earliest relatives in the Paleocene of Asia (especially China). Clearly, more fossils need to be found, especially in the Paleocene of China, India, and Pakistan, to help resolve this issue.

Theodor et al. (this volume) review the record of the earliest terrestrial artiodactyls (diacodexeids, dichobunids, and many other primitive forms). They are most diverse in the Eocene of Eurasia and only secondarily immigrated to North America, where they are represented in the Wasatchian by *Diacodexis, Bunophorus,* and *Simpsonodus.* Theodor et al. (this volume) recognize seven distinct families of primitive Eocene terrestrial artiodactyls: the Diacodexeidae, Dichobunidae, Homacodontidae, Leptochoeridae, Helohyidae, Cebochoeridae, and Raoellidae. Within these families, they outline the current generic and species-level systematics of all the valid taxa. They point out that "*Diacodexis*" *pakistanensis* of Thewissen et al. (1983) is now referable to *Gujaratia* Bajpai et al. (2005). Their chapter highlights the fact that many new taxa of European and Asian dichobunids have been described in recent years, yet they are poorly known to most paleontologists. This chapter provides a summary of the current state of their systematics.

ENDEMIC EUROPEAN PALEOGENE ARTIODACTYLS

Even more dramatic was the endemic radiation of artiodactyls in Europe during the Eocene, with most groups vanishing during the *Grande Coupure* except for a few Oligocene and early Miocene stragglers, such as cainotheres. Many different families (Cebochoeridae, Choeropotamidae, Mixtotheriidae, Cainotheriidae, Anoplotheriidae, Xiphodontidae, and Amphimerycidae) have been known since Cuvier first described *Anoplotherium* in 1804 and *Xiphodon* and *Choeropotamus* in 1822. Dozens of genera in these families are now recognized, with over a hundred species. Most were originally named and described in detail during the nineteenth century, culminating with the work of Depéret (1908) and Stehlin (1908, 1910a), but the past 30 years (since Sudre, 1978) has produced much new material and new taxa, which are summarized by Erfurt and Métais (this volume). Most of these groups are clearly restricted to Europe and have no apparent affinities with Eocene artiodactyls from other continents, although some (Dechaseaux, 1963, 1965, 1967; Gentry and Hooker, 1988; Norris, 1999) have tried to link xiphodonts with camelids. Viret (1961) and Erfurt and Métais (this volume) argue, however, that this appears to be convergence, and there is no evidence that xiphodonts and camelids are in fact closely related. Heissig (1993) argued that the anoplotheres and oreodonts might be closely related based on postcranial similarities. However, Gentry and Hooker (1988) related anoplotheres to dacrytheres and xiphodonts, combined in a clade with camels, oromerycids, and protoceratids. Gentry and Hooker (1988) considered mixtotheres to be closer to oreodonts.

One of the major controversies is the systematic position of the Amphimerycidae, which have long been placed as ancestors of the ruminants based their fused cubonaviculars. However, Viret (1961) and Webb and Taylor (1980) have argued that most other characters do not support this hypothesis, and Hooker and Weidmann (2000) have described further material that corroborates this. Erfurt and Métais (this volume) point out that the dental morphology of amphimerycids is more similar to that of dacrytheres or xiphodonts than it is to the primitive ruminants now known from the early Eocene of Asia (Vislobokova, 1998; Guo et al., 2000).

Erfurt and Métais (this volume) summarize the known occurrences of the endemic European Paleogene families (Erfurt and Métais, this volume, Fig. 5.1). The Cebochoeridae are restricted to the late middle and late Eocene and consist of just four genera: *Cebochoerus, Gervachoerus, Acotherulum,* and *Moiachoerus.* Choeropotamids consist of 11

genera, which range from the early Eocene to early Oligocene, mostly restricted to Europe except for one taxon from the middle Eocene of Turkey (Ducrocq and Sen, 1991). Mixtotheriids are monotypic, based on the middle Eocene genus *Mixtotherium* alone. Cainotheres, on the other hand, are represented by six genera from the middle and late Eocene, and three of them range into the Oligocene or even (in the case of *Cainotherium* itself) the early Miocene (MN 6). Six genera of anoplotheriids are known, mostly from the middle and late Eocene, although *Diplobune* extends into the early Oligocene. Xiphodonts are known from five genera, all restricted to the middle and late Eocene. Finally, the Amphimerycidae consists of two genera, *Amphimeryx* and *Pseudamphimeryx,* which range from the middle Eocene to the early Oligocene. As Erfurt and Métais (this volume) noted, most of these groups (along with many other European endemic groups) vanished at the *Grande Coupure,* when they were apparently replaced by immigrant groups from Asia such as enteledonts, anthracotheres, leptomerycids, gelocids, and primitive ruminants, although much of the extinction of native elements could have been caused by climatic changes at this time (Prothero, 1994, 2006). It should be noted that Erfurt and Métais (this volume) follow the common practice of placing the Eocene-Oligocene boundary at the *Grande Coupure,* between MP 20 and MP 21 (Headonian/Suevian boundary). However, Hooker (1992) showed quite clearly that the *Grande Coupure* occurred in the earliest Oligocene, consistent with the global cooling and climatic event in the earliest Oligocene. The formally recognized Eocene–Oligocene boundary, as defined since 1989 by the last appearance of *Hantkenina,* is a global nonevent with respect to extinction and climatic change.

THE RADIATION OF BUNODONT ARTIODACTYLS IN ASIA, AFRICA, AND NORTH AMERICA

In contrast to the endemicity of European middle and late Eocene artiodactyls, the contemporary groups (especially enteledonts and anthracotheres) among nonruminant artiodactyls show frequent ties across the Bering Corridor between Asia and North America. Foss (this volume) reviews the small, mostly North American family Helohyidae, whose contents were scattered among several groups by Simpson (1945). Gazin (1955) was the first to recognize a grouping that included *Helohyus* and *Achaenodon* (the latter often mistaken for an enteledont relative), and McKenna and Bell (1997) also recognized this grouping but included within it some Asian taxa (*Gobiohyus, Pakkokuhyus*) that Stucky (1998) and Foss (this volume) do not regard as helohyids. As Foss (this volume) notes, the helohyids were among the first taxa to develop teeth for an omnivorous diet and were eventually replaced by enteledonts with similar dentitions.

The Enteledontidae (Foss, this volume), by contrast, were a widespread and long-ranging group that arose in Asia during the middle Eocene, spread to Europe and North America in the late Eocene, and lasted until the early Miocene in North America and Asia. The more derived forms were generally piglike in appearance with massive skulls bearing large tusks and many tubercles on the jaws and on the broad flaring flangelike zygomatic arches. The largest forms, such as *Daeodon* (formerly *Dinohyus*), reached the size of some living rhinos. Their teeth were bunodont like those of most other primitive nonselenodont artiodactyls but show extensive evidence of apical wear (even on the canine tusks), suggesting a bone-crushing omnivorous diet. Foss (this volume) reviews their taxonomy, which was subjected to gross oversplitting based on minor differences in the highly variable jugals and mandibles and their tubercles, with as many as 56 species in 23 genera named. Foss (this volume) recognizes only seven genera, most of which are monotypic, based on much larger samples than available to previous authors and also based on a modern concept of the variability within species. Familiar genera such as *"Dinohyus"* are now referred to *Daeodon,* and *"Archaeotherium"* coarcta-tum from the Chadronian of Cypress Hills, Saskatchewan, is now referred to a new genus, *Cypretherium.* The entelodonts first appear in the late middle Eocene of China (*Eoentelodon*), and in the late Eocene *Brachyhyops* made its appearance in North America, while *Entelodon* and *Paraentelodon* appeared in Eurasia. *Cypretherium* occurs in the Chadronian and *Archaeotherium* in the Orellan, Whitneyan, and early Arikareean of North America. Entelodont evolution concluded in the early Miocene, when the huge *Daeodon* in North America and *Paraentelodon* in Asia were the last surviving species of the group, vanishing about 19 Ma. The relationships of entelodonts to other artiodactyls are still not well established. Gentry and Hooker (1988) regarded them as sister group of the Suoidea (pigs and peccaries), and Foss (this volume) prefers this hypothesis as well.

The other suoids were reviewed by Harris and Liu (this volume). In contrast to the entelodonts, the suoids were very diverse over most of Eurasia and North America since the middle Eocene, and seven genera and 12 species are still alive today. They are distributed between two families, the peccaries (Tayassuidae) and the pigs (Suidae). The oldest suoids are now reported from the late Eocene of China (Tong and Zhao, 1986; Liu, 2001) and Thailand (Ducrocq, 1994a, 1994b; Ducrocq et al., 1998), where the suids evolved in Asia until the Oligocene, then spread to Europe during the *Grande Coupure* and to Africa when the land bridge across the Arabian Peninsula opened in the middle Miocene. The peccaries, on the other hand, first appeared in North America in the late Eocene (Chadronian) and have been restricted to the New World ever since. Some authors have argued that a number of Old World fossils (the Palaeochoerinae and Doliochoerinae) are peccaries as well (Pickford, 1993; Van der Made, 1997a; McKenna and Bell, 1997). However, Wright (1998), Liu (2003), and Harris and Liu (this volume) showed that these specimens display no true derived characters of the Tayassuidae, and they have been redistributed to the Suidae or to Suoidea *incertae sedis.*

The Suidae is a large family with at least six subfamilies (seven if the sanitheres are demoted from separate family rank). At present, 44 different genera and dozens of species are recognized by McKenna and Bell (1997) and largely upheld by Harris and Liu (this volume). Some of the subfamilies of previous authors, such as the Palaeochoerinae and Hyotheriinae, are not monophyletic groups and so were not recognized by Liu (2003) and Harris and Liu (this volume). The first known true suid is *Eocenchoerus* Liu, 2001, from the late Eocene of China, although there are many "suoids" of uncertain taxonomic affinities from the late Eocene of China and Thailand that may prove to be sister taxa of the suids. Harris and Liu (this volume) review the evolutionary history of all the families of Suidae. This chapter is not the place to review all the complicated details of their evolution; for this discussion, the reader is referred to the excellent summary by Harris and Liu (this volume).

Harris and Liu (this volume) also briefly summarize the history of the North American peccaries (family Tayassuidae), which were last reviewed by Wright (1998). Unfortunately, Wright's (1998) review was just a brief summary chapter, with no detailed descriptions and hints at several unnamed new taxa that require description. Clearly, a complete monograph of the Tayassuidae is needed.

Although linked with the pigs and peccaries for a long time, recent evidence suggests that hippos are not suoids but closely related to whales in the "Whippomorpha." Boisserie et al. (2005a, 2005b) and Boisserie (this volume) and Lihoreau and Ducrocq (this volume) argued that anthracotheres are the sister group of hippos and thus also not "suoids" as long suggested (e.g., Gentry and Hooker, 1988; Kron and Manning, 1998). Anthracotheres were first described by Cuvier in 1822, and dozens of taxa have been described since then. Lihoreau and Ducrocq (this volume) provide a detailed review of the complex history of anthracotheres, which first appeared in Eurasia in the middle Eocene (*Siamotherium* from Thailand and Myanmar) and then spread to North America by the late middle Eocene (*Heptacodon* in the late Uintan) and to eastern Europe in the late Eocene of Bulgaria (*Bakalovia*). They did not reach western Europe until after the *Grande Coupure* in the Oligocene (when *Elomeryx* and *Anthracotherium* migrated in) but did reach Africa in the late Eocene (*Bothriogenys, Qatraniodon*). Thirty-seven genera were recognized by McKenna and Bell (1997), most of which were upheld by Lihoreau and Ducrocq (this volume). At the beginning of the Miocene, anthracotheres vanished from Europe, Africa, and North America and survived only in Asia. In the middle Miocene, anthracotheres reappeared in Africa after a local extinction where there were at least two radiations. According to Lihoreau and Ducrocq (this volume) and Boisserie et al. (2005a, 2005b), one of the lineages of bothriodontines in Africa gave rise to the hippopotamids. As Lihoreau and Ducrocq (this volume) point out, the commonly held notion that hippos drove the anthracotheres to extinction cannot be supported. The anthracotheres vanished in Europe and North America long before hippos appeared and overlapped with hippos for 4 million years in Asia and 8 million years in Africa, so it is unlikely that they competed and drove each other to extinction. Instead, if Boisserie et al. (2005a, 2005b) are correct, the anthracotheres are not extinct but survive today as their descendants, the hippos.

The Hippopotamidae are reviewed in this book by Boisserie (this volume). Such a review is long overdue because there has been no comprehensive systematic treatment of hippos in many decades. As Boisserie (this volume) indicates, hippos are the exception to the rule that most groups arose in the Eocene of Eurasia and spread elsewhere (Beard, 1998). Their oldest representatives are the poorly known Kenyapotaminae, dated between 16 and 14 Ma from Kenya, and from younger beds from Kenya, Tunisia, and Ethiopia. Boisserie (this volume) maintains hippos with "anatomically modern cheek teeth (or Hippopotaminae, excluding *Kenyapotamus*) burst into the fossil record in the uppermost Miocene and number among the most frequently collected mammals." They then diversified into a number of lineages in Africa, including the *Choeropsis-Saotherium* lineage, the *Archaeopotamus* lineage, and the *Hexaprotodon* lineage. They spread to Europe in the latest Miocene and by 5.9 Ma spread into Pakistan and reached Southeast Asia by 1.5 Ma (mostly specimens referred to *Hexaprotodon*). According to Boisserie (this volume), the first appearance of the genus *Hippopotamus* itself is not yet well constrained, but specimens from the Pliocene of West Turkana about 4 Ma may be referable to this genus. In the early Pleistocene, the large *Hippopotamus gorgops,* with elevated orbits and large sagittal crests, roamed all of Africa, to be replaced in the middle Pleistocene by *H. amphibius,* the living species. Several different species of hippos spread across Europe as well as the Middle East in the Pleistocene. Finally, hippos are well known for colonizing isolated islands such as Madagascar and many Mediterranean islands (Sicily, Malta, Crete, and Cyprus), where endemic dwarfed species evolved on each island during the Pleistocene.

TYLOPODS, OREODONTS, PROTOCERATIDS, AND OTHER NONRUMINANT SELENODONTS

The systematic position of oreodonts and their paraphyletic sister group, the agriochoeres, is still not well established. In their primitive skeletons, the oreodonts are somewhat suoid-like, so classifications such as those of Simpson (1945) and McKenna and Bell (1997) placed them in the Suiformes. However, this is basing relationships on symplesiomorphy and not on synapomorphy. Most researchers (e.g., Scott, 1940; Romer, 1966; Carroll, 1988; Gentry and Hooker, 1988; Norris, 1999) have placed oreodonts closer to the Tylopoda (particularly the camels and oromerycids). Thewissen et al. (2001b), Geisler (2001b), Geisler and Uhen (2003), and Theodor and Foss (2005), however, placed the oreodonts as sister taxon to Ruminantia plus Tylopoda, the group that Webb and Taylor (1980) had named Neoselenodontia. However, with the Neoselenodontia being broken up by the inclusion of the Whippomorpha, the position of oreodonts is again uncertain. Geisler and Uhen (2005) placed

them as immediate sister group to the ruminants and distant from the camels (which clustered with the Suoidea minus hippos). The larger data set of Geisler et al. (this volume), however, placed the oreodonts with European groups such as xiphodonts, amphmerycids, as well as the primitive protoceratids *Leptoreodon* and *Leptotragulus.* Together these formed a clade that was sister taxon to the rest of the protoceratids plus the ruminants. Clearly, the major switches of systematic affinities and the limited character sets employed on the few oreodont taxa sampled so far indicate a further need to expand the data matrices in order to find a stable hypothesis of relationships.

At a more detailed level, close comparisons of primitive agriochoerids such as *Protoreodon* have been made with the middle Eocene form *Asiohomacodon myanmarensis* (Tsubamoto et al., 2003) or with the homacodontids in general (Lander, 1998). However, this is based on a few dental similarities and not on the total evidence used by Geisler et al. (this volume).

Whatever their affinities, oreodonts were an endemic North American group that became extremely successful during the middle Eocene through Oligocene, then began diminishing by the early Miocene, and vanished by the end of the Miocene. In the Uintan, *Protoreodon* is by far the most common taxon; in the Orellan, the strata were long known as the "turtle-Oreodon beds" for their abundant *Merycoidodon* and *Miniochoerus.* Oreodonts continued to be diverse and abundant through the late Oligocene and early Miocene, so they have often been used as zonal index species for that interval. Ludtke (this volume) reviews the paraphyletic assemblage known as agriochoeres, which include the primitive Uintan form *Protoreodon,* and the strange long-tailed clawed forms known as *Agriochoerus* from the Chadronian through Arikareean. As Ludtke (this volume) points out, the agriochoeres were once grossly oversplit, with more than 10 genera and 35 species described. Lander (1998) and Ludtke (this volume) recognize only three of the named genera: *Protoreodon, Diplobunops,* and *Agriochoerus.* In addition, Lander (1998) claimed to recognize five more unnamed genera based on previously described species. The validity of these taxa cannot yet be determined because Lander has still not published full descriptions of these taxa in over 30 years of work on the group, and his published indications are inadequate for full diagnosis. Ludtke (this volume) also evaluates the longstanding controversy over whether the long-tailed, large-clawed *Agriochoerus* was arboreal (as suggested by Matthew, 1911). Interestingly, Wilhelm (1993) showed that limb morphology of *Agriochoerus* most closely resembles that of a tree kangaroo.

Arising from the agriochoeres was the monophyletic family Merycoidodontidae, the traditional oreodonts of most authors (called family Oreodontidae by some). This group is extremely abundant in the Oligocene and early Miocene and consequently has figured importantly in the biostratigraphy of those intervals (e.g., Prothero and Whittlesey, 1998). However, they have also suffered from the most incompetent and extreme taxonomic oversplitting by Schultz and Falkenbach (1940, 1941, 1947, 1949, 1950, 1954, 1956, 1968). As many authors have noted (summarized by Stevens and Stevens, 1996), Schultz and Falkenbach failed to make any useful statistical comparisons or take into account populational and sexual variability and often named taxa based purely on their stratigraphic level without any morphological differences. In many cases, the taxa are demonstrably based on postmortem deformation, so *"Stenopsochoerus"* is a *Miniochoerus* that has been laterally crushed, and *"Platyochoerus"* is the same taxon dorsoventrally crushed. The mess of oreodont systematics has been gradually cleared up by Stevens and Stevens (1996), with more work to come. Unfortunately, the systematics of Lander (1998) was published without adequate justification for his taxonomic decisions. Some of his decisions are clearly wrong, such as declaring the taxon *Merycoidodon* invalid because the type specimen is so poor (and resurrecting Schultz and Falkenbach's *"Prodesmatochoerus"* to replace it). As Stevens and Stevens (1996, this volume) clearly showed, Leidy (1851b, 1852, 1869) designated enough referred material to clearly show what he meant by *Merycoidodon,* so that taxon is still valid. Stevens and Stevens (this volume) provide a concise summary of their ongoing work, which is based on careful statistical analyses of large sample sizes and adequate consideration of postmortem deformation, and much of the Schultz and Falkenbach mess is gradually being cleared up. In their summary, Stevens and Stevens (this volume) recognize only 12 subfamilies and about 65 species, including the recent revisions to the leptauchenine oreodonts by Hoffman and Prothero (2004) and Prothero and Sanchez (in press). This is a considerable improvement over the taxonomy of Schultz and Falkenbach or Lander (1998).

The Tylopoda has always been based on a core taxon, the Camelidae, with various other sister groups, including Oromerycidae (once mistaken for early camels) and sometimes the Protoceratidae, Xiphodontidae, and Oreodontoidea included within the group (Gentry and Hooker, 1988). Traditionally, most systematists have regarded the camels as sister group to the ruminants (the Neoselenodontia of Webb and Taylor, 1980). As we have already discussed, the newer phylogenetic analyses place the camels in a more remote position within the artiodactyls, with the whales plus cetaceans as immediate sister group to the ruminants (Geisler et al., this volume; Marcot, this volume). This creates significant problems because camels have ruminating stomachs and other digestive similarities to the ruminants and also many similarities in the teeth and skull and limbs. Geisler (pers. comm.) suggests that these features may still be prone to convergence, as has been well demonstrated by the convergent evolution of fused and elongate metapodials independently in camelids and in many different ruminant groups. He also speculates that it seems unlikely that hippos, which are herbivorous and have the primitive hindgut fermentation system found in pigs and most other ungulates, secondarily lost these features that might have been shared by the common ancestor of camels, hippos, and ruminants. The resolution of this problem awaits fur-

ther analysis, particularly in the details of camelid anatomy and whether the details show it is identical to the condition found in ruminants or only similar enough to suggest convergence.

The Camelidae have a long and complex history in North America (with late emigration events in the Miocene and Pliocene to Eurasia and to South America), last reviewed by Honey et al. (1998). However, a complete systematic monograph of the Camelidae, utilizing the huge undescribed Frick Collection of the American Museum of Natural History, has still not been attempted. Because there was no point in rehashing the work in Honey et al. (1998) without any contribution from new discoveries, Honey (this volume) provides an interesting analysis of the protolabine camels *Protolabis* and *Michenia*. He shows that the larger *Protolabis* and the smaller *Michenia*, which have the exact same range in time and space, might actually be male and female morphs of the same genus, a conclusion that was originally rejected by Honey and Taylor (1978). If this is true, many of the taxa named in the Camelidae need to be reassessed as possibly reflecting sexual dimorphism.

The Protoceratidae are a bizarre group characterized (in the more advanced taxa) by peculiar combinations of cranial appendages, including large horns over the orbits and nasals and frontals as well and strange branched horns on the occiput (*Paratoceras*). They are relatively primitive in most skeletal features, with unfused metapodials and low-crowned bunoselenodont teeth, yet have many derived features as well. Endemic to North America and never very diverse (10 genera and about twice that many species), they ranged from the late middle Eocene (Uintan) to the very end of the Miocene (Hemphillian) but were usually quite rare elements through most of their history. During the Miocene, they became largely restricted to the more forested habitats of the Gulf Coastal Plain of North America (especially Texas and Florida, but also Central America), suggesting that they preferred more densely vegetated habitats. With their broad muzzles and cranial appendages, they have been compared to water-loving browsers such as the moose or bushbuck. Prothero and Ludtke (this volume) update the systematic review of Prothero (1998a) and provide the latest information on new discoveries that modify the systematics or expand the geographic range of the protoceratids.

As Prothero and Ludtke (this volume) show, the systematic position of protoceratids has fluctuated dramatically. Most authors up to and including Simpson (1945) have considered them to be related to ruminants, but Scott (1940), Stirton (1967), and Patton and Taylor (1971, 1973) have argued that they were tylopods, which was corroborated by Webb and Taylor (1980), Gentry and Hooker (1988), and Prothero (1998a). However, since then support for the tylopod hypothesis has eroded, with work by Joeckel and Stavas (1996), Norris (1999, 2000), and Geisler (2001b) placing them as sister group to the ruminants again or the ruminants plus tylopods. However, with the inclusion of hippos plus whales deeply nested within the Selendontia and the wide separation of camels and ruminants, the position of protoceratids must now be reconsidered. Geisler and Uhen (2005) placed them weakly with either the camels or the ruminants, as did Theodor and Foss (2005). However, as discussed above, Geisler et al. (this volume) scattered the contents of the Protoceratidae all over the cladogram, with the primitive forms *Leptotragulus* and *Leptoreodon* clustering with the xiphodonts and oreodonts as sister taxon to the ruminants, and *Protoceras* and *Heteromeryx* nested in the other clade as immediate sister group of the ruminants. As Geisler et al. (this volume) noted, their analysis does not take into account all the skeletal features used to diagnose and recognize the Protoceratidae, so this scattering of their contents is not well supported. As of this writing, the exact systematic affinities of the Protoceratidae are as much in the air as they have ever been, despite over a century of research.

EVOLUTION OF THE RUMINANTS

By far the largest, most widespread, and most diverse group of artiodactyls (or indeed of ungulates in general) are the ruminants, with their specialized four-chambered stomachs and many other adaptations for efficiently digesting low-quality vegetation. Although the modern groups are well known, their fossil history has long been difficult to interpret. This is largely because ruminant dentition became selenodont and relatively hypsodont very early in their evolution and soon became stereotyped as well, so the differences among most ruminant teeth are subtle and difficult to distinguish. Likewise, most ruminants have highly stereotyped limbs with fusion of the metapodials into a cannon bone, elongation of the phalanges, and fusion of the cubonavicular joint, so limbs are of limited use in most systematic studies. In some groups, such as the cervids, bovids, antilocaprids, and palaeomerycids, the cranial appendages are the most useful feature for classification, relegating the much more abundant but less diagnostic dental fossils and other stereotyped parts of the skeleton to a secondary role. Consequently, many specimens of ruminants cannot be identified without their cranial appendages, and much of the controversy over their relationships stems from the limited character sets available for analysis.

Most of the more primitive ruminant lineages, on the other hand, lacked any cranial appendages, as do living tragulids and moschids, but some compensate with larger canines in males. Recent work on the fossil record suggests that the earliest ruminants occur in middle Eocene (*Archaeomeryx* from the Irdinmanhan) of Asia (Métais and Vislobokova, this volume), and then slightly later in the late middle Eocene (late Uintan *Simimeryx, Leptomeryx,* and *Hendryomeryx*) of North America as well. Most authors regard the appearance of these taxa as an immigration event from Asia (Métais and Vislobokova, this volume), although Stucky (1998) posited that the American ruminants were descended from local homacodonts. A few ruminants (*Gelocus, Phaneromeryx*) occurred in Europe in the late Eocene (unless Amphimerycidae are considered ruminants—unlikely, according to Métais and Vislobokova, this volume), but the

Grande Coupure in the early Oligocene marked an invasion of many Asiatic ruminants, replacing the endemic artiodactyls of the Eocene of Europe (Métais and Vislobokova, this volume). Ruminants are not documented from Africa before the early Miocene (Barry et al., 2005) and did not spread into South America until the Great American Interchange in the middle Pliocene, where they formed a large radiation of native deer.

As Métais and Vislobokova (this volume) point out, many different phylogenetic hypotheses have been proposed for primitive ruminants, especially the fossil forms. Most of the fossils were placed in paraphyletic grades modeled on the living tragulids, forming wastebasket groups such as the Tragulina (Flower, 1883). The first comprehensive phylogeny was that of Webb and Taylor (1980), followed by the cladograms of Janis (1987), Janis and Scott (1987, 1988), Gentry and Hooker (1988), and Webb (1998). Families such as the Tragulidae, Hypertragulidae, Leptomerycidae, and Gelocidae have long been recognized, but recent work has added the Lophiomerycidae and Bachitheriidae of Janis (1987), the Praetragulidae of Vislobokova (2001), and the Archaeomerycidae as a subfamily raised to family rank, for a total of eight families (Métais and Vislobokova, this volume). These authors then cluster the eight families into two superfamilies: Traguloidea (Archaeomerycidae, Lophiomerycidae, Tragulidae, Leptomerycidae, Gelocidae, and Bachitheriidae) and Hypertraguloidea (Hypertragulidae including Hypisodontinae, and Praetragulidae). A detailed review of each fossil genus is then provided by Métais and Vislobokova (this volume). Their cladogram (Métais and Vislobokova, this volume: Fig. 15.2) first recovers the split between Traguloidea and Hypertraguloidea. Within the Traguloidea, the Lophiomerycidae and Tragulidae form one branch, and the Archaeomerycidae, Leptomerycidae, and Bachitheriidae form progressively more derived sister taxa to the Gelocidae. Thus, the cladogram seems to suggest that the earliest known ruminant, *Archaeomeryx,* is not the most primitive one but that the hypothetical common ancestor of the Traguloidea–Hypertraguloidea split, and the Tragulidae–Gelocidae split, must have preceded *Archaeomeryx.* The cladogram also suggests that the North American groups (Leptomerycidae, Hypertragulidae) each have separate origins nested within Asian clades, and each represents a separate immigration event in the late Uintan.

The details of the family Tragulidae are reviewed by Rössner (this volume). As the only living members of the family, the extant chevrotains or "mouse deer" are often used as models for the many extinct tiny hornless ruminants, and even considered "living fossils" by some (Janis, 1984; Thenius, 2000). Three genera (*Tragulus, Hyemoschus,* and *Moschiola*) with a total of 10 species are alive today (Groves and Grubb, 1987; Meijaard and Groves, 2004a), although all are highly endangered. All of the living taxa are restricted to tropical Asia, but the extinct forms were widespread over Eurasia and parts of Africa during the Miocene. Rössner (this volume) summarizes the biology and anatomy of the living tragulids, which are still not that well understood or studied despite their crucial importance. As many authors have noted, tragulids lack not only the cranial appendages but many other derived features found in most ruminants (especially the Pecora), such as only partially fused metapodials, incomplete distal keel on the metapodials, and an elongate relatively narrow astragalus with the distal articulation medially deflected. However, they have the fused cubonavicular and the incisiform lower canines found in nearly all ruminants (along with the ruminating stomach and other soft anatomical characters, although much more primitive than the condition found in other ruminants), so they are clearly members of the group. Rössner (this volume) discusses the possible origins of the family Tragulidae and points to *Archaeotragulus krabiensis* (Métais et al., 2000) from the late middle Eocene of Southeast Asia as the oldest known member of the group. According to Rössner (this volume), tragulids vanished from the fossil record for the rest of the Eocene and Oligocene, then suddenly reappeared in the early Miocene of Southeast Asia, Africa, and Europe. In the Pliocene and Pleistocene, they again became very rare in the fossil record, leaving only the living taxa as the relicts of their once widespread Old World distribution.

Another family of living ruminants without cranial appendages is the musk deer, or family Moschidae. Unlike the tragulids, however, they were diverse and abundant in North America as well as Eurasia during the Miocene. Today, they are represented only by the genus *Moschus* with seven valid species distributed throughout eastern Asia (Groves et al., 1995; Su et al., 2001). Like tragulids, their biology is poorly known because they are rare and elusive, and like tragulids, they are endangered and now being poached to extinction because the musk of the males is the most valuable substance in nature (US$45,000 a kilogram), more valuable than gold or cocaine or rhino horn. Prothero (this volume) reviews the systematics of the group. Traditionally, musk deer were placed within the cervids (Simpson, 1945) or as a sister group to the cervids (Prothero, this volume). The cytochrome *b* sequence (Su et al., 1999) also placed cervids and moschids as sister groups, as does the molecular-morphological-ethological analysis of Hernández Fernández and Vrba (2005). However, the molecular analysis of Hassanin and Douzery (2003) argued that moschids are sister group of the bovids, not cervids, and this conclusion is supported by Marcot (this volume). In addition to the living genus *Moschus,* there are at least eight other extinct Eurasian genera of moschids, which begin with *Dremotherium* and *Bedenomeryx* from the late Oligocene of Europe and spread across Eurasia by the early Miocene. By the middle Miocene, most of these taxa were extinct, leaving only the tiny *Micromeryx* in the middle and late Miocene faunas of Europe. *Moschus* is recorded from the late Miocene, Pliocene, and Pleistocene of Asia.

In addition to the Eurasian diversity of moschids, there was a North American radiation. Long misunderstood and misallocated by earlier authors, the moschids were combined into a group known as the blastomerycines by Frick (1937). Although the moschid affinities of the blastomerycines had been recognized early on, it was not until Webb and Taylor (1980) that they were formally united with the Moschidae. Frick (1937) also oversplit blastomerycines into numerous genera, subgenera, and species, mostly without adequate diagnoses or descriptions. From the 22 species that Frick (1937) recognized, only 8 species in six genera were recognized by Prothero (in press). Like the immigrant palaeomerycids (dromomerycids of some authors) and antilocaprids, the moschids (blastomerycines) first immigrated from Asia to North America in the late Arikareean (*Problastomeryx, Pseudoblastomeryx*) and reached moderate diversity in the early middle Miocene (Hemingfordian–Barstovian) before declining to just two genera, one of which (*Longirostromeryx*) survived until the end of the Miocene (Hemphillian).

Yet another group with Asian roots but a predominantly North American history consists of the pronghorns, or antilocaprids. They are often incorrectly called "antelopes"; they are not true antelopes (members of the family Bovidae) but, rather, their own unique and endemic North American family. Unlike bovids, their cranial appendages have both a bony core and a deciduous keratinous sheath, with at least one fork or tine, a feature found in no bovid. Like moschids and palaeomerycids, they appeared in North America as immigrants from Eurasia in the late Arikareean and diversified into at least 17 genera in the middle and late Miocene, with a wide variety of cranial appendages, including multiple prongs, long twisted horns, and many other shapes. Unlike moschids and palaeomerycids, they survived the extinctions at the end of the Miocene and survived in low diversities through the Pliocene and Pleistocene and today are represented by one remaining species, the American pronghorn, *Antilocapra americana*. Davis (this volume) reviews and updates the taxonomy and biogeography of antilocaprids, building on the work of Janis and Manning (1998b). As he notes, much of the species-level taxonomy of the group is based on Frick's (1937) work, which is grossly oversplit, so a detailed species-level revision using the large samples in the Frick Collection is long overdue and necessary to get a true picture of their diversity.

The origins of antilocaprids are still controversial. There are no obvious sister groups in Asia during the Oligocene or Miocene, although *Palaeohypsodontus* from the middle Oligocene Hsanda Gol Formation of Mongolia has been suggested as an ancestor. This taxon, however, is based on fragmentary material, and the teeth have no diagnostic features that would assign them to the antilocaprids, bovids, or some other ruminant group (Janis and Manning, 1998b). Whether we have the fossils or not, nearly all the candidates for closest relatives of antilocaprids (giraffids, bovids, and cervids) have their earliest representatives in the Eocene or Oligocene of Asia, so it likely the antilocaprids also originated there and then migrated to North America, where they diversified because there was no competition with bovids or cervids. Antilocaprids apparently performed the "antelope role" in North America, which lacked true antelopes. In recent years, antilocaprids were thought to be related to the cervoids (Leinders, 1979; Leinders and Heintz, 1980; Janis and Scott, 1987, 1988; Janis and Manning, 1998b), but earlier authors placed them with bovids based on convergent features of the cranial appendages and skeleton and dentition (Matthew, 1904; Pilgrim, 1941; Simpson, 1945; O'Gara, 1978; Gentry and Hooker, 1988; O'Gara, 1990). However, recent molecular evidence places them closer to the Giraffidae at the base of the pecoran radiation (Hassanin and Douzery, 2003; Hernández Fernández and Vrba, 2005). If this is true, the evolution of a horn with a keratinous sheath is convergent in bovids and antilocaprids.

The Palaeomerycidae is another cosmopolitan family that has long been misunderstood. Many of their members were shoehorned into the Cervidae by most early authors and only recognized as distinct in the past few decades (see Janis and Scott, 1987: 52–53, for a review). In the Old World, they were a common family throughout the Oligocene and Miocene. In addition, there were two separate immigration events of palaeomerycids to North America, forming a great radiation starting in the late Arikareean, peaking in the middle Miocene (Barstovian), and then declining to eventual extinction by the end of the Miocene (Hemphillian). These animals have been traditionally referred to a North American family, the Dromomerycidae, first recognized by Frick (1937) and placed in the European Palaeomerycidae by Stirton (1944). However, Prothero and Liter (in press) followed Janis and Scott's (1987) placement of the group as two separate lineages originating with the Eurasian Palaeomerycidae. Also, the palaeomerycids were grossly oversplit by Frick (1937), with many taxa having no diagnosis or description and apparently based on geographic locality rather than morphology. Prothero and Liter (in press) reduced Frick's (1937) 50 mostly invalid species to just 17 species in 11 genera (most of which are now monotypic), and no subgenera were recognized. This taxonomic work is reviewed by Prothero and Liter (this volume).

Like the antilocaprids, cervids, and bovids, the palaeomerycids evolved a wide variety of cranial appendages, although they were apparently made of bone with just a covering of skin, like giraffe ossicones, and were neither deciduous antlers as those in cervids, nor bone covered with keratin, as in bovids and antilocaprids. This complicates hypotheses of their relationships because the cranial appendages are largely nonhomologous with those in other horned ruminants. Frick (1937) placed the group within the Cervidae, but Janis and Scott (1987, 1988) and Janis and Manning (1998a) argued that they are a sister group of the cervids. Stirton (1944), on the other hand, thought they were closer to the Giraffidae, and this assignment was followed by some other authors (Crusafont, 1961; Viret, 1961; Hamilton, 1978b).

Whatever their relationships, by the late Miocene they had declined to one lineage (*Pediomeryx–Yumaceras*), and they vanished entirely at the end of the Miocene (latest Hemphillian), about the time that true cervids immigrated into North America. Janis and Manning (1998a) rule out di-rect competition between the groups because palaeomerycids were declining long before cervids appeared, and many of the late Miocene North American families (rhinocerotids, protoceratids, moschids, oreodonts, mylagaulids) also vanished from North America at the end of the Hemphillian, largely because of climate change rather than competition from similar ecomorphs.

Moving from deerlike forms to true deer, Groves (this volume) provides an updated review of the family Cervidae, the deer found on nearly every continent around the world. Most authors (e.g., Janis and Scott, 1987, 1988; Gentry and Hooker, 1988; Hernández Fernández and Vrba, 2005) place deer, palaeomerycids, and moschids in close relationship, but Hassanin and Douzery (2003) found that the molecular data support a close relationship of moschids with giraffids, not with cervids. As summarized by Janis and Scott (1987) and Groves (this volume), the earliest relatives of the Cervidae are probably the early Miocene genus *Lagomeryx* from Eurasia and *Procervulus dichotomus* from the early Miocene of Europe. Groves (this volume) summarizes the current status of the systematics of the many genera and species of living cervids and outlines the split between two clusters: the muntjacs and cervines (Plesiometacarpalia) and the odocoileines and hydropotines (Telemetacarpalia).

Like tragulids and moschids, giraffids comprise another family that has only a few living representatives (two species, the giraffe and okapi) but a long and diverse history in the Old World through most of the Neogene. Unlike most other groups of ruminants, giraffids have solid ossicones with just a covering of skin (possibly similar to those in palaeomerycids), rather than horns (found in bovids) or fully deciduous antlers (found in cervids) or partially deciduous horns (found in antilocaprids). However, as Solounias (this volume) points out, many of the fossil giraffids had bony horns growing out of the frontal bone rather than ossicones. Giraffids were widespread all over Eurasia and Africa by the early Miocene and moderately diverse (typically three or more taxa per locality) through much of the Miocene. The earliest known giraffids are the primitive families Climacoceridae and Canthumerycinae from the early Miocene, and by 16 Ma taxa such as *Georgiomeryx, Canthumeryx, Injanatherium,* and "*Palaeotragus*" *tungurensis* were widespread over Africa and Eurasia. Giraffid diversity was much lower at 14 Ma and known mainly from Africa and Pakistan ("*Palaeotragus*" *primaevus* and *Giraffokeryx*). In the late Miocene (7–9 Ma), giraffids became very diverse again, especially in Europe and Asia, although the late Miocene fossil record from Africa is poor, so giraffids may be underrepresented. By the Pliocene and Pleistocene, *Sivatherium* was widespread over Eurasia and Africa. Today giraffids are restricted to two relict species surviving in Africa, although they once roamed the entire Old World.

Solounias (this volume) provides a cladogram of over two dozen other extinct giraffid taxa. He regards the Palaeomerycidae as their sister group and places giraffids near the base of the ruminant radiation, along with tragulids. As discussed above, other authors have had varied interpretations of how giraffids fit into the ruminants. Pilgrim (1941), Stirton (1944), Simpson (1945), Crusafont-Pairó (1952), Viret (1961), Ginsburg and Heintz (1966), Hamilton (1973, 1978b), Webb and Taylor (1980), and Leinders (1983) clustered them with the cervids (usually including the palaeomerycids), as did Janis (1987; Janis and Scott, 1987, 1988). However, Gentry and Hooker (1988) placed giraffids as sister taxon to the bovids plus antilocaprids, with cervids, moschids, and palaeomerycids as more distant outgroups. Using molecular data, Hassanin and Douzery (2003) clustered them with antilocaprids as sister group to the cervid–bovid clade, as did Hernandez Fernández and Vrba (2005) using combined molecular-morphological-ethological data. Summarizing all the molecular phylogenies, however, Marcot (this volume) found that giraffids came out as a sister taxon to the cervid–moschid–bovid clade, and antilocaprids were a more distant outgroup. Clearly, much additional work is needed to reach consensus on this issue. As noted by Marcot (this volume), the difficulty with molecular analyses of both giraffids and antilocaprids is the low diversity of living species, resulting in long-branch attraction.

By far the largest family of artiodactyls is the Bovidae. At least 150 genera are recognized, of which 47 genera with about 158 species are still living (Nowak, 1999; McKenna and Bell, 1997). They show a wide range of body sizes from huge water buffaloes to tiny dik-diks and occupy nearly every habitat on earth. They have a number of distinguishing features, but the most diagnostic is their horns, which have a solid bony core covered by a nondeciduous keratinous sheath. Bovid diversity includes not only the more familiar varieties of cattle and oxen and buffaloes and bison but also the huge radiation of sheep and goats, many different families of true antelopes, and such tiny forms as dik-diks and duikers.

Solounias (this volume) provides a brief overview of this huge family. His chapter gives an extensive discussion of the key anatomical characteristics. The origin of bovids is still obscure. There are some Oligocene ruminants that have some bovid characteristics, such as *Andagameryx, Palaeohypsodontus, Hanhaicerus,* and *Hispanomeryx,* although these taxa are based on dental remains and lack any horn cores that would definitely assign them to the Bovidae. The earliest undisputed bovid is the early Miocene (18 Ma) *Eotragus,* which is found in Europe, Africa, the Middle East, and Pakistan. By the middle Miocene, bovids had diversified all over the Old World. However, they did not arrive in the New World until the Pleistocene with the immigration of bison and several other forms over the Bering Land Bridge.

Solounias (this volume) follows Gentry's (2000a) classification scheme of bovids into Boodontia (the primitive boselaphines, such as the nilgai, and the bovines, or cattle), and Aegodontia. Within the Aegodontia are the Antilopini (the true antelopes) and some Neotragini, and another cluster of the Caprinae (goats and sheep) plus the Alcelaphini (wildebeests and hartebeests). Using molecular data, Hassanin and Douzery (2003) also clustered the Bovini, Tragelaphini, and

Boselaphini in the Bovinae, the Alcelaphini and Hippotragini (sable antelopes) with the Caprinae, and all the rest of the tribes (Aepycerotini, Antilopini, Cephalophini, and Reduncini) in an unresolved polytomy. Hernández Fernández and Vrba (2005) used a combined morphological–molecular–behavioral approach and found that the Bovinae were the most primitive group within the Bovidae, followed by several clusters in an unresolved polytomy: Antilopinae plus Cephalophinae plus Reduncinae; *Oreotragus* and *Neotragus;* and the Aepycerotinae, Alcelaphinae, Hippotraginae, Pantholopinae, and Caprinae. Finally, Marcot (this volume) provides a consensus of all the different molecular analyses performed to date. The Bovinae (Bovini, Tragelaphini, and Boselaphini) once again emerge as one clade, and all the rest of the bovids are in the other. The most primitive clade in the nonbovine branch is the impalas (Aepycerotini), followed by a clade of Reduncini plus Cephalophini. The other major branch is the clade of Antilopini on one side, and a cluster of Hippotragini, Alcelphini, and Caprinae on the other. Thus, there is considerable consistency and congruence among all these approaches, and although there are still minor differences, these do not obscure the overall pattern of bovid diversification.

PALEOECOLOGICAL TRENDS IN ARTIODACTYLS

Finally, Janis (this volume) provides a paleoecological overview of artiodactyl evolution. She outlines a number of overarching features of their history. First, she reviews the well-known story of the replacement of perissodactyls (dominant in the middle Eocene) by artiodactyls (dominant ever since). She suggests that the development of selenodont teeth seen in the late middle Eocene and late Eocene artiodactyls also coincided with some sort of foregut fermentation in those groups, giving them an advantage in digesting vegetation. Meanwhile, the dramatic climatic change and loss of forests during the late Eocene and Oligocene took away much of the leaf- and fruit-based diets for many bunodont and lophodont herbivores, which vanished by the Oligocene, leaving the world to more specialized feeders such as ruminating artiodactyls and perissodactyls. In addition, the trend toward less carbon dioxide in the atmosphere meant that plants were more likely to produce higher quality but lower quantity vegetation, a condition that favors ruminants over hindgut-fermenting perissodactyls such as horses and rhinos.

Her second point concerns the radiation of omnivorous artiodactyls. She points out that although many of the early ungulates were bunodont, their descendants quickly developed into more specialized teeth, and the primitive bunodont taxa vanished by the late Eocene. In the artiodactyls, however, there has been a considerable radiation of bunodont forms (especially pigs and peccaries, but also hippos, anthracotheres, entelodonts, cebochoerids, and choeropotamids), and this radiation occurred when most archaic bunodont forms vanished. The "suiform" artiodactyls are the only group that still occupies this bunodont-omnivorous niche and have flourished despite their "primitive" dentitions. Janis (this volume) posits that no other group could reoccupy the bunodont omnivore niche because even the earliest perissodactyls were already more specialized toward lophodonty, and it is functionally unlikely that this kind of tooth could revert to bunodonty. This is another example of how artiodactyls have managed to outcompete perissodactyls in occupying niches that perissodactyls never did occupy and could not evolve to fill.

Janis (this volume) also speculates on the lack of ruminant megaherbivores. Megaherbivores are defined as weighing at least 1,000 kg (Owen-Smith, 1988). Only nonruminant hippos and possibly giraffes fall in this category among artiodactyls. By contrast, many species of rhinos, brontotheres, and proboscideans reached this size. Janis cites work by Clauss (Clauss et al., 2003; Clauss and Hummel, 2005) that might explain this. Rumination works best for smaller body sizes because the slow digestion delays the passage of food and maximizes cellulose breakdown. However, with very large body sizes, the digestion becomes even slower and thus gives no advantage to a megaherbivore to ruminate. In fact, there are some interesting arguments that suggest that a ruminating digestive tract is disadvantageous to a megaherbivore because it creates byproducts that larger animals would have trouble removing.

The radiation of Neogene artiodactyls (primarily pecorans) around the world is a topic that has fascinated many paleontologists. It coincided with the decline of many other artiodactyl groups that had dominated in the Oligocene, especially oreodonts, protoceratids, cainotheres, entelodonts, and anthracotheres. Most of the pecorans were more hypsodont than the other groups and tended to have more cursorial adaptations. This also coincided with the well-documented trends toward more open habitats and greater aridity by the early Miocene. It is not necessarily correlated with the first appearance of grasslands, which the phyloliths show occurred by the late Oligocene (Strömberg, 2006), or the spread of modern C4 grasslands, which did not occur until 7 Ma in the latest Miocene. Janis (this volume) has no clear answer to why these groups became dominant in the absence of clear evidence of vegetational change, although

the interval at 18 Ma is very interesting because it marks the beginning of the warm early-middle Miocene climatic optimum and also a period of tremendous faunal interchange among Africa, Eurasia, and North America. This may have promoted competition, diversification, and some extinction among ungulates, although it is clearly not the end of the story.

Finally, Janis (this volume) discussed the trends within the ruminants during the later Neogene. In the middle and late Miocene, there was still a great diversity of ruminant families, including giraffids, antilocaprids, moschids, camelids, and palaeomerycids, with bovids and cervids only beginning to diversify in the Old World. Most of these other families declined through the late Miocene and Pliocene, while bovids began their enormous radiation. There is also a parallel trend, where most of the lineages of browsing and mixed-feeding horses, rhinos, and proboscideans began their decline, and only the hypsodont obligate grazers in these groups (elasmotherine and dicerotine rhinos, elephantids) flourished. Janis (this volume) suggests that some of this may be a result of the rise of modern C4 grasslands in the Miocene, and the spread of these grasslands through the Pliocene, as many of the modern lineages of grazing bovids evolved at that time. Interestingly, even though cervids are also hypsodont, they have not exploited the grazing niche to the extent that bovids have; most are primarily browsers in more brushy and forested habitats. Janis (this volume) speculates that the cervids never occupied Sub-Saharan Africa, which has been dominated by hypsodont grazing bovids since the late Miocene, and provides much of the diversity of living bovids. By contrast, in tropical areas such as Southeast Asia, bovids and cervids are about equally diverse. Janis (this volume) does not have a complete explanation for why bovids were so successful at dominating the grazing niche but leaves this question open for further research.

ACKNOWLEDGMENTS

We thank C. Janis, J. Theodor, and J. Geisler for helpful review comments on this chapter.

REFERENCES

Abusch-Siewert, S. 1989. Bemerkungen zu den Anoplotherien (Artiodactyla, Mammalia) der Pariser Gipse. Münchner Geowissenschaftliche Abhandlungen, Reihe A 15:55–78.

Adachi, J., and M. Hasegawa. 1996. Instability of quartet analyses of molecular sequence data by the maximum likelihood method: the Cetacea/Artiodactyla relationships. Molecular Phylogenetics and Evolution 6:72–76.

Adams, C. G., R. H. Benson, R. B. Kidd, W. B. F. Ryan, and R. C. Wright. 1977. The Messinian salinity crisis and evidence of late Miocene eustatic changes in the world ocean. Nature 269:383–386.

Adkins, R. M., R. L. Honeycutt, and T. R. Disotell. 1996. Evolution of eutherian cytochrome *c* oxidase subunit II: heterogeneous rates of protein evolution and altered interaction with cytochrome *c*. Molecular Biology and Evolution 13:1393–1404.

Aguirre, E. 1963. *Hippopotamus crusafonti* n. sp. del Plioceno inferior de Arenas del Rey (Granada). Notas y Comunicaciones del Instituto Geologico y Minero de Espana 69:215–230.

Aguirre, E., F. Robles, L. Thaler, N. Lopez, M. T. Alberdi, and C. Fuentes. 1973. Venta del Moro, nueva fauna finimiocena de moluscos y vertebrados. Estudios Geologicos 29:569–578.

Agungpriyono, S., Y. Yamamoto, N. Kitamura, J. Jamada, K. Sigit, and T. Yamashita. 1992. Morphological study on the stomach of the Lesser Mouse Deer (*Tragulus javanicus*) with special reference to the internal surface. Journal of Veterinarian Medicine Science 54:1063–1069.

Agustí, J., and M. Anton. 2002. Mammoths, Sabertooths, and Hominids: 65 Million Years of Mammalian Evolution in Europe. Columbia University Press, New York, New York.

Agusti, J., L. Cabrera, M. Garces, W. Krijgsman, O. Oms, and J. M. Pares. 2001. A calibrated mammal scale for the Neogene of Western Europe. State of the art. Earth Science Reviews 52:247–260.

Alexeyev, A. (Alexjew). 1915. Animaux fossiles du village Novo-Elisavetovka. Odessa.

Allard, M. W., M. M. Miyamoto, L. Jarecki, F. Kraus, and M. R. Tennant. 1992. DNA systematics and evolution of the artiodactyl family Bovidae. Proceedings of the National Academy of Sciences, USA 89:3972–3976.

Allen, G. M. 1926. Fossil animals from South Carolina. Bulletin of the Museum of Comparative Zoology, Harvard University 67:447–467.

Alroy, J., P. L. Koch, and J. C. Zachos. 2000. Global climate change and North American mammalian evolution. Paleobiology 26 (suppl. to no. 4):259–288.

Alston, E. R. 1876. Zoological record for 1876, Zoological Society of London, London.

Alston, E. R. 1878. Mammalia. Zoological Record, Zoological Society of London, London 13:1–24.

Amato, G., M. G. Egan, and G. B. Schaller. 2000. Mitochondrial DNA variation in muntjac: evidence for discovery, rediscovery, and phylogenetic relationships; pp. 285–295 in E. S. Vrba and G. B. Schaller (eds.), Antelopes, Deer and Relatives. Yale University Press, New Haven, Connecticut, and London.

Amrine-Madsen, H., K.-P. Koepfli, R. K. Wayne, and M. S. Springer. 2003. A new phylogenetic marker, apolipoprotein B, provides compelling evidence for eutherian relationships. Molecular Phylogenetics and Evolution 28:225–240.

An, Z., J. E. Kutzbach, W. L. Prell, and C. Porter. 2001. Evolution of Asian monsoons and phased uplift of the Himalaya–Tibetan plateau since late Miocene times. Nature 411:62–66.

Andrews, C. W. 1899. Fossil mammals from Egypt. Geological Magazine 4:481–484

Andrews, C. W. 1902. Note on a Pliocene vertebrate fauna from the Wadi-Natrun, Egypt. Geological Magazine 9:432–439.

Andrews, C. W. 1906. A Descriptive Catalogue of the Tertiary Vertebrata of the Fayum, Egypt. Based on the Collection of the Egyptian Government in the Geological Museum, Cairo, and on the Collection in the British Museum (Natural History). British Museum (Natural History), London, 324 pp.

Andrews, C. W. 1914. On the lower Miocene vertebrates from British East Africa, collected by Dr. Felix Oswald. Quarterly Journal of the Geological Society, London 70:163–186.

Andrews, C. W., and H. J. L. Beadnell. 1902. A preliminary note of some new mammals from the upper Eocene of Egypt. Egypt Survey Department Publications, Works Mines 1:1–9.

Antunes, M. T., M. L. Casanovas, C. L. Cuesta, J. V. Santafé, and J. Agustí. 1997. Eocene mammals from Iberian Peninsula. Mémoires et travaux de l'Institut de Montpellier 21:337–352.

Arambourg, C. 1933. Mammifères miocènes du Turkana, Afrique orientale. Annales de Paléontologie 22:121–148.

Arambourg, C. 1944a. Les Hippopotames fossiles d'Afrique. Comptes Rendus de l'Académie des Sciences 218:602–604.

Arambourg, C. 1944b. Au sujet de l'*Hippopotamus hipponensis* Gaudry. Bulletin de la Société Géologique de France 14:147–153.

Arambourg, C. 1947. Mission Scientifique de l'Omo 1932–1933. Editions du Muséum, Paris, pp. 314–364.

Arambourg, C. 1963. Le genre *Bunolistriodon* Arambourg, 1933. Bulletin Société Géologique de France 4:107–109.

Arambourg, C., and J. Piveteau. 1929. Les Vertébrés du Pontien de Salonique. Annales de Paléontologie 18:57–140.

Arnason, U., and A. Janke. 2002. Mitogenomic analyses of eutherian relationships. Cytogenetic and Genome Research 96:20–32.

Arnason, U., A. Gullberg, and A. Janke. 2004. Mitogenomic analyses provide new insights into cetacean origin and evolution. Gene 333:27–34.

Arnason, U., A. Gullberg, S. Gretarsdottir, B. M. Ursing, and A. Janke. 2000. The mitochondrial genome of the sperm whale and a new molecular reference for estimating eutherian divergence dates. Journal of Molecular Evolution 50:569–578.

Astibia, H., and J. Morales. 1987. *Triceromeryx turiasonensis* nov. sp. (Palaeomerycidae, Artiodactyla, Mammalia) de Aragoniense medio de la cuenca del Ebro (Espagna). Palaeontologia i Evolució 21:75–115.

Astibia, H., J. Morales, and S. Moyà-Solà. 1998. *Tauromeryx,* a new genus of Palaeomerycidae (Artiodactyla, Mammalia) form the Miocene of Tarazona de Aragon (Ebro Basin, Aragon, Spain). Bulletin de la Societe Geologique de France 169:471–477.

Astibia, H., A. Aramburu, X. Pereda-Suberbiola, X. Murelaga, C. Sesé, M. A. Cuesta, S. Moyà-Solà, J. I. Baceta, A. Badiola, and M. Köhler. 2000. Un nouveau gisement à vertébrés continentaux de l'Èocène supérieur de Zambrana (Bassin de Miranda Treviño, Alava, Pays Basque). Géobios 33:233–248.

Astruc, J. G., M. Hugueney, G. Escarguel, S. Legendre, J.-C. Rage, R. Simon-Coincon, J. Sudre, and B. Sigé. 2003. Puycelci, nouveau site à vertébrés de la série molassique d'Aquitaine. Densité et continuité biochronologique dans la zone Quercy et bassins périphériques au Paléogène. Géobios 36:629–648.

Aubekerova, P. A. 1969. Novyy pedstavitel' semeystva Entelodontidae [New representative of the family Entelodontidae]. Izvestiya Akademiya Nauk Kazakhskoy SSR Seriya Biologicheskiy 4:47–52.

Aung Naing Soe. 2004. Géologie et paléontologie dans la Formation de Pondaung (Myanmar). Unpublished Ph.D. dissertation, Université Montpellier II, France, 304 pp.

Averianov, A. O. 1996. Artiodactyla from the early Eocene of Kyrgyzstan. Palaeovertebrata 25:359–369.

Axelrod, D. I. 1958. Evolution of the Madro-Tertiary Geoflora. Botanical Review 24:433–509.

Aymard, A. 1846 (1848). Essai monographique sur un nouveau genre de mammifère fossile trouvé dans la Haute-Loire, et nommé Entelodon. Annales de la Société d'Agriculture, Sciences, Arts et Commerce du Puy 12:227–267.

Aymard, A. 1855. Rapport sur les collections de M. Pichot-Dumazel. Congrégation scientifique de France 22:227–245.

Azanza, B. 1993a. Sur la nature des appendices frontaux des cervidés (Artiodacyla, Mammalia) du Miocène inférieur et moyen. Remarques sur leur systématique et leur phylogénie. Comptes Rendus de l'Académie des Sciences, Paris, II 316:1163–1169.

Azanza, B. 1993b. Sytématique et évolution du genre *Procervulus,* cervidé (Artiodacyla, Mammalia) du Miocène inférieur d'Europe. Comptes Rendus de l'Académie des Sciences Paris, II 316:717–723.

Azanza, B., and L. Ginsburg. 1997. A revision of the large lagomerycid artiodactyls of Europe. Palaeontology 40:461–485.

Azanza, B., and E. Menendez. 1990. Los ciervos fósiles del neógeno español. Paleontologia i Evolució 23:75–82.

Azanza, B., and J. Morales. 1994. *Tethytragus* nov. gen. et *Gentrytragus* nov. gen. Deux nouveaux Bovidés (Artiodactyla, Mammalia) du Miocène moyen. Proceedings Koninklijke Nederlandse Akademie van Wetenschappen Amsterdam 97:249–282.

Azanza, B., and B. Sanchez. 1990. Les cervidés du Pléistocène moyen d'Atapuerca (Burgos, Espagne). Quaternaire 3–4: 197–212.

Azanza, B., E. Menendez, and L. Alcalá. 1989. The middle-Upper Turolian and Ruscinian Cervidae in Spain. Bolletin Societa Paleontologia Italiana 28:171–182.

Azzaroli, A. 1981. On the Quaternary and recent cervid genera *Alces, Cervalces, Libralaces.* Bolletin Societa Paleontologia Italiana 20:147–154.

Azzaroli, A. 1982. Insularity and its effects on terrestrial vertebrates: Evolutionary and biogeographic aspects; pp. 193–213 in E. M. Gallitelli (ed.), Palaeontology, Essential of Historical Geology. Paleontologia come scienze geostorica. S.T.E.M. Mucchi, Modena, Italy.

Azzaroli, A. 1985. Taxonomy of Quarternary Alcini (Cervidae, Mammalia). Acta Zoologica Fennica 170:179–180.

Azzaroli, A. 1992. The cervid genus *Pseudodama* n.g. in the Villafranchian of Tuscany. Palaeontographica Italica 79: 1–41.

Azzaroli, A. 1994. Forest bed elks and giant deer revisited. Zoological Journal of the Linnaean Society 112:119–133.

Azzaroli, A., and P. Mazza. 1992a. On the possible origin of the Giant Deer genus *Megaceroides.* Rendiconti Fisiche Accademia dei Lincei 3:23–32.

Azzaroli, A., and P. Mazza. 1992b. The cervid genus *Eucladoceros* in the early Pleistocene of Tuscany. Palaeontographia Italica 79:43–100.

Azzaroli, A., and P. Mazza. 1993. Large early Pleistocene deer from Pietrafitta lignite mine, Central Italy. Palaeontographia Italica 80:1–24.

Azzaroli, A., C. De Giuli, G. Ficcarelli, and D. Torre. 1982. Table of the stratigraphic distribution of terrestrial mammalian faunas in Italy from the Pliocene to the early middle Pleistocene. Geographia Fisiche Dinamico Quaternario 5:55–58.

Bajpai, S., V. V. Kapur, D. P. Das, B. N. Tiwari, N. Saravanan, and R. Sharma. 2005. Early Eocene land mammals from Vastan lignite mine, District Surat (Gujarat), Western India. Journal of the Palaeontological Society of India 50:101–113.

Bakalov, P., and I. Nikolov. 1962. [Tertiary Mammals.] [Fossils of Bulgaria] 10:1–162. Bulgarian Academy of Science, Sofia. [Bulgarian].

Balakrishnan, C. N., S. L. Montfort, A. Gaur, L. Singh, and M. D. Sorenson. 2003. Phylogeography and conservation genetics of Eld's deer (*Cervus eldi*). Molecular Ecology 12:1–10.

Barbour, E. H. 1905. A new Miocene artiodactyl. Science 22: 797–798.

Barbour, E. H., and C. B. Schultz. 1934. A new antilocaprid and a new cervid from the late Tertiary of Nebraska. American Museum Novitates 273:1–4.

Barghoorn, S. F. 1985. Magnetic Polarity Stratigraphy of the Tesuque Formation, Santa Fe Group in the Española Valley, New Mexico, with a Taxonomic Review of the Fossil Camels. Ph.D. Dissertation, Columbia University, New York, New York, 430 pp.

Barnes, L. G. 1984. Whales, dolphins, and porpoises: origin and evolution of the Cetacea, in T. W. Broadhead (ed.), Mammals. Notes for a Short Course. University of Tennessee Department of Geology. Studies in Geology 8:139–154.

Barry, J. C., S. Cote, L. MacLatchy, E. H. Lindsay, R. Kityo, and A. R. Rajpar. 2005. Oligocene and early Miocene ruminants (Mammalia, Artiodactyla) from Pakistan and Ungunda. Palaeontologia Electronica 8:1–29.

Barry, J. C., M. L. E. Morgan, L. J. Flynn, D. Pilbeam, A. K. Behrensmeyer, S. M. Raza, I. A. Khan, C. Badgley, J. Hicks, and J. Kelley. 2002. Faunal and environmental change in the late Miocene Siwaliks of northern Pakistan. Paleobiology 18(suppl. 2):1–71.

Baudelot, S., and F. C. Crouzel. 1974. La faune burdigalienne des gisements d'Espira-du-Conflent (Pyrénées-Orientales). Bulletin de la Société d'Histoire naturelle de Toulouse 110:311–326.

Beard, K. C. 1998. East of Eden: Asia as an important center of taxonomic origination in mammalian evolution. Bulletin of Carnegie Museum of Natural History 34:5–39.

Beard, K. C., and M. R. Dawson. 1999. Intercontinental dispersal of Holarctic land mammals near the Paleocene / Eocene boundary: paleogeographic, paleoclimatic and biostratigraphic implications. Bulletin de la Société Géologique de France 170:697–706.

Beard, K. C., and B. Wang. 1991. Phylogenetic and biogeographic significance of the tarsiiform primate *Asiomomys changbaicus* from the Eocene of Jilin Province, People's Republic of China. American Journal of Physical Anthropology 85:159–166.

Beaumont, G. D. 1963. Deux importants restes d'Anoplotheriidae (Artiodactyla) des phosphorites du Quercy. Eclogae Geologicae Helvetiae 56:1169–1178.

Becker, D., G. E. Rössner, L. Picot, and J.-P. Berger. 2001. Early Miocene ruminants of Wallenried (USM, Aquitanian / Switzerland): sedimentology, biostratigraphy, and paleoecology. Eclogae Geologica Helvetica 94:547–564.

Behrensmeyer, A. K., A. L. Deino, A. Hill, J. D. Kingston, and J. J. Saunders. 2002. Geology and geochronology of the middle Miocene Kipsaramon site complex, Muruyur Beds, Tugen Hills, Kenya. Journal of Human Evolution 42: 11–38.

Bell, C. J., E. L. Lundelius, Jr., A. D. Barnosky, R. W. Graham, E. H. Lindsay, D. R. Ruez, Jr., H. A. Semken, Jr., S. D. Webb, and R. J. Zakrzewski. 2004. The Blancan, Irvingtonian, and Rancholabrean mammal ages; pp. 232–314 in M. O. Woodburne (ed.), Late Cretaceous and Cenozoic Mammals of North America: Biostratigraphy and Geochronology. Columbia University Press, New York, New York.

Bemmel, A. C. V. van. 1949. On the meaning of the name *Cervus javanicus* Osbeck 1765 (Tragulidae). Treubia 20:378–380.

Benton, M. J., M. A. Wills, and R. Hitchin. 2000. Quality of the fossil record through time. Nature 403:534–537.

Berger, F. E. 1959. Untersuchungen an Schädel- und Gebißresten von Cainotheriidae, besonders aus den oberoligocaenen Spaltenfüllungen von Gaimersheim bei Ingolstadt. Palaeontographica Abteilung A 112:1–58.

Berger, J.-P., B. Reichenbacher, D. Becker, M. Grimm, K. Grimm, L. Picot, A. Stornia, C. Pirkenseer, and A. Schäfer. 2005. Eocene–Pliocene time scale and stratigraphy of the Upper Rhine Graben (URG) and the Swiss Molasse Basin (SMB). International Journal of Earth Sciences on line.

Berggren, W. A., and D. R. Prothero. 1992. Eocene–Oligocene Climatic and Biotic Evolution: an Overview; in D. R. Prothero and W. A. Berggren (eds.), Eocene–Oligocene Climatic and Biotic Evolution. Princeton University Press, Princeton, New Jersey, p. 1–28.

Berggren, W. A., D. V. Kent, C. C. Swisher III, and M.-P. Aubry. 1995. A revised Cenozoic geochronology and chronostratigraphy. SEPM Special Publication 54:129–212.

Bergh, G. D. V. D., J. D. Vos, and P. Y. Sondaar. 2001. The late Quaternary palaeogeography of mammal evolution in the Indonesian Archipelago. Palaeogeography, Palaeoclimatology, Palaeoecology 171:385–408.

Bernard, J. B., S. R. DeBar, D. E. Ullrey, B. J. Schoeberl, J. Stromberg, and P. L. Wolff. 1994. Fiber utilization in the larger Malayan chevrotain (*Tragulus napu*). Proceedings of the American Association of Zoo Veterinarians 1994: 354–357.

Bernor, R. L., S. Bi, and J. Radovcic. 2004. A contribution to the evolutionary biology of *Conohyus olujici* n. sp. (Mammalia, Suidae, Tetraconodontinae) from the early Miocene of Luane, Croatia. Geodiversitas 26:509–534.

Bertram, B. C. R., and A. A. Biewener. 1990. Differential scaling of the long bones in the terrestrial Carnivora and other mammals. Journal of Morphology 20:157–169.

Bi, S., Y. Shen, D. Zhu, C. Zhu, L. Jia, and Z. Zhang. 1985. Study on ultrastructure of musk gland and musk secretion. Acta Theriologica Sinica 5:81–85.

Bininda-Emonds, O. R. P. 2004. The evolution of supertrees. Trends in Ecology and Evolution 19:315–322.

Biochrom. 1997. Synthèses et tableaux de corrélations; pp. 769–805 in J.-P. Aguilar, S. Legendre, and J. Michaux (eds.), Actes du Congrès BiochroM'97. Mémoires et Travaux de E.P.H.E. l'Institut de Montpellier, Montpellier.

Birungi, J., and P. Arctander. 2001. Molecular systematics and phylogeny of the Reduncini (Artiodactyla: Bovidae) inferred from the analysis of mitochondrial cytochrome *b* gene sequences. Journal of Mammalian Evolution 8:125–147.

Black, C. C. 1972. A new species of *Merycopotamus* (Artiodactyla: Anthracotheriidae) from the late Miocene of Tunisia. Notes du Service géologique de Tunisie 37:5–39.

Black, C. C. 1978a. Anthracotheriidae; pp. 423–434 in V. J. Maglio and H. B. S. Cooke (eds.), Evolution of African Mammals. Harvard University Press, Cambridge, Massachusetts.

Black, C. C. 1978b. Paleontology and geology of the Badwater Creek area, Central Wyoming, Part 14: The Artiodactyls. Annals of the Carnegie Museum 47:223–259.

Black, C. C., and M. R. Dawson. 1966. A review of late Eocene mammalian faunas from North America. American Journal of Science, 264:321–349.

Blainville, H. M. D. de. 1816. Prodrome d'une nouvelle distribution systematique du regne animal. Bulletin des Sciences de Societe Philomatique, Paris, series 3, 3:105–124.

Blainville, H. M. D. de. 1839. Recherches sur l'anciennete des edentes terrestres a la surface de la terre. Comptes Rendus Academie des Sciences, Paris, 8:65–69.

Blainville, H. M. D. de. 1849. Osteographie, ou description iconographique composée du squelette et du système dentaire de mammifères récents et fossiles pour servir de base à la zoologie ou à la geologie. Bailliere, Paris.

Bloch, J. I., D. C. Fisher, K. D. Rose, and P. D. Gingerich. 2001. Stratocladistic analysis of Paleocene Carpolestidae (Mammalia, Plesiadapiformes) with description of a new late Tiffanian genus. Journal of Vertebrate Paleontology 21: 119–131.

Blondel, C. 1997. Les ruminants de Pech Desse et de Pech du Fraysse (Quercy; MP 28); évolution des ruminants de l'Oligocène d'Europe. Géobios 30:573–591.

Blondel, C. 2005. New data on the Cainotheriidae (Mammalia, Artiodactyla) from the early Oligocene of south western France. Zoological Journal of the Linnean Society 144:145–166.

Blumenbach, J. F. 1779. Handbuch der Naturgeschichte. Johann Christian Dieterich, Göttingen.

Bobe, R., and Eck, G. G. 2001. Responses of African bovids to Pliocene climatic change. Paleobiology 27:1–47.

Bodenbender, B. E., and D. C. Fisher, 2001. Stratocladistic analysis of blastoid phylogeny. Journal of Paleontology 75:351–369.

Boekschoten, G. J., and P. Y. Sondaar. 1966. The Pleistocene of the Katharo Basin (Crete) and its hippopotamus. Bijdragen tot de Dierkunde, Aflevering 36(8):17–44.

Boekschoten, G. J., and P. Y. Sondaar. 1972. On the fossil Mammalia of Cyprus. Proceedings of the Koninklijke Nederlandse Akademie van Wetenschappen 75:306–338.

Bofill y Poch, A. 1897. Nota sobre la presencia del *Ancodus aymardi* en los lignitos de Calaf, provincia de Barcelona; su significación bajo los puntos de vista paleontológico y estratigráfico. Boletín de la Real Academia de Ciencias y Artes de Barcelona 1:332–337.

Bohlin, B. 1926. Die Familie Giraffidae. Palaeontologica Sinica Peking C 4:1–179.

Bohlin, B. 1935a. Cavicornier der *Hipparion*-fauna Nord-Chinas. Palaeontologia Sinica C 9:1–166.

Bohlin, B. 1935b. Kritische Bemerkungen über die Gatung *Tragocerus*. Nova Acta Regiae Society of Science of Uppsala Series 4 9:1–18.

Boisserie, J.-R. 2004. A new species of Hippopotamidae (Mammalia, Artiodactyla) from the Sagantole Formation, Middle Awash, Ethiopia. Bulletin de la Société Géologique de France 175:525–533.

Boisserie, J.-R. 2005. The phylogeny and taxonomy of Hippopotamidae (Mammalia: Artiodactyla): a review based on morphology and cladistic analysis. Zoological Journal of the Linnean Society 143:1–26.

Boisserie, J.-R. In press. Late Miocene Hippopotamidae from Lemudong'o, Kenya. Kirtlandia.

Boisserie, J.-R., and W. H. Gilbert. In press. Hippopotamidae; in W. H. Gilbert and B. Asfaw (eds.), Pleistocene Paleontology of the Daka Member, Middle Awash, Ethiopia. University of California Press, Berkeley, California.

Boisserie, J.-R., and Y. Haile-Selassie. In press. Hippopotamidae; in Y. Haile-Selassie and G. WoldeGabriel (eds.), Late Miocene Mammals of the Middle Awash, Ethiopia: Chronology, Paleoenvironment, and Evolution. University of California Press, Berkeley, California.

Boisserie, J.-R., and F. Lihoreau. 2006. Emergence of Hippopotamidae: new scenarios. Comptes Rendus Palévol 5: 749–756.

Boisserie, J.-R., and T. D. White. 2004. A new species of Pliocene Hippopotamidae from the Middle Awash, Ethiopia. Journal of Vertebrate Paleontology 24:464–473.

Boisserie, J.-R., M. Brunet, L. Andossa, and P. Vignaud. 2003. Hippopotamids from the Djurab Pliocene faunas, Chad, Central Africa. Journal of African Earth Sciences 36:15–27.

Boisserie, J.-R., F. Lihoreau, and M. Brunet. 2005a. Origins of Hippopotamidae (Mammalia, Cetartiodactyla): towards resolution. Zoologica Scripta 34:119–143.

Boisserie, J.-R., F. Lihoreau, and M. Brunet. 2005b. The position of Hippopotamidae within Cetartiodactyla. Proceedings of the National Academy of Sciences, USA 102:1537–1541.

Boisserie, J.-R., A. Likius, P. Vignaud, and M. Brunet. 2005c. A new late Miocene hippopotamid from Toros-Ménalla, Chad. Journal of Vertebrate Paleontology 25:665–673.

Boisserie, J.-R., A. Zazzo, G. Merceron, C. Blondel, P. Vignaud, A. Likius, H. T. Mackaye, and M. Brunet. 2005d. Diets of modern and late Miocene hippopotamids: evidence from carbon isotope composition and micro-wear of tooth enamel. Palaeogeography, Palaeoclimatology, Palaeoecology 221: 163–174.

Bonaparte, C. L. 1850. Conspectus systematis mastozoologiae (ornithologiae, etc.). Editio altera reformata; J. E. Brill, Leiden.

Bonarelli, G. 1947. Dinosauro fossile del Sahara Cirenaico. Rivista Biologica Colonial Roma 8:23–33.

Bonis, L. de. 1964. Étude de quelques mammifères du Ludien de la Débruge (Vaucluse). Annales de Palaeontologie 50:119–154.

Bonis, L. de, G. D. Koufos, and S. Sen. 1997. The sanitheres (Mammalia, Suoidea) from the middle Miocene of Chios Island, Aegean Sea, Greece. Revue de Paléobiologie Genève 16: 259–270.

Bonis, L. de, G. D. Koufos, and S. Sen. 1998. Ruminants (Bovidae and Tragulidae) from the middle Miocene (MN 5) of the island of Chios, Aegean Sea (Greece). Neues Jahrbuch für Geologie und Paläontologie Abhandlungen 210:399–420.

Borissiak, A. 1914. Les mammifères fossiles de Sebastopol. Mémoirs de Comite Geologique St. Petersburg Nouvelle Serie. 87:1–154.

Bouvrain, G. 1992. Antilopes à chevilles spiralées du Miocène supérieur de la province Gréco-Iranienne: nouvelles diagnoses. Annales de Paléontologie Paris 78:49–65.

Bouvrain, G. 1996. Les gazelles du Miocène supérieur de Macédoine, Grèce. Neues Jahrbuch für Geologie und Paläontologie Abhandlungen Stuttgart 199:111–132.

Bouvrain, G. 1997. Les gazelles du Miocène supérieur de Pentalophos (Macédoine Grèce). Münchener Geowissenschaftliche Abhandlungen A 34:5–22.

Bouvrain, G., D. Geraads, and Y. Jehenne. 1989. Nouvelles données relatives à la classification des Cervidae (Artiodactyla, Mammalia). Zoologischer Anzeiger 223:82–90.

Bouvrain, G., D. Geraads, and J. Sudre. 1986. Révision taxonomique de quelques ruminants oligocènes des phosphorites du Quercy. Comptes Rendus Hebdomadaire des Séances de l'Académie des Sciences, Paris. Série 2, 302:101–104.

Boyden, A., and D. Gemeroy. 1950. The relative position of Cetacea among the orders of Mammalia as indicated by precipitin tests. Zoologica 35:145–151.

Boyeskorov, G. 1999. New data on moose (*Alces,* Artiodactyla) systematics. Säugetierkundliche Mitteilung 44:4–14.

Bravard, A. 1828. Monographie de la montagne de Perrier près d'Issoire (Puy-de-Dome) et de deux espèces du genre *Felis,* découvert dans l'une de ces couches d'alluvion; Paris; pp. 1–147.

Bravard, A. 1835. Monographie du *Cainotherium,* nouveau genre fossile des Pachydermes, trouve dans les terrain tertiaires d'eau douce du departement du Puy-de-Dome; Paris; pp. 1–35.

Breda, M., and M. Marchetti. 2005. Systematical and biochronological review of Plio-Pleistocene Alceini (Cervidae: Mammalia) from Eurasia. Quaternary Science Reviews 24: 775–805.

Breukelman, H. J., P. A. Jekel, J.-Y. F. Dubois, P. P. M. F. A. Mulder, H. W. Warmels, and J. J. Beintema. 2001. Secretory ribonucleases in the primitive ruminant chevrotain (*Tragulus javanicus*). European Journal of Biochemistry 268: 3890–3897.

Bromage, T. G., W. Dirks, H. Erdjument, M. Huck, O. Kulmer, R. Öner, O. Sandrock, and F. Schrenk. 2001. Evolution and extinction of insular dwarfs: A North Cyprus case study. Journal of Morphology 248:211–212.

Brooke, V. 1878. On the classification of the Cervidae, with a synopsis of existing species. Proceedings of the Zoological Society of London 1878:883–928.

Brunet, M. M. 1973. Les Entelodontes des Phosphorites du Quercy. Palaeovertebrata. Montpellier 6:87–108.

Brunet, M. M. 1979. Les grands mammifères chefs de l'immigration Oligocène et le problème de la limite Eocène–Oligocène en Europe. [Large mammals, front rank of the Oligocene immigration and the problem of the Eocene–Oligocene boundary in Europe.] [French]. Editions de la Fondation Singer-Polignac, Paris, 281 pp.

Brunet, M., and Y. Jehenne. 1976. Un nouveau ruminant primitif des molasses oligocènes de l'Agenais. Bulletin de la Société Géologique de France 18:1659–1664.

Brunet, M., and Mission Paléoanthropologique Franco-Tchadienne. 2000. Chad: discovery of a vertebrate fauna close to the Mio-Pliocene boundary. Journal of Vertebrate Paleontology 20:205–209.

Brunet, M., and J. Sudre. 1980. Deux nouveaux dichobunidés (Artiodactyla, Mammalia) de l'Oligocène inférieur d'Europe. I, II. Proceedings of the Koninklijke Nederlandse Akademie Van Wetenschapen, Serie B 83:121–143.

Brunet, M., and J. Sudre. 1987. Evolution et systématique du genre *Lophiomeryx* Pomel 1853 (Mammalia, Artiodactyla). Münchner Geowissenschaftliche Abhandlungen 10:225–242.

Brunet, M., A. Beauvilain, D. Geraads, F. Guy, M. Kasser, H. T. Mackaye, L. M. Maclatchy, G. S. J. Mouchelin, and P. Vignaud. 1998. Tchad: découverte d'une faune de mammifères du Pliocène inférieur. Comptes Rendus de l'Académie des Sciences 326:153–158.

Bubenik, A. 1982. Taxonomy of the Pecora in relation to morphology of their cranial appendages; pp. 163–185 in R. Brown (ed.), Antler Development in the Cervidae. Caesar Kleberg Wildlife Research Institute, Kingsville, Texas.

Bubenik, A. B. 1990. Epigenetical, morphological, physiological, and behavioral aspects of evolution of horns, pronghorns, and antlers; pp. 3–113 in G. A. Bubenik and A. B. Bubenik (eds.), Horns, Pronghorns and Antlers. Springer-Verlag, New York, New York.

Buntjer, J. B., I. A. Hoff, and J. A. Lenstra. 1997. Artiodactyl interspersed DNA repeats in cetacean genomes. Journal of Molecular Evolution 45:66–69.

Burk, A., E. J. P. Douzery, and M. S. Springer. 2002. The secondary structure of mammalian mitochondrial 16S rRNA molecules: refinements based on a comparative phylogenetic approach. Journal of Mammalian Evolution 9:225–252.

Bush, L. P. 1903. Note on the dates of publication of certain genera of fossil vertebrates. American Journal of Science (ser. 4) 16:96–98.

Caloi, L., and M. R. Palombo. 1990. Gli erbivori pleistocenici delle isole del Mediteranneo: adattamenti. Hystrix 2: 87–100.

Caloi, L., and M. R. Palombo. 1994. Functional aspects and ecological implications in Pleistocene endemic herbivores of mediterranean islands. Historical Biology 8:151–172.

Caloi, L., and M. R. Palombo. 1996. Functional aspects and ecological implications in hippopotami and cervids of Crete; pp. 125–151 in D. S. Reese (ed.), Pleistocene and Holocene

Fauna of Crete and Its First Settlers. Monography in World Archaeology 28. Prehistory Press, Madison, Wisconsin.

Camp, C. L., and V. L. Van der Hoof. 1940. Bibliography of fossil vertebrates 1928–1933. Geological Society Special Publication 27:1–503.

Cap, H., S. Aulagnier, and P. Deleporte. 2002. The phylogeny and behaviour of Cervidae. Ethology, Ecology, and Evolution 14:199–216.

Capasso Barbato, L., and C. Petronio. 1983a. Considerazioni sistematiche e filogenetiche su *"Hippopotamus pentlandi"* Von Meyer, 1832. Atti della Societa italiana di Scienze naturali e del Museo civico di Storia naturale di Milano 124:229–248.

Capasso Barbato, L., and C. Petronio. 1983b. Considerazioni sistematiche e filogenetiche su *Hippopotamus melitensis* Major, 1902. Atti della Societa italiana di Scienze naturali e del Museo civico di Storia naturale di Milano 124:281–290.

Carlsson, A. 1926. Über die Tragulidae und ihre beziehungen zu den übrigen Artiodactyla. Acta Zoologica 6:69–99.

Caro, T. M., C. M. Graham, C. J. Stoner, and M. M. Flores. 2003. Correlates of horn and antler shape in bovids and cervids. Behavioral Ecology and Sociobiology 55:103–111.

Carr, S. M., S. W. Ballinger, J. N. Derr, L. H. Blankenship, and J. W. Bickham. 1986. Mitochondrial DNA analysis of hybridization between sympatric white-tailed deer and mule deer in west Texas. Proceedings of the National Academy of Sciences, USA 83:9576–9580.

Carroll, R. L. 1988. Vertebrate Paleontology and Evolution. W. H. Freeman and Company, New York, New York.

Casanovas, M. L. 1975a. *Choeropotamus sudrei* nva. sp. y algunas consideraciones sobre los restantes Artiodáctilos de Roc de Santa (prov. de Lérida). Geologica Acta 10:138–140.

Casanovas, M. L. 1975b. Estratigrafía y Paleontología del yacimiento ludiense de Roc de Santa (Área del Noguera-Pallaresa). Paleontologia i Evolució 10:1–158.

Cathey, J. C., J. W. Bickham, and J. C. Patton. 1998. Introgressive hybridization and nonconcordant evolutionary history of maternal and paternal lineages in North American deer. Evolution 52:1224–1229.

Cerling, T. E. 1992. Development of grasslands and savannas in East Africa during the Neogene. Palaeogeography, Palaeoclimatology, Palaeoecology 97:241–247.

Cerling, T. E., J. M. Harris, and M. G. Leakey. 2003. Isotope paleoecology of the Nawata and Nachukui Formations at Lothagam, Turkana Basin, Kenya; pp. 587–597 in J. M. Harris and M. G. Leakey (eds.), Lothagam. The Dawn of Humanity in Eastern Africa. Columbia University Press, New York, New York.

Cerling, T. E., J. M. Harris, S. H. Ambrose, M. G. Leakey, and N. Solounias. 1997. Dietary and environmental reconstruction with stable isotope analyses of herbivore tooth enamel from the Miocene locality of Fort Ternan, Kenya. Journal of Human Evolution 33:635–650.

Cerling, T. E., J. M. Harris, B. J. MacFadden, M. G. Leakey, J. Quade, V. Eisenmann, and J. R. Ehleringer. 1997. Global vegetation change through the Miocene/Pliocene boundary. Nature 389:153–158.

Chang, Y. 1974. Miocene suids from Kaiyuan, Yunnan, and Linchu, Shantung. Vertebrata PalAsiatica 12(2):118–123.

Chapman J. A., and G. A. Feldhamer. 1982. Wild Mammals of North America. The Johns Hopkins University Press, Baltimore, Maryland.

Checa, S. L. 2004. Revisión del género *Diacodexis* (Artiodactyla, Mammalia) en el Eoceno inferior del Noreste de España. Géobios 37:325–335.

Chen, G. F. 1988. Remakrs on *Oioceros* species (Bovidae, Artiodactyla, Mammalia) from the Neogene of China. Vertebrata PalAsiatica 26:169–172.

Chikuni, K., Y. Mori, T. Tabata, M. Saito, M. Monma, and M. Kosugiyama. 1995. Molecular phylogeny based on the k-casein and cytochrome *b* sequences in the mammalian suborder Ruminantia. Journal of Molecular Evolution 41:859–866.

Chow, B., and M. Shih.1978. A skull of *Lagomeryx* from middle Miocene of Linchu, Shantung. Vertebrata PalAsiatica 16(2):11–23.

Chow, M. 1957. On some Eocene and Oligocene mammals from Kwangsi and Yunnan. Vertebrata PalAsiatica 1:201–214.

Chow, M. 1958a. *Eoentelodon*—a new primitive entelodont from the Eocene of Lunan, Yunan. Vertebrata PalAsiatica 2:30–36.

Chow, M. 1958b. Some Oligocene mammals from Lunan, Yunnan. Vertebrata PalAsiatica 2:263–268 [in Chinese with English Abstract].

Chow, M. 1961. The first occurrence of fossil hippopotamus in China. Vertebrata PalAsiatica 5:39–40.

Chow, M. 1964. A lemuroid primate from the Eocene of Lantian, Shensi. Vertebrata PalAsiatica 8:257–262.

Chow, M., C.-K. Li, and Y.-P. Chang. 1973. Late Eocene mammalian faunas of Honan and Shansi with notes on some vertebrate fossils collected therefrom. Vertebrata PalAsiatica 11:165–181.

Churcher, C. S. 1970. Two new upper Miocene giraffids from Fort Ternan, Kenya, East Africa. *Paleotragus primaveus* n. sp. and *Samotherium africanum* n.sp.; pp. 1–109 in L. S. B. Leakey and D. J. G. Savage (eds.), Fossil Vertebrates of Africa, Volume 2. Academic Press, London.

Churcher, C. S. 1984. *Sangamona:* the furtive deer. Special Publications, Carnegie Museum of Natural History 8:317–221.

Churcher, C. S. 1991. The status of *Giraffa nebrascensis,* the synonymies of *Cervalces* and *Cervus,* and additional records of *Cervalces scotti.* Journal of Vertebrate Paleontology 11:391-397.

Churcher, C. S., and R. L. Peterson. 1982. Chronologic and environmental implications of a new genus of fossil deer from late Wisconsin deposits at Toronto, Canada. Quaternary Research 18:184–195.

Cifelli, R. L. 1981. Patterns of evolution among Artiodactyla and Perissodactyla (Mammalia). Evolution 35:433–440.

Cifelli, R. L. 1982. The petrosal structure of *Hyopsodus* with respect to that of some other ungulates, and its phylogenetic implications. Journal of Paleontology 56:795–805.

Clauss, M., and J. Hummel. 2005. The digestive performance of mammalian herbivores: why big may not be that much better. Mammal Review 35:174–187.

Clauss, M., R. Frey, B. Kiefer, M. Lechner-Doll, W. Loehlein, C. Polster, G. E. Rossner, and W. J. Streich. 2003. The maximum attainable body size of herbivorous mammals: morphophysiological constraints on foregut, and adaptations of hindgut fermenters. Oecologica 136:14–27.

Clyde, W. C., I. H. Khan, and P. D. Gingerich. 2003. Stratigraphic response and mammalian dispersal during initial India–Asia collision: Evidence from the Ghazij Formation, Baluchistan, Pakistan. Geology 31:1097–1100.

CoBabe, E. A. 1996. Leptaucheniinae, pp. 574–580 in D. R. Prothero and R. J. Emry (eds.), The Terrestrial Eocene–

Oligocene Transition in North America. Cambridge University Press, New York, New York.

Colbert, E. H. 1933. An upper Tertiary peccary from India. American Museum Novitates 635:1–9.

Colbert, E. H. 1935a. Siwalik Mammals in the American Museum of Natural History. Transactions of the American Philosophical Society 26:278–294.

Colbert, E. H. 1935b. The phylogeny of the Indian Suidae and the origin of the Hippopotamidae. American Museum Novitates 799:1–24.

Colbert, E. H. 1936a. *Palaeotragus* in the Tung Gur Formation of Mongolia. American Museum Novitates 871:1–17.

Colbert, E. H. 1936b. Was the extinct giraffe *Sivatherium* known to the early Sumerians? American Anthropologist 38: 605–608.

Colbert, E. H. 1937. Notice of a new genus and species of artiodactyl from the upper Eocene of Wyoming. American Journal of Science 33:473–474.

Colbert, E. H. 1938a. *Brachyhyops*, a new bunodont artiodactyl from Beaver Divide, Wyoming. Annals of the Carnegie Museum 27:87–108.

Colbert, E. H. 1938b. Fossil mammals from Burma in the American Museum of Natural History. Bulletin of the American Museum of Natural History 74:419–424.

Colbert, E. H. 1938c. The relationships of the *Okapi*. Journal Mammalogy 19:47–64.

Colbert, E. H. 1941. The osteology and relationships of *Archaeomeryx*, an ancestral ruminant. American Museum Novitates 1135:1–24.

Colbert, E. H., and R. G. Chaffee. 1939. A study of *Tetrameryx* and associated fossils from Papago Springs Cave, Sonoita, Arizona. American Museum Novitates 1034:1–21.

Collinson, M. E., and J. J. Hooker. 1987. Vegetational and mammalian faunal changes in the early Tertiary of southern England; pp. 259–303 in E. M. Fries, W. G. Chaloner, and P. R. Crane (eds.), The Origin of Angiosperms and Their Biological Consequences. Cambridge University Press, Cambridge.

Collinson, M. E., and J. J. Hooker. 1991. Fossil evidence of interactions between plants and plant-eating mammals. Philosophical Transactions of the Royal Society, London (B) 333: 197–208.

Collinson, M. E., and J. J. Hooker. 2003. Paleogene vegetation of Eurasia: framework for mammalian faunas. Deinsea 10:41–83.

Collinson, M. E., K. Fowler, and M. Boutler. 1981. Floristic changes indicate a cooling climate in the Eocene of southern England. Nature 291:315–317.

Cook, H. J. 1934. New artiodactyls from the Oligocene and lower Miocene of Nebraska. American Midland Naturalist 15:148–165.

Cooke, H. B. S. 1978. Suid evolution and correlation of African hominid localities: an alternative taxonomy. Science 201: 460–463.

Cooke, H. B. S., and S. C. Coryndon. 1970. Pleistocene mammals from the Kaiso Formation and other related deposits in Uganda; pp. 147–198 in L. B. S. Leakey and R. J. G. Savage (eds.), Fossil Vertebrates of Africa. Academic Press, London.

Cooke, H. B. S., and V. J. Maglio. 1972. Plio-Pleistocene stratigraphy in East Africa in relation to proboscidean and suid evolution; pp. 303–329 in W. W. Bishop and J. A. Miller (eds.), Calibration of Hominid Evolution. Scottish Academy Press, Edinburgh.

Coombs, M. C. 1983. Large mammalian clawed herbivores: a comparative study. Transactions of the American Philosophical Society 73(7):1–96.

Coombs, M. C., and W. P. Coombs, Jr. 1977. Dentition of *Gobiohyus* and a reevaluation of the Helohyidae (Artiodactyla). Journal of Mammalogy 58:291–308.

Coombs, M. C., and W. P. Coombs, Jr. 1982. Anatomy of the ear region of four Eocene artiodactyls: *Gobiohyus?, Helohyus, Diacodexis* and *Homacodon*. Journal of Vertebrate Paleontology 2:219–236.

Coombs, W. P., and M. C. Coombs. 1977. The origin of anthracotheres. Neues Jahrbuch für Geologie und Paläontologie Monatshefte 10:584–599.

Cooper, C. F. 1928. *Pseudamphimeryx hantonensis*, sp. n., with notes on certain species of artiodactyls from the Eocene deposits of Hordwell. Annals and Magazine of Natural History 2:49–55.

Cope, E. D. 1869. The Artiodactyla. American Naturalist 23: 111–136.

Cope, E. D. 1870. Fourth contribution to the history of the fauna of the Miocene and Eocene periods of the United States. Proceedings of the American Philosophical Society 11:285–294.

Cope, E. D. 1873a. Fourth notice of extinct Vertebrata from the Bridger and the Green River Tertiaries. Paleontological Bulletin 11:1–4.

Cope, E. D. 1873b. [On *Menotherium lemurinum, Hypisodus minimus, Hypertragulus calcaratus, Hypertragulus tricostatus, Protohippus*, and *Procamelus occidentalis*]. Proceedings of the Philadelphia Academy of Natural Sciences 25:410–420.

Cope, E. D. 1873c. On some Eocene mammals obtained by Hayden's Geological Survey of 1872. Palaeontological Bulletin 12:1–6.

Cope, E. D. 1873d. Third notice of extinct Vertebrata from the Tertiary of the Plains. Paleontological Bulletin 16:1–8.

Cope, E. D. 1874a. A horned *Elotherium*. American Naturalist 8: 437.

Cope, E. D. 1874b. Notes on the Santa Fe marls, and some of the contained vertebrate fossils. Proceedings of the Academy of Natural Sciences of Philadelphia 26:147–152.

Cope, E. D. 1874c. Report on the stratigraphy and Pliocene vertebrate paleontology of northern Colorado; pp. 9–28 in Bulletin of the United States Geological and Geographical Survey of the Territories, Series 1, 1(1).

Cope, E. D. 1875a. On fossil lemurs and dogs. Proceedings of the Academy of Natural Sciences of Philadelphia 1875:255–256.

Cope, E. D. 1875b. Systematic catalogue of Vertebrata of the Eocene of New Mexico, collected in 1874. Report to the Engineer Department, U.S. Army, in charge of Lieut. Geo. M. Wheeler, Washington, D.C., pp. 5–37.

Cope, E. D. 1876. On a new genus of Camelidae. Proceedings of the Academy of Natural Sciences of Philadelphia 28:144–147.

Cope, E. D. 1877. Report upon the extinct Vertebrata obtained in New Mexico by parties of the expedition of 1874. Wheeler Survey 4:1–370.

Cope, E. D. 1878a. A new genus of Oreodontidae. American Naturalist 12:129.

Cope, E. D. 1878b. Descriptions of new extinct Vertebrata from the upper Tertiary and Dakota formations. Bulletin, United States Geological and Geographic Survey, Territories 4:379–396.

Cope, E. D. 1878c. Descriptions of new Vertebrata from the upper Tertiary formations of the West. Proceedings, American Philosophical Society 17:219–231.

Cope, E. D. 1878d. On some of the characters of the Miocene fauna of Oregon. Proceedings of the American Philosophical Society, Philadelphia 18:63–78.

Cope, E. D. 1879a. Extinct Mammalia of Oregon. American Naturalist 13:131.

Cope, E. D. 1879b. Observations on the Fauna of the Miocene Territories of Oregon. Bulletin of the U.S. Geological and Geographical Survey of the Territories, 1880, 5:55–69.

Cope, E. D. 1879c. On some of the characters of the Miocene fauna of Oregon. Proceedings of the American Philosophical Society, Philadelphia 18:63–78.

Cope, E. D. 1881. The systematic arrangement of the Order Perissodactyla. Proceedings of the American Philosophical Society, Philadelphia 19:377–401.

Cope, E. D. 1882. The oldest artiodactyl. American Naturalist 16:71.

Cope, E. D. 1884a. Synopsis of the species of Oreodontidae. Proceedings, American Philosophical Society 21:503–572.

Cope, E. D. 1884b. The Vertebrata of the Tertiary formations of the West. Book I. Report U.S. Geological Survey of the Territories. F. V. Hayden, U.S. Geologist in Charge. Washington. 1009 pp.

Cope, E. D. 1885. The White River Beds of Swift Current River, Northwest Territory. American Naturalist 19:163.

Cope, E. D. 1887. The Perissodactyla. American Naturalist 21: 985–1007, 1060–1076.

Cope, E. D. 1889. Vertebrata of the Swift Current River. Nos. II, III. American Naturalist 23:151–155, 628–629.

Cope, E. D., and W. D. Matthew. 1915. Hitherto unpublished plates of Tertiary Mammalia and Permian Vertebrata. American Museum of Natural History Monograph Series Number 2, plates I–CLIV.

Corneli, P. S. 2002. Complete mitochondrial genomes and eutherian evolution. Journal of Mammalian Evolution 9:281–305.

Corvinus, G., and L. N. Rimal. 2001. Biostratigraphy and geology of the Neogene Siwalik group of the Surai Khola and Rato Khola areas in Nepal. Palaeogeography, Palaeoclimatology, Palaeoecology 165:251–279.

Coryndon, S. C. 1971. Evolutionary trends in East African Hippopotamidae. Bulletin de l'Association Française des Etudes sur le Quaternaire 1:473–478.

Coryndon, S. C. 1976. Fossil Hippopotamidae from Plio-Pleistocene successions of the Rudolf Basin; pp. 238–250 in Y. Coppens, F. C. Howell, G. L. Isaac, and R. E. F. Leakey (eds.), Earliest Man and Environments in the Lake Rudolf Basin. University of Chicago Press, Chicago, Illinois.

Coryndon, S. C. 1977. The taxonomy and nomenclature of the Hippopotamidae (Mammalia, Artiodactyla) and a description of two new fossil species. Proceedings of the Koninklijke Nederlandse Akademie van Wetenschappen 80(2):61–88.

Coryndon, S. C. 1978a. Hippopotamidae; pp. 483–495 in V. J. Maglio and H. B. S. Cooke (eds.), Evolution of African Mammals. Harvard University Press, Cambridge, Massachusetts.

Coryndon, S. C. 1978b. Fossil Hippopotamidae from the Baringo Basin and relationships within the Gregory Rift, Kenya; pp. 279–292 in W. W. Bishop (ed.), Geological Background to Fossil Man. Scottish Academic Press, Edinburgh.

Coryndon, S. C., and Y. Coppens. 1975. Une espèce nouvelle d'hippopotame nain du Plio-Pléistocène du bassin du lac Rodolphe (Ethiopie, Kenya). Comptes Rendus de l'Académie des Sciences 280:1777–1780.

Costeur, L., S. Legendre, and G. Escarguel. 2004. European large mammal palaeobiogeography and biodiversity during the Neogene. Palaeogeographic and climatic impacts. Revue de Paléobiologie volume spécial no. 9: 99-1-9.

Cronin, M. A., R. Stuart, B. J. Pierson, and J. C. Patton. 1996. κ-Casein phylogeny of higher ruminants (Pecora, Artiodactyla). Molecular Phylogenetics and Evolution 6:295–311.

Crusafont, M. 1961. Giraffoidea; pp. 1022–1037 in J. Pivetau (ed.), Traité de Paléontologie, Tome 6. Masson, Paris.

Crusafont, M., R. Adrover, and J. M. Golpe. 1964. Découverte dans le Pikermien d'Espagne du plus primitif des Hippopotames: Hippopotamus (Hexaprotodon) primaevus n. sp. Comptes Rendus de l'Académie des Sciences 258:1572–1575.

Crusafont-Pairó, M. 1952. Los Jiráfidos fósiles de España. Disp. Provincia Barcelona. Consejo Superior de Investigaciones Científicas, Mèmorias y Communicaciones del Instituto Geolólogico 8:1–239.

Crusafont-Pairó, M., and R. Lavocat. 1954. Schizochoerus un nuevo género de suidos del Pontiense inferior (Vallesiense) del Vallés Penedés. Notas y communicaciones Instituto Geologico y Minero de España 36:79–90.

Crusafont-Pairó, M., J. F. Villalta, and Y. J. Truyols. 1955. El Burdigaliense continental de la cuenca del Vallés-Penedés. Memoria y comunicaciones del Instituto Geológico Barcelona 12:1–272.

Cuesta, M. A. 1998. Presencia de Leptotheridium (Dacrytheriidae, Artiodactyla, Mammalia) en el yacimiento eocénico de Caenes (Cuenca del Duero, Salamanca, España). Studia Geologica Salmanticensia 34:69–78.

Cuesta, M. A., and E. Jiménez. 2000. Villamayor: nuevo yacimiento con artiodáctilos del Eoceno de la cuenca del Duero (Salamanca, Castilla y León, España). Studia Geologica Salmanticensia 36:3–12.

Cuesta, M. A., L. C. Soler, and M. L. Casanovas Cladellas. 2006. Artiodáctilos del yacimiento de Sossís (Eoceno superior, Cuenta Prepirenaica, Península Ibérica). Revista Española de Paleontología 21(2):123–144.

Cuvier, G. 1804–1805. Sur les espèces d'animaux dont provinnent les os fossiles répandus dans la pierre à plâtre des environs de Paris. Annales du Muséum d'Histoire Naturelle Paris 3:275–472.

Cuvier, G. 1822. Recherches sur les ossemens fossiles, où l'on rétablit les caractères de plusieurs animaux, dont les révolutions du globe ont détruit les espèces; pp. 1–412. G. Dufour et E. d'Ocagne, Paris.

Cuvier, G. 1824. Recherche sur les ossemens fossiles, tome V, 2ème partie, contenant les ossements de reptiles et le résumé général, 2nd edition. G. Dufour et E. d'Ocagne, Paris. 542 pp.

Dagg, I. A., and J. Bristol Foster. 1976. The Giraffe. Reinhold Co., New York, New York, 210 pp.

Dal Piaz, G. 1930. Nuovo genere e nuove specie di Artiodattili dell'oligocene veneto. Rendiconti della Reale Accademia Nazionale dei Lincei 12:61–64.

Dalquest, W. W. 1983. Mammals of the Coffee Ranch local fauna, Hemphillian of Texas. Texas Memorial Museum, Pearce-Sellards Series 38:1–41.

Dashzeveg, D. 1982. La faune de mammifères du Paléogène inférieur de Nuran-Bulak (Asie Centrale) et ses corrélations avec l' Europe et l'Amérique du Nord. Bulletin de la Société Géologique de France 24:275–281.

Daudt, W. 1898. Beiträge zur Kenntnis des Urogenitalapparates der Cetaceen. Jenaische Zeitschrift für Naturwissenschaft 32:231–312.

Davis, E. B. 2004. Multivariate analysis of Hemphillian antilocaprid astragali from northwestern Nevada reveals size partitioning between Ilingoceros and Sphenophalos. PaleoBios 24:2.

De Bruijn, H., R. Daams, G. Daxner-Höck, V. Fahlbusch, L. Ginsburg, P. Mein, and J. Morales. 1992. Report of the RCMNS working group on fossil mammals, Reisensburg 1990. Newsletters of Stratigraphy 26(2/3):65–118.

Dechaseaux, C. 1963. Une forme européenne du groupe des chameaux (Tylopodes): le genre *Xiphodon*. Comptes Rendus de l'Académie des Sciences 256:5607–5609.

Dechaseaux, C. 1965. Artiodactyles des Phosphorites du Quercy. 1. Étude sur le genre *Dichodon*. Annales de Paleontologie (Vertebres) 51:191–208.

Dechaseaux, C. 1967. Artiodactyles des Phosphorites du Quercy II—Etude sur le genre *Xiphodon*. Annales de Paleontologie (Vertebres) 53:27–47.

Dechaseaux, C. 1969. Moulages endocraniens d'Artiodactyles primitifs. Essai sur l'histoire du neopallium. Annales de Paleontologie (Vertebres) 55:198–248.

Dechaseaux, C. 1974. Artiodactyles primitifs des Phosphorites du Quercy. Annales de Paleontologie (Vertebres) 60:59–100.

Defen, H. 1986. Fossils of Tragulidae from Lufeng, Yunnan. Acta Anthropologica Sinica 5:68–78.

Dehm, R., and T. Öttingen-Spielberg. 1958. Paläontologische und geologische Untersuchungen im Tertiär von Pakistan. 2. Die mitteleocänen Säugetiere von Ganda Kas bei Basal, Nordwest-Pakistan. Abhandlungen der Bayerischen Akademie der Wissenschaften 91:1–54.

Delfortrie, M. 1874. Un pachyderme nouveau dans les phosphates de chaux du Lot. *Oltinotherium verdeaui*. Actes de la Société linéene de Bordeaux 29:261–263.

Demment, M. W., and P. Van Soest. 1985. A nutritional explanation of body-size patterns of ruminant and nonruminant herbivores. American Naturalist 125:641–672.

Deocampo, D. M. 2002. Sedimentary structures generated by *Hippopotamus amphibius* in a lake-margin wetland, Ngorongoro Crater, Tanzania. Palaios 17:212–217.

Déperet, C. 1887. Recherches sur les faunes de Vertébrés miocènes de la vallée du Rhône. Archives du Musée d'Histoire Naturelle de Lyon 4:45–313.

Depéret, C. 1895. Über die Fauna von miocänen Wirbelthieren aus der ersten Mediterranstufe von Eggenburg. Sitzungsberichten der Österreiche Akademie der Wissenschaften 4:395–416.

Depéret, C. 1906. Los vertebrados del Oligocenico inferior de Tárrega (prov. de Lerida). Memorias de la Academia de Ciencias y Artes de Barcelona 3:401–451.

Depéret, C. 1908. L'histoire géologique et la phylogénie des anthracothériidés. Comptes Rendus Hebdomadaires des Séances de l'Académie des Sciences 146:158–162.

Depéret, C. 1917. Monographie de la faune de mammifères fossiles du Ludien inférieur d'Euzet-les-Bains (Gard). Annales de l'Université de Lyon. I. Sciences, Médecine 40:1–290.

Deraniyagala, P. E. P. 1935. Some fossil animals from Ceylon. Journal of the Royal Asiatic Society (Ceylon Branch) 33(88):165–168.

Deraniyagala, P. E. P. 1969. Relationships of the fossil hippopotamus, *Hexaprotodon sinhaleyus*. Spolia Zeylanica 31:571–576.

Desmarest, A. G. 1822. Mammalogie ou description des espèces de mammifères. Seconde partie, contenant les ordres des rongeurs, des édentés, des pachydermes, des ruminans et des cétacés. Mme Veuve Agasse imprimeur, Paris, pp. 277–555.

De Zigno, A. 1888. Anthracoterio di Monteviale. Memorie dell' Istituto Veneto 23:1–35.

Diamond, J. M. 1992. Twilight of the pygmy hippos. Nature 359:15.

Dietrich, W. O. 1922. Beitrag zur Kenntniss der säugetierführenden Bohnerzformation in Schwaben. 1. Ein vergessenes, neu erschlossenes Höhlenvorkommen terrestrischen Eocäns auf der Ulmer Alb. Centralblatt für Mineralogie, Geologie und Paläontologie 19:209–224.

Dietrich, W. O. 1926. Fortschritte der Säugettierpaläontologie Afrikas. Forschungen und Fortschritte 15:121–122.

Dietrich, W. O. 1928. Pleistocäne Deutsch-Ostafrikanische Hippopotamus–reste; pp. 2–41 in H. Reck (ed.), Wissenschaftliche Ergebnisse des Oldoway Expedetion herausgeben von Prof. Dr. Reck, Neue Folge, Heft 3. G. Boerntraeger, Leipzig.

Dineur, H. 1981. Le genre *Brachyodus,* Anthracotheriidae (Artiodactyla, Mammalia) du Miocène inférieur d'Europe et d'Afrique. Ph.D. Dissertation, Université Paris 6, France, 186 pp.

Dingus, L. 1990. Systematics, stratigraphy, and chronology for mammalian fossils (late Arikareean to Hemingfordian) from the uppermost John Day Formation, Warm Springs, Oregon. PaleoBios 12(47 and 48):1–24.

Di Stefano, G., and C. Petronio. 2002. Systematics and evolution of the Eurasian Plio-Pleistocene tribe Cervini (Artiodactyla, Mammalia). Geologica Romana 36:311–334.

Dong, W., Y. Pan, and J. Liu. 2004. The earliest *Muntiacus* (Artiodactyla, Mammalia) from the late Miocene of Yuanmou, southwestern China. Comptes Rendus Palevol 3:379–386.

Douglass, E. 1899. The Miocene lake beds of western Montana. Ph.D. dissertation, University of Montana, Helena, Montana.

Douglass, E. 1901a. Fossil Mammalia of the White River Beds of Montana. Transactions of the American Philosophical Society 20:237–279.

Douglass, E. 1901b. New species of *Merycochoerus* in Montana, Pt. 2. American Journal of Science, series 4 11:73–83.

Douglass, E. 1902. Fossil Mammalia of the White River Beds of Montana. Transactions of the American Philosophical Society 20:237–279.

Douglass, E. 1903. New vertebrates from the Montana Tertiary. Annals of the Carnegie Museum 2:145–199.

Douglass, E. 1907. Some new merycoidodonts. Annals of the Carnegie Museum 4:99–109.

Douglass, E. 1909. *Dromomeryx,* a new genus of American ruminants. Annals of the Carnegie Museum 5:457–479.

Douzery, E. J. P., and E. Randi. 1997. The mitochondrial control region of Cervidae: evolutionary patterns and phylogenetic content. Molecular Biology and Evolution 14:1154–1166.

Driskell, A. C., C. Ané, J. G. Burleigh, M. M. McMahon, B. C. O'Meara, and M. J. Sanderson. 2004. Prospects for building the tree of life from large sequence databases. Science 306:1172–1174.

Dubost, G. 1964. Un ruminant à regime alimentaire partiellement carne: le chevrotaine aquatique (*Hyaemoschus aquaticus* Ogilby). Biologica Gabonica 1:21–23.

Dubost, G. 1965. Quelques traits remarquables du comportement de *Hyaemoschus aquaticus* (Tragulidae, Ruminantia, Artiodactyla). Biologica Gabonica 1:282–287.

Dubost, G. 1968. Les niches écologiques des forets tropicales sud-americaines et africaines, sources de convergences remarkquables entre rongeurs et artiodactyls. Terre Vie 22:3–28.

Dubost, G. 1975. Le comportement du Chevrotain africain, *Hyemoschus aquaticus* Ogilby (Artiodactyla, Ruminantia). Zeitschrift für Tierspsychologie 37:403–448.

Dubost, G. 1978. Un aperçu sur l'écologie du Chevrotain africain *Hyemoschus aquaticus* Olgilby, Artiodactyle Tragulidé. Mammalia 42:1–61.

Dubost, G. 1984. Comparison of the diets of frugivorous forest ruminants of Gabon. Journal of Mammalogy 65:298–316.

Dubost, G., and R. Terrade. 1970. La Transformation de la peau des Tragulidae en boucllier protecteur. Mammalia 34:505–513.

Ducrocq, S. 1994a. An Eocene peccary from Thailand and the biogeographical origins of the artiodactyl family Tayassuidae. Palaeontology 3:765–779.

Ducrocq, S. 1994b. Les anthracothères paléogènes de Thaïlande: paléogéographie et phylogénie. Comptes Rendus de l'Académie des Sciences, Paris 318:549–554.

Ducrocq, S. 1995. The contribution of Paleogene anthracotheriid artiodactyls in the paleobiogeographical history of southern Europe. Neues Jahrbuch für Geologie und Paläontologie Monatshefte 6:355–362.

Ducrocq, S. 1996. The Eocene terrestrial mammal from Timor, Indonesia. Geological Magazine 133:763–766.

Ducrocq, S. 1997. The anthracotheriid genus *Bothriogenys* (Mammalia, Artiodactyla) in Africa and Asia during the Paleogene: phylogenetical and paleobiogeographical relationships. Stuttgarter Beiträge zur Naturkunde (Geologie und Paläontologie) 250:1–44.

Ducrocq, S. 1999. The late Eocene Anthracotheriidae (Mammalia, Artiodactyla) from Thailand. Palaeontographica 252:93–140.

Ducrocq, S., and F. Lihoreau. 2006. The occurrence of bothriodontines (Artiodactyla, Mammalia) in the Paleogene of Asia with special reference to Elomeryx: paleobiogeographical implications. Journal of Asian Earth Sciences 27:885–891.

Ducrocq, S., and S. Sen. 1991. A new Haplobunodontidae (Mammalia, Artiodactyla) from the Eocene of Turkey. Neues Jahrbuch für Geologie und Paläontologie, Monatshefte 1:12–20.

Ducrocq, S., Y. Chaimanee, V. Suteethorn, and J.-J. Jaeger. 1996. An unusual anthracotheriid artiodactyl from the late Eocene of Thailand. Neues Jahrbuch für Geologie und Paläontologie Monatshefte 7:389–398.

Ducrocq, S., Y. Chaimanee, V. Suteethorn, and J.-J. Jaeger. 1997. First discovery of Helohyidae (Artiodactyla, Mammalia) in the late Eocene of Thailand: a possible transitional form for Anthracotheriidae. Comptes Rendus de l'Académie des Sciences, Paris (Sciences de la Terre) 325:367–372.

Ducrocq, S., Y. Chaimanee, V. Suteethorn, and J.-J. Jaeger. 1998. The earliest known pig from the upper Eocene of Thailand. Palaeontology 41:147–156.

Ducrocq, S., Y. Chaimanee, V. Suteethorn, and J.-J. Jaeger. 2003. Occurrence of the anthracotheriid *Brachyodus* (Artiodactyla, Mammalia) in the early middle Miocene of Thailand. Comptes Rendus Palévol 2:261–268.

Ducrocq, S., Y. Chaimanee, V. Suteethorn, S. Traimwichanon, and J.-J. Jaeger. 1997. The age of the Krabi mammal locality (South Thailand); pp. 177–182 in J. P. Aguilar, S. Legendre, and J. Michaux (eds.), The Age of the Krabi Mammal Locality (South Thailand). Mémoires et Travaux de E.P.H.E. l'Institut de Montpellier, Montpellier.

Ducrocq, S., B. Coiffait, P.-E. Coiffait, M. Mahboubi, and J.-J. Jaeger. 2001. The Miocene Anthracotheriidae (Artiodactyla, Mammalia) from the Nementcha, eastern Algeria. Neues Jahrbuch für Geologie und Paläontologie Monatshefte 3:145–156.

Ducrocq, S., A. N. Soe, A. K. Aung, M. Benammi, B. Bo, Y. Chaimanee, T. Tun, T. Thein, and J.-J. Jaeger. 2000. A new anthracotheriid artiodactyl from Myanmar, and the relative ages of the Eocene anthropoid primate-bearing localities of Thailand (Krabi) and Myanmar (Pondaung). Journal of Vertebrate Paleontology 20:755–760.

Duranthon, F., S. Moyà-Solà, H. Astiba, and M. Köhler. 1995. *Ampelomeryx ginsburgi* nov. gen., nov. sp. (Artiodactyla, Cervoidea) et la famille des Palaeomerycidae. Comptes Rendus Hebdomadaire Séances de l'Académie des Sciences, Paris 321:339–346.

Duwe, A. E. 1969. The relationship of the chevrotain, *Tragulus javanicus* to other Artiodactyla based on skeletal muscle antigens. Journal of Mammalogy 50:137–140.

Dyce, K. M., W. O. Sack, and C. J. G. Wensing. 1987. *Textbook of Veterinary Anatomy* (2nd edition) W. B. Saunders, Philadelphia.

Eaton, J. G. 1985. Paleontology and correlation of the Eocene Tepee Trail and Wiggins Formations in the north fork of Owl Creek area, southeastern Absaroka Range, Hot Springs County, Wyoming. Journal of Vertebrate Paleontology 5(4):345–370.

Eaton, J. G., J. H. Hutchison, P. A. Holroyd, W. W. Korth, and P. M. Goldstrand. 1999. Vertebrates of the Turtle Basin local fauna, Middle Eocene, Sevier Plateau, south-central Utah; pp. 463–468 in D. D. Gillette (ed.), Vertebrate Paleontology in Utah. Utah State Geological Survey Special Memoir 99-1.

Edwards, P. 1976. The subfamily Leptochoerinae (Artiodactyla, Dichobunidae) of North America (Oligocene). Contributions to Geology, University of Wyoming 14:99–113.

Effinger, J. A. 1987. Systematics and paleobiology of White River Group Entelodontidae (Mammalia, Artiodactyla). M.S. thesis, Geology, South Dakota School of Mines and Technology, Rapid City, South Dakota, 150 pp.

Effinger, J. A. 1998. Entelodontidae; pp. 375–380 in C. M. Janis, K. M. Scott, and L. L. Jacobs (eds.), Evolution of Tertiary Mammals of North America, Volume I: Terrestrial Carnivores, Ungulates, and Ungulatelike Mammals. Cambridge University Press, Cambridge.

Eisenberg, J. F. 1981. The Mammalian Radiations. University of Chicago Press, Chicago, Illinois.

Eisenberg, J. F., and G. M. McKay. 1974. Comparison of ungulate adaptations in the new world and old world tropical forests with special reference to Ceylon and the rainforests of Central America; in V. Geist and F. Walther (eds.), The Behaviour of Ungulates and Its Relation to Management. IUCN Publication new series No. 24. 585-602. Morges, Switzerland.

Ellerman, J. R., and T. C. S. Morrison-Scott. 1951. Checklist of Palaearctic and Indian Mammals 1758 to 1946. British Museum, London.

Ellsworth, D. L., R. L. Honeycutt, N. J. Silvy, J. W. Bickham, and W. D. Klimstra. 1994. Historical biogeography and contemporary patterns of mitochondrial DNA variation in white-tailed deer from the Southeastern United States. Evolution 48:122–136.

Eltringham, S. K. 1993a. The common hippopotamus (*Hippopotamus amphibius*); pp. 43–55 in W. L. R. Oliver (ed.), Pigs, Peccaries, and Hippos. IUCN, Gland.

Eltringham, S. K. 1993b. The pygmy hippopotamus (*Hexaprotodon liberiensis*); pp. 55–60 in W. L. R. Oliver (ed.), Pigs, Peccaries, and Hippos. IUCN, Gland.

Eltringham, S. K. 1999. The Hippos. Academic Press, London.

Emerson, B. C., and M. L. Tate. 1993. Genetic analysis of evolutionary relationships among deer (subfamily Cervidae). Journal of Heredity 84:266–273.

Emry, R. J. 1973. Stratigraphy and preliminary biostratigraphy of the Flagstaff Rim Area, Natrona County, Wyoming. Smithsonian Contributions to Paleobiology 18:1–44.

Emry, R. J. 1975. Revised Tertiary stratigraphy and paleontology of the western Beaver Divide, Fremont County, Wyoming. Smithsonian Contributions to Paleobiology 25:1–20.

Emry, R. J. 1978. A new hypertragulid (Mammalia, Ruminantia) from the Early Chadronian of Wyoming and Texas. Journal of Paleontology 52:1004–1014.

Emry, R. J. 1981. Additions to the mammalian fauna of the type Duchesnean, with comments on the status of the Duchesnean "age." Journal of Paleontology 55:563–570.

Emry, R. J. 1992. Mammalian range zones in the Chadronian White River Formation at Flagstaff Rim, Wyoming; pp. 106–115 in D. R. Prothero and W. A. Berggren (eds.), Eocene–Oligocene Climatic Evolution. Princeton University Press, Princeton, New Jersey.

Emry, R. J., and J. E. Storer. 1981. The hornless protoceratid *Pseudoprotoceras* (Tylopoda: Artiodactyla) in the early Oligocene of Saskatchewan and Wyoming. Journal of Vertebrate Paleontology 1:101–110.

Emry, R. J., S. G. Lucas, L. W.B. Tyutkova, and B. Wang. 1998. The Ergilian–Shandgolian (Eocene–Oligocene) transition in the Zaysan Basin, Kazakstan. Bulletin of the Carnegie Museum of Natural History 34:298–312.

Erfurt, J. 1988. Systematik, Paläoökologie und stratigraphische Bedeutung der Artiodactyla des Geiseltales. Ph.D. dissertation Martin-Luther-University Halle. 132 pp.

Erfurt, J. 1995. Taxonomie der eozänen Artiodactyla (Mammalia) des Geiseltales mit besonderer Berücksichtigung der Gattung *Rhagatherium*. Hallesches Jahrbuch für Geowissenschaften 17:47–58.

Erfurt, J. 2000. Rekonstruktion des Skelettes und der Biologie von *Anthracobunodon weigelti* (Artiodactyla, Mammalia) aus dem Eozän des Geiseltales. Hallesches Jahrbuch für Geowissenschaften 12:57–141.

Erfurt, J., and H. Altner. 2003. Habitus—Rekonstruktion von *Anthracobunodon weigelti* (Artiodactyla, Mammalia) aus dem Eozän des Geiseltales. Veröffentlichungen des Landesamtes für Archäologie 57:153–175.

Erfurt, J., and H. Haubold 1989. Artiodactyla aus den eozänen Braunkohlen des Geiseltales bei Halle (DDR). Palaeovertebrata 19:131–160.

Erfurt, J., and J. Sudre. 1995a. Revision der Gattung *Meniscodon* Rütimeyer 1888 (Artiodactyla, Mammalia) aus dem Mitteleozän Europas. Eclogae Geologicae Helvetiae 88:865–883.

Erfurt, J., and J. Sudre. 1995b. Un Haplobunodontidae nouveau *Hallebune krumbiegeli* nov. gen. nov. sp. (Artiodactyla, Mammalia) dans l' Eocene moyen du Geiseltal pres Halle (Sachsen-Anhalt, Allemagne). Palaeovertebrata 24:85–99.

Erfurt, J., and J. Sudre. 1996. Eurodexeinae, eine neue Unterfamilie der Artiodactyla (Mammalia) aus dem Unter- und Mitteleozän Europas. Palaeovertebrata 25:371–390.

Eriksson, O., E. M. Friis, and P. Lofgren. 2000. Seed size, fruit size, and dispersal systems in angiosperms from the Early Cretaceous to the late Tertiary. American Naturalist 156:47–58.

Erxleben, J. C. P. 1777. Systema regni animalis per classes, ordines, genera, species, varietates cum synonymia et historia animalium. Classis I. Mammalia; pp. I–XLVII, 1–636, index. Lipsiae. Weigand.

Escarguel, G. 1997. Implications phylétiques et applications biochronologiques de l' "Analyse Mandibulaire". Études de cas pour différents ordres: Marsupiaux, Insectivores, Rongeurs, Périssodactyles; pp. 83–96 in J. P. Aguilar, S. Legendre, and J. Michaux (eds.), Actes du Congrès BiochroM'97. Mémoires et Travaux de E.P.H.E. l'Institut de Montpellier, Montpellier.

Escarguel, G., B. Marandat, and S. Legendre. 1997. Sur l'áge numérique des faunes de mammifères du Paléogène d'Europe occidentale, en particulier celles de l'Éocène inférieur et moyen; pp. 443–460 in J.-P. Aguilar, S. Legendre, and J. Michaux (eds.), Actes du Congrès BiochroM'97. Mémoires et Travaux de E.P.H.E. l'Institut de Montpellier, Montpellier.

Estes, R. D. 1984. Social organization of the African Bovidae; pp. 166–205 in V. Geist and F. Walther (eds.), The Behaviour of Ungulates and Its Relation to Management. International Union for Conservation of Nature and Natural Resources, Morges, Switzerland.

Estravis, C., and D. E. Russell. 1989. Découverte d'un nouveau *Diacodexis* (Mammalia, Artiodactyla) dans l'Eocène inférieur de Silveirhina (Portugal). Palaeovertebrata 19:29–44.

Evans, H. E. 1993. Miller's Anatomy of the Dog, 3rd edition. W. B. Saunders, Philadelphia, Pennsylvania.

Fahlbusch, V. 1976. Report on the International Symposium on Mammalian Stratigraphy of the European Tertiary (München, April 11–14, 1975). Newsletters on Stratigraphy 5:160–167.

Fahlbusch, V. 1985. Säugerreste (*Dorcatherium, Steneofiber*) aus der miozänen Braunkohle von Wackersdorf/Oberpfalz. Mitteilungen der Bayerischen Staatssammlung für Paläontologie und Historische Geologie 25:81–94.

Fahlbusch, V., E. Martini, and G. E. Rössner. 2003. Ein *Pomelomeryx*-zahn (Mammalia, Ruminantia) aus den escheri-Schichten der Rhön (Miozän). Courier Forschung-Institut Senckenberg 241:313–317.

Falconer, H., and P. T. Cautley. 1836. Note on the fossil hippopotamus of the Siwalik Hills. Asiatic Research Calcutta 19(3):39–53.

Falconer, H., and P. T. Cautley. 1847. Fauna Antiqua Sivalensis, being the fossil zoology of the Siwalik Hills, in the north of India, Atlas. Smith, Elder, and Co., London, plates 25–80.

Fanning, J. C., and R. J. Harrison. 1974. The structure of the trachea and lungs in the South Australian *Tursiops truncates*; pp. 231–252 in R. J. Harrison (ed.), Functional Anatomy of Marine Mammals. Academic Press, New York, New York.

FAUNMAP Working Group. 1994. FAUNMAP: A Database Documenting Late Quaternary Distributions of Mammal Species in the United States. Illinois State Museum Scientific Papers 25:1–690.

Faure, M. 1981. Répartition des Hippopotamidae (Mammalia, Artiodactyla) en europe Occidentale. Implications stratigraphiques et plaéoécologiques. Géobios 14:191–200.

Faure, M. 1983. Les Hippopotamidae (Mammalia, Artiodactyla) d'Europe Occidentale. Ph.D. dissertation, Université Claude Bernard—Lyon I, Lyon.

Faure, M. 1984. *Hippopotamus incognitus* nov. sp., un Hippopotame (Mammalia, Artiodactyla) du Pléistocène d'Europe Occidentale. Géobios 17:427–434.

Faure, M. 1985. Les Hippopotames quaternaires non-insulaires d'Europe Occidentale. Nouvelles Archives du Muséum d'Histoire Naturelle de Lyon 23:13–79.

Faure, M. 1986. Les Hippopotamidés du Pléistocène ancien d'Oubeidiyeh (Israël). Mémoires et Travaux du Centre de Recherche Français de Jérusalem 5:107–142.

Faure, M. 1994. Les Hippopotamidae (Mammalia, Artiodactyla) du rift occidental (bassin du lac Albert, Ouganda). Etude Préliminaire; pp. 321–337 in B. Senut and M. Pickford (eds.), Geology and Paleobiology of the Albertine Rift Valley, Uganda-Zaïre. II: Paleobiology. Cifeg, Orléans.

Faure, M., and C. Guérin. 1990. *Hippopotamus laloumena* nov. sp., la troisième espèce d'hippopotame holocène de Madagascar. Comptes Rendus de l'Académie des Sciences 310:1299–1305.

Faure, M., and H. Méon. 1984. *L'Hippopotamus crusafonti* de la Mosson (près Montpellier). Première reconnaissance d'un Hippopotame Néogène en France. Comptes Rendus de l'Académie des Sciences 298(3):93–98.

Feibel, C. S., F. H. Brown, and I. McDougall. 1989. Stratigraphic context of fossil hominids from the Omo group deposits: Northern Turkana Basin, Kenya and Ethiopia. American Journal of Physical Anthropology 78:595–622.

Fejfar, O. 1987. A lower Oligocene mammalian fauna from Detan and Dverce NW Bohemia, Czechoslovakia. Münchner Geowissenschaftliche Abhandlungen 10:253–264.

Felsenstein, J. 1978. Cases in which parsimony or compatibility methods will be positively misleading. Systematic Zoology 27:401–410.

Felsenstein, J. 1985. Confidence limits on phylogenies: an approach using the bootstrap. Evolution 39:783–791.

Fernández, M. H., and E. S. Vrba. 2005. A complete estimate of the phylogenetic relationships in Ruminantia: a dated species-level supertree of extant ruminants. Biological Reviews 80:1–269.

Ferrusquía, I. 1967. Rancho Gaitan local fauna, early Chadronian, northeastern Chihuahua. Boletín de la Sociedad Geológica Mexicana 30:99–138.

Ferrusquia-Villafranca, I. 2003. The first Paleogene mammal record of middle America: Helohyidae new gen. and sp. Journal of Vertebrate Paleontology 23:49A–50A.

Ferrusquia-Villafranca, I. 2006. The first Paleogene mammal record of middle America: *Simojovelhyus pocitosense* (Helohyidae, Artiodactyla). Journal of Vertebrate Paleontology 26:989–1001.

Field, C. R. 1970. A study of the feeding habits of the hippopotamus (*Hippopotamus amphibius* Linn.) in the Queen Elizabeth National Park, Uganda, with some management implications. Zoologica Africana 5:71–86.

Filhol, H. 1876a. Mammiferes fossiles nouveaux provenant des dépôts de phosphate de chaux du Quercy. Comptes rendus hebdomadaires des séances de l'Académie des Sciences 82:288–289.

Filhol, H. 1876b. Recherches sur les phosphorites du Quercy. Etude des fossiles qu'on y rencontre et spécialement des mammifères. Annales des Sciences Géologiques de Paris 7(7):1–220.

Filhol, H. 1877. Recherches sur les Phosphorites du Quercy. Etudes des fossiles qu'on y rencontre et spécialment des mammifères. 2. Annales des Sciences Géologiques de Paris 8:1–338.

Filhol, H. 1880. Sur la découverte de mammifères nouveaux dans le dépots de phosphate de chaux du Quercy. Comptes Rendus de l'Académie des Sciences 90:1579–1580.

Filhol, H. 1882a. Découverte de nouveaux genres de mammifères fossiles dans les dépots de phosphate de chaux du Quercy. Comptes Rendus de l'Académie des Sciences 94:138–139.

Filhol, H. 1882b. Etude des mammifères fossiles de Ronzon (Haute-Loire). Annales des Sciences Géologiques de Paris 12:1–33.

Filhol, H. 1884. Oberservations relatives à des mammifères fossiles nouveaux provenant des depots de phosphate de chaux du Quercy. Bulletin de la Société des Sciences Physiques and Naturelles de Toulouse 5(2):159–203.

Filhol, H. 1890. Description d'un nouveau genre de mammifère. Bulletin de la Société Philomathique de Paris série 8 2:34–38.

Fisher, D. C., and B. E. Bodenbender. 2003. Blastoid stratocladistics—reply to Sumrall and Brochu. Journal of Paleontology 77:195–198.

Flerov, C. 1931. On the generic characters of the Family Tragulidae (Mammalia, Artiodactyla). Doklady Akademii Nauk SSSR 1931:75–79.

Flerov, C., 1952. [Musk deer and deer;] pp. 50–69 in Fauna of the U.S.S.R., Mammals, Volume 1, no. 2: Material of the Quaternary Period of the U.S.S.R., Academy of Sciences of the U.S.S.R., Moscow [Russian].

Flower, W. H. 1866. An Introduction to the Osteology of the Mammalia, 3rd ed. Asher and Co., Amsterdam, 382 pp.

Flower, W. H. 1883. On the arrangement of the orders and families of existing Mammalia. Proceedings of the Zoological Society of London 53:178–186.

Flower, W. H. 1884. Catalogue of the specimens illustrating the osteology and dentition of vertebrate animals, recent and extinct, contained in the Museum of the Royal College of Surgeons of England. Taylor and Francis, London, 779 pp.

Flower, W. H., and R. Lydekker. 1891. An introduction to the study of mammals living and extinct. Adam and Charles Black, London.

Forster-Cooper, C. 1915. New genera and species of mammals from the Miocene deposits of Baluchistan. Annals and Magazine of Natural History 8:404–410.

Forster-Cooper, C. 1928. On the ear region of certain of the Chrysochloridae. Philosophical Transactions of the Royal Society of London (B) 216:265–283.

Fortelius, M. 1985. Ungulate cheek teeth: developmental, functional, and evolutionary implications. Acta Zoologica Fennica 180:1–76.

Fortelius, M., and N. Solounias. 2000. Functional characterization of ungulate molars using the abrasion-attrition wear gradient: a new method for reconstructing paleodiets. American Museum Novitates 3301:1–26.

Fortelius, M., J. van der Made, and R. L. Bernor. 1996. Middle and late Miocene Suoidea of central Europe and the eastern Mediterranean: evolution, biogeography and paleoecology; pp. 348–379 in R. L. Bernor, V. Fahlbusch, and H. V. Mittmann (eds.), The Evolution of Western Eurasian Neogene Mammal Faunas. Columbia University Press, New York, New York.

Fortelius, M., M. Armour-Chelu, R. L. Bernor, and N. Fessaha. 2004. Systematics and paleobiology of the Rudubanya Suidae. Palaeontographia Italia 90:259–278.

Fortelius, M., J. Eronen, L. Liu, D. Pushkina, A. Tesakov, I. Vislobokova, and Z. Zhang. 2003. Continental-scale hypsodonty patterns, climatic paleobiogeography and dispersal of Eurasian Neogene large mammals; in J. W. F. Reumer and W. Wessels (eds.), Distribution and Migration of Tertiary Mammals in Eurasia. Deinsea 10:1–11.

Fortelius, M., L. Werdelin, P. Andrews, R. L. Bernor, A. Gentry, L. Humphrey, H.-W. Mittman, and S. Viranta. 1996. Provinciality, diversity, turnover, and paleoecology in land mammal faunas of the later Miocene of western Eurasia; pp. 414–448 in R. L. Bernor, V. Fahlbusch, and H.-W. Mittmann (eds.),

The Evolution of Western Eurasian Neogene Faunas. Columbia University Press, New York, New York.

Fortelius, M., J. Eronen, J. Jernvall, L. Liu, D. Pushkina, J. Rinne, A. Tesakov, I. Vislobokova, Z. Zhang, and L. Zhou. 2002. Fossil mammals resolve regional patterns of Eurasian climate change over 20 million years. Evolutionary Ecology Research 4:1005–1016.

Foss, S. E. 2001. Systematics and paleobiology of the Entelodontidae (Mammalia, Artiodactyla), Ph.D. Dissertation, Department of Biological Sciences, Northern Illinois University, DeKalb, Illinois, 222 pp.

Foss, S. E., and T. Fremd. 1998. A survey of the species of entelodonts (Mammalia, Artiodactyla) of the John Day Basin, Oregon. Dakoterra. South Dakota School of Mines and Technology 5:63–72.

Foss, S. E., and T. J. Fremd. 2001. Biostratigraphy of the Entelodontidae (Mammalia: Artiodactyla) from the John Day Basin, Oregon. PaleoBios 21:53.

Foss, S. E., and S. G. Lucas. 1997. Dental variability in a population sample of *Archaeotherium* (Mammalia, Entelodontidae) from the Chadronian of Colorado. Journal of Vertebrate Paleontology 17:47A.

Foss, S. E., and J. M. Theodor. 2003. New morphological support for the phylogenetic affinity of Artiodactyla and Cetacea. Journal of Vertebrate Paleontology 23:51A.

Foss, S. E., W. D. Turnbull, and L. Barber. 2001. Observations on a new specimen of *Achaenodon* (Mammalia, Artiodactyla) from the Eocene Washakie Formation of southern Wyoming. Journal of Vertebrate Paleontology 21:51A.

Foss, S. E., S. G. Lucas, T. J. Fremd, E. B. Lander, M. C. Mihlbachler, and C. B. Hanson. 2004. Reanalysis of the Hancock Mammal Quarry LF, Clarno Formation, Wheeler County, North-Central Oregon. Journal of Vertebrate Paleontology 24:59A.

Fourteau, R. 1918. Contribution à l'étude des vertébrés miocènes de l'Egypte. Geological Survey Department, Ministry of Finance, Egypt, Government Press, Cairo, 122 pp.

Fraas, E. 1904. Neue Zeuglodonten aus dem unteren Mitteleocän vom Mokattam bei Cairo. Geologische und paläontologische Abhandlungen 6:199–220.

Fraipont, J. 1907. *Okapia*. Annales du Musée Congo, Sér. II, Contributions a la Faune du Congo. Tome I. T. J. Musée Congo, Brussels, 116 pp.

Franzen, J. L. 1981. Das erste Skelett eines Dichobuniden (Mammalia, Artiodactyla), geborgen aus mitteleozänen Ölschiefern der "Grube Messel" bei Darmstadt (Deutschland, S-Hessen). Senckenbergiana Lethaea 61:299–353.

Franzen, J. L. 1983. Ein zweites Skelett von *Messelobunodon* (Mammalia, Artiodactyla, Dichobunidae) aus der "Grube Messel" bei Darmstadt (Deutschland, S-Hessen). Senckenbergiana Lethaea 64:403–445.

Franzen, J. L. 1988. Skeletons of *Aumelasia* (Mammalia, Artiodactyla, Dichobunidae) from Messel (M. Eocene, W. Germany). Courier Forschungs-Institut Senckenberg 107:309–321.

Franzen, J. L. 1994. Neue Säugetierfunde aus dem Eozän des Eckfelder Maares bei Manderscheid (Eifel). Mainzer naturwissenschaftliches Archiv, Beihefte 16:189–211.

Franzen, J. L. 2003. Mammalian faunal turnover in the Eocene of central Europe. Geological Society of America Special Paper 369:455–461.

Franzen, J. L., and H. Haubold. 1986. The middle Eocene of European mammalian stratigraphy. Definition of the Geiseltalian. Modern Geology 10:159–170.

Franzen, J. L., and G. Krumbiegel. 1980. *Messelobunodon ceciliensis* n.sp. (Mammalia, Artiodactyla)—ein neuer Dichobunide aus der mitteleozänen Fauna des Geiseltales bei Halle (DDR). Zeitschrift für Geologische Wissenschaften 8:1585–1592.

Franzen, J. L., and G. Richter. 1988. Die urtümlichen Paarhufer—Einzelgänger im Unterholz; pp. 249–256 in S. Schaal and W. Ziegler (eds.), Messel—Ein Schaufenster in die Geschichte der Erde und des Lebens. Kramer, Frankfurt am Main.

Franz-Odendaal, T., and N. Solounias. 2004. Dietary reconstruction of *Sivatherium hendeyi* (Mammalia Giraffidae, Sivatheriinae) from Langebaanweg, South Africa (early Pliocene). Geodiversitas 26:675–685.

Franz-Odendaal, T. A., J. A. Lee-Thorp, and A. Chinsamy. 2002. New evidence for the lack of C4 grassland expansions during the early Pliocene at Langebaanweg, South Africa. Paleobiology 28:378–388.

Fremd, T. J., E. A. Bestland, and G. J. Retallack. 1994. John Day Basin paleontology field trip guide and road log, 1994 Society of Vertebrate Paleontology Annual Meeting. John Day Fossil Beds National Monument 94. 80 pp.

Frick, C. 1937. Horned ruminants of North America. American Museum of Natural History Bulletin 59:1–669.

Frick, C., and B. E. Taylor. 1971. *Michenia*, a new protolabine (Mammalia, Camelidae) and a brief review of the early taxonomic history of the genus *Protolabis*. American Museum Novitates 2444:1–24.

Froehlich, D. J. 2002. Quo vadis *Eohippus*? The systematics and taxonomy of the early Eocene equids (Perissodactyla). Zoological Journal of the Linnean Society 134:141–256.

Furlong, E. L. 1927. The occurrence and phylogenetic status of Merycodus from the Mohave Desert Tertiary. University of California Publications, Bulletin of the Department of Geological Sciences 17:145–186.

Furlong, E. L. 1934. New merycodonts from the upper Miocene of Nevada. Contributions to Paleontology, Carnegie Institution of Washington 453:1–10.

Furlong, E. L. 1941. A new Pliocene antelope from Mexico with some remarks on some known antilocaprids. Publication of the Carnegie Institute of Washington 530:25–33.

Furlong, E. L. 1943. The Pleistocene antelope, *Stockoceros conklingi*, from San Josecito Cave, Mexico. Publication of the Carnegie Institution of Washington 551:1–8.

Gabunia, L. K. 1958. O cherepe rogatoi iskopaemoi svinia iz srednego miotsena Caucasia. Doklady Akademii Nauk SSSR 116:1187.

Gabunia, L. 1960. Kubanochoerinae, nouvelle sous-famille de porcs du Miocène moyen du Caucase. Vertebrata PalAsiatica 4:87–97.

Gabunia, L. K. 1964a. Benarskaya fauna oligotsenovykh pozvonochnykh [Benonafaun fauna of Oligocene Vertebrates]. Akademiya Nauk Gruzinskoy SSR Institut Paleobiologi Izdatelstvo "Metsnierba." 262 pp.

Gabunia, L. K. 1964b. The Oligocene mammalian fauna of Benara. Academy of Sciences of USSR, Tbilissi, 268 pp.

Gabunia, L. K. 1971. [On a new representative of condylarths (Condylarthra) from the Eocene of Zaisan Depression]. Bulletin of the Academy of Sciences of the Georgian SSR 61:233–235.

Gabunia, L. K. 1973. [On the presence of the Diacodexinae in the Eocene of Asia]. Bulletin of the Academy of Sciences of the Georgian SSR 71:741–744.

Gabunia, L. K. 1977. Contribution à la connaissance des mammifères Paléogènes du bassin de Zaissan (Kazakhstan Central). Géobios, Mémoire Spécial 1:29–37.

Gallagher, W. B., and D. C. Parris (eds.). 1996. Cenozoic and Mesozoic vertebrate paleontology of the New Jersey coastal plain. Society of Vertebrate Paleontology field trip. New Jersey State Museum, Trenton, New Jersey, 58 pp.

Gatesy, J. E. 1997. More DNA support for a Cetacea/Hippopotamidae clade: the blood-clotting protein gene y-fibrinogen. Molecular Biology and Evolution 14:537–543.

Gatesy, J. E. 1998. Molecular evidence for the phylogenetic affinities of Cetacea; pp. 63–111 in J. G. M. Thewissen (ed.), The Emergence of Whales. Plenum Press, New York, New York.

Gatesy, J. E., and P. Arctander. 2000a. Hidden morphological support for the phylogenetic placement of *Pseudoryx nghetinhensis* with bovine bovids: a combined analysis of gross anatomical evidence and DNA sequences from five genes. Systematic Biology 49:515–538.

Gatesy, J. E., and P. Arctander. 2000b. Molecular evidence for the phylogenetic affinities of Ruminantia; pp. 143–155 in E. S. Vrba and G. B. Schaller (eds.), Antelopes, Deer, and Relatives: Fossil Record, Behavioral Ecology, Systematics and Conservation. Yale University Press, New Haven, Connecticut.

Gatesy, J. E., and M. A. O'Leary. 2001. Deciphering whale origins with molecules and fossils. Trends in Ecology and Evolution 16:562–570.

Gatesy, J. E., P. O'Grady, and R. H. Baker. 1999a. Corroboration among data sets in simultaneous analysis: hidden support for phylogenetic relationships among higher level artiodactyl taxa. Cladistics 15:271–313.

Gatesy, J. E., C. Hayashi, M. A. Cronin, and P. Arctander. 1996. Evidence from milk casein genes that cetaceans are close relatives of hippopotamid artiodactyls. Molecular Biology and Evolution 13:954–963.

Gatesy, J. E., C. A. Matthee, R. DeSalle, and C. Hayashi. 2002. Resolution of a supertree/supermatrix paradox. Systematic Biology 51:652–664.

Gatesy, J. E., M. C. Milinkovitch, P. J. Waddell, and M. J. Stanhope. 1999b. Stability of cladistic relationships between Cetacea and higher-level artiodactyl taxa. Systematic Biology 48:6–20.

Gatesy, J. E., D. Yelon, R. DeSalle, and E. S. Vrba. 1992. Phylogeny of the Bovidae (Artiodactyla, Mammalia), based on mitochondrial ribosomal DNA Sequences. Molecular Biology and Evolution 9:433–446.

Gatesy, J. E., G. Amato, E. Vrba, G. B. Schaller, and R. DeSalle. 1997. A cladistic analysis of mitochondrial ribosomal DNA from the Bovidae. Molecular Phylogenetics and Evolution 7:303–319.

Gaubert, P., W. C. Wozencraft, P. Cordeiro-Estrela, and G. Veron. 2005. Mosaics of convergences and noise in morphological phylogenies: what's in a viverrid-like carnivoran? Systematic Biology 54:865–894.

Gaucher, E. A., L. G. Graddy, T. Li, R. C. M. Simmen, F. A. Simmen, D. R. Schreiber, D. A. Liberles, C. M. Janis, and S. M. Benner. 2004. The planetary biology of cytochrome P450 aromatases. BMC Biology 2:19–32.

Gaudry, A. 1862–1867. Animaux Fossiles et Géologie de l'Attique. Paris, 476 pp.

Gaudry, A. 1873. Sur l'*Anthracotherium* découvert à Saint-Menoux (Allier). Bulletin de la Société géologique de France 2:36–40.

Gaudry, A. 1876. Sur un hippopotame fossile découvert à Bone (Algérie). Bulletin de la Société Géologique de France 4:147–154.

Gaudry, A. 1961. Note sur les antilopes trouvées à Pikermi (Grèce). Bulletin de la Société Géologique de France 2(18):587–598.

Gaur, R. 1992. On *Dorcatherium nagrii* (Tragulidae, Mammalia)—with a review of Siwalik tragulids. Rivista Italiana di Paleontologia e Stratigrafia 98:353–370.

Gauthier, J. A., A. G. Kluge, and T. Rowe. 1988. Amniote phylogeny and the importance of fossils. Cladistics 4:105–209.

Gazin, C. L. 1935. A new antilocaprid from the Upper Pliocene of Idaho. Journal of Paleontology 9:390–393.

Gazin, C. L. 1952. The lower Eocene Knight Formation of western Wyoming and its mammalian faunas. Smithsonian Miscellaneous Collections 17(18):1–82.

Gazin, C. L. 1955. A review of the upper Eocene Artiodactyla of North America. Smithsonian Miscellaneous Collections 128:1–96.

Gazin, C. L. 1956. The geology and vertebrate paleontology of upper Eocene strata in the northeastern part of the Wind River Basin, Wyoming, Part 2. The mammalian fauna of the Badwater Area. Smithsonian Miscellaneous Collections 131(8):1–35.

Gazin, C. L. 1958. A new dichobunid artiodactyl from the Uinta Eocene. Breviora (Harvard Museum of Comparative Zoology) 96:1–6.

Gazin, C. L. 1965. A study of the early Tertiary condylarthran mammal *Meniscotherium*. Smithsonian Miscellaneous Collections 149:1–98.

Gazin, C. L. 1968. A study of the Eocene condylarthran mammal *Hyopsodus*. Smithsonian Miscellaneous Collections 153:1–90.

Gaziry, A. W. 1987a. *Hexaprotodon sahabiensis* (Artiodactyla, Mammalia): a new hippopotamus from Libya; pp. 303–315 in N. T. Boaz, A. El-Arnauti, A. W. Gaziry, J. De Heinzelin, and D. D. Boaz (eds.), Neogene Paleontology and Geology of Sahabi. Alan R. Liss, New York, New York.

Gaziry, A. W. 1987b. *Merycopotamus petrocchii* (Artiodactyla, Mammalia) from Sahabi, Libya; pp. 287–302 in N. T. Boaz, A. El-Arnauti, A. W. Gaziry, J. De Heinzelin, and D. D. Boaz (eds.), Neogene Paleontology and Geology of Sahabi. Alan R. Liss, New York, New York.

Geais, G. 1934. Le *Brachyodus borbonicus* des argiles de St. Henri (près Marseille). Travaux du Laboratoire de Géologie de la Faculté de Lyon 25:1–54.

Geisler, J. H. 2001a. Morphological and molecular evidence for the phylogeny of Cetacea and Artiodactyla: explaining incongruence between types of data. Ph.D. dissertation, Columbia University, New York, New York, 475 pp.

Geisler, J. H. 2001b. New morphological evidence for the phylogeny of Artiodactyla, Cetacea, and Mesonychidae. American Museum Novitates 3344:1–53.

Geisler, J. H., and Z. Luo. 1996. The petrosal and inner ear of *Herpetocetus* sp. (Mammalia: Cetacea) and their implications for the phylogeny and hearing of archaic mysticetes. Journal of Paleontology 70:1045–1066.

Geisler, J. H., and Z. Luo. 1998. Relationships of Cetacea to terrestrial ungulates and the evolution of cranial vasculature in Cete; pp. 163–212 in J. G. M. Thewissen (ed.), The Emergence of Whales. Plenum Press, New York, New York.

Geisler, J. H., and M. D. Uhen. 2003. Morphological support for a close relationship between hippos and whales. Journal of Vertebrate Paleontology 23:991–996.

Geisler, J. H., and M. D. Uhen. 2005. Phylogenetic relationships of extinct cetartiodactyls: results of simultaneous analyses of molecular, morphological, and stratigraphic data. Journal of Mammalian Evolution 12:145–160.

Geist, V. 1998. Deer of the World. Stackpole Books, Mechanicsburg, Pennsylvania.

Gentry, A. W. 1970. The Bovidae of the Fort Ternan fossil fauna; pp. 243-323 in L. S. B. Leakey and R. J. G. Savage (eds.), Fossil Vertebrates of Africa, Volume 2. Academic Press, London.

Gentry, A. W. 1971. The earliest goats and other antelopes from the Samos *Hipparion* fauna. Bulletin of the British Museum (Natural History) Geology, London 20:229–296.

Gentry, A. W. 1978a. Bovidae; pp. 540–581 in V. J. Maglio and H. B. S. Cooke (eds.), Evolution of African Mammals. Harvard University Press, Cambridge, Massachusetts.

Gentry, A. W. 1978b. Tragulidae and Camelidae; pp. 536–539 in V. J. Maglio and H. B. S. Cooke (eds.), Evolution of African Mammals. Harvard University Press, Cambridge, Massachusetts.

Gentry, A. W. 1990. Evolution and dispersal of African Bovidae; pp. 195–227 in G. A. Bubenik and A. B. Bubenik (eds.), Horns, Pronghorns and Antlers. Springer-Verlag, New York, New York.

Gentry, A. W. 1992. The subfamilies and tribes of the family Bovidae. Mammal Review 22:1–32.

Gentry, A. W. 1994. The Miocene differentiation of Old World Pecora (Mammalia). Historical Biology 7:115–158.

Gentry, A. W. 1999a. A fossil hippopotamus from the Emirate of Abu Dhabi, United Arab Emirates; pp. 271–289 in P. J. Whybrow and A. Hill (eds.), Fossil Vertebrate of Arabia. Yale University Press, New Haven, Connecticut.

Gentry, A. W. 1999b. Fossil pecorans from the Baynunah Formation, Emirate of Abu Dhabi, United Arab Emirates; pp. 290–316 in P. J. Whybrow and A. Hill (eds.), Fossil Vertebrates of Arabia. Yale University Press, New Haven, Connecticut.

Gentry, A. W. 2000a. The ruminant radiation; pp. 11–25 in E. S. Vrba and G. B. Schaller (eds.), Antelopes, Deer and Relatives: Fossil Record, Behavioral Ecology, Systematic, and Conservation. Yale University Press, New Haven, Connecticut.

Gentry, A. W. 2000b. Caprinae and Hippotragini (Bovidae Mammalia) in the upper Miocene; pp. 65–83 in E. S. Vrba and G. B. Schaller (eds.), Antelopes, Deer and Relatives: Fossil Record, Behavioral Ecology, Systematic, and Conservation. Yale University Press, New Haven, Connecticut.

Gentry, A. W. 2003. Ruminantia (Artiodactyla); pp. 332–379 in M. Fortelius, J. Kappelman, S. Sen, and R. L. Bernor (eds.), Geology and Paleontology of the Miocene Sinap Formation, Turkey. Columbia University Press, New York, New York.

Gentry, A. W., and E. P. J. Heizmann. 1996. Miocene ruminants of the Central Tethys and Eastern Tethys and Paratethys; pp. 378–391 in R. L. Bernor, V. Fahlbusch, and H.-W. Mittmann (eds.), The Evolution of the Western Eurasian Neogene Mammal Faunas. Columbia University Press, New York, New York.

Gentry, A. W., and J. J. Hooker. 1988. The phylogeny of the Artiodactyla; pp. 235–272 in M. J. Benton (ed.), The Phylogeny and Classification of the Tetrapods. Volume 2. Mammals. Systematics Association Special Volume 35B. Clarendon Press, Oxford.

Gentry, A. W., G. E. Rössner, and E. P. J. Heizmann. 1999. Suborder Ruminantia; pp. 225–258 in G. E. Rössner and E. P. J. Heizmann (eds.), The Miocene Land Mammals of Europe. Verlag Dr. Frierdich Pfeil, Munich.

Gentry, A. W., N. Solounias, J. C. Barry, and M. Raza. In press. Miocene Bovidae (Mammalia) from the Siwaliks.

Geoffroy Saint-Hilaire, É. F. 1833. Palaeontographie. Considérations sur des ossemens fossiles la plupart inconnus trouvés et observés dans les bassins de l'Auvergne. Revue Encyclopédique 54:76–95.

Geraads, D. 1977. Les Palaeotraginae (Giraffidae, Mammalia) du Miocène supéreur de la région de Thessalonique (Grèce). Géologie Méditerranéenne 5:269–276.

Geraads, D. 1979. Les Giraffidae (Ruminantia, Mammalia) du Miocène supéreur de la région de Thessalonique (Grèce). Bulletin Musee national Histoire naturelle, Paris, 4 séries 1:377–389.

Geraads, D. 1985. *Sivatherium maurusicum* (Pomel) (Giraffidae, Mammalia) du Pléistocène de la republique de Djibouti. Palaöntologische Zeitschrift 59:311–321.

Geraads, D. 1986. Remarques sur la systématique et phyloégnie des Les Giraffidae (Artiodactyla, Mammalia). Géobios 19: 465–477.

Geraads, D. 1989. Un nouveau Giraffidé du Miocène supérieur de Macédoine (Grèce). Bulletin de la Musee national Histoire naturelle, Paris, 4 séries 11:189–199.

Geraads, D. 1996. Le *Sivatherium* (Giraffidae, Mammalia) du Pliocène final d'Ahl al Oughalam (Casablanca Maroc), et l'évolution du genre en Afrique. Palaöntologische Zeitschrift 70:623–629.

Geraads, D., Z. Alemseged, and H. Bellon. 2002. The late Miocene mammalian fauna of Chorora, Awash Basin, Ethiopia: systematics, biochronology, and the ^{40}K-^{40}Ar ages of the associated volcanics. Tertiary Research 21: 113–122.

Geraads, D., G. Bouvrain, and J. Sudre. 1987. Relations phylétiques de *Bachitherium* Filhol, ruminant de l'Oligocène d'Europe occidentale. Palaeovertebrata 17:43–73.

Gervais, P. 1848–1852. Zoologie et Paléontologie françaises (animaux vertébrés) ou Nouvelles Recherches sur les Animaux Vivants et Fossiles de la France. Arthus Bertrand, Paris, 271 pp.

Gervais, P. 1849. Recherches sur les mammifères fossiles des genres *Palaeotherium* et *Lophiodon* et sur les autres animaux de la même classe que l'on a trouvés avec eux dans le midi de la France. Comptes Rendus de l'Académie des Sciences, Paris 29:381–384, 568–579.

Gervais, P. 1850. Nouvelles recherches relatives aux mammifères d'espèces éteintes qui sont enfouies auprès d'Apt, avec des *Palaeotherium* identiques à ceux de Paris. Comptes Rendus Hebdomadaires des Séances de l'Académie des Sciences 30:602–604.

Gervais, P. 1856. Sur les mammifères fossiles que l'on a recueillis dans le département du Gard. Comptes Rendus Hebdomadaires des Séances de l'Académie des Sciences 42:1159–1161.

Gervais, P. 1859. Zoologie et paléontologie françaises (animaux vertébrés) (2nd edition). Arthus Bertrand, Paris, 544 pp.

Gervais, P. 1873. Mammifères dont les ossements accompagnent les dépots de chaux phosphatée des départements de Tarn-et-Garonne et du Lot. Journal de Zoologie 2:356–380.

Gervais, P. 1876. Zoologie et Paléontologie générales, nouvelles recherches sur les Animaux Vertébrés dont on trouve les ossements enfouis dans le sol et sur leur comparasaison avec les espèces actuellement existantes. Série 2; pp. 1–72, Paris.

Gèze, R. 1980. Les Hippopotamidae (Mammalia, Artiodactyla) du Plio-Pléistocène de l'Ethiopie. Ph.D. dissertation, Université Pierre et Marie Curie—Paris VI.

Gèze, R. 1985. Répartition paléoécologique et relations phylogénétiques des Hippopotamidae (Mammalia, Artiodactyla) du néogène d'Afrique Orientale; pp. 81–100 in M. Beden, A. K. Berhensmeyer, N. T. Boaz, R. Bonnefille, C. K. Brain, B. Cooke, Y. Coppens, R. Dechamps, V. Eisenmann, A. Gentry, D. Geraads, R. Gèze, C. Guérin, J. Harris, J. Koeniguer, F. Letouzey, G. Petter, A. Vincens, and E. Vrba (eds.), L'environnement des hominidés au Plio-Pléistocène. Fondation Singer-Polignac, Masson, Paris.

Gheerbrant, E. 1987. Les vertebras continentaux de l'Adrar Mgorn (Maroc, Paléocène); une dispersion de mammifères transtéthysienne aux environs de la limite mésozoique/cénozoique? Geodinamica Acta 1:233–246.

Gheerbrant, E., J. Sudre, P. Tassy, M. Amaghzaz, and M. Iarochène. 2005. Nouvelles données sur *Phosphatherium escuillei* (Mammalia, Proboscidea) de l'Éocène inférieur du Maroc, apports à la phylogenie des Proboscidea et de ongulés lophodontes. Geodiversitas 27:239–333.

Gilbert, C., A. Ropiquet, and A. Hassanin. 2006. Mitochondrial and nuclear phylogenies of Cervidae (Mammalia, Ruminantia): systematics, morphology, and biogeography. Molecular Phylogenetics and Evolution 40:101–117.

Gill, T. 1872. Arrangement of the families of mammals and synoptical tables of characters of the subdivisions of mammals. Smithsonian Miscellaneous Collections 230(11):1–98.

Gingerich, P. D. 1976. Systematic position of the alleged primate *Lantianius xiehuensis* Chow, 1964, from the Eocene of China. Journal of Mammalogy 57:194–198.

Gingerich, P. D. 1986. Early Eocene *Cantius torresi*—oldest primate of modern aspect from North America. Nature 320: 319–321.

Gingerich, P. D. 1989. New earliest Wasatchian mammalian fauna from the Eocene of Northwestern Wyoming: composition and diversity in a rarely sampled high-floodplain assemblage. University of Michigan Papers in Paleontology 28:1–97.

Gingerich, P. D. 2003. Stratigraphic and micropaleontological constraints on the middle Eocene age of mammal-bearing Kuldana Formation of Pakistan. Journal of Vertebrate Paleontology 23:643–651.

Gingerich, P. D. 2005. Cetacea; pp. 234–252 in K. D. Rose and J. D. Archibald (eds.), The Rise of Placental Mammals. The Johns Hopkins University Press, Baltimore, Maryland.

Gingerich, P. D., and D. E. Russell. 1981. *Pakicetus inachus,* a new archaeocete (Mammalia, Cetacea) from the Early-Middle Eocene Kuldana Formation of Kohat (Pakistan). Contributions from the Museum of Paleontology University of Michigan 25:235–246.

Gingerich, P. D., and D. E. Russell. 1990. Dentition of the Early Eocene *Pakicetus* (Mammalia, Cetacea). Contributions from the Museum of Paleontology University of Michigan 28:1–20.

Gingerich, P. D., M. Arif, and W. C. Clyde. 1995. New archaeocetes (Mammalia, Cetacea) from the middle Eocene Domanda Formation of the Sulaiman Range; Punjab (Pakistan). Contributions from the Museum of Paleontology University of Michigan 29:291–230.

Gingerich, P. D., B. H. Smith, and E. L. Simons. 1990. Hind limbs of Eocene *Basilosaurus:* evidence of feet in whales. Science 249:154–157.

Gingerich, P. D., D. E. Russell, D. Sigogneau-Russell, and J. L. Hartenberger. 1979. *Chorlakkia hassani,* a new middle Eocene dichobunid (Mammalia, Artiodactyla) from the Kuldana Formation of Kohat (Pakistan). Contributions from the Museum of Paleontology University of Michigan 25:117–124.

Gingerich, P. D., M. U. Haq, I. S. Zalmout, I. H. Khan, and M. S. Malakani. 2001. Origin of whales from early artiodactyls: hands and feet of Eocene Protocetidae from Pakistan. Science 293:2239–2242.

Ginsburg, L. 1977. *Listriodon juba,* suidé nouveaux du Miocène de Bani Mellal (Maroc). Géologie Méditerranéenne 4:221–224.

Ginsburg, L. 1985. Systematique et evolution du genre Miocene *Palaeomeryx* (Artiodactyla, Giraffoidea) en Europe. Comptes Rendus de l'Academie des Sciences 301:1075–1078.

Ginsburg, L. 1989. Les mammifères des sables du Miocène inférieur des Beilleaux à Savigné-sur-Lathan (Indre-et-Loire). Bulletin du Musée national d'Histoire naturelle Paris, 4ème séries C 11:101–121.

Ginsburg, L., and B. Azanza. 1991. Présence de bois chez les femelles du cervidé miocène *Dicrocerus elegans* et remarques sur le problème de l'origine du dimorphisme sexuel sur les appendices frontaux des Cervidés. Comptes Rendus Hebdomadaires des Seances de l'Academie des Sciences Paris II 313:121–126.

Ginsburg, L., and C. Bulot. 1987. Les artiodactyles sélénodontes du Miocène de Bézian à La Romieu (Gers). Bulletin du Muséum national d' Histoire naturelle, Paris 9C:63–95.

Ginsburg, L., and E. Heintz. 1966. Sur les affinités du genre *Palaeomeryx* (ruminant du Miocène européen). Comptes Rendus Hebdomadaires des Seances de l'Academie des Sciences, Paris 301:1255–1257.

Ginsburg, L., and M. Hugueney. 1987. Les mammifères terrestres des sables stampiens du bassin de Paris. Annales de Palaeontologie 73:83–130.

Ginsburg, L., J. Huin, and J.-P. Locher. 1985. Les artiodactyles sélénodontes du Miocène inférieur des Beilleaux à Savigné-sur-Lathan (Indre-et-Loire). Bulletin du Muséum national d'Histoire Naturelle 7:285–304.

Ginsburg, L., J. Morales, and D. Soria. 1994. The ruminants (Artiodactyla, Mammalia) from the lower Miocene of Cetina de Aragon (Province of Zaragoza, Aragon, Spain). Proceedings of the Koninklijke Nederlandse Akademic van Wetenschappen 97:141–181.

Ginsburg, L., J. Morales, and D. Soria. 2001. Les Ruminantia (Artiodactyla, Mammalia) du Miocène des Bugti (Balouchistan, Pakistan). Estudios Geologica 57:155–170.

Godina, A. 1979. [History of the fossil giraffes of the genus *Palaeotragus*.] Trudy Paleontological Institute of the Academy Sciences. Russia 177:1–114 [Russian].

Godinot, M. 1978. Diagnoses de trois nouvelles espèces de mammifères du Sparnacien de Provence. Comptes Rendus Sommaire des Séances de la Société Géologique de France 6:286–288.

Godinot, M. 1982. Aspects nouveaux des echanges entre les faunes mammaliennes d'Europe et d'Amerique du Nord a la base de l'Eocene. Géobios, Mémoire Spécial 6:403–412.

Goloboff, P. 1994. NONA version 1.9. Computer program and documentation. Available at http://www.cladistics.com/aboutNona.htm.

Golpe-Posse, J. M. 1971. Suiformes del Terciario Español y sus yacimientos (Ph.D. disertación). Publicaciones des Instituto Provincial de Paleontologia de Sabadell, Paleontologia y Evolucion 2:1–197.

Golpe-Posse, J. M. 1972. Suiformes des Terciario español y sus yacimientos. Paleontologia y Evolución 8:1–87.

Golz, D. J. 1976. Eocene Artiodactyla of Southern California. Natural History Museum of Los Angeles County Science Bulletin 26:1–85.

González, S., F. Álvarez-Valin, and J. E. Maldonaldo. 2002. Morphometric differentiation of endangered Pampas deer (*Ozotoceros bezoarticus*), with description of new subspecies from Uruguay. Journal of Mammalogy 83:1127–1140.

Goodman, M., C. A. Porter, J. Czelusniak, S. L. Page, H. Schneider, J. Shoshani, G. Gunnell, and C. P. Groves. 1997. Toward a phylogenetic classification of Primates based on DNA evidence complemented by fossil evidence. Molecular Phylogenetics and Evolution 9:585–598.

Graham, A. 1993. History of the vegetation: Cretaceous (Maastrichtian)–Tertiary; pp. 57–70 in Flora of North America Editorial Committee (eds.), Flora of North America North of Mexico, Volume 1. Oxford University Press, New York, New York, 373 pp.

Graham, A. 1999. Cretaceous and Cenozoic history of North American Vegetation. Oxford University Press, New York, New York, 350 pp.

Graur, D., and D. G. Higgins. 1994. Molecular evidence for the inclusion of cetaceans within the order Artiodactyla. Molecular Biology and Evolution 11:357–364.

Gravlund, P., M. Meldgaard, S. Pääbo, and P. Arctander. 1998. Polyphyletic origin of the small-bodied, High-Arctic subspecies of tundra reindeer (*Rangifer tarandus*). Molecular Phylogenetics and Evolution 10:151–159.

Gray, J. E. 1821. On the natural arrangement of vertebrose animals. London Medical Repository 15:296–310.

Gray, J. E. 1843. List of the Specimens of Mammalia in the Collection in the British Museum. Trustees of the British Museum, London, 216 pp.

Gray, J. E. 1852. Catalogue of the specimens of Mammalia in the collection of the British Museum. Part 3: Ungulata Furcipeda. British Museum, London, pp. xvi + 286.

Gray, J. E. 1868. Synopsis of the pigs (Suidae) in the British Museum. Proceedings of the Zoological Society of London 1868:17–49.

Gray, J. E. 1869. On the bony dorsal shield of the male *Tragulus kanchil*. Proceedings of the Zoological Society in London 7:226–227.

Green, M. J. B. 1987. Scent-marking in the Himalayan musk-deer. Journal of Zoology, London, B 1:721–737.

Greene, E. C. 1935. Anatomy of the rat. Transactions of the American Philosophical Society 27:1–370.

Gregory, J. T. 1942. Pliocene vertebrates from Big Springs Canyon, South Dakota. University of California Publications, Bulletin of the Department of Geological Sciences 26:307–466.

Grey, J., and D. M. Harper. 2002. Using stable isotope analyses to identify allochthonous inputs to lake Naivasha mediated via the hippopotamus gut. Isotopes in Environmental and Health Studies 38:245–250.

Groves, C. P. 1976. The taxonomy of *Moschus* (Mammalia, Artiodactyla) with special reference to the Indian region. Journal of the Bombay Natural History Society 72:662–782.

Groves, C. P. 2001. Primate Taxonomy. Smithsonian Institution Press, Washington, D.C.

Groves, C. P. 2006. The genus *Cervus* in eastern Eurasia. European Journal of Wildlife Research 52:14–22.

Groves, C. P., and P. Grubb. 1987. Relationships of living deer; pp. 21–59 in C. M. Wemmer (ed.), Biology and Management of the Cervidae. Smithsonian Institution Press, Washington, D.C.

Groves, C. P., and P. Grubb. 1990. Muntiacidae; pp. 134–168 in G. A. Bubenik and A. B. Bubenik (eds.), Horns, Pronghorns and Antlers. Springer-Verlag, New York, New York.

Groves, C. P., and E. Meijaard. 2005. Interspecific variation in *Moschiola*, the Indian chevrotain. The Raffles Bulletin of Zoology, Supplement 12:413–421.

Groves, P., and G. F. Shields. 1996. Phylogenetics of the Caprinae based on cytochrome *b* sequence. Molecular Phylogenetics and Evolution 5:467–476.

Groves, P., and G. F. Shields. 1997. Cytochrome *b* sequences suggest convergent evolution of the Asian takin and Arctic muskox. Molecular Phylogenetics and Evolution 8:363–374.

Groves, C. P., Y. Wang, and P. Grubb. 1995. Taxonomy of musk deer, genus *Moschus* (Moschidae, Mammalia). Acta Theriologica Sinica 15:181–197.

Grubb, P. 2000. Valid and invalid nomenclature of living and fossil deer, Cervidae. Acta Theriologica 45:289–307.

Guldberg, G. A. 1883. Undersøgelser over en subfossil flodhest fra Madagascar. Videnskabs-selskabets forhandlinger, Christiania 6:1–24.

Gunnell, G. F. 1997. Wasatchian-Bridgerian (Eocene) paleoecology of the western interior of North America: changing paleoenvironments and taxonomic composition of omomyid (Tarsiiformes) primates. Journal of Human Evolution 32:105–132.

Gunnell, G. F., and W. S. Bartels. 2001. Basin margins, biodiversity, evolutionary innovation, and the origin of new taxa; pp. 404–430 in G. F. Gunnell (ed.), Eocene Biodiversity: Unusual Occurrences and Rarely Sampled Habitats. Kluwer Academic/Plenum Publishers, New York, New York.

Guo, J., M. R. Dawson, and K. C. Beard. 2000. *Zhailimeryx*, a new lophiomerycid artiodactyl (Mammalia) from the late middle Eocene of Central China and the early evolution of ruminants. Journal of Mammalian Evolution 7:239–258.

Guo, J., T. Qi, and H.-J. Sheng. 1999. A restudy of the Eocene ruminants from Baise and Yongle Basins, Guangxi, China, with a discussion of the systematic positions of *Indomeryx*, *Notomeryx*, *Gobiomeryx* and *Prodremotherium*. Vertebrata PalAsiatica 37:18–39.

Guthrie, D. A. 1966. A new species of dichobunid artiodactyl from the early Eocene of Wyoming. Journal of Mammalogy 47:487–490.

Guthrie, D. A. 1971. The mammalian fauna of the Lost Cabin Member, Wind River formation (Lower Eocene) of Wyoming. Annals of Carnegie Museum 43(4):48–113.

Hamilton, W. R. 1973. The Lower Miocene ruminants of Gebel Zelten, Libya. Bulletin of the British Museum of Natural History (Geology), London 21:76–150.

Hamilton, W. R. 1978a. Fossil giraffes from the Miocene of Africa and a revision of the phylogeny of Giraffoidea. Philosophical Transactions of the Royal Society (B) 283:165–229.

Hamilton, W. R. 1978b. Cervidae and Palaeomerycidae, pp. 496–508 in V. J. Maglio and H. B. S. Cooke (eds.), Evolution of African Mammals. Harvard University Press, Cambridge, Massachusetts.

Hammond, R. L., W. Macasero, B. Flores, O. B. Mohammed, T. Wacher, and M. W. Bruford. 2001. Phylogenetic reanalysis of the Saudi gazelle and its implications for conservation. Conservation Biology 15:1123–1133.

Han, D. 1986. Fossils of Tragulidae from Lufeng, Yunnan. Acta Anthropologica Sinica 5:68–78 [Chinese 68–77, English 77–78].

Hanson, C. B. 1996. Stratigraphy and vertebrate faunas of the Bridgerian-Duchesnean Clarno Formation, North-Central Oregon; pp. 206–239 in D. R. Prothero and R. J. Emry (eds.), The Terrestrial Eocene–Oligocene Transition in North America. Cambridge University Press, Cambridge.

Harrington, G. J. 2001. Impact of Paleocene/Eocene greenhouse warming on North American tropical forests. Palaios 16:266–278.

Harris, J. M. 1991. Family Hippopotamidae; pp. 31–85 in J. M. Harris (ed.), Koobi Fora Research Project. Clarendon Press, Oxford.

Harris, J. M., and T. E. Cerling. 2002. Dietary adaptations of extant and Neogene African suids. Journal of Zoology 256:45–54.

Harris, J. M., and M. G. Leakey. 2003. Lothagam Suidae; pp. 371–385 in M. G. Leakey and J. M. Harris (eds.), Lothagam: the Dawn of Humanity in Africa. Columbia University Press, New York, New York.

Harris, J. M., and T. D. White. 1979. Evolution of the Plio-Pleistocene African Suidae. Transactions of the American Philosophical Society 69(2):1–128.

Harris, J. M., F. H. Brown, and M. G. Leakey. 1988. Stratigraphy and paleontology of Pliocene and Pleistocene localities west of Lake Turkana, Kenya. Contributions in Science, Natural History Museum of the Los Angeles County, Los Angeles, California 399:1–128.

Harris, J. M., M. G. Leakey, and T. E. Cerling. 2004. Early Pliocene tetrapod remains from Kanapoi, Lake Turkana Basin, Kenya; pp. 39–113 in J. M. Harris and M. G. Leakey (eds.), Geology and Vertebrate Paleontology of the Early Pliocene Site of Kanapoi, Northern Kenya. Contributions in Science, Natural History Museum of the Los Angeles County, Los Angeles, California.

Harrison, D. L., P. J. J. Bates, and N. M. Thomas. 1995. The occurence of Acotherulum pumilum (Stehlin, 1908) (Mammalia, Artiodactyla, Cebochoeridae) in the Headonian (Upper Eocene) of England. Tertiary Research 15:139–143.

Harrison, T. 1997. The anatomy, paleobiology, and phylogenetic relationships of the Hippopotamidae (Mammalia, Artiodactyla) from the Manonga Valley, Tanzania; pp. 137–190 in T. Harrison (ed.), Neogene Paleontology of the Manonga Valley, Tanzania. Plenum Press, New York, New York.

Hartenberger, J.-L., B. Sigé, and J. Sudre. 1969. Les gisements de vertébrés de la région montpelliéraine. 1. Gisements éocènes. Bulletin du Bureau de recherches géologiques et minières 2:7–18.

Hasegawa, M., and J. Adachi. 1996. Phylogenetic position of cetaceans relative to artiodactyls: reanalysis of mitochondrial and nuclear sequences. Molecular Biology and Evolution 13:710–717.

Hassanin, A., and E. J. P. Douzery. 1999a. Evolutionary affinities of the enigmatic saola (Pseudoryx nghetinhensis) in the context of the molecular phylogeny of the Bovidae. Proceedings of the Royal Society of London, Series B 266:893–900.

Hassanin, A., and E. J. P. Douzery. 1999b. The tribal radiation of the family Bovidae (Artiodactyla) and the evolution of the mitochondrial cytochrome b gene. Molecular Phylogenetics and Evolution 13:227–243.

Hassanin, A., and E. J. P. Douzery. 2003. Molecular and morphological phylogenies of the Ruminantia and the alternative position of the Moschidae. Systematic Biology 52(2):206–228.

Hassanin, A., and A. Ropiquet. 2004. Molecular phylogeny of the tribe Bovini (Bovidae, Bovinae) and the taxonomic status of the Kouprey, Bos sauveli Urbain 1937. Molecular Phylogenetics and Evolution 33:896–907.

Hassanin, A., E. Pasquet, and J.-D. Vigne. 1998. Molecular systematics of the subfamily Caprinae (Artiodactyla, Bovidae) as determined from cytochrome b sequences. Journal of Mammalian Evolution 5:217–236.

Hatcher, J. B. 1901. Some new and little known fossil vertebrates. Annals of Carnegie Museum 1:128–144.

Haubold, H., and M. Hellmund. 1997. Contribution of the Geiseltal to the Paleogene biochronology and the actual perspective of the Geiseltal district. Mémoires et Travaux de l'Institut de Montpellier de l'Ecole Pratique des Hautes Etudes 21:353–359.

Hay, O. P. 1902. Bibliography and catalogue of the fossil Vertebrata of North America. United States Geological Survey Bulletin 179:1–868.

Heaton, T. H., and R. J. Emry. 1996. Leptomerycidae; pp. 581–608 in D. R. Prothero and R. J. Emry (eds.), The Terrestrial Eocene–Oligocene Transition in North America. Cambridge University Press, Cambridge.

Heinrich, R. E., and K. D. Rose. 1997. Postcranial morphology and locomotor behavior of two early Eocene miacoid carnivorans, Vulpavus and Didymictis. Palaeontology 40:279–305.

Heintz, E., and E. Aguirre. 1976. Le bois de Croizetoceros ramosus pueblensis, Cervidé de la faune villafranchienne de la Puebla de Valverde, Teruel (Espagne). Estudios Geológicos 32:569–572.

Heintz, E., and M. Dubar. 1981. Place et signification des dépôts villafranchiens de Moustiers-Ségriès et faune de Mammifères de Cornillet (Alpes de Haute-Provence). Bulletin Musee national Histoire naturelle, Paris 3C:363–397.

Heintz, E., and F. Poplin. 1981. Alces carnutorum (Laugel, 1862) du Pléistocéne de Saint-Prest (France). Systématique et évolution des Alcinés (Cervidae, Mammalia). Quartärpaläontologie 4:105–122.

Heissig, K. 1978. Fossil führende Spaltenfüllungen Süddeutschlands und die Ökologie ihrer oligozänen Huftiere. Mitteilungen der Bayerischen Staatssammlung für Paläontologie und Historische Geologie 18:237–288.

Heissig, K. 1990. Ein Oberkiefer von Anthracohyus (Mammalia, ?Artiodactyla) aus dem Eozän Jugoslawiens. Mitteilungen der Bayerischen Staatssammlung für Paläontologie und historische Geologie 30:57–64.

Heissig, K. 1993. The astragalus in anoplotheres and oreodonts, phylogenetical and paleogeographical implications. Kaupia 3:173–178.

Heissig, K. 2001. Anthracohyus (Artiodactyla, Mammalia), an Eurasian achaenodontid. Lynx 32:97–105.

Heizmann, E. P. J. 1983. Die Gattung *Cainotherium* (Cainotheriidae) im Orleanium und im Astaracium Sueddeutschlands. Eclogae Geologicae Helvetiae 76:781–825.

Heizmann, E. P. J. 1999. Family Cainotheriidae; pp. 217–220 in G. E. Rössner and K. Heissig (eds.), The Miocene Land Mammals of Europe. F. Pfeil, Munich.

Heller, F. 1934. *Anthracobunodon weigelti* n. g. et n. sp., ein Artiodactyle aus dem Mitteleozän des Geiseltales bei Halle a. S. Palaeontologische Zeitschrift 16:147–263.

Hellmund, M. 1991. Revision der europäischen Species der Gattung Elomeryx Marsh, 1894 (Anthracotheriidae, Artiodactyla, Mammalia)—odontologische Untersuchungen. Palaeontographica Abteilung A 220:1–101.

Hellmund, M. 1992. Schweineartige (Suina, Artiodactyla, Mammalia) aus oligo-miozanen Fundstellen Deutschlands, der Schweiz und Frankreichs. 2. Revision von *Palaeochoerus* Pomel 1847 und *Propalaeochoerus* Stehlin 1899 (Tayassuidae). Stuttgarter Beitraege zur Naturkunde Serie B (Geologie und Palaeontologie) 189:1–74.

Hellmund, M., and K. Heissig. 1994. Neuere Funde von *Prominatherium dalmatinum* (H. v. Meyer 1854) (Artiodactyla, Mammalia) aus dem Eozän von Dalmatien. Mitteilungen der Bayerischen Staatssammlung für Paläontologie und historische Geologie 34:273–281.

Hendey, Q. B. 1976. Fossil peccary from the Pliocene of South Africa. Science 192:787–789.

Hernández Fernández, M., and E. S. Vrba. 2005. A complete estimate of the phylogenetic relationships in Ruminantia: a dated species-level supertree of extant ruminants. Biological Reviews 80:269–302.

Hesse, C. J. 1935. New evidence on the ancestry of *Antilocapra americana*. Journal of Mammalogy 16:307–315.

Hessig, K. 1999. Chalicotheriidae; pp. 189–192 in G. E. Rössner and K. Heissig (eds.), Land Mammals of Europe. Verlag Dr. Friedrich Pfeil, Munich.

Hillis, D. M. 1999. SINEs of the perfect character. Proceedings of the National Academy of Sciences, USA 96:9979–9981.

Hillis, D. M., and J. J. Wiens. 2000. Molecules versus morphology in systematics: conflicts, artefacts and misconceptions; pp. 1–19 in J. J. Wiens (ed.), Phylogenetic Analysis of Morphological Data. Smithsonian Institution Press, Washington, D.C.

Hoffman, J. M., and D. R. Prothero. 2004. Revision of the late Oligocene dwarfed leptaucheniine oreodont *Sespia* (Mammalia: Artiodactyla). New Mexico Museum of Natural History and Science Bulletin 26:155–163.

Hofmann, A. 1893. Die Fauna von Göriach. Abhandlungen der Kaiserlich-königlichen geologischen Reichsanstalt 15(6):1–87.

Hofmann, R. R., and D. R. M. Stewart. 1972. Grazer or browser: a classification based on stomach-structure and feeding habits of East African mammals. Mammalia 36:227–240.

Holroyd, P. A. 2002. New record of Anthracotheriidae (Artiodactyla: Mammalia) from the middle Eocene Yegua Formation (Claiborne Group), Houston County, Texas. Texas Journal of Science 54:301–308.

Honey, J. G. 2004. Taxonomic utility of sequential wear patterns in some fossil camelids: comparison of three Miocene taxa. Bulletin of the Carnegie Museum of Natural History 36:43–62.

Honey, J. G., and B. E. Taylor. 1978. A generic revision of the Protolabidini (Mammalia, Camelidae), with a description of two new protolabidines. Bulletin of the American Museum of Natural History 161:369–425.

Honey, J. G., J. A. Harrison, D. R. Prothero, and M. S. Stevens. 1998. Camelidae; pp. 439–462 in C. M. Janis, K. M. Scott, and L. L. Jacobs (eds.), Evolution of Tertiary Mammals of North America. Volume 1: Terrestrial Carnivores, Ungulates, and Ungulatelike Mammals. Cambridge University Press, Cambridge.

Honeycutt, R. L., and R. M. Adkins. 1993. Higher level systematics of eutherian mammals: an assessment of molecular characters and phylogenetic hypotheses. Annual Review of Ecology and Systematics 24:279–305.

Hooijer, D. A. 1946. Notes on some Pontian mammals from Sicily, figured by Seguenza. Archives Néerlandaises de Zoologie 7:301–333.

Hooijer, D. A. 1950. The fossil Hippopotamidae of Asia, with notes on the recent species. Zoologische Verhandelingen 8:3–123.

Hooijer, D. A. 1958a. Fossil Bovidae from the Malay Archipelago and the Punjab. Zoologische Verhandelingen, Leiden 38: 1–122.

Hooijer, D. A. 1958b. Pleistocene remains of hippopotamus from the Orange Free State. Navorsinge van die Nasionale Museum, Bloemfontein 1(11):259–266.

Hooker, J. J. 1986. Mammals from the Bartonian (Middle/Late Eocene) of the Hampshire Basin. Bulletin of the British Museum (Natural History) Geology series 39:191–478.

Hooker, J. J. 1989. Character polarities in early perissodactyls and their significance for Hyracotherium and infraordinal relationships; pp. 79–101 in D. R. Prothero and R. M. Schoch (eds.), The Evolution of Perissodactyls. Oxford University Press, New York, New York.

Hooker, J. J. 1992. British mammalian paleocommunities across the Eocene–Oligocene transition and their environmental implications; pp. 494–515 in D. R. Prothero and W. A. Berggren (eds.), Eocene–Oligocene Climatic and Biotic Evolution. Princeton University Press, Princeton, New Jersey.

Hooker, J. J. 2000. Palaeogene mammals: crisis and ecological change; pp. 333–349 in S. J. Culver and P. F. Rawson (eds.), Biotic Response to Global Change: The Last 145 Million Years. Cambridge University Press, Cambridge.

Hooker, J. J., and K. M. Thomas 2001. A new species of *Amphirhagatherium* (Choeropotamidae, Artiodactyla, Mammalia) from the late Eocene Headon Hill Formation of southern England and phylogeny of endemic European "Anthracotherioids." Palaeontology 44:827–853.

Hooker, J. J., and M. Weidmann. 2000. The Eocene mammal faunas of Mormont, Switzerland: systematic revision and resolution of dating problems. Mémoires suisses de Paléontologie 120:1–141.

Hooker, J. J., M. E. Collinson, and N. P. Sille 2004. Eocene–Oligocene mammalian faunal turnover in the Hampshire Basin, UK: calibration to the global time scale and the major cooling event. Journal of the Geological Society London 16:161–172.

Hopwood, A. T. 1926. The Geology and Palaeontology of the Kaiso Bone Beds, Uganda. Part II, Palaeontology. Fossil Mammalia. Occasional Paper (Uganda Geological Survey) 2:13–36.

Hopwood, A. T. 1929. New and little-known mammals from the Miocene of Africa. American Museum Novitates 344:1–9.

Horwitz, L. K., and E. Tchernov. 1990. Cultural and environmental implications of hippopotamus bone remains in archaeological contexts in the Levant. Bulletin of American Schools of Oriental Research 280:67–76.

Houtemaker, J. L., and P. Y. Sondaar. 1979. Osteology of the fore limb of the Pleistocene dwarf hippopotamus from Cyprus with special reference to phylogeny and function. Palaeontology B 82:411–448.

Hu, C. K. 1963. A new Eocene anthracothere. Vertebrata Pal-Asiatica 7:315–316.

Huelsenbeck, J. P. 1991. When are fossils better than extant taxa in phylogenetic analysis? Systematic Zoology 40:458–469.

Huelsenbeck, J. P., and D. M. Hillis. 1993. Success of phylogenetic methods in the four-taxon case. Systematic Biology 42:247–264.

Huelsenbeck, J. P., and B. Rannala. 1997. Maximum likelihood estimation of phylogeny using stratigraphic data. Paleobiology 23:174–180.

Hulbert, R. C., Jr. 1998. Postcranial osteology of the North American middle Eocene protocetid *Georgiacetus;* pp. 235–267 in J. G. M. Thewissen (ed.), The Emergence of Whales. Plenum Press, New York, New York.

Hulbert, R. C., Jr., R. M. Petkewich, G. A. Bishop, D. Burky, and D. P. Aleshire. 1998. A new middle Eocene protocetid whale (Mammalia: Cetacea: Archaeoceti) and associated biota from Georgia. Journal of Paleontology 72:907–927.

Hundertmark, K. J., R. T. Bowyer, G. F. Shields, and C. C. Schwartz. 2003. Mitochondrial phylogeography of moose (*Alces alces*) in North America. Journal of Mammalogy 84:718–728.

Hünermann, K. A. 1968. Die Suidae (Mammalia, Artiodactyla) aus den Dinotheriernsandes (Unterpliozän: Pont) Rheinhessens (Sudwestdeutschland). Schweitzer Paläontologische Abhandlungen 86:1–96.

Hünermann, K. A. 1983. *Dorcatherium* (Mammalia, Artiodactyla, Tragulidae), das fossile Hirschferkel von Feuerthalen/Flurlingen (Kt. Zürich) bei Schaffhausen und seine Lagerstätte. Mitteilungen der Naturforschenden Gesellschaft Schaffhausen, Jahrgang 1981/84 32:1–17.

Hunt, R. M., Jr. 1990. Taphonomy and sedimentology of Arikaree (lower Miocene) fluvial, eolian, and lacustrine paleoenvironments, Nebraska and Wyoming; a paleobiota entombed in fine-grained volcaniclastic rocks. Geological Society of America, Special Paper 24:69–111.

Hunter, J. P., and M. Fortelius. 1994. Comparative dental occlusal morphology, facet development, and microwear in two sympatric species of Listrodon (Mammalia: Suidae) from the middle Miocene of western Anatolia (Turkey). Journal of Vertebrate Paleontology 14:105–126.

Hürzeler, J. 1936. Osteologie und Odontologie der Caenotheriden. Abhandlungen Schweizerischen Paläontologischen Gesellschaft 58:1–112.

Hürzeler, J. 1938. *Ephelcomenus* nov. gen., ein Anoplotheriide aus dem mittleren Stampien. Eclogae Geologicae Helvetiae 31:317–326.

Illiger, C. 1811. Prodromus systematis Mammalium et Avium additis terminis zoographicis utriusque classis, eorumque versione Germanica. Salfeld, Berlin, 301 pp.

International Commission on Zoological Nomenclature. 1999. International Code of Zoological Nomenclature, 4th edition. International Trust for Zoological Nomenclature, Natural History Museum, London, 306 pp.

Irwin, D. M., and Ú. Árnason. 1994. Cytochrome *b* gene of marine mammals: phylogeny and evolution. Journal of Mammalian Evolution 2:37–55.

Irwin, D. M., T. D. Kocher, and A. C. Wilson. 1991. Evolution of the cytochrome *b* gene of mammals. Journal of Molecular Evolution 32:128–144.

Isaac, G. L. 1977. Olorgesailie. Archeological Studies of a Middle Pleistocene Lake Basin in Kenya. The University of Chicago Press, Chicago, Illinois, 272 pp.

Jablonski, N. 2003. The hippo's tale: how the anatomy and physiology of Late Neogene *Hexaprotodon* shed light on Late Neogene environmental change. Quaternary International 117:119–123.

Jacobs, B. F., J. D. Kingston, and L. L. Jacobs. 1999. The origin of grass-dominated ecosystems. Annals of the Missouri Botanical Gardens 86:590–643.

Jaeckel, O. M. J. 1911. Die Wirbeltiere. Eine Übersicht über die fossilen und lebenden Formen. Gebrüder Bornträger, Berlin, 252 pp.

James, G. T., and B. H. Slaughter. 1974. A primitive new middle Pliocene murid from Wadi El Natrun, Egypt. Annals of the Geological Survey of Egypt 4:333–362.

Janis, C. M. 1976. The evolutionary strategy of the Equidae, and the origins of rumen and cecal digestion. Evolution 30:757–774.

Janis, C. M. 1982. Evolution of horns in ungulates: ecology and paleoecology. Biological Reviews of the Cambridge Philosophical Society 57:261–318.

Janis, C. M. 1984. Tragulids as living fossils; pp. 87–94 in N. Eldredge and S. Stanley (eds.), Living Fossils. Springer-Verlag, New York, New York.

Janis, C. M. 1987. Grades and clades in hornless ruminant evolution: the reality of the Gelocidae and the systematic position of *Lophiomeryx* and *Bachitherium.* Journal of Vertebrate Paleontology 7:200–216.

Janis, C. M. 1988. An estimation of tooth volume and hypsodonty indices in ungulate mammals and the correlation of these factors with dietary preferences; pp. 367–387 in D. E. Russell, J. P. Santorio, and D. Signogneu-Russell (eds.), Teeth Revisited: Proceedings of the VII International Symposium on Dental Morphology, series C, Volume 3. Muséum national de Histoire Naturelle Memoir, Paris.

Janis, C. M. 1989. A climatic explanation for patterns of evolutionary diversity in ungulate mammals. Palaeontology 32:463–481.

Janis, C. M. 1990. The correlation between diet and dental wear in herbivorous mammals, and its relationship to the determination of diets of extinct species; pp. 241–259 in A. J. Boucot (ed.), Evolutionary Paleobiology of Behaviour and Coevolution. Elsevier, Amsterdam, Toronto.

Janis, C. M. 1993. Tertiary mammal evolution in the context of changing climates, vegetation, and tectonic evens. Annual Review of Ecology and Systematics 24:467–500.

Janis, C. M. 1995. Correlation between craniodental morphology and feeding behaviour in ungulates: reciprocal illumination between living and fossil taxa; pp. 67–98 in T. Jeff (ed.), Functional Morphology in Vertebrate Paleontology. Cambridge University Press, Cambridge.

Janis, C. M. 2000a. Patterns in the evolution of herbivory of large terrestrial mammals: the Paleogene of North America; pp. 168–222 in H.-D. Sues (ed.), Evolution of Herbivory in Terrestrial Vertebrates. Cambridge University Press, Cambridge.

Janis, C. M. 2000b. The endemic ruminants of the Neogene of North America; pp. 26–37 in E. S. Vrba and G. B. Schaller (eds.), Antelopes, Deer, and Relatives: Fossil Record, Behavioral Ecology, Systematics, and Conservation. Yale University Press, New Haven, Connecticut.

Janis, C. M., and M. Fortelius. 1988. On the means whereby mammals achieve increased functional durability of their dentitions, with special reference to limiting factors. Biological Reviews 63:197–230.

Janis, C. M., and E. M. Manning. 1998a. Antilocapridae; pp. 491–507 in C. M. Janis, K. M. Scott, and L. L. Jacobs (eds.), Evolution of Tertiary Mammals of North America. Volume 1: Terrestrial Carnivores, Ungulates, and Ungulatelike Mammals. Cambridge University Press, Cambridge.

Janis, C. M., and E. M. Manning. 1998b. Dromomerycidae; pp. 477–490 in C. Janis, K. M. Scott, and L. Jacobs (eds.), Tertiary Mammals of North America, Volume 1: Terrestrial Carnivores, Ungulates, and Ungulatelike Mammals. Cambridge University Press, Cambridge.

Janis, C. M., and K. M. Scott. 1987. The interrelationships of higher ruminant families with special emphasis on the members of the Cervoidea. American Museum Novitates 2893:1–85.

Janis, C. M., and K. M. Scott. 1988. The phylogeny of the Ruminantia (Artiodactyla, Mammalia); pp. 273–282 in M. J. Benton (ed.), The Phylogeny and Classification of the Tetrapods, Volume 2: Mammals. Clarendon Press, Oxford.

Janis, C. M., J. Damuth, and J. M. Theodor. 2000. Miocene ungulates and terrestrial primary productivity: where have all the browsers gone? Proceedings of the National Academy of Sciences, USA 97:7899–7904.

Janis, C. M., J. Damuth, and J. M. Theodor. 2004. The species richness of Miocene browers, and implications for habitat type and primary productivity in the North American grassland biome. Palaeogeography, Palaeoclimatology, Palaeoecology 207:371–398.

Janis, C. M., M. R. Dawson, and L. J. Flynn. In press. Glires; in C. M. Janis, G. F. Gunnell, and M. D. Uhen (eds.), Evolution of Tertiary Mammals of North America. Volume 2: Small Mammals, Edentates, and Marine Mammals. Cambridge University Press, Cambridge.

Janis, C. M., P. Errico, and M. Mendoza. 2004. Morphological indicators of cursoriality in equids: legs fail to support the "arms race." Journal of Vertebrate Paleontology 24(3):75A.

Janis, C. M., I. M. Gordon, and A.W. Illius. 1994. Modeling equid/ruminant competition in the fossil record. Historical Biology 8:15–29.

Janis, C. M., K. M. Scott, and L. L. Jacobs (eds.). 1998. Evolution of Tertiary Mammals of North America. Volume 1: Terrestrial Carnivores, Ungulates and Ungulatelike Mammals. Cambridge University Press, Cambridge.

Janis, C. M., J. M. Theodor, and B. Boisvert. 2002. Locomotor evolution in camels revisited: a quantitative analysis of pedal anatomy and the acquisition of the pacing gait. Journal of Vertebrate Paleontology 22:110–121.

Janis, C. M., J. A. Effinger, J. A. Harrison, J. G. Honey, D. G. Kron, E. B. Lander, E. Manning, D. R. Prothero, M. S. Stevens, R. K. Stucky, S. D. Webb, and D. B. Wright. 1998. Artiodactyla; pp. 337–357 in C. M. Janis, K. M. Scott, and L. L. Jacobs (eds.), Evolution of Tertiary Mammals of North America. Volume 1: Terrestrial Carnivores, Ungulates, and Ungulatelike Mammals. Cambridge University Press, Cambridge.

Jehenne, Y. 1969. Étude du gisement de St.-Capraise d'Eymet en Dordogne. Bulletin des Sciences de la Terre de l'Université de Poitiers 10:1–42.

Jehenne, Y. 1977. Description du premier crâne de *Prodremotherium,* ruminant primitif de l'Oligocène eurasiatique. Géobios (special issue: Faunes de mammifères du Paléogène d'Eurasie) 1: 233–239.

Jehenne, Y. 1985. Les ruminants primitifs du Paléogène et du Néogène inférieur de l'Ancien Monde: systématique, phylogénie, biostratigraphie. Dissertation, Université de Poitiers, 288 pp.

Jehenne, Y. 1987. Intérêt biostratigraphique des ruminants primitifs du Paléogène et du Néogène inférieur d'Europe occidentale. Münchner Geowissenschaftliche Abhandlungen (A) 10:131–140.

Jehenne, Y. 1988. *Bedenomeryx,* un nouveau genre de ruminant primitif de l'Oligocene superieur et du Miocene inferieur d'Europe. Comptes Rendus de l'Academie des Sciences 307:1991–1996.

Jenner, R. A. 2004. Accepting partnership by submission? Morphological phylogenetics in a molecular millenium. Systematic Biology 53:333–342.

Jermann, T. M., J. G. Opitz, J. Stackhouse, and S. A. Benner. 1995. Reconstructing the evolutionary history of the artiodactyl ribonuclease superfamily. Nature 374:57–59.

Jernvall, J. 1995. Mammalian molar cusp patterns: Developmental mechanisms of diversity. Acta Zoologica Fennica 198: 1–61.

Jernvall, J., and M. Fortelius. 2002. Common mammals drive the evolutionary increase of hypsodonty in the Neogene. Nature 417:538–540.

Jernvall, J., J. P. Hunter, and M. Fortelius. 1996. Molar tooth diversity, disparity, and ecology in Cenozoic unglate radiations. Science 274:1489–1492.

Joeckel, R. M., and J. M. Stavas. 1996. Basicranial anatomy of *Syndyoceras cooki* (Artiodactyla, Protoceratidae) and the need for a reappraisal of tylopod relationships. Journal of Vertebrate Paleontology 16:320–327.

Joleaud, L. 1920. Contribution à l'étude des hippopotames fossiles. Bulletin de la Société Géologique de France 22: 13–26.

Jones, M. L. 1993. Longevity of ungulates in captivity. International Zoological Yearbook 32:159–169.

Kahlke, R. D. 1987. On the occurrence of *Hippopotamus* (Mammalia, Artiodactyla) in the Pleistocene of Achalkalaki (Gruzinian SSR, Soviet Union) and on the distribution of the genus in South-East Europe. Zeitschrift für Geologische Wissenschaften 15:407–414.

Kahlke, R. D. 1990. Zum stand der Erforschung fossiler Hippopotamiden (Mammalia, Artiodactyla). Eine übersicht. Quartärpaläontologie 8:107–118.

Kahlke, R.-D. 1997. Die Hippopotamus-Reste aus dem Unterpleistozän von Untermaßfeld; pp. 277–374 in R.-D. Kahlke (ed.), Das Pleistozän von Untermaßfeld bei Meiningen (Thüringen). Vol 1. Monographien des Römisch-Germanischen Zentralmuseums Mainz, 40(1). Verlag Dr. Rudolf Habelt, Bonn.

Kahlke, R.-D. 2001. Schädelreste von Hippopotamus aus dem Unterpleistozän von Untermaßfeld; pp. 483–500 in R.-D. Kahlke (ed.), Das Pleistozän von Untermaßfeld bei Meiningen (Thüringen). Volume 2. Monographien des Römisch-

Germanischen Zentralmuseums, 40(2). Verlag Dr. Rudolf Habelt, Bonn.

Kamis, A. 1981. Water metabolism in *Tragulus javanicus.* Malaysian Applied Biology 10:67–68.

Kangas, A. T., A. R. Evans, I. Thesleff, and J. Jernvall. 2004. Non-independence of mammalian dental characters. Nature 432:211–214.

Kaup, J.-J. 1833. Mitteilungen an Professor Bronn. Neues Jahrbuch für Mineralogie, Geognosie, Geologie und Petrefaktenkunde Jahrgang 1833:419–420.

Kaup, J.-J., and J. B. Scholl. 1934. Verzeichniss der Gypsabgüsse von den ausgezeichnetsten urweltlichen Thierresten des Grossherzoglichen Museums zu Darmstadt. Zweite Ausgabe. J. P. Diehl, Darmstadt, pp. 6–28.

Kay, R. N. B. 1987. The comparative anatomy and physiology of digestion in Tragulids and Cervids and its relation to food intake; pp. 214–222 in C. M. Wemmer (ed.), Biology and Management of the Cervidae. Smithsonian Institution Press, Washington, D.C., London.

Kellogg, A. R. 1936. A review of the Archaeoceti. Carnegie Institution of Washington Publication 482:1–366.

Kent, J. D. 1982. The domestication and exploitation of the South American Camelids: methods of analysis and their application to circum-lacustrine archaeological sites in Bolivia and Peru. Ph.D. dissertation, Washington University, St. Louis, Missouri, 348 pp. plus appendices.

Killgus, H. 1922. Unterpliozäne Säuer aus China. Palaeontologische Zeitschrift Volume 5, Berlin.

King, C. 1878. Systematic Geology. U.S. Geological Expedition along the 40th parallel; in Bulletin of the U.S. Geological Survey of the Territories. U.S. Government Printing Office, Washington, D.C.

Kingdon, J. 1979. East African mammals; pp. 180–277 in J. Kingdon (ed.), East African Mammals—An Atlas of Evolution in Africa. Academic Press, London.

Kingdon, J. 1997. The Kingdon Field Guide to African Mammals. Academic Press, London.

Kingdon, J., 1982a. East African Mammals (Bovids). Volume 3 Part C. University of Chicago Press, Chicago, Illinois.

Kingdon, J., 1982b. East African Mammals (Bovids). Volume 3 Part D. University of Chicago Press, Chicago, Illinois.

Kinkelin, F. 1884. Ueber Fossilien aus Braunkohlen der Umgebung von Frankfurt a/M. Berichte der Senckenbergische Naturforschende Gesellschaft 1884:165–182.

Kleineidam, R. G., G. Pesole, H. J. Breukelman, J. J. Beintema, and R. A. Kastelein. 1999. Inclusion of cetaceans within the order Artiodactyla based on phylogenetic analysis of pancreatic ribonuclease genes. Journal of Molecular Biology 48:360–368.

Kluge, A. G. 1989. A concern for evidence and a phylogenetic hypothesis of relationships among *Epicrates* (Boidae, Serpentes). Systematic Zoology 38:7–25.

Koerner, H. E. 1940. The geology and vertebrate paleontology of the Fort Logan and Deep River Formation of Montana. American Journal of Science 238:837–862.

Köhler, M. 1987. Boviden des turkischen Miozäns (Kanozoikum und Braunkohlen der Turkei 28). Paleontologia i Evolució Sabadell 21:133–246.

Köhler, M. 1993. Skeleton and habitat of recent and fossil ruminants. Münchner Geowissenschaftlichen Abhandlungen A 25:1–88.

Kondrashov, P. E., A. V. Lopatin, and S. G. Lucas. 2004. The oldest known Asian artiodactyl (Mammalia). New Mexico Museum of Natural History and Science Bulletin 26:205–208.

Korotkevich, E. L. 1981. Late Neogene tragocerines of the northern Black Sea region. Zoologia Institut Akademia Nauk. Ukranian SSR, Kiev. 155 pp.

Korth, W. W., and M. E. Diamond. 2002. Review of *Leptomeryx* (Artiodactyla, Leptomerycidae) from the Orellan (Oligocene) of Nebraska. Annals of Carnegie Museum 71:107–129.

Kostopoulos, D. S. 1996. The artiodactyls of the Plio-Pleistocene of Macedonia. Dissertation, Aristotelian University of Thessaloniki, Department of Geology [Greek].

Kostopoulos, D. S. 2004. Revision of some late Miocene spiral horned antelopes (Bovidae, Mammalia). Neues Jahrbuch für Geologie und Paläontologie Abhandlungen 231:167–190.

Kowalevsky, W. 1873. On the osteology of the Hyopotamidae. Philosophical Transactions of the Royal Society of London 163:19–94.

Kowalevsky, W. 1874. Monographie der Gattung *Anthracotherium* Cuv. und der Versuch einer natürlichen Classification der fossilen Huftiere (Fortsetzung). Palaeontographica, Neue Folge 22(4):211–290.

Kowalevsky, W. 1876. Monographie der Gattung *Anthracotherium*. Palaeontographica 22:287–347.

Kraus, F., and M. M. Miyamoto. 1991. Rapid cladogenesis among the pecoran ruminants: evidence from mitochondrial DNA sequences. Systematic Zoology 40:117–130.

Kraus, O. 2000. Internationale Regeln für die Zoologische Nomenklatur. Abhandlungen des Naturwissenschaftlichen Vereins Hamburg (NF) 34:1–232.

Krause, D. W., and M. C. Maas. 1990. The biogeographic origins of late Paleocene–early Eocene mammalian immigrants to the Western Interior of North America. Geological Society of America Special Paper 243:71–105.

Krishtalka, L., and R. K. Stucky. 1985. Revision of the Wind River Faunas, Early Eocene of Central Wyoming. Part 7. Revision of *Diacodexis* (Mammalia, Artiodactyla). Annals of Carnegie Museum 54:413–486.

Krishtalka, L., and R. K. Stucky. 1986. Early Eocene artiodactyls from the San Juan Basin, New Mexico, and the Piceance Basin, Colorado. Contributions to Geology University of Wyoming, Special paper 3:183–196.

Krommenhoek, W. 1969. Mammalian fossils from the Pleistocene of Lake Edward and the Kazinga Channel. Uganda Journal 33(1):79–84.

Krommenhoek, W. 1971. Further notes on the *Hippopotamus imaguncula* fossils from the Kazinga Channel area. Uganda Journal 35(2):211–213.

Kron, D. G., and E. Manning. 1998. Anthracotheriidae; pp. 381–388 in C. M. Janis, K. M. Scott, and L. L. Jacobs (eds.), Evolution of Tertiary Mammals of North America. Volume 1: Terrestrial Carnivores, Ungulates, and Ungulatelike Mammals. Cambridge University Press, Cambridge.

Kullmer, O. 1999. Evolution of African Plio-Pleistocene suids (Artiodactyla, Suidae) based on tooth pattern analysis. Kaupia 9:1–34.

Kumar, K., and A. Jolly. 1986. Earlist artiodactyl (*Diacodexis*, Dichobunidae: Mammalia) from the Eocene of Kalakot, North-Western Himalaya, India. Bulletin of the Indian Society of Geosciences 2:20–30.

Kumar, K., and A. Sahni. 1985. Eocene mammals from the Upper Subathu Group, Kashmir Himalaya, India. Journal of Vertebrate Paleontology 5(2):153–168.

Kumar, K., and A. Sahni. 1986. *Remingtonocetus harudiensis*, new combination, a middle Eocene archaeocete (Mammalia,

Cetacea) from Western Kutch, India. Journal of Vertebrate Paleontology 6:326–349.

Kurtén, B. 1975. A new Pleistocene genus of American mountain deer. Journal of Mammalogy 56:507–508.

Kurtén, B. 1979. The stilt-legged deer *Sangamona* of the North-American Pleistocene. Boreas, 8:313–321.

Kurtén, B., and E. Anderson. 1980. Pleistocene Mammals of North America. Columbia University Press, New York, New York, 443 pp.

Kuznetsova, M. V., and M. V. Kholodova. 2003. Molecular support for the placement of *Saiga* and *Procapra* in Antilopinae (Artiodactyla, Bovidae). Journal of Mammalian Evolution 9:271–280.

Lacomba, J. I., and J. Morales. 1987. Los mamíferos del Oligoceno superior de Carrascosa del Campo (Provincia Cuenca, España). Münchner Geowissenschaftliche Abhandlungen, Reihe A 10:289–299.

Lacomba, J. I., J. Morales, F. Robles, C. Santisteban, and M. T. Alberdi. 1986. Sedimentologia y paleontologia del yacimiento finimioceno de La Portera (Valencia). Estudios Geologicos 42:167–180.

Lalueza-Fox, C., J. Bertranpetit, J. A. Alcover, N. Shailer, and E. Hagelberg. 2000. Mitochondrial DNA from *Myotragus balearicus,* an extinct bovid from the Balearic Islands. Journal of Experimental Zoology (Molecular and Developmental Evolution) 288:56–62.

Lalueza-Fox, C., B. Shapiro, P. Bover, J. A. Alcover, and J. Bertranpetit. 2002. Molecular phylogeny and evolution of the extinct bovid *Myotragus balearicus*. Molecular Phylogenetics and Evolution 25:501–510.

Lambe, L. M. 1908. The vertebrata of the Oligocene of the Cypress Hills, Saskatchewan. Contributions to Canadian Paleontology 3:5–64.

Lander, E. B. 1998. Oreodontoidea; pp. 402–425 in C. M. Janis, K. M. Scott, and L. L. Jacobs (eds.), Evolution of Tertiary Mammals of North America. Volume 1: Terrestrial Carnivores, Ungulates, and Ungulatelike Mammals. Cambridge University Press, Cambridge, 691 pp.

Lander, E. B. In press. Oreodonts (Mammalia, Artiodactyla, Agriochoeridae and Oreodontidae) from strata of early Duchesnean to latest Hemingfordian age in the Clarno, John Day, and Mascall Formations, John Day Basin, North-Central Oregon. Paleobios.

Lander, E. B., and C. B. Hanson. In press. *Agriochoerus matthewi crassus* (Mammalia, Artiodactyla, Oreodontoidea, Agriochoeridae, Agriochoerinae) from the early Duchesnean Hancock Mammal Quarry local fauna, Clarno Formation, John Day Basin, North-Central Oregon. Paleobios.

Langer, P. 1974. Stomach evolution in the Artiodactyla. Mammalia 38:295–314.

Langer, P. 1988. The Mammalian Herbivore Stomach. Comparative Anatomy, Function and Evolution. Gustav Fischer, Stuttgart.

Langer, P. 2001. Evidence from digestive tract on phylogenetic relationships in ungulates and whales. Journal of Zoological Systematics and Evolutionary Research 39:77–90.

Langer, P. 2002. The digestive tract and life history of small mammals. Mammal Review 32:107–131.

Lankester, R. 1907a. The origin of the lateral horns of the giraffe in fetal life on the area of parietal bones. Proceedings of the Zoological Society, London 1907:100–125.

Lankester, R. 1907b. On the existence of rudimentary antlers in the okapi. Proceedings of the Zoological Society, London 1907:126–135.

Lankester, R. 1908. On certain points in the structure of the cervical vertebrae of the Okapi and Giraffe. Proceedings of the Zoological Society, London 1908:320–324.

Lartet, E. 1851. Notice sur la colline de Sansan—suivie d'une récapitulation de diverses espèces d'animaux vertébrés fossiles trouvés soit à Sansan, soit dans d'autres gisements du terrain tertiaire miocène dans le bassin sous-pyrénéen. J.-A. Portes, 45 pp.

Lavocat, R. 1951. Révision de la faune des mammifères Oligocènes d'Auvergne et du Velay. Science Avenir Editions, Paris, 154 pp.

Laws, R. M. 1968. Interactions between elephant and hippopotamus populations and their environments. East African Agricultural and Forestry Journal 33(Special):140–147.

Leakey, M. G., C. S. Feibel, R. L. Bernor, J. M. Harris, T. E. Cerling, I. McDougall, K. M. Stewart, A. Walker, L. Werdelin, and A. J. Winkler. 1996. Lothagam: a record of faunal change in the Late Miocene of East Africa. Journal of Vertebrate Paleontology 16:556–570.

Legendre, S. 1980. Etude du gisement de Port-la-Nouvelle; étude des Cainotheriidae d'Escamps; Master Thesis, University of Montpellier II, Montpellier.

Legendre, S. 1989. Les communautés de mammifères du Paléogène (Eocène supérior et Oligocène) d'Europe occidentale: structures, milieux et évolution. Müncher Geowissenshaftliche Abhandlungen (A) 16:1–110.

Legendre, S., B. Marandat, J. A. Remy, B. Sigé, J. Sudre, M. Vianey-Liaud, J. Y. Crochet, and M. Godinot. 1995. Coyrou 1–2, une nouvelle faune de mammifères des phosphorites du Quercy, niveau intermédiaire (MP 20–21) proche de la "Grande Coupure." Géologie de la France 1:63–68.

Lehmann, U., and H. Thomas. 1987. Fossil Bovidae from the Mio-Pliocene of Sahabi (Libya); pp. 323–335 in N. T. Boaz, A. El-Arnauti, A. W. Gaziry, J. de Heinzelin, and D. D. Boaz (eds.), Neogene Geology and Paleontology of Sahabi. Alan R. Liss, New York, New York.

Leidy, J. 1848. On a new fossil genus and species of ruminatoid, Pachyderma: *Merycoidodon culbertsonii*. Proceedings of the Academy of Natural Sciences, Philadelphia 4:47–50.

Leidy, J. 1850. [Observations on two new genera of mammalian fossils, *Eucrotaphus jacksoni* and *Archaeotherium mortoni*]. Proceedings of the Academy of Natural Sciences, Philadelphia 5:89–93.

Leidy, J. 1851a. Description of the genus *Arctodon*. Proceedings of the Academy of Natural Sciences, Philadelphia 5:277–278.

Leidy, J. 1851b. Descriptions of fossil ruminant ungulates from Nebraska. Proceedings of the Academy of Natural Sciences, Philadelphia, 5:237–239.

Leidy, J. 1851c. [Remarks on fossil mammals from Missouri.] Proceedings of the Academy of Natural Sciences, Philadelphia 5(6):121–122.

Leidy, J. 1852. Description of the remains of extinct Mammalia and Chelonia, from Nebraska Territory, collected during the geological survey under the direction of Dr. D. D. Owen; pp. 194–206, 534–572, pls. 9–25 in D. D. Owen (ed.), Report of a Geological Survey of Wisconsin, Iowa, Minnesota; and Incidentally of a Portion of Nebraska Territory. Lippincott, Grambo and Co., Philadelphia, 638 pp.

Leidy, J. 1853. Remarks on a collection of fossil Mammalia from Nebraska. Proceedings of the Academy of Natural Sciences, Philadelphia (ser. 2) 6:392–394.

Leidy, J. 1856a. Notice of remains of extinct Mammalia, discovered by Dr. F. V. Hayden, in Nebraska Territory. Proceedings of the Academy of Natural Sciences, Philadelphia 8:88–90.

Leidy, J. 1856b. Notice of some remains of extinct Mammalia recently discovered by Dr. F. V. Hayden in the badlands of Nebraska. Proceedings of the Academy of Natural Sciences, Philadelphia 8:1–59.

Leidy, J. 1856c. Notice of some remains of extinct vertebrated animals. Proceedings of the Academy of Natural Sciences, Philadelphia 8:163–165.

Leidy, J. 1858a. [Description of *Procamelus robustus* and *Procamelus gracilis*.] Proceedings of the Academy of Natural Sciences, Philadelphia 10:89–90.

Leidy, J. 1858b. Notice of remains of extinct Vertebrata, from the valley of the Niobrara River, collected during the exploring expeditions of 1857, in Nebraska. Proceedings of the Academy of Natural Sciences, Philadelphia 10:20–29.

Leidy, J. 1869. The extinct mammalian fauna of Dakota and Nebraska, including an account of some allied forms from other localities, together with a synopsis of the mammalian remains of North America. Proceedings of the Academy of Natural Sciences, Philadelphia (series 2) 7:1–472.

Leidy, J. 1870a. [Descriptions of *Palaeosyops paludosus, Microsus cuspidatus,* and *Notharctus tenebrosus*.] Proceedings of the Academy of Natural Sciences, Philadelphia 1870:111–114.

Leidy, J. 1870b. [Remarks on a collection of fossils from the western territories.] Proceedings of the Academy of Natural Sciences, Philadelphia 1870:109–110.

Leinders, J. J. M. 1979. On the osteology and function of the digits in some ruminants and their bearing on taxonomy. Zeitschrift Säugertieren 44:305–318.

Leinders, J. J. M. 1983. Hoplitomerycidae fam. nov. (Ruminantia, Mammalia) from Neogene fissure fillings in Gargano (Italy). Scripta Geologica 40:1–51.

Leinders, J. J. M., and E. Heintz. 1980. The configuration of the lacrimal orificesin pecorans and tragulids (Artiodactyla, Mammalia) and its significance for the distinction between Bovidae and Cervidae. Beaufortia 30:155–162.

Leinders, J. J. M., M. Arif, H. de Bruijn, S. T. Hussain, and W. Wessels. 1999. Tertiary continental deposits of northwestern Pakistan and remarks on the collision between the Indian and Asian plates. Jaarbericht van het Natuurmuseum Rotterdam 7:199–213.

Lemoine, V. 1878. Communication sur les ossements fossiles des terrains tertiaires inférieures des environs de Reims faite à la Sociéte d'Histoire Naturelle de Reims. Bulletin de la Société des Sciences Naturelles de Reims 2:90–113.

Lemoine, V. 1891. Étude d'ensemble sur les dents les mammifères fossiles des environs de Reims. Bulletin de la Société Géologique de France 19:263–290.

Lewis, G. E. 1939. A new *Bramatherium* skull. American Journal of Science 287:275–240.

Lewis, G. E., 1968. Stratigraphic paleontology of the Barstow Formation in the Alvord Mountain area, San Bernardino County, California. U.S. Geological Survey Professional Paper 600-C:C75–C79.

Lewis, P. O. 2001. Maximum likelihood phylogenetic inference: modeling discrete morphological characters. Systematic Biology 50:913–925.

Li, C., and S. Ting. 1983. The Paleogene mammals of China. Bulletin of Carnegie Museum of Natural History 21:9–93.

Li, M., and H. Sheng. 1998. MtDNA difference and molecular phylogeny among musk deer, Chinese water deer, munjak and deer. Acta Theriologica Sinica 18:184–191.

Lihoreau, F. 2003. Systématique et paléoécologie des Anthracotheriidae [Artiodactyla; Suiformes] du Mio-Pliocène de l'Ancien Monde: implications paléobiogéographiques. Ph.D. dissertation, Université de Poitiers, France. 395 pp.

Lihoreau, F., C. Blondel, J. C. Barry, and M. Brunet. 2004a. A new species of the genus *Microbunodon* (Anthracotheriidae, Artiodactyla) from the Miocene of Pakistan: genus revision, phylogenetic relationships and palaeobiogeography. Zoologica Scripta 33:97–115.

Lihoreau, F., J. C. Barry, C. Blondel, and M. Brunet. 2004b. A new species of Anthracotheriidae, *Merycopotamus medioximus* nov. sp. from the late Miocene of the Potwar Plateau, Pakistan. Comptes Rendus Palévol 3:653–662.

Lihoreau F., J. C. Barry, C. Blondel, Y. Chaimanee, J.-J. Jaeger, and M. Brunet. 2007. Anatomical revision of the genus Merycopotamus (Artiodactyla; Anthracotheriidae): its significance for late Miocene mammal dispersal in Asia. Palaeontology 50:503–524.

Lihoreau, F, J.-R. Boisserie, L. Viriot, Y. Coppens, A. Likius, H. T. Mackaye, P. Tafforeau, P. Vignaud, and M. Brunet. 2006. Anthracothere dental anatomy reveals a late Miocene Chado-Libyan bioprovince. Proceedings of the National Academy of Sciences, USA 103:8763–8767.

Lin, Y.-H., P. A. McLenachan, A. R. Gore, M. J. Phillips, R. Ota, M. D. Hendy, and D. Penny. 2002. Four new mitochondrial genomes and the increased stability of evolutionary trees of mammals from improved taxon sampling. Molecular Biology and Evolution 19:2060–2070.

Linné, C. [Linnaeus, C.] 1758. Systema naturae per regna tria naturae, secundum classes, ordines, genera, species, cum characteribus, differentiis, synonymis, locis. Laurentii Salvii Holmiae, Stockolm.

Lister, A. M. 1993. Patterns of evolution in Quaternary mammal lineages. Evolutionary Patterns and Processes. Linnean Society of London Symposium Series 14:71–93.

Lister, A. M. 1994. The evolution of the giant deer, *Megaloceros giganteus* (Blumenbach). Zoological Journal of the Linnean Society 112:65–100.

Lister, A. M., P. Grubb, and S. R. M. Sumner. 1998. Taxonomy, morphology and evolution of European roe deer; pp. 23–46 in R. Andersen, P. Duncan, and J. D. C. Linnell (eds.), The European Roe Deer: The Biology of Success. Oslo: Scandinavian University Press.

Lister, A. M., C. J. Edwards, D. A. W. Nock, M. Bunce, I. A. van Pijlen, D. G. Bradley, M. G. Thomas, and I. Barnes. 2005. The phylogenetic position of the "giant deer" *Megaloceros giganteus.* Nature 438:850–852.

Liu, G., and C. Zhang. 1993. Anthracothere found in Sihong, Jiangsu. Vertebrata PalAsiatica 31:111–116.

Liu, L. 2001. Eocene suoids (Artiodactyla, Mammalia) from Bose and Yongle basins, China, and the classification and evolution of the Paleogene suoids. Vertebrata PalAsiatica 39: 115–128.

Liu, L. 2003. Chinese fossil Suoidea—Systematics, Evolution, and Paleoecology. University of Helsinki, Department of Geology, Division of Geology and Paleontology, Doctoral dissertation, Printed: ISBN 952–10–1196–3 [PDF: ISBN 902–10–1197–1], 40 pp.

Liu, L., M. Fortelius, and M. Pickford. 2002. New fossil Suidae from Shanwant, Shandong, China. Journal of Vertebrate Paleontology 22:152–163.

Liu, L., D. S. Kostopoulos, and M. Fortelius. 2004. Late Miocene *Microstonyx* remains (Suidae, Mammalia) from Northern China. Géobios 37:49–64.

Liu, T. S., and Y. C. Lee. 1963. New species of *Listriodon* from Mioocene of Lantien, Shensi, China. Vertebrata PalAsiatica 7:300–304.

Lock, J. M. 1972. The effects of hippopotamus grazing on grasslands. Journal of Ecology 60:445–467.

Loomis, F. B. 1932. Two new Miocene entelodonts. Journal of Mammalogy 13:358–362.

Lucas, S. G. 1983a. Comments on two species of the Eocene artiodactyl *Homacodon* and the taxonomic status of *Nanomeryx caudatus* Marsh, 1894. New Mexico Journal of Sciences 23:48–56.

Lucas, S. G. 1983b. The Baca Formation and the Eocene–Oligocene boundary in New Mexico. New Mexico Geological Society Guidebook, 34th Field Conference, Socorro Region II:187–192.

Lucas, S. G., and R. J. Emry. 1996. Late Eocene entelodonts (Mammalia, Artiodactyla) from Inner Mongolia, China. Proceedings of the Biological Society of Washington 109:397–405.

Lucas, S. G., and R. J. Emry. 1999. Taxonomy and biochronological significance of *Paraentelodon,* a giant entelodont (Mammalia, Artiodactyla) from the late Oligocene of Eurasia. Journal of Vertebrate Paleontology 19:160–168.

Lucas, S. G., and R. J. Emry. 2004. The entelodont *Brachyhyops* (Mammalia, Artiodactyla) from the upper Eocene of Flagstaff Rim, Wyoming. New Mexico Museum of Natural History and Science Bulletin 26:97–100.

Lucas, S. G., R. J. Emry, and S. E. Foss. 1998. Taxonomy and distribution of *Daeodon,* an Oligocene–Miocene entelodont (Mammalia, Artiodactyla) from North America. Proceedings of the Biological Society of Washington 111:425–435.

Lucas, S. G., S. E. Foss, and M. C. Mihlbachler. 2004. *Achaenodon* (Mammalia, Artiodactyla) from the Eocene Clarno Formation, Oregon, and the age of the Hancock Quarry local fauna. New Mexico Museum of Natural History and Science Bulletin 26:89–96.

Luck, C. P., and P. G. Wright. 1964. Aspects of the anatomy and physiology of the skin of the hippopotamus (*H. amphibius*). Quarterly Journal of Experimental Physiology 49:1–14.

Luckett, W. P., and N. Hong. 1998. Phylogenetic relationships between the orders Artiodactyla and Cetacea: a combined assessment of morphological and molecular evidence. Journal of Mammalian Evolution 5:127–182.

Ludt, C. J., W. Schroeder, O. Rottmann, and R. Kuehn. 2004. Mitochondrial DNA phylogeography of red deer (*Cervus elaphus*). Molecular Phylogenetics and Evolution 31:1064–1083.

Ludtke, J. A., and D. R. Prothero. 2004. Taxonomic revision of the middle Eocene (Uintan-Duchesnean) protoceratid *Leptoreodon* (Mammalia: Artiodactyla). New Mexico Museum of Natural History and Science Bulletin 26:101–111.

Ludwig, A., and S. Fischer. 1998. New aspects of an old discussion—phylogenetic relationships of *Ammotragus* and *Pseudovis* within the subfamily Caprinae based on comparison of the 12S rDNA sequences. Journal of Zoological Systematics and Evolutionary Research 36:173–178.

Lull, R. S. 1920. New Tertiary artiodactyls. American Journal of Science 50:81–130.

Lull, R. S. 1921. Fauna of the Dallas Sand Pits. American Journal of Science, 5th series 11:159–176.

Lull, R. S. 1922. Primitive Pecora in the Yale Museum. American Journal of Science 5:111–119.

Luo, Z. 1998. Homology and transformation of cetacean ectotympanic structures; pp. 269–301 in J. G. M. Thewissen (ed.), The Emergence of Whales. Plenum Press, New York, New York.

Luo, Z., and P. D. Gingerich. 1999. Terrestrial Mesonychia to aquatic Cetacea: transformation of the basicranium and evolution of hearing in whales. University of Michigan Papers on Paleontology 31:1–98.

Luo, Z., and Marsh, K. 1996. Petrosal (periotic) and inner ear of a Pliocene kogiine whale (Kogiinae, Odontoceti): implications on relationships and hearing evolution of toothed whales. Journal of Vertebrate Paleontology 16:328–348.

Lydekker, R. 1876. Fossil mammalian faunae of India and Burma. Records of the Geological Survey of India 9:86–106.

Lydekker, R. 1877. Notices of new or rare mammals from the Siwaliks. Records of the Geological Survey of India 10:76–83.

Lydekker, R. 1883a. Catalogue of the fossil Mammalia in the British Museum (Natural History) 2:1–324.

Lydekker, R. 1883b. Indian Tertiary and post-Tertiary Vertebrata. Part 5: Siwalik selenodont Suina. Memoirs of the Geological Survey of India, Paleontologica Indica 10(5):143–177.

Lydekker, R. 1883c. Synopsis of the fossil Vertebrata of India. Records of the Geological Survey of India XVI:61–69.

Lydekker, R. 1884. Indian Tertiary and post-Tertiary Vertebrata. Siwalik and Narbada bunodont Suidae. Memoirs of the Geological Survey of India, Palaeontologica Indica 3(2):35–104.

Lydekker, R. 1885. Catalogue of the Fossil Mammalia in the British Museum (Natural History). British Museum, London, 324 pp.

Lydekker, R. 1889. On an apparently new species of *Hyracodontotherium.* Proceedings of the Zoological Society of London 18:67–69.

Macdonald, D. W. (ed.). 1984. The Encyclopedia of Mammals. Facts on File Publications, New York, New York, 1002 pp.

Macdonald, D. W. (ed.). 1999. The Encyclopedia of Mammals (2nd edition). Oxford University Press, Oxford, 830 pp.

Macdonald, J. R. 1951. Additions to the Whitneyan fauna of South Dakota. Journal of Paleontology 25:257–265.

Macdonald, J. R. 1955. The Leptochoeridae. Journal of Paleontology 29:439–459.

Macdonald, J. R. 1956. The North American anthracotheres. Journal of Paleontology 30:615–645.

Macdonald, J. R., and J. E. Martin. 1987. *Arretotherium fricki* (Artiodactyla, Anthracotheriidae) from the Hemingfordian (Miocene) Flint Hill local fauna in South Dakota; pp. 57–62 in J. E. Martin and G. E. Ostrander (eds.), Papers in Vertebrate Paleontology in Honor of Morton Greene. Dakoterra 3:57–62.

Macdonald, J. R., and C. B. Schultz. 1956. *Arretotherium fricki,* a new Hemingfordian anthracothere from Nebraska. Bulletin of the University of Nebraska State Museum 4:53–58.

MacFadden, B. J. 1980. An early Miocene land mammal (Oreodonta) from a marine limestone in northern Florida. Journal of Paleontology 54:93–101.

MacFadden, B. J. 2006. North American Miocene land mammals from Panama. Journal of Vertebrate Paleontology 26:720–734.

MacFadden, B. J., and R. M. Hunt, Jr. 1998. Magnetic polarity stratigraphy and correlation of the Arikaree Group, Arikareean (late Oligocene–early Miocene) of northwestern Nebraska. Geological Society of America Special Paper 325:143–166.

MacFadden, B. J., and G. S. Morgan. 2003. New oreodont (Mammalia, Artiodactyla) from the late Oligocene (early Arikareean) of Florida. Bulletin, American Museum of Natural History 279(15):368–396.

Macfarlane, W. V., B. Howard, and B. F. Good. 1974. Tracers in field measurements of water, milk and thyroxine metabolism in tropical ruminants; pp. 1–23 in Tracer Techniques in Tropical Animal Production. International Atomic Energy Agency, Vienna.

MacInnes, D. G. 1951. Fossil mammals of Africa: Miocene Anthracotheriidae from East Africa. Bulletin of the British Museum (Natural History) 4:1–24.

Mackie, C. 1976. Feeding habits of the hippopotamus on the Lundi River, Rhodesia. Arnoldia 7(34):1–16.

MacPhee, R. D. E. 1981. Auditory regions of primates and eutherian insectivores: morphology, ontogeny, and character analysis. Contributions to Primatology 18:1–282.

MacPhee, R. D. E. 1994. Morphology, adaptations, and relationships of *Plesiorycteropus,* and a diagnosis of a new order of eutherian mammals. Bulletin of the American Museum of Natural History 220:1–214.

MacPhee, R. D. E., and D. A. Burney. 1991. Dating of modified femora of extinct dwarf Hippopotamus from southern Madagascar: implications for constraining human colonization and vertebrate extinction events. Journal of Archaeological Science 18:695–706.

Maddison, W. P. 1989. Reconstructing character evolution on polytomous cladograms. Cladistics 5:365–367.

Maddison, W. P., and D. R. Maddison. 2000. MacClade, Vers. 4.0. Sinauer Associates, Sunderland, Massachusetts. Available at http://www.sinauer.com/.

Madsen, O., D. Willemsen, B. M. Ursing, U. Arnason, and W. W. de Jong. 2002. Molecular evolution of the mammalian alpha 2B adrenergic receptor. Molecular Biology and Evolution 19:2150–2160.

Madsen, O., M. Scally, C. J. Douady, D. J. Kao, R. W. DeBry, R. M. Adkins, H. Amrine-Madsen, M. J. Stanhope, W. W. de Jong, and M. S. Springer. 2001. Parallel adaptive radiations in two major clades of placental mammals. Nature 409:610–614.

Magallón, S., and M. J. Sanderson. 2001. Absolute diversification rates in angiosperm clades. Evolution 55:1762–1780.

Major, C. I. F. 1889. Sur un gisement d'ossements fossils dans l'ile de Samos, contemporains de l'âge de Pikermi. Comptes rendus des Séances de l'Académie, Paris 107:1178–1181.

Major, C. I. F. 1902. Some account of a nearly complete skeleton of *Hippopotamus madagascariensis,* Guld., from Sirabé, Madagascar, obtained in 1895. Geological Magazine 9:193–199.

Maniou, Z., O. C. Wallis, and M. Wallis. 2004. Episodic molecular evolution of pituitary growth hormone in Cetartiodactyla. Journal of Molecular Evolution 58:743–753.

Manlius, N. 2000. Biogéographie et écologie historique de l'hippopotame en Egypte. Belgian Journal of Zoology 130:59–66.

Marivaux, L., M. Benammi, S. Ducrocq, J.-J. Jaeger, and Y. Chaimanee. 2000. A new baluchimyine rodent from the late Eocene of the Krabi Basin (Thailand): paleobiogeographic and biochronologic implications. Comptes Rendus de l'Académie des Sciences, Paris 331:427–433.

Marsh, O. C. 1868. Geology of New Jersey. Proceedings of the Academy of Natural Sciences, Philadelphia 22:740.

Marsh, O. C. 1871. Notice of some new fossil mammals from the Tertiary formation. American Journal of Science (series 3) 2:35–41.

Marsh, O. C. 1872a. Notice of some remarkable fossil mammals. American Journal of Science (series 3) 3:343–344.

Marsh, O. C. 1872b. Preliminary description of new Tertiary mammals, Part I–IV. American Journal of Science and Arts 4:1–35.

Marsh, O. C. 1873. Notice of new Tertiary mammals. American Journal of Science (series 3) 5:407–410, 485–488.

Marsh, O. C. 1874. Notice of new Tertiary mammals. III. American Journal of Science (series 3) 7:531–534.

Marsh, O. C. 1875. Notice of new Tertiary mammals. IV. American Journal of Science (series 3) 8:239–250.

Marsh, O. C. 1876. Notice of new Tertiary mammals. V. American Journal of Science (series 3) 12:401–404.

Marsh, O. C. 1877. Introduction and succession of vertebrate life in America. American Journal of Science (series 3) 14:337–378.

Marsh, O. C. 1890. Notice of new Tertiary mammals. American Journal of Science 39:523–525.

Marsh, O. C. 1891. A horned artiodactyle (*Protoceras celer*) from the Miocene: American Journal of Science (series 2) 41:81–82.

Marsh, O. C. 1893. Description of Miocene Mammalia. American Journal of Science (series 3) 46:407–412.

Marsh, O. C. 1894a. A new Miocene mammal. American Journal of Science (series 3) 47:409.

Marsh, O. C. 1894b. Description of Tertiary artiodactyls. American Journal of Science (series 3) 48:259–274.

Marsh, O. C. 1894c. Miocene artiodactyls from the eastern *Miohippus* beds. American Journal of Science (series 3) 48:175–178.

Marsh, O. C. 1894d. Restoration of *Elotherium.* American Journal of Science (series 3) 47:407–408.

Martínez-Navarro, B., L. Rook, A. Segid, D. Yosieph, M. P. Ferretti, J. Shoshani, T. M. Tecle, and Y. Libsekal. 2004. The large fossil mammals from Buia (Eritrea). Rivista Italiana di Paleontologia e Stratigrafia 110(suppl.):61–88.

Matthee, C. A., and S. K. Davis. 2001. Molecular insights into the evolution of the family Bovidae: a nuclear DNA perspective. Molecular Biology and Evolution 18:1220–1230.

Matthee, C. A., and T. J. Robinson. 1999. Cytochrome *b* phylogeny of the family Bovidae: resolution within the Alcelaphini Antilopini, Neotragini, and Tragelaphini. Molecular Phylogenetics and Evolution 12:31–46.

Matthee, C. A., J. D. Burzlaff, J. F. Taylor, and S. K. Davis. 2001. Mining the mammalian genome for artiodactyl systematics. Systematic Biology 50:367–390.

Matthew, W. D. 1897. A revision of the Puerco Fauna. Bulletin of the American Museum of Natural History 9:59–110.

Matthew, W. D. 1899. A provisional classification of the Freshwater Tertiary of the West. Bulletin of the American Museum of Natural History 12:19–75.

Matthew, W. D. 1901. Fossil mammals of the tertiary of northeastern Colorado. Memoir, American Museum of Natural History 1:355–448.

Matthew, W. D. 1903. The fauna of the *Titanotherium* Beds at Pipestone Springs, Montana. Bulletin of the American Museum of Natural History 19:197–226.

Matthew, W. D. 1904. A complete skeleton of *Merycodus.* Bulletin of the American Museum of Natural History 20:101–129.

Matthew, W. D. 1905. Notice of two new genera of mammals from the Oligocene of South Dakota: Bulletin of the American Museum of Natural History 21:21–26.

Matthew, W. D., 1907. A lower Miocene fauna from South Dakota. Bulletin of the American Museum of Natural History, 23:169–220.

Matthew, W. D., 1908. Osteology of *Blastomeryx* and phylogeny of the American Cervidae: Bulletin of the American Museum of Natural History 23:535–562.

Matthew, W. D. 1909a. Faunal lists of the Tertiary Mammalia of the West. Bulletin of the United States Geological Survey 361:91–138.

Matthew, W. D. 1909b. Observations upon the genus *Ancodon*. Bulletin of the American Museum of Natural History 26:1–7.

Matthew, W. D. 1911. A tree climbing ruminant. American Museum Journal 11:162–163.

Matthew, W. D. 1918a. Contributions to the Snake Creek fauna with notes upon the Pleistocene of western Nebraska. Bulletin of the American Museum of Natural History 38: 183–229.

Matthew, W. D. 1918b. A revision of the Lower Eocene Wasatch and Wind River faunas, Part V. Insectivora (continued), Glires, Edentata. Bulletin of the American Museum of Natural History 33:565–657.

Matthew, W. D. 1924. Third contribution to the Snake Creek Fauna. Bulletin of the American Museum of Natural History 50:59–210.

Matthew, W. D. 1926. On a new primitive deer and two traguloid genera from the lower Miocene of Nebraska. American Museum Novitates 215:1–8.

Matthew, W. D. 1929a. Reclassification of the artiodactyl families. Bulletin of the Geological Society of America 40:403–408.

Matthew, W. D. 1929b. Tylopoda; pp. 641–642 in Encyclopedia Britannica. Cambridge University Press, Cambridge.

Matthew, W. D. 1934. A phylogenetic chart of the Artiodactyla. Journal of Mammalogy 15:207–209.

Matthew, W. D. 1937. Paleocene faunas of the San Juan Basin, New Mexico. Transactions of the American Philosophical Society 30:1–510.

Matthew, W. D., and H. J. Cook. 1909. A Pliocene fauna from western Nebraska (Sioux County). Bulletin of the American Museum of Natural History 26:361–414.

Matthew, W. D., and W. Granger. 1923. The fauna of the Houldjin Gravels. American Museum Novitates 97:1–6.

Matthew, W. D., and W. Granger. 1925a. New mammals from the Shara Murun Eocene of Mongolia. American Museum Novitates 196:1–11.

Matthew, W. D., and W. Granger. 1925b. New ungulates from the Ardyn Obo Formation of Mongolia. American Museum Novitates 195:1–12.

Matthew, W. D., and J. R. Macdonald. 1960. Two new species of *Oxydactylus* from the middle Miocene Rosebud Formation in western South Dakota. American Museum Novitates 2003: 1–7.

May-Collado, L., and I. Agnarsson. 2006. Cytochrome *b* and Bayesian inference of whale phylogeny. Molecular Phylogenetics and Evolution 38:344–354.

Mazza, P. 1991. Interrelations between Pleistocene hippopotami of Europe and Africa. Bolletino della Società Paleontologica Italiana 30:153–186.

Mazza, P. 1995. New evidence on the Pleistocene hippopotamuses of western Europe. Geologica Romana 31:61–241.

McCarthy, T. S., W. N. Ellery, and A. Bloem. 1998. Some observations on the geomorphological impact of hippopotamus (*Hippopotamus amphibius* L.) in the Okavango Delta, Botswana. African Journal of Ecology 36:44–56.

McDougall, I., and C. S. Feibel. 2003. Numerical age control for the Miocene–Pliocene succession at Lothagam, a hominoid-bearing sequence in the Northern Kenya Rift; pp. 45–63 in J. M. Harris and M. G. Leakey (eds.), Lothagam. The Dawn of Humanity in Eastern Africa. Columbia University Press, New York, New York.

McKenna, M. C. 1959. *Tapochoerus,* a Uintan dichobunid artiodactyl from the Sespe Formation of California. Bulletin Southern California Academy Science 58:125–132.

McKenna, M. C. 1972. Was Europe connected directly to North America prior to the Middle Eocene?; pp. 179–188 in T. Dobzhansky, M. Hecht, and C. W. Steeve (eds.), Evolutionary Biology. Appleton Century Crofts, New York, New York.

McKenna, M. C. 1975. Toward a phylogenetic classification of the Mammalia; pp. 21–46 in W. P. Luckett and F. C. Szalay (eds.), Phylogeny of Primates: A Multidisciplinary Approach. Plenum Press, New York, New York.

McKenna, M. C., and S. K. Bell. 1997. Classification of Mammals above the Species Level. Columbia University Press, New York, New York, 631 pp.

McKenna, M. C., and S. K. Bell. 1998. Classification of Mammals. Columbia University Press, New York, New York.

Medellín, R. A., A. L. Gardner, and J. M. Aranda. 1998. The taxonomic status of the Yucatán brown brocket, *Mazama pandora* (Mammalia: Cervidae). Proceedings of the Biological Society of Washington. 111:1–14.

Meijaard, E., and C. P. Groves. 2004a. A taxonomic revision of the *Tragulus* mouse-deer (Artiodactyla). Zoological Journal of the Linnean Society 140:63–102.

Meijaard, E., and C. P. Groves. 2004b. Morphometrical relationships between South-East Asian deer (Cervidae, tribe Cervini): evolutionary and biogeographic implications. Journal of the Zoological Society, London 263:179–196.

Mein, P. 1989. Updating of MN Zones; pp. 73–90 in E. H. Lindsay, V. Fahlbusch, and P. Mein (eds.), European Neogene Mammal Chronology. NATO ASI Series (A) 180, Plenum Press, New York, New York.

Mein, P., and L. Ginsburg. 1997. Les mammifères du gisement miocène inférieur de Li Mae Long, Thailand: systematique, biostratigraphie et palaéoenvironnement. Geodiversitas 19:783–844.

Meng, J., and M. C. McKenna. 1998. Faunal turnovers of Palaeogene mammals from the Mongolian Plateau. Nature 394: 364–367.

Merriam, J. C. 1909. The occurrence of strepsicerine antelopes in the Tertiary of northwestern Nevada. University of California Publications in Geological Sciences 5:319–330.

Merriam, J. C. 1911. Tertiary mammal beds of Virgin Valley and Thousand Creek in northwestern Nevada. Part II—Vertebrate faunas. University of California Publications in Geological Sciences 6:199–304.

Merriam, J. C. 1913. A peculiar horn or antler from the Mohave Miocene of California. University of California Publications, Bulletin of the Department of Geology 7:335–339.

Mertz, D. F., C. C. Swisher, J. L. Franzen, O. Neuffer, and H. Lutz. 2000. Numerical dating of the Eckfeld maar fossil site, Eifel, Germany: A calibration mark for the Eocene time scale. Naturwissenschaften 87:270–274.

Métais, G. 2006. New basal selenodont artiodactyls from the Pondaung Formation (Late Middle Eocene, Myanmar) and the phylogenetic relationships of early ruminants. Annals of the Carnegie Museum 75:51–67.

Métais, G., J. Guo, and K. C. Beard. 2004. A new small dichobunid artiodactyl from Shanghuang (middle Eocene, eastern China): implications for the early evolution of protoselenodonts in Asia. Bulletin Carnegie Museum of Natural History 36:177–197.

Métais, G., Aung Naing Soe, L. Marivaux, and K. C. Beard. In press. Artiodactyls from the Pondaung Formation (Myanmar): new data, and new interpretations on the South Asian faunal Province during the middle Eocene. Naturwissenschaften.

Métais, G., Y. Chaimanee, J.-J. Jaeger, and S. Ducrocq. 2001. New remains of primitive ruminants from Thailand: evidence of the early evolution of the Ruminantia in Asia. Zoologica Scripta 30:231–248.

Métais, G., T. Qi, J. Guo, and K. C. Beard. 2005. A new bunoselenodont artiodactyl from the Middle Eocene of China and the early record of selenodont artiodactyls in Asia. Journal of Vertebrate Paleontology 25:994–997.

Métais, G., P.-O. Antoine, L. Marivaux, J. L. Welcomme, and S. Ducrocq. 2003. New artiodactyl ruminant mammal from the late Oligocene of Pakistan. Acta Palaeontologica Polonica 48:375–382.

Métais, G., M. Benammi, Y. Chaimanee, J.-J. Jaeger, T. Tun, T. Thein, and S. Ducrocq. 2000. Discovery of new ruminant dental remains from the Middle Eocene Pondaung formation (Myanmar): reassessment of the phylogenetic position of *Indomeryx*. Comptes Rendus de l'Académie des Sciences de Paris, série IIa 330:805–811.

Meyer, H. von. 1832. Palaeologica, zur Geschichte der Erde und ihrer Geschöpfe. Frankfurt am Main, 560 pp.

Meyer, H. von. 1834. Die foccilen Zähne und Knochen und ihre Ablagerung in der Gegend von Georgensmünd in Bayern. Abhandlungen herausgegeben von der Senckenbergischen Naturforschenden Gesellschaft, Frankfurt am Main 1(VIII): 1–126.

Meyer, H. von. 1837. Mittheilungen, an Professor Bronn gerichtet. Neues Jahrbuch für Mineralogie, Geognosie, Geologie und Petrefaktenkunde 1837:557–562.

Meyer, H. von. 1846. Mittheilungen an Professor Bronn. Neues Jahrbuch für Mineralogie, Geologie, Geognosie und Petrefaktenkunde Jahrgang 1846:462–476.

Meyer, H. von. 1852. Mittheilungen, an Professor Bronn gerichtet. Neues Jahrbuch für Mineralogie, Geognosie, Geologie und Petrefaktenkunde 1852:831–833.

Meyer, H. von. 1854. *Anthracotherium dalmatinum* aus der Braunkohle des Monte Promina in Dalmatien. Palaeontographica 4:61–71.

Mihlbachler, M. C., and N. Solounias. 2006. Coevolution of tooth crown height and diet in oreodonts (Merycoidontidae, Artiodactyla) examined with phylogenetically independent constraints. Journal of Mammalian Evolution 13:11–36.

Mikkola, M. L., and I. Thesleff. 2003. Ectodysplasin signaling in development. Cytokine and Growth Factor Reviews 14:211–224.

Miller, E. R. 1999. Faunal correlation of Wadi Moghara, Egypt: implications for the age of *Prohylobates tandyi*. Journal of Human Evolution 36:519–533.

Miller, W. E., and T. Downs. 1974. A Hemphillian local fauna containing a new genus of antilocaprid from southern California. Contributions in Science, Natural History Museum of Los Angeles County 258:1–36.

Milne-Edwards, A. 1864. Recherches anatomiques, zoologiques et paléontologiques sur la famille des chevrotains. Annales de Science Naturelle Paris 5(2):1–167.

Milne-Edwards, A. 1868. Sur les découvertes zoologiques faites récemment à Madagascar par M. Alfred Grandidier. Comptes Rendus de l'Académie des Sciences 67:1165–1167.

Misonne, X. 1952. Quelques éléments nouveaux concernant *Hippopotamus imaguncula* Hopwood. Bulletin de l'Institut royal des Sciences naturelles de Belgique 28(3):1–12.

Miyamoto, M. M. 1999. Perfect SINEs of evolutionary history? Current Biology 9:R816–R819.

Miyamoto, M. M., and M. Goodman. 1986. Biomolecular systematics of eutherian mammals: phylogenetic patterns and classification. Systematic Zoology 35:230–240.

Miyamoto, M. M., F. Kraus, and O. A. Ryder. 1990. Phylogeny and evolution of antlered deer determined from mitochondrial DNA sequences. Proceedings of the National Academy of Sciences, USA 87:6127–6131.

Miyamoto, M. M., S. M. Tanhauser, and P. J. Laipis. 1989. Systematic relationships in the artiodactyl tribe Bovini (family Bovidae), as determined from mitochondrial DNA sequences. Systematic Zoology 38:342–349.

Miyamoto, M. M., F. Kraus, P. J. Laipis, S. M. Tanhauser, and S. D. Webb. 1993. Mitochondrial DNA phylogenies within Artiodactyla; pp. 268–281 in F. S. Szalay, M. J. Novacek, and M. C. McKenna (eds.), Mammal Phylogeny—Placentals. Springer-Verlag, New York, New York.

Molina, M., and J. Molinari. 1999. Taxonomy of Venezuelan white-tailed deer (*Odocoileus*, Cervidae, Mammalia), based on cranial and mandibular traits. Canadian Journal of Zoology 77:632–645.

Montgelard, C., F. M. Catzeflis, and E. Douzery. 1997. Phylogenetic relationships of artiodactyls and cetaceans as deduced from the comparison of cytochrome *b* and 12s rRNA mitochondrial sequences. Molecular Biology and Evolution 14:550–559.

Montgelard, C., S. Ducrocq, and E. J. P. Douzery. 1998. What is a suiforme (Artiodactyla)? Contribution of cranioskeletal and mitochondrial DNA data. Molecular Phylogenetics and Evolution 9:528–532.

Morales J. 1985. Nuevos datos sobre *"Decennatherium pachecoi"* (Crusafont, 1952) (Giraffidae, Mammalia): descripción del cráneo de Matillas. Coloquios de la Catedra de Paleontologia, Madrid 40:51–58.

Morales, J., Moyà-Solà, S., and Soria, D. 1981. Presencia de la familia Moschidae (Artiodactyla, Mammalia) en el Vallesiense de España: *Hispanomeryx duriensis* nov. gen. nov. sp. Estudios Geologicos 37:467–475.

Morales J., D. Soria, and M. Pickford. 1999. New stem giraffoid ruminants from the early and middle Miocene of Namibia. Geodiversitas (21) 2:229–253.

Morejohn, G. V., J. Hearst, and C. Dailey. 2005. Postcranial carpalial support for the extinct cervid genus *Bretzia* with associated antler. Journal of Mammalogy 86:115–120.

Morgan, J. K., and N. H. Morgan. 1995. A new species of *Capromeryx* (Mammalia: Artiodactyla) from the Taunton Local Fauna of Washington, and the correlation with other Blancan faunas of Washington and Idaho. Journal of Vertebrate Paleontology 15:160–170.

Moscarella, R. A., M. Aguilera, and A. A. Escalente. 2003. Phylogeography, population structure, and implications for conservation of white-tailed deer (*Odocoileus virginianus*) in Venezuela. Journal of Mammalogy 84:1300–1315.

Mottl, M. 1961. Die Dorcatherien (Zwerghirsche) der Steiermark. Mitteilungen des Museums für Bergbau, Geologie, Technik, Landesmuseum Joanneum Graz 22:21–71.

Moyà-Solà, S. 1983. Los Boselaphini (Bovidae Mammalia) del Neogeno de la península Ibérica. Publicaciones de Geologia, Universitat Autonoma de Barcelona 18:1–236.

Moyà-Solà, S. 1986. El genero *Hispanomeryx* Morales et al. (1981): posicion filogenetico y sistematica. Paleontologia i Evolució 20:267–287.

Moyà-Solà, S. 1987. Los ruminates (Cervoidea y Bovoidea, Artiodactyla, Mammalia) de Ageniense (Mioceno inferior) de Navarrete del Rio (Teruel, España). Paleontologia i Evolució 21:247–269.

Moyà-Solà, S. 1988. Morphology of lower molars of the ruminants (Artiodactyla, Mammalia): Phylogenetic implications. Paleontologia i Evolució 22:61–70.

Moyà-Solà, S., and M. Köhler. 1993. Middle Bartonian locality with *Anchomomys* (Adapidae, Primates) in the Spanish Pyrenees: preliminary report. Folia Primatologia 60: 158–163.

Müller, O. 1898. Untersuchungen über die Veränderungen, welche die Respirationsorgane der Säugetiere durch die anpassung an das Leben im Wasser erlitten haben. Jenaische Zeitschrift für Naturwissenschaft. 32:95–230.

Murphy, W. J., E. Eizirik, W. E. Johnson, Y. P. Zhang, O. A. Ryder, and H. P. O'Brien. 2001a. Molecular phylogenetics and the origins of placental mammals. Nature 409:614–618.

Murphy, W. J., E. Eizirik, S. J. O'Brien, O. Madsen, M. Scally, C. J. Douady, E. C. Teeling, O. A. Ryder, M. J. Stanhope, W. W. de Jong, and M. S. Springer. 2001b. Resolution of the early placental mammal radiation using Bayesian phylogenetics. Science 294:2348–2351.

Musakulova, L. T. 1963. *Gobiomeryx* from the Paleogene of Kazakhstan. Akademia Nauk Kazakh SSR, Institut Zoologia, Materialy Fauny i Flory 4:201–203.

Musakulova, L. T. 1971. Localities of fossils tragulids in Kazakhstan. Akademia Nauk Kazakh SSR, Institut Zoologia, Materialy Fauny i Flory 5:52–56.

Nakaya, H., M. Pickford, Y. Nakano, and I. Ishida. 1984. The late Miocene large mammal fauna from the Namurungule Formation, Samburu Hills, northern Kenya. African Study Monographs (supplementary issue) 2:87–131.

Nakaya, H., M. Pickford, K. Yasui, and Y. Nakano. 1986. Additional large mammalian fauna from the Namurungule Formation, Samburu Hills, northern Kenya. African Study Monographs (supplementary issue) 5:79–129.

Nanda, A. C., and A. Sahni. 1990. Oligocene vertebrates from the Ladakh Molasse Group, Ladakh Himalaya: palaeogeographic implications. Journal of Himalayan Geology 1:1–10.

Naylor, G. J. P., and D. C. Adams. 2001. Are the fossil data really at odds with the molecular data? Morphological evidence for Cetartiodactyla phylogeny reexamined. Systematic Biology 50:444–453.

Nesbit Evans, E. M., J. A. H. Van Couvering, and P. Andrews. 1981. Palaeoecology of Miocene sites in western Kenya. Journal of Human Evolution 10:99–116.

Nikaido, M., A. P. Rooney, and N. Okada. 1999. Phylogenetic relationships among cetartiodactyls based on insertions of short and long interspersed elements: hippopotamuses are the closest extant relatives of whales. Proceedings of the National Academy of Sciences, USA 96:10261–10266.

Nikaido, M., H. Hamilton, H. Makino, T. Sasaki, K. Takahashi, M. Goto, N. Kanda, L. A. Pastene, and N. Okada. 2006. Baleen whale phylogeny and a past extensive radiation event revealed by SINE insertion analysis. Molecular Biology and Evolution 23:866–873.

Nikolov, I. 1967. Neue obereozäne Arten der Gattung *Elomeryx*. Neues Jahrbuch für Geologie und Paläontologie Abhandlungen 128:205–214.

Nikolov, I., and K. Heissig. 1985. Fossile Säugetiere aus dem Obereozän und Unteroligozän Bulgariens und ihre Bedeutung für Palaeogeographie. Mitteilungen der Bayerischen Staatssammlung für Paläontologie und historische Geologie 25:61–79.

Nixon, K. C. 1999. The parsimony ratchet, a new method for rapid parsimony analysis. Cladistics 15:407–414.

Nixon, K. C., and J. M. Carpenter. 1996. On simultaneous analysis. Cladistics 12:221–241.

Noetling, F. 1901. Fauna of the Miocene Beds of Burma. Memoirs. Geological Survey India, Palaeontologia Indica 1(3):1–378.

Nolan, J. V., J. B. Liang, N. Abdullah, H. Kudo, H. Ismail, Y. W. Ho, and S. Jalaludin. 1995. Food intake, nutrient utilization and water turnover in the lesser mouse-deer (*Tragulus javanicus*) given lundai (*Sapium baccatum*). Comparative Biochemistry and Physiology 111A(1):177–182.

Nordin, M. 1978. Voluntary food intake and digestion by the lesser mouse deer. Journal of Wildlife Management 42:185–187.

Norell, M. A. 1992. Taxic origin and temporal diversity: the effect of phylogeny; pp. 89–118 in M. J. Novacek and Q. D. Wheeler (eds.), Extinction and Phylogeny. Columbia University Press, New York, New York.

Norris, C. A. 1999. The cranium of *Bunomeryx* (Artiodactyla: Homacodontidae) from the Upper Eocene Uinta deposits of Utah and its implications for tylopod systematics. Journal of Vertebrate Paleontology 19:742–751.

Norris, C. A. 2000. The cranium of *Leptotragulus*, a hornless protoceratid (Artiodactyla: Protoceratidae) from the middle Eocene of North America. Journal of Vertebrate Paleontology 20:341–348.

Noulet, J. B. 1870. Du chéropotame de Lautrec, espèce nouvelle des grès à palaeotheriums du bassin de l'Agout (Tarn). Mémoires de l'Académie des Sciences, Inscriptions et Belles Lettres de Toulouse 2:331–335.

Novacek, M. J. 1977. Aspects of the problem of variation, origin and evolution of the eutherian bulla. Mammal Reviews 7:131–149.

Novacek, M. J. 1980. Cranioskeletal features in tupaiids and selected Eutheria as phylogenetic evidence; pp. 35–93 in W. P. Luckett (ed.), Comparative Biology and Evolutionary Relationships of Tree Shrews. Plenum Press, New York, New York.

Novacek, M. J. 1982. Information for molecular studies from anatomical and fossil evidence on higher eutherian phylogeny. pp. 3–41 in M. Goodman (ed.), Macromolecular Sequences in Systematic and Evolutionary Biology. Plenum Press, New York, New York.

Novacek, M. J. 1986. The skull of lepticitid insectivorans and the higher-level classification of eutherian mammals. Bulletin of the American Museum of Natural History 183:1–112.

Novacek, M. J. 1992. Mammalian phylogeny: shaking the tree. Nature 356:121–125.

Novacek, M. J., and A. R. Wyss. 1986. Higher-level relationships of the Recent eutherian orders: morphological evidence. Cladistics 2:257–287.

Nowak, R. M. 1991. Walker's Mammals of the World (5th edition). The Johns Hopkins University Press, Baltimore, Maryland, 1629 pp.

Nowak, R. M. 1999. Walker's Mammals of the World (6th edition). The Johns Hopkins University Press, Baltimore, Maryland, 1936 pp.

Nowak, R. M., and J. L. Paradiso. 1983. Walker's Mammals of the World (4th edition). The Johns Hopkins University Press, Baltimore, Maryland.

O'Gara, B. W. 1978. *Antilocapra americana*. Mammalian Species 90:1–7.

O'Gara, B. W. 1990. The pronghorn (*Antilocapra americana*); pp. 231–264 in G. A. Bubenik and A. B. Bubenik (eds.), Horns, Pronghorns, and Antlers: Evolution, Morphology, Physiology, and Social Significance. Springer-Verlag, New York, New York.

Ogen-Odoi, A. A., and T. G. Dilworth. 1987. Effects of burning and hippopotamus grazing on savanna hare habitat utilization. African Journal of Ecology 27:47–50.

O'Leary, M. A. 1998a. Phylogenetic and morphometric reassessment of the dental evidence for a mesonychian and cetacean clade; pp. 133–161 in J. G. M. Thewissen (ed.), The Emergence of Whales. Plenum Press, New York, New York.

O'Leary, M. A. 1998b. Morphology of the humerus of *Hapalodectes* (Mammalia, Mesonychia). American Museum Novitates 3242:1–6.

O'Leary, M. A. 1999a. Whale evolution. Science 283:1641–1642.

O'Leary, M. A. 1999b. Parsimony analysis of total evidence from extinct and extant taxa and the cetacean–artiodactyl question (Mammalia, Ungulata). Cladistics 15:315–330.

O'Leary, M. A. 2001. The phylogenetic position of cetaceans: further combined data analyses, comparisons with the stratigraphic record and a discussion of character optimization. American Zoologist 41:487–506.

O'Leary, M. A., and J. H. Geisler. 1999. The position of Cetacea within Mammalia: Phylogenetic analysis of morphological data from extinct and extant taxa. Systematic Biology 48:455–490.

O'Leary, M. A., and K. D. Rose. 1995a. New mesonychian dentitions from the Paleocene of the Bighorn Basin, Wyoming. Annals of the Carnegie Museum 64:147–172.

O'Leary, M. A., and K. D. Rose. 1995b. Postcranial skeleton of the early Eocene mesonychid *Pachyaena* (Mammalia: Mesonychia). Journal of Vertebrate Paleontology 15:401–430.

O'Leary, M. A., S. G. Lucas, and T. E. Williamson. 2000. A new specimen of *Ankalagon* (Mammalia, Mesonychia) and evidence of sexual dimorphism in mesonychians. Journal of Vertebrate Paleontology 20:387–393.

Ord, G. 1815. North American zoology; pp. 292 and 308 in Guthrie's Geography (2nd American edition). Johnson and Warner, Philadelphia, Pennsylvania.

Orliac, M. J. 2006. Eurolistriodon tenarezensis sp.nov. from Montreal-du-Gers (France); implications for the systematics of the European Listriodontinae (Suidae, Mammalia). Journal of Vertebrate Paleontology 26:967–980.

Osborn, H. F. 1883. *Achaeonodon,* an Eocene bunodont. Contributions from the E. M. Museum of Geology and Archaeology of Princeton College, Bulletin 3:23–35, Fig. 6.

Osborn, H. F. 1895. Fossil mammals of the Uinta Basin, expedition of 1894. Bulletin of the American Museum of Natural History 7:71–105.

Osborn, H. F. 1909. Cenozoic mammal horizons of western North America. United States Geological Survey, Bulletin 361:1–90.

Osborn, H. F. 1910. The Age of Mammals in Europe, Asia, and North America. Macmillian, New York, New York.

Osborn, H. F. 1929. The titanotheres of ancient Wyoming, Dakota and Nebraska. United States Geological Survey Monograph 55. Washington, D.C.

Osborn, H. F., and J. L. Wortman. 1892. Characters of *Protoceras* (Marsh), the new artiodactyl from the Lower Miocene. Bulletin of the American Museum of Natural History 4:351–371.

Osborn, H. F., and J. L. Wortman. 1893. *Artionyx,* a new genus of Ancylopoda. Bulletin of the American Museum of Natural History 5:1–18.

Osborn, H. F., and J. L. Wortman. 1894. Fossil mammals of the lower Miocene White River Beds—collection of 1892. Bulletin of the American Museum of Natural History 6:199–228.

Osborn, H. F., W. B. Scott, and F. Speir, Jr. 1878. Palaeontological report of the Princeton scientific expedition of 1877. Contributions to the Museum of Geology and Archaeology, Princeton College 1:1–107.

Owen, R. 1841. Description of some fossil remains of *Chaeropotamus, Palaeotherium, Anoplotherium* and *Dichobune,* from the Eocene formation, Isle of Wight. Transactions of the Geological Society of London 6:41–45.

Owen, R. 1845. Odontography; or a treatise on the comparative anatomy of the teeth; their physiological relations, mode of development, and microscopic structure, in the vertebrate animals. Baillière, London, Paris, pp. 289–655.

Owen, R. 1848a. Description of teeth and proportion of jaws of two extinct Anthracotherioid quadrupeds (*Hyopotamus vectianus* and *Hyopotamus bovinus*) discovered by the Marchioness of Hastings in the Eocene deposits on the N.W. coast of the Isle of Wight: with an attempt to develop Cuvier's idea of the classification of pachyderms by the number of their toes. Quarterly Journal of the Geological Society of London 4:103–141.

Owen, R. 1848b. The Archetype and Homologies of the Vertebrate Skeleton. J. van Voorst, London, 203 pp.

Owen, R. 1857. Description of the lower jaw and teeth of an anoplotherioid quadruped (*Dichobune ovina,* Ow.) of the size of the *Xiphodon gracilis,* Cuv., from the upper Eocene Marl, Isle of Wight. Quarterly Journal of the Geological Society of London 13:254–260.

Owen-Smith, N. 1988. Megaherbivores. The Influence of Very Large Body Size on Ecology. Cambridge University Press, Cambridge.

Paden, M., and M. Nordin. 1978. Maximum food intake and passage of markers in the alimentary tract of the lesser mouse deer. Malaysian Applied Biology 7:11–17.

Page, D. M. 1999. http://taxonomy.zoology.gla.ac.uk/rod/NDE/manual.html.

Pagnac, D. 2005. New camels (Mammalia: Artiodactyla) from the Barstow Formation (middle Miocene), San Bernardino County, California. PaleoBios 25(2):19–31.

Palmer, T. S. 1897. Notes on the nomenclature of four genera of Tropical American mammals. Proceedings of the Biological Society, Washington 11:173–174.

Palmer, T. S. 1904. Index Generum Mammalium: A list of the genera and families of mammals. United States Department of Agriculture, Washington, D.C.

Panaretto, B. A. 1968. Body composition in vivo—III. The relation of body composition to the tritiated water spaces of ewes and wethers fasted for short periods. Australian Journal of Agricultural Research 19:272.

Pantanelli, D. 1878. Sugli strati mioceni del Casino (Siena) i considerazioni sul miocene superiore. Atti della Reale Accademia Nacionale dei Lincei, Memorie della Classe di Scienze Fisiche, Matematiche e Naturali 3:309–327.

Patnaik, R. 2003. Reconstruction of Upper Siwalik palaeoecology and palaeoclimatology using microfossil palaeocommunities. Palaeogeography, Palaeoclimatology, Palaeoecology 197:133–150.

Patton, T. H., and B. E. Taylor. 1971. The Synthetoceratinae (Mammalia, Tylopoda, Protoceratidae). Bulletin of the American Museum of Natural History 145:123–218.

Patton, T. H., and B. E. Taylor. 1973. The Protoceratinae (Mammalia, Tylopoda, Protoceratidae) and the systematics of the Protoceratidae. Bulletin of the American Museum of Natural History 150:351–413.

Pavlakis, P. P. 1990. Plio-Pleistocene Hippopotamidae from the Upper Semliki; pp. 203–223 in N. T. Boaz (ed.), Evolution of Environments and Hominidae in the African Western Rift Valley. Virginia Museum of Natural History Memoir 1.

Pavlov, M. 1900. Études sur l'histoire paléontologique des ongulés. VII. Artiodactyles anciens. Bulletin de la Société des Naturalistes de Moscou 13:266–328.

Pearson, H. S. 1927. On the skulls of early Tertiary Suidae, together with an account of the otic region in some other primitive Artiodactyla. Philosophical Transactions of the Royal Society, London, series B 215:389–462.

Pearson, H. S. 1928. Chinese fossil Suidae. Palaeontologica Sinica C5:1–75.

Pearson, P. N., and M. R. Palmer. 2000. Atmospheric carbon dioxide concentrations over the past 60 million years. Nature 206:695–699.

Pentland, J. B. 1828. Description of fossil remains of some animals from the North-East border of Bengal. Transactions of the Geological Society of London 2:393–394.

Pérez-Barbería, F. J., and I. J. Gordon. 2000. Differences in body mass and oral morphology between the sexes in the Artiodactyla: evolutionary relationships with sexual segregation. Evolutionary Ecology Research 2:667–684.

Peterson, O. A. 1904. Osteology of *Oxydactylus,* a new genus of camels from the Loup Fork of Nebraska, with descriptions of two new species. Annals of the Carnegie Museum 2:434–475.

Peterson, O. A. 1905a. A correction of the generic name (*Dinochoerus*) given to certain fossil remains from the Loup Fork Miocene of Nebraska. Science 22:719.

Peterson, O. A. 1905b. Preliminary note on a gigantic mammal from the Loup Fork Beds of Nebraska. Science 22:211–212.

Peterson, O. A. 1909. A revision of the Entelodontidae. Memoirs of the Carnegie Museum, Pittsburgh 4:41–156.

Peterson, O. A. 1911. A new camel from the Miocene of western Nebraska. Annals of the Carnegie Museum 7:260–266.

Peterson, O. A. 1919. II. Report upon the material discovered in the upper Eocene of the Uinta Basin by Earl Douglas in the years 1908–1909, and by O. A. Peterson in 1912. Annals of Carnegie Museum 12(2–4):40–168.

Peterson, O. A. 1931. Two new species of agriochoerids. Annals of Carnegie Museum 20:341–354.

Peterson, O. A. 1934. List of species and description of new material from Duchesne River Oligocene, Uinta Basin, Utah. Annals of the Carnegie Museum 23:373–389.

Petronio, C. 1986. Nuovi resti di ippopotamo del Pleistocene medio-inferiore del dintorni di Roma e problemi di tassonomia e filogenese del gruppo. Geologica Romana 25:63–72.

Pfeiffer, T. 1999. Die Stellung von Dama (Cervidae, Mammalia) im System plesiometacarpaler Hirsche des Pleistozäns. Courier Forschung-Institut Senckenberg 211:1–218.

Philippe, H., F. Delsuc, H. Brinkmann, and N. Lartillot. 2005. Phylogenomics. Annual Review of Ecology and Systematics 36:541–562.

Pickford, M. 1976. A new species of *Taucanamo* (Mammalia, Artiodactyla, Tayassuidae) from the Siwaliks of the Potwear Plateau, Pakistan. Pakistan Journal of Zoology 8:13–20.

Pickford, M. 1978. The taxonomic status and distribution of *Schizochoerus* (Mammalia, Tayassuidae). Tertiary Research 2:29–38.

Pickford, M. 1983. On the origins of Hippopotamidae together with descriptions of two new species, a new genus and a new subfamily from the Miocene of Kenya. Géobios 16:193–217.

Pickford, M. 1984. A revision of the Sanitheriidae, a new family of Suiformes (Mammalia, Artiodactyla). Géobios 16:133–154.

Pickford, M. 1986. A revision of the Miocene Suidae and Tayassuidae (Artiodactyla, Mammalia) of Africa. Tertiary Research, Special Paper No. 7, 82 pp.

Pickford, M. 1987. Révision des Suiformes (Artiodactyla, Mammalia) de Bugti (Pakistan). Annales de Paléontologie (Vertebrata-Invertebrata) 73:289–350.

Pickford, M. 1988a. Revision of the Miocene Suidae of the Indian subcontinent. Münchner Geowissenschaft Abhandlungen (A) 12:1–92.

Pickford, M. 1988b. Un étrange suidé nain du Néogène supérieur de Langebaanweg (Afrique du Sud). Annales de Paléontologie (Vertebrata–Invertebrata) 74:229–250.

Pickford, M. 1990. Découverte de Kenyapotamus en Tunisie. Annales de Paléontologie 76:277–283.

Pickford, M. 1991. Revision of the Neogene Anthracotheriidae of Africa; pp. 1491–1525 in M. J. Salem and M. T. Busrewil (eds.), The Geology of Libya. Academic Press, New York, New York.

Pickford, M. 1993. Old World suid systematics, phylogeny, biogeography and biostratigraphy. Paleontologia y Evolucion 26–27:237–269

Pickford, M. 1995. Suidae (Mammalia, Artiodactyla) from the early Middle Miocene of Arrisdrift, Namibia: *Namachoerus* (gen. nov.) *moruoroti,* and *Nguruwe kijivium.* Comptes Rendus de l'Academie des Sciences Sciences de la Terre et des Planetes 320:319–326.

Pickford, M. 1998. A new genus of Tayassuidae (Mammalia) from the middle Miocene of Uganda. Annales de Paléontologie 84(3–2):275–285.

Pickford, M. 2001a. Africa's smallest ruminant: a new tragulid from the Miocene of Kenya and the biostratigraphy of East African Tragulidae. Géobios 34:437–447.

Pickford, M. 2001b. New species of *Listriodon* (Suidae, Mammalia) from Bartule, Member A, Ngorora Formation (ca. 13 Ma), Tugen Hills, Kenya. Annales de Paléontologie 87:207–221.

Pickford, M. 2002. Ruminants from the Early Miocene of Napak, Uganda. Annales de Paléontologie 88:85–113.

Pickford, M. 2003. Early and middle Miocene Anthracotheriidae (Mammalia, Artiodactyla) from the Sperrgebiet, Namibia. Memoir of the Geology Survey of Namibia 19:283–289.

Pickford, M. 2004. Miocene Sanitheriidae (Suiformes, Mammalia) from Namibia and Kenya: systematic and phylogenetic implications. Annales de Paléontologie 90:223–278.

Pickford, M. 2006. Sexual and individual morphometric variation in *Libycosaurus* (Mammalia, Anthracotheriidae) from the Maghreb and Libya. Géobios 39:267–310.

Pickford, M., and C. Etürk. 1979. Suidae and Tayassuidae from Turkey. Bulletin of the Geological Society of Turkey 22: 141–154.

Pickford, M., and J. Morales. 2003. New Listriodontinae (Mammalia, Suidae) from Europe and a review of listriodont evolution, biostratigraphy and biogeography. Geodiversitas 25:347–404.

Pickford, M., and S. Moyà-Solà. 1995. *Eulistriodon* gen. nov. (Suoidea, Mammalia) from Els Casots, early Middle Miocene, Spain. Proceedings Koninklijke Nederlandse Akademie Wetenschappen 98:343–360.

Pickford, M., and A. Wilkinson. 1975. Stratigraphic and phylogenetic implications of new Listriodontinae from Kenya. Netherlands Journal of Zoology 25:128–137.

Pickford, M., B. Senut, and C. Mourer-Chauviré. 2004. Early Pliocene Tragulidae and peafowls in the Rift Valley, Kenya: evidence for rainforest in East Africa. Comptes Rendus Palevol 3:179–189.

Pictet, J. F. 1855–1857. Description des ossements fossiles trouvés au Mauremont. Seconde partie; pp. 27–120 in J. F. Pictet, C. Gaudin, and P. H. de Harpe (eds.), Mémoire sur les Animaux Vertébreés trouvés dans la terrain sidérolithique du canton de Vaud et appartemant à la fauna éocéne. C. J. Kessmann, Genève.

Pictet, J. F., and A. Humbert 1869. Mémoire sur les animaux vertébrés trouvés dans le terrain sidérolithique du Canton de Vaud et appartenant à la faune éocène. Supplément. Matériaux Paléontologie Suisse 5:121–197.

Pilgrim, G. E. 1907. Description of some new Suidae from the Bugti Hills, Baluchistan. Records of the Geological Survey of India 36:45–56.

Pilgrim, G. E. 1908. The Tertiary and post-Tertiary freshwater deposits of Baluchistan and Sind with notices of new vertebrates. Records of the Geological Survey of India 37:139–169.

Pilgrim, G. E. 1910. Notices of new mammalian genera and species from the Tertiaries of India. Records of the Geological Survey of India 40:63–71.

Pilgrim, G. E. 1915. The dentition of the tragulid genus *Dorcabune*. Records of the Geological Survey of India 45:226–238.

Pilgrim, G. E. 1928. The Artiodactyla of the Eocene of Burma. Palaeontologia Indica 13:1–39.

Pilgrim, G. E. 1934. Two new species of sheep-like antelope from the Miocene of Mongolia. American Museum Novitates 716:1–29.

Pilgrim, G. E. 1937. Siwalik antelopes and oxen in the American Museum of Natural History. Bulletin of the American Museum of Natural History 72:729–874.

Pilgrim, G. E. 1939. The fossil Bovidae of India. Memoirs of the Geological Survey of India, Palaeontologia Indica New Series 26:1–356.

Pilgrim, G. E. 1940. Middle Eocene mammals from North-West India. Proceedings of the Zoological Society of London 110:127–152.

Pilgrim, G. E. 1941. The dispersal of the Artiodactyla. Biological Reviews of the Cambridge Philosophical Society 16:134–163.

Pilgrim, G. E., and G. de P. Cotter. 1916. Some newly discovered Eocene mammals from Burma. Records of the Geological Survey of India 47:42–77.

Pitra, C., R. Furbass, and H.-M. Seyfert. 1997. Molecular phylogeny of the tribe Bovini (Mammalia: Artiodactyla): alternative placement of the anoa. Journal of Evolutionary Biology 10:589–600.

Pitra, C., J. Fickel, E. Meijaard, and C. P. Groves. 2004. Evolution and phylogeny of Old World deer. Molecular Phylogenetics and Evolution 33:880–895.

Pitra, C., R. A. Kock, R. R. Hofmann, and D. Lieckfeldt. 1998. Molecular phylogeny of the critically endangered Hunter's antelope (*Beatragus hunteri* Sclater 1889). Journal of Zoological Systematics and Evolutionary Research 36:179–184.

Ploeg, G. d., D. Dutheil, E. Gheerbrant, M. Godinot, A. Jossang, A. Nel, J.-L. Paicheler, D. Pons, and J.-C. Rage. 1998. Un nouveau gisement paleontologique konservat-lagerstaette a la base de l'Eocene dans le region de Creil (Oise). Strata série 1 9:108–110.

Poche, F. 1922. Zur Kenntnis der Amphilinidea. Zoologische Anzeiger 54:285.

Pocock, R. I. 1919. On the external characters of existing chevrotains. Proceedings of the Zoological Society London 1919:1–11.

Pol, D., and M. A. Norell. 2001. Comments on the Manhattan stratigraphic measure. Cladistics 17:285–289.

Polziehn, R. O., and C. Strobeck. 1998. Phylogeny of wapiti, red deer, sika deer, and other North American cervids as determined from mitochondrial DNA. Molecular Phylogenetics and Evolution 10:249–258.

Pomel, A. 1847a. Note critique sur les caractères et les limites du genre Palaeotherium. Archives des sciences physiques et naturelles 5:200–207.

Pomel, A. 1847b. Note sur un nouveau genre de pachydermes du bassin de la Gironde (*Elotherium magnum*). Bulletin de la Societe Geologique de France (2) 4:256–257, 1083–1085.

Pomel, A. 1847c. Sur un nouveau genre de Pachydermes fossiles (*Elotherium*) voisin des Hippopotames. Archives des Sciences Physiques et Naturelles 5:307–308.

Pomel, A. 1848. Sur la classification des mammifères ongulés. Bulletin de la Société Géologique de France 5:256–259.

Pomel, A. 1851. Nouvelles observations sur la structure des pieds dans les animaux de la famille des *Anoplotherium,* et dans le genre *Hyemoschus*. Comptes Rendus de l'Académie des Sciences, série IIa 33:16–18.

Pomel, A. 1853. Catalogue méthodique et descriptif des vertebrates fossils découverts dans le Basin Hydrographique Supérior de la Loire et surtout dans la Vallee de son Affluent Principal, l'Allier p. 88. Privately published

Pomel, A. 1890. Sur les Hippopotames fossiles de l'Algérie. Comptes Rendus de l'Académie des Sciences 110:1112–1116.

Prasad, K. N. 1970. The vertebrate fauna from the Siwalik Beds of Haritalyangar, Himachal Pradesh, India. Memoirs of the

Geological Survey of India, Palaeontologica Indica new series 34:1–55.

Prasad, K. N., and P. P. Satsangi. 1968. Fossil Tragulid from the Nagri beds of Haritalyan Gar, Himachal Pradesh. Records of the Geological Survey of India 95:538–540, pl. 39.

Prothero, D. R. 1985. North American mammalian diversity and Eocene–Oligocene extinctions. Paleobiology 11:389–405.

Prothero, D. R. 1993. Ungulate phylogeny: molecular vs. morphological evidence; pp. 173–181 in F. S. Szalay, M. J. Novacek, and M. C. McKenna (eds.), Mammal Phylogeny: Placentals. Springer-Verlag, New York, New York.

Prothero, D. R. 1994. The Eocene–Oligocene Transition: Paradise Lost. Columbia University Press, New York, New York, 291 pp.

Prothero, D. R. 1996. Camelidae; pp. 609–653 in D. R. Prothero and R. J. Emry (ed.), The Terrestrial Eocene–Oligocene Transition in North America. Cambridge University Press, Cambridge.

Prothero, D. R. 1998a. Protoceratidae; pp. 431–438 in C. Janis, K. M. Scott, and L. Jacobs (eds.), Evolution of Tertiary Mammals of North America. Cambridge University Press, Cambridge.

Prothero, D. R. 1998b. The chronological, climatic, and paleogeographic background to North American mammalian evolution; pp. 9–36 in C. M. Janis, K. M. Scott, and L. L. Jacobs (eds.), Evolution of Tertiary Mammals of North America. Cambridge University Press, Cambridge.

Prothero, D. R. 1998c. Oromerycidae; pp. 426–430 in C. M. Janis, K. M. Scott, and L. L. Jacobs (eds.), Evolution of Tertiary Mammals of North America. Cambridge University Press, Cambridge.

Prothero, D. R. 2006. After the Dinosaurs: The Age of Mammals. Indiana University Press, Bloomington, Indiana.

Prothero, D. R. In press. Systematics of the musk deer (Artiodactyla: Moschidae: Blastomerycinae) from the Miocene of North America. New Mexico Museum of Natural History and Science Bulletin.

Prothero, D. R., and R. J. Emry. 2004. The Chadronian, Orellan, and Whitneyan North American Land Mammal Ages; pp. 156–168 in M. O. Woodburne (ed.), Late Cretaceous and Cenozoic Mammals of North America. Columbia University Press, New York, New York.

Prothero, D. R., and Liter, M. R. In press. Systematics of the Dromomerycinae (Artiodactyla: Palaeomerycidae) from the Miocene of North America. New Mexico Museum of Natural History Bulletin.

Prothero, D. R., and F. Sanchez. In press. Systematics of the leptaucheniinae oreodonts (Mammalia: Artiodactyla) from the Oligocene and earliest Miocene of North America. New Mexico. Museum of Natural History and Science Bulletin.

Prothero, D. R., and R. M. Schoch. 2002. Horns, Tusks, and Flippers. The Johns Hopkins University Press, Baltimore, Maryland, 311 pp.

Prothero, D. R., and K. E. Whittlesey. 1998, Magnetostratigraphy and biostratigraphy of the Orellan and Whitneyan land mammal "ages" in the White River Group. Geological Society of America Special Paper 325:39–61.

Prothero, D. R., E. M. Manning, and M. Fischer. 1988. The phylogeny of the ungulates; pp. 201–234 in M. J. Benton (ed.), The Phylogeny and Classification of the Tetrapods, Mammals, Volume 2. Systematics Association Special Volume 35B. Clarendon Press, Oxford.

Pyrchodko, V. I., 2003. The musk deer: origins, systematics, ecology, behavior, and communications. Moscow, 443 pp. [in Russian with English summary].

Qiu, Z. 1977. Notes on the new species of *Anthracokeryx* from Guangxi. Vertebrata PalAsiatica 15:54–58.

Qiu, Z. 1978. Late Eocene Hypertragulids of Baise Basin, Kwangsi. Vertebrata PalAsiatica 16:7–12.

Qiu, Z., and Y. Gu. 1991. The middle Miocene vertebrate fauna from Xiacaowan, Sihong County, Jiangsu Province. 8. Dorcatherium (Tragulidae, Artiodactyla). Vertebrata PalAsiatica 29:21–37.

Qiu, Z., D. Yan, H. Jia, and B. Sun, B. 1985. Preliminary observations on newly found skeletons of *Palaeomeryx* from Shanwang, Shandong. Vertebrata PalAsiatica 23:173–200.

Qiu, Z., J. Ye, and F. Huo. 1988. Description of a *Kubanochoerus* skull from Tongxin, Ningxia. Vertebrata PalAsiatica 26:1–19.

Radinsky, L. B. 1965. Evolution of the tapiroid skeleton from *Heptodon* to *Tapirus*. Bulletin of the Museum of Comparative Zoology (Harvard) 34:1–106.

Radinsky, L. 1978. Evolution of brain size in carnivores and ungulates. American Naturalist 112:815–831.

Ralls, K., C. Barasch, and K. Minkowski. 1975. Behavior of Capitive Mouse Deer, *Tragulus napu*. Zeitschrift für Tierpsychologie 37:356–378.

Randi, E., V. Lucchini, and C. H. Diong. 1996. Evolutionary genetics of Suiformes as reconstructed using mtDNA sequencing. Journal of Mammalian Evolution 3:163–194.

Randi, E., M. Pierpaoli, and A. Danilkin. 1998a. Mitochondrial DNA polymorphism in populations of Siberian and European roe deer (*Capreolus pygargus* and *C. capreolus*). Heredity 80:429–437.

Randi, E., N. Mucci, M. Pierpaoli, and E. J. P. Douzery. 1998b. New phylogenetic perspectives on the Cervidae (Artiodactyla) are provided by the mitochondrial cytochrome *b* gene. Proceedings of the Royal Society of London, Series B 265:793–801.

Ranga Rao, A. 1971. New mammals from Murree (Kalakot Zone) of the Himalayan foot hills near Kalakot, Jammu and Kashmir State, India. Journal of the Geological Society of India 12:125–134.

Ranga Rao, A. 1972. New mammalian genera and species from the Kalakot zone of Himalayan foot hills near Kalakot, Jammu and Kashmir State, India; pp. 1–22 in Directorate of Geology, Oil and Natural Gas, Dehra Dun.

Rasmussen, D. T., and E. L. Simons. 1988. New Oligocene hyracoids from Egypt. Journal of Vertebrate Paleontology 8:67–83.

Rasmussen, D.T., J. Kappelman, W. J. Sanders, and K. M. Muldoon. In press. Middle Oligocene hyracoids and artiodactyls of Chilga, Ethiopia. Journal of Vertebrate Paleontology.

Rasmussen, D. T., G. C. Conroy, A. R. Friscia, K. E. Townsend, and M. D. Kinkel. 1999. Mammals of the Middle Eocene Uinta Formation; pp. 401–420 in D. D. Gillette (ed.), Fossil Vertebrates of Utah. Special Publication of the Utah Geological Survey, Salt Lake City, Utah.

Rebholz, W., and E. Harley. 1999. Phylogenetic relationships in the bovid subfamily Antilopinae based on mitochondrial DNA sequences. Molecular Phylogenetics and Evolution 12:87–94.

Reed, K. E. 1997. Early hominid evolution and ecological change through the Africa Plio-Pleistocene. Journal of Human Evolution 32:289–322.

Repelin, J. 1919. Sur les espèces ou mutations nouvelles du genre *Entelodon* Aymard. Bulletin de la Société Géologique de France 19:11–14.

Reyment, R. A. 1983. Palaeontological aspects of island biogeography: colonization and evolution of mammals on Mediterranean islands. Oikos 41:299–306.

Richard, M. 1942. Description et figuration du *Lophiobunodon minervoisensis*. Bulletin de la Société d'Histoire Naturelle de Toulouse 77:141–144.

Richards, G. D., and M. L. McCrossin. 1991. A new species of *Antilocapra* from the Late Quaternary of California. Géobios—Paléontologie, Stratigraphie, Paléoécologie 24:623–635.

Richter, G. 1981. Untersuchungen zur Ernährung von *Messelobunodon schaeferi* (Mammalia, Artiodactyla). Senckenbergiana Lethaea 61:355–370.

Richter, G. 1987. Untersuchungen zur Ernährung eozäner Säuger aus der Fossilfundstelle Messel bei Darmstadt. Courier Forschungs-Institut Senckenberg 91:1–33.

Rinnert, P. 1956. Die Huftiere aus dem Braunkohlenmiozän der Oberpfalz. Palaeontographica A 107:1–65.

Roberts, S. C. 1996. The evolution of hornedness in female ruminants. Behaviour, Leiden 133:399–442.

Robinson, P., G. F. Gunnell, S. L. Walsh, W. C. Clyde, J. E. Storer, R. K. Stucky, D. J. Froehlich, I. Ferrusquia-Villafranca, and M. C. McKenna. 2004. Wasatchian through Duschesnean biochronology; pp. 106–155 in M. O. Woodburne (ed.), Late Cretaceous and Cenozoic Mammals of North America. Columbia University Press, New York, New York.

Rodler, A., and K. A. Weitheofer. 1890. Die Wiederkäuer der Fauna von Maragha. Denkschriften der Kaiserlichten Akademie der Wissenschaften Wien 57:753–772.

Roger, J., M. Pickford, H. Thomas, F. Lapparent de Broin, P. Tassy, W. Van Neer, C. Bourdillon de Grissac, and S. Al-Busaidi. 1994. Découverte de vertébrés fossiles dans le Miocène de la région du Huqf au Sultanat d'Oman. Annales de Paléontologie 80:253–273.

Roger, O. 1896. Verzeichniss der bisher Bekännten fossilen Säugethiere. Bericht des Naturwissenschaftlichen Vereins für Schwaben und Neuburg, Augsburg 32:194, 204.

Roger, O. 1904. Wirbeltierreste aus dem Obermiocän der bayerisch-schwäbischen Hochebene. V. Bericht des Naturwissenschaftlichen Vereins für Schwaben und Neuburg, Augsburg 36:1–19.

Rögl, F. 1999. Mediterranean and Paratethys Palaeogeography during the Oligocene and Miocene; pp. 8–22 in J. Augusti, L. Rook, and P. Andrews (eds.), Hominoid Evolution and Climate Change in Europe. Volume 1. The Evolution of Neogene Terrestrial Ecosystems in Europe. Cambridge University Press, Cambridge.

Rokas, A., D. Krüger, and S. B. Carroll. 2005. Animal evolution and the molecular signature of radiations compressed in time. Science 310:1933–1938.

Romer, A. S. 1966. Vertebrate Paleontology, 3rd edition. University of Chicago Press, Chicago, Illinois, 468 pp.

Roosevelt, Q., and J. W. Burden. 1934. A new species of antilocaprine, *Tetrameryx onusrosagris*, from a Pleistocene cave deposit in southern Arizona. American Museum Novitates 754:1–4.

Ropiquet, A., and A. Hassanin. 2004. Molecular phylogeny of caprines (Bovidae, Antilopinae): the question of their origin and diversification during the Miocene. Journal of Zoological Systematics and Evolutionary Research 43:49–60.

Ropiquet, A., and A. Hassanin. 2005. Molecular evidence for the polyphyly of the genus *Hemitragus* (Mammalia, Bovidae). Molecular Phylogenetics and Evolution 36:154–168.

Rose, K. D. 1981. Composition and species diversity in Paleocene and Eocene mammal assemblages: an empirical study. Journal of Vertebrate Paleontology 1:367–388.

Rose, K. D. 1982. Skeleton of *Diacodexis*, oldest known artiodactyl. Science 216:621–623.

Rose, K. D. 1985. Comparative osteology of North American dichobunid artiodactyls. Journal of Paleontology 59:1203–1226.

Rose, K. D. 1987. Climbing adaptations in the early Eocene mammal *Chriacus*, and the origin of the Artiodactyla. Science 236:314–316.

Rose, K. D. 1990. Postcranial skeletal remains and adaptations in early Eocene mammals from the Willwood Formation, Bighorn Basin, Wyoming. Geological Society of America Special Paper 243:107–133.

Rose, K. D. 2001. Compendium of Wasatchian mammal postcrania from the Willwood Formation of the Bighorn Basin. University of Michigan Papers on Paleontology 33:157–183.

Rose, K. D., and M. A. O'Leary. 1995. The manus of *Pachyaena gigantea* (Mammalia: Mesonychia). Journal of Vertebrate Paleontology 15:855–859.

Rössner, G. 1995. Odontologische und schädelanatomische Untersuchungen an Procervulus (Cervidae, Mammalia). Münchner Geowissenschaftliche Abhandlungen, A: Geologie und Paläontologie 29:1–128.

Rössner, G. E. 2004. Community structure and regional patterns in late Early to Middle Miocene Ruminantia of Central Europe; pp. 91–100 in F. F. Steininger, J. Kovar-Eder, and M. Fortelius (eds.), The Middle Miocene Environments and Ecosystem Dynamics of the Eurasian Neogene (EEDEN). Courier Forschungs-Institut Senckenberg 249, Frankfurt am Main.

Rössner, G. E., and K. Heissig (eds.). 1999. Land Mammals of Europe. Verlag Dr. Friedrich Pfeil, Munich, 515 pp.

Rössner, G. E., and M. Rummel. 2001. *Pomelomeryx gracilis* (Pomel, 1853) (Mammalia, Artiodactyla, Moschidae) from the lower Miocene karstics fissure filling complex Rothenstein 10/14 (Germany, Bavaria). Lynx (Praha) 32:323–253.

Russell, D. E. 1964. Les mammiféres Paleocenes d'Europe. Memoir Musee National d'Histoire Naturelle Série C 13:1–324.

Russell, D. E., and R. J. Zhai. 1987. The Paleogene of Asia: mammals and stratigraphy. Mémoire du Museum Nationale d'Histoire Natürelle 53:1–488.

Russell, D. E., J. G. M. Thewissen, and D. Sigogneau-Russell. 1983. A new dichobunid artiodactyl (Mammalia) from the Eocene of North-West Pakistan, Part II: Cranial osteology. Proceedings of the Koninklijke Nederlandse Akademie van Wetenschappen Series B 86:285–299.

Russell, D. E., J. L. Hartenberger, C. Pomerol, S. Sen, N. Schmidt-Kittler, and M. Vianey-Liaud. 1982. Mammals and Stratigraphy: The Paleogene of Europe. Palaeovertebrata (Mémoire extraordinaire):1–77.

Russell, L. S. 1978. Tertiary mammals of Saskatchewan. Part IV: the Oligocene anthracotheres. Life Science Contributions Royal Ontario Museum 115:3–16.

Russell, L. S. 1980a. A new species of *Brachyhyops?* (Mammalia, Artiodactyla) from the Oligocene Cypress Hills Formation of Saskatchewan. Royal Ontario Museum, Life Sciences Occasional Paper No. 33:1–8.

Russell, L. S. 1980b. Tertiary mammals of Saskatchewan, Part V: the Oligocene entelodonts. Royal Ontario Museum, Life Sciences Contributions 122:1–42.

Rütimeyer, L. 1857. Ueber *Anthracotherium magnum* und *hippoideum*. Neue Denkschriften der schweizerischen Naturforschenden Gesellschaft 15:1–32.

Rütimeyer, L. 1862. Eocäene Säugetiere aus dem Gebiet des Schweizerischen Jura. Neue Denkschriften der Allgemeinen Schweizerischen Gesellschaft für die Gesammten Naturwissenschaften 19:1–98.

Rütimeyer, L. 1888. Sur la faune éocène d'Egerkingen (Soleure). Compte rendu des travaux de la Société Helvétique des Sciences Naturelles LXXII:46–48.

Rütimeyer, L. 1891. Die eocäne Säugetierwelt von Egerkingen. Gesamtdarstellung und dritter Nachtrag zu den "Eozänen Säugetieren aus dem Gebiet des Schweizerischen Jura (1862)." Abhandlungen der Schweizerischen Paläontologischen Gesellschaft 18:1–153.

Sahni, A., and A. Jolly. 1993. Eocene mammals from Kalakot, Kashmir Himalaya: community structure, taphonomy and palaeogeographical implications. Kaupia 3:209–222.

Sahni, A., and S. K. Khare. 1971. Three new Eocene mammals from Rajauri District, Jammu and Kashmir. Journal of the Palaeontological Society of India 16:41–53.

Sahni, A., S. Bal Bhatia, J.-L. Hartenberger, J.-J. Jaeger, K. Kumar, J. Sudre, and M. Vianey-Liaud. 1981. Vertebrates from the type section of the Subathu Formation and comments on the palaeobiography of the Indian subcontinent during the early Palaeogene. Bulletin of the Indian Geological Association 14:89–100.

Sánchez-Villagra, M. R., O. Aguilera, and I. Horovitz. 2003. The anatomy of the world's largest extinct rodent. Science 301: 1708–1710.

Sanderson, M. J., and A. C. Driskell. 2003. The challenge of constructing large phylogenetic trees. Trends in Plant Science 8:374–379.

Sanderson, M. J., A. Purvis, and C. Henze. 1998. Phylogenetic supertrees: assembling the trees of life. Trends in Ecology and Evolution 13:105–109.

Sanmartin, I., H. Enghoff, and F. Ronquist. 2001. Patterns of animal dispersal, vicariance and diversification in the Holarctic. Biological Journal of the Linnean Society 73:345–390.

Savage, D. E., and D. E. Russell. 1983. Mammalian Paleofaunas of the World. Addison-Wesley Publishing Company, London, 432 pp.

Scally, M., O. Madsen, C. J. Douady, W. W. de Jong, M. J. Stanhope, and M. S. Springer. 2001. Molecular evidence for the major clades of placental mammals. Journal of Mammalian Evolution 8:239–277.

Schaeffer, B. 1947. Notes on the origin of the artiodactyl tarsus. American Museum Novitates 1356:1–24.

Schaller, G. B. 1977. Mountain Monarchs. Wild Sheep and Goats of the Himalaya. University Press, Chicago, Illinois.

Scheele, W. E. 1955. The First Mammals. World Press, New York, New York.

Schiebout, J. A. 1979. An overview of the terrestrial early Tertiary of southern North America—fossil sites and paleopedology. Tulane Studies in Geology and Paleontology 15:75–94.

Schlaikjer, E. M. 1935a. Contributions to the stratigraphy and paleontology of the Goshen Hole area, Wyoming. 3. A new basal Oligocene formation. Bulletin of the Museum of Comparative Zoology, Harvard University 76:71–93.

Schlaijker, E. M. 1935b. New vertebrates and the stratigraphy and palcontology of thc Goshcn Holc arca, Wyoming. IV. New vertebrates and the stratigraphy of the Oligocene and early Miocene. Bulletin of the Museum of Comparative Zoology, Harvard University 76:97–189.

Schlosser, M. 1883. Über die Extremitäten des *Anoplotherium*. Jahrbuch für Mineralogie, Geologie und Paläontologie 2:141–152.

Schlosser, M. 1886. Beitrage zur kenntniss der Stammesgeschichte der Huftiere und versuch einer Systematik der Paar- und Unpaarhufer. Morphologische Jahrbuch 12:1–136.

Schlosser, M. 1901. Beiträge zur Kenntnis der Wirbeltierfauna der böhmischen Braunkohlenformatiom. Abhandlungen Naturwissenschaften Verlag, Prague 2:1–43.

Schlosser, M. 1902. Beiträge zur Kenntnis der Säugetierreste aus den süddeutschen Bohnerzen. Geologische und Paläontologische Abhandlungen, Neue Serie F 5(3):1–258.

Schlosser, M. 1904. Die fossilen Cavicornier von Samos. Beitrage zur Paläoetologie und Geologie Österreich-Ungarns und des Orients, Vienna 17:28–118.

Schmidt, M. 1913. Über Paarhufer der fluviomarinen Schichten des Fajum. Geologische und paläontologische Abhandlungen 11:1–111.

Schmidt-Kittler, N. 1971. Die Obermiozane Fossilaggerstatte—Sandelzhausen 3. Suidae (Artiodactyla, Mammalia). Mitteilungen der Bayerischen Staatssammlung für Paläontologie und Historische Geologie 11:129–170.

Schmidt-Kittler, N. 1987. International Symposium on Mammalian Biostratigraphy and Palaeoecology of the European Palaeogene. Münchner Geowissenschaftliche Abhandlungen, Reihe A 10:1–312.

Schrodt, A. K. 1980. Depositional environments, provenance, and vertebrate paleontology of the Eocene–Oligocene Baca Formation, Catron County, New Mexico. M.S. thesis, Geology, Louisiana State University, Baton Rouge, Louisiana, 174 pp.

Schultz, C. B., and C. H. Falkenbach. 1940. Merycochoerinae, a new subfamily of oreodonts. Bulletin of the American Museum of Natural History 77:213–306.

Schultz, C. B., and C. H. Falkenbach. 1941. Ticholeptinae, a new subfamily of oreodonts. Bulletin of the American Museum of Natural History 79:1–105.

Schultz, C. B., and C. H. Falkenbach. 1947. Merychyinae, a subfamily of oreodonts. Bulletin of the American Museum of Natural History 88:157–286.

Schultz, C. B., and C. H. Falkenbach. 1949. Promerycochoerinae, a new subfamily of oreodonts. Bulletin of the American Museum of Natural History 93:73–198.

Schultz, C. B., and C. H. Falkenbach. 1950. Phenacocoelinae, a new subfamily of oreodonts. Bulletin of the American Museum of Natural History 95:91–149.

Schultz, C. B., and C. H. Falkenbach. 1954. Desmatochoerinae, a new subfamily of oreodonts. Bulletin of the American Museum of Natural History 105:143–256.

Schultz, C. B., and C. H. Falkenbach. 1956. Miniochoerinae and Oreonetinae, two new subfamilies of oreodonts. Bulletin of the American Museum of Natural History 109: 377–482.

Schultz, C. B., and C. H. Falkenbach. 1968. The phylogeny of the oreodonts. Bulletin of the American Museum of Natural History 139:1–498.

Scopoli, G. A. 1777. Introductio ad historiam naturalem sistens genera lapidum, plantarum et animalium hactenus detecta, caracteribus essentialibus donata, in tribus divisa, subinde ad leges naturae. Gerie, Prague, 506 pp.

Scotland, R. W., R. G. Olmstead, and J. R. Bennett. 2003. Phylogeny reconstruction: the role of morphology. Systematic Biology 52:539–548.

Scott, K. M., and C. M. Janis. 1993. Relationships of the Rumi-nantia (Artiodactyla) and an analysis of the characters used in ruminant taxonomy; pp. 282–302 in F. S. Szalay, M. J. Novacek, and M. C. McKenna (eds.), Mammal Phylogeny. Springer-Verlag, New York, New York.

Scott, W. B. 1890. Beiträge zur kenntniss der Oreodontidae. Morphologisches Jahrbuch 16:319–395.

Scott, W. B. 1891. Osteology of *Poebrotherium:* a contribution to the phylogeny of the Tylopoda. Journal of Morphology 5:1–74.

Scott, W. B. 1893. The mammals of the Deep River beds. American Naturalist 27:659–662.

Scott, W. B. 1894a. On some new and little known creodonts. Journal of the Academy of Natural Sciences, Philadelphia 9:155–185.

Scott, W. B. 1894b. The structure and relationships of *Ancodus.* Journal of the Academy of Natural Sciences, Philadelphia 9:461–497.

Scott, W. B. 1895a. The Mammalia of the Deep River beds. Transactions of the American Philosophical Society, Philadelphia 18:55–183.

Scott, W. B. 1895b. The osteology and relations of *Protoceras.* Journal of Morphology 11:303–374.

Scott, W. B. 1898a. Preliminary note on the selenodont artiodactyls of the Uinta Formation. Proceedings of the American Philosophical Society 37:73–81.

Scott, W. B. 1898b. The osteology of *Elotherium.* Transactions of the American Philosophical Society, Philadelphia 19:273–324, pl. 17–18.

Scott, W. B. 1899. The selenodont artiodactyls of the Uinta Eocene. Transactions of the Wagner Free Institute of Science of Philadelphia 6:9–121.

Scott, W. B. 1913. A History of Land Mammals of the Western Hemisphere. Macmillan, New York, New York, 786 pp.

Scott, W. B. 1937. A History of Land Mammals in the Western Hemisphere (revised edition). Macmillan, New York, New York, 750 pp.

Scott, W. B. 1940. Artiodactyla, pp. 363–746 in W. B. Scott and G. L. Jepsen (eds.), The mammalian fauna of the White River Oligocene. Transactions of the American Philosophical Society, (new series) 28(4):1–980.

Scott, W. B. 1945. The mammalia of the Duchesne River Oligocene. Transactions of the American Philosophical Society, Philadelphia (new series) 34:209–253.

Scott, W. B., and G. L. Jepsen. 1940. The mammalian fauna of the White River Oligocene: part IV—Artiodactyla. Transactions of the American Philosophical Society, Philadelphia (new series) 28:363–746.

Scott, W. B., and H. F. Osborn. 1887. Preliminary report on the vertebrate fossils of the Uinta Formation, collected by the Princeton expedition of 1886. Proceedings of the American Philosophical Society 24:255–264.

Seguenza, L. 1907. Nuovi resti di mammiferi pontici di Gravitelli presso Messina. Bollettino della Società Geologica Italiana 26:106–119.

Semprebon, G., C. Janis, and N. Solounias. 2004a. The diets of the Dromomerycidae (Mammalia: Artiodactyla) and their response to Miocene vegetational change. Journal of Vertebrate Paleontology 24:427–444.

Semprebon, G. M., L. R. Godfrey, N. Solounias, M. R. Sutherland, and W. L. Jungers. 2004b. Can low-magnification stereomicroscopy reveal diet? Journal of Human Evolution 47:115–144.

Shedlock, A. M., and N. Okada. 2000. SINE insertions: powerful tools for molecular systematics. BioEssays 22:148–160.

Shedlock, A. M., M. C. Milinkovitch, and N. Okada. 2000. SINE evolution, missing data, and the origin of whales. Systematic Biology 49:808–817.

Shedlock, A. M., K. Takahashi, and N. Okada. 2004. SINEs of speciation: tracking lineages with retroposons. Trends in Ecology and Evolution 19:545–553.

Sherr, L. 1997. Tall Blondes. Andrews McMeel Publishing, Kansas City, Kansas, 167 pp.

Shimamura, M., H. Abe, M. Nikaido, K. Ohshima, and N. Okada. 1999. Genealogy of families of SINEs in cetaceans and artiodactyls: the presence of a huge superfamily of tRNAGlu-derived families of SINES. Molecular Biology and Evolution 16:1046–1060.

Shimamura, M., H. Yasue, K. Ohshima, H. Abe, H. Kato, T. Kishiro, M. Goto, I. Munechika, and N. Okada. 1997. Molecular evidence from retrotransposons that whales form a clade within even-toed ungulates. Nature 388:666–670.

Shoshani, J. 1986. Mammalian phylogeny: comparison of morphological and molecular results. Molecular Biology and Evolution 3:222–242.

Siddall, M. E. 1998. Stratigraphic fit to phylogenies: a proposed solution. Cladistics 14:201–208.

Sigogneau, D. 1968. Le genre *Dremotherium* (Cervoidea): anatomie du crane, denture et moulage endocranien. Annales Paleontologie (Vertebres) 54(1):3–100.

Simmons, A. H. 1988. Extinct pigmy hippopotamus and early man in Cyprus. Nature 333:554–557.

Simmons, A. H., and D. S. Reese. 1993. Hippo hunters of Akrotiri. Archaeology 46(5):40–43.

Simpson, C. D. 1984. Artiodactyls; pp. 563–587 in S. Anderson and J. K. Jones, Jr. (eds.), Orders and Families of Recent Mammals of the World. John Wiley & Sons, New York, New York.

Simpson, G. G. 1937. The Fort Union of the Crazy Mountain field, Montana, and its mammalian faunas: U.S. National Museum Bulletin 169:1–287.

Simpson, G. G. 1944. Tempo and Mode in Evolution. Columbia University Press, New York, New York.

Simpson, G. G. 1945. The principles of classification and a classification of mammals. Bulletin of the American Museum of Natural History 85:1–350.

Simpson, G. G. 1947. Holarctic mammalian faunas and continental relationships during the Cenozoic. Geological Society of America Bulletin 58:613–688.

Sinclair, W. J. 1905. New or imperfectly known rodents and ungulates from the John Day series. Bulletin of the Department of Geology, University of California, Berkeley 4:125–143.

Sinclair, W. J. 1914. A revision of the bunodont Artiodactyla of the middle and lower Eocene of North America. Bulletin of the American Museum of Natural History 33(21):267–295.

Sinclair, W. J. 1915. Additions to the fauna of the lower Pliocene Snake Creek beds. Proceedings of the American Philosophical Society 54:73–95.

Sinclair, W. J. 1921. Entelodonts from the Big Badlands of South Dakota in the geological museum of Princeton University. Proceedings of the American Philosophical Society, Philadelphia 60:467–495.

Sinclair, W. J. 1922. The small entelodonts of the White River Oligocene. Proceedings of the American Philosophical Society, Philadelphia 61:53–64.

Singh, A. 1988. History of aridland vegetation and climate: a global perspective. Biological Reviews 63:156–198.

Sisson, S. 1921. The Anatomy of the Domesticated Animals. W. B. Saunders, Philadelphia, Pennsylvania.

Skinner, M. F. 1942. The fauna of Papago Springs Cave, Arizona, and a study of Stockoceros, with three new antilocaprines from Nebraska and Arizona. Bulletin of the American Museum of Natural History 80:143–220.

Skinner, M. F., and B. E. Taylor. 1967. A revision of the geology and paleontology of the Bijou Hills, South Dakota. American Museum Novitates 2300:1–53.

Skinner, M. F., S. M. Skinner, and R. J. Gooris. 1968. Cenozoic rocks and faunas of Turtle Butte, South-Central South Dakota. Bulletin of the American Museum of Natural History 138:379–436.

Slijper, E. J. 1936. Die Cetaceen Vergleichend-Anatomische und Systematisch. Capita Zoologica 7:1–590.

Slijper, E. J. 1966. Functional morphology of the reproductive system in Cetacea; pp. 277–319 in K. S. Norris (ed.), Whales, Dolphins, and Porpoises. University of California Press, Berkeley, California.

Slijper, E. J. 1979. Whales. Cornell University Press, Ithaca, New York.

Sloan, R.E. 1969. Cretaceous and Paleocene terrestrial mammal communities of western North America. Proceedings of the North American Paleontological Convention 1E: 427–453.

Smith, A. B. 1988. Patterns of diversification and extinction in early Palaeozoic echinoderms. Palaeontology 31:799–828.

Smith, A. B. 1994. Systematics and the Fossil Record: Documenting Evolutionary Patterns. Blackwell Scientific Publications, London, 223 pp.

Smith, M. R., M. S. Shivji, V. G. Waddell, and M. J. Stanhope. 1996a. Phylogenetic evidence from the IRBP gene for the paraphyly of toothed whales, with mixed support for Cetacea as a suborder of Artiodactyla. Molecular Biology and Evolution 13:918–922.

Smith, N. D., and A. H. Turner. 2005. Morphology's role in phylogeny reconstruction: perspectives from paleontology. Systematic Biology 54:166–173.

Smith, R., T. Smith, and J. Sudre. 1996b. *Diacodexis gigasei* n. sp., le plus ancien Artiodactyle (Mammalia) belge, proche de la limite Paléocène-Eocène. Bulletin de l'Institut Royal des Sciences Naturelles de Belgique Sciences de la Terre 66: 177–186.

Smithers, H. N. 1983. The Mammals of the Southern African Subregion. University of Pretoria, Pretoria, South Africa.

Smit-van Dort, M. 1989. Skin, skull and skeleton characters of the mouse deer (Mammalia, Tragulidae), with keys to the species. Bulletin Zoölogisches Museum, Universiteit van Amsterdam 12(5):89–95.

Smuts, M. M. S., and A. J. Bezuidenhout. 1987. Anatomy of the Dromedary. Oxford University Press, New York, New York, 230 pp.

Sokolov, V. E. 1982. Mammal Skin. University of California Press, Berkeley, California.

Sokolov, V. E., M. Z. Kagan, V. S. Vasilieva, V. I. Prihodko, and E. P. Zinkevich. 1987. Musk deer (*Moschus moschiferus*): reinvestigation of main lipid components from preputial gland secretion. Journal of Chemical Ecology 13:71–83.

Solounias, N. 1981. The Turolian fauna from the island of Samos, Greece. Contributions to Vertebrate Evolution 6:1–232.

Solounias, N. 1988a. Evidence from horn morphology on the phylogenetic relationships of the pronghorn (*Antilocapra americana*). Journal of Mammalogy 69:140–143.

Solounias, N. 1988b. The prevalence of ossicones in Giraffidae (Artiodactyla, Mammalia). Journal of Mammalogy 69:845–848.

Solounias, N. 1990. New hypothesis uniting *Boselaphus* and *Tetracerus* with the Miocene *Boselaphini* (Bovidae, Mammalia) based on horn morphology. Annales Musei Goulandris, Kifisia, Greece 8:425–439.

Solounias, N. 1999. The remarkable anatomy of the giraffe's neck. Journal of Zoology, London 247:257–268.

Solounias, N. 2000. Giraffe. McGraw-Hill Yearbook of Science and Technology 2001:174–175.

Solounias, N., and B. Dawson-Saunders. 1988. Dietary adaptations and paleoecology of the late Miocene ruminants from Pikermi and Samos in Greece. Palaeogeography, Palaeoclimatology, Palaeoecology 65:149–172.

Solounias, N., and L.-A. C. Hayek. 1993. New methods of tooth microwear analysis and application to dietary determination of two extinct antelopes. Journal of Zoology, London 229: 421–445.

Solounias, N., and S. M. C. Moelleken. 1991. Evidence for the presence of ossicones in *Giraffokeryx punjabiensis*. Journal of Mammalogy 72:215–217.

Solounias, N., and S. M. C. Moelleken. 1992a. Tooth microwear analysis of *Eotragus sansaniensis* (Mammalia: Ruminantia), one of the oldest known bovids. Journal of Vertebrate Paleontology 12(1):113–121.

Solounias, N., and S. M. C. Moelleken. 1992b. Dietary adaptation of two goat ancestors and evolutionary considerations. Géobios 6(25):797–809.

Solounias, N., and S. M. C. Moelleken. 1993a. Determination of dietary adaptations of extinct ruminants through premaxillary analysis. Journal of Mammalogy 74:1059–1074.

Solounias, N., and S. M. C. Moelleken. 1993b. Tooth microwear and premaxillary shape of an archaic antelope. Lethaia 26: 261–268.

Solounias, N., and S. M. C. Moelleken. 1994. Dietary differences between two archaic ruminant species from Sansan, France. Historical Biology 7:203–220.

Solounias, N., and S. M. C. Moelleken. 1999a. Dietary determination of extinct bovids through cranial foraminal analysis, with radiographic applications. Annales Musei Goulandris, Greece 10:267–290.

Solounias, N., and S. M. C. Moelleken. 1999b. The Miocene gazelle from Greece as a model for detecting Darwinian evolutionary change. Annales Musei Goulandris, Greece 10:291–308.

Solounias, N., and G. M. Semprebon. 2002. Advances in the reconstruction of ungulate ecomorphology and application to early fossil equids. American Museum Novitates 3366:1–47.

Solounias, N., and N. Tang. 1990. The two types of cranial appendages in *Giraffa camelopardalis* (Mammalia, Artiodactyla). Journal of Zoology, London 222:293–302.

Solounias, N., M. Fortelius, and P. Freeman. 1994. Molar wear rates in ruminants: a new approach. Annales Zoologici Fennici 31:219–227.

Solounias, N., S. M. C. Moelleken, and J. M. Plavcan. 1995. Predicting the diet of extinct bovids using masseteric morphology. Journal of Vertebrate Paleontology 15:795–805.

Solounias, N., M. Teaford, and A. Walker. 1988. Interpreting the diet of extinct ruminants: the case of a non-browsing giraffid. Paleobiology 14:287–300.

Solounias, N., W. S. McGraw, L.-A. C. Hayek, and L. Werdelin. 2000. The paleodiet of the Giraffidae; pp. 84–95 in E. E. Vrba and G. B. Schaller (eds.), Antelopes, Deer and Relatives: Fossil Record, Behavioral Ecology, Systematics and Conservation. Yale University Press, New Haven, Connecticut.

Solounias, N., J. M. Plavcan, J. Quade, and L. Witmer. 1999. The paleoecology of the Pikermian Biome and the savanna myth; pp. 436–453 in J. Agusti, L. Rook, and P. Andrews (eds.), The Evolution of Neogene Terrestrial Ecosystems in Europe. Cambridge University Press, Cambridge.

Sondaar, P. Y. 1977. Insularity and its effect on mammal evolution; pp. 671–707 in M. K. Hecht, P. C. Goody, and B. M. Hecht (eds.), Major Patterns in Vertebrate Evolution. Plenum Press, New York, New York.

Spaan, A. 1996. *Hippopotamus creutzburgi*: the case of the Cretan hippopotamus; pp. 99–110 in D. S. Reese (ed.), Pleistocene and Holocene Fauna of Crete and Its First Settlers. Monography in World Archaeology 28. Prehistory Press, Madison, Wisconsin.

Spencer, L. M. 1997. Dietary adaptations of Plio-Pleistocene Bovidae: implications for hominid habitat use. Journal of Human Evolution 32:201–228.

Spinage, C. A. 1968a. The Book of the Giraffe. Collins, London, 191 pp.

Spinage, C. A. 1968b. Horns and other bony structures of the skull of the giraffe, and their functional significance. East African Wildlife Journal 6:53–61.

Spinage, C. A. 1993. The median ossicone of *Giraffa camelopardalis*. Journal of Zoology, London 230:1–5.

Springer, M. S., and E. J. P. Douzery. 1996. Secondary structure and patterns of evolution among mammalian mitochondrial 12S rRNA molecules. Journal of Molecular Evolution 43:357–373.

Springer, M. S., W. J. Murphy, E. Eizirik, and S. J. O'Brien. 2003. Placental mammal diversification and the Cretaceous–Tertiary boundary. Proceedings of the National Academy of Sciences, USA 100:1056–1061.

Springer, M. S., W. J. Murphy, E. Eizirik, and S. J. O'Brien. 2005. Molecular evidence for major placental clades; pp. 37–49 in K. D. Rose and J. D. Archibald (eds.), The Rise of Placental Mammals: Origins and Relationships of Major Clades. The John Hopkins University Press, Baltimore, Maryland.

Springer, M. S., M. J. Stanhope, O. Madsen, and W. W. de Jong. 2004. Molecules consolidate the placental mammal tree. Trends in Ecology and Evolution 19:430–438.

Springer, M. S., A. Burk, J. R. Kavanagh, V. G. Waddell, and M. J. Stanhope. 1997. The interphotoreceptor retinoid binding protein gene in therian mammals: implications for higher level relationships and evidence for loss of function in the marsupial mole. Proceedings of the National Academy of Sciences, USA 94:13754–13759.

Springer, M. S., E. C. Teeling, O. Madsen, M. J. Stanhope, and W. W. de Jong. 2001. Integrated fossil and molecular data reconstruct bat echolocation. Proceedings of the National Academy of Sciences, USA 98:6241–6246.

Steensma, K. J., and S. T. Hussain. 1992. *Merycopotamus dissimilis* (Artiodactyla, Mammalia) from the Upper Siwalik Subgroup and its affinities with Asian and African forms. Proceedings of the Koninklijke Nederlandse Akademie van Wetenschappen 95:97–108.

Stehlin, H. G. 1899. Ueber der Geschichte des Suiden-Gebisses. I. Abhandlungen der Schweizerischen Paläontologischen Gesellschaft 26:1–336.

Stehlin, H. G. 1900. Ueber der Geschichte des Suiden-Gebisses. II. Abhandlungen der Schweizerischen Paläontologischen Gesellschaft 27:337–527.

Stehlin, H. G. 1906. Die Säugetiere des schweizerischen Eocaens. Teil 4. Abhandlungen der Schweizerischen Paläontologischen Gesellschaft 33:597–690.

Stehlin, H. G. 1908. Die Säugetiere des schweizerischen Eocaens. Teil 5. Abhandlungen der Schweizerischen Paläontologischen Gesellschaft 35:691–837.

Stehlin, H. G. 1909. Remarques sur les faunules de mammifères de l'Eocène et de l'Oligocène du Bassin de Paris. Bulletin de la Société Geologique de France 4:488–520.

Stehlin, H. G. 1910a. Die Säugetiere des schweizerischen Eocaens. Teil 6. Abhandlungen der Schweizerischen Paläontologischen Gesellschaft 36:839–1164.

Stehlin, H. G. 1910b. Zur Revision der europäischen Anthracotherien. Verhandlungen der Naturforschenden Gesellschaft in Basel 21:165–185.

Stevens, M. S. 1970. *Merychyus verrucomalus*, a new species of oreodont (Mammalia, Artiodactyla) from the middle Miocene Runningwater Formation. American Museum Novitates 2425:1–11.

Stevens, M. S., and J. B. Stevens. 1996. Merycoidodontinae and Miniochoerinae; pp. 498–573 in D. R. Prothero and R. J. Emry (eds.), The Terrestrial Eocene–Oligocene Transition in North America. Cambridge University Press, Cambridge. 688 pp.

Stevens, M. S., J .B. Stevens, and M. R. Dawson. 1969. New early Miocene formation and vertebrate local fauna, Big Bend National Park, Brewster County, Texas. The Pearce-Sellards Series, Texas Memorial Museum 15:3–52.

Stirton, R. A. 1929. Artiodactyla from the fossil beds of Fish Lake Valley, Nevada. University of California Publications. Bulletin of the Department of Geological Sciences 18(11):291–302.

Stirton, R. A. 1932a. An association of horn cores and upper molars of the antelope *Sphenophalos nevadanus* from the lower Pliocene of Nevada. American Journal of Science 24:46–51.

Stirton, R. A. 1932b. A new genus of Artiodactyla from the Clarendon lower Pliocene of Texas. University of California Publications in Geological Sciences 21:147–168.

Stirton, R. A. 1936. A new ruminant from the Hemphill middle Pliocene of Texas. Journal of Paleontology 10:644–647.

Stirton, R. A. 1938. Notes on some late Tertiary and Pleistocene antilocaprids. Journal of Mammalogy 19:366–370.

Stirton, R. A. 1944. Comments on the relationships of the Palaeomerycidae. American Journal of Science 242: 633–653.

Stirton, R. A. 1967. Relationships of the protoceratid artiodactyls, and a description of a new genus. University of California Publications in Geological Sciences 72:30–43.

Stock, C. 1930a. Oreodonts from the Sespe deposits of South Mountain, Ventura County, California. Publications, Carnegie Institute of Washington 404:27–42.

Stock, C. 1930b. Quaternary antelope remains from a second cave deposit in the Organ Mountains, New Mexico. Los Angeles County Museum Publications 2:1–18.

Stock, C. 1934a. A hypertragulid from the Sespe uppermost Eocene, California. Proceedings of National Academy of Sciences, USA 20:625–629.

Stock, C. 1934b. Microsyopinae and Hyopsodontidae in the Sespe upper Eocene, California. Proceedings of the National Academy of Sciences, USA 20:349–354.

Stock, C. 1949. Mammalian fauna from the Titus Canyon Formation, California. Publications, Carnegie Institution of Washington 584:229–244.

Stoner, C. J., T. M. Caro, and C. M. Graham. 2003. Ecological and behavioral correlates of coloration in artiodactyls: systematic analyses of conventional hypotheses. Behavioral Ecology 14:823–840.

Storer, J. E. 1975. Tertiary mammals of Saskatchewan, Part III: The Miocene fauna. Life Sciences Contributions, Royal Ontario Museum 103:1–134.

Storer, J. E. 1983. A new species of the artiodactyl *Heptacodon* from the Cypress Hills Formation, Lac Pelletier, Saskatchewan. Canadian Journal of Earth Sciences 20:1344–1347.

Storer, J. E. 1984. Mammals of the Swift Current Creek local fauna (Eocene: Uintan), Saskatchewan. Natural History Contributions, Saskatchewan Museum of Natural History 7:1–158.

Storer, J. E. 1996. Eocene–Oligocene Faunas of the Cypress Hills Formation, Saskatchewan; pp. 240–261 in D. R. Prothero and R. J. Emry (eds.), The Terrestrial Eocene–Oligocene Transition in North America. Cambridge University Press, Cambridge and New York, New York.

Strahl, H. 1905. Zur Kenntnis der Placenta von *Tragulus javanicus*. Anatomischer Anzeiger 26:425–428.

Strömberg, C. A. E. 2006. Evolution of hypsodonty in equids: testing a hypothesis of adaptation. Paleobiology 32: 236–258.

Stucky, R. K. 1990. Evolution of land mammal diversity in North America during the Cenozoic; pp. 375–432 in H. H. Genoways (ed.), Current Mammalogy. Plenum Press, London.

Stucky, R. K. 1992. Mammalian faunas in North America of Bridgerian to early Arikareean "ages" (Eocene and Oligocene); pp. 464–493 in D. R. Prothero and W. A. Berggren (eds.), Eocene–Oligocene climatic and biotic evolution. Princeton University Press, Princeton, New Jersey.

Stucky, R. K. 1998. Eocene bunodont and bunoselenodont Artiodactyla ("dichobunids"); pp. 358–374 in C. M. Janis, K. M. Scott, and L. L. Jacobs (eds.), Evolution of Tertiary Mammals of North America, Volume I: Terrestrial Carnivores, Ungulates, and Ungulatelike Mammals. Cambridge University Press, Cambridge.

Stucky, R. K., and L. Krishtalka. 1990. Revision of the Wind River faunas, early Eocene of central Wyoming. Part. 10. *Bunophorus* (Mammalia, Artiodactyla). Annals of the Carnegie Museum 59:149–171.

Stucky, R. K., and M. C. McKenna. 1993. Mammalia; pp. 739–777 in M. J. Benton (ed.), The Fossil Record 2. Chapman and Hall, London.

Stucky, R. K., D. R. Prothero, W. G. Lohr, and J. R. Snyder. 1996. Magnetic stratigraphy, sedimentology, and mammalian faunas of the early Uintan Washakie Formation, Sand Wash Basin, Northwestern Colorado; pp. 40–51 in D. R. Prothero and R. J. Emry (eds.), The Terrestrial Eocene–Oligocene Transition in North America. Cambridge University Press, Cambridge and New York, New York.

Stuenes, S. 1989. Taxonomy, habits, and relationships of the subfossil Madagascan Hippopotami *Hippopotamus lemerlei* and *H. madagascariensis*. Journal of Vertebrate Paleontology 9:241–268.

Su, B., Y.-X. Wang, and Q.-S. Wang. 2001. Mitochondrial DNA sequences imply Anhui musk deer a valid species in genus *Moschus*. Zoological Research 22:169–173.

Su, B., Y.-X. Wang, H. Lan, W. Wang, and Y. Zhang. 1999. Phylogenetic study of the complete cytochrome *b* genes in musk deer (genus *Moschus*) using museum samples. Molecular Phylogenetics and Evolution 12:241–249.

Sudre, J. 1969. Les gisements de Robiac (Eocène supérieur) et leurs faunes de Mammiféres. Palaeovertebrata 2:95–156.

Sudre, J. 1972. Révision des Artiodactyles de l'Eocène moyen de Lissieu (Rhône). Palaeovertebrata 5:111–156.

Sudre, J. 1973. Un *Dichodon* géant de La Débruge et une nouvelle interprétation phylétique du genre. Bulletin du Muséum national d'Histoire naturelle, section C 25:73–78.

Sudre, J. 1974. D'important restes de *Diplobune minor* (Filhol) à Itardies (Quercy). Palaeovertebrata 6(1–2):47–54.

Sudre, J. 1977. L'évolution du genre Robiacina SUDRE 1969 et l'origine de Cainotheriidae; implications systématiques. Géobios, Mémoire Spécial 1:213–231.

Sudre, J. 1978. Les Artiodactyles de l'Eocène moyen et supérieur d'Europe Occidentale (Systématique et évolution). Mémoires et Travaux de l'Institut de Montpellier de l'Ecole Pratique des Hautes Etudes 7:1–229.

Sudre, J. 1980. *Aumelasia gabineaudi* n.g. n.sp. nouveau dichobunidae (Artiodactyla, Mammalia) du gisement d'Aumelas (Hérault) d'age Lutétien terminal. Palaeovertebrata (Mémoire Jubil. R. Lavocat):197–211.

Sudre, J. 1983. Interprétation de la denture et description des éléments du squelette appendiculaire de l'espèce Diplobune minor (Filhol 1877): apports à la connaissance de l'anatomie des Anoplotheriinae Bonaparte 1850; pp. 439–458 in E. Buffetaut, J. M. Mazin, and E. Salmon (eds.), Actes du Symposium Paléontologique. G. Cuvier, Montbeliard.

Sudre, J. 1984. *Cryptomeryx* Schlosser, 1886, Tragulidé de l'Oligocène d'Europe—relations du genre et considérations sur l'origine des ruminants. Palaeovertebrata 14:1–31.

Sudre, J. 1986. Le genre *Bachitherium* Filhol 1882 (Mammalia, Artiodactyla): diversité spécifique, phylogénie, extension chronologique. Comptes Rendus de l'Académie des Sciences de Paris, série IIa 303:749–754.

Sudre, J. 1988a. Apport à la connaissance du *Dichobune robertiana* Gervais, 1848–1852 (Mammalia, Artiodactyla) du Lutétien: considérations sur l'évolution des Dichobunidés. Courier Forschungs-Institut Senckenberg 107:409–418.

Sudre, J. 1988b. Le gisement du Bretou (Phosphorites du Quercy, Tarn-et-Garonne, France) et sa faune de vertébrés de l'Eocène supérieur. VI. Artiodactyles. Palaeontographica Abteilung A 205:129–154.

Sudre, J. 1995. Le Garouillas et les sites contemporains (Oligocène, MP 25) des phosphorites du Quercy (Lot, Tarn-et-Garonne, France) et leur faunes de vertébrés. 12. Artiodactyla. Palaeontographica Abteilung A 236:205–256.

Sudre, J. 1997. Les remplissages karstiques polyphasés (Éocène, Oligocène, Pliocène) de Saint-Maximin (Phosphorites du Gard) et leur apport à la connaisance des faunes européennes, notamment pour l'Éocène moyen (MP 13). Mémoires et Travaux de l'Institut de Montpellier de l'Ecole Pratique des Hautes Etudes 21:759–765.

Sudre, J., and C. Blondel. 1995. Le tarse des Amphimerycidae d'Europe—Paraphylie des Ruminantia. Journées spécialisées de la Société Francaise d'Ecologie et de l'Association Francaise de Mammalogie 2.

Sudre, J., and C. Blondel. 1996. Sur la présence de petits gélocidés (Artiodactyla) dans l'Oligocène inférieur du Quercy (France); considérations sur les genres *Pseudogelocus* Schlosser 1902,

Paragelocus Schlosser 1902 et *Iberomeryx* Gabunia 1964. Neues Jahrbuch für Geologie und Paläontologie. Monatshefte 3:169–182.

Sudre, J., and J. Erfurt. 1996. Les artiodactyles du gisement Ypresien terminal de Prémontré (Aisne, France). Palaeovertebrata 25:391–414.

Sudre, J., and L. Ginsburg. 1993. La faune de mammifères de La Défense (Calcaire grossier; Lutétien supérieur) à Puteaux près Paris; artiodactyles et *Lophiodon parisiense* Gervais, 1848–1852. Bulletin du Muséum National d'Histoire Naturelle, section C 15(4):155–181.

Sudre, J., and G. Lecomte. 2000. Relations et position systématique du genre *Cuisitherium* Sudre et al., 1983, le plus dérivé des artiodactyles de l'Éocène inférieur d'Europe. Geodiversitas 22:415–432.

Sudre, J., and B. Marandat. 1993. First discovery of an Homacodontinae (Artiodactyla, Dichobunidae) in the Middle Eocene of Western Europe: *Eygalayodon montenati* new genus, new species. Considerations on the evolution of primitive artiodactyls. Kaupia Darmstädter Beiträge zur Naturgeschichte 3:157–164.

Sudre, J., D. E. Russell, P. Louis, and D. E. Savage. 1983a. Les Artiodactyles de l'Éocène inférieur d'Europe. Première partie. Bulletin du Muséum National d'Histoire Naturelle, section C 5:281–333, 339–365.

Sudre, J., B. Sigé, J. A. Remy, B. Marandat, J. L. Hartenberger, M. Godinot, and J. Y. Crochet. 1990. Une faune du niveau d'Egerkingen (MP14; Bartonien inferieur) dans les Phosphorites du Quercy (Sud de la France). Palaeovertebrata 20:1–32.

Sudre, J., L. De Bonis, M. Brunet, J.-Y. Crochet, F. Duranthon, M. Godinot, J.-L. Hartenberger, Y. Jehenne, S. Legendre, B. Marandat, J. A. Remy, M. Ringeade, B. Sigé, and M. Vianey-Liaud. 1992. La biochronologie mammalienne du Paléogène au Nord et au Sud des Pyrénées: état de la question. Comptes Rendus de l'Academie des Sciences, Série II 314:631–636.

Sumrall, C. D., and C. A. Brochu. 2003. Resolution, sampling, higher-taxa, and assumptions in stratocladistic analysis. Journal of Paleontology 77:189–194.

Suteethorn, V., E. Buffetaut, R. Helmcke-Ingavat, J.-J. Jaeger, and Y. Jongkanjanasoontorn. 1988. Oldest known Tertiary mammal from South-East Asia: Middle Eocene primate and anthracotheres. Neues Jahrbuch für Geologie und Paläontologie Monatshefte 9:563–570.

Swofford, D. L. 1984. Phylogenetic analysis using parsimony (PAUP). Illinois Natural History Survey, Champaign, Illinois.

Swofford, D. L. 2002. PAUP*, Phylogenetic analysis using parsimony (*and other methods). Sinauer Associates, Sunderland, Massachusetts. Vers. 4.0β10. Available at http://paup.csit.fsu.edu/.

Szalay, F. S. 1969. The Hapalodectinae and a phylogeny of the Mesonychidae (Mammalia, Condylarthra). American Museum Novitates 2361:1–26.

Tabrum, A. R., D. R. Prothero, and D. Garcia. 1996. Magnetostratigraphy and biostratigraphy of the Eocene–Oligocene transition, southwestern Montana; pp. 278–311 in D. R. Prothero and R. J. Emry (eds.), The Terrestrial Eocene–Oligocene Transition in North America. Cambridge University Press, Cambridge and New York, New York.

Tang, Y. 1978. Two new genera of Anthracotheriidae from Kwangsi. Vertebrata PalAsiatica 16:13–21.

Tarling, D. H. 1982. Land bridges and plate tectonics. Géobios, Mémoire Spécial 6:361–374.

Tassy, P. 1996. Dental homologies and nomenclature in the Proboscidea; pp. 21–25 in J. Shoshani and P. Tassy (eds.), The Proboscidea: Evolution and Palaeoecology of Elephants and Their Relatives. Oxford University Press, Oxford.

Taylor, B. E., and S. D. Webb. 1976. Miocene Leptomerycidae (Artiodactyla, Ruminantia) and their relationships. American Museum Novitates 2596:1–22.

Tchernov, E. 1987. The age of the 'Ubeidiya Formation, an early Pleistocene hominid site in the Jordan Valley, Israel. Israel Journal of Earth Sciences 36:3–30.

Tedford, R. H., M. F. Skinner, R. S. Fields, J. M. Rensberger, D. P. Whistler, R. Galusha, B. E. Taylor, J. R. McDonald, and S. D. Webb. 1987. Faunal succession and biochronology of the Arikareean through Hemphillian (late Oligocene through earliest Pliocene epochs) in North America; pp. 153–210 in M. O. Woodburne (ed.), Cenozoic Mammals of North America. University of California Press, Berkeley, California.

Tedford, R. H., L. B. Albright III, A. D. Barnosky, I. Ferrusquia-Villafranca, R. M. Hunt, Jr., J. E. Storer, C. C. Swisher III, M. R. Voorhies, S. D. Webb, and D. P. Whistler. 2004. Mammalian biochronology of the Arikareean through Hemphillian interval (late Oligocene through early Pliocene Epochs); pp. 169–231 in M. O. Woodburne (ed.), Late Cretaceous and Cenozoic Mammals of North America. Columbia University Press, New York, New York.

Teeling, E. C., M. S. Springer, O. Madsen, P. Bates, S. J. O'Brien, and W. J. Murphy. 2005. A molecular phylogeny for bats illuminates biogeography and the fossil record. Science 307:580–584.

Teller, F. 1884. Neue Anthracotherienreste aus der Südsteiermark und Dalmatien. Beiträge zur Paläontologie Oesterreich-Ungarns 4:45–134.

Telles-Antunes, M. 1986. *Anoplotherium* (mammalia, artiodactyla) et *Geochelone* (reptilia, testudines) à côja: les vertebras fossils et l'Éocène supérieur au Portugal. Ciências da Terra 8:99–110.

Thenius, E. 1952. Die Säugetierfauna aus dem Torton von Neudorf an der March (SR). Neues Jahrbuch für Geologie und Paläontologie. Abhandlungen 96:27–136.

Thenius, E. 1979. Das Genus *Xenochoerus* Zdarsky 1909, ein aberranter Tayassuide (Artiodactyla, Mammalia) aus dem Miozan Europas. Anzeiger der Oesterreichischen Akademie der Wissenschaften Mathematisch-Naturwissenschaftliche Klasse 116:1–8.

Thenius, E. 2000. Lebende Fossilien. Oldtimer der Tier- und Pflanzenwelt. Zeugen der Vorzeit. Verlag Dr. Friedrich Pfeil, Munich, 228 pp.

Theodor, J. M. 1996. Phylogeny, locomotor evolution and diversity patterns in Eocene Artiodactyla. Ph.D. dissertation, University of California at Berkeley, Berkeley, California, 177 pp.

Theodor, J. M. 1999. *Protoreodon walshi,* a new species of agriochoerid (Oreodonta, Artiodactyla, Mammalia) from the late Uintan of San Diego County, California. Journal of Paleontology 73:1179–1190.

Theodor, J. M. 2001. Clock estimates and the artiodactyl–whale relationship. PaleoBios 21:125.

Theodor, J. M., and S. E. Foss. 2005. Deciduous dentitions of Eocene cebochoerid artiodactyls and cetartiodactyl relationships. Journal of Mammalian Evolution 12:161–181.

Theodor, J. M., and M. J. Mahoney. 1998. Why do molecules and morphology conflict? Examination of the artiodactyls–cetacean relationship. Journal of Vertebrate Paleontology Supplement 17(3):80A.

Theodor, J. M., K. D. Rose, and J. Erfurt. 2005. Artiodactyla; pp. 215–233 in K. D. Rose and J. D. Archibald (eds.), The Rise of Placental Mammals: Origins and Relationships of Major Clades. The Johns Hopkins University Press, Baltimore, Maryland.

Thewissen, J. G. M. 1990. Evolution of Paleocene and Eocene Phenacodontidae (Mammalia, Condylarthra). University of Michigan Papers in Paleontology 29:1–107.

Thewissen, J. G. M. 1994. Phylogenetic aspects of cetacean origins: A morphological perspective. Journal of Mammalian Evolution 2:157–183.

Thewissen, J. G. M. 1998. The Emergence of Whales: Evolutionary Patterns in the Origin of Cetacea. Plenum Press, New York, New York, 491 pp.

Thewissen, J. G. M., and Domning, D. P. 1992. The role of phenacodontids in the origin of the modern orders of ungulate mammals. Journal of Vertebrate Paleontology 12: 494–504.

Thewissen, J. G. M., and S. T. Hussain 1990. Postcranial osteology of the most primitive artiodactyl *Diacodexis pakistanensis* (Dichobunidae). Anatomy, Histology, Embryology 19:37–48.

Thewissen, J. G. M., and S. T. Hussain. 1993. Origin of underwater hearing in whales. Nature 361:444–445.

Thewissen, J. G. M., and S. T. Hussain. 1998. Systematic review of Pakicetidae, early and middle Eocene Cetacea (Mammalia) from Pakistan and India. Bulletin of the Carnegie Museum of Natural History 34:220–238.

Thewissen, J. G. M., and S. I. Madar. 1999. Ankle morphology of the earliest cetaceans and its implications for the phylogenetic relations among ungulates. Systematic Biology 48:21–30.

Thewissen, J. G. M., P. D. Gingerich, and D. E. Russell. 1987. Artiodactyla and Perissodactyla (Mammalia) from the early-middle Eocene Kuldana Formation of Kohat (Pakistan). Contributions from the Museum of Paleontology University of Michigan 27:247–274.

Thewissen, J. G. M., S. I. Madar, and S. T. Hussain. 1996. *Ambulocetus natans,* an Eocene cetacean (Mammalia) from Pakistan. Courier Forschunginstitut Senckenberg 191:1–86.

Thewissen, J. G. M., E. M. Williams, and S. T. Hussain. 2001a. Eocene Mammal faunas from Northern Indo-Pakistan. Journal of Vertebrate Paleontology 21:347–366.

Thewissen, J. G. M., D. E. Russell, P. D. Gingerich, and S. T. Hussain. 1983. A new dichobunid artiodactyl (Mammalia) from the Eocene of North-West Pakistan. Part I: Dentition and classification. Proceedings of the Koninklike Nederlandse Akademie Van Wetenschapen, Serie B 86:153–180.

Thewissen, J. G. M., E. M. Williams, L. J. Roe, and S. T. Hussain. 2001b. Skeletons of terrestrial cetaceans and the relationship of whales to artiodactyls. Nature 413:277–281.

Thomas, H. 1981. Les Bovidés miocènes de la formation de Ngorora du Bassin de Baringo (Kenya). Proceedings Koninklijke Nederlandse Akademie van Wetenschappen B 84: 335–409.

Thomas, H. 1984a. Les Bovidae (Artiodactyla : Mammalia) du Miocène du Sous-Continent Indien, de la Peninsule Arabique et de L'Afrique: Biostratigraphie, Biogéographie et Écologie. Palaeogeography, Palaeoclimatology, Palaeoecology 84: 251–299.

Thomas, H. 1984b. Les bovidés anté-hipparions des Siwaliks inféreurs (Plateau du Potwar, Pakistan). Mémoires de la Société Géologique de France, Paris NS 145:1–68.

Thomas, H., L. Ginsburg, C. Hintong, and V. Suteethorn. 1990. A new tragulid, *Siamotragulus sanyathanai* n. gen., n. sp. (Artiodactyla, Mammalia) from the Miocene of Thailand (Amphoe Pong, Phayao Province). Comptes Rendues de l'Académie des Sciences, série II 310:989–995.

Thomasson, J. R., R. J. Zakrzewski, H. E. Lagarry, and D. E. Mergen. 1990. A late Miocene (late early Hemphillian) biota from northwestern Kansas. National Geographic Research 6:231–244.

Thorpe, M. R. 1921. John Day eporeodonts, with descriptions of new genera and species. American Journal of Science, Series 5 6:93–111.

Thorpe, M. R. 1937. The Merycoidodontidae: An extinct group of ruminant mammals. Memoirs of the Peabody Museum of Natural History 3(4):1–428.

Ting, S., and C. Li. 1987. The skull of *Hapalodectes* (?Acreodi, Mammalia), with notes on some Chinese Paleocene mesonychids. Vertebrata PalAsiatica 25:161–186.

Tobien, H. 1980. Ein anthracotherioider Paarhufer (Artiodactyla, Mammalia) aus dem Eozän von Messel bei Darmstadt (Hessen). Geologisches Jahrbuch Hessen 108:11–22.

Tobien, H. 1985. Zur Osteologie von Masillabune (Mammalia, Artiodactyla, Haplobunodontidae) aus dem Mitteleozän der Fossilfundstätte Messel bei Darmstadt (S-Hessen, Bundesrepublik Deutschland). Geologisches Jahrbuch Hessen 113:5–58.

Todd, N. B. 1975. Chromosomal mechanisms in the evolution of artiodactyls. Paleobiology 1:175–188.

Tong, Y., and J. Wang 1998. A preliminary report of the early Eocene mammals of the Wutu fauna, Shandong Province, China. Bulletin of Carnegie Museum of Natural History 34:186–193.

Tong, Y., and J. Wang. 2006. Fossil mammals from the Early Eocene Wutu Formation of Shandong Province. Palaeontologia Sinica 192:1–223.

Tong, Y. S., and Z.-R. Zhao. 1986. *Odiochoerus,* a new suoid (Artiodactyla, Mammalia) from the early Tertiary of Guangxi. Vertebrata PalAsiatica 24:129–138.

Tong Y., S. Zheng, and Z. Qiu. 1995. Cenozoic mammal ages of China. Vertebrate PalAsiatica 33:307–314.

Torres, V. R. P., A. A. Enciso, and E. G. Porras. 1986. The osteology of South American camelids. Archaeological Research Tools 3:1–32.

Townsend, K. E. 2004. Stratigraphy, paleoecology, and habitat change in the middle Eocene of North America. Ph.D. Thesis. Department of Anthropology, Washington University, St. Louis, Missouri.

Trofimov, B. A. 1952. Novyye entelodontidy iz Mongoli I Kazakhstana. Akademiya Nauk SSSR Trudy Paleontologicheskiy Institut, Moscow 41:144–154.

Trofimov, B. A. 1957. Nouvelles données sur les Ruminantia les plus anciens d'Asie. Cursillos y Conferencies del Instituto "Lucas Mallada" 4:137–141.

Trouessart, E. L. 1898. Catalogus Mammalium tam viventium quam fossilium. Nova editio, Berolini 8:665–1264.

Troxell, E. L. 1920. Entelodonts in the Marsh collection. American Journal of Science (series 4) 50:243–255, 361–386, 431–445.

Troxell, E. L. 1921. The American bothriodonts. American Journal of Science 3:325–339.

Tsubamoto, T., T. Masanaru, and N. Egi. 2004. Quantitative analysis of biogeography and faunal evolution of middle to late Eocene mammals in East Asia. Journal of Vertebrate Paleontology 24:657–667.

Tsubamoto, T., N. Egi, M. Takai, C. Sein, and M. Maung. 2005. Middle Eocene ungulate mammals from Myanmar: A review with description of new specimens. Acta Palaeontologica Polonica 50:117–138.

Tsubamoto, T., M. Takai, N. Egi, N. Shigehara, S. T. Tun, A. K. Aung, A. N. Soe, and T. Thein. 2002. The Anthracotheriidae (Mammalia; Artiodactyla) from the Eocene Pondaung Formation (Myanmar) and comments on some other anthracotheres from the Eocene of Asia. Paleontological Research 6:363–384.

Tsubamoto, T., S. T. Tun, N. Egi, M. Takai, N. Shigehara, A. N. Soe, A. K. Aung, and T. Thein. 2003. Reevaluation of some ungulate mammals from the Eocene Pondaung Formation, Myanmar. Paleontological Research 7:219–243.

Turner, H. N. 1849. On the evidences of affinity afforded by the skull in the ungulate Mammalia. Proceedings of the Zoological Society of London 17:147–158.

Turner, W. 1876. On the structure of the diffused, the polycotyledonary and zonary forms of placenta. Journal of Anatomy and Physiology 10:127–177.

Tyson, E. 1683. Anatomy of the Mexican musk hog. Philosophical Transactions 13:359–385.

Uhen, M. D. 1998. Middle to late Eocene basilosaurines and dorudontines; pp. 29–61 in J. G. M. Thewissen (ed.), The Emergence of Whales. Plenum Press, New York, New York.

Ursing, B. M., and U. Arnason. 1998. Analyses of mitochondrial genomes strongly support a hippopotamus–whale clade. Proceedings of the Royal Society of London B 265:2251–2255.

Ursing, B. M., K. E. Slack, and U. Arnason. 2000. Subordinal artiodactyl relationships in the light of phylogenetic analysis of 12 mitochondrial protein-coding genes. Zoologica Scripta 29:83–88.

Van Couvering, J. A. H. 1980. Community evolution in East Africa during the late Cenozoic; pp. 272–282 in A. K. Behrensmeyer and A. H. Hill (eds.), Fossils in the Making. University of Chicago Press, Chicago, Illinois.

Van der Made, J. 1989–1990a. Iberian Suoidea. Paleontologia i Evolució 23:83–97.

Van der Made, J. 1989–1990b. A range chart for European Suidae and Tayassuidae. Paleontologia i Evolució 23:99–104.

Van der Made, J. 1992. African Lower and Middle Miocene Suoidea (pigs and peccaries). VIII Jornadas de paleontologia, Barcelona, 8–10 de Octubre de 1992, Resúmenes, 87–97.

Van der Made, J. 1994. Suoidea from the Lower Miocene of Cetina de Aragón (Spain). Revista Española de Paleontologia 9(1):1–23.

Van der Made, J. 1996a. Albanohyus, a small Miocene pig. Acta Zoologica Cracoviensia 39:293–303.

Van der Made, J. 1996b. Listriodontinae (Suidae, Mammalia), their evolution, systematics and distribution in time and space. Contributions to Tertiary and Quaternary Geology 33(1–4):3–254.

Van der Made, J. 1997a. Systematics and stratigraphy of the genera Taucanamo and Schizochoerus and a classification of the Palaeochoeridae. Proceedings of the Koninklijke Nederlandse Akademie Wetenschappen 100:127–139.

Van der Made, J. 1997b. On Bunolistriodon (= Eurolistriodon) and kubanochoeres. Proceedings of the Koninklijke Nederlandse Akademie Wetenschappen 100:141–160.

Van der Made, J. 1997c. Pre-Pleistocene land mammals from Crete; pp. 69–79 in D. S. Reese (ed.), Pleistocene and Holocene Fauna of Crete and Its First Settlers. Monographs in World Archaeology 28. University of Wisconsin Press, Madison, Wisconsin.

Van der Made, J. 1999a. Intercontinental relationship Europe-Africa and the Indian subcontinent; pp. 457–472 in G. Rössner and K. Heissig (eds.), The Miocene Land Mammals of Europe. Dr. Friedrich Pfeil, Munich.

Van der Made, J. 1999b. Biometrical trends in the Tetraconodontinae, a subfamily of pigs. Transactions of the Royal Society of Edinburgh: Earth Sciences 89:199–225.

Van der Made, J. 1999c. Superfamily Hippopotamoidea; pp. 203–208 in G. E. Rössner and K. Heissig (eds.), The Miocene Land Mammals of Europe. Dr. Friedrich Pfeil, Munich.

Van der Made, J., and D. Han. 1994. Suoidea from the Upper Miocene hominoid locality of Lufeng, Yunnan province, China. Proceedings of the Koninklijke Nederlandse Akademie van Wetenschappen 97:27–82.

Van der Made, J., and J. Morales. 1999. Aureliachoerus (Suidae, Mammalia) from Agreda and other Miocene localities of Spain. Géobios 32:901–914.

Van der Made, J., and S. Moyà-Solà. 1989. European Suinae (Artiodactyla) from the Late Miocene onwards. Bolletino della Società Paleontologica Italiana 28(2–3):329–339.

Van Soest, P. J. 1982. Nutritional Ecology of the Ruminant. Cornell University Press, Ithaca, New York.

Van Soest, P. 1994. Nutritional Ecology of the Ruminant (2nd ed.). Cornell University Press, Ithaca, New York, 528 pp.

Van Valen, L. 1965. Paroxyclaenidae, an extinct family of Eurasian mammals. Journal of Mammalogy 46:388–397.

Van Valen, L. 1966. Deltatheridia, a new order of mammals. Bulletin of the American Museum of Natural History 132: 1–126.

Van Valen, L. 1971. Toward the origin of artiodactyls. Evolution 25:523–529.

Van Vuuren, B. J., and T. J. Robinson. 2001. Retrieval of four adaptive lineages in duiker antelope: evidence from mitochondrial DNA sequences and flourescence in situ hybridization. Molecular Phylogenetics and Evolution 20:409–425.

Vekua, A. 1959. O gippopotame iz ni neplejtocevych otlo enij Gruzij. Soobshcheniia Akademii nauk Gruzii 23:561–566.

Vekua, A. 1986. The lower Pleistocene mammalian fauna of Akhalkalaki (Southern Georgia, USSR). Palaeontographia Italica 74:63–96.

Verheyen, R. 1954. Monographie éthologique de l'Hippopotame (Hippopotamus amphibius Linné). Institut des Parcs Nationaux du Congo Belge, Morges.

Vidyadaran, M. K., S. Vellayan, and R. Kuruppiah. 1983. Muscle weight distribution of the Malaysian less mousedeer. Peranika 6:63–69.

Vignaud, P., P. Duringer, H. T. Mackaye, A. Likius, C. Blondel, J. R. Boisserie, L. de Bonis, V. Eisenmann, M. E. Etienne, D. Geraads, F. Guy, T. Lehmann, F. Lihoreau, N. Lopez-Martinez, C. Mourer-Chauviré, O. Otero, J. C. Rage, M. Schuster, L. Viriot, A. Zazzo, and M. Brunet. 2002. Geology and palaeontology of the upper Miocene Toros-Ménalla hominid locality, Chad. Nature 418:152–155.

Villalta, J., M. Crusafont, and R. Lavocat.1946. Primer ballazgo en Europa de Ruminantes fosiles tricornes. Publicaciones Museo de Sabadell, Paleontologia, Sabadell, España.

Viret, J. 1961. Artiodactyla; pp. 887–1021, 1038–1084 in J. Piveteau (ed.), Traité de Paléontologie. Masson, Paris.

Vislobokova, I. A. 1997. Eocene–early Miocene ruminants in Asia; in J.-P. Aguilar, S. Legendre, and J. Michaux (eds.), Actes du Congrès BiochroM'97. Mémoires et Travaux de l'E.P.H.E., Institut de Montpellier 21:213–215.

Vislobokova, I. A. 1998. A new representative of the Hypertraguloidea (Tragulina, Ruminantia) from the Khoer-Dzan locality in Mongolia, with remarks on the relationships of the Hypertragulidae. American Museum Novitates 3225:1–24.

Vislobokova, I. A. 2001. Evolution and classification of Tragulina (Ruminantia, Artiodactyla). Paleontological Journal 35(Suppl. Issue 2):69–145.

Vislobokova, I. A. 2004a. A new representative of the family Raoellidae (Suiformes) from the middle Eocene of Khaichin-Ula 2. Paleontological Journal 38:102–107.

Vislobokova, I. A. 2004b. Artiodactyls from the middle Eocene of Khaichin-Ula II, Mongolia. Paleontological Journal 38: 90–96.

Vislobokova, I. A., and G. Daxner-Höck. 2002. Oligocene–early Miocene ruminants from the Valley of Lakes (Central Mongolia). Annalen des Naturhistorischen Museums in Wien 103A:213–235.

Vislobokova, I. A., and E. L. Dmitrieva. 2000. Changes in enamel ultrastructure at the early stages of ruminant evolution. Paleontological Journal 34:242–249.

Vislobokova, I. A., and B. A. Trofimov. 2000. The family Archaeomerycidae (Tragulina): classification and role in the evolution of the Ruminantia. Paleontological Journal 4:92–99.

Vislobokova, I. A., and B. A. Trofimov. 2002. *Archaeomeryx* (Archaeomerycidae, Ruminantia): Morphology, Ecology, and Role in the Evolution of the Artiodactyla. Paleontological Journal 35(Supplement Issue 5):429–523.

Von Koenigswald, G. H. R. 1967. An upper Eocene mammal of the Family Anthracotheriidae from the Island of Timor, Indonesia. Proceedings of the Koninklijke Nederlandse Akademie van Wetenschappen 70:529–533.

Voorhies, M. R. 1990. Vertebrate biostratigraphy of the Ogallala Group in Nebraska; pp. 115–151 in T. C. Gustavson (ed.), Geologic Framework and Regional Hydrology: Upper Cenozoic Blackwater Draw and Ogallala Formations, Great Plains. University of Texas at Austin, Bureau of Economic Geology, Austin, Texas.

Vrba, E. S. 1995. The fossil record of African antelopes (Mammalia, Bovidae) in relation to human evolution and paleoclimate; pp. 385–424 in E. S. Vrba, G. H. Denton, T. C. Partrige, and L. D. Burckle (eds.), Palaeoevolution and Climate with Emphasis on Human Origins. Yale University Press, New Haven, Connecticut.

Vrba, E. S. 1997. New fossils of Alcelaphini and Caprinae (Bovidae: Mammalia) from Awash, Ethiopia, and phylogenetic analysis of Alcelaphini. Paleontologia Africana Johannesburg 34:127–198.

Vrba E. S., and G. B. Schaller (eds.). 2000a. Antelopes, Deer and Relatives: Fossil Record, Behavioral Ecology, Systematics, and Conservation. Yale University Press, New Haven, Connecticut, 356 pp.

Vrba, E. S., and G. B. Schaller. 2000b. Phylogeny of Bovidae based on behavior, glands, skulls, and postcrania; pp. 203–222 in E. S. Vrba and G. B. Schaller (eds.), Antelopes, Deer and Relatives: Fossil Record, Behavioral Ecology, Systematics, and Conservation. Yale University Press, New Haven, Connecticut, 356 pp.

Waddell, P. J., and S. Shelly. 2003. Evaluating placental interordinal phylogenies with novel sequences including RAG1, g-fibrinogen, ND6, and mt-tRNA, plus MCMC-driven nucleotide, amino acid, and codon models. Molecular Phylogenetics and Evolution 28:197–224.

Waddell, P. J., N. Okada, and M. Hasegawa. 1999. Towards resolving the interordinal relationships of placental mammals. Systematic Biology 48:1–5.

Wagner, P. J. 1998. A likelihood approach for evaluating estimates of phylogenetic relationships among fossil taxa. Paleobiology 24:430–449.

Wagner, P. J. 2000. Phylogenetic analyses and the fossil record: tests and inferences, hypotheses and models; pp. 341–371 in D. H. Erwin and S. L. Wing (eds.), Deep Time: Paleobiology's Perspective. The Paleontological Society, Lawrence, Kansas.

Walker, A. C. 1969. Lower Miocene fossils from Mount Elgon, Uganda. Nature 223:591–596.

Wall, W. P., and M. J. Shikany. 1993. Comparison of feeding mechanisms in Oligocene Agriochoeridae and Merycoidodontidae, from Badlands National Park; pp. 41–54 in V. L. Santucci (ed.), National Park Service, Paleontological Research, National Park Service, Technical Report, 2, NPS/NRPO/NRTR-93/11.

Wallis, O. C., and M. Wallis. 2001. Molecular evolution of growth hormone (GH) in Cetartiodactyla: cloning and characterization of the gene encoding GH from a primitive ruminant, the chevrotain (*Tragulus javanicus*). General and Comparative Endocrinology 123:62–72.

Walsh, S. L. 1996. Middle Eocene mammal faunas of San Diego County, California; pp. 75–119 in D. R. Prothero and R. J. Emry (eds.), The Terrestrial Eocene–Oligocene Transition in North America. Cambridge University Press, New York, New York.

Walsh, S. L. 2000. Bunodont artiodactyls (Mammalia) from the Uintan (Middle Eocene) of San Diego County, California. Proceedings of the San Diego Society of Natural History 37:1–27.

Wang, B. 1992. The Chinese Oligocene: a preliminary review of mammalian localities and local faunas; pp. 529–547 in D. R. Prothero and W. A. Berggren (eds.), Eocene–Oligocene Climatic and Biotic Evolution. Princeton University Press, Princeton, New Jersey.

Wang, B., and Y. Zhang. 1983. New finds of fossils from Paleogene of Qujing, Yunnan. Vertebrata PalAsiatica 21:119–128.

Wang, J. 1985. A new species of *Anthracokeryx* from Shanxi. Vertebrata PalAsiatica 23:58–59.

Wang, X. 1994. Phylogenetic systematics of the Hesperocyoninae (Carnivora: Canidae). Bulletin of the American Museum of Natural History 221:1–207.

Wang, X., R. H. Tedford, and B. E. Taylor. 1999. Phylogenetic systematics of the Borophaginae (Carnivora: Canidae). Bulletin of the American Museum of Natural History 243:1–391.

Ward, T. J., R. L. Honeycutt, and J. N. Derr. 1997. Nucleotide sequence evolution at the k-casein locus: evidence for positive selection within the family Bovidae. Genetics 147:1863–1872.

Webb, S. D. 1969. The Burge and Minnichaduza Clarendonian mammalian faunas of North-Central Nebraska. University of California Publications, Bulletin of the Department of Geological Sciences 78:1–191.

Webb, S. D. 1973. Pliocene pronghorns in Florida. Journal of Mammalogy 54:203–221.

Webb, S. D. 1977. A history of the savannah vertebrates of the New World. Part I, North America. Annual Review of Ecology and Systematics 8:355–380.

Webb, S. D. 1981. *Kyptoceras amatorum,* new genus and species from the Pliocene of Florida, the last protoceratid artiodactyls. Journal of Vertebrate Paleontology 1:357–365.

Webb, S. D. 1983. A new species of *Pediomeryx* from the late Miocene of Florida, and its relationships within the subfamily Cranioceratinae (Ruminantia: Dromomerycidae). Journal of Mammalogy 64:261–272.

Webb, S. D. 1998. Hornless ruminants; pp. 463–476 in C. M. Janis, K. M. Scott, and L. L. Jacobs (eds.), Evolution of Tertiary Mammals of North America. Volume 1: Terrestrial Carnivores, Ungulates, and Ungulatelike Mammals. Cambridge University Press, Cambridge.

Webb, S. D. 2000. Evolutionary history of the New World Cervidae; pp 38–64 in E. S. Vrba and G. B. Schaller (eds.), Antelopes, Deer and Relatives: Fossil Record, Behavioral Ecology, Systematics, and Conservation. Yale University Press, New Haven, Connecticut.

Webb, S. D., and S. C. Perrigo. 1984. Late Cenozoic vertebrates from Honduras and El Salvador. Journal of Vertebrate Paleontology 4:237–254.

Webb, S. D., and B. E. Taylor. 1980. The phylogeny of hornless ruminants and a description of the cranium of *Archaeomeryx*. Bulletin of the American Museum of Natural History 16:121–157.

Webb, S. D., B. L. Beatty, and G. Poinar, Jr. 2003. New evidence of Miocene Protoceratidae including a new species from Chiapas, Mexico. Bulletin of the American Museum of Natural History 279:348–367.

Welcomme, J.-L., P.-O. Antoine, F. Duranthon, P. Mein, and L. Ginsburg. 1997. Nouvelles découvertes de Vertébrés miocènes dans le synclinal de Dera Bugti (Balouchistan, Pakistan). Comptes Rendus de l'Académie des Sciences de Paris 325:531–536.

West, R. M. 1980a. A minute new species of *Dorcatherium* (Tragulidae, Mammalia) from the Chinji Formation near Daud Khel, Mianwali District, Pakistan. Milwaukee Public Museum Contributions in Biology and Geology 1980:1–6.

West, R. M. 1980b. Middle Eocene large mammal assemblage with Tethyan affinities, Ganda Kas region, Pakistan. Journal of Paleontology 54:508–533.

West, R. M. 1982. Fossil mammals from the lower Buck Hill Group, Eocene of Trans-Pecos Texas: Marsupicarnivora, Primates, Taeniodonta, Condylarthra, bunodont Artiodactyla, and Dinocerata. Pearce-Sellards Series Texas Memorial Museum 35:1–20.

West, R. M. 1984. Paleontology and geology of the Bridger Formation, Southwestern Wyoming. Part 7. Survey of Bridgerian Artiodactyla, including description of a skull and partial skeleton of *Antiacodon pygmaeus*. Contributions in Biology and Geology of the Milwaukee Public Museum 56:1–47.

West, R. M., and E. G. Atkins 1970. Additional middle Eocene (Bridgerian) mammals from Tabernacle Butte, Sublette County, Wyoming. American Museum Novitates 2404:1–26.

Westgate, J. W. 1990. Uintan land mammals (excluding rodents) from estuarine facies of the Laredo Formation (Middle Eocene, Claiborne Group) of Webb County, Texas. Journal of Paleontology 64:454–468.

Westgate, J. W. 1994. A new leptochoerid from middle Eocene (Uintan) deposits of the Texas coastal plain. Journal of Vertebrate Paleontology 14:296–299.

Westgate, J. W., and R. J. Emry. 1985. Land mammals of the Crow Creek local fauna, Late Eocene, Jackson Group, St. Francis County, Arkansas. Journal of Paleontology 59:242–248.

Weston, E. M. 1997. A biometrical analysis of evolutionary change within the Hippopotamidae. Ph.D. dissertation, Cambridge University, Cambridge.

Weston, E. M. 2000. A new species of hippopotamus *Hexaprotodon lothagamensis* (Mammalia: Hippopotamidae) from the late Miocene of Kenya. Journal of Vertebrate Paleontology 20:177–185.

Weston, E. M. 2003. Fossil Hippopotamidae from Lothagam; pp. 380–410 in J. M. Harris and M. G. Leakey (eds.), Lothagam. The Dawn of Humanity in Eastern Africa. Columbia University Press, New York, New York.

Wheeler, J. C. 1995. Evolution and present situation of the South American Camelidae. Biological Journal of the Linnean Society 54:271–295.

Wheeler, J. C., and E. J. Reitz. 1987. Allometric prediction of live weight in the Alpaca. Archaeozoologia 1:31–46.

Wheeler, P. E. 1992. The thermoregulatory advantages of large body size for hominids foraging in savanna environments. Journal of Human Evolution 23:351–362.

Whistler, D. D. 1984. An early Hemingfordian (early Miocene) fossil vertebrate fauna from the western Mohave Desert. Contributions in Science, Natural History Museum of Los Angeles County 355:1–36.

White, T. D., and J. M. Harris. 1977. Suid evolution and correlation of African hominid localities. Science 198:13–21.

White, T. E. 1941. Additions to the fauna of the Florida Pliocene. Proceedings of the New England Zoölogical Club 18:67–70.

White, T. E. 1942. Additions to the fauna of the Florida phosphates. Proceedings of the New England Zoölogical Club 21:87–91.

Whittow, G. C., C. A. Scammell, M. Leong, and D. Rand. 1977. Temperate regulation in the smallest ungulate, the lesser mousedeer (*Tragulus javanicus*). Comparative Biochemistry and Physiology 56A:23–26.

Whitworth, T. 1958. Miocene ruminants of East Africa. British Museum of Natural History, Fossil Mammals of Africa 15:1–50.

Wible, J. R. 1987. The eutherian stapedial artery: Character analysis and implications for superordinal relationships. Zoological Journal of the Linnean Society 91:107–135.

Wible, J. R. 1990. Petrosals of Late Cretaceous marsupials from North America, and a cladistic analysis of the petrosal in therian mammals. Journal of Vertebrate Paleontology 10:183–205.

Wiens, J. J. 2001. Character analysis in morphological phylogenetics: problems and solutions. Systematic Biology 50:689–699.

Wiens, J. J. 2004. The role of morphological data in phylogeny reconstruction. Systematic Biology 53:653–661.

Wiens, J. J., P. T. Chippindale, and D. M. Hillis. 2003. When are phylogenetic analyses misled by convergence? A case study in Texas cave-dwelling salamanders. Systematic Biology 52:501–514.

Wilhelm, P. B. 1993. Morphometric analyses of the limb skeleton of generalized mammals in relationship to locomotor behavior, with applications to fossil mammals. Unpublished Ph.D. dissertation, Brown University, Providence, Rhode Island, 210 pp.

Wilkinson, A. F. 1976. The Lower Miocene Suidae of Africa. Fossil Vertebrates of Africa, 4:173–282.

Williams, S. H., and R. F. Kay. 2001. A comparative test of adaptive explanations for hypsodonty in ungulates and rodents. Journal of Mammalian Evolution 8:207–229.

Williamson, T. E., and S. G. Lucas. 1992. *Meniscotherium* (Mammalia, "Condylarthra") from the Paleocene–Eocene of Western North America. New Mexico Museum of Natural History Bulletin 1:1–75.

Wilson, J. A. 1971a. Early Tertiary vertebrate faunas, Vieja Group, Trans-Pecos Texas: Agriochoeridae and Merycoidodontidae. Bulletin of the Texas Memorial Museum 18:1–83.

Wilson, J. A. 1971b. Early Tertiary vertebrate faunas, Vieja Group, Trans-Pecos Texas: Entelodontidae. Pearce-Sellards Series, Texas Memorial Museum, Austin, Texas 17:1–17.

Wilson, J. A. 1974. Early Tertiary vertebrate faunas, Vieja Group and Buck Hill Group, Trans-Pecos Texas: Protoceratidae, Camelidae, Hypertragulidae. Bulletin of the Texas Memorial Museum 23:1–34.

Wilson, J. A. 1986. Stratigraphic occurrence and correlation of early Tertiary vertebrate faunas, Trans-Pecos Texas: Agua Fria-Green Valley areas. Journal of Vertebrate Paleontology 6:350–373.

Wilson, J. A., and M. S. Stevens. 1986. Fossil vertebrates from the latest Eocene, Skyline Channels, Trans-Pecos Texas; pp. 221–235 in K. M. Flanagan and J. A. Lillegraven (eds.), Vertebrates, phylogeny, and philosophy. University of Wyoming Contributions to Geology Special Paper 3.

Wing, E. S. 1988. Use of animals by the Inca as seen at Huanuco Pampa: pp. 167-179 in E. S. Wing and J. C. Wheeler (eds.), Economic Prehistory of the Central Andes. BAR International Series 427.

Wing, S. L. 1998. Tertiary vegetation of North America as a context for mammalian evolution; pp. 37–60 in C. M. Janis, K. M. Scott, and L. L. Jacobs (eds.), Evolution of Tertiary Mammals of North America. Volume 1: Terrestrial Carnivores, Ungulates, and Ungulatelike Mammals. Cambridge University Press, Cambridge.

Wolanski, E., and E. Gereta. 1999. Oxygen cycle in a hippo pool, Serengeti National Park, Tanzania. African Journal of Ecology 37:419–423.

Wolfe, J. A. 1978. A paleobotanical interpretation of Tertiary climates in the northern hemisphere. American Scientist 66:694–703.

Wolfe, J. A. 1985. Distribution of major vegetational types during the Tertiary. Geophysical Monographs 32:357–375.

Woodburne, M. O. 1969. Systematic, biogeography and evolution of *Cynorca* and *Dyseohyus* (Tayassuidae). Bulletin of the American Museum of Natural History 19:52–70.

Woodburne, M. O. 1987. Cenozoic Mammals of North America. University of California Press, Berkeley, California.

Woodburne, M. O. 2004a. Global events and the North American mammalian biochronology; pp. 315–343 in M. O. Woodburne (ed.), Late Cretaceous and Cenozoic Mammals of North America. Columbia University Press, New York, New York.

Woodburne, M. O. 2004b. Late Cretaceous and Cenozoic Mammals of North America. Columbia University Press, New York, New York.

Woodburne, M. O., and C. C. Swisher III. 1995. Land mammal high-resolution geochronology, intercontinental overland dispersals, sea level, climate, and vicariance. SEPM Special Publication 54:335–364.

Woodburne, M. O., R. H. Tedford, and C. C. Swisher III. 1990. Lithostratigraphy, biostratigraphy, and geochronology of the Barstow Formation, Mojave Desert, southern California. Geological Society of America Bulletin 102:459–477.

Wortman, J. L. 1898. The extinct Camelidae of North America and some associated forms. Bulletin of the American Museum of Natural History 10(7):93–142.

Wortman, J. L. 1901. Studies of the Eocene Mammalia in the Marsh collection, Peabody Museum. American Journal of Science 11:1–90.

Wright, D. B. 1998. Tayassuidae; pp. 389–401 in C. M. Janis, K. M. Scott, and L. L. Jacobs (eds.), Evolution of Tertiary Mammals of North America. Volume 1: Terrestrial Carnivores, Ungulates, and Ungulatelike Mammals. Cambridge University Press, Cambridge.

Xu, Y. 1961. Some Oligocene mammals from Chuching, Yunnan. Vertebrata PalAsiatica 5:315–325.

Xu, Y. 1962. Some new anthracotheres from Shansi and Yunnan. Vertebrata PalAsiatica 6:243–250.

Yang, Q., J. Hu, and J. Peng. 1990. A study on the population ecology of forest musk deer (*Moschus berezovskii*) at the north of Hengduan Mountains. Acta Theriologica Sinica 10:255–262.

Ye, J., Z. Qiu, and G. Zhang. 1992. *Bunolistriodon intermedius* (Suidae, Artiodactyla) from Tongxi, Ningxia. Vertebrata PalAsiatica 30:135–145.

Zachos, J., M. Pagani, L. Sloan, E. Thomas, and K. Billups. 2001. Trends, rhythms, and abberations in global climate 65 Ma to present. Science 292:686–693.

Zdansky, O. 1930. Die alttertiären Säugetiere Chinas nebst stratigraphischen Bemerkungen. Palaeontologia Sinica 6:5–84.

Zhang, Y., Y. Long, H. Ji, and S. Ding. 1983. The Cenozoic deposits of the Yunnan Region, Professional Papers on Stratigraphy and Paleontology, Volume 7. Geological Publishing House, Peking, China, 21 pp.

Zhanxiang, Q., and G. Yumin. 1991. The Aragonian vertebrate fauna of Xiacaowan—8. *Dorcatherium* (Tragulidae, Artiodactyla). Vertebrata PalAsiatica 29:21–37 [Chinese 21–31; English 32–37].

Zhao, Z. 1981. The vertebrate fossils and lower Tertiary from Nanning Basin. Vertebrata PalAsiatica 19:218–227.

Zhao, Z. 1983. A new species of anthracothere from Nanning Basin, Guangxi. Vertebrata PalAsiatica 21:266–270.

Zheng, J. 1978. Description of some late Eocene mammals from Lian-Kan Formation of Turfan Basin, Sinkiang. Memoirs, Institute of Vertebrate Paleontology and Paleoanthropology, Academia Sinica 13:115–125.

Zhou, X., R. Zhai, P. D. Gingerich, and L. Chen. 1995. Skull of a new mesonychid (Mammalia, Mesonychia) from the Late Paleocene of China. Journal of Vertebrate Paleontology 15:387–400.

Zittel, K. A. von. 1893. Handbuch der Palaeontologie, Abtheilung I. Palaeozoologie. IV Band, 1ste Lief. Vertebrata (Mammalia). R. Oldenbourg, Munich and Leipzig, 799 pp.

INDEX

363